# Handbook of Communication for Development and Social Change

Jan Servaes
Editor

# Handbook of Communication for Development and Social Change

Volume 1

With 83 Figures and 60 Tables

Springer

*Editor*
Jan Servaes
Department of Media and Communication
City University of Hong Kong
Hong Kong, Kowloon, Hong Kong

Katholieke Universiteit Leuven
Leuven, Belgium

ISBN 978-981-15-2013-6     ISBN 978-981-15-2014-3 (eBook)
ISBN 978-981-15-2015-0 (print and electronic bundle)
https://doi.org/10.1007/978-981-15-2014-3

© Springer Nature Singapore Pte Ltd. 2020
This work is subject to copyright. All rights are reserved by the Publisher, whether the whole or part of the material is concerned, specifically the rights of translation, reprinting, reuse of illustrations, recitation, broadcasting, reproduction on microfilms or in any other physical way, and transmission or information storage and retrieval, electronic adaptation, computer software, or by similar or dissimilar methodology now known or hereafter developed.
The use of general descriptive names, registered names, trademarks, service marks, etc. in this publication does not imply, even in the absence of a specific statement, that such names are exempt from the relevant protective laws and regulations and therefore free for general use.
The publisher, the authors, and the editors are safe to assume that the advice and information in this book are believed to be true and accurate at the date of publication. Neither the publisher nor the authors or the editors give a warranty, expressed or implied, with respect to the material contained herein or for any errors or omissions that may have been made. The publisher remains neutral with regard to jurisdictional claims in published maps and institutional affiliations.

This Springer imprint is published by the registered company Springer Nature Singapore Pte Ltd.
The registered company address is: 152 Beach Road, #21-01/04 Gateway East, Singapore 189721, Singapore

# Introduction

This *Handbook of Communication for Development and Social Change* provides a single reference resource regarding communication for development and social change. Increasingly, one considers communication crucial to effectively tackle the major problems of today. Hence, the question being addressed in this handbook is, "Is there a right communication strategy?" Perspectives on sustainability, participation, and culture in communication have changed over time in line with the evolution of development approaches and trends, and in response to the need for effective applications of communication methods and tools to new issues and priorities.

In essence, the coherence of Communication for Development and Social Change (CDSC) is expressed in its different common underlying premises. In all its diversity, such as in theory, method, medium, and region, it is characterized by a number of underlying common values and starting points.

Therefore, the *Handbook of Communication for Development and Social Change* starts from the following premises and assumptions:

- *The use of a culturalist viewpoint*
  By means of such a viewpoint, specific attention is given to communication in social change processes. By putting culture centrally from a user's perspective, other social science disciplines can significantly contribute to the field of Communication for Development and Social Change.
- *The use of an interpretative perspective*
  Participation, dialogue, and an active vision of human beings as the interpreters of their environments are of the utmost importance. A highly considered value is the showing of respect and appreciation for the uniqueness of specific situations and identities in social change environments.
- *The preference for a transdisciplinary approach*
  While interdisciplinary collaborations create new knowledge synthesized from existing disciplines, a transdisciplinary approach relates all disciplines into a coherent whole. Transdisciplinarity combines interdisciplinarity with a participatory approach.
  The field of "sustainability" is, in essence, a transdisciplinary one.

- *The use of integrated methods and theories*
  In the field of Communication for Development and Social Change, it is considered important that the chosen methods should be connected with the used theoretical perspective. This implies that openness, diversity, and flexibility in methods and techniques are valued. In practice, it generally means triangulation and a preference for qualitative methods. This does not mean, however, that quantitative methods are excluded, and indeed an emphasis is placed on evidence-based scientific methodologies.
- *To show mutual understanding and attach importance to formal and informal intercultural teaching, training, and research*
  Tolerance, consciousness-raising, acceptance, and respect can only be arrived at when members of different cultures not only hear but also understand each other. This mutual understanding is a condition for development and social change. In order to prevent all forms of miscommunication, intercultural awareness, capacity building, and dialogue is deemed very important.

A short-hand definition of Communication for Development and Social Change (CDSC) reads as follows:

*Communication for development and social change is the nurturing of knowledge aimed at creating a consensus for action that takes into account the interests, needs and capacities of all concerned. It is thus a social process, which has as its ultimate objective sustainable development/change at distinct levels of society.*
*Communication media and ICTs are important tools in achieving social change but their use is not an end in itself. Interpersonal communication and traditional, group, and social media must also play a fundamental role.*

Divided into prominent themes comprising relevant chapters written by experts in the field and reviewed by renowned editors, the book addresses topics where communication and social change converge in both theory and praxis. Specific concerns and issues include climate change, poverty reduction, health, equity and gender, sustainable development goals (SDGs), and information and communication technologies (ICTs).

The *Handbook of Communication for Development and Social Change* integrates 11 clusters of concepts and practices. The parts are as follows:

1. An historic cluster which looks at the origins of the Communication for Development and Social Change field both from a theoretical as well as political/economic perspective
2. A *normative cluster of concepts* (e.g., on sustainability, democracy, participation, empowerment, equity, social inclusion, human rights, and accountability)
3. A *cluster of concepts that sets an important context for communication activities for developmen*t (e.g., globalization, gender, social movements, cultural diversity, and Sustainable Development Goals (SDGs))

4. A *cluster of strategic and methodological concepts* (e.g., diffusion of innovations, social marketing, advocacy, entertainment-education, social mobilization, knowledge management)
5. A *cluster of concepts that relate to methods, techniques, and tools* (e.g., (digital) storytelling, (participatory) mapping, monitoring and evaluation, intercultural value differences, empowerment evaluation, social media, oral history, film, video, drama, and art)
6. A *cluster of cases that focus on climate change and sustainable development* (e.g., on youth participation in Indonesia, emergency communication during the 2015 Nepal earthquake, innovative education in Kenya)
7. A *cluster of cases that focus on Information and Communication Technologies (ICTs) for development* (e.g., on ICTs for learning in rural communication, strategic social media management for NGOs, online activism in Central Asia)
8. A *cluster of cases that focus on health communication* (e.g., on sustainable health, maternal health in Africa, countering sexual harassment in Bangladesh)
9. A *cluster of cases that focus on participatory communication* (e.g., on natural resource management, conflict and power, capacity building, participatory video)
10. A *set of regional cases and overviews* (e.g., on Africa, Europe, Western journalism education, ASEAN)
11. A *set of specific case studies* (e.g., on entertainment-education in radio in Africa, participatory communication in Afghanistan, Egyptian media, protest in South Africa)

This book shows how communication is essential at all levels of society. It helps readers understand the processes that underlie attitude change and decision-making and the work uses powerful models and methods to explain the processes that lead to sustainable development and social change.

This handbook is published in multiple formats: print, static electronic, and live electronic.

It is essential reading for academics and practitioners, students and policy makers alike.

World Rainforest Day
22 June 2019                                                                Jan Servaes

# Acknowledgments

The *Handbook of Communication for Development and Social Change* is the collective effort of many distinguished researchers, scholars, and professionals in the field of development communication or communication for social change.

At the editorial level, I was assisted by two excellent associate editors: Dr. Rico Lie, assistant professor at Wageningen University in The Netherlands, and Dr. Patchanee Malikhao, director of the consultancy agency Fecund Communication in Chiang Mai, Thailand.

A number of renowned experts assisted with the selection and peer review of the manuscripts. The nature of their work requires that they remain anonymous. However, I wish to single-out Professor emeritus Royal Colle from Cornell University in Ithaca NY, USA, who has kindly assisted me with the section on ICTD, Information and Communication Technologies for Development, and Dr. Jessica Noske-Turner, lecturer at Loxbourough University London, who assisted as external reviewer.

We also wish to acknowledge and thank Ms. Jayanthie Krishnan, then Publishing Editor at Springer Singapore, who initially invited me to coordinate this project. Once the project was on rails Ms. Alexandra Campbell, Editor for Social Sciences and Humanities at Springer Nature Singapore, took over. The capable Springer Reference Editorial Team was coordinated by Ms. Mokshika Gaur and her hardworking assistants Dr. Lijuan Wang and Ms. Sunaina Dadhwal.

Special thanks go to all the contributing authors, researchers, and students who have made this handbook possible.

**Important Note**

In all cases where gender is not implicit, we have attempted to combine feminine and masculine pronouns. However, as English is not the mother tongue of the editors and many authors in this handbook, we apologize to the reader if she/he still finds errors.

# Contents

## Volume 1

**Part I  Introduction** .................................... 1

1. Terms and Definitions in Communication for Development and Social Change .................................... 3
   Jan Servaes

2. Communication for Development and Social Change: In Search of a New Paradigm .................................... 15
   Jan Servaes

3. Key Concepts, Disciplines, and Fields in Communication for Development and Social Change .................................... 29
   Jan Servaes and Rico Lie

**Part II  Historic Cluster** .................................... 61

4. Communication for Development and Social Change: Three Development Paradigms, Two Communication Models, and Many Applications and Approaches .................................... 63
   Jan Servaes and Patchanee Malikhao

5. Family Tree of Theories, Methodologies, and Strategies in Development Communication .................................... 93
   Silvio Waisbord

6. A Changing World: FAO Efforts in Communication for Rural Development .................................... 133
   Silvia Balit and Mario Acunzo

7. Daniel Lerner and the Origins of Development Communication .................................... 157
   Hemant Shah

8   The Pax Americana and Development .................... 167
    P. Eric Louw

**Part III   Normative Concepts** .............................. **193**

9   Media and Participation ............................... 195
    Nico Carpentier

10  Empowerment as Development: An Outline of an Analytical
    Concept for the Study of ICTs in the Global South ........... 217
    Jakob Svensson

11  The Theory of Digital Citizenship ........................ 237
    Toks Dele Oyedemi

12  Co-creative Leadership and Self-Organization: Inclusive
    Leadership of Development Action ....................... 257
    Alvito de Souza and Hans Begeer

13  Communication for Development and Social Change Through
    Creativity ........................................... 269
    Arpan Yagnik

14  The Relevance of Habermasian Theory for Development and
    Participatory Communication ........................... 287
    Thomas Jacobson

15  The Importance of Paulo Freire to Communication for
    Development and Social Change ......................... 309
    Ana Fernández-Aballí Altamirano

16  De-westernizing Alternative Media Studies: Latin American
    Versus Anglo-Saxon Approaches from a Comparative
    Communication Research Perspective .................... 329
    Alejandro Barranquero

**Part IV   Context for Communication Activities for Development
and Social Change** ...................................... **341**

17  A Threefold Approach for Enabling Social
    Change: Communication as Context for Interaction, Uneven
    Development, and Recognition .......................... 343
    Gloria Gómez Diago

18  Shifting Global Patterns: Transformation of Indigenous
    Nongovernmental Organizations in Global Society ........... 359
    Junis J. Warren

19  Women's Empowerment in Digital Media: A Communication
    Paradigm ............................................. 379
    Xiao Han

20  Development Communication and the Development Trap ....... 395
    Cees J. Hamelink

21  The Soft Power of Development: Aid and Assistance as Public
    Diplomacy Activities ................................. 407
    Colin Alexander

22  Asian Contributions to Communication for Development and
    Social Change ........................................ 421
    Cleofe S. Torres and Linje Manyozo

23  Development Communication in Latin America .............. 443
    José Luis Aguirre Alvis

24  Development Communication in South Africa ............... 469
    Tanja Bosch

25  Development Communication as Development Aid for
    Post-Conflict Societies .............................. 481
    Kefa Hamidi

26  Glocal Development for Sustainable Social Change ........ 501
    Fay Patel

27  Communication Policy for Women's Empowerment: Media
    Strategies and Insights .............................. 519
    Kiran Prasad

**Part V  Strategic and Methodological Concepts** ............... **531**

28  Three Types of Communication Research Methods:
    Quantitative, Qualitative, and Participatory ......... 533
    Jan Servaes

29  Visual Communication and Social Change .................. 555
    Loes Witteveen and Rico Lie

30  Multidimensional Model for Change: Understanding Multiple
    Realities to Plan and Promote Social and Behavior Change .... 579
    Paolo Mefalopulos

31  Broadcasting New Behavioral Norms: Theories Underlying the
    Entertainment-Education Method ....................... 595
    Kriss Barker

| 32 | Protest as Communication for Development and Social Change | 615 |

Toks Dele Oyedemi

| 33 | Political Engagement of Individuals in the Digital Age | 633 |

Paul Clemens Murschetz

| 34 | Family and Communities in Guatemala Participate to Achieve Educational Quality | 647 |

Antonio Arreaga

| 35 | Digital Communication and Tourism for Development | 667 |

Alessandro Inversini and Isabella Rega

## Part VI  Methods, Techniques, and Tools .................. 679

| 36 | A Community-Based Participatory Mixed-Methods Approach to Multicultural Media Research | 681 |

Rukhsana Ahmed and Luisa Veronis

| 37 | Digital Stories as Data | 693 |

Valerie M. Campbell

| 38 | Participatory Mapping | 705 |

Logan Cochrane and Jon Corbett

| 39 | Evaluations and Impact Assessments in Communication for Development | 715 |

Lauren Kogen

| 40 | Transformative Storywork: Creative Pathways for Social Change | 733 |

Joanna Wheeler, Thea Shahrokh, and Nava Derakhshani

| 41 | Recollect, Reflect, and Reshape: Discoveries on Oral History Documentary Teaching | 755 |

Yuhong Li

| 42 | Differences Between Micronesian and Western Values | 767 |

Tom Hogan

## Volume 2

## Part VII  Climate Change and Sustainable Development ........ 793

| 43 | Communicating Climate Change: Where Did We Go Wrong, How Can We Do Better? | 795 |

Emily Polk

44  Bottom-Up Networks in Pacific Island Countries: An Emerging
    Model for Participatory Environmental Communication  . . . . . . .  815
    Usha S. Harris

45  Youth Voices from the Frontlines: Facilitating Meaningful
    Youth Voice Participation on Climate, Disasters, and
    Environment in Indonesia  . . . . . . . . . . . . . . . . . . . . . . . . . . . . . . . .  833
    Tamara Plush, Richard Wecker, and Swan Ti

46  Key SBCC Actions in a Rapid-Onset Emergency: Case Study
    From the 2015 Nepal Earthquakes  . . . . . . . . . . . . . . . . . . . . . . . . .  847
    Rudrajit Das and Rahel Vetsch

47  Importing Innovation? Culture and Politics of Education in
    Creative Industries, Case Kenya  . . . . . . . . . . . . . . . . . . . . . . . . . .  861
    Minna Aslama Horowitz and Andrea Botero

**Part VIII  ICTs for Development**  . . . . . . . . . . . . . . . . . . . . . . . . . . . . .  **871**

48  ICTs for Learning in the Field of Rural Communication  . . . . . . .  873
    Rico Lie and Loes Witteveen

49  How Social Media Mashups Enable and Constrain Online
    Activism of Civil Society Organizations  . . . . . . . . . . . . . . . . . . . .  891
    Oana Brindusa Albu and Michael Andreas Etter

50  Strategic Social Media Management for NGOs  . . . . . . . . . . . . . .  911
    Claudia Janssen Danyi and Vidhi Chaudhri

51  ICTs and Modernization in China  . . . . . . . . . . . . . . . . . . . . . . . . .  929
    Song Shi

52  Online Social Media and Crisis Communication in China:
    A Review and Critique  . . . . . . . . . . . . . . . . . . . . . . . . . . . . . . . . . .  939
    Yang Cheng

53  Diffusion and Adoption of an E-Society: The Myths and Politics
    of ICT for the Poor in India  . . . . . . . . . . . . . . . . . . . . . . . . . . . . . .  953
    Ravindra Kumar Vemula

54  Online Activism in Politically Restricted Central Asia:
    A Comparative Review of Kazakhstan, Kyrgyzstan, and
    Tajikistan  . . . . . . . . . . . . . . . . . . . . . . . . . . . . . . . . . . . . . . . . . . . . .  961
    Bahtiyar Kurambayev

55  New Media: The Changing Dynamics in Mobile Phone
    Application in Accelerating Health Care Among the Rural
    Populations in Kenya  . . . . . . . . . . . . . . . . . . . . . . . . . . . . . . . . . . .  977
    Alfred Okoth Akwala

56  ICTs for Development: Building the Information Society by
    Understanding the Consumer Market ...................... 989
    Shahla Adnan

**Part IX   Health Communication ............................ 1013**

57  Health Communication: Approaches, Strategies, and Ways to
    Sustainability on Health or Health for All ................... 1015
    Patchanee Malikhao

58  Health Communication: A Discussion of North American and
    European Views on Sustainable Health in the Digital Age ....... 1039
    Isabell Koinig, Sandra Diehl, and Franzisca Weder

59  Millennium Development Goals (MDGs) and Maternal
    Health in Africa ......................................... 1063
    Alfred Okoth Akwala

60  Impact of the Dominant Discourses in Global Policymaking
    on Commercial Sex Work on HIV/STI Intervention Projects
    Among Commercial Sex Workers ........................... 1075
    Satarupa Dasgupta

61  Designing and Distribution of Dementia Resource Book to
    Augment the Capacities of Their Caretakers ................. 1091
    Avani Maniar and Khyati Deesawala

62  Strategic Communication to Counter Sexual Harassment in
    Bangladesh ............................................. 1119
    Nova Ahmed

63  Multiplicity Approach in Participatory Communication: A Case
    Study of the Global Polio Eradication Initiative in Pakistan ..... 1131
    Hina Ayaz

**Part X   Participatory Communication ....................... 1139**

64  Participatory Development Communication and Natural
    Resources Management ................................... 1141
    Guy Bessette

65  Participatory Communication in Practice: The Nexus to Conflict
    and Power .............................................. 1155
    Saik Yoon Chin

66  Capacity Building and People's Participation in e-Governance:
    Challenges and Prospects for Digital India .................. 1177
    Kiran Prasad

| | | |
|---|---|---|
| 67 | **Fifty Years of Practice and Innovation Participatory Video (PV)** .................................................. Tony Roberts and Soledad Muñiz | 1195 |
| 68 | **Reducing Air Pollution in West Africa Through Participatory Activities: Issues, Challenges, and Conditions for Citizens' Genuine Engagement** .................................................. Stéphanie Yates, Johanne Saint-Charles, Marius N. Kêdoté, and S. Claude-Gervais Assogba | 1213 |
| 69 | **Community Radio in Ethiopia: A Discourse of Peace and Conflict Reporting** .................................................. Mulatu Alemayehu Moges | 1231 |

## Part XI   Regional Overviews .................................................. 1241

| | | |
|---|---|---|
| 70 | **Political Economy of ICT4D and Africa** .................................................. Tokunbo Ojo | 1243 |
| 71 | **Mainstreaming Gender into Media: The African Union Backstage Priority** .................................................. Bruktawit Ejigu Kassa and Katharine Sarikakis | 1257 |
| 72 | **Idiosyncrasy of the European Political Discourse Toward Cooperation** .................................................. Teresa La Porte | 1277 |
| 73 | **The Challenge of Promoting Diversity in Western Journalism Education: An Exploration of Existing Strategies and a Reflection on Its Future Development** .................................................. Rozane De Cock and Stefan Mertens | 1293 |
| 74 | **Institutionalization and Implosion of Communication for Development and Social Change in Spain: A Case Study** .................................................. Víctor Manuel Marí Sáez | 1311 |
| 75 | **A Sense of Community in the ASEAN** .................................................. Pornpun Prajaknate | 1325 |

## Part XII   Case Studies .................................................. 1341

| | | |
|---|---|---|
| 76 | **Entertainment-Education in Radio: Three Case Studies from Africa** .................................................. Kriss Barker and Fatou Jah | 1343 |
| 77 | **The Role of Participatory Communication in Strengthening Solidarity and Social Cohesion in Afghanistan** .................................................. Hosai Qasmi and Rukhsana Ahmed | 1355 |

78   Sinai People's Perceptions of Self-Image Portrayed by the
     Egyptian Media: A Multidimensional Approach .............. 1365
     Alamira Samah Saleh

79   Protest as Communication for Development and Social Change
     in South Africa ....................................... 1381
     Elizabeth Lubinga

80   Case Study of Organizational Crisis Communication: Oxfam
     Responds to Sexual Harassment and Abuse Scandal .......... 1399
     Claudia Janssen Danyi

81   Communication and Culture for Development: Contributions to
     Artisanal Fishers' Wellbeing in Coastal Uruguay ............ 1413
     Paula Santos and Micaela Trimble

82   Fostering Social Change in Peru Through Communication:
     The Case of the Manuani Miners Association ................ 1429
     Sol Sanguinetti

83   Communicative Analysis of a Failed Coup Attempt in Turkey ... 1439
     Zafer Kıyan and Nurcan Törenli

84   Plurality and Diversity of Voices in Community Radio: A Case
     Study of Radio Brahmaputra from Assam ................... 1455
     Alankar Kaushik

**Part XIII   Conclusion** ...................................... **1469**

85   Communication for Development and Social Change:
     Conclusion ............................................ 1471
     Jan Servaes

**Index** ...................................................... 1483

# About the Editor

**Jan Servaes** (Ph.D.) was UNESCO Chair in Communication for Sustainable Social Change. He has taught International Communication and Communication for Social Change in Australia, Belgium, China, Hong Kong, The Netherlands, Thailand, and the United States, in addition to several teaching stints at about 120 universities in 55 countries.

Servaes is Editor of the Lexington Book Series Communication, Globalization and Cultural Identity (https://rowman.com/Action/SERIES/LEX/LEXCGC) and the Springer Book Series Communication, Culture and Change in Asia (http://www.springer.com/series/13565) and was Editor-in-Chief of the Elsevier journal *Telematics and Informatics: An Interdisciplinary Journal on the Social Impacts of New Technologies"* (http://www.elsevier.com/locate/tele).

Servaes has been President of the European Consortium for Communications Research (ECCR, later www.ecrea.eu) and Vice President of the International Association of Media and Communication Research (IAMCR, www.iamcr.org), in charge of Academic Publications and Research, from 2000 to 2004. He chaired the Scientific Committee for the World Congress on Communication for Development (Rome, 25–27 October 2006),organized by the World Bank, FAO, and the Communication Initiative.

Servaes has undertaken research, development, and advisory work around the world and is the author of more than 500 journal articles and 25 books/monographs, published in Chinese, Dutch, French, German, Indonesian, Portuguese, Russian, Spanish, and Thai, on topics such as as international and development communication, ICT and media policies, intercultural communication, participation and social change, and human rights and conflict management. He is known for his "multiplicity paradigm" in *Communication for Development: One World, Multiple Cultures* (1999).

Some of his recent book titles include: (2017) Servaes, Jan (ed.) *The Sustainable Development Goals in an Asian Context*, Singapore: Springer; (2016) Servaes, Jan & Oyedemi, Toks (Eds.) *The Praxis of Social Inequality in Media: A Global Perspective*, Lanham, MD: Lexington Books, Rowman and Littefield; (2016) Servaes, Jan & Oyedemi, Toks (Eds.) *Social Inequalities, Media, and Communication: Theory and Roots*, Lanham, MD: Lexington Books, Rowman and Littefield; (2014) Jan Servaes (Ed.). *Technological Determinism and Social Change*, Lanham: Lexington Books; (2014) Jan Servaes and Patchanee Malikhao. *Communication for Social Change* (in Chinese), Wuhan: Wuhan University Press; (2013) Jan Servaes (Ed.). *Sustainable Development and Green Communication: African and Asian Perspectives*, London/New York: Palgrave/MacMillan; (2013) J. Servaes (Ed.). *Sustainability, Participation and Culture in Communication. Theory and Praxis*, Bristol-Chicago: Intellect-University of Chicago Press; (2008) J. Servaes. *Communication for Development and Social Change*, Los Angeles, London, New Delhi, Singapore: Sage; (2007) J. Servaes & Liu S. (Eds.). *Moving Targets: Mapping the Paths Between Communication, Technology and Social Change in Communities*, Penang: Southbound; (2006) P. Thomas & J. Servaes (Eds.). *Intellectual Property Rights and Communications in Asia*, New Delhi: Sage; (2006) J. Servaes & N. Carpentier (Eds.). *Towards a Sustainable European Information Society*, ECCR Book Series, Bristol: Intellect; (2005) Shi -Xu, Kienpointner M. & J. Servaes (Eds.). *Read the Cultural Other: Forms of Otherness in the Discourses of Hong Kong's Decolonization*, Berlin: Mouton De Gruyter; (2003) J. Servaes (Ed.). *The European Information Society: A Reality Check*, ECCR Book Series, Bristol: Intellect; and (2003) J. Servaes (Ed.). *Approaches to Development: Studies on Communication for Development*, Paris: UNESCO Publishing House.

# About the Associate Editors

**Rico Lie** (Ph.D. 2000, Catholic University of Brussels, Belgium) is a social anthropologist working at the research group Knowledge, Technology and Innovation, Wageningen University & Research (WUR), The Netherlands. He previously worked at the University of Brussels in Belgium and the Universities of Nijmegen and Leiden in The Netherlands. At WUR, he is an assistant professor in international communication with an interest in the areas of communication for development and intercultural learning.

**Patchanee Malikhao** (Ph.D. 2007, The University of Queensland, Australia) is a Sociologist with competencies in mass communication research and graphic arts research. She received her education in two prestigious universities in Thailand, Chulalongkorn and Thammasat universities, and higher education in the United States (RIT, Rochester, NY) and Australia (UQ). Dr. Malikhao has worked in Australia, the United States, Belgium, the Netherlands, and Thailand in the fields of Communication for Social Change research, Health Communication research, teaching and research in the graphic arts, journalism, and communication. She has been an academic writer since the 1980s. Dr. Malikhao has a wide range of research interests from communication for sustainable social change, over health related areas such as HIV/AIDS to mindful journalism, Thai Buddhism and social communication, globalization and Thai culture, intercultural communication, and Thai culture and communication. Currently she is the director of Fecund Communication Consultancy in Thailand.

She was the recipient of a Fulbright Scholarship, the Australian Postgraduate Award Scholarship, and a distinguished award in printing studies from the Tab Nilanidhi Foundation, Thailand.

She is the single author of the following books:

Malikhao, P. (2017). *Culture and Communication in Thailand*. Singapore: Springer. (https://www.springer.com/gp/book/9789811041235)

Malikhao, P. (2016). *Effective Health Communication for Sustainable Development*. New York: Nova Publishers, Inc. (https://www.novapublishers.com/catalog/product_info.php?products_id=58305&osCsid=98458b3794e19851cf28ac915a8c232b).

Malikhao, P. (2012). *Sex in the Village. Culture, Religion and HIV/AIDS in Thailand.* Southbound & Silkworm Publishers: Penang-Chiang Mai, 238 pp. (ISBN: 978-983-9054-55-2).

# Contributors

**Mario Acunzo**  FAO Communication for Development Team, Rome, Italy

**Shahla Adnan**  Department of Communication and Media Studies, Faculty of Social Sciences, Fatima Jinnah Women University, Rawalpindi, Pakistan

**Nova Ahmed**  North South University, Dhaka, Bangladesh

**Rukhsana Ahmed**  Department of Communication, University at Albany, SUNY, Albany, NY, USA

**Alfred Okoth Akwala**  Department of Language and Communication Studies, Faculty of Social Sciences and Technology, Technical University of Kenya, Nairobi, Kenya

**Oana Brindusa Albu**  Department of Marketing and Management, University of Southern Denmark, Odense M, Denmark

**Colin Alexander**  School of Arts and Humanities, Nottingham Trent University, Nottingham, UK

**José Luis Aguirre Alvis**  Department of Social Communication, Universidad Católica Boliviana "San Pablo", La Paz, Bolivia

**Antonio Arreaga**  Juarez & Associates, Inc., Guatemala City, Guatemala

**S. Claude-Gervais Assogba**  Faculté d'Agronomie, Université de Parakou, Abomey-Calavi, Benin

**Hina Ayaz**  Berlin, Germany

**Silvia Balit**  Rome, Italy

**Kriss Barker**  International Programs, Population Media Center, South Burlington, VT, USA

**Alejandro Barranquero**  Universidad Carlos III de Madrid, Madrid, Spain

**Hans Begeer**  BMC Consultancy, Brussels, Belgium

**Guy Bessette**  Gatineau, Canada

**Tanja Bosch** Centre for Film and Media Studies, Cape Town, South Africa

**Andrea Botero** Interact Research Group, University of Oulu, Oulu, Finland

**Valerie M. Campbell** University of Prince Edward Island, Charlottetown, PE, Canada

**Nico Carpentier** Uppsala University, Uppsala, Sweden
Vrije Universiteit Brussel (VUB), Brussels, Belgium
Charles University, Prague, Czech Republic

**Vidhi Chaudhri** Department of Media and Communication, Erasmus University Rotterdam, Rotterdam, The Netherlands

**Yang Cheng** Department of Communication, North Carolina State University, Raleigh, NC, USA

**Saik Yoon Chin** Southbound, George Town, Penang, Malaysia

**Logan Cochrane** International and Global Studies, Carleton University, Ottawa, ON, Canada

**Jon Corbett** University of British Columbia, Vancouver, BC, Canada

**Rudrajit Das** UNICEF, Kathmandu, Nepal

**Satarupa Dasgupta** Communication Arts, School of Contemporary Arts, Ramapo College of New Jersey, Mahwah, NJ, USA

**Rozane De Cock** University of Leuven, Leuven, Belgium

**Alvito de Souza** Co-Creative Communication, Brussels, Belgium

**Khyati Deesawala** Department of Extension and Communication, The Maharaja Sayajirao University of Baroda, Vadodara, India

**Nava Derakhshani** Cape Town, South Africa

**Gloria Gómez Diago** Department of Communication Sciences and Sociology, Rey Juan Carlos University, Fuenlabrada, Madrid, Spain

**Sandra Diehl** Department of Media and Communications, Alpen-Adria-Universitaet Klagenfurt, Klagenfurt, Austria

**Bruktawit Ejigu Kassa** Department of Communication, University of Vienna, Vienna, Austria
Department of Journalism and Mass communication, Haramaya University, Dire Dawa, Ethiopia

**P. Eric Louw** School of Communication and Arts, University of Queensland, Brisbane, QLD, Australia

**Michael Andreas Etter** Marie Curie Research Fellow, Faculty of Management, Cass Business School, City, University of London, London, UK

**Ana Fernández-Aballí Altamirano** Department of Communication, Universitat Pompeu Fabra, Barcelona, Spain

**Cees J. Hamelink** University of Amsterdam, Amsterdam, The Netherlands

**Kefa Hamidi** Institute communication and media Studies, Leipzig University, Leipzig, Germany

**Xiao Han** Mobile Internet and Social Media Centre, Communication University of China, Beijing, China

**Usha S. Harris** Macquarie University, Sydney, Australia

**Tom Hogan** Sydney, NSW, Australia

**Minna Aslama Horowitz** St. John's University, New York City, NY, USA
University of Helsinki, Helsinki, Finland

**Alessandro Inversini** Henley Business School, University of Reading, Greenlands Henley-on-Thames, UK

**Thomas Jacobson** Department of Media Studies and Production, Lew Klein College of Media and Communication, Temple University, Philadelphia, PA, USA

**Fatou Jah** International Programs, Population Media Center, South Burlington, VT, USA

**Claudia Janssen Danyi** Department of Communication Studies, Eastern Illinois University, Charleston, IL, USA

**Alankar Kaushik** EFL University, Shillong Campus, Shillong, India

**Marius N. Kêdoté** Institut Régional de Santé Publique, Comlan Alfred Quenum, Université d'Abomey-Calavi, Abomey Calavi, Benin

**Zafer Kıyan** Department of Journalism, Ankara University, Ankara, Turkey

**Lauren Kogen** Department of Media Studies and Production, Temple University, Philadelphia, PA, USA

**Isabell Koinig** Department of Media and Communications, Alpen-Adria-Universitaet Klagenfurt, Klagenfurt, Austria

**Bahtiyar Kurambayev** Department of Media and Communications, College of Social Sciences, KIMEP University, Almaty, Kazakhstan

**Teresa La Porte** International Political Communication, University of Navarra, Pamplona, Navarra, Spain

**Yuhong Li** Hong Kong International New Media Group, Hong Kong, China

**Rico Lie** Research Group Knowledge, Technology and Innovation, Wageningen University, Wageningen, The Netherlands

**Elizabeth Lubinga** Department of Strategic Communication, School of Communication, University of Johannesburg, Johannesburg, South Africa

**Patchanee Malikhao** Fecund Communication, Chiang Mai, Thailand

**Avani Maniar** Department of Extension and Communication, The Maharaja Sayajirao University of Baroda, Vadodara, India

**Linje Manyozo** School of Media and Communication, RMIT University, Melbourne, Australia

**Víctor Manuel Marí Sáez** Faculty of Communication and Social Sciences, Universidad de Cádiz, Jerez de la Frontera, Spain

**Paolo Mefalopulos** UNICEF, Montevideo, Uruguay

**Stefan Mertens** University of Leuven, Leuven, Belgium

**Mulatu Alemayehu Moges** School of Journalism and Communication, Addis Ababa University, Addis Ababa, Ethiopia

**Soledad Muñiz** InsightShare, London, UK

**Paul Clemens Murschetz** Faculty of Digital Communication, Berlin University of Digital Sciences, Berlin, Germany

**Tokunbo Ojo** Department of Communication Studies, York University, Toronto, ON, Canada

**Toks Dele Oyedemi** Communication and Media Studies, University of Limpopo, Sovenga, South Africa

**Fay Patel** International Higher Education Solutions, Sydney, Australia

**Tamara Plush** Royal Roads University, Victoria, BC, Canada

**Emily Polk** Stanford University, Stanford, CA, USA

**Pornpun Prajaknate** Graduate School of Communication Arts and Management Innovation, National Institute of Development Administration, Bangkok, Thailand

**Kiran Prasad** Sri Padmavati Mahila University, Tirupati, India

**Hosai Qasmi** Institute of Feminism and Gender Studies, University of Ottawa, Ottawa, ON, Canada

**Isabella Rega** Centre of Excellence in Media Practice, Faculty of Media and Communication, Bournemouth University, Bournemouth, UK

**Tony Roberts** Institute of Development Studies, University of Sussex, Sussex, UK

**Johanne Saint-Charles** Département de communication sociale et publique, axe santé environnementale, CINBIOSE, Université du Québec à Montréal, Montréal, QC, Canada

**Alamira Samah Saleh** Faculty of Mass Commination, Cairo University, Giza, Egypt

**Sol Sanguinetti** Universidad de Lima, Lima, Peru

Programme, Environment and Technology Institute, Lima, Peru

**Paula Santos** Universidad Católica del Uruguay, Montevideo, Uruguay

**Katharine Sarikakis** Department of Communication, University of Vienna, Vienna, Austria

**Jan Servaes** Department of Media and Communication, City University of Hong Kong, Hong Kong, Kowloon, Hong Kong

Katholieke Universiteit Leuven, Leuven, Belgium

**Hemant Shah** School of Journalism and Mass Communication, University of Wisconsin-Madison, Madison, WI, USA

**Thea Shahrokh** Centre for Trust Peace and Social Relations at Coventry University, Coventry, UK

**Song Shi** Department of East Asian Studies, McGill University, Montreal, QC, Canada

**Jakob Svensson** School of Arts and Communication (K3), Malmö University, Malmö, Sweden

**Swan Ti** PannaFoto Institute, Jakarta, Indonesia

**Nurcan Törenli** Department of Journalism, Ankara University, Ankara, Turkey

**Cleofe S. Torres** College of Development Communication, University of the Philippines Los Baños, Laguna, Philippines

**Micaela Trimble** South American Institute for Resilience and Sustainability Studies (SARAS), Bella Vista-Maldonado, Uruguay

**Ravindra Kumar Vemula** Deptartment of Journalism and Mass Communication, The English and Foreign Languages University, Shillong, Meghalaya, India

**Luisa Veronis** Department of Geography, Environment and Geomatics, University of Ottawa, Ottawa, ON, Canada

**Rahel Vetsch** UNICEF, Kathmandu, Nepal

**Silvio Waisbord** School of Media and Public Affairs, George Washington University, Washington, DC, USA

**Junis J. Warren** York College, City University of New York, New York, USA

**Richard Wecker** UNICEF Indonesia, Jakarta, Indonesia

**Franzisca Weder** Department of Media and Communications, Alpen-Adria-Universitaet Klagenfurt, Klagenfurt, Austria

**Joanna Wheeler** University of Western Cape, Cape Town, South Africa

**Loes Witteveen** Research Group Communication, Participation and Social Ecological Learning, Van Hall Larenstein University of Applied Sciences, Velp, The Netherlands

Research Group Environmental Policy, Wageningen University, Wageningen, The Netherlands

**Arpan Yagnik** Department of Communication, Penn State, Erie, Erie, PA, USA

**Stéphanie Yates** Département de communication sociale et publique, Université du Québec à Montréal, Montréal, QC, Canada

# Part I
# Introduction

# Terms and Definitions in Communication for Development and Social Change

Jan Servaes

## Contents

| | | |
|---|---|---|
| 1.1 | Development Communication | 4 |
| 1.2 | Communication | 5 |
| 1.3 | Development and Social Change | 6 |
| 1.4 | Policies | 6 |
| 1.5 | Culture | 7 |
| 1.6 | Ideology and Power | 8 |
| 1.7 | Identity | 9 |
| 1.8 | Sustainability and Resilience | 9 |
| 1.9 | Communication for Development and Social Change | 11 |
| References | | 12 |

### Abstract

In this introductory chapter, a number of important and frequently used terms and definitions in the field of communication for development and social change are being explained briefly: development communication, communication, development and social change, policies, culture, ideology and power, identity, sustainability and resilience, and communication for development and social change.

### Keywords

Development communication · Social change · Policies · Culture · Ideology and power · Identity · Sustainability · Resiliency

---

J. Servaes (✉)
Department of Media and Communication, City University of Hong Kong, Hong Kong, Kowloon, Hong Kong

Katholieke Universiteit Leuven, Leuven, Belgium
e-mail: jan.servaes@kuleuven.be; 9freenet9@gmail.com

© Springer Nature Singapore Pte Ltd. 2020
J. Servaes (ed.), *Handbook of Communication for Development and Social Change*,
https://doi.org/10.1007/978-981-15-2014-3_109

Though most social scientists reckon that the concepts of power, culture, communication, development, and social change are essential for an understanding of the social reality, the concepts are often not explicitly defined and therefore interpreted in different ways. This is mainly due to the complexity and multidimensionality of these concepts. Therefore, by way of introduction, a few words on words we use frequently in this handbook.

## 1.1 Development Communication

The term *development communication* was first coined in 1971 by Nora C. Quebral (Manyozo 2006). Quebral (1971: 69) defined the field as "the art and science of human communication applied to the speedy transformation of a country and the mass of its people from poverty to a dynamic state of economic growth that makes possible greater social equality and the larger fulfillment of the human potential". However, Quebral (1988) acknowledges that the term borrows from contributions made by Alan Chalkley (1968) and Juan Jamias (1975) on development journalism, and Erskine Childers and Mallica Vajrathon's (1968) work on development support communication. Erskine Childers started the first Development Support Communication unit at UNDP in Bangkok in the 1960s (see Fraser and Restrepo-Estrada 1998, or Balit & Acunzo for more details).

Throughout the past century, in separate regions of the globe, and both in the professional and academic world, different definitions of development communication have emerged. We present a brief selection:

> Development Communication is the study of social change brought about by the application of communication research, theory, and technologies to bring about development ... Development is a widely participatory process of social change in a society, intended to bring about both social and material advancement, including greater equality, freedom, and other valued qualities for the majority of people through their gaining greater control over their environment. (Everett Rogers 1976)

> Communication for development is a social process, designed to seek a common understanding among all the participants of a development initiative, creating a basis for concerted action. (UN FAO 1984)

> The planned use of communication techniques, activities and media gives people powerful tools both to experience change and actually to guide it. An intensified exchange of ideas among all sectors of society can lead to the greater involvement of people in a common cause. This is a fundamental requirement for appropriate and sustainable development. (Colin Fraser and Jonathan Villet 1994)

> Development communication involves creating mechanisms to broaden public access to information on reforms; strengthening clients' ability to listen to their constituencies and negotiate with stakeholders; empowering grassroots organizations to achieve a more

participatory process; and undertaking communication activities that are grounded in research. (World Bank, http://sitresources.worldbank.org, 2006)

"Communication for Development is a social process based on dialogue using a broad range of tools and methods. It is also about seeking change at different levels including listening, building trust, sharing knowledge and skills, building policies, debating and learning for sustained and meaningful change. *It is not public relations or corporate communication" (emphasis added)* (http://www.devcomm-congress.org/worldbank/macro/2.asp). The Rome Consensus agreed at the World Congress on Communication for Development (Rome, 25–27 October 2006)

These definitions have been interpreted and applied in different ways by organizations working at distinct societal and geographic levels. Both at theory and research levels, as well as at the levels of policy and planning-making and implementation, divergent perspectives were and still are on offer. They are based on different ontological and epistemological assumptions and therefore originate and relate to different worldviews, disciplinary perspectives, and methodological and case-based applications (Servaes 1999, 2008).

## 1.2 Communication

*Communication*, as a social process, is constituted in a specific spatial and temporal framework. How people communicate, where and why they communicate, with whom they communicate, what and why they communicate, is a function of historical processes. Communication is nothing more, and nothing less, than the articulation of social relations between people. Consequently, a concern with the communication process necessitates the identification of the web of social relations within which the process of communication is interwoven. Therefore, "a definition of 'communication' is totally subject to the theoretical framework that is to be placed on social interaction" (Thomas 1982: 80).

Several perspectives on communication for development can be identified.

A first perspective could be of communication as a process, often seen in metaphor as the fabric of society. It is not confined to the media or to messages but to their interaction in a network of social relationships. By extension, the reception, evaluation, and use of media messages, from whatever source, are as important as their means of production and transmission.

A second perspective is of communications media as a mixed system of mass communication and interpersonal channels, with mutual impact and reinforcement. In other words, the mass or social media should not be seen in isolation from other conduits.

One could, for instance, examine the role and benefits of radio versus the internet for development and democracy. Both the Internet and the radio are characterized by their interactivity. However, if, as many believe, better access to information, education, and knowledge would be the best stimulant for development, the

Internet's primary development potential is as a point of access to the global knowledge infrastructure. The danger, now widely recognized, is that access to knowledge increasingly requires a telecom infrastructure that is inaccessible to the poor and is controlled by a few powerful technology giants. Therefore, the digital divide is not about technology, it is about the widening gaps between the developed and developing worlds and the info-rich and the info-poor.

While the benefits offered by the Internet are many, its dependence on a telecom infrastructure means that they are only available to a few. Radio is much more pervasive, accessible, and affordable. Blending the two could be an ideal way of ensuring that the benefits accruing from the internet have wider reach.

Another perspective of communications in the development process is from an intersectoral and interagency concern. This view is not confined to information or broadcasting organizations and ministries but extends to all sectors, and its success in influencing and sustaining development depends to a large extent on the adequacy of mechanisms for integration and co-ordination.

## 1.3 Development and Social Change

The same can be said of the concept of *development*. "The idea of development, if not the word, is as old as the expansion of Europe, and, interestingly enough, in all the interpretations it has been given, there has been present some notion of duty. What has changed has been the answer to the question to whom the duty is owned" (Mair 1984: 1). Therefore, some authors have proposed to get rid of the concept of development altogether: "The term, development, in its current usage lacks the rigorous epistemological, conceptual, and at times, methodological power to be of much value in enhancing our understanding of some of the most complex historical, social, political, economic, and spiritual occurrences of our times" (Mowlana 1986: 1).

Though we have great sympathy with this argument, we don't see how another term could easily replace it in practice. Some suggest to replace "development" with the more neutral "social change." In general, Development or *Social Change* can be described as a significant change of structured social action or of the culture in a given society, community, or context. Such a broad definition could be further specified on the basis of a number of "dimensions" of social change: *space* (micro, meso-, macro), *time* (short, medium, long-term), *speed* (slow, incremental, evolutionary versus fast, fundamental, revolutionary), *direction* (forward or backward), *content* (sociocultural, psychological, sociological, organizational, anthropological, economic, and so forth), and *impact* (peaceful versus violent) (for more details, see Servaes 2011).

## 1.4 Policies

Instead of going for a dispute on terminologies, we rather examine the very policies in order to find out what *communication for development or social change* actually is, while being aware that these policies may express overtly the latent aspirations of the societies. Therefore, we have distinguished between development "from above" versus "from below" as in essence a consideration of the nature of development itself. One aims to produce a common understanding among all the participants in a development initiative by implementing a policy or a development project, that is, the *top-down* model. The other emphasizes engaging the grassroots in making decisions that enhance their own lives, or the *bottom-up* model. Because in an ultimate sense development is a reflection of personal values, conditioned by the societal framework in which one lives. The value a society holds, which themselves change over time, are the ultimate standard by which development or lack of it will be judged.

Development is shaped and done by people – not for people. In order for people to be able to do so, they need to understand "how the system works." Therefore, development or social change should be equated with *empowerment*: the ability of people to influence the wider system and take control of their lives (see Svensson).

## 1.5 Culture

This discussion is also directly related to the definition of *culture*. Raymond Williams (1981) once said that it is one of the two or three words that are the most difficult to define. One of the first critical reviews of concepts and definitions of culture cited 164 different definitions (Kroeber and Kluckhohn 1952). According to Thompson et al. (1990) two families of definitions vie for supremacy. Culture is a complex phenomenon that can be interpreted in a narrow or broad sense. One views culture as composed of values, beliefs, norms, rationalizations, symbols, ideologies, i.e., mental products. The other sees culture as referring to the total way of life as a people, their interpersonal attitudes as well as their attitudes.

Therefore, again, we don't attempt to start with a definition, rather a description of how we perceive "culture." In other words, what exactly constitutes a culture, or different cultures? Culture is the collective equivalent of personality, and consequently is not amenable to simplistic classification or "pigeonholing." Cultures have indistinct peripheries, and they shade off into one another in a quite indefinite way. We do not always recognize a culture when we see one. Cultures can overlap, absorb, encompass, and blend. They can be differentiated according to environment, custom, social class, world-view, or *Weltanschauung*. The tendency is to think of another culture as somewhat foreign or exotic, as existing outside of one's national borders. However, some intranational communication can be far more cross-cultural than international communication. Often, for instance, there exists an easily discernable cultural gap between the ruling elite and the masses in many developing nations. In other words, culture varies with the parameters through which we choose to look at it.

The meaning of concepts and symbols, as well as the use of language, as such, is culture bound.

We reject the viewpoint that conceives of culture as only a partial phenomenon of the social reality and thus considers cultural sociology a subdiscipline of the social sciences. Cultural sociology derives its special character not so much from a specific object field but rather from a specific perspective on society.

In this sense, cultures can be defined as social settings in which a certain reference framework has taken concrete form or has been institutionalized (further elaborated in Servaes 1999). It orients and structures the interaction and communication of people within this historical context. The classic distinction between structure and culture as an empirical duality becomes meaningless. All structures are cultural products and all culture gives structure. This intrinsic bond with a society in which actions are full of value makes all social facts cultural goods. Social facts, like institutions, behavioral patterns, norm systems, structures, and societal models are construed and cultivated in the light of certain values, preferences, or options that have developed in a society in response to certain common needs or problems. Culture has material and immaterial aspects which are part of a certain way of life, passed on, and corroborated via socialization processes (e.g., school, media, and religion) to the members of that society.

## 1.6 Ideology and Power

The reproduction of any social organization entails a basic correspondence between processes of "subjection" and "qualification." This basic social functioning of subjection/qualification involves three fundamental modes of ideological interpolation. *Ideologies* impact and qualify subjects by expressing to them, relating them to, and making them recognize: (a) what exists and what does not exist (i.e., a sense of identity); (b) what is good and bad (i.e., normalization); and (c) what is possible and impossible (i.e., a logic of conservation versus a logic of change) (Therborn 1980). Ideological interpolations are made all the time, everywhere, and by everybody. However, ideological interpolations tend to cluster at nodal points in the ongoing social process, which one could call ideological institutions, or apparatuses, and which are both discursive and non-discursive.

This process through which meaning is transmitted is never linear. It is linked to *power* in conscious and unconscious ways; it is sporadic and ubiquitous and transcends spatial and cultural boundaries: "Culture is the deposit of knowledge, experiences, beliefs, values, attitudes, meanings, hierarchies, religion, timing, roles, spatial relations, concepts of the universe, and material objects and possessions acquired by a large group of people in the course of generations through individual and group striving" (Samovar and Porter 1995: 19). It not only concerns decisions about good and evil and so on but also the way we eat, live, or dress. Ideologies do function in rational as well as irrational, in conscious as well as unconscious forms. The latter, unconscious aspects are, in our opinion, more important though often overlooked. Joseph Campbell (1988) called them "myths" or "dreams" in the sense that a dream is a personal myth, and a myth is the public dream of a society. Myths are therefore culture-bound creations of the human mind and spirit:

> National cultures are structured around myths which explain the origins of the particular grouping, their specific national identities and their concepts of national destiny. Such

national mythologies seek in grounding in broader cosmic myths, and thus gain a sacred, timeless character. Myths function more at the unintentional, symbol level, defining that what a national society is trying to become. Mythological functions are likely to be especially strong at times of national crisis, rapid change or external threat (White 1988: 19–20).

## 1.7 Identity

Culture can be taken as the way we perceive and interact with the world, and those with whom we share similar perceptions. It is precisely such shared, often unarticulated and sometimes implicit patterns of perception, communication, and behavior which are referred to as "a culture."

Therefore, various cultures also manifest different identities. Four empirical dimensions can be distinguished in such frameworks of reference: a world view (Weltanschauung), a value system, a system of symbolic representation, and social organization. Worldviews are not static any more than the cultures that enshrine them are static even though values and customs within worldviews can be and are inherited from one generation to another, as are various values and customs. Ideological institutions or apparatuses fulfill a key role here. They are forms of behavior that are crystalized on the basis of social acceptance into more-or-less standardized self-evident routines and which can work as both negative-repressing and as positive-liberating. They exist in strategies of relations of forces supporting, and supported by, types of knowledge which are both discursive and nondiscursive. They form clusters of institutions that have an impact on and influence each other and that are distinct from others by their own identity. The term cultural identity refers to two complementary phenomena: on the one hand, an inward sense of association or identification with a specific culture or subculture; on the other hand, an outward tendency within a specific culture to share a sense of what it has in common with other cultures and of what distinguishes it from other cultures. Like all social processes, these processes are not purely rational or preplanned events. We therefore live, as Benedict Anderson (1983) eloquently puts it, in "imagined communities." In such communities culture must be seen as the unintended result of an interweaving of the behavior of a group of people who interrelate and interact with each other.

## 1.8 Sustainability and Resilience

*Sustainability and resilience* are currently two of the many "buzz-words" in the academic community, especially with regard to how we understand processes of lasting social change. Indeed, although there is no formal definition of "sustainability," it remains popular in various political, social, and economic discourses, particularly those of environmental groups as a call to action to raise awareness around the current depletion of finite natural resources (for recent overviews, see Senaratne 2017; Servaes 2013, 2017).

*Sustainable Development* is seen as a means of enhancing decision-making so that it provides a more comprehensive assessment of the many multidimensional problems society faces. What is required is an evaluation framework for categorizing programs, projects, policies, and/or decisions as having sustainability potential.

The word is most often associated with being able to meet the needs of the present (socially, economically, environmentally), without compromising the ability of future generations to meet their needs (World Commission on Environment and Development 1987).

Phra Dhammapidhok (Payutto 1998) points out that sustainable development in a Western perspective lacks the human development dimension. He states that the Western ideology emphasizes "competition." Therefore the concept of "compromising" is used in the WCED definition. Compromising means lessen the needs of all parties. If the other parties do not want to compromise, you have to compromise your own needs and that will lead to frustration. Development won't be sustained if people are not happy.

He consequently reaches the conclusion that the western perception of and road to sustainability, based on Western ethics, leads development into a *cul-de-sac*.

From a Buddhist perspective, sustainability concerns ecology, economy, and *evolvability*. The concept "evolvability" means the potential of human beings to develop themselves into less selfish persons. The main core of sustainable development is to encourage and convince human beings to live in harmony with their environment, not to control nor destroy it. If humans have been socialized correctly, they will express the correct attitude towards nature and the environment and act accordingly. He argues that: "A correct relation system of developed mankind is the acceptance of the fact that human-being is part of the existence of nature and relates to its ecology. Human-being should develop itself to have a higher capacity to help his fellows and other species in the natural domain; to live in a harmonious way and lessen exploitations in order to contribute to a happier world" (Payutto 1998: 189).

This holistic approach of human relates to cultural development in three dimensions:

- Behaviors and lifestyles which do not harm nature.
- Minds in line with (Eastern) ethics, stability of mind, motivation, etc., to see other creatures as companions.
- Wisdom includes knowledge and understanding, attitude, norms, and values in order to live in harmony with nature.

Four dimensions are generally recognized as the "pillars" of sustainable development: economic, environmental, social, and cultural. "The essence of sustainability therefore, is to take the contextual features of economy, society, and environment – the uncertainty, the multiple competing values, and the distrust among various interest groups – as givens and go on to design a process that guides concerned groups to seek out and ask the right questions as a preventative approach to environmentally and socially regrettable undertakings" (Flint 2007: IV).

Recently, the term resiliency has been used by a variety of researchers, policy makers, and community organizers as a more relevant supplant to the term sustainability in the context of development and social change. Hopkins (2008: 54) defines resilience as "the capacity of a system to absorb disturbance and reorganize while undergoing change, so as to retain essentially the same function, structure, identity and feedbacks." According to Zolli & Healy (2012: 10), a "truly resilient system is able to ensure continuity by dynamically reorganizing both the way it serves its purpose and the scale at which it operates." Thus, strategies of resilience might be developed for economic, social, and ecologic systems. There is no doubt that current global problems such as climate change and the economic crises make the call for "sustainability" and "resiliency" from different groups more urgent.

## 1.9  Communication for Development and Social Change

In essence, the coherence of Communication for Development and Social Change is expressed in its different common underlying premises. In all its diversity, such as in theory, method, medium, and region, it is characterized by a number of underlying common values and starting-points. These premises, values, and starting-points are (Servaes 2007):

- *The use of a culturalist viewpoint*
  By means of such a viewpoint, specific attention is given to communication in social change processes. By putting culture centrally from a user's perspective, other social science disciplines can significantly contribute to the field of Communication for Development and Social Change.
- *The use of an interpretative perspective*
  Participation, dialogue and an active vision of human beings as the interpreters of their environments are of the utmost importance. A highly considered value is the showing of respect and appreciation for the uniqueness of specific situations and identities in social change environments.
- *The use of integrated methods and theories*
  In the field of Communication for Development and Social Change, it is considered important that the chosen methods should be connected with the used theoretical perspective. This implies that openness, diversity, and flexibility in methods and techniques are valued. In practice it generally means triangulation and a preference for qualitative methods. This does not mean, however, that quantitative methods are excluded, and indeed an emphasis is placed on evidence-based scientific methodologies.
- *To show mutual understanding and attach importance to formal and informal intercultural teaching, training, and research*
  Tolerance, consciousness-raising, acceptance, and respect can only be arrived at when members of different cultures not only hear but also understand each other. This mutual understanding is a condition for development and social change. In order to prevent all forms of miscommunication, intercultural awareness, capacity building, and dialogue is deemed very important.

Therefore, a short-hand definition of Communication for Development and Social Change could read as follows:

*Communication for development and social change is the nurturing of knowledge aimed at creating a consensus for action that takes into account the interests, needs and capacities of all concerned. It is thus a social process, which has as its ultimate objective sustainable development/change at distinct levels of society.*
*Communication media and ICTs are important tools in achieving social change but their use is not an end in itself. Interpersonal communication, traditional and group media must also play a fundamental role.*

## References

Anderson B (1983) Imagined communities: reflections on the origin and spread of nationalism. Verso, London
Campbell J (1988) The power of myths. Doubleday, Garden City
Chalkley A (1968) A manual of development journalism. Press Foundation of Asia, Manila
Childers E, Vajrathon M (1968) Development support communication for project support. Paper. UNDP, Bangkok
FAO (1984) Expert consultation on communication for development. FAO, Rome
Flint W (2007) Sustainability manifesto. Exploring sustainability: getting inside the concept. http://www.eeeee.net/sd_manifesto.htm. Accessed 21 July 2015
Fraser C, Restrepo-Estrada S (1998) Communicating for development. Human change for survival. I.B. Tauris, London
Fraser C, Villet J (1994) Communication – a key to human development. FAO, Rome
Hopkins R (2008) The transition handbook: from oil dependency to local resilience. UK, Cambridge: Green Books
Jamias J (ed) (1975) Readings in development communication. University of the Philippines, Los Baños
Kroeber A, Kluckhohn C (1952) Culture. A critical review of concepts and definitions. Vintage Books, New York
Mair L (1984) Anthropology and development. Macmillan, London
Manyozo L (2006) Manifesto for development communication: Nora Quebral and the Los Banos school of development communication. Asian J Commun 16(1):79–99
Mowlana H (1986) Development. A field in search of itself. American University, Washington, DC
Payutto P (1998) Sustainable development. Buddhadham Foundation, Bangkok
Quebral NC (1971) Development communication in the agricultural context. In: Search of breakthroughs in agricultural development. University of the Philippines, Los Baños, December 9–10
Quebral NC (1988) Development communication. University of the Philippines, Los Baños
Rogers EM (1976) Communication and development: critical perspectives. Sage, Beverly Hills
Samovar LA, Porter RE (1995) Communication between cultures. Wadsworth, Belmont
Senaratne M (2017) The transition from MDGs to SDGs: rethinking buzzwords. In: Servaes J (ed) Sustainable development goals in the Asian context, communication, culture and change in Asia. Springer Nature, Singapore
Servaes J (1999) Communication for development: one world, multiple cultures. Hampton, Cresskill
Servaes J (ed) (2007) Communication for development. Making a difference – a WCCD background study. In: World congress on communication for development: lessons, challenges and the way forward. World Bank, Washington, DC, pp 209–292
Servaes J (ed) (2008) Communication for development and social change. Sage, London

Servaes J (2011) Social change. Oxford Bibliographies Online (OBO), Oxford University Press, New York, 58 pp. http://www.oxfordbibliographiesonline.com/display/id/obo-9780199756841-0063
Servaes J (ed) (2013) Sustainability, participation and culture in communication. Theory and praxis. Intellect-University of Chicago Press, Bristol/Chicago
Servaes, J (ed.) (2017) The Sustainable Development Goals in an Asian context. Singapore: Springer, 174pp. ISBN 978-981-10-2814-4 http://www.springer.com/in/book/9789811028144
Therborn G (1980) The ideology of power and the power of ideology. Verso, London
Thomas S (1982) Some problems of the paradigm in communication theory. In: Whitney DC, Wartella E (eds) Mass Communication Review Yearbook 3. Sage, London
Thompson M, Ellis R, Wildavsky A (1990) Cultural theory. Westview Press, Boulder
White R (1988) Media, politics and democracy in the developing world. Center for the Study of Communication and Culture, London, April
Williams R (1981) Culture. Fontana, Glasgow
World Bank (2006) Global monitoring report 2006. World Bank, Washington, DC
World Commission on Environment and Development (1987) Our common future. Published as annex to general assembly document A/42/427, development and international co-operation: environment. UN, New York
Zolli A, Healy M (2012) Resilience: why things bounce back. Free Press, New York

# Communication for Development and Social Change: In Search of a New Paradigm

Jan Servaes

## Contents

| | | |
|---|---|---|
| 2.1 | Three Paradigms on Communication for Development | 17 |
| 2.2 | Two Communication Models | 19 |
| 2.3 | The Policy Options of the Three Paradigms | 20 |
| 2.4 | Globalization and Localization | 22 |
| 2.5 | The Right to Communicate | 22 |
| 2.6 | The Constraints of the Framework Under Study | 23 |
| 2.7 | Conclusion | 25 |
| References | | 26 |

### Abstract

This chapter outlines three development paradigms which are still being applied in theory and praxis: modernization, dependency, and multiplicity. It then argues that these paradigms determine the way two dominant communication models have emerged. Each development paradigm and communication model leads to divergent policy options at different levels of societal change: local, national, international, global.

The right to communication is presented as a translation of the communication policy of the multiplicity paradigm. This principle, as a fundamental human right, clearly indicates that another communication model necessitates democratization and thus a redistribution of the power on all levels. The point of departure is not an elitist position but development from the grass roots.

---

J. Servaes (✉)
Department of Media and Communication, City University of Hong Kong, Hong Kong, Kowloon, Hong Kong

Katholieke Universiteit Leuven, Leuven, Belgium
e-mail: jan.servaes@kuleuven.be; 9freenet9@gmail.com

**Keywords**

Modernization · Dependency · Multiplicity · Diffusion model · Participatory communication · Policies · Right to communicate

Cultural and communication dimensions of development have long been given short shrift: only in the last 40 years or so has it been realized that culture and communication could well have a fundamental impact on the entire question of development or social change.

It is generally agreed that thinking about development and communication as a distinct discipline emerged after the Second World War. Since that time this new field has become increasingly complicated, mainly due to changes in the societal and world system on one hand and the interdisciplinary nature of development and communication studies on the other hand.

Besides this interdisciplinary character of doing communication for development, Bjorn Hettne (1990) reports three more reasons for its complexity. First, more than conventional economics, sociology, or political science, development theory is concerned with structural change. Second, development theory has been closely related to development strategy in order to find solutions. And, third, development theory is more explicitly normative than social sciences in general. In our opinion, the same is true for communication theory analyzed at the macro level. That is the level of communication as a social function in society.

At the risk of oversimplification, we would like to take Karl Erik Rosengren's typology (1981, 2000: 57) as our starting point. Rosengren returns to the old but still basic question of the relationship between culture and social structure. He distinguishes between two extremes: "social structure influences culture" versus "culture influences social structure." For our purposes, we read social structure as social change or development and culture as communication (both software and hardware, thus including technology). Next, Rosengren puts these basic propositions in a cross-tabulation in order to classify the four main types of the relationship: cultural autonomy, culture creates social structure (idealism), social structure creates culture (materialism), and the interdependence or mutual influence between both (cf. Table 1).

Far too often the discussion on the relationship between development and communication concerns mainly the axis materialism-idealism. But, the focus of attention has somewhat shifted more in favor of interdependence-autonomy and from a holistic, overall perspective to a more differentiated one.

As we will elaborate later, most of the communication approaches toward modernization can be placed under the heading of idealism in Rosengren's typology. According to the modernization theory, the developed world played an important role in modernizing and facilitating economic development in the developing world. The mass media were seen as being instrumental in achieving this goal, as it was believed that media messages had a powerful impact in these underdeveloped societies.

**Table 1** Types of relations between culture (communication) and society (development)

|  |  | Development influences communication | |
|---|---|---|---|
|  |  | YES | NO |
| Communication influences development | YES | Interdependence | Idealism |
|  | NO | Materialism | Autonomy |

## 2.1 Three Paradigms on Communication for Development

Most authors who are concerned with the problem of development set two views, schools, or paradigms next to or in opposition to each other: "modernization and growth" versus "dependency and underdevelopment." While the first paradigm can be considered the oldest and most deeply rooted view in Western thought, the dependency theory came from Latin America. However, since paradigms in the social sciences tend to build on each other rather than to reject each other radically, a new vision has emerged over the last few decades. It focuses on the elements neglected by the previous paradigms. Therefore, it can be put forward that social sciences are in a pluri-paradigmatic state, thus emphasizing that the multiplicity of viewpoints and associated methodological approaches are the norm rather than the exemption. We have described this new paradigm as another and multidimensional development or in short the "multiplicity" paradigm (Servaes 1999, 2008).

During the late 1940s and 1950s, most development thinkers stated that the problem of underdevelopment or "backwardness" could be solved by a more or less mechanical application of the economic and political system in the West to countries in the Third World, under the assumption that the differences were of degree and not of kind. This mainly economic-oriented view, characterized by endogenism and evolutionism that occurred in the industrialized countries, ultimately resulted in the modernization and growth theory. Development in a given country can be stimulated by importing factors – capital, technology, expertise, and skills – from industrialized nations and by imposing internal measures that support modernized sectors and modernize traditional sectors.

Development in the modernization paradigm is seen as inevitable. It is a spontaneous and an irreversible process inherent in every single society. This need for a more global perspective has been strengthened as the present worldwide crisis shows the degree to which the world economy has become a reality. Therefore, the need for a more contextual analysis becomes apparent. Progressing through discernible, predetermined stages assessed primarily by industrial base and economic criteria, development of this genre implies structural differentiation and functional specialization – capital-centered modes of production. The concept is defined by and for those nations that are developed. And it was also accepted by the elites in developing countries.

In a word, development is the state of being or becoming "westernized." The process of development can be divided into distinct stages showing the level of development achieved by each society. Whereas the cause of underdevelopment in this paradigm lies in the "backwardness" of the developing society, the stimulus to change and development or growth is exogenous. Hence, culture and social structure, the essence of society, are seen as the primary impediments to its "progress."

As a result of the general intellectual "revolution" that took place in the mid-1960s, this Euro- or ethnocentric view on development, which is generally referred to as modernization, was challenged by Latin American social scientists, and a theory dealing with dependency and underdevelopment was born. This dependency approach formed part of a general structuralistic reorientation in the social sciences. The so-called dependistas were primarily concerned with the effects of dependency in peripheral countries, but implicit in their analysis was the idea that development and underdevelopment must be understood in the context of the world system.

The most basic criticism of the modernization paradigm was that it did not work. Rather than underdevelopment being a stage on the road to development, the dependistas saw them as two sides of the same coin; an underdeveloped periphery was prerequisite for the existence of the developed center. Due to the fact that the periphery is deprived of its surplus, development in the center implies underdevelopment in the periphery.

Therefore, the process of development is analyzed in terms of relations between regions, central and peripheral. Dependency theory asserts that the causes of underdevelopment are external to the underdeveloped nation, that is, global economic structures, while the cure could be found in the nation itself, chiefly through economic and cultural dissociation from world markets and self-reliance.

The dependency paradigm, both in theory and policy, focused almost exclusively on analysis at the international level and bypassed analogous relations of exploitation and inequity intra-nationally, serving to enhance the positions of the many Third World elites handsomely, as well as failing to prescribe any concrete solution.

Both the modernization and dependency paradigms employ an overwhelmingly economic, technological framework of analysis, ignoring cultural, aesthetic, environmental, or other more holistic considerations. Dependency theory is much more a description of structures than a prescription for a sustainable growth or significant social change.

Contrary to the more economic- and political-oriented views of the modernization and dependency theories, the central idea in the multiplicity paradigm is that there is no universal path to development and that development must be conceived as an integral, multidimensional, and dialectic process which can differ from one country to another. In other words, every society must define development for itself and find its own strategy.

At the same time, this also implies that the problem of development is relative. Therefore, according to this paradigm, no part of the world can claim to be developed in all respects. Furthermore, the discussion on the degree and scope of inter(in) dependence is connected with the content of development. According to this view, "another" development could be defined as need-oriented, endogenous, self-reliant, and ecologically sound and based on participatory democracy and structural transformations.

All nations, in one way or the other, are dependent upon one another. Consequently, internal and external factors inevitably influence the development process. Development has to be studied in a global context, in which both center and periphery, as well as their interrelated subdivisions, have to be taken into consideration.

More attention is also being paid to the content of development, which implies a more normative approach. Another development questions whether "developed" countries are in fact developed and whether this genre of progress is sustainable or desirable. It favors a multiplicity of approaches based on the context and the basic, felt needs and the empowerment of the most oppressed sectors of various societies at divergent levels. A main thesis is that change must be structural and occur at multiple levels in order to achieve these ends.

In order to provide a brief overview of the historical development of these three major development theories, we can identify the following number of shifts in scientific thinking:

1. From a more positivistic, quantitative, and comparative approach to a normative, qualitative, and structural approach
2. From a universal, formally descriptive model to a more substantial, change-oriented, and less predictable model
3. From a Euro- or ethnocentric view to an indigenistic view and then to a contextual and polycentric view
4. From endogenism to exogenism and then to globalism and localism
5. From an economic interest to more universal and interdisciplinary interests
6. From a primarily national frame of reference to an international perspective and then to combined levels of analysis
7. From segmentary to holistic approaches and then to more problem-oriented approaches
8. From an integrative and reformist strategy to revolutionary options and then to an integral vision on revolutionary and evolutionary change

## 2.2 Two Communication Models

In line with linear models of communication formulated in the West, communication was understood as a straight line, from the developed "source" to the underdeveloped "receiver." Congruent with the modernization philosophy, the diffusion and development support communication approaches tend to assign responsibility for the problem of underdevelopment to peoples residing in those societies.

Development as modernization and communication as one-way persuasion reached their zenith through the diffusion of innovations, the two-step flow, and other "social marketing" strategies of attitude and behavior change directed at "underdeveloped" peoples. Mass media play the preeminent role in the campaign of development through communication, and early predictions were of great effects.

Bi-directional models and strategies such as feedback were added to render the initial message more effective.

Contrary to the top-down communication model, the participatory model sees people as the controlling actors or participants for development. Development means lifting up the spirits of a local community to take pride in its own culture, intellect, and environment. Development aims to educate and stimulate people to be active in self and communal improvements while maintaining a balanced ecology. Authentic participation, though widely espoused in the literature, is not in everyone's interest. Such programs are not easily implemented, highly predictable, or readily controlled. People will have self-appreciation instead of self-depreciation. Development is meant to liberate and emancipate people. Local culture is respected.

The participatory model emphasizes on the local community rather than the nation state, on universalism rather than nationalism, on spiritualism rather than secular humanism, on dialogue rather than monologue, and on emancipation rather than alienation.

Participation involves the redistribution of power. Participation aims at redistributing the elites' power so that a community can become a full-fledged democratic one. As such, it directly threatens those whose position and/or very existence depends upon power and its exercise over others. Reactions to such threats are sometimes overt, but most often are manifested as less visible, yet steady and continuous resistance.

## 2.3 The Policy Options of the Three Paradigms

The principle of free flow of information can be considered the communication policy application of the modernization paradigm. After a fascistic and authoritarian period of war, it took little effort in 1945 for the Free West, led by the United States, to have this principle accepted as a universal value within the United Nations. This principle of freedom was initially interpreted in a rather individualistic and liberal manner and formed the basis of the so-called free press theory that still determines the international communication policy of many Western government and communication transnationals, as well as of the Third World elites oriented to the West. It took considerable time before this extreme liberal vision would be provided with a more social explanation particularly in the so-called social responsibility or social-liberal theory. The major difference between the free press and social responsibility theory is related to the question whether the freedom of information principle can or should only be guaranteed by private competition or also by public authorities and institutionalized groups of media workers and media consumers. In everyday life one observes that publishers, both in Western and Southern countries, advocate extreme liberal interpretations of the above principle. Governments, on the other hand, tend to take more ambivalent and varied positions, which are more in line with the latter interpretations of the free flow principle. However, one observes differences between Western European and US governmental policies. While the former attempts to regulate the market, the United States is, at least theoretically, in favor of total liberalization and deregulation policies.

With the development of the dependency viewpoint, this free and unhindered flow of information, grafted into the free flow doctrine, was challenged. From the Third World, arguments were put forth for a free and balanced flow of information, a principle that was backed particularly by the non-aligned nations in the debate on the New International Information Order (Nordenstreng and Schiller 1993). At the same time, it was contended that this free and balanced flow could be better guaranteed and organized by governments than by private enterprises. These viewpoints are explicitly formulated in the so-called development media theory (McQuail 2005) and are also implicitly there in social-authoritarian and social-centralistic philosophies (Gordon and Merrill 1988).

This position was strongly contested by the defenders of a free press, who charged that it could lead to governmental censure and curbs on the press. And, indeed, in reality, this did often appear to be the case. In Latin America, the mother continent of the dependency paradigm, the process of capitalistic state intervention brought authoritarian, initially military governments into power that tried to centralize the decision-making and opinion formation. These governments control the production and distribution of communication and used the media for their own legitimation purposes.

These countries have a contradictory communication policy. Abroad, they support a free and balanced flow of information, while they do as much as they can to keep it under control within their own borders. The same applies, moreover, for many Western nations. The US government, for example, supports free export of American communication products but tries to stop the import of such products from abroad as much as possible by protectionist measures. Even Ithiel de Sola Pool (1983), one of the fierce propagandists of the free flow principles, had to admit that "in rhetoric, the United States government favors diversity of voices and seeks to break up communications monopolies. The reality, however, is more ambiguous" (Pool 1983: 241).

In general, the governments of these countries hold rather contradictory views with regard to external versus internal communication policy principles. They support the demands for an expansion of the free and balanced flow between and among countries, but not within the borders of their individual nations. Therefore, the policy options of the two paradigms have one fundamental trait in common: they are elitist in the sense that they only want to increase the power of their respective elites and certainly do not strive to achieve universal social development. While the modernization paradigm legitimates the interests of Western political and economic interest groups and their "bridgeheads" in the Global South, the dependency theory meets the economic and political needs of those Global South elites who want to play an autonomous role. While the first group thus strives for international integration, the second group wants to turn back the international dependency relationship by means of a radical and dissociative policy. In both cases, however, little is done to alter the internal power relationships and dependency structures. We can therefore distinguish between an international or extern-national and intra-national or domestic policy level (see Table 2).

**Table 2** Policy options of the three paradigms

|  | Modernization | Dependency | Multiplicity |
|---|---|---|---|
| External (International) | Integration | Dissociation | Selective participation |
| Internal (Domestic) | Integration | Integration | Dissociation |

## 2.4 Globalization and Localization

Globalization has not fundamentally altered the above situation. It is fair to say that communication has been facing new challenges in the last decade, as a consequence of globalization, media liberalization, rapid economic and social change, and the emergence of new information and communication technologies (ICTs). Liberalization has led not only to greater media freedom but also to the emergence of an increasingly consumer-led and urban-centered communication infrastructure, which is less and less interested in the concerns of the poor and rural people. Women and other vulnerable groups continue to experience marginalization and lack of access to communication resources of all kinds (Brown 2016; Hafez 2007; Sparks 2007; Servaes and Oyedemi 2016; Scholte 2005).

The issue of ensuring access to information and the right to communication as a precondition for empowering marginalized groups has been addressed by several meetings and international conferences (for instance, at the World Summit on the Information Society in 2005). The debates in the general field of international and intercultural communication have shifted and broadened. They have shifted in the sense that they are now focusing on issues related to "global culture," "local culture," "(post)modernity," and "multiculturalism" instead of their previous concern with "modernization," "synchronization," and "cultural imperialism." With these "new" discussions, the debates have also shifted from an emphasis on homogeneity toward an *emphasis on differences* (De Cuellar 1995; Lie 2003; Servaes 2015).

Diversity not only exists *between* cultures but also *within* cultures. All cultures are plural, creole, hybrid, and multicultural from within. There are no (more) authentic, pure, traditional, and isolated cultures in the world, if they ever existed at all (Hopper 2007). In relation to the media, this means that perspectives need to be no longer monistic but from a multiplicity approach, and media have to be looked at in combined ways, such as in multimedia approaches. Within this multiplicity perspective, then, community media need to be appreciated as a major form of media from a people's perspective for development (Carpentier et al. 2012).

## 2.5 The Right to Communicate

The right to communication is offered as a translation of the communication policy of the multiplicity paradigm. This principle, as a fundamental human right, clearly indicates that another communication model necessitates democratization and thus a redistribution of the power on all levels. The point of departure is not an elitist position but development from the grass roots. Even the well-known MacBride

**Table 3** A comparison between a dominance and a pluralism model

|  | Dominance Top to bottom | Pluralism Bottom to top |
| --- | --- | --- |
| Societal source | Ruling class or dominant elite | Competing political, socio-cultural interest groups |
| Media | Under concentrated ownership and of uniform type | Many and independent of each other |
| Structure | Hierarchical, bureaucratic | Horizontal, egalitarian |
| Production | Standardized, routinized, controlled, synchronic | Creative, free, original, free, diachronic |
| Content and ideology | Selective and coherent, decided from "above" | Diverse and competing views, decided from "below" |
| Audience | Dependent, passive, organized on large scale | Fragmented, selective, reactive, and active |
| Effects | Strong and confirmative of established social order | Numerous, without consistency or predictability of direction |

report suggests that the right to communicate "promises to advance the democratization of communication on all levels – international, national, local, individual" (MacBride 1980: 171).

Fundamental here is the other vision of the role of the authorities in processes of social change. Unlike the confidence in and respect for the role of the state, which is characteristic of the modernization and dependency paradigms, the multiplicity paradigm has a rather reserved attitude toward the authorities. Policies should be built on more selective participation strategies of dissociation and association (Servaes 1999). Communication rights are closely related to human rights (Hamelink 2004; Servaes 2017).

The points of difference between the two options and the ideal-typical political consequences following from them are set side by side in Table 3, which makes a political division between a dominant and a pluralistic model (adapted from McQuail 2005: 88). Along with this "political" distinction goes an implicit distinction between two models of the process by which the top of society is related, by way of mass communication, to the various groups at the bottom. This dominance model could stand for the "elitist" aims of both the modernization and dependency paradigms, while the pluralist model is more suitable for the "basic" needs of the third perspective.

These two models have fundamentally different policy and planning implications. It makes a great difference whether policy priorities and planning projects are designed and worked out from the perspective of the top or of the grass roots. This is shown by Table 4 (adapted from Tehranian 1979: 123), which places the two positions next to each other for a number of development and communication objectives.

## 2.6 The Constraints of the Framework Under Study

First, the chronological approach applied here has a certain bias in creating the impression that later theoretical contributions replace earlier ones. Although new innovations often were stimulated by shortcomings of previous theorizing on the

**Table 4** Development and communication objectives viewed from a top versus bottom perspective

| Top perspective | Bottom perspective |
| --- | --- |
| National power and security | Individual choice and freedom |
| Social mobilization | Social mobility |
| National identity, integration, and unity | Sub-national (ethnic or group interest) identity and solidarity |
| Economic growth (rise in national income) | Income distribution and social justice |
| Political socialization | Political participation |
| Property and business rights | Public and consumer rights |
| Educational and professional advance | Educational and professional opportunities |
| Information control | Information access |
| Majority rule (where there is electoral democracy) | Minority rights |
| Central control and direction | Regional and local autonomy |
| Cultural and artistic direction (sometimes censorship) | Cultural and artistic creativity (sometimes subversive) |
| Ideological and cultural control | Intellectual and artistic freedom |
| Mass media centered (one-way communication) | Multiple communication use (social media and interpersonal communication) |
| Emphasis on solutions | Emphasis on questions and problems |
| Product oriented | Process oriented |

one hand and according to Kuhn (1962) theoretical changes in the social sciences are different from positive sciences on the other hand, these three paradigms still do find support among academics, policymakers, as well as the general public. Both the modernization and dependency paradigms today still have their proponents and opponents and are thus used in specific circles, by policymakers and by the public at large. In general, in academic circles, the modernization vision was dominant until the second half of the 1960s. Nowadays, this dominance remains discerned in Western policymaking institutions and public opinion. The majority of the Western public opinion sticks to the ethnocentric and paternalistic views which are present in the modernization theory. Because in the social sciences "paradigms" tend to accumulate rather than to replace each other, we conceive paradigms as "frames of meaning" which are mediated by others: "The process of learning a paradigm or language-game as the expression of a form of life is also a process of learning what that paradigm is not : that is to say, learning to mediate it with other, rejected, alternatives, by contrast to which the claims of the paradigm in question are clarified. The process is itself often embroiled in the struggles over interpretation which result from the internal fragmentation of frames of meaning, and from the fragility of the boundaries that separate what is 'internal' to the frame from what is 'external' to it, that is, belongs to discrete or rival meaning frames" (Giddens 1976: 144). Therefore, Kuhn (1970: 172) puts it nicely that a paradigm is what the members of a scientific community share, and, by the same token, a scientific community consists of men and women who share a paradigm.

Consequently, in each time period, one can make a distinction between a dominant and an alternative perspective on development on the one hand, and, on the other hand, each paradigm can be further subdivided in what can be called its "mainstream" of thinking and its "counterpoints." As already has been pointed out, the modernization paradigm continues to be influential at the political level although it no longer enjoys the widespread theoretical endorsement which it did up until the mid-1960s. However, broadly speaking, the beliefs upon which it is built – that is, economic growth, centralized planning, and the explanation for the state of underdevelopment as sought in chiefly internal causes which can be solved by external (technological) "aid" – are still shared by many governments, development agencies associated with the United Nations, the World Bank and so on, and, of course, transnational companies. Also public opinion is sensitive to the paternalism with respect to non-Western cultures that characterize the modernization theory.

Another observation concerns the evolution in thinking of certain theorists and researchers. As has been argued already, paradigms in the social sciences build on one another rather than breaking fundamentally with previous theories. As a consequence, many individual thinkers deepen and widen their views in an evolutionary, sometimes dialectic way. So it might occur that their earlier work can be viewed as an eloquent evidence of the modernization or dependency theory, while their later publications are more in line with the multiplicity paradigm.

However, a related and less likely aspect, which can be considered to be one of the major flaws in many well-intended policy recommendations, concerns the mixing of these paradigms in a rather contradictory fashion. Very often reports of academics and policymakers are heaped with all sorts of fashionable recommendations. This is, for instance, the case with much of the technology debate.

Another related topic deals with the problem of power and the legitimization of power relationships. Each social order can be characterized by an interrelated division between an (economic) base and an (ideological and symbolic) superstructure. According to Pierre Bourdieu (1979, 1980), call the dominant classes upon an ideological and symbolic preponderance not only to maintain their position in the social hierarchy but also to justify it. This symbolic system has a "symbolic power" because it is capable of construing reality in a directed manner. Its symbolic power does not lie in the symbolic system itself but in the social relationships between those who exercise the power and those who are subject to it. Symbolic power functions mainly "unconsciously" as the legitimation criterion for the existing social and economic power relationships and creates "myths" and "ways of life." So, in reality, not only normative but also, and especially so, power factors play a role in policy and planning making, especially when it comes to confirming and carrying out policy recommendations.

## 2.7 Conclusion

There have historically been two dominant paradigms informing the communication for development sector. In recent decades, a third paradigm emerged: multiplicity or one world, multiple cultures. The paradigm shifts can be observed among individual scientists as well as in particular research projects and publications. Despite some reorientations and shifts, older paradigms have not "vanished" as theories, and in fact there are still quite vocal applications of these.

The shifts in the area of scholarly research are also visible in policy and planning making. Policy and planning making is based on a number of assumptions, usually provided and supported by social-scientific research. These assumptions serve as a guidance for policymakers of various social sectors to justify their policy.

In the domain of the freedom of expression and the freedom of the press, one can observe a double evolution in the postwar period. Whereas originally the active right of the so-called sender-communicator to supply information without externally imposed restrictions was emphasized, nowadays the passive as well as active right of the receiver to be informed and to inform gets more attention. Therefore the principle of the right to communicate was introduced as it contains both the passive and active right of the receiver to inform and be informed. It is worth noting that development broadcasting, by applying a "from below" approach, plays an important role in the development communication process. We believe that it is the multiplicity paradigm which brings together some past approaches in order to better serve the need for communication for development. Adequate policies for the public to be well-informed must be put in place, but just as much attention must be given to the rights of those audiences to not just receive information but also to actively engage in the communication process.

## References

Bourdieu P (1979) La distinction: critique sociale du jugement. Éditions de Minuit, Paris
Bourdieu P (1980) Questions de sociologie. Minuit, Paris
Brown K (2016) Resilience, development and global change. Routledge, London
Carpentier N, Lie R, Servaes J (2012) Multitheoretical approaches to community media: capturing specificity and diversity. In: Fuller L (ed) The power of global community media. Palgrave Macmillan, New York, pp 219–236
De Cuellar JP (1995) Our creative diversity. Report of the World Commission on Culture and Development. UNESCO, Paris
Giddens A (1976) New rules of sociological method. Hutchinson, London
Gordon D, Merrill J (1988) A power theory of press freedom, Paper ICA conference, New Orleans, May
Hafez K (2007) The myth of media globalization. Polity Press, Cambridge, UK
Hamelink CJ (2004) Human rights for communicators. Hampton Press, Cresskill
Hettne B (1990) Development theory and the three worlds. Longman, New York
Hopper P (2007) Understanding cultural globalization. Polity, Cambridge, UK
Kuhn T (1962) The structure of scientific revolutions. University of Chicago Press, Chicago
Kuhn TS (1970) The Structure of Scientific Revolutions. Enlarged (2nd ed.). University of Chicago Press
Lie R (2003) Spaces of intercultural communication. Hampton Press, Cresskill
MacBride S (ed) (1980) Many voices, one world. Communication and society. Today and tomorrow. UNESCO, Paris
McQuail D (2005) McQuails mass communication theory, 5th edn. Sage, London
Nordenstreng K, Schiller HI (red) (1993) Beyond national sovereignty: international communication in the 1990s. Ablex, Norwood

Pool IDS (1983) Technologies of freedom. Belknap Press, Cambridge
Rosengren KE (1981) Mass media and social change: some current approaches. In: Katz E, Szecsko T (eds) Mass media and social change. Sage, Beverly Hills
Rosengren KE (2000) Communication. An introduction. Sage, London
Scholte J (2005) Globalization: a critical introduction. Palgrave, New York
Servaes J (1999) Communication for development. One world, multiple cultures. Hampton Press, Cresskill
Servaes J (ed) (2008) Communication for development and social change. Sage, London
Servaes J (2015) Studying the global from within the local. Commun Res Pract 1(3):242–250
Servaes J (2017) All human rights are local. The resiliency of social change. In: Tumber H, Waisbord S (eds) The Routledge companion to media and human rights. Routledge, London, pp 136–146
Servaes J, Oyedemi T (eds) (2016) Social inequalities, media and communication. Theory and roots. Lexington, Lanham
Sparks C (2007) Globalization, development and the mass media. Sage, London
Tehranian M (1979) Development theory and communication policy. The changing paradigm. In: Voigt M, Hanneman G (eds) Progress in communication sciences, vol 1. Ablex, Norwood

# Key Concepts, Disciplines, and Fields in Communication for Development and Social Change

## Jan Servaes and Rico Lie

## Contents

3.1  Introduction .................................................................... 30
3.2  Key Concepts and Practices ................................................... 32
    3.2.1  Normative Concepts ..................................................... 32
    3.2.2  Contextual Concepts .................................................... 36
    3.2.3  Strategies and Methodologies ........................................... 38
    3.2.4  Methods, Techniques, and Tools ........................................ 39
    3.2.5  Advocacy and Impact Assessment ....................................... 39
3.3  Sub-disciplines ................................................................. 40
    3.3.1  Strategic Communication and Participatory Communication ............. 41
    3.3.2  Crisis Communication and Risk Communication ......................... 42
    3.3.3  Journalism and International Communication ............................ 43
3.4  Thematic Sub-disciplines ...................................................... 43
    3.4.1  Health Communication .................................................. 44
    3.4.2  Agricultural Extension and Rural Communication ...................... 45
    3.4.3  Environmental Communication ......................................... 46
3.5  Fields and Areas ............................................................... 47
    3.5.1  Right to Communicate .................................................. 48
    3.5.2  Education and Learning ................................................. 48
    3.5.3  Innovation, Science, and Technology ................................... 49
    3.5.4  Natural Resource Management ......................................... 49
    3.5.5  Food Security ........................................................... 50
    3.5.6  Inequality and Poverty Reduction ...................................... 51
    3.5.7  Peace and Conflict ...................................................... 51

J. Servaes (✉)
Department of Media and Communication, City University of Hong Kong, Hong Kong, Kowloon, Hong Kong

Katholieke Universiteit Leuven, Leuven, Belgium
e-mail: jan.servaes@kuleuven.be; 9freenet9@gmail.com

R. Lie
Research Group Knowledge, Technology and Innovation, Wageningen University, Wageningen, The Netherlands
e-mail: Rico.Lie@wur.nl

© Springer Nature Singapore Pte Ltd. 2020
J. Servaes (ed.), *Handbook of Communication for Development and Social Change*,
https://doi.org/10.1007/978-981-15-2014-3_113

3.5.8 Children and Youth, Women, and Senior Citizens .............................. 52
3.5.9 Tourism ................................................................................ 52
3.6 Conclusions ................................................................................ 53
References ...................................................................................... 55

**Abstract**

We first identify five clusters of concepts and practices that are currently actively circulating and determine the activities and approaches in the field of Communication for Sustainable Development and Social Change: (1) a *normative* cluster of concepts; (2) a cluster of concepts that sets an important *context* for communication activities for development; (3) a cluster of *strategic and methodological* concepts; (4) a cluster of concepts that relate to *methods, techniques*, and *tools*; and (5) a cluster of concepts that addresses the practices of *advocacy, (participatory) monitoring and evaluation*, and *impact assessment*.

Then we examine *sub-disciplines* and *fields and areas* of aspects of Communication for Sustainable Development and Social Change. We make a distinction between *non-thematic sub-disciplines*, which cover a domain within communication science and *thematic sub-disciplines*, which cover a life science theme in the development sector. The thematic sub-disciplines are (a) *health communication*, (b) *agricultural extension and rural communication*, and (c) *environmental communication (including climate change communication)*.

The fields and areas that are discussed are (a) right to communicate; (b) *education and learning*; (c) *innovation, science*, and *technology*; (d) *natural resource management*; (e) *food security*; (f) *poverty reduction*; (g) *peace and conflict*; (h) *children and youth, women*, and *senior citizens*; and (i) *tourism*.

**Keywords**

Normative concepts · Context of development · Strategic and methodological concepts · Methods, techniques, and tools · Practices of advocacy, (participatory) monitoring and evaluation, and impact assessment · Health communication · Agricultural extension · Rural communication · Environmental communication · Climate change communication · Journalism · International communication

## 3.1 Introduction

Most of the available historical accounts on development communication are related to distinctively different paradigms: the modernization, the dependency, and the multiplicity paradigm (see, for instance, Melkote and Steeves 2001; Servaes 1999; Waisbord 2018). Development within these paradigms stemmed from the Western drive to develop the world after the colonization period and the Second World War. A rethink of the "power" of development within which communication plays an important role has been caused by a number of events: the collapse of the Soviet

Union that resulted in the United States as the only remaining "superpower" in the late 1980s, the birth of the European Union and the rising power of China; and the gradual coming to the foreground of the regional powers such as Brazil, Russian, India, and South Africa in the BRICS countries. The traditional ways of thinking about doing development as a copycat of the Western way are being challenged by a more cultural, religious, and ethnic base. The dichotomous way of looking at development as "developed" as the Western World and "underdeveloped" (or developing) has been diverted to pay more attention to the "hearts and minds" of the stakeholders.

In this chapter, which builds on Lie and Servaes (2015) and Servaes and Lie (2015), we take a different approach. It starts with the current existing situation in daily development practices and frames this situation from a communication science perspective. It first identifies five clusters of concepts and practices that are currently actively circulating and determining the activities and approaches in the field. These clusters are (1) a *normative* cluster of concepts (e.g., sustainability, democracy, participation, empowerment, equity, social inclusion, human rights, and accountability); (2) a cluster of concepts that sets an important *context* for communication activities for development (e.g., globalization, gender, social movements, cultural diversity, and the Sustainable Development Goals (SDGs)); (3) a cluster of *strategic and methodological* concepts (e.g., social marketing, entertainment-education, social mobilization, knowledge management, knowledge sharing, and co-creation); (4) a cluster of concepts that relate to *methods*, *techniques*, and *tools* (e.g., (digital) storytelling, the most significant change approach, community/village mapping, empowerment evaluation, mobile phones, social media, film, video, drama, and art); and (5) a cluster of concepts that addresses the practices of *advocacy, (participatory) monitoring and evaluation*, and *impact assessment*.

After having discussed these five clusters, we examine *sub-disciplines* and *fields and areas* of aspects of Communication for Development and Social Change. The sub-disciplines we discuss are more or less established sub-disciplines within the discipline of communication science and at the same time have established a community of interest within the field of Communication for Development and Social Change. We make a distinction between *non-thematic sub-disciplines*, which cover a domain within communication science (strategic communication, participatory communication, crisis communication, risk communication, journalism and international communication), and *thematic sub-disciplines*, which cover a life science theme in the development sector. The thematic sub-disciplines are (a) *health communication*, (b) *agricultural extension and rural communication*, and (c) *environmental communication (including climate change communication)*. Fields and areas are seen as currently of importance and concern within development studies and development interventions and have demonstrated an interest in communication. The fields and areas that are discussed are (a) right to communicate; (b) *education and learning*; (c) *innovation, science*, and *technology*; (d) *natural resource management*; (e) *food security*; (f) *poverty reduction*; (g) *peace and conflict*; (h) *children and youth, women*, and *senior citizens*; and (i) *tourism*. The sections that follow address the role of communication within these areas.

## 3.2 Key Concepts and Practices

This section makes a distinction between concepts: normative ones; contextual ones; strategic and methodological ones; ones that relate to methods, techniques, and tools; and concepts and practices that relate to advocacy, (participatory) monitoring and evaluation (PM&E), and impact assessment. Practices are seen as activities that use particular approaches. Normative concepts deal with good and bad, right and wrong, morality, ethics, norms, and values and make claims about how things should or ought to be. The discussed contextual concepts are important to better understand the role of the environment in which development communication interventions take place. Strategies, methodologies, methods, techniques, and tools are important to know how to design interventions in the field of communication for development. Strategies and methodologies are systematic ways of approaching complex problems. They state the philosophy and vision behind the way things are done and imply a set of systematic procedures. They include an epistemological and ontological positioning. Methods, techniques, and tools refer to the ways the strategies and methodologies are actually used. Methods employ techniques and thus include considerations of appropriate tools. Finally, advocacy is discussed as a communication practice that aims to bring about change in governance and policies. Monitoring and evaluation of projects is then discussed as related to assessing the impact of interventions.

### 3.2.1 Normative Concepts

*Sustainable development* has emerged as one of the most prominent development paradigms. In 1987, the World Commission on Environment and Development (WCED) – in short Brundtland Commission – concluded that "sustainable development is development that meets the needs of the present without compromising the ability of future generations to meet their own needs." Sustainable development is seen as a means of enhancing decision-making so that it provides a more comprehensive assessment of the many multidimensional problems society faces. What is required is an evaluation framework for categorizing programs, projects, policies, and/or decisions as having sustainability potential (Lennie and Tacchi 2013).

Four dimensions are generally recognized as the "pillars" of sustainable development: economic, environmental, social, and cultural. "The essence of sustainability therefore, is to take the contextual features of economy, society, and environment – the uncertainty, the multiple competing values, and the distrust among various interest groups – as givens and go on to design a process that guides concerned groups to seek out and ask the right questions as a preventative approach to environmentally and socially regrettable undertakings" (Flint 2007: IV).

In addition to the "Western" perspective represented by the Brundtland Commission, one could also refer to "Eastern" or Buddhist perspectives, as presented by the Thai philosophers and social critics Sulak Sivaraksa (2010) and Phra Dhammapidhok (Payutto 1998). They point out that sustainable development in a Western perspective lacks the human development dimension and that the Western ideology emphasizes "competition." Therefore the concept of "compromising" is used in the above WCED definition. Compromising means lessen the needs of all parties. If the other parties do not want to compromise, you have to compromise your own needs, and that will lead to frustration. Development won't be sustained if people are not happy.

From a Buddhist perspective, sustainability concerns ecology, economy, and *evolvability*. The concept "evolvability" means the potential of human beings to develop themselves into less selfish persons. The main core of sustainable development is to encourage and convince human beings to live in harmony with their environment, not to control or destroy it. If humans have been socialized correctly, they will express the correct attitude toward nature and the environment and act accordingly. Therefore, Payutto states that: "A correct relation system of developed mankind is the acceptance of the fact that human-being is part of the existence of nature and relates to its ecology. Human-being should develop itself to have a higher capacity to help his fellows and other species in the natural domain; to live in a harmonious way and lessen exploitations in order to contribute to a happier world" (Payutto 1998: 189).

Different perspectives (such as the TERMS approach developed in Thailand that builds on Buddhist principles and the "efficiency economy" concept outlined by the late King Bhumibol – see Servaes and Malikhao 2007; Malikhao 2019) have, over the years, influenced this holistic and integrated vision of sustainable development as a viable way to sustainable development and happiness. Sivaraksa (2010: 66) lists the following indicators of happiness:

- The degree of trust, social capital, cultural continuity, and social solidarity
- The general level of spiritual development and emotional intelligence
- The degree to which basic needs are satisfied
- Access to and the ability to benefit from health care and education
- The level of environmental integrity, including species loss or gain, pollution, and environmental degradation

It is relevant to emphasize that this "Eastern" perspective is not "uniquely" Eastern as it has been promoted in other parts of the world as well. For instance, in the late 1970s, the Dag Hammarskjold Foundation advocated three foundations for "another" or sustainable development: (a) Another Development is geared to the satisfaction of needs, beginning with the eradication of poverty; (b) Another Development is endogenous and self-reliant; and (c) Another Development is in harmony with the physical and cultural ecology (Nerfin 1977).

Also the World Commission on Culture and Development, chaired by Javier Pérez de Cuéllar (1995), started from similar assumptions. It stressed that "development," which is divorced from its human or cultural context, is like "growth without a soul." This means culture is as essential as economic growth when thinking about development. The report states that people's culture cannot be compelled by a government but the government is partly influenced by the culture (De Cuéllar 1995: 15).

The basic principle should be the fostering of respect for all cultures that are tolerant to others' values. Respect goes beyond tolerance, and that implies having a positive attitude toward and rejoicing in cultures. Social peace is necessary for human development; it requires that we should not regard differences among cultures as neither something alien, nor unacceptable, nor hateful. Rather, we should look at those differences as the experiments of ways of living together that we all gain valuable lessons and information from (De Cuéllar 1995: 25).

The Human Development Report 2004 and the *United Nations Millennium Declaration* (2000) advocate these principles of cultural liberty and cultural respect in today's diverse world for similar reasons: "The central issue in cultural liberty is the capability of people to live as they would choose, with adequate opportunity to consider other options" (UNDP 2004: 17).

Therefore, in contrast to the more economically and politically oriented approach in the traditional perspectives on sustainable development, the central idea in alternative, more culturally oriented versions is that there is *no universal development model* which leads to sustainability at all levels of society and the world. Development is an integral, multidimensional, and dialectic process that can differ from society to society, community to community, and context to context. In other words, each society and community must attempt to delineate its own strategy for sustainable development (Servaes 1999). This implies that development problems are relative and there is no single society that can proclaim it is "developed" in every aspect.

In other words, sustainable rural development is the kind of change that takes the current as well as the future generations of humanity and nature into account. It aims to improve the quality of life and the integrity of the environment for all. Communication plays a vital role in this kind of change (for more details, see Servaes 2013). Communication for Development and Social Change is therefore never neutral. Communication takes side with human development with respect to one another and for the environment in which we live. Taking such a normative stand illuminates some concepts as more valuable than others. These concepts are especially normative political ones: democracy, participation, human rights, empowerment, equity, accountability, leadership, champions, and resilience.

Thinking about *participation* is more balanced today as compared to the 1970s and 1980s when different development organizations started questioning existing theories and practices of development and demanded more participatory approaches to be introduced. Nowadays, not everything needs to be "bottom-up" anymore and not everybody needs to participate in everything all of the time. Participatory communication, participatory action research, participatory rural appraisals (PRA), participatory learning and action (PLA), participatory needs assessments, and participatory budgeting found their way into the mainstream of development activities, and we seem to have found a new balance, at least in theory. Nevertheless, participation remains one of the key concepts in development studies and interventions, and many other concepts relate in a direct or indirect way to participation (Carpentier). *Empowerment* and *giving people a voice* still address the de-marginalization of particular groups in society. In the context of democratization and civil society movements, scholars value *involvement of all stakeholders*, *engagement of stakeholders*, and *stakeholder analyses and consultations* (Wallerstein et al. 2017). They also started to address the concepts of equity and social inclusions. They gradually add (human) rights-based approaches on to the area of development and social change activities (Van Hemelrijck 2013). Rights-based approaches stress that everyone has the right to engage in decision-making processes that affect their own lives.

Much of the political economy and political science literature continues viewing *democracy* in one-dimensional terms, primarily in terms of political rights. This is particularly pronounced in the empirical literature, especially in the recent strand that seeks to identify the determinants of democracy. BenYishay and Betancourt (2013) expand on this view of democracy by incorporating the role of civil liberties, noting that these are conceptually at the core of modern democracy. They offer a conceptual framework that identifies five sources of potential differences in the evolution of political rights and civil liberties and investigate the empirical evidence on this differential evolution using cross-national panel data based on the Freedom House measures of political rights and civil liberties. Perhaps the most important policy implication of their analysis is that in promoting democracy, it makes sense to emphasize the provision of civil liberties. "Our empirical findings suggest that at least some civil liberties are necessary for political rights but the reverse is not the case. Free and fair elections are necessary for democracy, but they are far from sufficient. Part of their value is that by conferring legitimacy on the winners, they prevent the loss of life and property associated with other mechanisms for transferring power inter-temporally. By the same token, however, they provide no legitimacy for the trampling of civil liberties associated with illiberal democracies" (BenYishay and Betancourt 2013: 31).

*Accountability*, transparency, (corporate) social responsibility, accessibility, and efficiency are concepts that directly relate to how organizations should operate. The concept of accountability is relatively new to the field of Communication for Development and Social

Change. Being able to manifest accountability and social responsibility reflects good governance. In its basic form, accountability is about taking explicit responsibility for one's actions. Mutual accountability in partnerships is increasingly becoming the norm. Other related concepts that have recently entered the arena of development discourse are *leadership*, *champions*, and *resilience*. Centralizing the issue of leadership came with the changed view on participation. Leadership is different from management. It not only deals with executing tasks, it also addresses establishing direction, aligning resources, generating motivation, and providing inspiration (Kotter 2001). Leaders are change agents, but champions or "animators" are another particular type of change agents who have been given quite some consideration in change processes. Champions, animators, or facilitators often play a role in an early stage of transformation and in a context of resistance to change. They combine specific personality characteristics, behavior, knowledge, and power (see, for instance, Kennedy 2008; Quarry and Ramírez 2009). Resilience refers to the ability of people to absorb, adapt, and reorganize. It is about overcoming vulnerability to change and creating the potential for sustainability (Zolli and Healy 2012). Building resilient communities as a normative objective has now become a key issue in the field of Communication for Development and Social Change (Polk 2015).

### 3.2.2 Contextual Concepts

This section addresses contextual concepts that are part of the enabling environment in which Communication for Development and Social Change operates. Among these concepts are globalization, gender, social movements, cultural diversity, and the Sustainable Development Goals (SDG). These contextual concepts are actively shaping the circumstances that determine existence, direction, success, and impact of communication interventions in the development sector.

Communication for Development and Social Change is not an isolated process. It is embedded in different contexts, which highly influence the kind of development and change one would strive for. "Think globally, act locally" might be a bit outdated slogan as economic and political globalization still has major consequences for developing countries (see, for instance, Goldin and Reinert 2012). *Gender* also has an important place on the development agenda. Gender and globalization both address power and discourse (see, for instance, Benería et al. 2003). In the 1980s, feminism and modernization found each other and merged in what was called a women in development (WID) approach. Women and development (WAD) is and was grounded in the dependency paradigm. Today we talk about gender and development (GAD). Different from the other two approaches, the focus with GAD is not on women but on gender and thus on different roles, identities, discourses, responsibilities, and power positions between men and women in a specific sociocultural context. The UNESCO Global Forum on Media and Gender, held in Bangkok on 2–4 December 2013, took the first historic steps toward a gender-equal media (http://www.un.org/sg/statements/?nid=7318).

*Social movements* (and other civil society actors) set another context for Communication for Development and Social Change (Nash 2005). Social movements connect to social change in an active way. Examples of such movements are the anti-globalization movement, the fair trade movement, and the indigenous people movement. They set an important context for development and especially for communication in development, as these movements ventilate alternative voices. The way these movements behave, communicate, and organize themselves has changed somewhat with the arrival of the Internet and social media. The media and social media consumption has made people behave differently than before. Media audiences have more channels to respond and interact with the media organizations and one another more, as the media are being used to record artifacts and portray the society at a particular point in time. It is not wrong to say these new ways of media consumption have changed the culture in a particular society. In turn, culture determines the performance of communication and of development. Culture influences the way people act and communicate (and the other way around). *Cultural diversity* is one of the driving forces of development, and UNESCO sees it as a key dimension of sustainable development. "Cultural diversity must be seen as a cross-cutting dimension (rather than as a separate, fourth pillar of sustainability), with an important role to play in all development projects, from poverty eradication and the safeguarding of biodiversity to resource management and climate change" (UNESCO 2009: 189). As we have outlined earlier, the other three pillars of sustainability are economic viability, social responsiveness, and respect for the environment.

In the development sector, the *Millennium Development Goals* (MDGs), followed by the Sustainable Development Goals (SDGs) (Servaes 2017), provide the most influential guidelines for making decisions on interventions. Eight MDGs were drawn from the Millennium Declaration that was adopted by 189 nations and signed by 147 heads of state and governments during the UN Millennium Summit in September 2000. In September 2015, the 70th session of the United Nations General Assembly in New York set the post-2015 development agenda in the form of 17 SDGs and 169 associated targets, as successors to the MDGs. The aim was to overcome the compartmentalization of technical and policy work by promoting integrated approaches to the interconnected economic, social, and environmental challenges confronting the world.

This universal set of 17 goals, 169 targets, and 304 indicators was approved by all UN member states in the UN General Assembly on 25 September 2015. It is meant to be implemented for the next 15 years starting January 2016 (UN 2014; SDSN 2015).

The SDGs go further than the MDGs, which aside from having an increased set of agenda to work upon have more demanding targets (such as the elimination of poverty, instead of reducing its occurrence), as well as closely related and interdependent goals. In general, it is hoped that by 2030, poverty and hunger will be eliminated; quality of life will be greatly improved. All forms of capital are intact and functioning under ideal climatic situations. Lastly, peace and prosperity will be shared by all.

Implementing the SDGs follows the Global Partnership for Sustainable Development framework, as outlined in the outcome document of the 3rd International Conference on Financing for Development held in Addis Ababa, Ethiopia, on July 16, 2015 (UNSC 2015). Importantly, the numerous targets indicated in the SDGs require huge capital investments, not just for the SDGs as a whole but also for its individual components. On a global scale, crude estimates have put the cost of providing a social safety net to eradicate extreme poverty at about 66 billion USD a year, while annual investments in improving essential infrastructure could reach 7 trillion USD (Ford 2015).

The complexity and interconnectivity of the SDGs require the mobilization of global knowledge operating across many sectors and regions in order to identify frameworks and systems that would help realize the goals by 2030 (Sachs 2012). Specifically, effective accountability measures need to be in place, such as annual reviews of specific goals, monitoring systems which gather needed information to support and update conducted projects, and follow-up statuses of precursor programs essential for meeting a specific SDG as initiated by certain stakeholders (SDSN 2015). The current framework for achieving and monitoring SDGs which rely on triple bottom line parameters and statistics still needs to be improved through robust conceptual and methodological work (Hak et al. 2016).

### 3.2.3 Strategies and Methodologies

Communication strategies and methodologies that are currently used in the field of sustainable development interventions can be divided in two methodological schools. The first school has a strong position in many development circles. *Communication campaigns* are often used within this school to sensitize people and to influence attitudes and behavior, especially within the health sector and/or related to environmental issues. General marketing techniques are used to sell ideas and to raise awareness, increase knowledge, and influence attitudes. Social marketers often use the distinction between sender, message, and receiver to get a grip on the communication situation. Although sometimes not used in the classic "Shannon and Weaver" way (Weaver and Shannon 1963) that presumes direct linear effects, this school continues in this line of thinking and argues in terms of persuading target audiences to change their behavior. One of the objectives is that change within target populations can be measured by quantitative research methods and techniques. *Entertainment-education* is a much used communication strategy to achieve such behavior change. *Social (community) mobilization* goes a step further and aims to mobilize people into action. It differs from marketing in the sense that it involves the community and not solely aims for individual behavior change. It takes the social context into account while seeking diffusion and community ownership of ideas and innovations.

The second methodological school is dynamically constructed by practitioners and academics working in different sectors and disciplines of development and social change. This school explicitly centralizes the normative concepts presented above. Scholars working within this school attach importance to processes of participation, empowerment, equity, and democratization. This is not to say that the first school completely disregards these concepts. However, the first school does not take these processes as point of departure. Instead, scholars working in that line of thought centralize targeted behavior and attitude change. Within the second school, there are different platforms and communities of practice discussing various topics and producing different kinds of output. This field often uses the terms *knowledge management*, *knowledge sharing*, and *co-creation* in referring to the role of knowledge and interaction in communication processes. This field sometimes prefers to centralize the concept of knowledge in order to address the often existing knowledge gap between stakeholders and to connect to issues of governance and management in the context of the *knowledge economy* and *knowledge society*.

## 3.2.4 Methods, Techniques, and Tools

The first methodological school mentioned in the previous section uses traditional, long established, widely accepted, and often quantitative, measurable methods and techniques such as communication campaigns, social marketing techniques, surveys, broadcast media methods, and mass entertainment-education techniques in development communication interventions. The other school uses so-called creative methods and techniques such as (digital) storytelling and the most significant change approach. The methods and techniques that are often used in this second school are more qualitative and participatory in nature. Techniques such as appreciative inquiry, world café, open space, brown-bag lunch gatherings, knowledge fairs, and mind-mapping exercises are just a few among many others that could be mentioned. Tools that are often used here include film and video, drama, sport, and art. General tool kits for communication for development methods and techniques are widely available on the Internet (e.g., http://www.kstoolkit.org/; http://www.participatorymethods.org/).

Using *(digital) storytelling, film and (participatory) video*, but also increasingly *mobile phones* seems to be very popular in present-day sustainable development interventions. Storytelling is a method that has been recognized and developed as a meaningful intervention method in social change activities in and outside the development sector (e.g., Bell 2010; Lundby 2008; Zingaro 2009). Stories and narratives in audio, print, still images, and film can be used in multiple ways. In all cases stories challenge our linear way of thinking and can have a real enabling power. Digital storytelling refers to people telling stories through means of digital media such as film, web blogs, and social media. The use of film and video has been dominated by the term *participatory video* (Campbell). Participatory video often refers to the process of handing over the camera to ordinary people who can then film each other's stories. The empowerment of the process itself is valued more than the end product. However, this is only one particular form of using video. Other methods (e.g., Kennedy 2008; Witteveen 2009; Lie and Witteveen 2019) do value the end product and are in need of professional production. Lie and Mandler made an inventory of all kinds of uses of film in 2009. The use of new ICTs also changed its focus. It shifted from using telecenters and information kiosks to using mobile phones. The mobile phone seems to be the new tool of interest as it rapidly spreads around the world. Especially in countries in the Global South, this diffusion provides a huge potential for development interventions and social change activities.

## 3.2.5 Advocacy and Impact Assessment

*Advocacy* is another communication practice that needs to be mentioned here as advocacy communication is a key action term in development discourse (Servaes and Malikhao 2012). Advocacy is a very specific communication process that aims to bring about change in governance and policies. Advocacy is about persuasion and targets to influence the specific audience of decision-makers. Advocacy is a decision-making process that has the assent of the community as a whole. In this process the community and the decision-maker and the analysts are involved. Therefore, three streams of action are important:

- Media must be activated to build public support and upward pressure for policy decisions.
- Interest groups must be involved and alliances established for reaching a common understanding and mobilizing societal forces. This calls for networking with influential individuals and groups, political forces and public organizations, professional and academic institutions, religious- and cause-oriented groups, and business and industry.
- Public demand must be generated and citizens' movements activated to evoke a response from national leaders.

Therefore, advocacy can be defined as "speaking and/or acting on behalf of people to secure the services they need and the rights to which they are entitled. Advocacy aims to ensure that people's opinions, wishes or needs are expressed and listened to" (Suffolk County Council 2008). Put another way, it aims to convince people with the power (e.g., policy-makers) to address the urgent concerns of a particular group of people (see also Lie and Mandler 2009: 10–11). Advocacy is high on the agenda of many organizations, including the agenda of the UN. Advocacy tool kits are various and widely available. Video advocacy is recently added to many tool boxes and refers to the use of video in advocacy communication activities.

*Monitoring, evaluation*, and *impact assessment* are other issues that need to be mentioned when talking about strategic or methodological practices in the development sector. Specific guides and tool kits on how to do *(participatory) monitoring and evaluation* and also reflexive monitoring in action (RMA) are widely available on the Internet. Lennie and Tacchi (2013) provide a comprehensive framework for evaluating communication for development in a complex theory and systems approach context and also have a tool kit online (Lennie et al. 2011). There is a growing body of literature addressing *impact assessment* of development projects (comprehensive overviews include Bessette (2004), Chambers (2008), or Guijt (2008)). In one way or the other, we need to assess the impact of development (communication) interventions. This is not only important for donor organizations but also for being able to improve the quality of interventions. Servaes et al. (2012a, b) review existing impact assessment practices and provide a framework of sustainability indicators for Communication for Development and Social Change projects.

## 3.3 Sub-disciplines

This section reviews some sub-disciplines within communication science that are related to development communication or in some way address development communication within their scope. Sub-disciplines that are ramifications of both communication science and development communication are strategic communication, participatory communication, crisis communication, risk communication, journalism, and international communication. It is worth noting that the sub-discipline of political communication has not, or only in a very marginal way, engaged with development communication. The same comment seems to hold true for the sub-discipline of intercultural communication. The fact that these two abovementioned sub-disciplines could have engaged with development communication should be well noted. Before we go into each sub-discipline, we need to know the different levels at which communication operates.

To distinguish between levels of communication, it is essential to better understand the different kinds of change. Change can occur at the individual level, the group level, the community level, the regional level, the national level, the global level, or anywhere in-between. In this regard, sociologists like to make a distinction between (a) *individual and collective levels* on the one hand and (b) *global, macroregional, national, intranational, and local levels* on the other hand.

The first distinction is made to be able to differentiate between *behavior change* and *social change*. Looking at *desired or expected outcomes*, one could think of (a) approaches that attempt to *change attitudes* (through information dissemination, public relations, etc.); (b) *behavioral change* approaches (focusing on changes of individual behavior, interpersonal behavior, and/or community and societal behavior); (c) *advocacy* approaches (primarily targeted at policy-makers and decision-makers at all levels and sectors of society); and (d) *communication for structural and sustainable change* approaches (which could be either top-down, horizontal, or bottom-up).

The first three approaches, though useful by themselves, are in isolation not capable of creating sustainable development. Therefore, *sustainable social change can only be achieved in combination with and incorporating aspects of the wider environment that influences (and constrains) structural and sustainable change.* These aspects include (see also McKee et al. 2000) structural and conjunctural factors (e.g., history, migration, conflicts); policy and legislation; service provision; education systems; institutional and organizational factors (e.g., bureaucracy, corruption); cultural factors (e.g., religion, norms, and values); sociodemographic factors (e.g., ethnicity, class); sociopolitical factors; socioeconomic factors; and the physical environment.

This brings us to the second distinction, which is often made in *economic, political, and governance arenas* to distinguish between different kinds of institutions operating at different societal levels. Advocacy communication can, for instance, be situated within this second distinction. The two types of distinctions have relevance to the field of development communication as will be shown in the following sections.

### 3.3.1 Strategic Communication and Participatory Communication

Strategic communication deals with the organizational planning of communication and relates to persuasive communication, marketing, and public relations. Strategic communication is planned communication with a strategic, intentional goal. Being strategic means thinking in terms of executing a stakeholder analysis, a risk analysis, and a SWOT analysis, setting objectives, identifying target audiences, developing key messages, and designing an effective communication plan. The social psychological line within communication science has always been strong within the sub-discipline of strategic communication. Strategic communication in the field of development communication is often applied in the thematic sub-discipline of health communication (see later under health communication), but is not.

Participatory communication had and still has a strong presence in development communication (for an overview see, for instance, Srampickal 2006). Especially in the 1980s and 1990s, when participation became the buzz word in the development sector, participatory communication established itself as a sub-discipline of communication science. It is often associated with the new multiplicity paradigm in development communication and builds on widely formulated critiques on linear, top-down, diffusionist, and modernist perspectives on the one hand and the thinking of Latin American scholars such as Augusto Boal, Jesús Martín-Barbero, Luis Ramiro Beltrán, and Paolo Freire on the other hand (Barranquero; Fernandez). Several books carry participatory communication explicitly in the title. Different UN agencies have been active in advocating a participatory communication approach (Mefalopulos). The International Association for Media and Communication Research (IAMCR) has a section on participatory communication, which has since its inception always demonstrated a strong connecting to the field of Communication for Development and Social Change.

### 3.3.2 Crisis Communication and Risk Communication

Crisis communication and risk communication are often linked. It involves (a) assessment and management of risk and crisis situations and (b) communication with the public. Crisis communication and risk communication have both become sub-disciplines of communication science. They address all *assessments*, *managements*, and *communications* related to unexpected events that could have negative impacts. Several books are available on crisis communication (e.g., Coombs 2012) and risk communication (e.g., Lundgren and McMakin 2013) or both sub-disciplines together (Heath and O'Hair 2010).

In the development sector, crises and risks often refer to *natural disasters*, such as earthquakes, hurricanes, floods, and tsunamis, on the one hand, and *health issues*, such as those related to HIV/AIDS, avian flu, and foodborne illnesses, on the other hand. Organizations, including governments, are increasingly recognizing the importance of crisis and risk communication. Crisis and risk communication have close connections with the thematic sub-disciplines of communication for development, especially with environmental communication and health communication. The United Nations Office for Disaster Risk Reduction (UNISDR) defines disaster risk as "the potential disaster losses in lives, health status, livelihoods, assets and services, which could occur to a particular community or a society over some specified future time period" (UNISDR 2009: 9). Issues that are being dealt with within this sub-discipline are global warming and related climate risks such as water scarcity, droughts, intense rainfall, river floods, and early warning systems. ICTs can play an important role in disaster risk reduction, in building and increasing resilience, and in improving early warning systems.

### 3.3.3 Journalism and International Communication

Journalism studies are a long-established sub-discipline of communication science. Journalism: Theory, Practice and Criticism (Sage) and Journalism Studies (Taylor & Francis) are leading academic journals. Journalism has several connections to sustainable development and social change. Journalists deliver news to the public by making it public. In this sense it relates to the second part of the definition of crisis and risk communication given above. In the same manner, journalism has a strong connection to health communication and environmental communication (see below). Normative and critical journalism studies take position in how journalism ought to work in democratic societies and what role journalists play in societal change processes. *Citizen journalism* is a specific field that relates to development and change (see, for instance, Allan and Thorsen 2009). *Peace journalism* is another area where journalism connects to sustainable change (see, for instance, Lynch et al. 2011; Terzis and Vassiliadou 2008). In the field of *journalism education*, UNESCO plays an important role in the development sector. "UNESCO recognizes the fact that sound journalism education contributes towards professional and ethical practice of journalism. Such journalism is better suited to foster democracy, dialogue and development" (UNESCO n.d.). UNESCO also publishes a series on journalism education.

International communication, according to the term itself, deals with communication that crosses international borders (see, for instance, Thussu 2014). It is sometimes explicitly connected to intercultural communication or to development communication, but in the majority of cases, this sub-discipline is mainly concerned with international media, global (tele)communication infrastructures, international communication governance, policy-making, and the political economy of global communication. In this sense it mainly deals with development communication by addressing questions of communication infrastructures and telecommunications in or in relation to developing countries. It also addresses policy issues and the regulation of information flows in the South.

## 3.4 Thematic Sub-disciplines

This section addresses thematic sub-disciplines that can be distinguished within the field of Communication for Development and Social Change. Thematic sub-disciplines cover a life science theme. The thematic sub-disciplines are

(a) *health communication*, (b) *agricultural extension and rural communication*, and (c) *environmental communication (including climate change communication)*.

### 3.4.1 Health Communication

The sub-discipline of health communication seems to have a close link with social marketing and with the discipline of social psychology within the social sciences. Whereas Rogers' theory of diffusions of innovation had and to a certain extent still has a dominant presence within the sub-discipline of agricultural extension, it is especially the theory of planned behavior (Fishbein and Ajzen 1975) that occupies a firm place within health communication. Other popular theories are the health belief model (Rosenstock 1974) and the elaboration likelihood model (Petty and Cacioppo 1986). Behavior change is the dominant form of change that is focused on here. Research on social change gets less attention. Health promotion is sometimes used to refer to the branch within health communication that studies the persuasive use of communication messages and media to promote public health.

Lines of thinking within the strands of Information, Education and Communication (IEC) and Knowledge, Attitude and Practice (KAP) dominate the other strands in health communication. IEC is used in strategic communication intervention programs aiming at inducing quantifiable changes in behavior and attitude by, most of the time, means of knowledge transfer to a specific target group. KAP often refers to a specific quantitative survey research to measure any change in knowledge, attitude, and behavior. Jargons used in these sub-disciplines are rooted in social psychology and behavior change. Scholars often refer to jargons such as communication strategies, communication campaigns, social marketing and mobilization, persuasive communication, and awareness raising. Risk communication has an important stance in health communication as well as in environmental communication.

Different from extension and rural agricultural communication, health communication has a firm place within communication science. Academic organizations such as the National Communication Association (NCA), the International Communication Association (ICA), and the International Association for Media and Communication Research (IAMCR) all have divisions or working groups that address health communication in the North as well as in the South. Different universities around the world offer a variety of educational programs in health communication.

The United Nations, WHO, and UNFPA are leading users of the above terminology and accompanying approach to communication for sustainable development. Health communication in a development context, within and outside the UN, seems to be directly linked to specific pandemics, diseases, and risks. HIV/AIDS, malaria, and polio are among those specifically addressed. Emerging infectious diseases such as SARS, avian flu, and Ebola for which cures have not been developed and where the main instrument of control in the event of a pandemic is timely and effective crisis communication are also paid appropriate attention. The field of family planning, reproductive health, and sexual and reproductive health rights is another area where communication theory is often applied. Besides the traditional communication and awareness-raising campaigns and prevention activities, entertainment-education is an often used methodology in this sub-discipline (Malikhao 2012, 2016).

Leading journals are Health Communication and the Journal of Health Communication. The academic communication associations listed above organize annual or biannual gatherings.

### 3.4.2 Agricultural Extension and Rural Communication

Agricultural extension is one of the oldest sub-disciplines of development communication. It has a peculiar history within development communication, which started with main concepts like diffusions of innovations (Rogers 1962) and knowledge and technology transfer (see, for instance, Chambers et al. 1989). In the 1960s, the 1970s, and through the 1980s, it was generally agreed upon that "proper extension" would directly and immediately profit the increase of crop yields for farmers. Rogers (1962) and later Van den Ban and Hawkins (1988) published leading textbooks on agricultural extension. Extension, then, was perceived in those early days as mechanistic and linear knowledge transfer. However, that way of thinking became obsolete. In the 1970s, the 1980s, and through the 1990s, the FAO, first through its Development Support Communication (DSC) branch and later through other channels, has played a key role in changing the old interpretation by promoting communication as the focal and crucial factor for decision-making in development interventions (Balit and Acunzo 2019). Worth mentioning is that FAO was the first UN specialized agency, besides UNESCO, which has a mandate for communication.

The sub-discipline moved on to the studies of Agricultural Knowledge and Information Systems (AKIS) (FAO and World Bank 2000) and Agricultural Innovation Systems and adopted Farmer Field Schools (FFS) as new methods to better fit the new way of thinking extension. By far scholars have taken another step by renewing and adjusting focus on communication and (agricultural) innovation (see, for instance, Leeuwis and Aarts 2011, for a theoretical rethinking of communication in the field of innovations). They have also improved on knowledge and brokering in complex systems and the role of champions (see, for instance, Klerkx and Aarts 2013; Quarry and Ramírez 2009). Innovations are now seen as an assimilation of technical innovations with new social and organizational arrangements. An innovation is a combination of software (i.e., new knowledge and modes of thinking), hardware (i.e., new technical devices and practices), and orgware (i.e., new social institutions and forms of organization), whereas communication is regarded as a phenomenon in which those involved construct meanings in interaction (Leeuwis and Aarts 2011).

This new thinking and focus have consequences for how we approach communication in the agricultural and rural sector. According to Leeuwis and Aarts (2011), networks, power, and social learning are the key issues in this new thinking about agricultural and rural communication. Rural communication engulfs the concept of agricultural extension; therefore, it offers a wider scope. It includes health issues, education, and sociocultural, political, and other issues that are not directly related to agriculture. Therefore, agricultural extension officers can also play an initiating role in rural communication and can be real champions in a broader sense. These officers are often the first to be in direct contact with rural households, and so these people can signal rural household problems. The media seem to be secondary in addressing the improvement of rural livelihood strategies.

The media field seems to be alive and kicking and has flourished after the emergence of new media. The media and ICT field within agricultural extension and communication has also developed in a specific way. Films and mobile phones have replaced television, radio, and even telecenters or information kiosks. Many cases are available for review, learning, and upscaling. For film see, for instance, the work of Digital Green (http://www.digitalgreen.org). The mobile phone seems to be the most popular ICT at the moment, and many interventions focus on their use.

The journal of Agricultural Education and Extension (JAEE) is addressing this sub-discipline. Other journals occasionally address issues of agricultural extension and rural communication. The Association for International Agricultural and Extension Education (AIAEE) is a professional association for agricultural and extension educators who share the common goal of strengthening agricultural and extension education programs and institutions worldwide.

### 3.4.3 Environmental Communication

Environmental communication addresses all interactions of humans with the environment and is a relatively new sub-discipline within communication science. Environmental communication not only involves the management of the environment but also the study of public opinion and perceptions (Cox 2013). As within the other two sub-disciplines, communication is often closely associated with education. Environmental communication and environmental education sometimes overlap, especially when the terms are used outside academic circles in the public domain.

The sub-discipline of environmental communication seems to be dominated by the issue of climate change, which has been on the agenda for a few years now. Climate change communication even seems to become a field in itself (Kelly 2012; Servaes 2013b). Different from agricultural extension, but maybe similar to areas within health communication, environmental communication, especially climate change communication, often focusses on risk, public engagement, and public opinion (see, for instance, http://www.climatechangecommunication.org). Agricultural communication is often far more concerned with specific target group communication activities. Communication theories used in the areas of public environmental communication and public health communication overlap. In this context environmental communication also has a close link with journalism. Different universities around the world offer courses and programs in environmental communication, sometimes linked with health communication or with other areas within the life sciences.

In the development sector, it is often stressed that again the poor are among those who feel the consequences of climate change the most. This also applies to other areas within environmental communication such as energy security, biodiversity, deforestation, overexploitation of natural resources, and extreme weather conditions. It is often in this context that environmental communication is linked to sustainability. Communication for sustainable development emphasizes the sustainability aspects of interactions: human-human and human-nature. Since the Brundtland report *Our Common Future* (1987) and the United Nations Conference on Environment and Development in Rio de Janeiro (1992), sustainable development is now on many agendas. The 2013 report by the Intergovernmental Panel on Climate Change (IPCC) says there is now 95 percent certainty that humans have caused most of the warming of the planet's surface that has occurred since the 1950s. A balance must be sought between economic growth, social equity, and the natural environment. Communication plays a decisive role in creating this balance. Sustainability communication is a term used for responsible interaction with the natural and social environment. One of the tasks of sustainability communication is to critically evaluate social discourse of the human-environment relationship. Manfred Max-Neef (1991) and the Dag Hammarskjold Foundation call this a transdisciplinary model for human scale development with self-reliance among human beings, nature, and technology; the personal and the social; the micro and the macro; planning and autonomy; and the state and civil society as central to empower groups and social actors: "The fundamental issue is to enable people from their many small and heterogeneous spaces to set up, sustain and develop their own projects" (Max-Neef 1991, p. 85).

The communication science associations provide several homes for environmental communication scientists. The NCA has a division named Environmental Communication, and the ICA has an interest group with the same name. The IAMCR has a working group called Environment, Science and Risk Communication, and the European Communication Research and Education Association (ECREA) is home to the section Science and Environmental Communication. There is also a specific international association for environmental communication, the International Environmental Communication Association (IECA). *Environmental Communication: A Journal of Nature and Culture* is their official journal. IECA also organizes a biannual Conference on Communication and Environment (COCE). *Applied Environmental Education and Communication* and the *Journal of Environmental Education* are other leading journals.

## 3.5 Fields and Areas

This section addresses fields and areas that can be identified within the field of Communication for Development and Social Change. The fields and areas considered must be of importance at present, be concerned with development studies, and pay an interest in the role of communication. The fields and areas that are discussed are (a) *the right to communicate*; (b) *education and learning*, (c) *innovation, science, and technology*; (d) *natural resource management*; (e) *food security*; (f) *inequality and poverty reduction*; (g) *peace and conflict*; (h) *children and youth (and senior citizens)*; and (i) *tourism*. The sections that follow address the role of communication within these fields and areas.

### 3.5.1 Right to Communicate

Liberalization has led not only to greater media freedom but also to the emergence of an increasingly consumer-led and urban-centered communication infrastructure, which is less and less interested in the concerns of poor and rural people. Women and other vulnerable groups continue to experience marginalization and lack of access to communication resources of all kinds. According to Nobel Prize winner Amartya Sen (1999), development should be measured by how much freedom a country has, because without freedom people cannot make the choices that allow them to help themselves and others. He defines freedom as an interdependent bundle of political freedom and civil rights, economic freedom, social opportunities (arrangements for health care, education, and other social services), interactions with others, including the government, and protective security (which includes unemployment benefits, famine and emergency relief, and general safety nets). Cultural respect and the right to communicate are essential in this regard (Dakroury et al. 2009; De Cuéllar 1995). Ensuring free and equal access to information and the right to communicate is a precondition for empowering marginalized groups, as has been addressed by several meetings and international conferences (World Summit on the Information Society and the World Social Forum).

The issue of equal access to knowledge and information and the right to communicate is becoming one of the key aspects of sustainable development. Vulnerable groups in the rural areas of developing countries are on the wrong side of the digital divide and risk further marginalization. In the rush to "wire" developing countries, little attention has been paid to the design of ICT programs for the poor. The trend ignores many lessons learned over the years by communication for development approaches, which emphasize communication processes and outcomes over the application of media and technologies. There needs to be a focus on the needs of communities and the benefits of the new technologies rather than the quantity of technologies available. Local content and languages are critical to enable the poor to have access to the benefits of the information revolution and to be able to actively participate. The creation of local content requires building on existing and trusted traditional communication systems and methods for collecting and sharing information. However, access is only the start of this process. Full participation implies the provision of capacity training and the development of competencies.

### 3.5.2 Education and Learning

The existence of Information, Education and Communication (IEC) and the organization UNESCO itself are proof of the close connection between communication and education. This area ranges from establishing educational infrastructures, "training of trainers"(ToT) activities, vocational education and training (VET) to adult learning, and addressing theories of learning, such as social learning and transformative learning.

In many processes of human development, social learning is seen as an important key process, and, in fact, social learning is intrinsically linked to communication. Social learning and transformative learning are two important theories about how people learn. They directly connect to communication and play an important role in sustainable development and social change. Social learning refers to observational learning in a social context and imitating the actions of others (basic historical text: Bandura 1977). Transformative learning is the process of transforming frames of reference of adult learners (basic historical text: Mezirow 1991). This theory of transformative learning is considered uniquely adult, that is, grounded in human communication, where "learning is understood as the process of using a prior interpretation to construe a new or revised interpretation of the meaning of one's experience in order to guide future action" (Mezirow 1991: 162).

### 3.5.3 Innovation, Science, and Technology

Communication and innovation studies and the field of science and technology studies (STS) are closely related fields. Innovation studies and STS link to communication and development in several ways. Combing the areas of communication, innovation, and development initiated from the work of Rogers, but the study of these concepts also forms disciplines in themselves. Communication studies, innovation studies (see Fagerberg and Verspagen 2009), and development studies are now established disciplinary fields as is STS. Each discipline has its own way of looking at and incorporating the other concepts. Emerging technologies such as nanotechnology, biotechnologies, and sustainable energy technologies are among the technologies that must somehow merge with societal change. It is this merging of technology and society where knowledge sharing and communication are vital processes. The flow of knowledge in a societal and developmental context is the focus of this area.

Communication within and between knowledge and policy networks is analyzed to gain a better understanding of innovation and policy processes. The central theme is the role of communication in innovation processes. Due attention is paid to the integration of knowledge and perspectives of various stakeholders and disciplines. Knowledge and policy networks increasingly involve institutions and persons to facilitate the interaction between the various parties with a stake in the innovation and policy process. Among the sub-themes dealt with are interactive (policy) design processes, transdisciplinary collaboration, social learning and negotiation processes, process and system innovation, and the organization of knowledge and policy networks.

### 3.5.4 Natural Resource Management

Natural resource management is the area that deals with the organization, the control, and the administration of water, land, animals, and plants. Sustainable natural resource management seeks a balance between economic growth and the quality of life and the environment. Adaptive management includes issues such as information and knowledge management, monitoring and evaluation, and risk management. Integrated natural resource management (INRM) brings adaptive management together with participatory planning and community participation and firmly grounds it in sustainability. Community-based development, co-management, and stakeholder analysis are other processes of importance to INRM.

Natural resource management is often related to issues of stakeholder groups, multi-stakeholder actions, and processes of social learning and thus to communication (see, for instance, Muro and Jeffrey 2008; Röling 1994). "In the natural resource management literature, the relevance of communication to cooperation has mainly been based on general theories and formal models of cooperation. Of late, investigators of natural resource management have started using theories and models from social network analysis to argue for the importance of communication and network structure. Network analysis promises to be a productive approach because interpersonal communication is a natural and appealing example of a social network relation" (De Nooy 2013: 44).

FAO has been active in this field. The organization has published *Information and Communication for Natural Resource Management in Agriculture: A Training Sourcebook* in 2006.

### 3.5.5 Food Security

Food security closely relates to natural resource management and of course to the sub-discipline of agricultural extension and rural communication. Food security was in 1996 defined by the World Food Summit as "when all people at all times have access to sufficient, safe, nutritious food to maintain a healthy and active life" (WHO 1996). It thus connects to sustainable livelihoods, and disruptions in affordable, locally produced food supplies can cause widespread food insecurity. In addressing hunger and malnutrition but also food safety and healthy diets, it directly relates to health communication. FAO published a communications tool kit in this area (FAO 2011). The kit provides detailed guidelines for food security professionals to develop a communication strategy and to communicate more effectively with target audiences.

The area of food security also touches upon a new information and communication technology that has not been discussed till now, namely, Geographic Information System (GIS). GIS can help in an indirect way by visualizing spatial data at early stages of crises. Google Maps and Bing Maps are popular GIS applications. ArcGIS is a more sophisticated, professional one. GIS can help with short-term food emergencies and long-term food insecurities by combining geographically referenced data gathered by GIS technologies and combining them with other types of data. The visualization of spatial data opens up new possibilities for food security analyses and can produce valuable new insights. The same technology is also used in other fields and areas, for instance, to monitor the outbreak of health pandemics such as the avian flu. Geotagging, participatory mapping, crisis mapping, and map-based storytelling are just a few among many new initiatives to be found within the wider field of development communication.

## 3.5.6 Inequality and Poverty Reduction

The relationship between communication and poverty is complex. Poor and marginalized people do not only have unequal access to land, livestock, and food but also to information and communication. The knowledge gap theory informed us that in many countries in the South, there is a lot of information available for the relatively few rich people and politically powerful elite, but only little information and marginal access to communication is available for the relatively large poor population. In visualized economic development theory, they form reversed triangles. It is difficult for the poor to participate in decision-making processes and to have their voice heard in distant economic and political arenas. It is in this area that issues of participation and empowerment, engagement and dialogue, equity, democratization, and rights are most pressing.

Furthermore, economic growth is only a limited measure of progress if we consider happiness and well-being of the people as necessary qualitative measures as well. Wilkinson and Pickett (2011) convincingly report that societies with more equal distribution of incomes have better health, have fewer social problems, and are more cohesive than ones in which the gap between the rich and poor is greater. Therefore, poverty reduction often directly relates to livelihood strategies and to governance issues. The Department for International Development (DFID) rural livelihoods model, which distinguishes between several capitals, has been dominant in better understanding livelihood strategies and communications. Poverty is also often directly related to high rates of illiteracy and lack of education. ICTs can help in several ways, and different cases demonstrate how (see, for instance, Harris and Rajora 2006). It is difficult to replicate and scale up the use of ICTs as serious evaluations are often lacking. The World Bank analyzed the role of communication in poverty reduction strategy (Mozammel 2011). The report focuses, for instance, on the communication and governance challenges facing three stakeholder groups: government, donors, and civil society.

## 3.5.7 Peace and Conflict

Conflict prevention, conflict resolution, conflict management, conflict transformation, humanitarian relief, peacebuilding, and the consolidation of peace (peacekeeping) all address the undesirable situation of disharmony, disagreement, or controversy. These situations include among others civil wars, armed conflict and genocide, ethnic disputes, and cultural and identity clashes on the one hand and peace negotiations and mediations on the other hand. A stable social system includes having conflicts. The existence of conflicts is in itself not necessarily problematic. In fact, conflicts form an essential part of all healthy relationships (Raven 2008). The crux lies in managing the existence of the conflict and the magnitude and impact of the conflict. Needless to say that communication plays a crucial role in that process of management.

Conflict management connects to the area of conflict communication. Within the discipline of communication science, much of the work done in the thematic sub-discipline of conflict communication relates to interpersonal communication and studies the skills and competencies that are necessary for dealing effectively with those conflicts (Servaes and Malikhao 2012; Terzis and Vassiliadou 2008). Besides its focus on interpersonal conflicts, it is also concerned with organizational and community conflict. "Most communication theorists prefer the term conflict management to conflict resolution because the former suggests an ongoing communication process focusing attention on interaction, whereas the latter suggests episodes that must be dealt with as they occur, focusing attention on the discrete content of each episode" (Nicotera 2009: 164). The area of conflict management frequently makes reference to different *negotiations styles*, such as competing, collaborating, compromising, avoiding, and accommodating.

### 3.5.8 Children and Youth, Women, and Senior Citizens

Youth stands out and seems to be an emerging area of interest in the development sector. It has overlap with all other fields and areas that have been distinguished and discussed, but communicating with children and youth is a specialized field and is increasingly seen as an important area in itself to address. Cross-cutting themes are, among others, youth sexuality, HIV/AIDS, sexual and reproductive health and rights, formal and informal education, and media literacy. Specific youth problems are, among others, a reduced interest in studying agriculture, unemployment, bullying, identity formation, and a lack of young people's engagement in community issues.

However, not only the needs of children and youth but also those of women (Malik 2013) and senior citizens (De Cuéllar 1995) should be recognized. UNICEF has a global mandate and according to their own saying "supports an environment that guarantees the participation of children and women in social development programmes through raising awareness and mobilizing communities, developing capacity, and strengthening partnerships among key allies and stakeholders" (UNICEF 2013).

In the area of *youth development* and communication, creative techniques such as using film and (performing) arts are often used with children and youth. Moreover, young adults are pioneer users of new media. Nordicom and Ørecomm have been active in addressing the theme of media, youth, and social change.

### 3.5.9 Tourism

International tourism deals to a large extent with *intercultural communication*. Tourism is for many countries in the South an important area for development. It incorporates significant social changes. Cultural diversity and identity are issues that especially come to the fore in this area of international tourism. Sustainability is also an important issue in the field of tourism.

In 2006, the World Bank Development Communication Division, the USAID Development Communication and Sustainable Tourism Unit, and the United Nations World Tourism Organization (UNWTO) organized a conference on the role of development communication in sustainable tourism. The peer-reviewed academic journal Tourism Culture & Communication addresses aspects of this area.

## 3.6 Conclusions

The field of Communication for Development and Social Change is active and dynamic. The field matured, settled in sub-disciplines, and found accommodation in different fields and areas of development and social change. Still, there are future imperatives to identify.

These future imperatives are:

- One way of *mainstreaming communication for development* is firmly grounding the field of Communication for Development and Social Change in thematic and non-thematic sub-disciplines of communication science. These sub-disciplines provide a foundation by underpinning the work of development communication professionals and academics and giving them a solid basis to work from.
- Human and environmental *sustainability* must be central in development and social change activities. Sustainable interventions are necessary to ensure a world worth living in for future generations. Besides political-economic approaches, we need sociocultural approaches to guarantee acceptable and integrated levels of sustainability and to build resilience. *Building resilient communities* should be a priority issue in the field of Communication for Development and Social Change: "...common themes concerning the maintenance of ecological balance, a move away from environmentally unfriendly modernization, and an emphasis on local systems that shift from solely Western-led development and focus on local culture and participation are crucial to an understanding of sustainable development" (Servaes et al. 2012b: 117); and "...the scope and degree of sustainability must be studied in relationship with the local concept of development contingent upon the cultural values of each community" (Servaes et al. 2012b: 118).
- It is essential to recognize that development problems are complex. *Complex or so-called wicked problems*, such as the existence of climate change, conflict and war, HIV/AIDS, and malaria, are problems that do not have one single solution that is right or wrong, good or bad, or true or false. These are problems in which many stakeholders are involved, all of them framing the problems and issues in a different way. Therefore, solutions need to be negotiated, for instance, in multi-stakeholder platforms. Such types of negotiating or "social dialogue" are promoted for concrete purposes, such as reclaiming indigenous knowledge or monitoring and evaluation, but increasingly also from *a rights-based perspective* that all people have a right to be heard.
- Though participatory approaches have gained some visibility, and sometimes even recognition, among mainstream development agencies, an interesting

alliance could be forged at the level of *participatory budgeting*. Many of the steps in participatory budgeting could be seen as participatory communication processes to deepen democracy (Fung and Wright 2001). The fiscal focus of this work offers the potential to cut across all facets of a community's life – climate change, agriculture, health, gender, etc. may be addressed via a community's budgetary process. The need to effectively process budgetary debates via an integrated "weighing" of alternative public spending offers and requires a holistic approach in communication that is particularly vital in addressing climate change (Yoon 2013).

- There is a need for *transdisciplinarity*. We need to rethink and reorder the relationships between communication academics, communication professionals (e.g., extension agents, health communication specialists, intermediaries, knowledge brokers, change agents, M&E specialists), technical field-specific professionals (technical ICT specialists, agronomists, medical doctors), policy-makers (international, national, intranational), civil society members (e.g., NGOs, social movements, societal agents), and local people (e.g., farmers, fishermen, households, audiences, clients). *Linkages and dialogues* need improvement. There is a demand for building knowledge and communication networks and to attach importance to stakeholder interactions and knowledge system approaches. Climate change adaptation and livelihood adaptations require *multi-stakeholder actions* and processes of *social learning*.
- *New creative techniques and methodologies* need further attention. New questions need to be addressed such as: What is the role of *creativity* in development and social change interventions? What actually is "out-of-the-box" thinking and is everybody willing and able to think in that way? And should everybody think in that way? Especially the use of (digital) storytelling, film, (participatory) video, and mobile phones has huge potential. The new ICTs do have potential, but centralizing them incorporates the danger of *technological determinism*. We can learn from history and not make the same mistakes that we made with the introduction of, for instance, broadcast television.
- There is a need to connect communication to *learning*, *education*, and *knowledge exchange*. Focusing on processes of transformative learning, social learning, experiential learning, reflexive learning, organizational learning, and double-loop learning is essential to better understand processes of change by looking at how people learn. For interventions to be successful, it seems necessary to invest in double-loop (second order) learning. Double-loop learning involves learning about methodologies and understanding why things are learnt and why certain knowledge is needed. In this context we acknowledge, with Ingie Hovland, that we need a shift from instrumental change to conceptual change. "The current focus is on instrumental change through immediate and identifiable change in policies, and less on conceptual change in the way we see the world and the concepts we use to understand it" (Hovland 2003:viii, 15–16).
- The 2013 Human Development Report identifies four specific areas of focus for sustaining development momentum: *enhancing equity*, including on the gender dimension; enabling greater voice and *participation of citizens*, including youth;

*confronting environmental pressures*; and *managing demographic change*. For the first time in 150 years, the combined output of the developing world's three leading economies – Brazil, China, and India – is about equal to the combined GDP of the long-standing industrial powers of the North: Canada, France, Germany, Italy, the United Kingdom, and the United States. This represents a dramatic rebalancing of global economic power.

The middle class in the South is growing rapidly in size, income, and expectations. The South is now emerging alongside the North as a breeding ground for technical innovation and creative entrepreneurship. Not only the larger countries, such as Brazil, China, India, Indonesia, Mexico, South Africa, Thailand, and Turkey, but also smaller economies, such as Bangladesh, Chile, Ghana, Mauritius, Rwanda, and Tunisia, have made rapid and substantial progress.

However, "unless people can participate meaningfully in the events and processes that shape their lives, national human development paths will be neither desirable nor sustainable," the report claims (Malik 2013: 18). In that regard – see also one of the Millennium Development Goals – *educating women* through adulthood is the closest thing to a "silver bullet" formula for accelerating human development.

## References

Allan S, Thorsen E (2009) Citizen journalism: global perspectives. Peter Lang, New York
Balit S, Acunzo M (2019) A Changing World: FAO Efforts in Communication for Rural Development. In: Servaes J. (eds) Handbook of Communication for Development and Social Change. Springer, Singapore
Bandura A (1977) Social learning theory. General Learning Press, Englewood Cliffs
Bell L (2010) Storytelling for social justice: connecting narrative and the arts in antiracist teaching. Routledge, New York
Benería L, Berik G, Floro M (2003) Gender, development, and globalization. Economics as if all people mattered. Routledge, New York
BenYishay A, Betancourt R (2013) Unbundling democracy: Tilly Trumps Schumpeter. Available at SSRN: http://ssrn.com/abstract=2282729. https://doi.org/10.2139/ssrn.2282729
Bessette G (2004) Involving the community: a guide to participatory development communication. Southbound, Penang
Chambers R (2008) Revolutions in development inquiry. Earthscan, London
Chambers R, Pacey A, Thrupp LA (1989) Farmer first: farmer innovation and agricultural research. Intermediate Technology Publications, London
Coombs WT (2012) Ongoing crisis communication: planning, managing, and responding. Sage, Thousand Oaks
Cox R (2013) Environmental communication and the public sphere. Sage, Thousand Oaks
Dakroury A, Eid M, Yahya K (eds) (2009) The right to communicate. Historical hopes, global debates, and future premises. Kendall Hunt, Dubuque
De Cuéllar JP (ed) (1995) Our creative diversity. Report of the world commission on culture and development. UNESCO, Paris
De Nooy W (2013) Communication in natural resource management: agreement between and disagreement within stakeholder groups. Ecol Soc 18(2):44. https://doi.org/10.5751/ES-05648-180244

Fagerberg J, Verspagen B (2009) Innovation studies. The emerging structure of a new scientific field. Res Policy 38(2):218–233

FAO (2011) Food security communications toolkit. FAO, Rome

FAO, World Bank (2000) Agricultural knowledge and information systems for rural development (AKIS/RD). In: Strategic vision and guiding principles. FAO/World Bank, Rome/Washington, DC

Fishbein M, Ajzen I (1975) Belief, attitude, intention, and behavior: an introduction to theory and research. Addison-Wesley, Reading

Flint W (2007) Sustainability manifesto. Exploring Sustainability: getting inside the concept. http://www.eeeee.net/sd_manifesto.htm. Accessed 17 Dec 2011

Ford L (2015) Sustainable development goals: all you need to know. The Guardian. Retrieved from: http://www.theguardian.com/global-development/2015/jan/19/sustainable-development-goals-united-nations

Fung A, Wright EO (eds) (2001) Deepening democracy. Institutional Innovations in Empowered Participatory Governance. http://www.ssc.wisc.edu/~wright/DeepDem.pdf. Retrieved 5 Feb 2014

Goldin I, Reinert KA (2012) Globalization for development: meeting new challenges. Oxford University Press, Oxford

Guijt I (2008) Critical readings on assessing and learning for social change: a review. IDS development bibliographies 21. IDS, Sussex

Hak T, Janouskova S, Moldan B (2016) Sustainable development goals: a need for relevant indicators. Ecol Indic 60:565–573

Harris R, Rajora R (2006) Empowering the poor. Information and communications technology for governance and poverty reduction. A study of rural development projects in India. The Asia-Pacific Development Information Programme (APDIP), Bangkok

Heath RL, O'Hair HD (eds) (2010) Handbook of risk and crisis communication. Routledge, New York

Hovland I (2003) Communication of research for poverty reduction: a literature review. Working paper 227. Overseas Development Institute (ODI), UK Department for International Development (DFID), London

Kelly MR (2012) Climate change communication research here and now: a reflection on where we came from and where we are going. Appl Environ Educ Commun 11(3–4):117–118. https://doi.org/10.1080/1533015X.2012.777289

Kennedy T (2008) Where the rivers meet the sky: a collaborative approach to participatory development. Southbound, George Town

Klerkx L, Aarts N (2013) The interaction of multiple champions in orchestrating innovation networks: conflicts and complementarities. Technovation 33(6–7):193–210. https://doi.org/10.1016/j.technovation.2013.03.002

Kotter JP (2001) What leaders really do. Harvard Business Review, December

Leeuwis C, Aarts N (2011) Rethinking communication in innovation processes: creating space for change in complex systems. J Agric Educ Ext 17(1):21–36. https://doi.org/10.1080/1389224X.2011.536344

Lie R, Witteveen L (2019) ICTs for Learning in the Field of Rural Communication. In: Servaes J. (eds) Handbook of Communication for Development and Social Change. Springer, Singapore

Lennie J, Tacchi J (2013) Evaluating communication for development. A framework for social change. Routledge Earthscan, London

Lennie J, Tacchi J, Koirala B, Wilmore M, Skuse A (2011) Equal access participatory monitoring and evaluation toolkit. Retrieved from http://betterevaluation.org/toolkits/equal_access_participatory_monitoring

Lie R, Mandler A (2009) Video in development: filming for rural change. Centre for Agricultural and Rural Cooperation (CTA), GTZ (Deutsche Gesellschaft fur Technische Zusammenarbeit), Food and Agriculture Organization of the United Nations (FAO), and Wageningen UR, Wageningen

Lie R, Servaes J (2015) Disciplines in the field of communication for development and social change. Commun Theory 25(2):244–258

Lundby K (ed) (2008) Digital storytelling, mediatized stories: self-representations in new media. Peter Lang, New York
Lundgren RE, McMakin AH (2013) Risk communication: a handbook for communicating environmental, safety, and health risks. Wiley, New York
Lynch J, Shaw IS, Hackett RA (2011) Expanding peace journalism: comparative and critical approaches. Sydney University Press, Sydney
Malik K (ed) (2013) The rise of the south: human progress in a diverse world. Human development report. UNDP, New York
Malikhao P (2012) Sex in the village. Culture, religion and HIV/AIDS in Thailand. Southbound/Silkworm Publishers, Penang/Chiang Mai
Malikhao P (2016) Effective health communication for sustainable development. Nova Publishers, New York
Malikhao P (2019) Sustainability and communication from a Thai Buddhist perspective. Paper international conference on "Media and communication in sustainable development", February 2019, Visva-Bharati, Santiniketan
Max-Neef M (1991) Human scale development. Conception, application and further reflections. The Apex Press, New York
McKee N, Manoncourt E, Chin SY, Carnegie R (eds) (2000) Involving people evolving behavior. Southbound Publishers, Penang
Melkote SR, Steeves HL (2001) Communication for development in the third world: theory and practice for empowerment. 2nd ed. Sage, London
Mezirow J (1991) Transformative dimensions of adult learning. Jossey-Bass, San Francisco
Mozammel M (ed) (2011) Poverty reduction with strategic communication: moving from awareness raising to sustained citizen participation. The World Bank. Communication for Governance and Accountability Program (CommGAP), Washington, DC
Muro M, Jeffrey P (2008) A critical review of the theory and application of social learning in participatory natural resource management processes. J Environ Plan Manag 51(3):325–344. https://doi.org/10.1080/09640560801977190
Nash J (ed) (2005) Social movements: an anthropological reader. Blackwell, Malden
Nerfin M (ed) (1977) Another development. Approaches and strategies. Dag Hammerskjold Foundation, Uppsala
Nicotera A (2009) Conflict communication theories. In: Littlejohn S, Foss K (eds) Encyclopedia of communication theory. Sage, Thousand Oaks, pp 165–171
Payutto P (1998) Sustainable development. Buddhadham Foundation, Bangkok
Petty RE, Cacioppo JT (1986) From communication and persuasion: central and peripheral routes to attitude change. Springer, New York
Polk E (2015) Communicating global to local resiliency. A case study of the transition movement. Lexington Books, Lanham
Quarry W, Ramírez R (2009) Communication for another development: listening before telling. Zed Books, London
Raven BH (2008) The bases of power and the power/interaction. Model of interpersonal influence. Anal Soc Issues Public Policy 8(1):1–22
Rogers EM (1962) Diffusion of innovations. Free Press, New York
Röling NG (1994) Communication support for sustainable natural resource management. Special issue: knowledge is power? The use and abuse of information in development. IDS Bull 25(2):125–133
Rosenstock I (1974) Historical origins of the health belief model. Health Educ Monogr 2(4):328
Sachs JD (2012) From millennium development goals to sustainable development goals. Lancet 379(9832):2206–2211
SDSN (2015) Indicators and a monitoring framework for the sustainable development goals. Launching a data revolution for the SDGs. In: A report by the leadership council of the sustainable development solutions network. Revised working draft (June 22, 2015)
Sen A (1999) Development as freedom. Random House, New York
Servaes J (1999) Communication for development: one world, multiple cultures. Hampton, Cresskill

Servaes J (2013a) Sustainability, participation & culture in communication: theory and praxis. Intellect, Bristol
Servaes J. (ed.) (2013b) Sustainable Development and Green Communication. African and Asian Perspectives, London/New York: Palgrave/MacMillan
Servaes J (ed) (2017) Sustainable development goals in the Asian context. Springer, Singapore
Servaes J, Lie R (2015) New challenges for communication for sustainable development and social change. A review essay. J Multicult Discourses 10(1):124–148
Servaes J, Malikhao P (2007) Communication and sustainable development. In: Servaes J, Liu S (eds) Moving targets. Mapping the paths between communication, technology and social change in communities. Southbound, Penang, pp 11–42
Servaes J, Malikhao P (2012) Advocacy communication for peace building. Dev Pract 22(2):229–243
Servaes J, Polk E, Shi S, Reilly D, Yakupitijage T (2012a) Sustainability testing for development projects. Dev Pract 22(1):18–30. https://doi.org/10.1080/09614524.2012.634177
Servaes J, Polk E, Shi S, Reilly D, Yakupitijage T (2012b) Towards a framework of sustainability indicators for 'communication for development and social change' projects. Int Commun Gaz 74(2):99–123. https://doi.org/10.1177/1748048511432598
Sivaraksa S (2010) The wisdom of sustainability. Buddhist economics for the 21st century. Silkworm Books, Chiang Mai
Srampickal SJ (2006) Development and participatory communication. Commun Res Trends 25:2:1–25
Suffolk County Council (2008) What is advocacy? Retrieved from http://www.suffolk.gov.uk/CareAndHealth/CustomerRights/Advocacy/Advocacy.htm
Terzis G, Vassiliadou M (2008) Working with media in areas affected by ethno-political conflict. In: Servaes J (ed) Communication for development and social change. Sage, Thousand Oaks
Thussu DK (2014) International communication: continuity and change. Bloomsbury Academic, New York
UN (2014) Millennium development goals report 2014. United Nations, New York
UNDP (2004) Human development report 2004: cultural liberty in today's diverse world. United Nations Development Programme and Oxford University Press, New York
UNESCO (2009) UNESCO world report. Investing in cultural diversity and intercultural dialogue. UNESCO, Paris
UNESCO (n.d.) Journalism education and training. Retrieved from http://www.unesco.org/new/en/communication-and-information/media-development/journalism-education-and-training/browse/7/
UNICEF (2013) Communication for development support to public health preparedness and disaster risk reduction in East Asia and the Pacific: a review. Young Child Survival and Development Section, UNICEF East Asia and Pacific Regional Office, New Delhi
UNISDR (2009) Terminology on disaster risk reduction. United Nations International Strategy for Disaster Reduction, Geneva
UNSC (2015) Technical report by the Bureau of the United Nations Statistical Commission on the process of the development of an indicator framework for the goals and targets of the post-2015 development agenda (Working draft). Retrieved from: https://sustainabledevelopment.un.org/content/documents/6754Technical%20report%20of%20the%20UNSC%20Bureau%20%28final%29.pdf
Van den Ban AW, Hawkins HS (1988) Agricultural extension. Longman Scientific & Technical, London
Van Hemelrijck A (2013) Powerful beyond measure? Measuring complex systemic change in collaborative settings. In: Servaes J (ed) Sustainability, participation & culture in communication: theory and praxis. Intellect, Bristol
Waisbord S (2018) Family Tree of Theories, Methodologies, and Strategies in Development Communication. In: Servaes J. (eds) Handbook of Communication for Development and Social Change. Springer, Singapore
Wallerstein N, Duran B, Oetzel JG, Minkler M (eds) (2017) Community-based participatory research for health: advancing social and health equity, 3rd edn. Jossey-Bass/Wiley, New York

Weaver W, Shannon C (1963) The mathematical theory of communication. University of Illinois Press, Urbana

WHO (1996) Food security. Retrieved from http://www.who.int/trade/glossary/story028/en/

Wilkinson R, Pickett K (2011) The spirit level: why greater equality makes societies stronger. Bloomsbury, New York

Witteveen L (2009) The voice of the visual. Visual learning strategies for problem analysis, social dialogue and mediated participation. Eburon, Delft

World Commission on Environment and Development (WCED) (1987) Our common future. Published as annex to General Assembly document A/42/427, Development and International Co-operation: Environment. UN, New York

Yoon CS (2013) Future imperatives of practice: the challenges of climate change. In: Servaes J (ed) Sustainable development and green communication. Palgrave, Basingstoke, pp 58–77

Zingaro L (2009) Speaking out: storytelling for social change. Left Coast Press, Walnut Creek

Zolli A, Healy M (2012) Resilience: why things bounce back. Free Press, New York

# Part II
# Historic Cluster

# Communication for Development and Social Change: Three Development Paradigms, Two Communication Models, and Many Applications and Approaches

**4**

Jan Servaes and Patchanee Malikhao

## Contents

| | | |
|---|---|---|
| 4.1 | Introduction | 64 |
| | 4.1.1 What Is Communication for Development and Social Change (CDSC)? | 65 |
| 4.2 | Summarizing the Past | 66 |
| | 4.2.1 Development Paradigms | 66 |
| | 4.2.2 Communication Paradigms | 68 |
| | 4.2.3 Research Priorities | 73 |
| 4.3 | Mapping the Future | 76 |
| | 4.3.1 Interdisciplinarity | 76 |
| | 4.3.2 The Power of Culture in Homogeneity and Diversity | 76 |
| | 4.3.3 A New Form of Modernization? | 77 |
| | 4.3.4 The Sustainability of Social Change Processes | 77 |
| | 4.3.5 Nation-States and National Cultures | 80 |
| | 4.3.6 The Place of Civil Society and the Role of New Social Movements | 81 |
| | 4.3.7 Linking the Global and the Local | 81 |
| 4.4 | Directions for Future Research | 82 |
| | 4.4.1 The Transformation of Society | 82 |
| | 4.4.2 The Cosmopolitan Challenge | 83 |
| | 4.4.3 The Content of Development Agendas | 83 |
| | 4.4.4 More Participatory Communication Research Needed | 84 |
| | 4.4.5 Measurement and Evaluation for "Social Usefulness" | 88 |
| 4.5 | Concluding Remarks | 89 |
| References | | 89 |

---

J. Servaes (✉)
Department of Media and Communication, City University of Hong Kong, Hong Kong, Kowloon, Hong Kong

Katholieke Universiteit Leuven, Leuven, Belgium
e-mail: jan.servaes@kuleuven.be; 9freenet9@gmail.com

P. Malikhao
Fecund Communication, Chiang Mai, Thailand
e-mail: pmalikhao@gmail.com

© Springer Nature Singapore Pte Ltd. 2020
J. Servaes (ed.), *Handbook of Communication for Development and Social Change*,
https://doi.org/10.1007/978-981-15-2014-3_110

**Abstract**

All those involved in the analysis and application of Communication for Development and Social Change would probably agree that in essence communication for development and social change is the sharing of knowledge aimed at reaching a consensus for action that takes into account the interests, needs, and capacities of all concerned. It is thus a social process. Communication media are important tools in achieving this process, but their use is not an aim in itself – interpersonal communication and traditional forms of communication too must play a fundamental role.

The basic consensus on development communication has been interpreted and applied in different ways throughout the past century. Both at theory and research levels, as well as at the levels of policy- and planning-making and implementation, divergent perspectives are on offer.

In this chapter it summarizes the past of Communication for Development and Social Change; identify the roadmap for the future of Communication for Development and Social Change; and look at some of the key purposes, functions, and approaches needed to steer communication for development and social change.

**Keywords**

Development paradigms · Modernization · Dependency · Multiplicity · Social change · Diffusion of innovations · Participatory communication · Information rights

## 4.1 Introduction

The collapse of the Soviet Union in the late 1980s, together with the rise of China as the new "competitor" to the USA (the remaining "superpower"); the tensions within the European Union; the gradual coming to the fore of regional powers, such as Brazil, Russia, India, China, and South Africa (the so-called BRICS countries); the surge of populism and illiberal or authoritarian regimes; global warming; and the 2008 meltdown of the world financial system with its disastrous lingering consequences for people everywhere, necessitates a rethink of the "power of development."

The study of communication for development and social change has been through several paradigmatic changes during the past decades. From modernization and growth theory to the dependency approach and the participatory model, the new traditions of discourse are now characterized by a turn toward local communities as targets for research and debate on the one hand and the search for an understanding of the complex relationships between globalization and localization on the other. Our present-day "globalized" world as a whole and its distinct regional and national entities are confronted with multifaceted crises from the economic and financial to those relating to social, cultural, ideological, moral, political, ethnic, ecological, and security issues. Previously held traditional modernization and dependency perspectives have become more difficult to support because of the growing interdependency of regions, nations, and communities in our globalized world.

### 4.1.1 What Is Communication for Development and Social Change (CDSC)?

Today almost all of those involved in the analysis and application of Communication for Development and Social Change – or what was broadly termed *development communication* or *devcom* – would probably agree that in essence communication for social change is the sharing of knowledge aimed at reaching a consensus for action that takes into account the interests, needs, and capacities of all concerned. It is thus a social process, which has as its ultimate objective *sustainable development* at distinct levels of society. Communication media and Information and Communication Technologies (ICTs) are important tools in achieving social change, but their use is not an aim in itself – interpersonal communication and traditional and group media must also play a fundamental role.

Such a basic consensus on what constitutes *development communication*, though still being debated, is relatively new. Summarizing the history of development communication is not easy because it is considered to have different origins and "founding fathers." Some explain it as the logical offspring of the Western drive to develop the world after colonization and the Second World War. Staples (2006), for instance, explains that after 1945 the West considered development as an international obligation, the beginning of a broad international civil service and the start of the continuing effort to find a way of promoting the well-being of the earth's people as a whole (see also ▶ Chap. 8, "The Pax Americana and Development"). Others (e.g., Habermann and De Fontgalland 1978; Jayaweera and Amunugama 1987) position it as a regional (mainly Asian) reply to modernization. There are also some who look for founding "fathers" and "mothers" in the academic world (Nora Quebral from The Philippines or Luis Ramiro Beltran in Latin America are likely candidates) or in the world of development agencies.

The latest attempt to present a unified and agreeable definition was made during the first World Congress on Communication for Development (Rome, 25–27 October 2006). The so-called Rome Consensus states that "Communication for Development is a social process based on dialogue using a broad range of tools and methods. It is also about seeking change at different levels including listening, building trust, sharing knowledge and skills, building policies, debating and learning for sustained and meaningful change. *It is not public relations or corporate communication*" (http://www.devcomm-congress.org/worldbank/macro/2.asp, emphasis added).

However, major aspects of many projects and programs currently being promoted and implemented are, we believe, nothing but "public relations or corporate communication." We are joined by Adam Rogers (2005), former Head of Communications and Information at UNCDF and now with the UN Development

Group (UNDG), who aptly summarizes the major "brand" of devcom approaches as "Participatory diffusion or semantic confusion": "Many development practitioners are avoiding the semantic debates ... in order to harness the benefits of both approaches. For them, what is most important is not what an approach is called, the origins of an idea or how it is communicated. What is critical is that we find the most effective and efficient tools to achieve the noble objectives outlined in the Millennium Declaration" (Rogers 2005).

To state that development communication is a good case to illustrate the importance and relevance for comparative and interdisciplinary research is a no-brainer. By its very nature, development communication would not exist without comparative research methods. However, one of the major problems in comparative research is that the data sets in different contexts may not use the same categories or define categories differently (e.g., by using different definitions of development, communication, growth, poverty, inequality, participation, etc.). That, indeed, is the situation in development communication.

## 4.2 Summarizing the Past

There are at least three ways of summarizing the past at different levels: by identifying the different (1) development paradigms and (2) communication paradigms and (3) by looking at the research priorities in different time periods (see also ▶ Chap. 5, "Family Tree of Theories, Methodologies, and Strategies in Development Communication").

### 4.2.1 Development Paradigms

1. After the Second World War, the founding of the United Nations stimulated relations among sovereign states, especially the North Atlantic Treaty nations and the so-called developing nations, including the new states emerging from colonial rule. During the Cold War period, the superpowers – the USA and the former Soviet Union – tried to expand their own interests to the so-called Third World or developing countries. In fact, the USA was defining development and social change as the replica of its own political-economic system and opening the way for the transnational corporations. At the same time, the developing countries saw the "welfare state" of the North Atlantic Treaty nations as the ultimate goal of development. These nations were attracted by the new technology transfer and the model of a centralized state with careful economic planning and centrally directed development bureaucracies for agriculture, education, and health as the most effective strategies to catch up with those industrialized countries (Fraser and Restrepo 1998; McMichael 2008; Nederveen Pieterse 2010; see also ▶ Chap. 6, "A Changing World: FAO Efforts in Communication for Rural Development").

   This mainly economic-oriented view, characterized by endogenism and evolutionism, ultimately resulted in the *modernization and growth theory*. It sees development as a unilinear, evolutionary process and defines the state of underdevelopment in terms of observable quantitative differences between the so-called poor and rich countries on the one hand and traditional and modern societies on the other (for more details, see Servaes 1999, 2003, 2008).

2. As a result of the general intellectual "revolution" that took place in the mid-1960s, this Euro- or ethnocentric perspective on development was challenged by Latin-American social scientists, and a theory concerning *dependency and underdevelopment* was born (see one of the "classics": Cardoso and Falletto 1969). The dependency approach formed part of a general structuralist reorientation in the social sciences. The *dependistas* were primarily concerned with the effects of dependency in peripheral countries, but implicit in their analysis was the idea that development and underdevelopment must be understood in the context of a world system (Chew and Denemark 1996).

   This dependency paradigm played an important role in the movement for a New International Economic Order (NIEO) and New World Information and Communication Order (NWICO) from the late 1960s to the early 1980s (Mowlana 1986). At that time, the new states in Africa and Asia and the success of socialist and popular movements in Cuba, China, Chile, and other countries provided the goals for political, economic, and cultural self-determination within the international community of nations. These new nations shared the ideas of being independent from the superpowers and moved to form the Non-Aligned Nations. The Non-Aligned Movement defined development as a political struggle.

3. Since the demarcation of the First, Second, and Third Worlds has broken down and the crossover center-periphery can be found in every region, there was a need for a new concept of development which emphasized *cultural identity and multidimensionality* (further discussed in Martin-Barbero 1993; Canclini 1993).

   From the criticism of the two paradigms above, particularly that of the dependency approach, a new viewpoint on development and social change came to the forefront. The common starting point here is the examination of the changes from "bottom-up," from the self-development of the local community. The basic assumption was that there are no countries or communities that function completely autonomously and that are completely self-sufficient, nor are there any nations of which development is exclusively determined by external factors. Every society is dependent in one way or another, both in form and in degree. Thus, a framework was sought within which both the center and the periphery could be studied separately and in their mutual relationship, at global, national, and local levels.

   More attention was also being paid to the *content of development*, which implied a more normative, holistic, and ecological approach. "Another development" questions whether "developed" countries are in fact developed and whether this genre of progress is sustainable or desirable. It favors a multiplicity of approaches based on the context and the basic, felt needs, and the empowerment of the most oppressed sectors of various societies at divergent levels. A main thesis is that change must be structural and occur at multiple levels in order to achieve sustainable ends.

4. The modernization paradigm was avowed on both ends of the ideological spectrum, both by the classic liberal and neoliberal theorists like Keynes and by the classic Marxist thinkers. Both ideologies have a lot in common. The differences in approach lie on the level of the means, the relative role that is assigned to the market versus the state. But the objective is the same: development on the basis of the Western vision of growth and progress. The obstacles for development are indicated only in the traditional sectors and are initially only attacked with economic means. Where liberals try to achieve development by means of a massive transfer of capital technology to the Third World, the classic Marxists strive for state intervention, the stimulation of the public sector, and the establishment of the heavy industry as an initial step in the development process – in other words, development according to the Soviet model (Galtung 1980; Sparks 2007). Some scholars (Bartolovich and Lazarus 2007; Chang 2008) claim that even in Mao's China, this view on development was of great importance. Since the revolution, there have been two ideological "lines." The first stands for a highly centralized, technocratic guidance of society toward modernization; the second line is based upon the elimination of the so-called Three Great Differences (i.e., city vs. country, mental vs. manual labor, worker vs. peasant), a collective functioning on the basis of mass democracy and self-reliance. These authors state that these lines continue to guide the Chinese development process in apparently "nonantagonistic" contradiction.

The characteristics of the different development paradigms have been summarized in Servaes (1999) (Table 1).

### 4.2.2 Communication Paradigms

1. The above typology of the so-called development paradigms can also be found at the communication and culture levels. The communication media are, in the context of development, generally used to support development initiatives by the dissemination of messages that encourage the public to support development-oriented projects. Although development strategies in developing countries diverge widely, the usual pattern for broadcasting and the press has been predominantly the same: informing the population about projects, illustrating the advantages of these projects, and recommending that they be supported. A typical example of such a strategy is situated in the area of family planning, where communication means like posters, pamphlets, radio, and television attempt to persuade the public to accept birth control methods. Similar strategies are used on campaigns regarding health and nutrition, agricultural projects, education, HIV/AIDS prevention, and so on.

    This model sees the communication process mainly as a message going from a sender to a receiver. This hierarchic view on communication can be summarized in Laswell's classic formula – "Who says What through Which channel to Whom with What effect?" – and dates back to (mainly American) research on (political)

**Table 1** Main characteristics of the modernization, dependency, and multiplicity paradigms

| | | Modernization | Dependency | Multiplicity |
|---|---|---|---|---|
| Theory | **Period origins** | 1940–1960<br>Western<br>neo-liberalism (Keynes) –<br>functionalism behaviorism | 1960–1980<br>Latin-American<br>neo-marxism – structuralism | Since 1980<br>divers origins<br>postmodernism – anthropology –<br>psychoanalysis – pedagogy |
| | Level of analysis | Nation-state<br>*Internal* (traditional society) | International: Center-Periphery<br>*External* (First World) | Multidimensional: Inter-dependency<br>global-regional-national-local<br>*Internal* and *external cultural variables* |
| | Cause problem | *Economic variables* | *Economic and political variables* | |
| | Emphasis on economic | endless growth in linear stages<br>(Rostow) – capital intensive – GNP | development and underdevelopment<br>(Frank) global exploitation class<br>struggle – holism – subjectivism | End of growth – world system analysis<br>– coupling of modes of production –<br>new social movements – empowerment |
| | Socio-cultural | dualism (traditional-modern) –<br>structural differentiation – functional<br>specialization – socio-psychology –<br>harmony model Western democratic | conflict model – regional contradictions | – participation – ecology – culture<br>normative approach participatory<br>democracy |
| | Political | | Third World elitism – Technology<br>creates dependency technological | |
| | technological Communication | system – ethnocentric technology<br>transfer – value-free technological<br>determinism two-step flow – diffusion<br>of innovations – emphasis on mass<br>media – communicator – opinion leader<br>– individual attitude change empathy –<br>medium is the message – media are<br>mobility multipliers Lerner – Pool –<br>Rogers – Schramm | appropriateness cultural imperialism/<br>synchronization – institutional and<br>structural constraints – professional<br>values – imbalance of flow and content | Ideology and culture – modes of<br>communication. cultural identity –<br>participatory communication receiver<br>oriented – exchange of meanings<br>hybridization – cultural proximity |
| | Scholars | | Beltran – Bordenave – Hamelink –<br>Mattelart – Schiller – Somavia | Barbero – Dissanayake – White<br>Canclini – Wang |
| Research | Methodology | Quantitative – positivism – effects | Quantitative + qualitative | Quantitative + qualitative +<br>participatory |
| | Assumptions | cause-effect – short-term – explicit –<br>a-contextual – individual | quantity equals quality | Contextual analysis participatory<br>observation Part. Action Research |

(continued)

**Table 1** (continued)

| | | Modernization | Dependency | Multiplicity |
|---|---|---|---|---|
| Policies | Communautes | Top down | Top down | Bottom up |
| | 'Solution' coming | From 'outside' (west) integration/assimilation | From 'inside' (third world) Integration | No universal model available Dissociation |
| | Internal Domestic | Explicit Integration | Implicit Dissociation/self-reliance | Explicit Selective participation |
| | External International | Implicit Free flow of information | Explicit Free and balanced flow | Explicit Right to communicate |
| | Principles | Free press theory social Responsibility theory | Media dependency Theory authoritarian/Marxist-Leninist theory | Democratic-participatory theory |
| | Normative media theories | | | |

campaigns and diffusions in the late 1940s and 1950s (Lerner 1958; Schramm 1964). The most important hypothesis of Lerner's (1958) study was that a high level of empathy is "the predominant personal style only in modern society, which is distinctively industrial, urban, literate and participant" (p. 50). Empathic persons have a higher degree of mobility, meaning a high capacity for change, and were more future oriented and rational than so-called traditional people. According to Lerner, the general psychological conditions captured by the concept of empathy stimulated mobility and urbanization which, in turn, increases literacy and consequently economic and political participation – all essential to the modernization process. The media would serve to stimulate, in direct and indirect ways, the conditions of "psychological mobility" that are so crucial to economic development (see also ▶ Chap. 7, "Daniel Lerner and the Origins of Development Communication").

Schramm stressed the importance of developing modern communication infrastructures and, based on the weakness of the private sector and the lack of economic resources in most developing countries, he advocated that the state, or governments, should play a leading role in infrastructural development. While this obviously enhanced the role of the state vis-à-vis the media, since the government under Schramm's proposals would own and control the communications media, other theorists suggested the training of media professionals and journalists in the traditions of Western media theory. As such, the overriding interest was in gaining access to the technology as quickly as possible, even if this meant turning a blind eye to the power relations that would shape the transfer and deployment of technology, as in the case of Schramm's proposal for government development of national media systems.

Referring to the agricultural extension work done in the USA in the first half of the past century and borrowing concepts from rural sociology, the American scholar Everett Rogers (1962, 1986) is said to be the person who introduced the *diffusion theory* into the context of development. Modernization is here conceived as a process of diffusion, whereby individuals move from a traditional way of life to a different, more technically developed and more rapidly changing way of life. Building primarily on sociological research in agrarian societies, Rogers stressed the adoption and diffusion processes of cultural innovation. This approach is therefore concerned with the process of *diffusion and adoption of innovations* in a systematic and planned way (Hoffmann 2007). Mass media are important in spreading awareness of new possibilities and practices, but at the stage where decisions are being made about whether to adopt or not to adopt, personal communication is far more likely to be influential. Therefore, the general conclusion of this line of thought is that *mass communication is less likely than personal influence to have a direct effect on social behavior.*

2. Newer perspectives on development communication claim that the diffusion and modernization perspective is a limited view of development communication. Juan Diaz Bordenave (1977) and Luis Ramiro Beltran (1976) were among the first Latin-American scholars who fundamentally questioned the capacity of the diffusionist communication model to modernize Latin-American societies (see

also ▶ Chap. 16, "De-westernizing Alternative Media Studies: Latin American Versus Anglo-Saxon Approaches from a Comparative Communication Research Perspective").

They argued that this diffusion model is a vertical or one-way perspective on communication and that development will accelerate mainly through active involvement in the process of the communication itself. Research has shown that, while the public can obtain information from impersonal sources like radio, television, and nowadays the Internet, this information has relatively little effect on behavioral changes. Development, furthermore, envisions precisely such change. Similar research has led to the conclusion that more is learned from interpersonal contacts and from mass communication techniques that is based on them. On the lowest level, before people can discuss and resolve problems, they must be informed of the facts – information that the media provide nationally as well as regionally and locally. At the same time, the public, if the media are sufficiently accessible, can make its information needs known.

Communication theories such as the "diffusion of innovations," the "two-step flow," or the early "extension" approaches are quite congruent with the above modernization theory. The elitist, *vertical or top-down orientation of the diffusion model* is obvious (for more details, see Servaes 2003, 2008).

One of the more useful attempts to theorize the means by which *cultural/media dependency* works through the modern forms of electronic communication is that offered by Oliver Boyd-Barrett (1977). According to Boyd-Barrett, the international communication process consists of *four major interrelated components*: (a) the shape of the communication vehicle, involving a specific technology for the distribution and consumption of media products; (b) a set of industrial arrangements that organize the relations between media finance, production, distribution, and consumption; (c) a body of values about ideal practice; and (d) specific media contents.

Herb Schiller (1976), one of the most prolific American communication scholars on dependency issues, called it "cultural imperialism." He studied the links between the US military-industrial complex and the media industry and its consequences for culture and communication worldwide. Schiller's (1976) definition of cultural imperialism is broad: "The sum of the processes by which a society is brought into the modern world system and how its dominating stratum is attracted, pressured, forced, and sometimes bribed into shaping social institutions to correspond to, or even promote, the values and structures of the dominating centre of the system" (p. 9).

3. The *participatory model*, on the other hand, incorporates the concepts in the framework of multiplicity. It stresses the importance of cultural identity of local communities and of *democratization and participation at all levels* – global, international, national, local, and individual. It points to a strategy, not merely inclusive of, but largely emanating from, the traditional "receivers." Paulo Freire (1983) refers to this as the right of all people to individually and collectively speak for themselves: "This is not the privilege of some few men, but the right of every (wo)man. Consequently, no one can say a true word alone – nor can he say

it for another, in a prescriptive act which robs others of their words" (p. 76) (see also ▶ Chap. 15, "The Importance of Paulo Freire to Communication for Development and Social Change").

In order to share information, knowledge, trust, commitment, and a right attitude in development projects, participation is very important in any decision-making process for development. Therefore, the *International Commission for the Study of Communication Problems*, chaired by the late Sean MacBride, illuminated that "this calls for a new attitude for overcoming stereotyped thinking and to promote more understanding of diversity and plurality, with full respect for the dignity and equality of peoples living in different conditions and acting in different ways" (MacBride 1980, p. 254). This model stresses reciprocal collaboration throughout all levels of participation (for more details, see Jacobson and Servaes 1999; see also ▶ Chaps. 9, "Media and Participation," and ▶ 10, "Empowerment as Development: An Outline of an Analytical Concept for the Study of ICTs in the Global South").

Also, these newer approaches posit that the *point of departure must be the community* (see, for instance, Geertz 1973; Servaes and Liu 2007). It is at the community level that the problems of living conditions are discussed and interactions with other communities are elicited. The most developed form of participation is self-management. This principle implies the rights to participation in the planning and production of media content. However, not everyone wants to or must be involved in its practical implementation. More important is that participation is made possible in the decision-making regarding the subjects treated in the messages and regarding the selection procedures.

One of the fundamental hindrances to the decision to adopt the participation strategy is that it threatens existing hierarchies. However, participation does not imply that there is no longer a role for development specialists, planners, and institutional leaders. It only means that the viewpoint of the local groups of the public is considered before the resources for development projects are allocated and distributed and that suggestions for changes in the policy are taken into consideration.

### 4.2.3 Research Priorities

1. Development communication in the *1958–1986 period* was generally greeted with enthusiasm and optimism. In her PhD thesis, Jo Ellen Fair (1988; summarized in Gazette, 1989) examined 224 studies of communication and development published between 1958 and 1986 and found that models predicting either powerful effects or limited effects informed the research:

    > Communication has been a key element in the West's project of developing the Third World. In the one-and-a-half decades after Lerner's influential 1958 study of communication and development in the Middle East, communication researchers assumed that the introduction of media and certain types of educational, political, and economic

information into a social system could transform individuals and societies from traditional to modern. Conceived as having fairly direct and powerful effects on Third World audiences, the media were seen as magic multipliers, able to accelerate and magnify the benefits of development. (p. 145)

Three directions for future research were suggested: (a) to examine the relevance of message content, (b) to conduct more comparative research, and (c) to conduct more policy research.

2. As a follow-up to this research, Jo Ellen Fair and Hemant Shah (1997) studied 140 journal articles, book chapters, and books published in English *between 1987 and 1996*. Their findings are quite illuminating:

*In the 1987–1996 period, Lerner's modernization model completely disappears.* Instead, the most frequently used theoretical framework is participatory development, an optimist postmodern orientation, which is almost the polar opposite of Lerner who viewed mass communication as playing a top-down role in social change. Also vanishing from research in this latter period is the two-step flow model, which was drawn upon by modernization scholars. (p. 10)

3. *Both periods* do make use of theories or approaches such as knowledge gap, indirect influence, and uses and gratifications. However, research appearing in the years from 1987 to 1996 can be characterized as much more theoretically diverse than that published between 1958 and 1986.
In the 1987–1996 study, the most frequent suggestion was "the need to conduct more policy research, including institutional analysis of development agency coordination. This was followed by the need to research and develop indigenous models of communication and development through participatory research" (Fair and Shah 1997, p. 19). Therefore, nobody was making the optimistic claims of the early years any longer.
4. In 2007, Hemant Shah completed an analysis of 167 items (123 journal articles, 38 book chapters, and 6 books) covering the *1997–2005 period*, and Christine Ogan et al. (2009) analyzed 211 on- and off-line peer-reviewed articles on communication and development between *1998 and 2007*. Though the meta-research technique may still need some fine-tuning, some of the findings are noteworthy:
- *Most authors work at Western institutions* (mainly North America, then Western Europe), rather than in the non-Western world, although the majority of authors actually come from developing countries.
- Surveys, *secondary data analysis, content analysis, and meta-research* were the most popular quantitative methods used in 1997–2007, whereas on the qualitative side were interviews, case studies, observation, focus groups, and ethnography.
- On the content side, the trend to conduct a theoretical research in this field persists. Three-quarters of all the studies used no theory to define their work. In those studies which build on theories, *modernization theories remain dominant*, followed by participatory development, dependency, and feminist

development. Globalization did not feature prominently in most studies. Lerner's model of media and development has reappeared in the 1997–2005 time period after totally disappearing in the 1987–1996 period. Only two other theories from the traditional US-based behavioral science approach, social learning theory and knowledge gap, appear in the 1997–2005 period. Shah (2007) explains the persistence of "old" ideas from a technological deterministic perspective:

> Each new technological innovation in the postcolonial world since 1958 – television, satellites, microwave, computers, call centers, wireless technology – has been accompanied by determined hope that Lerner's modernization model will increase growth and productivity and produce modern cosmopolitan citizens. (p. 24)

- This point is echoed by Ogan et al. (2009) who conclude that studies have moved away from mass communication and *toward ICTs' role in development* that they infrequently address development in the context of globalization and often continue to embrace a modernization paradigm despite its many criticisms:

  > We believe that the more recent attention to ICTs has to do with the constant search for the magic solution to bringing information to people to transform their lives, allowing them to improve their economic condition, educate their children, increase literacy and the levels of education and spread democracy in their countries. (p. 667)

- The *consequences of development communication* are very much associated with the more traditional views on modernization; that is, media activate modernity, and media raise knowledge levels. This more traditional perspective makes a strong return, as it was less pronounced during the 1987–1996 timeframe. The three other consequences listed are more critical to modernization: media create participatory society, media benefit certain classes, and media create development problems.
- The *optimistic belief* that there are overall positive impacts of development communication on individuals, dominant in 1958–1986, has consistently dropped from 25% (in 1958–1986) to 6% (in 1997–2005). Increasingly, however, it is pointed out that *more attention needs to be paid to theory and research*.
- *Suggestions for future research* prioritize the development of new development communication models and the examination of content relevance (both 27%), the need for indigenous models (24%), the study of new technologies (21%), more comparative research (18%), the need for more policy research (8%), and the development of a new normative framework (5%).

The findings by Fair, Shah, and Ogan present us with a clear but at the same time complex picture of the devcom field. The implicit assumptions on which the so-called dominant modernization paradigm is built do still linger on and continue to influence the policy- and planning-making discourse of major actors in the field of Communication for Development and Social Change, both at theoretical and applied levels.

## 4.3 Mapping the Future

The substantial components and major issues currently being discussed as the future of Communication for Development and Social Change can be listed under seven headings: (1) the interrelated processes of the emergence of interdisciplinarity, (2) the increasing role of the power of culture, (3) the birth of a new form of modernization, (4) the sustainability of social change processes, (5) the changing role of the nation-state, (6) the role of new social movements, and (7) the emerging attempts to address the link between the global and the local. Rico Lie (2003) listed five of them in this so-far most comprehensive roundup of the future of Communication for Development and Social Change.

### 4.3.1 Interdisciplinarity

Because of the complexity of societies and cultures, especially in a "world-system" perspective, the future of the social sciences seems to lie in interdisciplinarity. Theory on the impact of culture on globalization and localization has become a truly interdisciplinary academic field of study (Baumann 1999; Hopper 2007). Marxists, anthropologists, philosophers, political scientists, historians, sociologists, economists, communication specialists, and scholars in the field of cultural studies are attempting to integrate the field. The "meeting" of different perspectives, or in Geertz's terms "blurred genres" (Geertz 1983), on sociocultural phenomena seems to be the most adequate way to grasp the complexity. It is these united attempts that can provide fruitful insights and shed new light on old and new emerging problems.

### 4.3.2 The Power of Culture in Homogeneity and Diversity

Culture has long been regarded as only *context*, but culture is increasingly becoming *text* (Kroeber and Kluckhohn 1952). At the same time, it looks as if culture is also the concept that constitutes the common interests of the different disciplines and is as such responsible for interdisciplinarity. Robertson (1992) termed this increasing interest in culture "the cultural turn":

> For not merely is culture increasingly visible as a topic of specialized concern, it is evidently being taken more seriously as a relatively 'independent variable' by sociologists working in areas where it had previously been more or less neglected. (p. 32)

More focus on culture, and the increased attention that is being given to culture and cultural identity, has changed the debate from its inception with modernization, dependency, and "world-system" theory. In the Latin-American context, authors like Lawrence Harrison (1985) or Carlos Rangel (1977) started arguing in line with a Weberian perspective that *culture is the principle determinant of development*:

> In the case of Latin America, we see a cultural pattern, derivative of traditional Hispanic culture, that is anti-democratic, anti-social, anti-progress, anti-entrepreneurial, and, at least among the elite, anti-work. (p. 165)

In order to change this cultural pattern, Harrison proposes that at least seven issues need to be resolved: leadership, religious reform, education and training, the media, development projects, management practices, and child-rearing practices (Harrison 1985, pp. 169–177). Therefore, Anne Deruyttere (2006), former Chief of the Indigenous Peoples and Community Development Unit at the Inter-American Development Bank, pleads for *development with identity*:

> Development with identity refers to a process that includes strengthening of indigenous peoples, harmony and sustained interaction with their environment, sound management of natural resources and territories, the creation and exercise of authority, and respect for the rights and values of indigenous peoples, including cultural, economic, social and institutional rights, in accordance with their own worldview and governance. (p. 13)

### 4.3.3 A New Form of Modernization?

Globalization represents a new form of modernization that no longer equates Westernization (Hannerz 1996). However, in the current age of modernism, postmodernism, late or high modernism, or whatever new prefix is used, to grasp the current modern state of the world, it is important to, once again, point out the linear implications of this thinking. Globalization, which is highly associated with modernisms as a process of the changing cultural state of the world, is quite linear in its conceptualization (Held 2000; Sparks 2007). Although the process is less American oriented and is no longer equivalent to Westernization as those theories on cultural and media imperialism in the 1970s crudely referred to, it does not *fundamentally* change the line of thinking that the world has a modern end state determined by external forces (Hafez 2007; Lie 1998).

### 4.3.4 The Sustainability of Social Change Processes

In Servaes (2008), we subdivided communication strategies for development and social change at five levels: (a) *behavior change communication* (BCC) (mainly interpersonal communication), (b) *mass communication* (MC) (community media, mass media, and ICTs), (c) *advocacy communication* (AC) (interpersonal and/or mass communication), (d) *participatory communication* (PC) (interpersonal communication and community media), and (e) *communication for structural and sustainable social change* (CSSC) (interpersonal communication, participatory communication, and mass communication).

Interpersonal communication and mass communication form the bulk of what is being studied in the mainstream discipline of communication science. Behavior change communication is mainly concerned with short-term individual changes in attitudes and behavior. It can be further subdivided in perspectives that explain individual behavior, interpersonal behavior, and community or societal behavior.

Behavioral change communication (BCC), mass communication (MC), and advocacy communication (AC), though useful in itself, will not be able to create sustainable development. Therefore, participatory communication (PC) and communication for structural and sustainable social change (CSSC) are more concerned about long-term sustained change at different levels of society.

Looking at *desired or expected outcomes*, one could think of four broad categories: (a) approaches that attempt to *change attitudes* (through information dissemination, public relations, etc.), (b) *behavioral change* approaches (focusing on changes of individual behavior, interpersonal behavior, and/or community and societal behavior), (c) *advocacy* approaches (primarily targeted at policy-makers and decision-makers at all levels and sectors of society), and (d) *communication for structural and sustainable change* approaches (which could be either top-down, horizontal, or bottom-up). The first three approaches, though useful by themselves, are in isolation not capable of creating sustainable development. Sustainable social change can only be achieved in combination with incorporating aspects of the wider environment that influences (and constrains) structural and sustainable change. These aspects include structural and conjunctural factors (e.g., history, migration, conflicts), policy and legislation, service provision, education systems, institutional and organizational factors (e.g., bureaucracy, corruption), cultural factors (e.g., religion, norms, and values), sociodemographic factors (e.g., ethnicity, class), sociopolitical factors, socioeconomic factors, and the physical environment.

For an overview and more details, see McKee et al. (2000), Papa et al. (2006), and Tremblay (2007). McKee et al. (2000) provide a multi-sectoral and interdisciplinary synthesis of field experiences and lessons learned in the context of behavior development and change. Its aim is to challenge traditional approaches to program design, implementation, and monitoring, with a view to increase the impact and sustainability of international development programs such as the ones coordinated by UNICEF. Omoto's (2005) interdisciplinary academic volume addresses a variety of topics related to service learning, social movements, political socialization, civil society, and especially volunteerism and community development. Papa et al. (2006) address four dialectic tensions which are considered essential to the process of organizing social change: control and emancipation, oppression and empowerment, dissemination and dialogue, and fragmentation and unity. They advocate for a dialectic approach.

As part of an ongoing research project (see Servaes et al. 2012; Servaes 2011; Servaes 2013a, b, 2014), we have developed a framework for the assessment of sustainability. Based on a review of the literature, our indicators were designed for four sectors of development or social change: *health, education, environment, and governance*. We selected eight indicators for each of the sectors of development: *actors* (the people involved in the project, which may include opinion leaders, community activists, tribal elders, youth, etc.), *factors* (structural and conjunctural), *level* (local, state, regional, national, international, global), *development communication approach* (behavioral change, mass communication, advocacy, participatory communication, or communication for sustainable social change – which is likely a mix of all of the above), *channels* (radio, ICT, TV, print, Internet, etc.), *message* (the content of the project, campaign), *process* (diffusion-centered, one-way, and information-persuasion strategies or interactive and dialogical), and *method* (quantitative, qualitative, participatory, or in combination) (Table 2).

For each indicator, we developed a set of questions designed to specifically measure the sustainability of the project. We defined "sustainability," for example,

**Table 2** Testing sustainability framework: categories and indicators

| Indicators | | Health | Education | Governance | Environment |
|---|---|---|---|---|---|
| **Actors** | | | | | |
| **Factors** | Structural | | | | |
| | Conjuctural | | | | |
| **Level** | Local | | | | |
| | National | | | | |
| | Regional | | | | |
| **Type communication** | Behavioral | | | | |
| | Mass communication | | | | |
| | Advocacy | | | | |
| | Participatory communication | | | | |
| | Communication for social change | | | | |
| **Channels** | Face to face | | | | |
| | Print | | | | |
| | Radio | | | | |
| | Television | | | | |
| | ICT | | | | |
| | Telephone/cellular phone | | | | |
| **Process** | Persuasion strategies | | | | |
| | One-way transmission | | | | |
| | Interactive dialogue | | | | |
| **Methods** | Quantitative | | | | |
| | Qualitative | | | | |
| | Participatory | | | | |
| | Mixed methods | | | | |
| **Message** | Was it developed by the community? | | | | |
| | Was it received? | | | | |
| | Was it understood? | | | | |

by analyzing whether the channels are compatible with both the capacity of the actors and the structural and conjunctural factors. If they are, the project will have a higher likelihood of being sustainable in the long run. We asked to what extent was the process participatory and consistent with the cultural values of the community. Was the message developed by local actors in the community and how was it understood? Our research shows that the more local and interactive the participation – in levels, communication approaches, channels, processes, and methods – the more sustainable the project will be. We analyzed each of the indicators in the context of sustainability further for each project.

### 4.3.5 Nation-States and National Cultures

Nation-states are seen by most scholars, especially Marxists (Hirst and Thompson 1996), as the basic elements in a world system and the main actors in the process of globalization; but is this also true for cultural globalization? Does the globalization thesis automatically imply that national cultures are the main elements or actors in a "global culture"? Are the nation-states and national cultures the central points of convergence and main actors in globalization (Anderson 1983; Sunkel and Fuenzalida 1980)?

On the one hand, the globalization thesis centers the nation-state as the main actor. This has become even more apparent since the financial meltdown in 2008. Many national governments have not only moved to forms of protectionism unheard of a decade ago, but they also tend to argue that this might be the only realistic way out of the crisis. On the other hand, different scholars try to escape from the limitations of state-centrism. This problematizing of state-centrism is in essence what the globalization thesis is all about. According to Delcourt (2009) and Scholte (2005), we must go beyond the nation-state and develop sociology of the global system. The same seems to be true for the cultural field. Discussions on global and local culture seem to go beyond discussions that centralize the nation-state and thus center national culture, national identity, and nationalism. The nation-state might be the most significant political-economic unit into which the world is divided, but a cultural discussion on globalization must include other levels, because the nation-state is not the only cultural frame used for the construction of a cultural identity (Servaes and Lie 1998). Tomlinson (1999) also showed us by analyzing the discourse within UNESCO, where cultural identity seems to be seen as national identity equal, that this statement cannot stand ground because cultural identity transcends national identity (Tomlinson 1999, pp. 70–75).

If culture at the national level is identified as being only one level that structures and frames the construction of identity, we need to initiate a discussion on other levels that play roles in the process of identity construction. There is little discussion possible about what the global level incorporates. There is no bigger sociocultural or economic-political analytical frame possible. But there seems to be some different interpretations possible about what "local" actually is. Is it an extended family, a village, a tribe, a neighborhood in a town, a city, a county, a region, an island, or even a nation-state? Or should it go beyond a nation-state and become associated with "local" global communities, real or virtual or both? These are important issues that also need to be incorporated in the discussions on macro-micro linkages in the social sciences (Elliott and Lemert 2006).

### 4.3.6 The Place of Civil Society and the Role of New Social Movements

New identities and the need for new meanings arise from and are a consequence of the encounters between individualization, privatization, globalization, the extensions of control, and disjunctures between established forms of representation and emerging needs. These struggles for identity and meaning have been carried out by a variety of civil society and new social movements. These local, national, regional, and transnational projects and struggles have been powered by new media networks that have proven invaluable for organizing, recruiting, campaigning, lobbying, and experimenting with the planning, organization, implementation, and delivery of innovative, participatory projects.

Moreover, culture is also increasingly seen as an important factor in international communication, social processes, *social movements, and civil society* (de Sousa Santos 2007).

Each from their specific vantage point provides unique insights and perspectives for a better understanding of what the future has in store: Braman and Sreberny-Mohammadi (1996, on globalization and localization), Cimadevilla and Carniglia (2004, on sustainability and rural development), Friedmann (1992, on empowerment), Downing (2001, on radical communication and social movements), Kennedy (2008, one of the best researched and documented cases on participatory development with the Native Alaskans), Kronenburg (1986, compares two development projects in Kenya), and Nash (2005, analyzes social movements from an anthropological perspective).

### 4.3.7 Linking the Global and the Local

Globalization and localization are seen as interlinked processes, and this marks a *radical change in thinking about change and development*. As Anthony Giddens (1995) observed, "Globalization does not only concern the creation of large-scale systems, but also the transformation of the local, and even personal, context of social experience" (pp. 4–5). Potentially, it integrates global dependency thinking, world-system theory, and local, grassroots, interpretative, participatory theory and research on social change (Bauman 1998; Berger and Huntington 2002).

Obviously, the debates in the general field of "international and intercultural communication for development and social change" have shifted and broadened. They have shifted in the sense that they are now focusing on issues related to "global culture," "local culture," "(post)modernity," and "multiculturalism" instead of their previous concern with "modernization," "synchronization," and "cultural imperialism" (Grillo et al. 1998). Therefore, in contrast to mainstream views on globalization, which center on the political economy, the global industry, and have a capitalist-centered view of the world, here, the focus is *on situating the field of globalization in the local* (Harindranath 2006).

At the same time, the debates have shifted from an emphasis on homogeneity toward an *emphasis on differences*. Therefore, the total conglomerate of changes accounts for something new, and especially the last issue of linking the global with the local can be identified as a central point of change. But how can this conglomerate of global changes be linked to development and political-economic and social change at local levels and from within local levels?

## 4.4 Directions for Future Research

Today almost nobody would dare to make the optimistic claims of the early years any longer. The experience of the past 50 years has demonstrated that development is possible, but not inevitable. But, as mentioned earlier, the findings by Jo Ellen Fair, Christine Ogan, and Hemant Shah present us with a clear but at the same time complex picture of our field: the re-emergence of *modernization theories* with a technological-deterministic twist, followed by participatory development, dependency, feminist development, and globalization.

### 4.4.1 The Transformation of Society

As many others, we are in search of a new paradigm for development and social change, one which looks at development as a *transformation of society*. In such a perspective, "change is not an end in itself, but a means to other objectives. The changes that are associated with development provide individuals and societies more control over their own destiny. Development enriches the lives of individuals by widening their horizons and reducing their sense of isolation. It reduces the afflictions brought on by disease and poverty, not only increasing lifespans, but improving the vitality of life" (Stiglitz 1998, p. 3). We have referred to this paradigm as *multiplicity* (Servaes 1999).

This perspective argues that considerations of communication need to be explicitly built into development plans to ensure that a mutual sharing/learning process is facilitated. Such communicative sharing is deemed the best guarantee for creating successful transformative projects. Therefore, as already mentioned at the outset, we define Communication for Development and Social Change as:

> *Communication for social change is the nurturing of knowledge aimed at creating a consensus for action that takes into account the interests, needs and capacities of all concerned. It is thus a social process, which has as its ultimate objective sustainable development at distinct levels of society.*

> Communication media and ICTs are important tools in achieving social change but their use is not an end in itself. Interpersonal communication, traditional and group media must also play a fundamental role.

The new starting point is examining the processes of "bottom-up" change, focusing on self-development of local communities. The basic assumption is that there are no countries or communities that function completely autonomously and that are completely self-sufficient, nor are there any nations whose development is exclusively determined by external factors. Every society is dependent in one way or another, both in form and in degree.

### 4.4.2 The Cosmopolitan Challenge

This also implies, as argued by Southern scholars like Kwame Anthony Appiah (2006), Wimal Dissanayake (2006), Shelton Gunaratne (2005), or Majid Tehranian (2007), that a cultural perspective has to be fully embraced. Wimal Dissanayake (2006), for instance, argues for a *new concept of humanism*:

> Humanism as generally understood in Western discourse ... places at the center of its interest the sovereign individual – the individual who is self-present, the originator of action and meaning, and the privileged location of human values and civilizational achievements. However, the concept of the self and individual that is textualized in the kind of classical works that attract the attention of Asian communication theorists present a substantially different picture. The ontology and axiology of selfhood found in Buddhism differs considerably from those associated with European humanism. What these differences signpost is that there is not one but many humanisms. (p. 6)

Many humanisms may lead to what Appiah (2006) calls *the cosmopolitan challenge*:

> If we accept the cosmopolitan challenge, we will tell our representatives that we want them to remember those strangers. Not because we are moved by their suffering – we may or may not be – but because we are responsive to what Adam Smith called 'reason, principle, conscience, the inhabitant of the beast.' The people of the richest nations can do better. This is a demand of simple morality. But it is one that will resonate more widely if we make our civilization more cosmopolitan. (p. 174)

### 4.4.3 The Content of Development Agendas

Attention is also needed to critically analyze the *content of development agendas*. An understanding of the way in which development projects both encounter and transform power relationships within (and between) the multiple stakeholders who are impacted by such projects and an understanding of the way in which communication plays a central part in building (or maintaining or changing) power relationships are needed. Therefore, *three streams of action* are important:

- Media must be activated to build public support and upward pressure for policy decisions.
- Interest groups must be involved and alliances established for reaching a common understanding and mobilizing societal forces. This calls for networking with influential individuals and groups, political forces and public organizations, professional and academic institutions, religious and cause-oriented groups, business, and industry.
- Public demand must be generated and citizens' movements activated to evoke a response from national leaders. It may not always be easy to build up a strong public movement around development issues – but even a moderate display of interest and effort by community, leaders could stimulate the process for policy decisions and resource allocation for combating the problem.

### 4.4.4 More Participatory Communication Research Needed

Research that addresses the achievement and sustainability of processes and outcomes of communication for development should be encouraged. This requires a participatory approach, a shared framework between development agencies and local stakeholders, and community involvement in design, implementation, and dissemination (Servaes and Malikhao 2010).

1. There are two major approaches to participatory communication that everybody today accepts as common sense. The first is the dialogical pedagogy of Paulo Freire (1970, 1973, 1983, 1994), and the second involves the ideas of access, participation, and self-management articulated in the UNESCO debates of the 1970s (Berrigan 1977, 1979). Every communication project that calls itself participatory accepts these principles of democratic communication. Nonetheless there exist today a wide variety of practical experiences and intentions. Before moving on to explore these differences, it is useful to briefly review the common ground.

    The *Freirian argument* works by a dual theoretical strategy. He insists that subjugated peoples must be treated as fully human subjects in any political process. This implies dialogical communication. Although inspired to some extent by Sartre's existentialism – a respect for the autonomous personhood of each human being – the more important source is a theology that demands respect for otherness, in this case that of another human being. The second strategy is a moment of utopian hope derived from the early Marx that the human species has a destiny which is more than life as a fulfillment of material needs. Also from Marx is an insistence on collective solutions. Individual opportunity, Freire stresses, is no solution to general situations of poverty and cultural subjugation.

    These ideas are deeply unpopular with elites, including elites in the Third World, but there is nonetheless widespread acceptance of Freire's notion of dialogic communication as a normative theory of participatory communication. One problem with Freire is that his theory of dialogical communication is based on group dialogue rather than such amplifying media as radio, print, and television. Freire also gives little attention to the language or form of communication, devoting most of his discussion to the intentions of communication actions.

The second discourse about participatory communication is the UNESCO language about *self-management, access, and participation* from the 1977 meeting in Belgrade, the former Yugoslavia. The final report of that meeting defines the terms in the following way:

- Access refers to the use of media for public service. It may be defined in terms of the opportunities available to the public to choose varied and relevant programs and to have a means of feedback to transmit its reactions and demands to production organizations.
- Participation implies a higher level of public involvement in communication systems. It includes the involvement of the public in the production process and also in the management and planning of communication systems.
  Participation may be no more than representation and consultation of the public in decision-making.
- On the other hand, self-management is the most advanced form of participation. In this case, the public exercises the power of decision-making within communication enterprises and is also fully involved in the formulation of communication policies and plans.

Access by the community and participation of the community are to be considered key defining factors, as Berrigan eloquently summarizes: "[Community media] are media to which members of the community have access, for information, education, entertainment, when they want access. They are media in which the community participates, as planners, producers, and performers. They are the means of expression of the community, rather than for the community" (Berrigan 1979, p. 8). Referring to the 1977 meeting in Belgrade, Berrigan (1979, p. 18) (partially) links access to the reception of information, education, and entertainment considered relevant by/for the community: "[Access] may be defined in terms of the opportunities available to the public to choose varied and relevant programs, and to have a means of feedback to transmit its reactions and demands to production organizations."

Others limit access to mass media and see it as "the processes that permit users to provide relatively open and unedited input to the mass media" (Lewis 1993, p. 12) or as "the relation to the public and the established broadcasting institutions" (Prehn 1991, p. 259). Both the production and reception approaches of "access" can be considered relevant for an understanding of "community media."

These ideas are important and widely accepted as a normative theory of participatory communication: it must involve access and participation (Pateman 1972). However, one should note some differences from Freire. The UNESCO discourse includes the idea of a gradual progression. Some amount of access may be allowed, but self-management may be postponed until sometime in the future. Freire's theory allows for no such compromise. One either respects the culture of the other or falls back into domination and the "banking" mode of imposed education. The UNESCO discourse talks in neutral terms about "the public."

Freire talked about "the oppressed." Finally, the UNESCO discourse puts the main focus on the institution. Participatory or community radio means a radio station that is self-managed by those participating in it.
2. Participation involves the more equitable sharing of both political and economic power, which often decreases the advantage of certain groups. Structural change involves the redistribution of power. In mass communication areas, many communication experts agree that structural change should occur first in order to establish participatory communication policies. Mowlana and Wilson (1987, p. 143), for instance, state: "Communications policies are basically derivatives of the political, cultural and economic conditions and institutions under which they operate. They tend to legitimize the existing power relations in society, and therefore, they cannot be substantially changed unless there are fundamental structural changes in society that can alter these power relationships themselves." Therefore, the development of a participatory communication model has to take place in relation with overall societal emancipation processes at local, national as well as international levels. Several authors have been trying to summarize the criteria for such a communication model. The Latin-American scholar Juan Somavia (1977, 1981) sums up the following (slightly adapted) components as essential for it:

(a) *Communication is a human need.* The satisfaction of the need for communication is just as important for a society as the concerns for health, nutrition, housing, education, and labor. Together with all the other social needs, communication must enable the citizens to emancipate themselves completely from being oppressed. The rights to inform and to be informed and the rights to communicate are thus essential human rights at an individual and a collective level.

(b) *Communication is delegated human rights.* Within its own cultural, political, economic, and historical context, each society has to be able to define the concrete form in which it wants to organize its social communication process independently. Because of varieties of culture, various organizational structures arise. Whatever the form in which the social communication function is embodied, priority must be given to the principles of participation and accessibility.

(c) *Communication is a facet of the societal conscientization, emancipation, and liberation process.* The social responsibility of the media in the process of social change is very large. Indeed, after the period of formal education, the media are the most important educational and socialization agents. They are capable of informing or misinforming, exposing or concealing important facts, interpreting events positively or negatively, and so on.

(d) *The communication task involves rights and responsibilities/obligations.* Since the media in fact provide a public service, they must carry it out in a framework of social and juridical responsibility that reflects the social consensus of the society. In other words, there are no rights without obligation.

The freedom and right to communicate, therefore, must be approached from a threefold perspective: *firstly*, it is necessary for the public to participate effectively in the communication field; *secondly*, there is the design of a framework in which this can take place; and, *thirdly*, the media must enjoy professional autonomy, free of economic, political, or whatever pressure.

In other words, participatory communication for social change sees people as the nucleus of development. Development means lifting up the spirits of a local community to take pride in its own culture, intellect, and environment. Development aims to educate and stimulate people to be active in self and communal improvements while maintaining a balanced ecology. Authentic participation, though widely espoused in the literature, is not in everyone's interest. Due to their local concentration, participatory programs are, in fact, not easily implemented, nor are they highly predictable or readily controlled.

3. In spite of the widespread acceptance of the ideas of Freire and UNESCO by development organizations and communication researchers, there is still a very wide range of projects calling themselves "participatory communication projects." There is an evident need for clarification in descriptive and normative theories of participatory media. What does it mean to be participatory? It is necessary to make further distinctions and arguments to deal with a wide variety of actually existing experiences and political intentions.

   A review of the literature turns up the following *types* (Berrigan 1979):
   1. Participatory media are internally organized on democratic lines (as worker cooperatives or collectives).
   2. Participatory media are recognized by their opposition to cultural industries dominated by multinational corporations.
   3. Participatory media may be traced to the liberation of linguistic and ethnic groups following a major social transformation.
   4. The strong existence of participatory media may be explained in terms of class struggle within the society.
   5. Participatory media may be identified as "molecular" rather than "molar" (a collectivity of individual autonomous units rather than one that is homogenized and one-dimensional).
   6. Participatory media (like the montage of Eisenstein and the theater of Brecht) by design require a creative and varied reception from its audience.

Reyes-Matta (1986) states that participatory communication is first and foremost an alternative to media dominated by transnational corporations. This is the context in which any alternative must operate. To succeed is to have won against the culture industries that are dominated by multinational corporations. The line of thought developed by CINCO (1987) is a development of this because it involves above all a structural analysis of communicative institutions. For the CINCO researchers, a medium is alternative, if it has a democratic institutional structure. Here the issue is one of ownership and control that is external to the community against access and participation in the media organization.

Legitimacy and political credibility can be fostered by the establishment of what is called *participatory democracy*, the building in of actual participation from the public. This is only possible when the communication system is decentralized. The

**Table 3** Positioning the four theoretical approaches on community media (CM)

|  | Media-centered | Society-centered |
|---|---|---|
| **Autonomous identity of CM** (Essentialist) | *Approach I*: Serving the community | *Approach III*: Part of civil society |
| **Identity of CM in relation to other identities** (Relationalist) | *Approach II*: An alternative to mainstream | *Approach IV*: Rhizome |

control over communication and information may not be monopolized by one or a few segments of the society. Unfortunately, most of the time, structural aspects stand in the way of the ideal of democracy. In most developing countries, the first stone for bridging the gap between the ruling elite and the masses has still to be laid. For the establishment of participatory democracy, therefore, dialogue must be made possible between the authorities and the public, nationally, regionally, and locally. In the political sector, this can be done through political parties, pressure groups, civil action groups, environmental movements, and the like. Thus political credibility as well as social and cultural identity of the population and an awareness and support of the development goals are needed.

The concept of *community media* (CM) has shown to be, in its long theoretical and empirical tradition, highly elusive. The multiplicity of media organizations that carry this name has caused most mono-theoretical approaches to focus on certain characteristics, while ignoring other aspects of the identity of community media. This theoretical problem necessitates the use of different approaches toward the definition of community media (Table 3), which will allow for a complementary emphasis on different aspects of the identity of "community media" (for an elaboration, see Carpentier et al. 2001, 2012).

### 4.4.5 Measurement and Evaluation for "Social Usefulness"

Evaluation and impact assessments should include participatory baseline formulations and communication needs assessments. They should also include self-evaluation by the communities themselves and the concept of "social usefulness." They should be used to feed back at the policy level. There is a need for effective and convincing evaluation models and data to show evidence of the impact of communication for development. Sustainability indicators based on qualitative dimensions of development need to be emphasized, involving the potential of ICTs to collect feedback interactively. Research should also be reinforced in order to better identify communication needs.

While many successful small-scale examples of communication for development exist, these need to be scaled up, thus improving practice and policy at every level. A focus on small-scale projects (pilot projects) is acceptable, but evidence-based and properly researched benchmarks need to be set.

## 4.5 Concluding Remarks

In this chapter, it summarizes the past of Communication for Development and Social Change; identified the roadmap for the future of Communication for Development and Social Change; and looked at some of the key purposes, functions, and approaches needed to steer communication for development and social change. In the other chapters published in this handbook, specific aspects and themes will be further detailed, explored, and discussed in order to hopefully arrive at a more comprehensive and holistic perspective on the increasingly complex issue of Communication for Development and Social Change.

## References

Anderson B (1983) Imagined communities: reflections on the origin and spread of nationalism. Verso, London
Appiah KA (2006) Cosmopolitanism: ethics in a world of strangers. Allen Lane, London
Bartolovich C, Lazarus N (eds) (2007) Marxism, modernity, and postcolonial studies. Peking University Press, Peking
Bauman Z (1998) Globalization: the human consequences. Columbia University Press, New York
Baumann G (1999). The multicultural riddle. Rethinking National, Ethnic, and Religious Identities. Routledge, London
Beltran LR (1976) TV etchings in the minds of Latin Americans: conservatism, materialism and conformism. Paper presented to the IAMCR conference, Leicester, September
Berger P, Huntington S (eds) (2002) Many globalizations: cultural diversity in the contemporary world. Oxford University Press, New York
Berrigan FJ (1977) Access: some Western models of community media. UNESCO, Paris
Berrigan FJ (1979) Community communications. The role of community media in development. UNESCO, Paris
Bordenave JD (1977) Communication and rural development. UNESCO, Paris
Boyd-Barrett O (1977) Media imperialism: towards an international framework for the analysis of media systems. In: Curran J, Gurevitch M, Woollacott J (eds) Mass communication and society. Arnold, London
Braman S, Sreberny-Mohammadi A (eds) (1996) Globalization, communication and transnational civil society. Hampton Press, Cresskill
Canclini NG (1993) Culturas hibridas. Estrategias para entrar y salir de la modernidad. Grijalbo
Cardoso FH, Falletto E (1969) Dependencia y desarrollo en América Latina. Siglo XXI, Mexico D.F.
Carpentier N, Lie R, Servaes J (2001) Making community media work. Report prepared for UNESCO, Paris, 50 pp + CD-Rom
Carpentier N, Lie R, Servaes J (2012) Multitheoretical approaches to community media: capturing specificity and diversity. In: Fuller L (ed) The power of global community media. Palgrave Macmillan, New York, pp 219–236
Chang P-C (2008) Does China have alternative modernity? An examination of Chinese modernization discourse. J Commun Dev Soc Change 2:4
Chew SC, Denemark RA (eds) (1996) The underdevelopment of development: essays in honor of Andre Gunder Frank. Sage, Newbury Park
Cimadevilla G, Carniglia E (eds) (2004) Comunicacion, Ruralidad y Desarrollo. Mitos, paradigmas y dispositivos del cambio. Instituto Nacional de Tecnologia Agropecuaria, Buenos Aires
Cinco. (1987) Comunicacion Dominante y Comunicacion Alternativa en Bolivia. Cinco/IDRC, La Paz

de Sousa Santos B (ed) (2007) Democratizing democracy: beyond the liberal democratic canon. Verso, London
Delcourt L (2009) Retour de l'Etat. Pour quelles politiques sociales? Alternatives Sud, vol XVI, 2. Centre Tricontinental, Louvain-la-Neuve
Deruyttere A (ed) (2006) Operational policy on indigenous peoples and strategy for indigenous development. Inter-American Development Bank, Washington, DC
Dissanayake W (2006) Postcolonial theory and Asian communication theory: towards a creative dialogue. China Media Res 2(4):1–8
Downing J (2001) Radical media: rebellious communication and social movements. Sage, Thousand Oaks
Elliott A, Lemert C (2006) The new individualism: the emotional costs of globalization. Routlege, London
Fair JE (1988) A meta-research of mass media effects on audiences in developing countries from 1958 through 1986. Unpublished doctoral dissertation, Indiana University, Bloomington
Fair JE (1989) 29 years of theory and research on media and development: the dominant paradigm impact. Gazette 44:129–150
Fair JE, Shah H (1997) Continuities and discontinuities in communication and development research since 1958. J Int Commun 4(2):3–23
Fraser C, Restrepo-Estrada S (1998) Communicating for development. Human change for survival. I.B. Tauris, London
Freire P (1970) Cultural action for freedom. Penguin, Harmondsworth
Freire P (1973) Extension o comunicacion? La concientizacion en el medio rural. Siglo XXI, Mexico
Freire P (1983) Pedagogy of the oppressed. Continuum, New York
Freire P (1994) Pedagogy of hope. Reliving pedagogy of the oppressed. Continuum, New York
Friedmann J (1992) Empowerment: the politics of alternative development. Blackwell, Cambridge, MA
Galtung J (1980) The true worlds. A transnational perspective. Free Press, New York
Geertz C (1973) The interpretation of cultures. Basic Books, New York
Geertz C (1983) Local knowledge: further essays in interpretative anthropology. Basic Books, New York
Giddens A (1995) Modernity and self-identity: self and society in the late modern age. Polity, Cambridge
Grillo M, Berti S, Rizzo A (1998) Discursos locales. Universidad Nacional de Rio Cuarto, Rio Cuarto
Gunaratne S (2005) The Dao of the press: a humanocentric theory. Hampton Press, Cresskill
Habermann P, de Fontgalland G (eds) (1978) Development communication–rhetoric and reality. AMIC, Singapore
Hafez K (2007) The myth of media globalization. Polity, Cambridge, UK
Hannerz U (1996) Transnational Connections. Routledge, London
Harindranath R (2006) Perspectives on global cultures. Open University Press, Maidenhead
Harrison L (1985) Underdevelopment is a state of mind. The Latin American case, 2nd edn. Madison Books, Lanham
Held D (ed) (2000) A globalizing world? Culture, economic, politics. Routledge, London
Hirst P, Thompson G (1996) Globalization in question: the international economy and the possibilities of governance. Polity, London
Hoffmann V (2007) Five editions (1962–2003) of Everett Rogers's diffusion of innovations. J Agric Educ Ext 13(2):147–158
Hopper P (2007) Understanding cultural globalization. Polity, Cambridge, UK
Jacobson T, Servaes J (eds) (1999) Theoretical approaches to participatory communication. Hampton, Cresskill
Jayaweera N, Amunugama S (eds) (1987) Rethinking development communication. AMIC, Singapore

Kennedy T (2008) Where the rivers meet the sky: a collaborative approach to participatory development. Southbound, Penang
Kroeber A, Kluckhohn C (1952) Culture. A critical review of concepts and definitions. Vintage Books, New York
Kronenburg J (1986) Empowerment of the poor: a comparative analysis of two development endeavors in Kenya. Third World Center, Nijmegen
Lerner D (1958) The passing of traditional society: modernizing the Middle East. Free Press, New York
Lewis P (ed.) (1993) Alternative Media: Linking Global and Local, Reports and Papers on Mass Communication, no. 107. UNESCO, Paris
Lie R (1998) What's new about cultural globalization? Linking the global from within the local. In: Servaes J, Lie R (eds) Media and politics in transition: cultural identity in the age of globalization. ACCO, Leuven, pp 141–155
Lie R (2003) Spaces of intercultural communication: an interdisciplinary introduction to communication, culture, and globalizing/localizing identities. Hampton, Cresskill
MacBride S (ed) (1980) Many voices, one world: communication and society. Today and tomorrow. UNESCO, Paris
Martin-Barbero J (1993) Communication, culture and hegemony, from the media to mediations. Sage, London
McKee N, Manoncourt E, Saik Yoon C, Carnegie R (2000) Involving people evolving behaviour. Southbound, Penang
McMichael P (2008) Development and social change: a global perspective. Pine Forge Press, Los Angeles
Mowlana H (1986) Development. A field in search of itself. American University, Washington, DC
Mowlana H, Wilson L (1987) Communication and development: a global assessment. UNESCO, Paris
Nash J (ed) (2005) Social movements: an anthropological reader. Blackwell, Malden, MA
Nederveen Pieterse J (2010) Development theory, 2nd edn. Sage, Los Angeles
Ogan CL, Bashir M, Camaj L, Luo Y, Gaddie B, Pennington R, Rana S, Salih M (2009) Development communication: the state of research in an era of ICTs and globalization. Int Commun Gaz 71(8):655–670
Omoto A (2005) Processes of community change and social action. Lawrence Erlbaum, Mahwah
Papa MJ, Singhal A, Papa W (2006) Organizing for social change. A dialectic journey of theory and praxis. Sage, New Delhi
Pateman C (1972) Participation and democratic theory. Cambridge University Press, Cambridge
Prehn O (1991) From small scale utopianism to large scale pragmatism. In: Jankowski N, Prehn O, Stappers J (eds) The people's voice. Local radio and television in Europe. John Libbey, London/Paris/Rome, pp 247–268
Rangel C (1977) The Latin Americans: their love-hate relationship with the United States. Harcourt Brace Jovanovich, New York
Reyes-Matta F (1986) Alternative communication: solidarity and development in the face of transnational expansion. In: Atwoord R, Mcanany E (eds) Communication and Latin American society. Trends in critical research 1960–1985. University of Wisconsin Press, Madison, pp 190–214
Robertson R (1992) Globalization: social theory and global culture. Sage, London
Rogers EM (1962) The diffusion of innovations. The Free Press, New York
Rogers EM (1986) Communication technology: the new media in society. The Free Press, New York
Rogers A (2005) Participatory diffusion or semantic confusion. In: Harvey M (ed) Media matters: perspectives on advancing governance & development from the Global Forum for Media Development. Internews Europe, London, pp 179–187
Schiller HI (1976) Communication and cultural domination. International Arts and Sciences Press, New York

Scholte J (2005) Globalization: a critical introduction. Palgrave, New York
Schramm W (1964) Mass media and national development: the role of information in the developing countries. Stanford University Press, Stanford
Servaes J (1999) Communication for development: one world, multiple cultures. Hampton, Creskill
Servaes J (ed) (2003) Approaches to development: studies on communication for development. UNESCO, Paris
Servaes J (ed) (2008) Communication for development and social change. Sage, London
Servaes J (2011) Social change. Oxford bibliographies online (OBO). Oxford University Press, New York, 58 pp. http://www.oxfordbibliographiesonline.com/display/id/obo-9780199756841-0063
Servaes J (ed) (2013a) Sustainable development and green communication. African and Asian perspectives. Palgrave Macmillan, London/New York
Servaes J (ed) (2013b) Sustainability, participation and culture in communication. Theory and praxis. Intellect-University of Chicago Press, Bristol/Chicago
Servaes J (ed) (2014) Technological determinism and social change. Communication in a tech-mad world. Lexington Books, Lanham
Servaes J, Lie R (eds) (1998) Media and politics in transition: cultural identity in the age of globalization. Acco, Louvain-la-Neuve
Servaes J, Liu S (eds) (2007) Moving targets: mapping the paths between communication, technology and social change in communities. Southbound, Penang
Servaes J, Malikhao P (2010) Comunicacion participativa. El Nuevo paradigma? In: Thornton R, Cimadevilla G (eds) Usos y Abusos del Participare/Usos e abusos do participar. Instituto Nacional de Tecnología Agropecuaria (INTA), Buenos Aires, pp 67–90
Servaes J, Polk E, Shi S, Reilly D, Yakupitijage T (2012) Towards a framework of sustainability indicators for 'communication for development and social change' projects. Int Commun Gaz (Sage) 74(2):99–123
Shah H (2007) Meta-research of development communication studies, 1997–2005: patterns and trends since 1958. Paper presented to ICA, San Francisco, May
Somavia J (1977) Third world participation in international communication. Perspective after Nairobi. Paper symposium 'International communication and third world participation', Amsterdam, September
Somavia J (1981) The democratization of communication: from minority social monopoly to majority social representation. Dev Dialogue 2:13
Sparks C (2007) Globalization, development and the mass media. Sage, London
Staples A (2006) The birth of development: how the World Bank, Food and Agriculture Organization, and World Health Organization changed the world, 1945–1965. The Kent State University Press, Kent
Stiglitz J (1998) Towards a new paradigm for development: strategies, policies, and processes. Prebisch lecture at UNCTAD, Geneva, 19 October 1998
Sunkel O, Fuenzalida E (1980) La transnacionalizacion del capitalismo y el desarrollo nacional. In: Sunkel O, Fuenzalida E, Cardoso F et al (eds) Transnacionalizacion y dependencia. Cultura Hispania, Madrid
Tehranian M (2007) Rethinking civilization: resolving conflict in the human family. Routledge, London
Tomlinson J (1999) Globalisation and culture. Polity, Cambridge
Tremblay S (ed.) (2007) Developpement durable et communications. Au-dela des mots, pour un veritable engagement. Presses de l'Universite du Quebec, Quebec

# Family Tree of Theories, Methodologies, and Strategies in Development Communication

## 5

Silvio Waisbord

## Contents

5.1 Introduction ........................................................................... 94
5.2 Development Communication .......................................................... 94
5.3 The Dominant Paradigm ................................................................ 95
5.4 Theories in the Tradition of the Dominant Paradigm ................................... 98
    5.4.1 Social Marketing ............................................................... 99
    5.4.2 Health Promotion and Health Education ......................................... 103
    5.4.3 Entertainment-Education ....................................................... 105
5.5 Critiques of the Dominant Paradigm .................................................... 108
    5.5.1 Dependency Theory ............................................................. 108
    5.5.2 Participatory Theories and Approaches .......................................... 110
    5.5.3 Media Advocacy ................................................................ 116
    5.5.4 Social Mobilization ............................................................ 118
5.6 Toward a Theoretical and Empirical Convergence? ...................................... 120
    5.6.1 General Remarks ................................................................ 120
    5.6.2 Points of Convergence .......................................................... 122
References ................................................................................ 129

## Abstract

This chapter presents a family tree of theories, concepts, methodologies, and strategies for change in the field of development communication. It presents a chronological evolution and comparison of approaches and findings. The goal of this report is to clarify the understandings and the uses of the most influential theories, strategies, and techniques. Theory refers to sets of concepts and propositions that articulate relations among variables to explain and predict situations and results. Theories explain the nature and causes of a given problem and

S. Waisbord (✉)
School of Media and Public Affairs, George Washington University, Washington, DC, USA
e-mail: waisbord@gwu.edu

© Springer Nature Singapore Pte Ltd. 2020
J. Servaes (ed.), *Handbook of Communication for Development and Social Change*,
https://doi.org/10.1007/978-981-15-2014-3_56

provide guidelines for practical interventions. Diagnoses of problems translate into strategies, that is, specific courses of action for programmatic interventions that use a variety of techniques.

> **Keywords**
>
> Communication for social change (CSC) · Convergence points · Dependency theory · Entertainment-education · Family planning programs · Health education · Media advocacy · Participatory theories · Social mobilization

## 5.1 Introduction

Since the 1950s, a diversity of theoretical and empirical traditions has converged in the field of development communication. Such convergence produced a rich analytical vocabulary but also conceptual confusion. The field has not experienced a unilinear evolution in which new approaches superseded and replaced previous ones. Instead, different theories and practices that originated in different disciplines have existed and have been used simultaneously. This report identifies the main theoretical approaches and their practical applications, traces their origins, draws comparisons, and indicates strengths and weaknesses. It also analyzes the main understandings of development communication that express the outlook of the main "trunks" and "branches" of the family tree.

## 5.2 Development Communication

Development communication has its origins in post-war international aid programs to countries in Latin America, Asia, and Africa that were struggling with poverty, illiteracy, poor health, and a lack of economic, political, and social infrastructures. Development communication commonly refers to the application of communication strategies and principles in the developing world. It is derived from theories of development and social change that identified the main problems of the post-war world in terms of a lack of development or progress equivalent to Western countries.

Development theories have their roots in mid-century optimism about the prospects that large parts of the postcolonial world could eventually "catch up" and resemble Western countries. After the last remains of European empires in Africa and Asia crumbled in the 1950s and 1960s, a dominant question in policy and academic quarters was how to address the abysmal disparities between the developed and underdeveloped worlds. Development originally meant the process by which Third World societies could become more like Western developed societies as measured in terms of political system, economic growth, and educational levels (Inkeles and Smith 1974). Development was synonymous with political democracy, rising levels of productivity and industrialization, high literacy rates, longer life expectancy, and the like. The implicit assumption was that there was one form of development as expressed in developed countries that underdeveloped societies needed to replicate.

Since then, numerous studies have provided diverse definitions of development communication. Definitions reflect different scientific premises of researchers as well as interests and political agendas of a myriad of foundations and organizations in the development field. Recent definitions state that the ultimate goal of "development communication" is to raise the quality of life of populations, including increase income and well-being, eradicate social injustice, promote land reform and freedom of speech, and establish community centers for leisure and entertainment (Melkote 1991, 229). The current aim of development communication is to remove constraints for a more equal and participatory society.

Although a multiplicity of theories and concepts emerged during the past 50 years, studies and interventions have fundamentally offered two different diagnoses and answers to the problem of underdevelopment. While one position has argued that the problem was largely due to lack of information among populations, the other one suggested that power inequality was the underlying problem. Because the diagnoses were different, recommendations were different, too. Running the risk of overgeneralization, it could be said that theories and intervention approaches fell in different camps on the following points:

- Cultural versus environmental explanations for underdevelopment
- Psychological versus sociopolitical theories and interventions
- Attitudinal and behavior models versus structural and social models
- Individual versus community-centered interventions development
- Hierarchical and sender-oriented versus horizontal and participatory communication models
- Active versus passive conceptions of audiences and populations
- Participation as means versus participation as end approaches

These divergences are explored in the examination of theories and approaches below.

## 5.3 The Dominant Paradigm

Behavior change models have been the dominant paradigm in the field of development communication. Different theories and strategies shared the premise that problems of development were basically rooted in lack of knowledge and that, consequently, interventions are needed to provide people with information to change behavior.

The early generation of development communication studies was dominated by modernization theory. This theory suggested that cultural and information deficits lie underneath development problems and therefore could not be resolved only through three economic assistance (a la Marshall Plan in post-war Europe). Instead, the difficulties in Third World countries were at least partially related to the existence of a traditional culture that inhibited development. Third World countries lacked the necessary culture to move into a modern stage. Culture was viewed as the "bottleneck" that prevented the adoption of modern attitudes and behavior. McClelland (1961) and Hagen (1962), for example, understood that personalities determined social structure. Traditional personalities, characterized by authoritarianism, low self-esteem, and resistance to innovation, were diametrically different from modern personalities and, consequently, anti-development.

These studies best illustrated one of modernization's central tenets: ideas are the independent variable that explains specific outcomes. Based on this diagnosis, development communication proposed that changes in ideas would result in transformations in behavior. The underlying premise, originated in classic sociological theories, was that there is a necessary fitness between a "modern" culture and economic and political development. The low rate of agricultural output, the high rate of fertility and mortality, or the low rates of literacy found in the underdeveloped world were explained by the persistence of traditional values and attitudes that prevented modernization. The goal was, therefore, to instill modern values and information through the transfer of media technology and the adoption of innovations and culture originated in the developed world. The Western model of development was upheld as the model to be emulated worldwide.

Because the problem of underdeveloped regions was believed to be an information problem, communication was presented as the instrument that would solve it. As theorized by Daniel Lerner (1958) and Wilbur Schramm (1964), communication basically meant the transmission of information. Exposure to mass media was one of the factors, among others (e.g., urbanization, literacy), that could bring about modern attitudes. This knowledge-transfer model defined the field for years to come. Both Lerner's and Schramm's analyses and recommendations had a clear pro-media, pro-innovation, and pro-persuasion focus. The emphasis was put on media-centered persuasion activities that could improve literacy and, in turn, allow populations to break free from traditionalism.

This view of change originated in two communication models. One was the Shannon-Weaver model of sender-receiver, originally developed in engineering studies that set out to explain the transmission of information among machines. It became extremely influential in communication studies. The other was the propaganda model developed during World War II; according to which the mass media had "magic bullet" effects in changing attitudes and behavior.

From a transmission/persuasion perspective, communication was understood as a linear, unidirectional process in which senders send information through media channels to receivers. Consequently, development communication was equated with the massive introduction of media technologies to promote modernization, and the widespread adoption of the mass media (newspapers, radio, cinemas, and later television) was seen as pivotal for the effectiveness of communication interventions. The media were both channels and indicators of modernization: they would serve as the agents of diffusion of modern culture and, also, suggested the degree of modernization of the society.

The emphasis on the diffusion of media technologies meant that modernization could be measured and quantified in terms of media penetration. The numbers of television and radio sets and newspaper consumption were accepted as indicators of modern attitudes (Lerner 1958; Inkeles and Smith 1974). Statistics produced by the United Nations Educational, Scientific and Cultural Organization (UNESCO) showing the penetration of newspapers, radio, and television sets became proxy of development. Researchers found that in countries where people were more exposed to modern media, there are more favorable attitudes toward modernization and development. Based on these findings, national governments and specialists agreed to champion the media as instruments for the dissemination of modern ideas that would improve agriculture, health, education, and politics. So-called "small" media such as publications, posters, and leaflets were also recommended as crucial to the success of what became known as Development Support Communication, that is, the creation of the human environment necessary for a development program to succeed (Agunga 1997).

The "diffusion of innovations" theory elaborated by Everett Rogers (1962, 1983) became one of the most influential modernization theories. It has been said that Rogers' model has ruled development communication for decades and became the blueprint for communication activities in development. Rogers' intention was to understand the adoption of new behaviors. The premise was that innovations diffuse over time according to individuals' stages. Having reviewed over 500 empirical studies in the early 1960s, Rogers posited five stages through which an individual passes in the adoption of innovations: awareness, knowledge and interest, decision, trial, and adoption/rejection. Populations were divided in different groups according to their propensity to incorporate innovations and timing in actually adopting them. Rogers proposed that early adopters act as models to emulate and generate a climate of acceptance and an appetite for change, and those who are slow to adopt are laggards. This latter category was assumed to describe the vast majority of the population in the Third World.

For Rogers, the subculture of the peasantry offered important psychological constraints on the incorporation of innovations and, consequently, development. His view on development reflected the transmission bias also found in Lerner and Schramm. According to Rogers, development communications entailed a "process by which an idea is transferred from a source to a receiver with the intent to change his behavior. Usually the source wants to alter the receiver's knowledge of some idea, create or change his attitude toward the idea, or persuade him to adopt the idea as part of his regular behavior" (Rogers 1962).

However, diverging from the media centrism and "magic bullet" theory of effects that underpinned earlier analyses, Rogers and subsequent "diffusion" studies concluded that the media had a great importance in increasing awareness but that interpersonal communication and personal sources were crucial in making decisions to adopt innovations. This revision incorporated insights from the opinion leader theory (Katz and Lazarsfeld 1955); according to which there are two steps in information flow: from the media to opinion leaders and from leaders to the masses. Media audiences rely on the opinions of members of their social networks rather than solely or mainly on the mass media. In contrast to powerful media effects models that suggested a direct relation between the mass media and the masses, Lazarsfeld and Katz found that interpersonal relations were crucial in channeling and shaping opinion. This insight was incorporated in diffusion studies, which proposed that both exposure to mass media and face-to-face interaction were necessary to induce effective change. The effectiveness of field workers in transmitting information in agricultural development projects also suggested the importance of interpersonal networks in disseminating innovations. Consequently, a triadic model of communication was recommended that included change agents, beneficiaries, and communicators.

Confirming Lerner's and Schramm's ideas, another important finding of diffusion research was that what motivates change is not economics but communication and culture. This is what studies on how farmers adopted new methods showed. Such studies were particularly influential because a substantial amount of early efforts targeted agricultural development in the Third World (Rogers 1983). Other applications targeted literacy programs and health issues, mainly family planning and nutrition.

In the mid-1970s, main representatives of modernization/diffusion theories considered it necessary to review some basic premises (Rogers 1976, 1983). In a widely quoted article, Rogers admitted "the passing of the dominant paradigm." Schramm and Rogers recognized that early views had individualistic and psychological biases. It was necessary to be sensitive to the specific sociocultural environment in which "communication" took place, an issue that was neglected in early analyses. To a large extent, these revisions resulted from the realization that the "trickle down" model that was originally championed was not proven to be effective in instrumenting change. The stages model remained, but the top-down perspective according to which innovations diffuse from above needed modification.

Other positions suggested that the traditional model needs to integrate a process orientation that was not only focused on the results of intervention but also to content and address the cognitive dimensions (not just behavior). Many of these observations were integrated into the diffusion approach. By the mid-1970s, Rogers' definition of communication showed important changes that partially responded to criticisms. Development was theorized as a participatory process of social change intended to bring social and material advancement. Communication was no longer focused on persuasion (transmission of information between individuals and groups), but was understood as a "process by which participants create and share information with one another in order to reach a mutual understanding" (Rogers 1976).

## 5.4 Theories in the Tradition of the Dominant Paradigm

In the early 1970s, modernization theory was the dominant paradigm of development communication. The climate of enthusiasm and "missionary zeal," as Wilbur Schramm (1997) described it, that had existed a decade earlier had notably receded, but the notion that the diffusion of information and innovations could solve problems of underdevelopment prevailed.

### 5.4.1 Social Marketing

Social marketing has been one of the approaches that has carried forward the premises of diffusion of innovation and behavior change models. Since the 1970s, social marketing has been one of the most influential strategies in the field of development communication.

The origins of social marketing hark back to the intention of marketing to expand its disciplinary boundaries. It was clearly a product of specific political and academic developments in the United States that were later incorporated into development projects. Among various reasons, the emergence of social marketing responded to two main developments: the political climate in the late 1960s that put pressure on various disciplines to attend to social issues and the emergence of nonprofit organizations that found marketing to be a useful tool (Elliott 1991). Social marketing was marketing's response to the need to be "socially relevant" and "socially responsible." It was a reaction of marketing as both discipline and industry to be sensitive to social issues and to strive toward the social good. But it was also a way for marketing to provide intervention tools to organizations whose business was the promotion of social change.

Social marketing consisted of putting into practice standard techniques in commercial marketing to promote pro-social behavior. From marketing and advertising, it imported theories of consumer behavior into the development communication. The analysis of consumer behavior is required to understand the complexities, conflicts, and influences that create consumer needs and how needs can be met (Novelli 1990). Influences include environmental, individual, and information processing and decision-making. At the core of social marketing theory is the exchange model; according to which individuals, groups, and organizations exchange resources for perceived benefits of purchasing products. The aim of interventions is to create voluntary exchanges.

In terms of its place on the "family tree" of development communication, social marketing did not come out of either diffusion or participatory theories, the traditions that dominated the field in the early 1970s. Social marketing was imported from a discipline that until then had little to do with modernization or dependency theories, the then-dominant approaches in development communication. Social marketing grew out of the disciplines of advertising and marketing in the United States. The central premise of these disciplines underlies social marketing strategies: the goal of an advertising/marketing campaign is to make the public aware about the existence, the price, and the benefits of specific products.

Social marketing's focus on behavior change, understanding of communication as persuasion ("transmission of information"), and top-down approach to instrument change suggested an affinity with modernization and diffusion of innovation theories. Similar to diffusion theory, it conceptually subscribed to a sequential model of behavior change in which individuals cognitively move from acquisition of knowledge to adjustment of attitudes toward behavior change. However, it was not a natural extension of studies in development communication.

What social marketing brought was a focus on using marketing techniques such as market segmentation and formative research to maximize the effectiveness of interventions. The use of techniques from commercial advertising and marketing to promote social/political goals in international issues was not new in the 1970s. Leading advertising agencies and public relations firms had already participated in support of US international policies, most notably during the two wars in drumming up domestic approval and mobilization for war efforts. Such techniques, however, had not been used before to "sell" social programs and goals worldwide.

One of the standard definitions of social marketing states that "it is the design, implementation, and control of programs calculated to influence the acceptability of social ideas and involving consideration of product planning, pricing, communication, distribution, and marketing research" (Kotler and Zaltman 1971, 5). More recently, Andreasen (1994, 110) has defined it as "the adaptation of commercial marketing technologies to programs designed to influence the voluntary behavior of target audiences to improve their personal welfare and that of the society of which they are a part." Others have defined it as the application of management and marketing technologies to pro-social and nonprofit programs (Meyer and Dearing 1996).

Social marketing suggested that the emphasis should be put not so much on getting ideas out or transforming attitudes but influencing behavior. For some of its best-known proponents, behavior change is social marketing's bottom line, the goal that sets it apart from education or propaganda. Unlike commercial marketing, which is not concerned with the social consequences of its actions, the social marketing model centers on communication campaigns designed to promote socially beneficial practices or products in a target group.

Social marketing's goal is to position a product such as condoms by giving information that could help fulfill, rather than create, uncovered demand. It intends to "reduce the psychological, social, economic and practical distance between the consumer and the behavior" (Wallack et al. 1993, 21). The goal would be to make condom use affordable, available, and attractive (Steson and David 1999). If couples of reproductive age do not want more children but do not use any contraceptive, the task of social marketing is to find out why and what information needs to be provided so they can make informed choices. This requires sorting out cultural beliefs that account for such behavior or for why people are unwilling to engage in certain health practices even when they are informed about their positive results. This knowledge is the baseline that allows a successful positioning of a product. A product needs to be positioned in the context of community beliefs.

In the United States, social marketing has been extensively applied in public information campaigns that targeted a diversity of problems such as smoking, alcoholism, seat-belt use, drug abuse, eating habits, venereal diseases, littering, and protection of forests. The Stanford Three-Community Study of Heart Disease is frequently mentioned as one of the most fully documented applications of the use of marketing strategies. Designed and implemented as a strictly controlled experiment, it offered evidence that it is possible to change behavior through the use of marketing methodologies. The campaign included television spots, television programming, radio spots, newspaper advertisements and stories, billboard messages, and direct mail. In one town the media campaign was supplemented by interpersonal communication with a random group of individuals at risk of acquiring heart disease. Comparing results among control and experimental communities, the research concluded that media could be a powerful inducer of change, especially when aligned with the interpersonal activities of community groups (Flora et al. 1989).

Social marketing has been used in developing countries in many interventions such as condom use, breast-feeding, and immunization programs. According to Chapman Walsh and associates (1993, 107–108), "early health applications of social marketing emerged as part of the international development efforts and were implemented in the Third World during the 1960s and 1970s. Programs promoting immunization, family planning, various agricultural reforms, and nutrition were conducted in numerous countries in Africa, Asia and South America during the 1970s...The first nationwide contraceptive program social marketing program, the Nirodh condom project in India, began in 1967 with funding from the Ford Foundation." The substantial increase in condom sales was attributed to the distribution and promotion of condoms at a subsidized price. The success of the Indian experience informed subsequent social marketing interventions such as the distribution of infant-weaning formula in public health clinics.

According to Fox (n.d.), "problems arose with the social marketing approach, however, over the motives of their sponsors, the effectiveness of their applications, and, ultimately, the validity of their results. The social marketing of powdered milk products, replacing or supplementing breastfeeding in the third world, provides an example of these problems. In the 1960s multinational firms selling infant formulas moved into the virgin markets of Asia, Africa and Latin America. Booklets, mass media, loudspeaker vans, and distribution through the medical profession were used in successful promotion campaigns to switch traditional breastfeeding to artificial products. Poor people, however, could not afford such products, and many mothers diluted the formula to make it last longer or were unable to properly sterilize the water or bottle. The promotion of breast milk substitutes often resulted in an erosion of breastfeeding and led to increases in diarrheal diseases and malnutrition, contributing to the high levels of infant mortality in the third world."

Critics have lambasted social marketing for manipulating populations and being solely concerned with goals without regard for means. For much of its concerns about ethics, critics argue, social marketing subscribes to a utilitarian ethical model that prioritizes ends over means. In the name of achieving certain goals, social marketing justifies any nine methods. Like marketing, social marketing deceives and manipulates people into certain behaviors (Buchanan et al. 1994).

Social marketers have responded by arguing that campaigns inform publics and that they use methods that are not intrinsically good or bad. Judgments should be contingent on what goals they are meant to serve, they argue. Moreover, the widely held belief that marketing has the ability to trick and make people do what otherwise they would not is misinformed and incorrect. The reluctance of people to tailor behavior to the recommendations of social marketing campaigns and the fact that campaigns need to be adjusted to sociocultural contexts and morals are evidence that social marketing lacks the much-attributed power of manipulating audiences. If a product goes against traditional beliefs and behavior, campaigns are likely to fail.

Social marketing needs to be consumer oriented and knowledgeable of the belief systems and the communication channels used in a community (Maibach 1993). Products need to be marketed according to the preferences and habits of customers. Market research is necessary because it provides development specialists with tools to know consumers better and, therefore, to prevent potential problems and pitfalls in behavior change. This is precisely marketing's main contribution: systematic, research-based information about consumers that is indispensable for the success of interventions. Marketing research techniques are valuable for finding out thoughts and attitudes about a given issue that help prevent possible failures and position a product.

For its advocates, one of the main strengths of social marketing is that it allows to position products and concepts in traditional belief systems. The inclination of many programs to forgo in-depth research of targeted populations for funding or time considerations, social marketers suggest, reflects the lack of understanding about the need to have basic research to plan, execute, and evaluate interventions. They argue that social marketing cannot manipulate populations by positioning a product with false appeals to local beliefs and practices. If the desired behavior is not present in the local population, social marketing cannot deceive by wrapping the product with existing beliefs. When a product is intended to have effects that are not present in the target population, social marketers cannot provide false information that may resonate with local belief systems but, instead, need to provide truthful information about its consequences. For example, if "dehydration" does not exist as a health concept in the community, it would be ethically wrong for social marketing to position a dehydration product by falsely appealing to existing health beliefs in order to sell it. That would be deceptive and manipulative and is sure to backfire. The goal should be long-term health benefits rather than the short-term goals of a given campaign (Kotler and Roberto 1989).

Theorists and practitioners identified with participatory communication have been strong critics of social marketing. For them, social marketing is a non-participatory strategy because it treats most people as consumers rather than protagonists. Because it borrows techniques from Western advertising, it shares it premises, namely, a concern with selling products rather than participation. To critics, social marketing is concerned with individuals, not with groups or organizations. They also view social marketing as an approach that intends to persuade people to engage in certain behaviors that have already decided by agencies and planners. It does not involve communities in deciding problems and courses of action. The goal should be, instead, to assist populations in changing their actions based on critical analysis of social reality (Beltrán 1976; Diaz-Bordenave 1977). According to participatory approaches, change does not happen when communities are not actively engaged in development projects and lack a sense of ownership.

Social marketers have brushed aside these criticisms, emphasizing that social marketing is a two-way process and that it is genuinely concerned about community participation. As Novelli (1990, 349) puts it, "the marketing process is circular." This is why input from targeted communities, gathered through qualitative methods such as focus groups and in-depth interviews, is fundamental to design campaign activities and content. Social marketing is premised on the idea of mutual exchange between agencies and communities. Marketing takes a consumer orientation by assuming that the success of any intervention results from an accurate evaluation of perceptions, needs, and wants of target markets that inform the design, communication, pricing, and delivery of appropriate offerings. The process is consumer-driven, not expert-driven.

Also, social marketing allows communities to participate by acting upon health, environmental, and other problems. Without information, there is no participation, and this is what social marketing offers. Such participation is voluntary: Individuals, groups, and organizations are not forced to participate but are offered the opportunity to gain certain benefits. Such explanation is not satisfactory to participatory communication advocates who respond that social marketing does not truly involve participation. More than a narrow conception of participation, they argue, social marketing offers the appearance of it to improve interventions that are centralized. Social marketing's conception of participation basically conceives campaigns' targets are "passive receivers," subjects from whom information is obtained to change products and concepts.

After three decades of research and interventions, the lessons of social marketing can be summarized as follows (Chapman Walsh et al. 1993):

- Persistence and a long-term perspective are essential. Only programs with sustainable support and commitment have proven to have impact on diffusion of new ideas and practices, particularly in cases of complex behavior patterns.
- Segmentation of the audience is central. Some researchers have identified different lifestyle clusters that allow a better identification of different market niches.
- Mapping target groups is necessary. Designers of interventions need to know where potential consumers live, their routines, and relations vis-à-vis multiple messages.
- Incentives foster motivation among all participants in interventions.
- The teaching of skills is crucial to support behavior change.
- Leadership support is essential for program success.
- Community participation builds local awareness and ownership. Integrating support from different stakeholders sets apart social marketing from commercial advertising as it aims to be integrated with community initiatives.
- Feedback makes it possible to improve and refine programs.

## 5.4.2 Health Promotion and Health Education

The trajectory of health promotion in development communication resembles the move of social marketing and diffusion of innovation, from originally gaining influence in the United States to being introduced in interventions in developing countries. The same approaches that were used to battle chronic diseases, high-fat diets, and smoking in the United States in the 1970s and 1980s were adopted in development interventions such as child survival and other programs that aimed to remedy health problems in the Third World.

As it crystallized in the Lalonde report in Canada in 1974 and the US Surgeon General's 1979 Healthy People report, health promotion was dominated by the view that individual behavior was largely responsible for health problems and, consequently, interventions should focus on changing behavior. It approached health in terms of disease problems (rather than health generally), namely, the existence of lifestyle behaviors (smoking, heavy drinking, poor diet) that had damaging consequences for individual and, by extension, social health (Terris 1992).

The prevalent view was that changes in personal behaviors were needed to have a healthier population. Although the idea that institutional changes were also necessary to achieve that goal made strides, health promotion remained focused on personal change at the expense of community actions and responsibility. A substantial number of studies were offered as conclusive evidence that personal choices determined changes in health behavior and were positively related with new developments that indicated the decrease of unhealthy practices.

This highly individualistic perspective was initially criticized in the context of developed countries for "blaming the victim" and ignoring social conditions that facilitated and encouraged unhealthy behaviors. It gave a free ride to larger social and political processes that were responsible for disease and essentially depoliticized the question of health behavior. To its critics, individual-centered health promotion ignores the surrounding social context (poverty, racism) in which individual health behaviors take place as well as the fact that certain unhealthy behaviors are more likely to be found among certain groups (Minkler 1999; Wallack and Montgomery 1992). They pointed out that the overall context needs to be considered both as responsible and as the possible target of change.

Recent understandings of health promotion such as the one promoted by the World Health Organization have moved away from individualistic views by stressing the idea that individual and social actions need to be integrated. The goal of health promotion is to provide and maintain conditions that make it possible for people to make healthy choices.

Health education is an important component of health promotion. It refers to learning experiences to facilitate individual adoption of healthy behaviors (Glanz et al. 1990). The evolution of health education somewhat mirrored the evolution of the field of development communication. Health education was initially dominated by conventional educational approaches that, like modernization/diffusion models, were influenced by individual behaviorist models that emphasized knowledge transmission and acquisition as well as changes in knowledge, attitudes, and beliefs. Later, theories and strategies that stressed the importance of social and environmental changes gained relevance. This meant that both health education and health promotion became more broadly understood. Health education includes different kinds of interventions such as conventional education, social marketing, health communication, and empowerment actions (Steston and Davis 1999). Consequently, a vast range of activities such as peer education, training of health workers, community mobilization, and social marketing are considered examples of health education interventions.

Health promotion became no longer understood as limited to educational efforts and individual changes. It also includes the promotion of public policies that are responsible for shaping a healthy environment. The goal of health promotion is to facilitate the environmental conditions to support healthy behaviors. Individual knowledge, as conceived in traditional approaches, is insufficient if groups lack basic systems that facilitate the adoption of healthy practices. The mobilization of a diversity of social forces including families and communities is necessary to shape a healthy environment (Bracht 1990; Rutten 1995).

The emphasis on social mobilization to improve general conditions does not mean that behavior change models are absent in health promotion but, rather, that they need to be integrated among other strategies. Still, the behavior change model has incorporated the idea that interventions need to be sensitive to the education and the choices of receivers (Valente et al. 1998), understanding the interests at stake, using social marketing technique to know individuals better, and the role of the community in interventions.

### 5.4.3 Entertainment-Education

Entertainment-education is another strategy that shares behavior change premises with the aforementioned theories and strategies. Entertainment-education is a communication strategy to disseminate information through the media. As applied in development communication, it was originally developed in Mexico in the mid-1970s and has been used in 75 countries, including India, Nigeria, the Philippines, Turkey, Gambia, and Pakistan. Paradigmatic examples of this approach have been soap operas in Latin America (telenovelas) and in India that were intended to provide information about family planning, sexual behavior, and health issues. Literacy and agricultural development have also been central themes of several entertainment-education efforts.

Entertainment-education is not a theory but a strategy to maximize the reach and effectiveness of health messages through the combination of entertainment and education. The fact that its premises are derived from socio-psychology and human communication theories places entertainment-education in the modernization/diffusion theory trunk. It subscribes to the Shannon-Weaver model of communication of sender-channel-message-receiver. Like diffusion theory, it is concerned with behavior change through the dissemination of information. It is based on Stanford Professor Albert Bandura's (1977) social learning theory, a framework currently dominant in health promotion. Entertainment-education is premised on the idea that individuals learn behavior by observing role models, particularly in the mass media. Imitation and influence are the expected outcomes of interventions. Entertainment-education telenovelas were based on Bandura's model of cognitive subprocesses: attention, retention, production, and motivational processes that help understand why individuals imitate socially desirable behavior. This process depends on the existence of role models in the messages: good models, bad models, and those who transition from bad to good. Besides social learning, entertain-education strategies are based on the idea that expected changes result from self-efficacy, the belief of individuals that they can complete specific tasks (Maibach and Murphy 1995).

Entertainment-education refers to "the process of purposely designing and implementing a media message to both entertain and educate, in order to increase audience knowledge about an educational issue, create favorable attitudes, and change overt behavior" (Singhal and Rogers 1999, xii). Like social marketing and health promotion, it is concerned with social change at individual and community levels. Its focus is on how entertainment media such as soap operas, songs, cartoons, comics, and theater can be used to transmit information that can result in pro-social behavior. Certainly, the use of entertainment for social purposes is not new as they have been used for centuries. What is novel is the systematic research and implementation of educational, pro-social messages in entertainment media in the developed world.

One of the starting points of entertainment-education is that populations around the world are widely exposed to entertainment media content. The heavy consumption of media messages suggests that the media have an unmatched capacity to tell people how to dress, talk, and think. The problem is, as numerous studies document, that entertainment messages are rarely positive. In the attempt to maximize audiences by appealing to the lowest common denominator, the media are filled with antisocial messages such as violence, racism, stereotyping, and sexual promiscuity. However, the pervasiveness of the media provides numerous opportunities to communicate messages that can help people in solving a myriad of problems that they confront.

Another central premise is that education does not necessarily need to be dull but it can incorporate entertainment formats to generate pro-social attitudes and behavior. This could solve the problem that audiences find social messages uninteresting and boring and prefer to consume entertainment media. What characterizes the latter is the intention of the messages (to divert rather than to educate) and to capture audiences' interest. These characteristics should not be dismissed as superficial and mindless but need to be closely examined to analyze the potential of entertainment to educate the public in an engaging manner. Moreover, because they are entertaining and widely popular, entertainment-education messages can also be profitable for television networks and other commercial ventures.

*Simplemente María*, a 1969 Peruvian telenovela, has been often mentioned as having pioneered entertainment-education even though it was not intended to have pro-social effects. The protagonist was a maid who attended night sewing classes. The program has been credited with having turned sewing into a craze among poor, migrant women as well as increasing the purchase of sewing machines and contributing to higher enrollment numbers in literacy classes. This example and subsequent ones were deemed to be important in two regards. The programs contribute to self-efficacy (an individual's belief that he or she is able to take action and control specific outcomes) and social learning (individuals not only learn through their own experiences but also by observing and imitating the behavior of other individuals as role models).

Besides television entertainment, entertainment-education interventions were also implemented in music and music videos promoting sexual control and radio soap operas that promoted women's issues, AIDS and sex education, and family planning. In the mid-1980s, a campaign was implemented to promote sexual restraint among Mexican teenagers. It consisted of songs and music videos featuring a male and female singer as well as public service announcements. Evaluation analysis concluded that the campaign had a number of positive consequences: teenagers felt freer to talk about sex and became more sensitized about the relevance of sex, messages reinforced teenagers who already practiced abstinence, and demand for family planning services modestly increased (Singhal and Rogers 1999).

Comparable findings were documented in a similar intervention in the Philippines. The campaign also featured songs, video, live presentations of the performers, and PSAs. It resulted in positive changes in knowledge, attitude, and behavior. Other less effective campaigns suggested that appeals that may work in some cultural contexts could fail in others. Performers need to be credible, that is, audiences need to believe that they truly represent the values promoted.

Some studies have concluded that entertainment-education strategies are successful in attracting large audiences, triggering interpersonal communication about issues and lessons from interventions, and in engaging and motivating individuals to change behavior and support changes among their peers. Rogers et al. (1999) concluded that a soap-opera radio broadcast in Tanzania played an important role in fertility changes. The broadcast increased listeners' sense of self-efficacy, ideal age at marriage of women, approval contraceptive use, interspousal communication about family planning, and current practice of family planning. Similarly, Piotrow et al. (1992) report that the "Male Motivation Project" in Zimbabwe, which involved a radio drama intended to influence men's decisions in opting for different reproductive choices, resulted in changes in beliefs and attitudes. Also, Valente et al. (1994) found that individuals who listened to a radio drama in the Gambia have better knowledge, attitudes, and practices than the control group. The study also concluded that substantial changes in the use of contraceptive methods existed after the broadcast. Both studies concluded that audiences incorporate language presented in the programming, talk to others, and introduce behavior changes. A hierarchy of effects was observed in interventions in Mexico, Nigeria, and the Philippines. In decreasing order, campaigns contributed to audience recall, comprehension, agreement, and talking with others about the messages promoted in the campaigns.

In contrast, other studies have found little evidence that entertainment-education strategies have resulted in such effects (Yoder et al. 1996). Yoder and coauthors have argued that the changes in behavior reported in the Zimbabwe and Gambia studies were not statistically significant. An analysis of the impact of a radio drama in Zambia suggested that improvement in knowledge and awareness about AIDS could not be directly attributed to the intervention. Significant changes in condom use were not associated with exposure to the radio drama as there was a substantial amount of information and public debate about HIV/AIDS during the time the drama was broadcast. Exposure to discrete radio programs per se did not account for changes between target and control groups. Moreover, a return to previous behavior after the broadcast suggested the lack of evidence of long-term impact and attributed the findings to the timing when the data were collected. Exposure to entertainment-education messages was positively associated with the use of modern contraceptive methods, but the data did not allow a direct causal inference. It was not clear whether the campaign had influenced knowledge and practices. Studies did not reject the possible counter-explanation that people more predisposed to family planning were more likely to be exposed and recall media content (Westoff and Rodriguez 1995).

Rather than discounting the possibility of any media effects, Yoder and associates concluded that it is problematic to reach comprehensive conclusions about the effectiveness of entertainment education. In contrast to more optimistic evaluations that suggest that the task ahead is to measure what works better, they recommended a more cautionary approach. Entertainment-education projects are effective in stimulating people predisposed to change behavior to engage in a new behavior (e.g., use contraceptive methods). They provide the push for those already inclined to act to behave differently. Media interventions catalyze latent demand into contraceptive use among ready-to-act populations (see Freedman 1997; Zimicki et al. 1994).

## 5.5 Critiques of the Dominant Paradigm

Beginning in the late 1960s, the field of development communication split in two broad approaches: one that revised but largely continued the premises and goals of modernization and diffusion theories and another that has championed a participatory view of communication in contrast to information- and behavior-centered theories. Both approaches have dominated the field. Although in recent years there have been attempts to incorporate insights from both traditions, no comprehensive view has evolved (Servaes 1996a). Integrative attempts are analyzed in the last section of this report.

### 5.5.1 Dependency Theory

One of the most powerful critiques of modernization/diffusion theories came from the dependency paradigm. Originally developed in Latin America, dependency analysis was informed by Marxist and critical theories; according to which the problems of the Third World reflected the general dynamics of capitalist development. Development problems responded to the unequal distribution of resources created by the global expansion of Western capitalism.

Against modernization theories, dependency theorists argued that the problems of underdevelopment were not internal to Third World countries but were determined by external factors and the way former colonies were integrated into the world economy. It forcefully stated that the problems of the underdeveloped world were political rather than the result of the lack of information. What kept Third World countries underdeveloped were social and economic factors, namely, the dominated position that those countries had in the global order. Underdevelopment, they argued, was the flip side and the consequence of the development of the Western world. The latter concentrated economic power and political decisions that maintained underdevelopment and dependency. Third World countries were politically and culturally dependent on the West, particularly on the United States.

Asides from external problems, internal structures were also responsible for the problems of underdevelopment. Dependency positions charged development programs for failing to address structures of inequality and targeting individual rather than social factors. Unequal land distribution, lack of credit for peasants, and poor health-care services strongly limited the possibilities for an overall improvement in social conditions. Interventions were doomed when basic conditions that could make it possible for people to adopt new attitudes and behaviors were missing.

Also, innovations promoted by development programs were adopted by individuals from higher socioeconomic strata living in cities rather than by rural and poor populations. In singling out the mass media as having a central role in introducing innovations, modernization theories ignored the issue of media ownership and control. Urban and powerful interests controlled the media that was supposed to promote development. The media were not interested in championing social goals or helping underprivileged populations but in transmitting entertainment and trivial information. The relation between media structure and content was virtually ignored in modernization theories. Only a small percentage of programming was devoted to development issues, and in regions such as Latin America, the media were commercially run and their central goal was profit-making not social change.

For dependency theorists, modernization theories were driven by behaviorist, positivist, and empiricist approaches in the mold of the "scientific model" that prevailed in the US universities and research centers. These particular biases accounted for why structural factors were ignored and for why interventions were focused on behavior changes at the individual level rather than on social causes of poverty and marginalization. Modernization theories as applied in the Third World featured, to quote Bolivian communication researcher Luis Ramiro Beltrán (1976), "alien premises, objects and methods." The solution to underdevelopment problems was essentially political, rather than merely informational. What was required was social change in order to transform the general distribution of power and resources. Information and media policies were necessary to deal with communication problems. Solutions to underdevelopment required major changes in media structures that were dominated by commercial principles and foreign interests. Policies needed to promote national and public goals that could put the media in the service of the people rather than as pipelines for capitalist ideologies. Such positions were expressed in a number of international fora, particularly during the UNESCO-sponsored debates about the New World Information and Communication Order in the 1970s and 1980s. Representatives from Third World countries proposed "national communication policies" that emphasized the need for governments to control media structures and oppose domestic and foreign elites and business interests.

### 5.5.2 Participatory Theories and Approaches

Participatory theories also criticized the modernization paradigm on the grounds that it promoted a top-down, ethnocentric, and paternalistic view of development. They argued that the diffusion model proposed a conception of development associated with a Western vision of progress. Development communication was informed by a theory that "became a science of producing effective messages" (Hein in Quarmyne 1991). After decades of interventions, the failure to address poverty and other structural problems in the Third World needed to be explained on the faulty theoretical premises of the programs. Any intervention that was focused on improving messages to better reach individuals or only change behavior was, by definition, unable to implement social change.

Development theories also criticized traditional approaches for having been designed and executed in the capital cities by local elites with guidance and direction from foreign specialists. Local people were not involved in preparing and instrumenting development interventions. Interventions basically conceived of local residents as passive receivers of decisions made outside of their communities and, in many cases, instrumented ill-conceived plans to achieve development. Governments decided what was best for agricultural populations, for example, without giving them a sense of ownership in the systems that were introduced (see Mody 1991; Servaes 1989; White 1994).

The top-down approach of persuasion models implicitly assumed that the knowledge of governments and agencies was correct and that indigenous populations either did not know or had incorrect beliefs. Because programs came from outside villages, communities felt that innovations did not belong to them but to the government and thus expected the latter to fix things when they went wrong. The sense of disempowerment was also rooted in the fact that "targeted" populations did not have the choice to reject recommendations or introduce modifications to interventions.

For participatory theorists and practitioners, development communication required sensitivity to cultural diversity and specific context that were ignored by modernization theories. The lack of such sensitivity accounted for the problems and failures of many projects. Experts learned that development was not restricted to just building roads, piping water, and distributing electricity. Nor was it limited to efforts to increase farm yields nor switching farmers over to cash crops. Many of the agricultural projects failed because farmers were reluctant to abandon their traditional ways for foreign and unknown methods. As McKee (1999) writes, "they were also nervous about planting exotic crops that they could not eat but had to sell for money with which to buy food from the market." Modernization projects undermined the importance of local knowledge and the consequences of the interaction between local cultures and foreign ideas. When piped water arrived, it was frequently used for washing rather than drinking and cooking because the people disliked its flavor. Persuading people of the benefits of healthy practices on the basis of scientific reasons was a tough sell. People were asked to change time-old practices on the basis of a foreign form of knowledge that dismissed their local traditions in the name of "true" knowledge (McKee 1999).

The lack of local participation was viewed as responsible for the failure of different programs. In the case of agricultural programs, it was concluded that the issue at stake was not the transmission of information to increase output but rather the low prices of agricultural goods in the market or the absence of a more equal distribution of land ownership. In explaining the failures of family planning programs, it was suggested that mothers were disinclined to follow instructions because fathers believed that having more children meant having more hands to work in the fields and carry out other tasks.

Participatory theories considered a redefinition of development communication necessary. One set of definitions stated that it meant the systematic utilization of communication channels and techniques to increase people's participation in development and to inform, motivate, and train rural populations mainly at the grassroots. For others, development communication needed to be human – rather than media centered. This implied the abandonment of the persuasion bias that development communication had inherited from propaganda theories and the adoption of a different understanding of communication.

Communication means a process of creating and stimulating understanding as the basis for development rather than information transmission (Agunga1997). Communication is the articulation of social relations among people. People should not be forced to adopt new practices no matter how beneficial they seem in the eyes of agencies and governments. Instead, people needed to be encouraged to participate rather than adopt new practices based on information.

This understanding of communication was central to the ideas developed by Brazilian educator Paulo Freire (1970), whose writings and experiences became an influential strand in participatory communication. Freire's work in northeastern Brazil in the 1960s and early 1970s challenged dominant conceptions of development communication, particularly as applied to literacy training. He argued that development programs had failed to educate small farmers because they were interested in persuading them about the benefits of adopting certain innovations. Development programs tried to domesticate foreign concepts, to feed information, and to force local populations to accept Western ideas and practices without asking how such practices fit existing cultures. The underlying premise of such programs was an authoritarian conception of communication that stood against the essence of communication understood as community interaction and education.

Freire offered the concept of liberating education that conceived communication as dialogue and participation. The goal of communication should be conscientization, which Freire defined as free dialogue that prioritized cultural identity, trust, and commitment. His approach has been called "dialogical pedagogy" which defined equity in distribution and active grassroots participation as central principles. Communication should provide a sense of ownership to participants through sharing and reconstructing experiences. Education is not transmission of information from those "who have it" to those "who lack it," from the powerful to the powerless, but the creative discovery of the world.

Freire's ideas ran against fundamental principles in the diffusion model, namely, the sender-focus and behavioral bias that it inherited from persuasion models in the United States. He diagnosed the problems in the Third World as problems of communication, not information as persuasion theories proposed. Solutions, then, are needed to have an understanding of communication that was not limited to the application of Western ideas. Freire also challenged the value judgment in early development theories that viewed agricultural and health practices in the Third World as backward and obstacles to modernization.

Freire's model and participatory models in general proposed a human-centered approach that valued the importance of interpersonal channels of communication in decision-making processes at the community level. Studies in a variety of Third World rural settings found that marginal and illiterate groups preferred to communicate face-to-face rather than through mass media or other one-way sources of communication (Okunna 1995). The recommendation was that development workers should rely more on interpersonal methods of communication rather than national media and technologies and that they should act as facilitators of dialogue.

Because media and technologies were perceived as foreign to local communities, they should be used to supplement instead of dominate interpersonal methods. The notion of "group media" drew from Freire to call the media that are means for small groups to develop a critical attitude toward the reality of self, the group, community, and society through participation in group interaction. Group media has helped marginal groups to speak to one another and to articulate their thoughts and feelings in the process of community organizing (Hamelink 1990). Community-based forms of communication such as songs, theater, radio, video, and other activities that required group intervention are needed to be promoted. More than mechanisms to disseminate information, they could provide opportunities to identify common problems and solution, to reflect upon community issues, and to mobilize resources. Community members, rather than "professionals," should be in charge of the decision and production processes. This is precisely what "small" media offer: an opportunity for media access in countries where the mass media are usually controlled by governments and urban elites.

The value of participatory media is not in being instruments of transmission but of communication, that is, for exchanging views and involving members. Community media dealt with various subjects: literacy, health, safety, agricultural productivity, land ownership, gender, and religion.

There have been a number of paradigmatic examples. In Latin America, miners' and peasants' radio in Bolivia, grassroots video in peasant and indigenous movements in Brazil, tape recorders in Guatemala, small-scale multimedia in Peru, and other cases of low-powered media based in unions and churches were offered as concrete examples of participatory communication development (Beltrán 1993a). Canada's "Fogo process" was another experience informed by similar principles in which populations living in remote areas actively produced videos to discuss community issues of people living in remote areas and to communicate with outsiders about their concerns and expectations (Williamson 1991). In Africa, popular theater has been successfully used to increase women's participation and ability to deal with primary care problems. Through songs and storytelling, women were able to raise awareness and attention to issues and address problems, something that had not been achieved through "modern" media such as television and newspapers (Mlama 1991). Community participation through popular theater motivated rural communities to become involved in health care. Participation was credited for the reduction of preventable diseases such as cholera and severe diarrhea after communities constructed infrastructure that helped to improve sanitary condition situation (Kalipeni and Kamlongera 1996).

In stressing the relevance of "other" media and forms of communication, participatory theories lifted development communication out of the "large media" and "stimulus-response" straitjacket and opened new ways of understanding interventions. They expanded the concept of participation that in modernization theories was limited to voting in party and electoral politics and championed a view of democracy that implied different forms of participation at different levels.

They also removed professionals and practitioners from having a central role as transmitters of information who would enlighten populations in development projects. People, not change agents, were central to community participation. It downplayed the role of expert and external knowledge while stressing the centrality of indigenous knowledge and aspirations in development. Communication was a horizontal process, diametrically different from the vertical model that placed knowledge in the domain of modern experts.

Participatory communication identified encouraging participation, stimulating critical thinking, and stressing process, rather than specific outcomes associated with modernization and progress, as the main tasks of development communication (Altafin 1991). Participation should be present in all stages of development projects. Communities should be encouraged to participate in decision-making, implementation, and evaluation of projects. This would give a sense of involvement in their lives and communities and provide them with a sense of ownership and skills that they can use beyond the timetable of development projects (Kavinya et al. 1994). Community empowerment has become one of the main contributions of participatory theories to development communication. Empowerment is possible only if community members critically reflect on their experiences and understand the reasons for failure and success of interventions (Bradford and Gwynne 1995; Purdey et al. 1994).

Certainly, participatory communication has not lacked critics. Even though vindicating some tenets of participatory theories, other positions argued that they were elaborated at a theoretical level and did not provide specific guidelines for interventions.

One problem in participatory models was that it was not clear that communities needed to be involved for certain results to be achieved. In some cases such as epidemics and other public health crises, quick and top-down solutions could achieve positive results. Participation communication ignores that expediency may also positively contribute to development. Belaboring through grassroots decision-making process is slower than centralized decisions and thus is not advisable in cases that require prompt resolutions. Participation might be a good long-term strategy but has shortcomings when applied to short-term and urgent issues.

Another problem was that participation in all stages does not have similar relevance. It was not clear what participation entailed. If decisions were made outside of the community and the latter was assigned the role of implementing and evaluating results, some positions argued, participation was limited to instances that depended on decisions previously made (McKee 1999). It was not true participation and, therefore, maintained power inequalities.

Another problem was that the focus on interpersonal relations underplayed the potential of the mass media in promoting development as participation and process. Little attention was paid to the uses of mass media in participatory settings, an issue that is particularly relevant considering that populations, even in remote areas, are constantly exposed to commercial media messages that stand in opposition to the goals set by programs. This lack was particularly evident in Freire's theory of dialogical communication that is based on group interactions and underplays the role of the mass media.

Participatory approaches usually avoided the issue that people who lived in nondemocratic societies might be wary to participate out of fear of retaliation. Moreover, people can be manipulated into participating. This would violate local autonomy and the possibility that members might not be interested in taking an active role. Critics argued that participatory communication, like social marketing, could also be seen as foreign, pushing for certain goals and actions that have not resulted from inside communities. Participatory communication did not offer the chance not to participate and implicitly coerced people to adopt a certain attitude.

Social marketers charged that participatory approaches were too idealistic, falling short from offering specific practical guidelines and offering recommendations with limited impact. These shortcomings are particularly pronounced when funds for development communication are short and funding agencies are interested in obtaining cost-effective results not just at the local but also the national level.

Other critics, particularly in Asia, thought that participatory models were premised on Western-styled ideas of democracy and participation that do not fit political cultures elsewhere. Individualism rather than community and conflict rather than consensus lies at the heart of participatory models developed in the West. Participation can also promote division, confusion, and disruption that do little to solve problems. It may privilege powerful and active members of the community at the expense of the community as a whole. Education and decision-making skills, rather than participation for its own sake, should be promoted.

To these criticisms, advocates of participatory models admitted that divisions and conflicts might result, but, they argued, the answer should be teaching negotiation and mediation skills rather than opting for interventions that disempower people in the name of consensus building. Although advocates of participatory theories viewed their critics as favoring government centralization and leaving power inequalities intact, they admitted that some original premises needed to be revised (White 1994).

Participatory approaches needed to:

- Be sensitive to the potential convenience of short-term and rapid solutions.
- Recognize that recommendations for participation could also be seen as foreign and manipulative by local communities (just like modernization theories).
- Translate participatory ideas into actual programs.
- Be aware that the communities may be uninterested in spending time in democratic processes of decision-making and, instead, might prefer to invest their time on other activities.
- Recognize that communities are not necessarily harmonious and that participation may actually deepen divisions. Servaes (1996, 23) admits that "participation does not always entail cooperation nor consensus. It can often mean conflict and usually poses a threat to existent structures ... Rigid and general strategies for participation are neither possible nor desirable."

To prevent some of these problems, it was suggested that it was preferable that projects be carried out in communities where agencies already had linkages (McKee 1999). Previous knowledge of problems and characteristics of a given community was fundamental to identify activities and define projects. Existing linkages could also provide agents that were familiar with (or even were from) the community who could assist in creating organizations and networks to stimulate participation. No previously determined set of activities was advisable if the interests and dynamics of communities were not known. Workers would also provide important feedback information about the progress of projects through regular, face-to-face contact with participants. These practices function as a sort of transmission belt for making sure that community issues are addressed and that members have a voice in deciding future courses. The peril is to focus solely on professional technicians and leaders without consideration of involving the community at large.

Against criticisms that participatory communication leads to the existence of a myriad, disconnected projects carried out by multiple NGOs, coordination plans were deemed necessary. Providing a sense of orientation and organization was required to prevent that development efforts become too fragmented and thus weaker. Because NGOs are closer to communities than governments and funding agencies, they have the capacity to respond relatively quickly to demands and developments. But without a more encompassing vision, projects may only obtain, at best, localized results that fail to have a larger impact.

It was also recommended that relying on grassroots media was not sufficient. Populations needed media education to develop skills to be critical of commercial media and to develop alternatives that would help them gain a sense of empowerment and counter other messages. Yet, it was undeniable that local media provided a sense of ownership and participation that was key to sustainable development and could not be replaced by any other strategies.

Responding to critics who were impatient with obtaining "results," participatory approaches suggested that development communication requires a long-term perspective that is usually missing among funding agencies and governments interested in getting quick results and knowing whether efforts pay off. Participatory theorists turned the criticisms about "timing" and "impact" onto their critics, arguing that the so-called problems of participatory approaches in "showing results" did not originate in the model but in how organizations approach development communication (Melkote 1991). Short-term projects that are prone to be terminated according to different considerations make it difficult to promote participation and examine the results of interventions in the long run. The interests of funders and politicians, who were urged to prove effectiveness of investments, ran against the timing of participatory development communication projects. For the latter to be possible, NGOs, funding agencies, and other actors involved should be sensitive to the fact that grassroots projects cannot be expected to "produce results" in the manner of top-down interventions. Neither community development nor empowerment fits the timetables of traditional programs.

### 5.5.3 Media Advocacy

Media advocacy is another approach that questions central premises of the traditional paradigm. Media advocacy is the strategic use of mass media to advance social or public policy initiatives (Wallack et al. 1993). Its goals are to stimulate debate and promote responsible portrayals and coverage of health issues. Advocacy requires the mobilization of resources and groups in support of certain issues and policies to change public opinion and decisions. It consists of the organization of information for dissemination through various interpersonal and media channels toward gaining political and social acceptance of certain issues.

Like education-entertainment strategies, media advocacy rejects the idea that the media can be a source of only antisocial messages and, instead, proposes to include socially relevant themes in entertainment. Both share the perspective that because the media are the main source of information about health issues, interventions need to focus on the media. Both also believe in the capacity of the media to transmit information that can result in changes. Unlike education-entertainment, which has been mostly concerned with directly influencing audiences, media advocacy centers on shaping the public debate about public health. It is not information-centered but aims to incorporate social themes in entertainment content in order to influence public agendas. It takes a political and social approach that differs from the social-psychological premises and diagnoses found in education-entertainment. And, in contrast to education-entertainment, it is less convinced about the power of the media to be extremely effective in changing attitudes and behaviors.

Because it locates problems in political and social conditions, social advocacy promotes social, rather than individual and behavioral, changes to health issues. It approaches health not as a personal issue but as a matter of social justice. It is explicitly set against the individualistic assumptions of mainstream approaches found in the dominant paradigm of development communication that fault individuals for unhealthy and antisocial behaviors and propose individual solutions based on the idea that health is primarily a question of individual responsibility. Instead, it advocates changes in the social environment that legitimize certain behaviors. For example, it sees tobacco and alcohol companies rather than individual smokers and drinkers as responsible for unhealthy behavior. Therefore, those companies should be the targets of advocacy and communication activities. Actions should target, for example, access to unhealthy products by involving communities in implementing policy changes (Holder and Treno 1997).

Here the contrast with behavior-centered health approaches is clear as media advocacy proposes that social conditions should be the target of interventions. Such interventions entail fundamentally a political process of changing conditions and redressing social inequalities rather persuading individuals about the benefits of certain lifestyles and behavior change. Health is a matter of social justice and partnering with interested parties rather than providing information to change individual behavior (Brawley and Martinez-Brawley 1999).

These premises set media advocacy apart from social marketing. Media advocacy criticizes social marketing for having an individualistic, behaviorist approach to health and social problems that narrows interventions to public information campaigns. Media advocacy espouses a community-level model of intervention in health issues. Development, defined to be the well-being of communities, can be achieved through promoting structures and policies that support healthy lifestyles. Community organization is the process by which community groups are helped to identify common problems or goals, mobilize resources, and develop and implement strategies for reaching their goals (Glanz and Rimer 1995).

According to media advocacy theory, campaigns are not the panacea not only because their effectiveness is questionable but also because they ignore the social causes of unhealthy behavior. Public service announcements have shown limited success in stimulating change and fail to address the social and economic environment that ultimately determines health risk factors. Social marketing does not face head-on the fundamental structures that sustain unhealthy behavior. Social advocacy does not minimize the importance of individual changes, but, instead, it strongly argues that the latter require changes in social conditions. Because external conditions are responsible for health, the strategy should target those conditions instead of centering on lifestyle behaviors. Promoting individual health habits in developing countries without, for example, advocating for clean water supplies underplays the factors responsible for disease.

Media advocacy adopts a participatory approach that emphasizes the need of communities to gain control and power to transform their environments. It assigns the media a pivotal role in raising issues that need to be discussed and putting pressure on decision-makers. However, advocacy is not solely concerned with media actions. Because it concludes that health problems are fundamentally rooted in power inequalities, it promotes a dual strategy to build power that includes the formation of coalitions and grassroots actions coupled with media actions and lobbying.

Media advocacy theory assumes that the media largely shape public debate and, consequently, political and social interventions. To be politically effective, then, influencing news agendas is mandatory. AIDS and tobacco control coalitions and groups in the United States have been successful in their use of the mass media that has resulted in support, funding, and the implementation of public policies. Media savviness is necessary to get widespread coverage of certain health issues and to shape how stories are presented. Here again social advocacy differs from social marketing. Social advocacy is not about putting in action centralized actions to relay information to consumers but, rather, providing skills to communities so they can influence media coverage. It approaches the media not in terms of "health messages" but as agenda setters of policy initiatives. Placing messages is not only insufficient to correct problems, but it is also the wrong strategy: the target of media interventions should be news divisions rather than the advertising departments of media organizations (Wallack 1989).

Moreover, the media might be willing to feature public service announcements for a variety of reasons to further its own goals. Lobbying the media to feature PSAs does not necessarily result in an examination of structural conditions responsible for health problems, however. Media interest in participating in health promotion activities by donating free airtime fall short from moving away from the individualistic view that dominates behavior change models. Such contributions by media organizations do not deal with external factors, unequal access and structures, and the political environment that is ultimately responsible for public health problems. This is why public health needs to incorporate a broader view that conceives actions in terms of community participation and mobilization to transform public opinion and change health policies.

In summary, advocacy consists of a large number of information activities, such as lobbying with decision-makers through personal contacts and direct mail; holding seminars, rallies, and newsmaking events; ensuring regular newspaper, magazine, television, and radio coverage; and obtaining endorsements from known people. The goal of advocacy is to make the innovation a political or national priority that cannot be swept aside with a change in government. In the context of development programs, media advocacy may be carried out by key people in international agencies, as well as special ambassadors, but is gradually taken over by people in national and local leadership positions and the print and electronic media.

### 5.5.4 Social Mobilization

Social mobilization is a term used by the United Nations International Children's Emergency Fund (UNICEF) to describe a comprehensive planning approach that emphasizes political coalition building and community action (UNICEF 1993; Wallack 1989). It is the process of bringing together all feasible and practical inter-sectoral social allies to raise people's awareness of and demand for a particular development program, to assist in the delivery of resources and services, and to strengthen community participation for sustainability and self-reliance. A successful mobilization must be built on the basis of mutual benefits of partners and a decentralized structure. The more interested the partners are, the more likely that a project of social mobilization can be sustained over time. This approach does not require that partners abandon their own interests and perceptions on a given issue but are willing to coalesce around a certain problem.

One of the basic requisites is that groups carefully consider the best-suited groups to partner for a specific program. A child survival and development program in Ghana, for example, started with an analysis to identify individuals and organizations with the potential to serve as partners in a social mobilization project. The study included three sub-studies: interviews with members of governmental institutions, trade unions, revolutionary organizations, and traditional leaders, among others, media content analysis that suggested the need for collective efforts between journalists and health workers, and the assessment of health information sources among parents. It concluded that partners included religious organizations, women's groups, and school teachers (Tweneboa-Kodua et al. 1991).

Mobilization is a process through which community members become aware of a problem, identify the problem as a high priority for community action, and decide steps to take action (Thompson and Pertschuk 1992). It starts with problem assessment and analysis at the community level and moves to action on chosen courses, involving many strategic allies at all levels in a wide range of support activities. Central to social mobilization interventions is empowerment or the process through which individuals or communities take direct control over their lives and environment (Minkler 1990).

Social mobilization suggests that wide community participation is necessary for members to gain ownership so innovations would not be seen as externally imposed. Community mobilization is one of the main resources in implementing behavior change. Social mobilization differs from traditional social marketing approaches that are largely based on appeals to individuals. When there is no individual interest in adopting innovations and, particularly in the context of developing countries, reaching people only through social marketing techniques is not effective, interpersonal channels stimulated by social mobilization allow the wide diffusion of concepts and innovations and increasing demand. Social mobilization is closely interlinked with media advocacy. To McKee, social mobilization "is the glue that binds advocacy activities to more planned and researched program communication activities." It strengthens advocacy efforts and relates them to social marketing activities. It makes it possible to add efforts from different groups to reach all levels of society by engaging in different activities: service delivery, mobilizing resources, providing new channels for communication; providing training and logistical support for field workers, and managing field workers.

Examples of social mobilization interventions include World Bank (1992) nutrition and family planning projects in Bangladesh that also used a social mobilization approach by assigned non-governmental organizations (NGOs) the role of mobilizing communities. It defined community mobilization as "the process of involving and motivating interested stakeholders (general public, health workers, policymakers, etc.) to organize and take action for a common purpose. Mobilization of communities should focus on building confidence, trust and respect, increasing knowledge base, and enabling community members to participate, and become more proactive with regard to their own health behavior." The implementation is required to identify and utilize village communication networks, train field workers, locate and mobilize opinion leaders, activate link persons, establish rotating peer group discussions, provide information and supplies at meetings.

McKee (1999) states that social mobilization programs require that government agencies, NGOs, and donor agencies need to meet and review the objectives and methodology of the research, follow its progress through periodic briefings, and give feedback on the final report. These activities have proven to strengthen the sense of ownership among different stakeholders which ultimately results in a more successful intervention.

## 5.6 Toward a Theoretical and Empirical Convergence?

Can the two broad approaches that dominated the field of development communication, diffusion and participatory models, converge around certain principles and strategies? Or is that unthinkable given that their underlying premises and goals are still essentially different? The last section of this report reviews attempts to bring together different "branches of the family tree," agreements, and disagreements among different approaches based on lessons learned in the last decades. The fact that some coincidences can be identified does not imply that old differences have been completely bridged. This is impossible because different theoretical premises and diagnoses continue to inform approaches and strategies. The fundamental issue continues to be that definitions of the problem are different and, expectedly, theories, strategies, and techniques still offer essentially opposite analyses and recommendations. To identify points of convergence does not imply that a specific value judgment is made about the desirability or necessity of the process. The intention is to map out trends and directions that attest to the richness and complexity of the field rather than to pass judgment on them.

### 5.6.1 General Remarks

Since the 1950s, the meaning of development communication has changed. Changes should not be surprising considering that "development," a concept that together with "modernization" and "Third World" emerged and dominated academic and policy debates in the 1950s, has lost much of its past luster. New concepts have been coined and gained popularity but have not displaced the broad notion of development communication. Despite its multiple meanings, development communication remains a sort of umbrella term to designate research and interventions concerned with improving conditions among people struggling with economic, social, and political problems in the non-Western world. Like "development," "communication" has also undergone important transformations in the past five decades that reflected the ebbs and flows of intellectual and political debates as well as the changing fortunes of theoretical approaches.

The absence of a widespread consensus in defining "development" and "communication" reflects the larger absence of a common vocabulary in the field (Gibson n.d.). This conceptual ambiguity and confusion should not be surprising considering that different disciplines and theories have converged in the field of development communication. There has been a confluence of overlapping traditions from a variety of disciplines that imported vocabularies that had little in common. For example, do concepts such as "empowerment," "advocacy engagement of communities," and "collective community action" refer to fundamentally different ideas? Not really. The presence of different terminologies does not necessarily reflect opposite understandings but, mainly, the existence of different trunks in the family tree. In a fragmented field, diverse programs and strategies are rooted in a myriad of intellectual fields that were rarely in fluid contact.

Despite the diversity of origins, however, it is remarkable that there has been a tendency toward having a more comprehensive understanding of "development communication." The historic gap between approaches has not been bridged, but, certainly, there have been visible efforts to integrate dissimilar models and strategies. Consider Jan Servaes' (1996b) definition of development as a multidimensional process that involves change in social structures, attitudes, institution, economic growth, reduction of inequality, and the eradication of poverty. For him, development is a "whole change for a better life." This notion comes close to the idea of "another development" that emphasizes the satisfaction of needs, endogenous self-reliance, and life in harmony with the environment (Melkote 1991). We would be hard-pressed to find approaches and interventions that essentially disagree with such encompassing idea of development.

Similarly, different approaches have gradually adopted an understanding of communication that is not reduced to the idea of information transmission but includes the idea of process and exchange. Certainly, the persuasion model of communication maintains a towering presence in the field. Sociopsychological models of behavior and perspectives grounded in stimulus-response communication theories continue to dominate, arguably because some premises of the "dominant paradigm" remain widely accepted. The model of top-down, sender-receiver communication has been revised, however.

The idea of "communication as process" has gained centrality in approaches informed by both behavior change and participatory models. Moemeka's (1994, 64) words illustrate a widespread sentiment in the field: "Communication should be seen both an independent and dependent variable. It can and does affect situations, attitudes, and behavior, and its content, context, direction, and flow are also affected by prevailing circumstances. More importantly, communication should be viewed as an integral part of development plans – a part whose major objective is to create systems, modes, and strategies that could provide opportunities for the people to have access to relevant channels, and to make use of these channels and the ensuing communication environment in improving the quality of their lives."

This perspective is somewhat akin to "ritualistic" models of communication that prioritize the Latin roots of the word (as in "making common" through the exchange of meaning) that gained currency in the field of communication in the last decades (Carey 1989). Communication is understood as communities and individuals engaging in meaning-making. It is a horizontal, deinstitutionalized, multiple process in which senders and receivers have interchangeable roles, according to participatory theorist Jan Servaes (1996a). From a perspective rooted in behavior change models, Kincaid (1988) has similarly argued that all participants are senders and receivers.

The difference lies in the fact that whereas approaches largely informed by the dominant paradigm continue to think of communication as a process that contributes to behavior change, participatory models are not primarily concerned with "behavior" but with transforming social conditions.

Another salient feature of recent studies in development communication is the increasing influence of theories and approaches that were originated or have been widely used in health communication. Health communication has received more attention than education or agriculture, issues that were central in early projects of development communication. Certainly, issues such as literacy, agricultural productivity, and violence are included in many contemporary development plans. Behavior change, social marketing, and health promotion models have become increasingly influential in development communication, however. In a way, the growing centrality of health issues should not be surprising considering that family planning and nutrition, for example, have been dominant in the agenda of development communication since the 1960s. Additionally, attention to HIV/AIDS since the 1980s further contributed to the ascendancy of health and health-related approaches in the field.

On the one hand, this shift could be interpreted as a reflection of the priorities of funding agencies. Although further research is needed to support this finding, it seems that the presence of health issues and, consequently, the influence of health communication approaches express the agenda of development organizations. On the other hand, it can also be interpreted as a result of the emergence of a broader approach to health issues. The definition of health as "a state of well-being," widely cited in contemporary studies, allows a more comprehensive approach that includes issues such as illiteracy and poverty that were not integrated in early development communication projects.

### 5.6.2 Points of Convergence

Notwithstanding important persistent differences among theories and approaches, it is possible to identify several points of convergence that suggest possible directions in the field of international communication.

#### 5.6.2.1 The Need of Political Will

One point of convergence is that political will is necessary in order to bring about change. Development communication should not only be concerned with instrumenting specific outcomes as defined in the traditional paradigm but also with the process by which communities become empowered to intervene and transform their environment. Community empowerment should be the intended outcome of interventions. This requires coming up with a set of indicators that measure the impact of interventions in terms of empowerment.

Empowerment lacks a single definition, however. It can refer to communities making decisions for themselves and acquiring knowledge (e.g., about health issues). Whereas for participatory/advocacy approaches empowerment involves changes in power distribution, behavior models use empowerment to represent ways for communities to change behavior, for example, discontinuing unhealthy practices. Advocates of social marketing suggest that marketing empowers people by proving information and having constant feedback from consumers so they can be responsible for their well-being.

Because understandings of empowerment are different, expectations about interventions are different, too. If development requires redressing power inequalities, then, it conceivably takes longer time than interventions that aim to change knowledge, attitudes, and practices. The pressures for relatively quick results and short-term impact of interventions are better suited for a particular understanding of empowerment (and thus development communication) which is more aligned with behavior change than participatory approaches. The slowness of policy and political changes required for more equal distribution of resources and decision-making, as advocated by participatory models, does not fit short-term expectations.

The problems of measuring results, however, are not unique to participatory strategies. Many observers have indicated that behavior change models have not satisfactorily answered the question of long-term effects. The lack of longitudinal studies that document changes over time makes it difficult to know the extent of the influence of interventions and environmental factors that could help reach solid conclusions about the long-term impact of communication strategies.

The fact that the debate over "results indicators" has not concluded and that no easy resolution seems in sight reflects the persistence of disagreements over measuring development. Answers to questions such as "what are the right results?" are expected to be different given that, notwithstanding a growing consensus on the issues of community empowerment and horizontal communication as central to development communication, behavior change and participatory models still define the task of interventions in different terms. In other words, there continues to be a tension between approaches that are oriented to achieving results as measured in behavior change and those that prioritize the building of sustainable resources as the goal of programs.

### 5.6.2.2 A "Tool-Kit" Conception of Strategies

Another important point of convergence is the presence of a "tool-kit" conception of approaches within the behavior change tradition. Practitioners have realized that a multiplicity of strategies is needed to improve the quality of life of communities in developing countries. Rather than promoting specific theories and methodologies regardless of the problem at stake, there has been an emerging consensus that different techniques are appropriate in different contexts in order to deal with different problems and priorities. Theories and approaches are part of a "tool kit" that is used according to different diagnoses. There is the belief that the tools that are used to support behavior change depend on the context in which the program is implemented, the priorities of funders, and the needs of the communities.

For example, conventional educational interventions might be recommended in critical situations such as epidemics when large masses of people need to be reached in a short period of time. Such strategies, however, would be unlikely to solve structural, long-term health problems.

Social marketing could be useful to address certain issues (e.g., increase rates of immunization) but is inadequate to address deeper problems of community participation that are ultimately responsible for permanent changes. It also can result in the problem that interventions conclude when public information campaigns are terminated. One of the problems is that such interventions create a dependency on media programs; the alternative, then, is focusing on self-maintaining resources that are responsible for the sustainability of programs. Another problem is that even when social marketing strategies are successful at raising awareness, they do not last forever and, therefore, other support systems are necessary to maintain participation and communication.

Because of the limitations of social marketing, other strategies are needed to address the problem of empowering and politically involving different groups. Social mobilization, for example, offers a way to deal with certain issues such as education, sanitation, nutrition (including breastfeeding), family planning, respiratory problems, AIDS, and diarrheal diseases. Still, the mobilization of a vast array of partners is necessary, but this does not exclude the uses of media advocacy and social marketing to target specific problems. A breastfeeding program in Brazil successfully integrated social mobilization and social marketing (Fox n.d.). As a result of the program, there was an increase in the median duration of breastfeeding and a reduction in infant mortality. Ministries and professional medical and nutrition groups participated in elaborating plans and stimulating actions at the national and state levels among their employees, members, and associated institutions. At the community level, mothers' groups were formed and breastfeeding was promoted through extension workers, university students, the church, and other voluntary groups.

Family planning programs in Egypt have been another example of successful integration of different approaches (Wisensale and Khodair 1998). After the intervention, the use of contraceptives doubled and the birthrate dropped from 39.8 to 27.5 percent in 10 years. The achievements of the program have been attributed to fact that the Information, Education & Communication Center of the State Information Service used five tools, including the mass media, interpersonal communication, and entertainment-education. The participation of the government, health organizations, and religious groups was also considered to be responsible for the success of the program.

The application of any prescriptive theory and methods might not work everywhere. Because of political and religious reasons, it is difficult to bring together a wide spectrum of forces to rally behind issues such as breastfeeding, family planning, and AIDS education in some countries. Under these circumstances, searching for a broad coalition is not recommended. In cases where governments strictly control the mass media or believe that they should be the only actors involved in public information campaigns, then, social marketing interventions confront many problems.

There has been a growing sensitivity to the problems of the universal application of strategies that were successful in specific contexts. In countries where political and cultural factors limit participation and maintain hierarchical relationships, participatory approaches might be difficult to implement as they require a long-term and highly political process of transformation. This does not mean that participation should be abandoned as a desirable goal but that interventions that aim to mobilize communities necessarily adopt different characteristics in different circumstances. Public service announcements may be perceived as contradicting official power and policies. When access to national media is limited or extremely conditional, grassroots strategies whether community participation and local media could offer an alternative. But if populations are afraid of participating for fear of repression or because of past frustrations, then, participatory approaches face clear obstacles and may not be advisable.

**Integration of "Top-Down" and "Bottom-Up" Approaches**
Faced with different scenarios and choices, the growing consensus is that a multiple approach that combines "top-down" and "bottom-up" interventions is recommended. Here it becomes evident that development communication has gone beyond transmission models focused on implementing behavior changes through communication activities.

The Iringa Nutrition Improvement Program in Tanzania has been mentioned as a successful example of integrating media advocacy, social mobilization, and social marketing (FPRI 1994). The Program included the mobilization of different groups at different levels, community participation, media advocacy to popularize the goals of fighting malnutrition and child mortality, and social marketing to raise awareness among all sectors of the populations. It included "child growth monitoring, strengthening the health infrastructure, health education, and women's activities." Government commitment, long-term sustainability of the program, and anti-poverty efforts have been mentioned as decisive in contributing to its success in reducing malnutrition despite larger economic problems. Environmental factors such as a tradition of grassroots participation and national policies that dramatically increased literacy were crucial for the success of the program.

### 5.6.2.3 Integration of Multimedia and Interpersonal Communication

Much with the current thinking is that successful interventions combine media channels and interpersonal communication. Against arguments of powerful media effects that dominated development communication in the past, recent conclusions suggest that blending media and interpersonal channels is fundamental for effective interventions (Flay and Burton 1990; Hornik 1989). The media are extremely important in raising awareness and knowledge about a given problem (Atkin and Wallack 1990).

The media are able to expose large amounts of people to messages and generate conversation among audiences and others who were not exposed (Rogers 1998). But it would be wrong to assume that development mainly or only requires media channels. Because social learning and decision-making are not limited to considering media messages but listening and exchanging opinions with a number of different sources, as Bandura (1989) suggested, interventions cannot solely resort to the mass media. Although television, radio, and other media are important in disseminating messages, social networks are responsible for the diffusion of new ideas (Rogers and Kincaid 1981; Valente et al. 1994). Entertainment-education programming is one way, for example, to activate social networks and peer communication in the diffusion of information (Rogers et al. 1999). Similarly, information given through the media is also important in raising awareness and knowledge as integrated into peer conversations and in contacts with field workers (Mita and Simmons 1995; Ogundimu 1994).

According to McKee (1999), interpersonal communication and the actions of community workers account for much of the success of several projects. Nothing can replace community involvement and education in the effective dissemination of information. Media-centered models are insufficient for behavior change. McKee argues that the most successful strategies in family planning, HIV/AIDS, and nutritional and diarrhea programs have involved multiple channels, including strong, community-based programming, networks, peer counseling, and government and NGO field workers. Successful initiatives attest to the fact that redundancy and multiple channels should be used. The media has powerful effects only indirectly by stimulating peer communication and thus making possible for messages to enter social networks and become part of everyday interactions.

Without disputing the value of interpersonal communication, McDivitt et al. (1997) have stressed the importance of the mass media in behavior change (also see Hornik (1988). In an evaluation of the impact of a vaccination campaign in the Philippines, they concluded that the media, rather than interpersonal channels, was responsible for changes in vaccination knowledge. Media exposure was sufficient to generate more knowledge about the specifics of the campaign and change in vaccinations without the intervention of social networks. It would be wrong, according to the researchers, to ignore the unmatched reach of the media, particularly among certain groups, in getting the message out. Mass media messages per se, however, do not explain the success of the campaign. The campaign provided specific information that mothers needed in order to engage in expected behaviors, and other conditions (access to health centers, sufficient vaccine supplies) were also fundamental in making behavior change possible.

**Personal and Environmental Approaches Should Be Integrated**
The revision of traditional health promotion strategies and then integration of social marketing and social mobilization are examples of the tendency to integrate personal and environmental approaches.

Consider the "ecological approach" as an example in that direction as used in the North Karelia Project in Finland, the Minnesota Health program, and the Stanford Three Community study (Bowes 1997). It espouses organizational and environmental interventions and aims to be more comprehensive that efforts directed only at individuals or social action (McLeroy et al. 1988; Glanz and Rimer 1995). Non-behavioral factors such as unemployment, poverty, and lack of education are included as part of the broad view that ecological approaches encompass. Health promotion should be integrated into existing social systems such as schools, health delivery systems, and community organizations. The mentioned projects required coordination among a variety of intermediate agencies that acted as liaisons between developers of health promotion innovation and potential adopters. The focus is still on behavior change, but programs feature environmental supports to encourage individuals to adopt and maintain changes. Similarly, community participation approaches have recognized the need to promote a "holistic approach" that integrates the contributions of both personal behavior change and broader environmental changes in facilitating health improvement (Minkler 1999).

"Communication for social change" (CSC) is another example of recent efforts to integrate different theories and approaches in development communication (Rockefeller Foundation 1999). Whereas traditional interventions were based on behavior change models, CSC relies on participatory approaches in emphasizing the notion of dialogue as central to development. Development is conceived as involving work to "improve the lives of the politically and economically marginalized" (1998, 15). In contrast to the sender-receiver, information-based premises of the dominant paradigm, it stresses the importance of horizontal communication, the role of people as agents of change, and the need for negotiating skills and partnership. Another important contribution of CSC is to call attention to the larger communication environment surrounding populations.

In contrast to behavior change and participatory theories that, for different reasons, pay little if any attention to the wide organization of information and media resources, CSC calls attention to the relevance of ongoing policy and structural changes in providing new opportunities for communication interventions. Unlike neo-dependency theories that negatively view worldwide changes in media and information industries as stimulating a process of power concentration, CSC offers a mixed evaluation. It recognizes that transformations open possibilities for community-based, decentralized forms of participation, but also admits that some characteristics of contemporary media are worrisome in terms of the potential for social change. CSC views changes in health and in quality of life in general in terms of citizens' empowerment, a notion that became more relevant in behavior change models (Hornik 1997).

But unlike participatory theories, CSC stresses the need to define precise indicators to measure the impact of interventions. It is particularly sensitive to the expectations of funding agencies to find results of interventions and to the needs of communities to provide feedback and actively intervene in projects. Here accountability, a concept that is also fundamental in contemporary global democratic projects, is crucial to development efforts. Projects should be accountable to participants in order to improve and change interventions and involve those who are ultimately the intended protagonists and beneficiaries. Because the intended goals are somewhat different from behavior change approaches, then, it is necessary to develop a different set of indicators that tell us whether changes are achieved (although certainly some measurements traditionally used in health interventions are useful too). The goals are not only formulated in terms that could perfectly fit health promotion/social marketing/behavior change theories (e.g., elimination of HIV/AIDS, lower child and maternal mortality) but also in broader social terms such as eradicating poverty and violence and increasing employment and gender equality. These goals express a more comprehensive understanding of development that is not limited to "better health and well-being" but is aware of the need to place traditional approaches in larger social and environmental contexts.

Despite the cross-pollination of traditions and a multi-strategy approach to interventions, the rift between behavior change and participatory approaches and theories still characterizes the field. The divisions are less pronounced than a few decades ago given the integration of different strategies discussed in the previous section but are still important.

For participatory and advocacy approaches, behavior change models are still associated with a certain scientific paradigm that is questionable on several grounds. Behavior change models are based on premises that do not necessarily translate to developing countries (Stetson and Davis 1999). From a perspective influenced by recent theoretical developments in the social sciences, particularly postcolonial and post-modernist thinking, critics have challenged Western models of rationality and knowledge that inform behavior models. What is necessary is to change the traditional perspective according to which "traditional cultures" are backward and antithetical to development interventions. Because what populations know is considered wrong, local knowledge is viewed as obstacle and unnecessary in development interventions. Overcoming ethnocentric conceptions is crucial. It requires recognition that understandings of information and knowledge are different. Interventions also need to be sensitive to the fact that local cultures do not necessarily fit philosophical assumptions about individual rationality that are embedded in traditional models. Sense-making practices that are found in the developing world contradict key premises of behavior change models. Behavior models assume that individuals engage in certain actions after weighing costs and benefits of the action. Whereas individual interest and achievement are the underlying premises of those models, non-rationalistic forms of knowledge as well as community values are central to non-Western cultures.

Critics charge behavior change models for being focused on individual changes while underplaying (or minimizing) the need to instrument larger political transformations that affect the quality of life. They call attention to the organizational structures that inhibit the successful implementation of projects for social change (Wilkins 1999). The concentration of information resources worldwide, the growing power of advertising in media systems, and the intensification of inequalities that underlie the persistence of development problems require more than ever to examine structural-political factors. Media systems have changed dramatically in the last decades. These changes, however, have been particularly revolutionary in the non-Western world as privatization and liberalization of media systems radically transformed the production, distribution, and availability of information resources.

Behavior change models have recognized the merits of insights from participatory approaches as well as the need to be sensitive to media access and new technologies (Piotrow et al. 1997). They continue to be mainly concerned with refining analytical and evaluation instruments and measuring the success of different intervention strategies. One of the main tasks is to identify the impact of communication/information campaigns in the context of other factors that affect behavior (Hornik 1997). The integration of social mobilization and social marketing strategies has been found to be successful and a positive referent for future interventions (McKee 1999).

The realization that communities should be the main actors of development communication may constitute a starting point for further integration. Likewise, efforts to integrate theories and strategies that recognize that media campaigns are insufficient without community participation, that social marketing efforts are weak without environmental changes, and that community empowerment might be the ultimate goal to guarantee sustainable development are encouraging to promote dialogue among different theories and traditions.

## References

Agunga RA (1997) Developing the Third world. A communication approach. Nova Science, Commack

Altafin I (1991) Participatory communication in social development evaluation. Comm Develop J 26(4):312–314

Andreasen AR (1994) Social marketing: its definition and domain. J Public Policy Market 13(1):108–114

Atkin C, Wallack L (eds) (1990) Mass communication and public health: complexities and conflicts. Sage Publications, Newbury Park

Bandura A (1977) Social learning theory. Prentice Hall, Englewood Cliffs, NJ

Bandura A (1989) Perceived self-efficacy in the exercise of control over AIDS infection. In: Mays VM, Albee GW, Schneider SS (eds) Primary prevention of AIDS: psychological approaches. Sage, Newbury Park, pp 128–141

Beltrán LR (1976) Alien premises, objects, and methods in Latin American communication research. In: Rogers EM (ed) Communication and development: critical perspectives. Sage, Beverly Hills, pp 15–42

Beltrán LR (1993a) Communication for development in Latin America: a forty-year appraisal. In: Nostbakken D, Morrow C (eds) Cultural expression in the global village. Southbound, Penang, pp 10–11

Beltrán LR (1993b) The quest for democracy in communication: outstanding Latin American experiences. Development 45-47(38):3

Bowes JE (1997) Communication and community development for health information: constructs and models for evaluation. www.nnlm.nlm.nih.gov/pnr/eval/bowes/

Bracht N (ed) (1990) Health promotion at the community level. Sage, Newbury Park

Bradford B, Gwynee MA (1995) Down to earth: community perspectives on health, development, and the environment. Kumarian Press, West Hartford

Brawley EA, Martinez-Brawley EE (1999) Promoting social justice in partnership with the mass media. J Sociology Social Welfare 26(2):63–86

Buchanan DR, Reddy S, Hossian Z (1994) Social marketing: a critical appraisal. Health Promot Int 9(1):49–57

Carey JW (1989) Communication as culture: essays on media and society. Unwin Hyman, Boston

Chapman Walsh D, Rudd RE, Moeykens BA, Moloney TW (1993) Social marketing for public health. Health Affairs 12(2):104–119

Diaz-Bordenave J (1977) Communication and rural development. UNESCO, Paris

Elliott BJ (1991) A re-examination of the social marketing concept. Elliott & Shanahan Research, Sydney

FPRI Report (1994) Seminar series focuses on successful nutrition programs, 16, 2, www.cgiar.org/ifpri/reports/0694RPT/0694e.htm

Flay BR, Burton D (1990) Effective mass communication strategies for health campaigns. In: Atkin C, Wallack L (eds) Mass communication & public health, pp 129–145
Flora JA, Maccoby N, Farquhar JW (1989) Communication campaigns to prevent cardiovascular disease: the Stanford community studies. In: Atkin C, Rice R (eds) Public communication campaigns. Sage, Newbury Park, pp 233–252
Fox E (n.d.) Conductismo y Comunicación Social Hacia Dónde Nos Llevó?
Freedman R (1997) Do family planning programs affect fertility preferences? A literature review. Stud Fam Plan 28(1):1–13
Freire P (1970) Pedagogy of the oppressed. Herder & herder, New York
Gibson C (n.d.) Strategic communications for health and development. Typescript
Glanz K, Lewis FM, Rimer BK (eds) (1990) Health behavior and health education: theory, research and practice. Jossey-Bass Publishers, San Francisco
Glanz K, Rimer BK (1995) Theory at a glance. National Institute of Health, Washington
Hagen E (1962) On the theory of social change. University of Illinois Press, Urbana
Hamelink C (1990) Integrated approaches to development communication: a study and training kit. J Develop Comm 1(1):77–79
Holder HD, Treno AJ (1997) Media advocacy in community prevention: news as a means to advance policy change. Addiction 92:189–199
Hornik RC (1989) Channel effectiveness in development communication programs. In: Rice RE, Atkin CK (eds) Public information campaigns, 2nd edn. Sage, Newbury Park, pp 309–330
Hornik RC (1997) Public health education and communication as policy instruments for bringing about changes in behavior. In: Goldberg M, Fishbein M, Middlestadt S (eds) Social marketing. Lawrence Erlbaum, Mahwah, pp 45–60
Inkeles A, Smith DH (1974) Becoming modern. Harvard University Press, Cambridge, MA
Kalipeni E, Kamlongera C (1996) The role of 'theatre for development' in mobilizing rural communities for primary health care: the case of Liwonde PHC unit in southern Malawi. J Soc Dev Afr 11(1):53–78
Katz E, Lazarsfeld PF (1955) Personal influence: the part played by people in the flow of mass communications. Free Press, New York
Kavinya A, Alam S, Decock A (1994) Applying DSC methodologies to population issues: a case study in Malawi. FAO, Rome
Kincaid L (1988) The convergence theory of communication: its implications for intercultural communication. In: Kim YY (ed) Theoretical perspectives on international communication. Sage, Beverly Hills
Kotler P, Zaltman G (1971) Social marketing: an approach to planned social change. J Mark 35:3–12
Kotler P, Roberto E (1989) Social marketing: strategies for changing public behavior. Free Press, New York
Lerner D (1958) The passing of traditional society. Free Press, New York
Maibach E (1993) Social marketing for the environment: using information campaigns to promote environmental awareness and behavior change. Health Promot Int 3(8):209–224
Maibach E, Murphy DA (1995) Self-efficacy in health promotion research and practice: conceptualization and measurement. Health Educ Res 10(1):37–50
McClelland D (1961) The achieving society. Van Nostrand, New York
McDivitt JA, Zimicki S, Hornik RC (1997) Explaining the impact of a communication campaign to change vaccination knowledge and coverage in the Philippines. Health Commun 9(2):95–118
McKee N (1999) Social Mobilization & Social Marketing in developing communities: lessons for communicators. Southbound, Penang
McLeroy KR, Bibeau D, Steckler A, Glanz K (1988) An ecological perspective on health promotion programs. Health Educ Q 15(4):351–377
Melkote SR (1991) Communication for development in the Third world. Sage, Newbury Park
Meyer G, Dearing JW (1996) Respecifying the social marketing model for unique populations. Soc Mark Q 3(1):44–52

Minkler M (1990) Improving health through community organization. In: Glanz K, Lewis FM, Rimer BK (eds) Health behavior and health education: theory, research, and practice. Jossey-Bass Publishers, San Francisco, pp 257–287

Minkler M (1999) Personal responsibility for health? A review of the arguments and the evidence at century's end. Health Educ Behav 26(1):121–140

Mita R, Simmons R (1995) Diffusion of the culture of contraception: program effects on young women in rural Bangladesh. Stud Fam Plan 26(1):1–13

Mlama PM (1991) Women's participation in "communication for development": the popular theater alternative in Africa. Res Afr Lit 22(3):41–53

Mody B (1991) Designing messages for development communication: an audience participation-based approach. Sage, Newbury Park

Moemeka AA (ed) (1994) Communicating for development: a new pan-disciplinary perspective. State University of New York Press, Albany

Novelli W (1990) Applying social marketing to health promotion and disease prevention. In: Glanz K, Lewis FM, Rimer BK (eds) Health behavior and health 41 education: theory, research, and practice. Jossey-Bass Publishers, San Francisco, pp 324–369

Ogundimu F (1994) Communicating knowledge of immunization for development: a case study from Nigeria. In: Moemeka AA (ed) Communicating for development, pp 219–243

Okunna CS (1995) Small participatory media technology as an agent of social change in Nigeria: a non-existent option? *Media*. Culture Society 17(4):615–627

Piotrow PT, Kincaid DL, Hindin MJ, Lettenmaier CL, Kuseka I, Silberman T, Zinanga A, Ikim YM (1992) Changing men's attitudes and behavior: the Zimbabwe male motivation project. Stud Fam Plan 23(6):365–375

Piotrow PT, Kincaid DL, Rimon JG, Rinehart W (1997) Health communication: lessons from family planning and reproductive health. Praeger, Westport

Purdey AF, Adhikari GB, Robinson SA, Cox PW (1994) Participatory health development in rural Nepal: clarifying the process of community empowerment. Health Educ Q 21(3):329–343

Quarmyne W (1991) Towards a more participatory environment: cross-linking establishment and alternative media. In: Boafo K (ed) Communication processes: alternative channels and strategies for development support. IDRC, Ottawa

Rockefeller Foundation (1999) Communication for social change: a position paper and conference report. Rockefeller Foundation, New York

Rogers EM (1962) Diffusion of innovations, 1st edn. Free Press, New York

Rogers EM (1976) Communication and development: the passing of the dominant paradigm. Commun Res 3(2):213–240

Rogers EM (1983) Diffusion of innovations, 3rd edn. Free Press, New York

Rogers EM (1998) When the mass media have strong effects: intermedia processes. In: Trent J (ed) Communication: views from the helm for the twenty-first century. Allyn and Bacon, Boston, pp 276–285

Rogers EM, Kincaid DL (1981) Communication networks: a paradigm for new research. Free Press, New York

Rogers EM, Vaughan PW, Swalehe RMA, Rao N, Svenkerud P, Sood S (1999) Effects of an entertainment-education radio soap opera on family planning behavior in Tanzania. Stud Fam Plan 30(3):193–211

Rutten A (1995) The implementation of health promotion: a new structural perspective. Social Science Medicine 41(12):1627–1637

Schramm W (1964) Mass media and national development. Stanford University Press, Stanford

Schramm W (1997) The beginnings of communication study in America. Sage, Thousand Oaks

Servaes J (1989) One world, multiple cultures: a new paradigm on communication for development. Acco, Leuven, Belgium

Servaes J (1996a) Introduction: participatory communication and research in development settings. In: Servaes J, Jacobson T, White SA (eds) Participatory communication for social change. Sage, Thousand Oaks

Servaes J (1996b) Communication for development in a global perspective: the role of governmental and non-governmental agencies. Communications 21(4):407–418

Singhal A, Rogers EM (1999) Entertainment-education: a communication strategy for social change. Lawrence Erlbaum, Mahwah

Stetson V, Davis R (1999) Health education in primary health care projects: a critical review of various approaches. Core group, Washington, DC

Terris M (1992) Concepts of health promotion: dualities in public health theory. J Public Health Policy 13:267–276

Thompson B, Pertschuck M (1992) Community intervention and advocacy. In: Ockene JK, Ockene JS (eds) Prevention of coronary heart disease. Little, Brown, Boston, pp 493–515

Tweneboa-Kodua A, Obeng-Quaidoo I, Abu K (1991) Ghana social mobilization analysis. Health Educ Q 18(1):25–134

UNICEF (1993) We will never go back: social mobilization in the child survival and development programme in the United Republic of Tanzania. UNICEF, New York

Valente T, Kim YM, Lettenmaier C, Glass W, Dibba Y (1994) Radio promotion of family planning in the Gambia. Int Fam Plan Perspect 20(3):96–104

Valente T, Paredes P, Poppe P (1998) Matching the message to the process: the relative ordering of knowledge, attitudes, and practices in behavior change research. Hum Commun Res 24(3):366–385

Wallack L (1989) Mass communication and health promotion: a critical perspective. In: Rice RE, Atkin C (eds) Public communication campaigns, 2nd edn. Sage, Newbury Park

Wallack L, Montgomery K (1992) Advertising for all by the year 2000: public health implications for less developed countries. J Public Health Policy 13(2):76–100

Wallack L, Dorfman L, Jernigan D, Themba M (1993) Media advocacy and public health: power for prevention. Sage, Newbury Park

Westoff C, Rodriguez G (1995) The mass media and family planning in Kenya. Int Fam Plan Perspect 21(1):26–31, 36

White SA (1994) The concept of participation: transforming rhetoric to reality. In: White SA et al (eds) Participatory communication: working for change and development. Sage, New Delhi

Wilkins KG (1999) Development discourse on gender and communication in strategies for social change. J Commun 49(1):46

Williamson HA (1991) The Fogo process: development support communications in Canada and the developing world. In: Casmir FL (ed) Communication in development. Ablex Publishing Corporation, Norwood, pp 270–287

Wisensale SK, Khodair AA (1998) The two-child family: the Egyptian model of family planning. J Comp Fam Stud 29(3):503–516

World Bank (1992) The determinants of reproductive change, population and health sector study. World Bank, South Asia Region, Health, Population, and Nutrition Unit, Washington, DC

Yoder PS, Robert RC, Chirwa BC (1996) Evaluating the program effects of a radio drama about AIDS in Zambia. Stud Fam Plan 27(4):188–203

Zimicki S, Hornik RC, Verzosa CC, Hernandez JR, de Guzman E, Dayrit M, Fausto A, Lee MB, Abad M (1994) Improving vaccination coverage in urban areas through a health communication campaign: the 1990 Philippine experience. Bull World Health Organ 72(3):409–422

# A Changing World: FAO Efforts in Communication for Rural Development

## Silvia Balit and Mario Acunzo

## Contents

| | | |
|---|---|---|
| 6.1 | The Landscape | 135 |
| 6.2 | Telling the Story | 136 |
| 6.3 | The Good Times: Reasons for Success | 139 |
| 6.4 | What Followed? Current Perspectives | 140 |
| | 6.4.1 ComDev Assistance | 142 |
| | 6.4.2 Building Capacities and Partnerships | 145 |
| | 6.4.3 Evidence-Based Approaches | 146 |
| 6.5 | Lessons Learned: Some Essential Guiding Principles | 147 |
| | 6.5.1 Starting Upstream | 147 |
| | 6.5.2 A Social Process | 147 |
| | 6.5.3 Sharing of Knowledge and Two-Way Communication | 147 |
| | 6.5.4 Listening to People | 148 |
| | 6.5.5 In Line with National Policies | 148 |
| | 6.5.6 Promoting Policy Change | 149 |
| | 6.5.7 Holistic Approach | 149 |
| | 6.5.8 Preserving Indigenous Knowledge, Values, and Culture | 149 |
| | 6.5.9 Respect for Culture | 149 |
| | 6.5.10 Shared Interests and Trust | 150 |

S. Balit (✉)
Rome, Italy
e-mail: silvia.balit@gmail.com

M. Acunzo
FAO Communication for Development Team, Rome, Italy
e-mail: Mario.Acunzo@fao.org

© Springer Nature Singapore Pte Ltd. 2020
J. Servaes (ed.), *Handbook of Communication for Development and Social Change*,
https://doi.org/10.1007/978-981-15-2014-3_29

6.5.11 Flexibility and Duration .................................................. 150
6.5.12 Ensuring Gender Sensitivities ........................................... 150
6.5.13 A ComDev Approach to the Use of the ICTs ...................... 151
6.5.14 Multimedia Approach ..................................................... 152
6.5.15 Political Space and Context ............................................. 152
6.5.16 Toward Inclusive Rural Communication Services ............... 152
6.5.17 No-Off-the-Shelf Solution ............................................... 153
6.6 Conclusion ............................................................................. 153
References ...................................................................................... 155

### Abstract

The chapter describes relevant approaches, trends, and challenges of applying Communication for Development to agricultural and to rural development in a changing world. It focuses on the experience generated by the United Nations (UN) Food and Agriculture Organization (FAO), starting with the first years when FAO was considered a leading agency in the discipline, followed by an analysis of the present situation, the perspectives, and the challenges for the future. The authors argue that there are new and favorable conditions for Communication for Development in the context of the changing social and political scene and that new economic and environmental challenges affecting rural areas demand new communication approaches and practice to be blended with the principles and participatory methods that have inspired the discipline.

Today more than ever, there is a need for mainstreaming Communication for Development to contribute managing the structural transformation affecting rural areas and to bridge the divide of opportunities ensuring equal rights to communication through demand-led communication services dedicated to rural dwellers. This changing landscape also requires new policy and financial commitments on the part of development institutions and governments, as well as policy dialogue with interested parties, including producer and indigenous peoples' organizations, representatives of women and youth, as well as the private sector. By facilitating the dialogue, the definition of priorities, and options for inclusive rural communication services, Communication for Development will contribute significantly to meeting the challenges of our changing world.

### Keywords

Communication · Communication for Rural development · ComDev · C4D · Rural development · Family farming · Information and communication technologies · Participation · Rural communication services · Sustainable Development Goals · FAO

No innovation, however brilliantly designed becomes development unless it is communicated. Erskine Childers

# 6 A Changing World: FAO Efforts in Communication for Rural Development

## 6.1 The Landscape

We live in a fast changing world, conditioned by the information revolution. And the continuous explosion of new technologies and media are a powerful force to share knowledge and influence values and behavior. The benefits of the information revolution can have positive repercussions for economic and social development but also negative results.

These technologies can be used for enlightenment but also for manipulation (Ansah 1994). The three Internet giants Amazon, Facebook, and Google influence the way billions of people think, access information, and act. No political figure can aspire to win an election without spin and often fake news. Financial centers operate in real time on a global scale, bypassing national states and creating super-national powers. The nation state is in crisis as new technologies have amplified people's voices and created aspirations for autonomy. At the same time, the incredible spread of the mobile phone has meant that people in rural areas throughout the developing world now use their mobile phones and send SMS messages to gain information, sell their products, and keep in touch with relatives abroad.

But, despite progress in some areas, how are information and communication technologies (ICTs) helping to eliminate poverty, hunger, and malnutrition that affect nearly one billion people, mainly rural areas of developing countries?

We live in uncertain times characterized by unprecedented climate change amid widening social, political, and economic inequalities. While agriculture is increasingly knowledge-intensive, the rural population tends to be not only poor resource-wise but also information-wise. In particular, smallholder and family farmers face significant social, economic, and environmental challenges which demand improved knowledge and information and their active engagement to foster suitable solutions. Today more than ever, smallholder family farmers and rural communities require access to information and communication to make their voices heard and change their lives for the better (FAO 2016a). Thus communication is key and central to these processes.

To be able to respond and adapt to the new conditions, farmers and rural communities need access to reliable sources of information making available relevant contents in languages and formats they can easily use. Nevertheless, while dedicated information and communication services addressing the need of family farmers and their organizations are rarely available in rural areas, many economic and social barriers affect their access to these services and the active participation in decision-making especially for women, youth, or minority groups. Participatory communication processes, community media, and local appropriations of ICTs have proven essential to overcoming such limitations and increasing the self-reliance of millions of poor farmers worldwide.

Within this framework national and international development institutions have yet to fully recognize the potential of communication processes to contribute to eliminating poverty and promoting sustainable rural development. Neither the Sustainable Development Goals include the right to communication as one of the fundamental conditions to fight poverty and hunger and to provide equal development opportunities to both rural and urban people. The solution for these inequalities is generically associated to the access to the ICTs.

In general terms, our societies value more the power of media and of technologies, jointly with pervasive forms of communication such as publicity, public relations, corporate communication, and spin. There is a rhetorical emphasis on the "positive impact" of media and technology, rather than a deep understanding of the role that communication plays in the structural transformations in rural areas that would allow to generate adequate policies and services accordingly. The lack of Communication for Development efforts in turn hampers the possibility of ensuring the engagement of rural stakeholders, while limiting the efficient use of the resources destined to development.

Over the years billions have been spent for development assistance with limited results. Recognizing this failure development aid is now changing, moving away from donor-driven approaches to country-led systems. The emphasis now is on country programs concentrating on participation, community ownership, capacity building, and effectiveness. In principle, there is a new enabling environment for Communication for Development, with new opportunities and challenges.

## 6.2 Telling the Story

There was a different landscape when the Food and Agriculture of the United Nations initiated activities in Development Support Communication, as the discipline was originally called, in 1969. Top-down approaches were prevalent. Development efforts concentrated on economic growth and transferring western, industrialized technology to developing countries. Development specialists were convinced they had all the answers and that it would be sufficient to spread the word for the knowledge to trickle down and for people to accept new ideas and put them into practice. The role of the media was overestimated, and in particular the emphasis was on the use of mass media. There were few governments in developing countries willing to establish a true dialogue with rural people and allow de-centralized decision-making to take place. The role of communication was primarily limited to supporting transfer of technical knowledge (Fugelsang 1987).

Development Support Communication was established as a discipline within the UN system at the end of the 1960s. The first Development Support Communication (DSC) Service was established in Bangkok in 1967 with financing from UNDP and UNICEF. Development Support Communication owed its vision to Erskine Childers and Mallica Vajrathon of UNDP and Jack Ling of UNICEF. Erskine Childers convinced the Administrator of UNDP Paul Hoffman of the importance of communication in development, and UNDP sent out a circular to all UN agencies requesting them to include communication components in all UNDP-financed projects. And in 1969 units were established in various agencies with different names, but with a common consensus that communication should become a component in development programs.

It made sense to start a DSC program in FAO since it dealt with agriculture, rural development, and extension. It also already had a farm radio program. Erskine Childers visited FAO, and interest came from the Director of the Public Information program Ted Kaghan, rather than from the extension services. The extension services were suspicious of the emerging discipline that they considered to be a threat, and did not promote the development of an integrated and collaborative relationship. They saw the role of communication being limited to the production of media material and the packaging of information from a central source to transfer technology to rural audiences.

The FAO Development Support Communication Branch (DSC) was established in 1969, under the leadership of Colin Fraser. It made use of the media resources of the Public Information Division but did not deal with public relations or corporate communication. It was entirely field oriented. Objectives were to improve project planning, participation, and capacity development.

From the beginning it was difficult to receive recognition within an Organization that was very technical and especially in the first years entirely devoted to transfer of technology and the hard sciences. Many of the specialists came from colonial backgrounds. There was little understanding of what communication was all about, and many understood it as dealing with telephones, telegraphs, and roads. It was a new frontier, and little was known in those days about how to communicate with illiterate people in rural areas of developing countries. As Colin Fraser used to say, it was like painting on an empty canvas. No one in the group had an academic background, although some had backgrounds in media or learning methodologies. The group started learning by doing, learning from mistakes as well as from the first successes. Many requests for assistance were confined to production of media materials for the field, in line with predominant approaches. The partners were primarily government representatives and projects. The media applied were simple and appropriate to conditions in rural areas in developing countries such as radio, slide sets, and portable video. But even the use of portable video was contested by many government representatives as being too sophisticated for communities in rural areas.

With the evolution of development paradigms and more emphasis being given to the human aspects of development and participatory processes, communication started being recognized as a social process to improve the lives of rural people and increase participation, also in FAO. The practical experience in the field led to the development of a number of innovative communication systems and approaches, thanks also to the presence of a number of "champions" as project managers (Quarry and Ramirez 2009).

In Peru an audio visual training methodology based on the use of video was developed for large-scale training of illiterate communities. A holistic approach was applied with programs produced on all the themes of interest to the farm family. Attention was paid to peasant culture and traditional technologies.

In Mexico the rural communication system in Proderith, an integrated rural development program, participatory video was applied to consult with communities and involve them in local development planning. Sharing of knowledge was then based on needs as identified by them. Young farmers, both men and women, were trained to produce and use communication materials with their communities. Messages were produced by the farmers, with the farmers and for the farmers (Fraser 1987).

Based on the oral tradition, in Africa radio was considered the most effective medium to reach large numbers of rural audiences over vast distances. A participatory approach to rural radio was developed which gave a voice to rural people (Balit and Ilboudo 1996). Rural radio services were brought to the villages and programs produced in local languages with their participation, in accordance with traditional customs and values, at the same time providing them with relevant and essential information.

Another successful approach was the use of multimedia campaigns to support a clearly defined and particular development strategy, such as the elimination of rinderpest in Africa or pest control in Asia. Campaigns combined interpersonal and multimedia channels in an orchestrated and mutually reinforcing manner. They also used techniques such as audience research, pretesting of materials, and message fine-tuning on the basis of feedback and evaluation.

The use of traditional media such as local artists, music, storytellers, theater groups, and puppets combined with qualitative research, group discussions, and other participatory tools were successfully used in several countries in Africa to provoke change in life styles on sensitive subjects such as population issues.

After some 20 years of pragmatic experience, it was considered important to have a more strategic approach. For this purpose an Expert Consultation was organized in 1987, and participants included academics as well as communication specialists from all regions, the so-called champions. The consultation was truly a milestone in FAO's experience. It contributed to a better understanding of the role of communication in agricultural and rural development. It provided directions for future work and defined a strategic framework and guidelines that are still a reference and are continuously being updated according to the evolution of FAO work (FAO 2014a).

The first Expert Consultation defined communication as "a social process designed to seek a common understanding among all the participants of a development initiative, creating a basis for concerted action. It promotes knowledge sharing and participation. Sharing is not a one way transfer of information; it implies rather an exchange between communication equals: on the one hand, technical specialists learn about people's needs and their knowledge, and on the other hand, people learn of the techniques and proposals of the specialists. Communication technology and the media are useful tools to facilitate this process, but should not be considered an end in itself. What is important is the process, rather than the product or the technology/media used."

The experts also defined DSC as part science, part art, and part craft:

- It is part science because it draws heavily on social science theory and methodology.
- It is part art because it incorporates artistic talents and skills such as graphics, photography, radio, video, instructional technology, and so on.
- It is part craft because it employs a wide variety of tools and technology such as cameras, projectors, computers, and broadcasting and telecommunication

equipment for preparing, projecting, and disseminating messages (Report of FAO Expert Consultation 1987).

Another landmark in the history of Communication for Development was the World Congress held in Rome in October 2006 and organized by FAO, the World Bank, and the Communication Initiative. "Without communication there is no development" was the primary message that aimed at demonstrating to policymakers that Communication for Development is an essential tool for meeting today's most pressing development challenges and therefore should be more fully integrated into development policies and practices (World Bank 2007).

The meeting brought together some 900 participants from around the world to share experiences and best practices. For the first time on such a large-scale communication specialists gathered together with policymakers to discuss how to increase the success of development programs through more widespread use of communication. The many voices at the Congress were evidence of how the discipline has evolved, the variety of people and institutions now working in this area and the richness of practices in the field.

The Rome Consensus adopted by the Congress provided a definition of Communication for Development which is now commonly adopted by all UN agencies, along with guiding principles and a vision for the future:

> Communication for Development (ComDev) is a social process based on dialogue using a broad range of tools and methods. It is about seeking change at different levels including listening, establishing trust, sharing knowledge and skills, building policies, debating and learning for sustained and meaningful change. The process goes beyond information dissemination to facilitate active participation and stakeholder dialogue. It highlights the importance of raising awareness, the cultural dimensions of development, local knowledge, experiential learning, information sharing and the active participation of rural people and other stakeholders in decision making. (World Bank 2007)

It was hoped that the Congress would be a major tool for advocacy among policymakers and development institutions. But instead no major action followed immediately after and FAO downgraded the discipline during the reform process of a renewed Organization, as did UNESCO. The World Bank highlighted corporate communication. The only UN organization that today is giving major emphasis to the discipline is UNICEF.

## 6.3 The Good Times: Reasons for Success

In the 1980s and 1990s, FAO was recognized internationally as a pioneer and leader in the field of communication for development. There were a number of reasons for this recognition in the past. These included:

- After the first few years of pragmatic experiences, a strategic approach was applied. This was the result of the 1987 Expert Consultation. The strategic

approach ensured that FAO communication activities at headquarters and in the field followed a common vision.
- Although initially in the Public Information Division, Communication for Development was integrated into the Organization's policy, programs, and budget. The discipline was an officially recognized subprogram of the FAO Economic and Social Program. The activities of the then DSC Branch were presented to the Program and Finance Committees for supervision and review of its program of work and budget.
- The Branch had a large field program, with projects that were able to supply evidence of the added value of Communication for Development. Projects lasted 5–10 years even, so it was possible to develop innovative communication methodologies and document them with case studies that were distributed widely. The program had skilled and committed project managers, "champions" for Communication for Development. Over the years, three communication experts received the B.R. Sen Award as best field expert. This was important for advocacy within the Organization (Balit 2012).
- The results of projects and programs were systematized and presented regularly to divisions and departments within FAO. Over the years the Branch produced a number of supporting multimedia materials for advocacy including booklets, slide sets, a professional video "sharing knowledge," and power point presentations. These materials were presented to FAO senior management and widely distributed to other UN agencies, member governments, and NGOs.
- The Branch had a critical mass of staff, up to nine professionals as well as general service supporting staff, with expertise in different aspects of communication ranging from audio visual media, rural radio, and ICTs to planning, research, and evaluation. Staff traveled frequently to field programs to ensure not only supervision and dialogue but to formulate new projects in consultation with local staff. The staff at headquarters together with the field experts created a community of practitioners, sharing experiences and methodologies.
- The field experts had capacity building and ownership as their main objective, supporting local and national institutions that would be able to sustain ComDev activities even when the international projects finished. Respect for local traditions, language, and culture was paramount, with always a gender perspective. In the majority of projects, half of the national staff were women. This was the case also for the program at headquarters. Many of the national staff trained by the FAO experts went on to initiate and implement similar projects in other developing countries. Today they are still carrying on the legacy.

## 6.4 What Followed? Current Perspectives

After the positive experience in the 1980s and 1990s, why was the discipline not fully recognized during the reforms of the different development agencies including FAO that occurred later on? There are a number of factors, some related to context, and some missed opportunities, that have limited the potential of Communication for Development as a strategic element for renewing FAO and the UN.

In the last 20 years, the global and rural development context changed considerably. Structural adjustment programs were prevalent, with less attention to social issues and assistance to poor and vulnerable groups. Governments and traditional rural institutions withdrew from certain functions that were taken over by civil society and the private sector. Globalization shaped the world economy, and privatization of public services, free markets, and international trade agreements created new scenarios for development, with serious effects on governments, local communities, and vulnerable groups. The vertical structure of institutions and the operational mode of most organizations based on results-based management and logical frameworks were not conducive to participatory approaches. The financial crisis did the rest: with fewer resources available for development, competition for funding of projects became tough. In addition, there was less attention given to participatory models on the part of governments and donors. In this scenario, Communication for Development is like the fifth wheel on the cart, the moment there is a budget cut communication is the first to be eliminated.

As for lost opportunities, during the FAO reform process, the Communication for Development Group (ComDev) was not able to effectively advocate for the discipline with senior management. In fact, the downgrading within the organizational structure from branch to group hampered the possibility for the theme to be represented at program committee level. Furthermore, another limiting factor for the lack of internal advocacy can be associated with a weak research capacity to document results and generate evidence. In particular, the group lacked sufficient quantitative data on impact to be able to prove the value of ComDev to policymakers with backgrounds in hard sciences.

Finally, during the reform process the term "communication" was used for relabeling public information officers, deepening the misunderstanding that still exists today between Communication for Development, corporate communication, and public information.

In general terms, in the last two decades, Communication for Development has been facing new issues and challenges, as a consequence of globalization media liberalization, rapid economic and social changes, and the emergence of new information and communication technologies (ICTs). Furthermore, the ICTs set up around the world, including in very poor countries have multiplied, and donors have heavily invested in hardware and software, but not with a strategic vision (Gumucio-Dagron 2011). Technologies are often used as silver bullets and the Communication for Development dimension is absent in many projects. Communities rarely participate in the decision-making process, and content is often prepackaged.

The WCCD was one of the many attempts to advocate for Communication for Development mainstreaming at the policy and institutional level (Servaes et al. 2007). UN agencies such as UNESCO, FAO, and UNICEF have been among the main international referents in this area, because they have traditionally supported communication for development as a programmatic strategy for sustainable development and promoted civil society participation (Gumucio-Dagron 2011). Several attempts were systematically made by FAO and partner agencies through international events such as the IX UN Roundtable on Communication for Development (UNRT) in 2004, the XI UNRT in 2009, and the FAO Expert Consultation on Communication for Development in 2011. In particular, the 2011 FAO Expert Consultation served as a visioning exercise for FAO and partners, which did in fact initiate processes to implement a mainstreaming strategy within the Organization.

Notwithstanding these efforts, in the last 10 years, different agencies went through in-depth reform processes that resulted in downsizing of themes and technical areas including Communication for Development, among others. In the case of FAO, ComDev functions were shifted several times due to the long institutional reform process resulting in changes both in terms of priorities and focus of the work. For example, the FAO_ComDev_Team moved in several rounds from the Economic and Social Affairs to the Natural Resource Department and from the Research and Extension Division to its present location as part of the Family Farming and Partnerships with Civil Society Unit in the Partnerships Division.

Today the work of the ComDev Team comprises both normative and field activities combining the technical assistance to projects with the systematization of lessons learned, methods and tools, and the advocacy at the policy level. Three main lines of work have been prioritized in the last years to support FAO member countries and partners: (a) assistance in ComDev to FAO strategic objectives and programs; (b) capacity building, partnerships, and networking; and (c) evidence, advocacy, and policy dialogue for mainstreaming rural communication services. From these activities several experiences and lessons have emerged.

### 6.4.1 ComDev Assistance

Notwithstanding limiting factors such as the downsizing in terms of staff and resources and the reduced "visibility" of the theme, recent institutional changes and the strategic programs of the Organization provide new opportunities and challenges for the FAO work in ComDev. Presently FAO activities include as an integral part capacity development of individuals and organizations at the local, national, and regional level and the institutionalization of ComDev services within national agricultural policies (FAO 2014c). Mainstreaming ComDev is therefore one of the priorities jointly with the support to program delivery.

The FAO project cycle identifies Communication for Development as an integral element that has to be considered in any project design (FAO 2014a). Along the years several ComDev components and projects have been designed, implemented, and evaluated on different topics such as food security; family farming; agricultural innovation systems; governance of tenure; rural livelihoods; natural resource management (Acunzo and Thompson 2006); emergency relief, community-based climate adaptation, and resilience; and poverty reduction, among others.

The FAO Communication for Sustainable Development Initiative (CSDI, 2007–2012) has been one of the first global projects dedicated to the application of Communication for Development to climate change adaptation and sustainable natural resource management. The CSDI piloted and documented participatory ComDev strategies in several countries as a mean to promote community engagement and collaborative change toward climate adaptation and mitigation (Acunzo and Protz 2010).

The CSDI was also one of the last FAO projects exclusively focused on piloting ComDev, building capacities in different countries and facilitating the sharing of lessons learned in this field. One of the key features of the project has been enhancing rural institutions and service provider in their capacity to deliver ComDev assistance. Technical support to project activities has been provided in the field by local experts, NGOs, and community media networks, reinforcing the role of ComDev entities often at the regional level. This has been also in line with efforts to improve local capacities and encouraging south-south collaboration to advance sustainability.

In recent years, the priority of FAO ComDev Team has been supporting family farming initiatives and projects in the field through regional ComDev plans. Special attention is being given to ComDev strategies leveraging community media and ICTs for resilient family farming, youth employment, and inclusive food systems. An important dimension is the monitoring of results and the involvement of rural institutions and producer organizations to seek upscaling and sustainability of ComDev interventions.

### 6.4.1.1 Mainstreaming Rural Communication Services

Several UN Roundtables highlighted the need to strengthen and mainstream Communication for Development at the policy level, within development agencies and in the context of the UN. In fact, when treated only as an ad hoc project component, ComDev activities have a limited duration risking not to be fully integrated into development programs and policies (FAO 2014c). As indicated by the second Expert Consultation on Communication for Development, the only way to achieve sustainability is mainstreaming ComDev as part of sectoral policies (FAO 2012).

Institutionalization can take different forms at different levels depending on the existing organizational structures and, of course, policy frameworks. In the case of the agriculture and rural development sector, the institutionalization of ComDev is presently relatively poor due to lack of understanding of its potential and the limited capacity of the agricultural sector to influence telecommunication or media laws. This is, for example, the case of the Universal Access Funds (and Universal Service Funds) that are underutilized to ensure dedicated communication services for rural people (FAO 2014b).

Within the framework of a renewed FAO and the new mission to support advocacy and multi-stakeholder engagement on corporate priorities, FAO ComDev Team was tasked to contribute to the activities of the 2014 International Year of Family Farming (IYFF) that included regional dialogues among governments, development agencies, NGOs, and farmers' organizations, such as the World Rural Forum (WRF), the World Farmers' Organization (WFO), and La Via Campesina (LVC).

The Forum on Communication for Development and Community Media for Family Farming (FCCM, FAO, Rome – October 2014) provided the opportunity to share experiences and showcase evidence of the contribution of communication, ICTs, and community media to family farming. It addressed opportunities for promoting rural communication services as sustained, inclusive, and demand-led communication processes involving family farmers and the rural population (FAO 2017). The FCCM brought together more than 100 participants representing various countries from all regions. This was the first specialized ComDev event that gave voice all together to farmers; civil society organizations, including community media and communication networks; academia; and specialized ComDev entities and stimulated a dialogue among these constituencies. It was a unique space for grass roots organizations to dialogue with subject matter specialists, government regulators, practitioners, research institutions, and academia (FAO 2017).

The agenda FCCM included interactive session on relevant issues like the contribution of communication to family farming, needs and priorities for communication policies, enabling policy and institutional frameworks and investments, and partnership opportunities. Specific attention was also given to the appropriation of communication by farmers and to enhancing the capacities of rural stakeholders.

Participants recognized communication as an asset for farmers and the need to integrate it as a dimension of family farming policies, the challenge being promoting institutional and policy frameworks for inclusive rural communication services, to facilitate access to information and the active participation of smallholders and the rural population in development initiatives, while fostering economic opportunities and the sustainable use of natural resources (FAO 2017).

The FCCM also recalled the WCCD Rome Consensus and its pledge to move toward a right-based approach in communication for development. The Forum endorsed a working definition of "Rural Communication Services" (RCS), as a wide range of demand-led communication processes, activities, media applications, and institutional arrangements to respond in a sustained and inclusive manner to the communication need of the rural population (FAO 2017).

Furthermore, the background document of the FCCM Farming for the Future presents a thorough analysis of experiences of rural communication services – understood as public goods – that seek to enhance rural livelihoods by facilitating equitable access to knowledge and information along with social inclusion in decision-making and stronger links between rural institutions and local communities (FAO 2014b).

The FCCM represents another landmark in the present work of FAO in ComDev. It allowed to establish a dialogue among ComDev practitioners, producer organizations, community media, and the academia. It also allowed to initiate a new path for mainstreaming ComDev as part of FAO support to family farming policies and field projects, through the lens of the RCS and a specific road map including collection of evidence, capacity building, and promotion of partnerships among rural stakeholders, a major asset in these efforts being the claim by producer organizations for their right to inclusive RCS. This path will be continued by FAO promoting ComDev as a key dimension of the activities of the UN Decade of Family Farming (UNDFF 2019–28).

## 6.4.2 Building Capacities and Partnerships

Mainstreaming ComDev goes hand in hand with building capacities, both at the field level and within development organizations, and establishing effective partnerships. It implies improving knowledge sharing and collaboration to strengthen national institutions, organizations, and people according to their priorities and characteristics (FAO 2007b).

The resolution of the second FAO Expert Consultation on Communication for Development emphasized the need for relevant agencies, organizations, funds, and programs of the United Nations to develop a systematic approach to capacity building in Communication for Development, particularly with respect to the training of field and development workers, technicians, as well as communication planners and specialists, especially in developing countries" (FAO 2012).

In recent years FAO prepared a large number of field guides, manuals, and case studies to help building capacities on ComDev methods and tools. In particular, the Communication for Development Sourcebook (FAO 2014e) and the e-learning Communication for Rural Development (FAO 2016b) available on the web in different languages for free are important references for practitioners involved in agriculture and rural development (FAO 2018). To enhance experience sharing, advocacy, and collaboration among practitioners, projects, academicians, and rural stakeholders at national and international level, three regional web-based platforms have been supported in Africa (YenKasa), Asia (ComDevAsia), and Latin America (Onda Rural) in partnership with communication networks and grass root organizations. The platforms aim at building regional communities of practice and knowledge networks in ComDev. Furthermore, FAO has joint efforts with the College of Development Communication, University of Philippines Los Baños, to spearhead the Collaborative Change Communication initiative through a web-based platform named CCComDev that facilitates exchanges on research activities, trainings, and learning resources (FAO 2018).

Horizontal communication is critical to enhance ComDev capacities of farmers, indigenous peoples, and producer organizations stimulating peer-to-peer sharing and learning. This is why experiences of the appropriation of communication systems and ICTs by indigenous peoples and farmers have been documented, shared, and promoted through the regional ComDev platform, especially in Latin America (FAO 2014d).

In order to build capacities and promote RCS as part of agricultural and rural development policies, FAO has catalyzed strategic partnerships in ComDev with relevant entities including farmers' organizations (e.g., World Rural Forum, La Via Campesina), communication networks (e.g., World Association of Community Media Broadcasters, AMARC), research centers, and the six universities of the Global Research Initiative on Rural Communication (GRI/RC), the Centro Internacional de Estudios Superiores de Comunicacion para America Latina, CIESPAL, and the Mercosur Specialized Meeting on Family Farming (REAF). Each entity is actively involved in animating a regional ComDev platform promoting research, learning, or advocacy activities in a specific context and contributing to the consolidation of communities of practices and regional partnerships to enhance south-south and triangular cooperation in ComDev.

### 6.4.3 Evidence-Based Approaches

One of the main findings of the 2014 FCCM is the need for evidence-based approaches to orient the institutionalization of RCS as part of family farming, poverty reduction, and rural development policies and programs. It also requires to research the role of communication (processes, media, ICTs, telecommunications, laws, etc.) as part of the structural transformations happening in rural areas. The analysis of present communication trends may help informing policies about how to reduce the urban/rural communication divide and contribute to achieving the SDGs also in rural areas (FAO 2017). Specific studies should also lead to a clearer understanding of the appropriations by people of communication processes, media, and ICTs in rural areas and their influence on local economies and in relation to development opportunities.

The increasing interest of the research community and FAO to combine the analysis of changing communication environment in rural areas with the need to build evidence of the results of RCS motivated researchers from six universities to partner under the Global Research Initiative on Rural Communication and to establish an ad hoc working group on rural communication as part of the International Association of Media and Communication Research (IAMCR). As results of this collaboration, the study "Inclusive rural communication services. Building evidence, informing policy" was published by FAO with contributions of the universities of Queensland, Guelph, Reading, Wageningen, Van Hall Larenstein, and the University of the Philippines Los Baños. This scoping study is the first research aimed at compiling existing evaluation cases based on proven methodologies to assess and document evidence-based approaches in the field of Communication for Development that may be used for designing rural communication services as part of agricultural and rural development policies (FAO 2017).

The study compares evaluation approaches applied to RCS and provides recommendations on how to build evidence that can orient governments, institutions, and organizations in establishing and operationalizing inclusive RCS. The conclusions highlight the need for policies to take into account all the dimensions of RCS initiatives, including the focus on investing in stakeholders' capacities and strengthening rural knowledge institutions and farmer organizations. It also recommends streamlining the planning, implementation, and evaluation of rural communication service, with focus on learning-based approaches. In all, the systematic evaluation approaches and the evidences demonstrated in the cases can inform policy and can be used for establishing rural communication services.

Following the results of the FCCM and the GRI-RC publication, FAO launched a series of regional studies on trends, experiences, and opportunities for institutionalizing evidence-based RCS with a view to mainstream this topic in the policy dialogues that is being promoted in the context of the UN Decade of Family Farming and the SDGs.

## 6.5 Lessons Learned: Some Essential Guiding Principles

Notwithstanding institutional ups and downs, the results of practice in the field and evaluations carried out by communication scholars across the globe in the last 40 years have confirmed a series of essential guiding principles. These should not be forgotten, even though new challenges require new approaches and adaptation to a changing world.

### 6.5.1 Starting Upstream

To be effective, communication should be built into development programs and policies from the start and integrated into development policies and poverty reduction strategies together with the participants and all stakeholders. A ComDev strategy planned with participants from the start will ensure their active involvement and ownership and give development programs better chances of being successful and sustainable.

### 6.5.2 A Social Process

Communication for Development is essentially "a social process, designed to seek a common understanding among all the participants of a development initiative, creating a basis for concerted action." Communication technology and the media are useful tools to facilitate this process, but should not be considered as an end in itself. What is important is the process, rather than the product or the technology/media used (FAO 1987). Interpersonal communication and the participation of a community change agent as a facilitator or broker with the rural community remain essential. Whenever possible, communication programs and projects should be designed and implemented with and by rural people themselves, and not just by outsiders. The professional quality of the products is not as important as guaranteeing local contents based on people's needs and cultural codes.

### 6.5.3 Sharing of Knowledge and Two-Way Communication

The first FAO Expert Consultation indicated that "people oriented and sustainable development can only realize its full potential if rural people are involved, and if information and knowledge is shared. Sharing is not a one transfer of information; it implies rather an exchange between communication equals" (Balit 1999). One of the most important methodological findings of the PRODERITH project can be resumed as follows: "The main challenge faced by a good communication system in the field is not, as one might generally think, filling a social space with words; it consists in the establishment of an initial silence, where the actors present recognize each other as equals, with the same rights and possibilities for generating the new knowledge required to improve the quality of life and working conditions" (Funes 1996).

### 6.5.4 Listening to People

Thus, Communication for Development efforts should begin by listening to people, learning about their perceived needs, and taking into account their knowledge and experience, their culture and traditions, and the reality of rural areas (Anyaegbunam et al. 1998). Listening is an important skill developed by oral cultures that rely exclusively on oral communication. Dialogue also requires the capacity to listen and to be silent. However there is still much to be done in terms of training in participatory communication methods and changing the attitudes of field staff and development agents who have been educated to apply top-down, authoritarian methods and tell people what to do. Fortunately today there are many participatory research and diagnostic methods that have been developed to listen and learn from communities, such as Knowledge, Attitudes, and Practices (KAP) surveys, Participatory Rapid Appraisal (PRA), Participatory Rural Communication Appraisal (PRCA), focus group discussions, ethnographic research, etc. These methods ensure that the needs of beneficiaries are taken into account in the planning of projects and that monitoring and evaluation make use of qualitative indicators identified with the communities themselves. These methods integrated under Communication for Development approach should serve to guide the use of the ICTs for development.

### 6.5.5 In Line with National Policies

ComDev policies and strategies should be designed in accordance with national policies for poverty reduction and sustainable development and help to promote them (FAO 2007a). Otherwise, these strategies and programs will remain isolated and not receive the support of national policymakers and decision-makers. For example, a key characteristic of the PRODERITH rural communication system in Mexico was its placement within a public policy where participation was prevalent. This helped its development and is crucial to understanding the causes of its success (Funes 1996). National communication policies have been introduced by FAO in a number of countries in the West and Central Africa region to articulate and regulate the role of communication in development. They have proposed different institutional frameworks ComDev interventions, as well as for establishing rural communication services, bringing together the variety of ministries and stakeholders involved in poverty reduction and development programs.

### 6.5.6 Promoting Policy Change

Communication for Development also has an important role to play in enacting and promoting national policy change, since there is a tendency to move away from top-down policy planning toward more participatory and interactive processes of policy planning, adjustment, and implementation. Often today country programs concentrate on community participation and ownership, and communication can foster the required dialogue between policymakers and communities, like in the case of family farming policies in several Latin American countries. There is also the need to build governmental receptivity for sustainable development and poverty reduction and to engage civil society and the private sector. In this respect, policy dialogue and advocacy can play a key role for the establishment of new legislation or for promoting the implementation and/or enforcement of an existing one.

### 6.5.7 Holistic Approach

Communication is a multidisciplinary and cross-cutting discipline. Thus, efforts should cover all the multifaceted aspects of life that concern poor people and their livelihoods. Communication activities should deal with agriculture, health, habitat, education, nutrition, the environment, population, gender, and all issues related to development in an integrated manner. Furthermore, communication can facilitate the necessary linkages to achieve improved coordination and multidisciplinary efforts to involve a number of development sectors and concerned institutions.

### 6.5.8 Preserving Indigenous Knowledge, Values, and Culture

It is important to incorporate indigenous knowledge and practice in the planning of Communication for Development projects: "The outcome of useful sharing of knowledge is not so much the replacement of traditional techniques by modern ones, as a merging of modern and traditional systems to produce a more appropriate hybrid, one that befits the economic and technical capacities of rural populations as well as their cultural values" (FAO 1987). Indigenous knowledge can provide an understanding and analysis of a situation that will ensure that programs are formulated in harmony with the environment and relevant cultural issues. Indigenous groups have access to a large volume of traditional knowledge about their environment and are highly efficient users of available resources that have been crucial for their survival.

### 6.5.9 Respect for Culture

Communities in rural areas contain a wealth of traditional cultural resources, a rich but fragile heritage which risks being lost with the advent of modern technology. The tendency to cultural homogenization that comes with globalization processes can erase cultural identities and diversities. To be successful, communication efforts must take into account the values of rural people rather than borrowing communication strategies from outside that promote change without due consideration for culture (Decock 2000). This also implies supporting local appropriation of ICT by rural people for their own purpose.

### 6.5.10 Shared Interests and Trust

Shared interests and trust are also essential elements of the communication process. Shared interests are the cohesive force in social networks and can explain the evolution of such formations. The ideal sharing of interests is expressed by the achievement of a social consensus and a state of mutual trust. Trust and credibility are also closely associated. It is well known that the effectiveness of a medium or a communicator hinges on their credibility and trust. For rural people, communication often is about building personal relationships, and when it comes to building trust and collaboration, face-to-face communication remains the preferred channel (Fugelsang 1987). In the case of rural youth, the use of cell phones and social media is introducing new patterns for socialization that have to be considered to avoid generational divide and to facilitate new mechanisms for social cohesion and economic inclusion for rural communities, limiting migration trends.

### 6.5.11 Flexibility and Duration

Social change and participatory communication processes take much more time than the typical 3–5-year project cycle during which governments and donors usually wish to obtain quick results. The most successful communication programs have had a duration of 7–10 years, with a long-term perspective. The development context is dynamic, with unanticipated events and variables that are difficult to predict. Communication can produce lasting and successful results when it is part of a flexible and inclusive approach in designing projects, where the whole endeavor is planned, executed, and evaluated with the participants, and when there is sufficient time and flexibility to allow for changes in direction as the project develops according to the feedback provided by the stakeholders

### 6.5.12 Ensuring Gender Sensitivities

The empowerment of rural women through the sharing of information, knowledge, and communication is crucial for achieving development goals. However, women are often underestimated and overlooked in development strategies and remain the "invisible" partners in development. They belong to the culture of silence. Communication activities, services, and materials should reflect the special needs and preferences of women and address the social, economic, and cultural constraints they face in the design of communication messages, choice of appropriate channels, and best timing and locations for delivery. Women in rural areas have communication requirements that are different from those of men. They do not have equal access to information, due to factors such as restricted mobility outside the home, lack of education, and sometimes men's control over information or media technology. Because women are involved in many aspects of rural life, their traditional knowledge systems are holistic. For generations they have relied on the spoken word and traditional forms of communication as a means of sharing knowledge and information. They have made use of the oral tradition to develop a rich communication environment, transmitting culture, knowledge, customs, and history through traditional forms of communication such as poetry, proverbs, songs, stories, dances, and plays. Within their communities they are active participants in social communication networks, using indigenous communication methods to disseminate knowledge and strategies for mutual assistance and survival (Balit 1999). More generally, communication programs should always include a gender perspective also in relation to women access to rural communication services and the use of ICT. A new issue on the development agenda is the increasing violence that women are subject to in situations of conflict and post-conflict.

### 6.5.13 A ComDev Approach to the Use of the ICTs

Communication is central to improving traditional farming systems and promoting sustainable agriculture by facilitating the appropriation and management of processes and technologies by farmers (Acunzo and Santucci 2012). The spread of digital technology in rural areas, especially cell phones, has made communication services an increasingly cost-effective option for providing basic information to the rural population even in remote areas. Furthermore, unlike other mass media, ICTs offer opportunities for two-way communication and for opening up new, non-traditional communication channels for rural communities, producer, and development organizations.

A Communication for Development approach is central to the use of the ICTs in order to bridge the rural digital divide and to provide universal access to information to the rural populations (Richardson 1997). This should support the creation of local contents and ownership, taking full advantage of the convergence between "conventional media," such as radio and the information and communication technologies.

Nevertheless, the focus must remain on the needs of the rural audiences, rather than on the communication technologies and the media utilized, with the aim of enhancing the capacity of local stakeholders to manage communication processes, develop local content, and appropriate the use of media and tools for their own purposes, including the ICTs (FAO 2006).

### 6.5.14 Multimedia Approach

Communication initiatives should make use of all media channels available, both modern and traditional, in an integrated manner. The advent of new technologies implies that new mixes can be made, to develop more effective communication programs with vulnerable groups. And, in some cases, using traditional communication processes and media that rural people are familiar with and are already available, to be integrated at their own passes with new media, can provide the most effective solution to promote information sharing and inclusive dialogue.

### 6.5.15 Political Space and Context

Communication for Development is contextual. There is no universal formula capable of addressing all situations, and therefore ComDev should be applied according to the cultural, social, and economic context. As a process of empowerment, participatory communication approaches can only take place if there is political space and will on the part of government and local authorities (Balit et al. 2011). Otherwise, the local people will pay the price. An example of best practice in this sense is again the PRODERITH program, which flourished when participatory approaches were part of the government policy. When the government changed its policies, the project was forced to implement conventional communication activities for the transfer of technology (Fraser and Restrepo 1996). Authentic participation directly addresses power and its distribution in society, and usually authorities do not want to upset the status quo, even if they pay lip service to participation. Good governance, transparency, respect for freedom of expression, and human rights are essential prerequisites for successful participatory communication approaches. Shared values, commitment, and will on the part of the communities involved are also essential prerequisites.

### 6.5.16 Toward Inclusive Rural Communication Services

Today more than ever, there is a need for mainstreaming ComDev as part of national policies and programs to ensure dedicated communication services for rural people. This also requires new policy and financial commitments on the part of development institutions and governments, as well as policy dialogue with producer and indigenous peoples organizations, representatives of women and youth, and the private sector. To be effective, mainstreaming ComDev activities require to be grounded on solid evidence, communication research, and political commitment. It should be framed within sectoral policies and take advantage of international processes such as the UN Decade of Family Farming (2019–2028) and the SDGs.

The definition of Rural Communication Services (RCS), understood as a public good, provides a framework to promote policy dialogue, social inclusion, participatory decision-making, appropriation of the ICTs by rural people, and stronger links between rural institutions, farmer organizations, and communities. Evidence-based approaches are needed to orient the design of RCS jointly with capacity building, advocacy initiatives, and partnerships among universities, research centers, community media, and civil society organizations, especially in the Global South.

### 6.5.17 No-Off-the-Shelf Solution

There is no single model for planning successful communication strategies and programs, since they deal with people, and not only with technology. Every situation is different – different audience groups, their location and distribution, their languages, their characteristics and needs, their levels of education, their access to the media, and their existing attitudes and behavior. Each group of users has a different cultural approach to information, whether they are from different regions, countries, or neighborhoods. One must tailor an information/communication activity to each group's needs and preferences. The communication objectives will be different, as well as the media that are available or most appropriate. Every situation calls for an analysis, audience research, and a strategic communication plan to implement a successful communication program.

## 6.6 Conclusion

What can we expect in future years? Is there an enabling environment for the discipline? The international context has changed considerably since the first specialized Unit was established in FAO. More recently the information revolution has provided communication practitioners with a wealth of new technologies and media to increase the potential of sharing knowledge in rural areas and giving people a voice. In addition the development community is seeking new approaches to overcoming the failures of the last decades, and there are many voices calling for "another development."

Apart from the overriding commitment to eradicating poverty and hunger, agricultural and rural development, natural resource management, and climate change are again priorities on the development agenda as indicated in the Sustainable Development Goals (SDGs). The challenges and emergencies are so great that the international community is supporting public agricultural policies in developing countries and discussing new investments to promote the development of rural areas.

Donors and development agencies are coordinating their efforts and demanding that ownership, dialogue, participation, and transparency become essential components of programs they contribute to. Contrary to the past, small holders and family farmers are now considered protagonists in fighting hunger, the deterioration of ecosystems, and climate change (FAO 2015). They have to be involved also in defining the type of rural communication services they should have access to, providing opportunities also for the youth to relay on those services to shorten the information divide that pushes them out of their communities.

These are all approaches that FAO has privileged when working with rural communities. And today the marriage of new information and communication technologies with other media such as radio has increased the potential and suitability of communication media outlet to reach distant audiences and work with the disadvantaged. Social media and communication networks are enabling people to voice their demands and promoting major changes in developing countries that governments can no longer ignore.

Communication for Development has many more practitioners locally, nationally, and on the international field. There are more universities and academic programs that provide courses in communication (CDC 2018). FAO has published a comprehensive e-learning Communication for Rural Development and a series of valuable manuals and tools accessible for free online (FAO 2018). There are specialized networks and platforms globally and regionally with ComDev practitioners and multiple stakeholders that facilitate exchange and experience and collaboration across regions (FAO 2018). Some of these networks publicize consultancies and assignments in the discipline. In addition to governments, partners include representatives of civil society, NGOs, and rural communities. Thus these are new opportunities for Communication for Development practitioners to contribute to the emerging development challenges.

According to Chin Saik Yoon, "The severity of the challenge to communication practitioners comes from the complexity of the problems posed by these 'new' issues, many of which have no ready solutions for communication and action, and are without precedence. They will be the big challenges of practice in the generations to come." Future generations of communication practitioners will need to develop new processes of practice to address the challenges these issues pose. The solutions will require involving multiple stakeholders and civil society, in addition to traditional partners (Saik Yoon 2010).

In the search for new approaches, it is important not to forget the many successful experiences accumulated until now that can still contribute to finding solutions to these many issues. FAO's accumulated experience with participatory communication with the poor and disadvantaged contains many principles and methodologies that are still valid for meeting new challenges. Communication practitioners will need to draw upon the best practices and lessons learned in the past and merge them with new communication methods and approaches to meet the changing needs of the of the plurality of societies we recognize in our dynamic world. Essential elements for successful and sustainable efforts will continue to be dialogued, ownership on the part of rural communities and integration with existing indigenous knowledge and communication systems.

There is a need for mainstreaming Communication for Development. UN Inter-Agency Roundtables have identified the priorities for mainstreaming the discipline (FAO 2014c). They include advocacy with policymakers, improved methods for monitoring and evaluation, and capacity building at all levels. These have been priorities since the beginning but still require action. The changing landscape also requires new commitment on the part of development institutions and governments but also the direct involvement of farmer and indigenous peoples organizations, as well as social movements and community media and the academia to ensure the adoption of the right-based approach to communication proclaimed during the WCCD (World Bank et al. 2007). All together we need to break the barriers to participatory communication and make the case that this is the only path to ensure inclusiveness, ownership, and appropriation of development process, thus ensuring sustainability and contributing to the SDGs. Only with real commitment on the part of all stakeholders Communication for Development can finally become an essential element in development and assist in meeting the challenges of this changing world.

**Disclaimer** The views expressed in this publication are those of the author(s) and do not necessarily reflect the views or policies of the Food and Agriculture Organization of the United Nations.

## References

Acunzo M, Protz M (2010) Collaborative change: a communication framework for climate change adaptation and food security. Rome, FAO

Acunzo M Santucci F (2012) Knowledge generation and sharing for organic agriculture in Costa Rica. J Ext Syst 28(1):15–27, Mumbai, India

Acunzo M, Thompson JS (2006) Building communication capacity for natural resource management in Cambodia. In: Bessette G (ed) People, land and water, participatory development communication for natural resource management. IDRC/Earthscan, Ottawa/London

Ansah P (1994) An African perspective. In: Nostbakken D, Morrow C (eds) Cultural expression in the global village. Southbound, Penang

Anyaegbunam C, Mefalopoulos P, Moetsabi T (1998) Participatory rural communication appraisal: starting with the people. SADC Center of Communication for Rural Development, Harare

Balit S (1999) Voices for change: rural women and communication. FAO, Rome

Balit S (2012) Communication for development in good and bad times. Nordicom Rev Commun Media Develop Prob Perspect 33(Special Issue):105–120. Gothenberg University Press

Balit S, Ilboudo JP (1996) Development of rural radio in Africa. FAO, Rome

Balit S, Pafumi M, Vertiz V (2011) New directions for mainstreaming communication for development in FAO, background paper presented to FAO expert consultation, communication for development, meeting today's agriculture and rural development challenges, 14–16 September 2011, FAO, Rome

College of Development Communication – CDC (2018) Collaborative Change Communication (CCComDev) – Web Site: http://cccomdev.org//. University of the Philippines, Los Baños

Decock A (2000) Tapping local cultural resources for development. In: Servaes J et al (eds) Walking on the other side of the information highway: communication, culture and development in the 21st century. Penang, Southbound

FAO (1987) Report of development support communication expert consultation. FAO, Rome

FAO (2006) Communication for Sustainable Development, Background Paper of the first World Congress on Communication for Development, Rome

FAO (2007a) Communication and sustainable development: selected papers, Rome

FAO (2007b) A compendium of regional perspectives in communication for development. FAO, Rome

FAO (2012) FAO expert consultation. Communication for Development. Meeting Today's Agriculture and Rural Development Challenges, Rome

FAO (2014a) Communication for rural development. Guidelines for planning and project formulation. FAO, Rome

FAO (2014b) Farming for the future. Communication Efforts to Advance Family Farming, Rome

FAO (2014c) Mainstreaming ComDev in policies and programmes. Enabling social inclusion to support food security and resilient livelihoods and family farming. FAO, Rome

FAO (2014d) Regional ComDev platforms. Strengthening partnerships and actions in communication for rural development. FAO, Rome

FAO (2014e) Communication for development sourcebook. FAO, Rome

FAO (2015) Planning communication for agricultural disaster risk management. A field guide. FAO, Rome

FAO (2016a) Rural communication services for family farming. Contributions, evidence and perspectives, Results of the forum on communication and community media for family farming. FAO, Rome (FCCM).

FAO (2016b) Communication for rural development – E-learning. http://www.fao.org/elearning/#/elc/en/course/COMDEV

FAO (2018) Communication for development – Web Page. http://www.fao.org/communication-for-development/en/

FAO and The University of Queensland (2017) Inclusive rural communication services. Building evidence, informing policy. FAO, Rome

Fraser C (1987) Pioneering a new approach to communication in rural areas: the Peruvian experience with video for training at grassroots level. FAO, Rome

Fraser C, Restrepo-Estrada S (1996) Communication for rural development in Mexico: in good times and in bad. FAO, Rome

Fugelsang A (1982) About understanding – ideas and observations on cross-cultural communication. Dag Hammarskold Foundation, Uppsala

Fugelsang A (1987) The paradigm of communication in development: from knowledge transfer to community participation – lessons from the Grameen Bank, Bangladesh. FAO, Rome

Funes S (1996) Foreword. In: Fraser C, Restrepo-Estrada S (eds) Communication for rural development in Mexico: in good times and in bad. FAO, Rome

Gumucio-Dagron (2011) Communication for development: meeting today's agriculture and rural development challenges. – Background paper of the FAO Expert Consultation on Communication for Rural Development (Rome, 14-16/09/2011), FAO

Quarry W, Ramírez R (2009) Communication for another development, listening before telling. Zed Books, London

Richardson D (1997) The Internet and agricultural and rural development. FAO, Rome

Saik Yoon C (2010) Future imperatives. Paper presented at the international conference on future imperatives of communication and information for development and social change, Bangkok, December 20–22

Servaes J et al (eds) (2007) Communication for development: making a difference, a WCCD background study. The Communication Initiative, FAO and The World Bank. Washington DC

World Bank (2007) The International Bank for Reconstruction and Development. The World Bank Washington DC

World Bank, Food and Agriculture Organization of the United Nations, and the Communication Initiative (2007) World congress on communication for development: lessons, challenges, and the way forward. World Bank, Washington, DC

# Daniel Lerner and the Origins of Development Communication

Hemant Shah

## Contents

| | | |
|---|---|---|
| 7.1 | Introduction | 158 |
| 7.2 | The Production of *Passing of Traditional Society* | 158 |
| 7.3 | The Reception to *Passing of Traditional Society* | 159 |
| 7.4 | Lerner's Influence on Development Communication Research | 161 |
| 7.5 | Why Does "Lerner" Persist? | 164 |
| References | | 165 |

### Abstract

Daniel Lerner's 1958 book *Passing of Traditional Society* was a foundational text for mass media and modernization theory. The theory, in turn, was the basis for early approaches to development communication. Lerner's ideas were embraced by some and challenged by others. Eventually, Lerner himself engaged in a series of revisions to his theory, leaving a distillation of ideas emphasizing (a) the universality of modernization and (b) faith in technology to solve social problems. These ideas continue, in important ways, to influence the field of development communication.

### Keywords

Daniel Lerner · Development communication · Development-support communication · Modernization

H. Shah (✉)
School of Journalism and Mass Communication, University of Wisconsin-Madison, Madison, WI, USA
e-mail: hgshah@wisc.edu

## 7.1 Introduction

While it might be an exaggeration to say that American political scientist Daniel Lerner is the father of development communication, there is a little doubt that Lerner's ideas about the role of mass media in modernization are a clear and direct antecedent of development communication. Lerner presented his ideas about mass media and modernization in a 1958 book, *The Passing of Traditional Society: Modernizing the Middle East* (Lerner 1958). Lerner's theory of modernization posited a model of societal transformation made possible by embracing western manufacturing technology, political structures, values, and systems of mass communication. As a policy initiative, modernization was the centerpiece of Cold War efforts to thwart the spread of Soviet communism in Asia, Africa, and the Middle East. A package of western industrial organization, patterns of governance, and general lifestyles was conceived and offered up as an irresistible and obviously superior path to entering the modern postwar world.

Development communication is the art and science of using communication technology to inculcate promodernization values. As Shah (2011) notes, the phrase "development communication" largely replaced "mass media and modernization" by the 1970s. The field was further legitimized with the publication of the *Journal of Development Communication* beginning in 1990. The concept of development communication originally thought of messages as moving "top-down" – from west to non-west, elite to nonelite, expert to nonexpert, capital city to outlying areas. The purpose of these messages was to ensure a social and culture climate in which people would favorably receive ideas from outside their systems of thought.

In the 1980s and 1990s, the concept of "development-support communication" was coined to capture new emphases, initiatives, and theoretical approaches to development that were rooted in concepts of participation and empowerment of people at the grass-roots. "Development-support communication" signaled the use of horizontally networked information flow (as opposed to top-down models) not simply as an input into modernization but as a way of supporting people's aspiration for control over their resources and basic needs (Shah 2011: 7). Development-support communication was not much different from development communication (Melkote and Steeves 2001). Shah (2008) further noted that development-support communication, while it was not top-down, was frequently "outside-in," meaning that western experts with particular values and institutional biases were still highly influential. In any case, several reviews of the literature have shown that some elements of Lerner's model remained touchstones for scholars, policy makers, and practitioners alike, albeit less and less frequently as time passed (Fair 1989; Fair and Shah 1997; Shah 2007).

## 7.2 The Production of *Passing of Traditional Society*

Since *Passing of Traditional Society* is central to the origins of development communication as a field of study, it is informative to examine origins of the book itself. The book was a "production," Shah (2011) notes, in the sense that it is an outcome of a research process embedded in a particular postwar political and cultural economy. The specific methods, conceptual frameworks, and theoretical linkages described in the book, as well as the specific policy implications thereof, were possible – imaginable – only at *that* particular historical moment. However, the general idea that the west is superior to the "rest" can be traced much further back in history than 1958. The perspective that people outside the Euro-American fold are inferior goes back at least to the fifteenth century "voyages of discovery" undertaken by the sailing fleets from Portugal, Spain, and England. A discourse that counterpoised the superiority of Europeans and the inferiority of non-Europeans emerged during this time and justified conquest, plunder, and murder of peoples throughout Africa, Asia, and the "new world" of the Americas. This discourse, which Edward Said labeled "Orientalism," persists to this day in many ways and continues to be employed to justify newer – and older – forms of western colonialism and imperialism (Said 1979).

During the Second World War, academics from a variety of disciplines served in various arms of government and military service (Cravens 2004; Rogers 1994; Smith 1994). They deployed their skills to address the problems associated with winning a war. They also formed social networks based on their shared intellectual interests. These networks transferred quite easily into civilian life after the war when the academics went back to their positions as professors and administrators at leading American universities. Some of these academics became heavily involved in discussions about crafting postwar foreign policy. Indeed, in the late 1940s and early 1950s, a handful of these academics were central to the first articulations of a nascent modernization theory, which helped American leaders legitimize Cold War policies to compete geopolitically in strategic locations such as the Middle East. The intellectual network for modernization thinking was interdisciplinary, spanning a number of fields including sociology, political science, economics, psychology, anthropology, ethnic and racial studies, and mass communication. As modernization emerged as a theoretical construct from discussions in conferences, seminars, and consultancies, a scholarly consensus emerged over a relatively short period of time as to the meaning and relevance of the idea. The production of *Passing of Traditional Society* played a significant part in the formation of this consensus.

Lerner's book is a production in a second sense, also. As Lerner moved through life – as a college student, military officer, researcher, and esteemed Ivy League professor – he was shaped by a number of personal experiences and intellectual influences. For example, Lerner learned statistical methods, embraced comparative cross-national approaches, and identified Harold Lasswell and Talcott Parsons as his "theoretical heroes." Lerner also found his gift for languages and literary analysis useful as he moved toward the opportunity that resulted in the writing of *Passing of Traditional Society*. The book is a product, Shah (2011) has shown, of a personal assemblage of specific research skills, disciplinary dogma, a system of values, and a vision of America's superiority and liberal altruism.

## 7.3 The Reception to *Passing of Traditional Society*

Lerner's book received positive and negative reviews. On the one hand, the book was praised at the time of its publication for its "solid social science" (Badeau 1959) and its impact on all meaningful studies of the Middle East and other modernizing cultures as well (Berger 1958). Even more recently, the book has been called the "seminal work" (Posusney 2004) and the "locus classicus" (Levy and Banerjee 2008) in the field of development communication.

However, reviews of *Passing of Traditional Society* were not uniformly positive. Criticism came from many sides, as Shah (2011) notes. Conservative thinker Edward Banfield (1959) said some societies were fated to remain backward: "The American Indian, for example, has had extensive aid for decades, but he is still in most cases very far from belonging to the modern world." Another conservative thinker, Samuel Huntington (1965), thought that rather than promoting political stability, modernization could just as easily create instability in the postcolonial world, which was not in the interest of the United States. Rather than promoting modernization, Huntington argued, the United States ought to be preventing modernization. The sharpest criticisms came from scholars in the postcolonial world. In a representative comment, the Pakistani economist Inayatullah (1967: 100) said: "[Modernization] presupposes that because the 'traditional' societies have not risen to the higher level of technological development in comparison to the Western society, they are sterile, unproductive, uncreative, and hence worth liquidating."

Responding to criticism of *Passing of Traditional Society* and the course of historical events, Lerner made some minor and some not-so-minor adjustments to his theory. Following a military coup in Turkey in 1960, he acknowledged that too-rapid urbanization may lead to social and political instability in the form of a "revolution of rising frustration" in which aspirations outruns achievements (Lerner and Robinson 1960). Other revisions included revising the "traditional-modern" dichotomy (Shah 2011: 135), accepting that the modernization process was neither universal nor linear (Shah 2011: 135), and acknowledging that urbanization, the first step in his theory of modernization, could be counterproductive (Shah 2011: 140).

One quite important revision Lerner made in his thinking about modernization occurred in 1975. At a conference at the East-West Center in Honolulu, Lerner prepared a paper titled "Research in Group Communication: A Retrospective Note." In the paper, Lerner linked for the first time his theory of modernization to Tönnies's society (*Gesselschaft*) and community (*Gemeinschaft*) dichotomy. Lerner saw the connection between Tönnies' descriptions of the rural migrants who left their community behind, became urban industrial workers, changed their ways of life and became part of society, and the modernization-theory idea of traditional to transitional to modern transformation in postcolonial nations. "But," Lerner wrote, "Tönnies misled us. By focusing our concern on *Gesselschaft* . . . he diverted our attention away from the continuing life of the *Gemeinschaft*" (Lerner 1975). Lerner cited anthropologists Robert Redfield (1930) and Oscar Lewis (1951) to argue that it was clear now that "Gemeinschaft did not simply pass out of history." Community continued to exist with and even develop important new roles within the larger society. Lerner's argument represented a most radical departure from his modernization theory. By acknowledging that community did not pass away, that the community survived *in* society, he was saying that traditional society *does not pass*. Lerner was directly undermining the very premise – and the title – of the book that put him on the modernization map.

## 7.4 Lerner's Influence on Development Communication Research

Twenty years after Lerner published *Passing of Traditional Society*, he still believed passionately that the western model of modernization was superior to all others and the western media and western experts could diffuse the model worldwide if only they could better communicate the virtues of western values and institutions. But as to the specific elements of his model, Lerner had revised his views, rethought central concepts, and refined variables not only to account for critiques from fellow academics but also the reality of rising frustration with and opposition to western-style modernization in postcolonial regions. How potent were the remaining ideas? How attractive have they been for social scientists still experimenting with social change in the postcolonial world? Partial answers were provided in a set of meta-research studies focused on the published research literature on development communication.

Shah (2011) reviewed three separate meta-research projects, collectively spanning from 1958 to 2006, on development communication. The first of the three projects examined 224 studies published between 1958 and 1986 (Fair 1989); the second focused on 140 items published between 1987 and 1996 (Fair and Shah 1997); and the third examined 183 studies published between 1997 and 2006 (Shah 2007). Each of the three projects focused on English-language books, book chapters, and academic journal articles that dealt with the role of radio, television, newspapers, magazines, and new communication technology in the process of development (primarily social and economic improvement, broadly conceived). Though the three time periods are not equivalent and the historical contexts each study covers are different, the data and analyses in these papers speak to the question about the extent to which and the way in which Lerner's ideas have been used by scholars in the area of development communication since the publication of *Passing of Traditional Society*. Presented below are some of the relevant results from the three research projects and some discussion of their significance.

Table 1 shows how researchers defined development in two time periods (no data was available on this topic for the 1958–1986 time period). In the 1987–1996 period, we see what may be the influence of Lerner's theory, as more than 55% of the studies defined development in terms of being or becoming "modern." In the 1997–2006 time period, there are more definitions of development and only about one-quarter of the studies refer directly to becoming "modern," though raising living standards and democratic development might well be considered consistent with Lerner's ideas of economic participation and political participation.

Table 2 shows the extent to which Lerner's theory is used to think about mass media impact on development. Lerner's modernization model was used in nearly one-third of the studies conducted between 1958 and 1986, and in about 14% of the studies conducted between 1997 and 2006. In the middle time period, Lerner's theory did not receive any attention, at least by direct reference. It is conceivable, however, that "agenda-setting" or "social learning" may have been in reference to the empathy process. The clear trend, however, is that Lerner's modernization model per se is invoked less frequently over time as alternative models of thinking about

**Table 1** Conceptual definitions of development

| 1987–1996 ($n = 126$) | | 1997–2006 ($n = 115$) | |
|---|---|---|---|
| Traditional-to-modern transformation | 34.1% | Raise living standards | 28.7% |
| Modern attitudes | 22.2 | Modern attitudes | 20.9 |
| Meeting basic needs | 15.9 | Health education | 13.0 |
| Self-reliance | 15.1 | Social change | 7.8 |
| Raise living standards | 6.3 | Democratic development | 6.1 |
| National integration | 6.3 | Meeting basic needs | 5.2 |
| | | Traditional-to modern transformation | 3.4 |
| | | Self-reliance | 3.4 |
| | | National integration | 3.4 |

Totals add up to more than 100% because of rounding

**Table 2** Media theories about media impact on development

| 1958–1986 ($n = 80$) | | 1987–1996 ($n = 58$) | | 1997–2006 ($n = 86$) | |
|---|---|---|---|---|---|
| Lerner model | 30.8% | Participatory communication | 37.8% | Participatory communication | 20.9 |
| Indirect effect | 16.9 | Information environment | 11.1 | Social Learning | 19.7 |
| Knowledge gap | 16.9 | Indirect effects | 6.7 | Lerner model | 13.9 |
| Uses and gratifications | 13.8 | Knowledge gap | 6.7 | Conscientization | 9.3 |
| Two-step flow | 13.8 | Agenda setting | 6.7 | Public sphere | 6.9 |
| | | Uses and gratifications | 4.4 | Digital divide | 6.9 |
| | | Social learning | 4.4 | Hegemony | 6.9 |
| | | Cultivation | 2.2 | Knowledge gap | 4.6 |
| | | Direct effects | 2.2 | Media imperialism | 4.6 |
| | | Feminist | 2.2 | Dependency | 3.5 |
| | | Public sphere | 2.2 | | |

Totals add up to more than 100% because of multiple coding

modernization or development had been formulated. In the second edition of *Passing of Traditional Society,* Lerner hoped new models would render his model obsolete. One wonders, though, to what extent Lerner would have welcomed theoretical models arising (in the 1997–2006 time period) from a neo-Marxist tradition (hegemony, media imperialism, and dependency) and from various critical perspectives (such as participatory communication, conscientization, public sphere, and digital divide).

As time passed, researchers examined more and more mediating variables in their studies of modernization, including those that had been suggested by Lerner in his writings after *Passing of Traditional Society.* Table 3 shows the increasing number of

**Table 3** Mediating variables included in development communication research

| 1958–1986 ($n = 224$) | | 1987–1996 ($n = 129$) | | 1997–2006 ($n = 80$) | |
|---|---|---|---|---|---|
| Interpersonal communication | 21.7% | Interpersonal communication | 13.1% | Gender | 25.6% |
| Literacy | 21.6 | Government policies | 12.1 | Local culture | 21.2 |
| Educational level | 10.2 | Political economy | 8.5 | Infrastructure | 10.6 |
| Socioeconomic status | 7.4 | Local culture | 7.2 | Literacy | 6.2 |
| Gender | 3.5 | Message credibility | 6.9 | Media access | 6.2 |
| Infrastructure | 2.4 | Infrastructure | 6.2 | Interpersonal communication | 5.3 |
| Cost of media | 2.0 | Media access | 6.2 | Spirituality | 3.5 |
| | | Literacy | 5.2 | Race and ethnicity | 3.5 |
| | | Socioeconomic status | 4.3 | Local politics | 3.5 |
| | | Capitalism | 3.9 | Language | 2.7 |
| | | Educational level | 3.0 | Cost of media | 1.8 |
| | | Gender | 1.0 | Technology | 1.8 |
| | | | | Socioeconomic status | 1.8 |

Totals add up to more than 100% because of multiple coding

mediating variables considered by these researchers even as literacy, socioeconomic status, and media-related variables, which were the core of Lerner's original model, continue to be included in later studies.

In Table 4, we see that the consequences of development communication are somewhat in line with Lerner's predictions – that is, media activate the modernization process, raise knowledge levels, and create a more participatory society. These three outcomes of development communication account for significant proportions of consequences mentioned in all three time periods, indicating that researchers were thinking about the impact of development communication in ways consistent with Lerner's general ideas about mass media and modernization. It should be noted also that the two other consequences listed are associated with views critical of modernization – that media create development problems and that media benefit certain classes. The rate with which these negative consequences are mentioned increased dramatically over time, indicating recognition among researchers of the complexity and ambivalence of the modernization process, which Lerner also acknowledged though he likely would not have framed the problem in the language of class conflict.

Shah (2011) summarized the meta-research projects as follows: "The[y] show that the study of development communication is much more conceptually and theoretically complex than when Lerner contemplated writing *Passing of Traditional Society* and Lerner's ideas about a broadly universal and teleological model of societal change based on western history were no longer widely supported in development communication research. And yet, some of the variables, descriptions,

**Table 4** Consequences of development communication

|  | 1958–1986 ($n = 224$) | 1987–1996 ($n = 140$) | 1997–2006 ($n = 148$) |
| --- | --- | --- | --- |
| Media activate modernization | 36.4% | 27.4% | 44.0% |
| Media raise knowledge and awareness | – | 25.0 | 41.9 |
| Media create participatory society | – | 10.7 | 27.8 |
| Media create development problems | 13.3 | 15.5 | 23.0 |
| Media benefit only certain classes | 15.4 | 31.0 | 21.0 |

Totals add up to more than 100% because of multiple coding

vocabulary related to Lerner's modernization theory continue to persist in more recent studies."

## 7.5 Why Does "Lerner" Persist?

By the end of his life, Lerner had revised his theory to the extent that only the basic propositions remained intact: that modernization was a psychosocial system of change, that the core variables were literacy, mass media, and participation (urbanization had diminished in importance) and that western media could help change the lifeways of traditional societies. The linear mechanism of modernization that Lerner had precisely outlined had given way to multiple paths and the explanatory power of the concepts "modern" and "traditional" had diminished considerably. However, Lerner remained fairly confident in his view that the Western model of democratic capitalist society was superior to any other system. Thus, when we say Lerner persists, we ought to say "Lerner" persists, to indicate that it was a certain distillation of his thought that continues to appear in development communication research and remains influential in some ways on the field of development communication. Though it would be a stretch to say "Lerner" *is* modernization, his ideas certainly were foundational to the growing literature on modernization from the 1960s onward.

One reason "Lerner" persists is because it helps modernizers define themselves first, as exceptional and special and, second, as benevolent and altruistic. How is this self-conception sustained, Gilbert Rist (1997) has asked, in the face of heavy criticism of modernization theory and evidence that the modernization in practice only occasionally followed the anticipated singular path? It is because, said Rist, for modernizers, modernization is the truth and "the Truth cannot lie." Thus, when failings are acknowledged they are always attributed to faulty implementations (e.g., the postcolonial countries urbanized too fast), human failings (e.g., some people in the postcolonial world do not have sufficient capacity to empathize), or lack of information (e.g., we were not good enough at communicating how to avoid a revolution of rising frustration). Most postwar American development communication specialists viewed their message as correct. The core of that message was the "Truth" that American values and institutions were the epitome of postwar modernity. As Lerner (1965) put it, American modernity was the new Enlightenment, rescued from the horrors of world wars, concentration camps, and atomic bombs. American modernizers stood ready to teach backward people how to become modern. That mass media would take a central role in this tutelage is not surprising given that Lerner worked as a propaganda specialist during World War II.

For the true believers, the obstacles to modernization resided in the realm of technique and technology, not in the realm of ideas, and even less in the realm of ideology. In *Passing of Traditional Society*, Lerner identified radio and cinema as key vehicles for teaching people to be modern. As new technologies of communication emerged and diffused around the world – television, satellites, computers – Lerner was confident that they could revitalize modernization theory and practice. Lerner's faith in technology to solve a social problem places him in a long line of scholars who "all think that the imperfections of human interchange can be redressed by improved technology or techniques" (Peters 1999: 11).

The reason Lerner's dictum that mass communication is a key input into the modernization process continues to have such power, it seems, is that despite the somewhat limited explanatory power of modernization theory, development communicators have faith that if technical problems are solved – if more people can be reached, if access can be improved, if connectivity can be enhanced, if service can be provided for longer periods of time, – then mass media can surely, eventually, fulfill its potential as a multiplier of predicted economic, social, and political benefits of modernization. And this vision is also seductively open ended; a fact that ensures continued innovation and experimentation in the technique and technology of message delivery alongside an unshakable Lernerian confidence in the message itself.

## References

Badeau J (1959) Untitled review of *Passing of Traditional Society*. Am Polit Sci Rev 53:1135
Banfield E (1959) American foreign aid doctrines. In: Goldwin R (ed) Why foreign aid: two messages by President Kennedy and essays. Rand McNally, Chicago, pp 10–31
Berger M (1958) Untitled review of *Passing of Traditional Society*. Public Opin Q 22:427
Cravens H (2004) The social sciences go to Washington: the politics of knowledge in the postmodern age. Rutgers University Press, New Brunswick
Fair JE (1989) 29 years of theory and research on media and development: the dominant paradigm impact. Gazette 44:129–150
Fair JE, Shah H (1997) Continuities and discontinuities in communication and development research since 1958. J Int Commun 4:3–23
Huntington S (1965) Political development and political decay. World Polit 17:391–393
Inayatullah (1967) Toward a non-western model of development. In: Lerner D, Schramm W (eds) Communication and change in the developing countries. East-West Center Press, Honolulu, pp 98–102
Lerner D (1958) Passing of traditional society: modernizing the Middle East. Free Press, New York

Lerner D (1965) Communication and Enlightenment. Paper presented at the symposium on "Comparative theories of social change," Ann Arbor

Lerner D (1975) Research in group communication: a retrospective note. (Unpublished manuscript, January 1975). Daniel Lerner Collection, Box 13, Folder 2 (MC336), Institute Archives and Special Collections, MIT Libraries, Cambridge, MA

Lerner D, Robinson RD (1960) Swords and ploughshares: the Turkish army as a modernizing force. World Polit 13:33–35

Levy M, Banerjee I (2008) Urban entrepreneurs, ICTs, and emerging theories: a new direction for development communication. Asian J Commun 18:304–317

Lewis O (1951) Life in a Mexican village: Tepoztlán revisited. University of Illinois Press, Urbana

Melkote S, Steeves HL (2001) Communication for development in the Third World: theory and practice for empowerment. Sage, New Delhi

Peters JD (1999) Speaking into the air: a history of the idea of communication. University of Chicago Press, Chicago

Posusney MP (2004) Enduring authoritarianism: Middle East lessons for comparative theory. Comp Polit 36:127–138

Redfield R (1930) Tepotzlan: a Mexican village. University of Chicago Press, Chicago

Rist G (1997) The history of development: from western origin to global faith. Zed Books, London

Rogers EM (1994) A history of communication study: a biographical approach. Free Press, New York

Said E (1979) Orientalism. Vintage Books Edition, New York

Shah H (2007) Meta-research of development communication research, 1997–2006: patterns and trends since 1958. Paper presented at the annual meeting of the International Communication Association, San Francisco

Shah H (2008) Communication and marginal sites: the Chipko movement and the dominant paradigm of development communication. Asian J Commun 18:32–46

Shah H (2011) The production of modernization: Daniel Lerner, mass media and the passing of traditional society. Temple University Press, Philadelphia

Smith MC (1994) Social science in the crucible: the American debate over objectivity and purpose, 1918–1941. Duke University Press, Durham

# The Pax Americana and Development

P. Eric Louw

## Contents

| | | |
|---|---|---|
| 8.1 | Differences Between Informal and Formal Empires | 168 |
| 8.2 | The Origins of the Post-World War II "Development" Idea | 171 |
| 8.3 | Some Difference Between American and British Conceptions of "Development" | 173 |
| 8.4 | Assumptions Underpinning Pax Americana "Development" | 176 |
| | 8.4.1 Jefferson's "Empire of Liberty" | 176 |
| | 8.4.2 Universalizing "Democratic Capitalism" | 177 |
| | 8.4.3 A Wilsonian Informal Empire | 178 |
| | 8.4.4 Continuities and Discontinuities Between American and British Models | 180 |
| 8.5 | The Mutating "Development Industry" | 182 |
| | 8.5.1 Development as Modernization | 182 |
| | 8.5.2 Critics of Modernization | 185 |
| | 8.5.3 Modernization Revived | 188 |
| 8.6 | And What of the Future? | 190 |
| References | | 191 |

### Abstract

Development in its many iterations was born from the needs of the Pax Americana to create reliable and competent compradors, plus viable trading partners in order that its post-1945 informal empire would function. This chapter will examine the way the Pax Americana replaced Europe's formal empires, the reason informal empires need comprador partners, and the difficulties encountered by the Pax Americana with identifying and appointing satisfactory comprador partners. The relationship between this "comprador problem" and the founding of a development industry and aid industry will be unpacked, as will the core assumption underpinning Pax Americana "development." The

P. Eric Louw (✉)
School of Communication and Arts, University of Queensland, Brisbane, QLD, Australia
e-mail: e.louw@uq.edu.au

© Springer Nature Singapore Pte Ltd. 2020
J. Servaes (ed.), *Handbook of Communication for Development and Social Change*,
https://doi.org/10.1007/978-981-15-2014-3_35

differences between Pax Americana and Pax Britannica governance and development will also be examined, as will the continuities and discontinuities between these two varieties of imperialism.

The idea of "development" emerged in the 1950s as one of the consequences of how World War II transformed the world by replacing British global hegemony with the Pax Americana (see Louw PE Roots of the Pax Americana. Manchester University Press, Manchester, 2010). Before this war European empires (especially the British Empire) straddled the globe, while the USA found itself in the frustrating position of being a new rising global power which kept finding its opportunities for expanding its trade thwarted by the fact the Europeans had gotten there first. And so, even before America had entered World War II, the US State Department had established working committees charged with conceptualizing how to terminate European imperialism through decolonization (O'Sullivan CD Sumner Welles, postwar planning, and the quest for a new world order. Columbia University Press, New York, 2008: chapter 4), plus how to replace Europe's formal empires with an American informal empire. After 1945, American power was used to implement this State Department planning such that Europe's empires were deconstructed and a Pax American trading empire constructed in their place. This shift from formal empire to informal empire is significant because it created conditions that produced the "development" phenomenon.

### Keywords

Asian decolonization · Dependency theory · Human rights-driven foreign policy · Jeffersonian "cloning" model · Modernization · New International Economic Order (NIEO) · Pax Americana development

## 8.1 Differences Between Informal and Formal Empires

The building of a Pax Americana represented more than a global shift of power across the Atlantic; it also signaled a new way of organizing global governance because the Americans exercise their global hegemony quite differently to how the British exercised theirs. At heart, the difference between pre- and post-1945 global governance is due to the fact that Britain ran a formal empire and America runs an informal empire. A summary of the differences between these two empires can be found in Table 1.

A key to understanding American hegemony is how decolonization was undertaken after 1945 and how this necessarily led to the growth of "development." Three features of this decolonization process are noteworthy.

Firstly, the Pax Americana retained the same administrative units created by the European imperialists (e.g., Iraq, Nigeria, Indonesia) and gave them independence as Woodrow Wilson's self-determination model (Louw 2010: 29–33) was rolled out globally. Most of these administrative entities made no "political" sense because they were not based on linguistic, ethnic, or religious bonds. Generally they were artificial

**Table 1** Differences between the British and American empires

| Pax Britannica (formal empire) | Pax Americana (informal empire) |
|---|---|
| British annexed foreign territory and ran up the British flag | Foreign territory not annexed |
| Dispatched British occupation army | Independent states have their own armies which America often helps train. America uses an alliance system which it dominates (e.g., NATO). American-led (multilateral) occupation armies only rarely dispatched (e.g., Germany, Japan, Afghanistan, Iraq) |
| British-ran police systems in each occupied territory | Independent states have their own police force which America often helps train |
| British-ran judicial systems in each occupied territory (with privy Council in London as final court of appeal) | Independent states run their own judiciaries. But power of international court to impose "international will" is growing |
| British governor oversaw political process in each occupied territory. (political processes had various levels of "local participation" and various levels "autonomy" from Britain) | No American governors in the informal empire. Instead, America encourages "comprador elites" within independent states |
| Local economies were restructured to suit British economic interests | Local economies restructured to suit American economic interests, but this is achieved through intervention of American corporations, American aid, and World Bank (rather than direct American government policymaking) |
| British merchants emigrated to colonies (to network these colonies with London) | American corporations dispatch some staff to foreign states, but control mostly exercised via telecommunications, short-term visits, or comprador managers |
| British built infrastructures such as railways, ports, roads, hospitals, schools, universities, irrigation, electricity/water/sewerage, and telecommunications. This involved building British-run bureaucracies in each colony to administer and develop these territories | America builds no foreign infrastructure and runs no overseas bureaucracy. However, a system of multilateral organizations (UN, World Bank, etc.) operates as a "neo-imperial bureaucracy" |
| Colonial economies tied to London (often via system of imperial preferences) | Foreign economies tied to New York through American corporations and American dominance of global financial system |
| Britain ran a global system of maritime trade routes (protected by globally dispersed naval bases) and a global telegraph system | America runs a global system of maritime and aviation trade routes (protected by globally dispersed naval and air force bases) and a global telecommunications system |

constructs created when Europe's imperialists randomly drew boundaries on a map. The new states often threw together people with no desire to inhabit the same state and who often harbored historical animosities toward one other.

However, these administrative units were useful for the Pax Americana because they had been developed by European imperialists as economic entities. And the Pax Americana, an informal (trading) empire, was fundamentally interested in these states' economies (as trading partners), not in their political difficulties. So what America wanted was for the "economic entities" created by Europeans (e.g., Iraq or Indonesia) to keep functioning. But within the new world order, they were to be tied to New York (rather than London, Amsterdam, Paris, etc.). What the Pax Americana needed was effectively run sovereign states since such states formed the basis for functioning liberal-capitalism (Mandelbaum 2002: 77). Americans naively believed one could simply transform imperial administrative/economic entities into new functioning sovereign nation states. Herbst (2000) has shown this was not the case.

Secondly, in each administrative entity created by European imperialism, a Westernized middle class had emerged. These Westernized middle classes lived in the cities and were a minority of their country's population. To maintain their Western middle-class lifestyles, the economies built by the colonials needed to be maintained. America saw these Westernized middle classes as "natural allies" (comprador partners) of the Pax Americana and advocated power be transferred to them. Where European colonial possessions did not have "native" middle-class populations large enough to run independent states, America pushed the colonial powers to actively build such middle-class populations. Successfully rolling out the American informal empire model meant transferring power to local Westernized elites, who stepped into the shoes of departing colonials (Fanon 1968: 152) – effectively becoming administrators of new independent states serving American economic interests. The model was based on the idea of a synergy between America (which needed comprador partners in its far-flung informal empire) and Westernized minorities in each new state (who needed powerful overseas allies to help them maintain their affluent lifestyles and often precarious hold on power). Consequently, the Pax Americana came to be characterized by comprador partnerships based on mutual need. America ruled "indirectly" and "informally" through comprador allies. Thus, independence often brought not "liberty," but rule by compradors who were often corrupt, brutal, and inept. But not all these Westernized populations turned into the "natural allies" America had anticipated. Instead, some African, Asian, and Latin American middle classes became radicalized and aligned themselves with the Soviet Union. This resulted in America slowing down its push for decolonization during the 1950s. But this slowdown did not end America to push for the development of middle-class comprador-building – instead it simply saw the European colonial powers pushed into the running of "development" programs which were to build moderate middle-class elites and thereby prepare backward colonies to become independent as reliable trading partners.

Thirdly, the ruling elites of these new states were allowed to have their flags and symbols of "independence" as long as they conformed to the rules of the game – i.e., power effectively shifted from European empires to (non-state) multilateral financial organizations (e.g., IMF, WTO, World Bank), with American control of the global financial and monetary system constituting the new locus of global governance (Underhill 2000). Within the Pax Americana, the role of independent states is to "administer" the "linkages" of a global trading system (e.g., foreign affairs, finance, and defense) (Cox 1987: 228), thereby facilitating a global market (Underhill 2000: 131). The rulers of these states operate as de facto functionaries of America's globalized trading system. The Pax Britannica's colonial administrators fulfilled similar roles within Britain's globalized trading system. But whereas the "rules" Britons administered were openly (and proudly) "imperial," the Pax Americana disguises its "rules" through multilateralism.

But, of course, before the Pax American could be established, the European empires had to go; and that was achieved through decolonization.

## 8.2    The Origins of the Post-World War II "Development" Idea

Franklin Roosevelt strongly pushed for decolonization; so during his presidency, anticolonialism was placed firmly onto America's foreign policy agenda. Domestically Roosevelt's progressive liberalism created the New Deal. Significantly, Roosevelt's team wanted to internationalize this New Deal – i.e., develop interventionist foreign policies that would "socially engineer" and "change" foreign societies to make them conform to Democratic Party's dreams of "progress." This shifted America toward a human rights-driven foreign policy – and important part of which was promoting decolonization (which would conveniently remove those European empires blocking the creation of a Pax Americana). Roosevelt effectively flagged the link between decolonization and "development" during a Press Conference for the Negro Newspaper Publishers Association when he spoke about a stopover in Britain's colony of Gambia. In this important statement about America's "development idea," Roosevelt said:

> It's the most horrible thing I have ever seen in my life. The natives are five thousand years back of us. Disease is rampant...for every dollar that the British...have put into Gambia, they have taken out ten. It's just plain exploitation of those people. There is no education whatsoever...the agriculture there is perfectly pitiful...the natives grow a lot of peanuts...and they still use a pointed stick. Nobody ever saw a plow in Gambia. The British have never done a thing about it. Now, as I say, we have got to realize that in a country like Gambia – and there are a lot of them down there – the people, who are in the overwhelming majority, have no possibility of self-government for a long time. But we have got to move...to teach them self-government. That means education, it means sanitation. (Roosevelt 1944)

Importantly, this Roosevelt statement encodes America's decolonization vision of "development," "trusteeship," plus how "development" will culturally transform/Westernize primitive people through "education." The State Department built both its 1950–1960 decolonization and development policies upon this Rooseveltian vision.

Once World War II ended, "decolonization" shifted from being "an idea" to a "political project." But decolonization-as-project became a complex affair because the US State Department had to deal with complex new post-war realities. The key issue necessitating some rethinking was the emergence of a Soviet Union powerful enough to block a full global rollout of a Pax Americana. Instead the world was divided into a Soviet-aligned communist bloc and a USA-aligned Western bloc. This meant that as European empires were deconstructed after World War II, the Americans could not guarantee new independent states became part of their informal empire. Instead Cold War competition meant decolonized states might end up as Soviet allies. This produced America's new core concern of containing the spread of communism – which meant reconceptualizing "decolonization." As a result, between 1945 and 1965, three iterations of decolonization emerged:

- Rapid Asian self-determination (1940s)
- Cold War gradualism and trusteeship (1950–1957)
- Abandoning gradualism for rushed independence (from 1958)

The Asian decolonization phase happened before Cold War concerns caused America to reconsider their original Rooseveltian plan. However, from 1950 to 1957, a gradualist approach to decolonization was applied. This 1950s gradualism was associated with three themes: firstly, the idea of "trusteeship" and Third World "development," secondly, constructing Third World nationalist elites to govern decolonized states, and thirdly, blocking Third World communists from gaining control of decolonized states. Faced with the threat the Soviets might breach Truman's containment lines by winning allies among new Third World ruling elites, Washington decided to rigorously manage future decolonizations to ensure outcomes favorable to America. Effectively, Washington decided the continuance of European empires was preferable to creating independent Third World states ruled by communists. This did not mean America abandoned its long-term vision of a thoroughly decolonized world wherein new states were tied into America's informal empire through a system of comprador partners. But it did mean reaching this goal would now be more difficult and drawn out, might necessitate short-term Washington support for maintaining Europe's empires (as long as these empires were liberalizing themselves), and might even involve American support for counterinsurgency wars against communist-aligned anticolonial activists. So under Truman a new decolonization model emerged that placed much greater emphasis on "development" – a form of "development" that America pushed the European colonial powers to carry out. Essentially, America wanted to ensure that, at independence, power was handed over to moderate ruling elites who functioned as reliable Pax Americana comprador partners. This new Pax Americana position took some pressure off Europe's empires, because America was now open to European arguments that some colonies were not ready for independence – after all Roosevelt's Gambia speech had acknowledged "the natives are five thousand years back of us...[and]... have no possibility of self-government for a long time" (Roosevelt 1944). Consequently, in the early 1950s, the Europeans and Americans reached agreement that African colonies were too backward to be granted independence; and US State Department Africa policymakers began making statements like: "premature independence for primitive, uneducated people can do them more harm than good" (Metz 1984: 521–522). A widespread feeling grew in the State Department that European empires were keeping Africa as a "stable and secure" area for the West (Metz 1984: 521), that colonialism in Africa was "progressive" because it was developing backward areas into economically viable states and modernizing Africa (Metz 1984: 521), and that it would take a long time to "fully prepare" Africa for independence and it would be detrimental to grant independence before this development process was completed (Metz 1984: 521). Consequently, for much of the 1950s, the State Department held the view that "the only obvious result of severing the colonial relationship before the sufficient development of the economy and political system would be chaos, a situation conducive to communist penetration" (Metz 1984: 521). So this was the context within which the post-World War II development industry was born.

For the British, French, Portuguese, and Belgians, this shift in US thinking was a godsend. They all concluded this new American "realism" gave them the opportunity to "salvage" their African empires as long as three conditions were met – America insisted upon (liberalized) "development" of the colonies (to build viable trading partners for the Pax Americana); maintaining open trading/no imperial tariff barriers was nonnegotiable (i.e., America insisted upon trading access to these colonies while they were being developed); sociopolitical stability had to be maintained in these African colonies while future nationalist ruling elites were being nurtured into existence through "development."

But in 1958 the ground shifted again when US policymakers suddenly panicked about their 1950s policy of gradualism. This policy shift seems to have been triggered by Nkrumah hosting the All-African People's Congress in 1958. Thereafter the State Department concluded that if decolonization did not happen rapidly, the West was in danger of losing Africa to the Soviets through the rise of radicals and revolution, because as Joseph Satterthwaite, the first assistant secretary of state for African affairs, said "1958 was a landmark year when things changed because the political situation in Africa became 'vibrant if not effervescent'" (Metz 1984: 524). But far from ending the need for "development," this shift to rapid decolonization actually increased the need for "development" because it meant independence was now going to be granted to ruling groups not deemed to have enough skills to run these states and economies. As British Colonial Secretary Ian MacLeod said: "Were the counties fully ready for independence? Of course not" (Shepherd 1994: 256). Consequently, the model shifted after 1959 to providing post-independence "development" which meant instead of "development" being provided by colonial administrations, a new "development industry" had to grow up outside the old colonial administrative arrangements. This shift also ended any residual European empire notions of "development" (which had still been evident in the 1950s) and generated a development model that more fully conformed Pax Americana conceptualizations.

## 8.3 Some Difference Between American and British Conceptions of "Development"

Because there were differences between America's (informal) and Britain's (formal) approaches to empire, there were also differences between how these two different empires understood what they were trying to do and hence how they conceptualized "development." Understanding these different visions is instructive.

Firstly, because both formal and informal empires were geared to building trading partners, both had a vested interest in "development." However, when the British encountered weak states and people whom they deemed unable to govern themselves well enough to be viable trading partners, they took control themselves by annexing these areas into their empire and then actively intervening to establish what they deemed satisfactory governance and infrastructures for commerce and trade to flourish (Gallagher 1982: 7). So Britons took direct responsibility for "development" by going into the field to impose ("British") order, build infrastructures, and establish businesses. America, on the other hand, prefers not to take such direct responsibility for establishing the governance and infrastructures needed for trade. Instead, Americans prefer local comprador partners to do this for them. This means the Pax Americana faces difficulties the Pax Britannica did not – i.e., because the British appointed their own governors and officials they could ensure quality control over governance. In the Pax Americana, if compradors turn out to be corrupt, incompetent, or simply demagogues (as has often been the case), Washington cannot fire them the way London could fire its governors (instead Washington has to mount expensive "regime change" operations). This is one of the reasons the Pax Americana has generated so many instances of poor governance, corruption, and failed states.

Secondly, there are differences between the Pax Americana and Pax Britannica notions of "development." Lord Lugard (who had a major influence on British Colonial Office policy from the 1920s) argued the British Empire had a "dual mandate" to develop British trading interests and to simultaneously progress the "native" peoples materially and morally (Lugard 1922). Lugard's vision of development was, like America's, grounded in the promotion of global trade. He believed Britain's imperial interventions (driven by trade) were not only good for Britain but were also good for natives of the colonial possessions, because Britain's efforts to build trade generated employment/wealth and diffused skills in Africa and Asia. Lugard's key development argument was that Britain should train "native" middle managers. These native middle managers (rather than white colonial officials) would become the "face" of British governance with whom local people actually interacted on a daily basis. He saw this as part of a civilizing mission in which the wealth derived from trade was used to train locals in the art of good governance. Significantly, Lugard aimed to leave existing native authorities largely intact. Training would empower the traditional authorities by giving them the ability to govern effectively and thereby the ability to protect their local native producers (and their culture) against the intrusion of white planter economies. Within this Lugardian vision, Britain's empire was a "trustee" for development and a long-term "training academy in liberal democracy" (Owen 1999: 193). It is important to note the Lugardian "development" model differs from the Pax Americana "development" model, because Pax Americana development is geared to producing Third World middle-class elites who are Westernized (i.e., "clones" of middle-class America), whereas Lugard's model was geared to actually preserving existing local authorities and cultures, but then grafting onto these the techniques of good governance. So in the British model, the local middle managers would not resemble the British middle class, but would rather be "translaters" between their own local culture and the culture of the British Colonial Office for whom they worked. In other words, British Empire "development" did not promote a middle-class social engineering model of radically transforming society (into new middle-class-led "nation states"). Rather, it was a gradualist development model promising to incrementally transform the empire into a multiracial Commonwealth based on confederalism and cultural pluralism (an outcome quite different to the Pax Americana model of secularized independent nation states governed by Westernized compradors).

After World War II, the British Colonial Office necessarily adjusted itself to the reality of the Pax Americana and abandoned the Lugardian approach in favor of America's preference for developing (Westernized) middle-class elites prepared to run secular-liberal nation states and serve as comprador partners of America's empire. This policy shift to "modernization" was put into effect by the Colonial Development and Welfare Act 1945. This produced a "development industry" geared to preparing Europe's empires for independence by "developing" urban middle classes with the necessary political, economic, and administrative skills to lead each imperial possession to independence. Hence, "development" effectively meant "liberalization" – i.e., an attempt to promote liberal democracy and capitalism in these new Third World states. Because this was essentially an exercise in Westernization, the provision of Western education to Asians and Africans was a central feature of this development process that ran from the mid-1940s to the late 1950s. Unfortunately for the Pax Americana, this form of development had the unintended consequence of producing radicalized middle classes who demanded immediate independence (Gallagher 1982: 148). And refusing to grant independence threatened to produce (Soviet-funded) armed rebellion and communist governments. To halt this radicalization process, the USA insisted that independence be granted to many Third World states before the "development" process had been completed – i.e., before they were deemed "ready" for independence (Goldsworthy 1971: 363). For the Pax Americana, this generated the problem of underdeveloped states ill-prepared to be valuable members of America's trading network. As a result, Westerners concluded "underdeveloped" states (who prefer to call themselves "the developing world") needed to be taught, through "development" programs, the appropriate skills to govern themselves and run Western economies (so they could become reliable trading partners within the Pax Americana). And this produced a new type of "development industry" – one focused on already-independent states deemed poorly organized – i.e., "underdeveloped" states that still need to learn how to organize themselves economically and politically.

These underdeveloped states have presented Washington with a particular set of challenges. They are weak, fragile, or even failed states characterized by poor governance, corruption, poor infrastructures, and a lack of the cultural capital required to organize or develop themselves into the sort of "nation states" preferred by Washington. Their weakness and poor governance make them precisely the sort of places the British Empire would have annexed in order to establish satisfactory conditions for commerce and trade. But what can Washington do to create satisfactory conditions for commerce and trade given that America (unlike Britain) does not run a formal empire, claims to be morally opposed to "imperialism," and does not possess the sort of ruling class capable and prepared to govern distant colonies? The Pax Americana's solution has been the growth of a "development industry" (geared to "teaching" good governance and building cultural capital and functioning infrastructures). In effect, the "development industry" was deployed as an alternative to formal imperialism (where imperialists directly intervened to establish "satisfactory conditions"). The informal empire approach involves trying to "improve conditions" through pedagogy, persuasion, and consultation.

So what is the nature of the "development" being undertaken? Answering this question requires unpacking some key assumptions of the Pax Americana's style of "development."

## 8.4 Assumptions Underpinning Pax Americana "Development"

Americans came to see themselves as "exceptional." And once they came to believe that the American way of life, social order, and political system was self-evidently superior, it was easy to reach the conclusion the whole world would benefit from adopting America's sociopolitical and economic system. This belief in the superiority of "the American way" underpins America's preferred model of "development" and social change. At heart, this model can be traced back to Jefferson's conceptualization of spreading American "liberty" westward across the continent to the Pacific.

### 8.4.1 Jefferson's "Empire of Liberty"

Jefferson's "empire of liberty" was premised upon the notion that American "democracy," "freedom," and state formation were simply superior. Jefferson advocated spreading the American model across the American continent within an "Enlightenment vision of a benign imperial order" (Onuf 2000: 53). Spreading this model across North America necessarily involved taking control of territory and then reconfiguring that territory politically, economically, and socially until a new "state" was deemed ready ("developed" enough) for incorporation into the Union. Each of these new states replicated America's model of state-building. Hence, rolling out the 50 states across the continent involved a cloning process. It was also an intensely imperialist process of incorporating other people's territory into America's own political formation. The result was America's first empire – an empire based on conquest, colonization, and cloning. In this way Jefferson's republican union spread irresistibly westward through a proliferation (cloning) of "free republican states" (Onuf 2000: 53). The irony is, this process of westward expansion and state formation ultimately produced an American political culture at war with itself – i.e., those Americans morally opposed to imperialism and colonialism must now either obfuscate their own history or oppose American villainy. On the other hand the other half of America celebrates the building of a transcontinental nation through the development of the West.

The Jeffersonian model of cloning "democratic states" did not die with the ending of America's first empire. Instead, a modified version of this Jeffersonian "cloning" model now effectively underpins today's American vision of how (liberalized) "development" should be rolled out across the "underdeveloped" world – i.e., a key feature of the Pax Americana involves the idea of developing "democratic states" across the world. Within America's post-1945 informal empire, these new states are not created or developed by American settlers. Instead, they are run by their own indigenous people. However, since they are not deemed to be well run, the West operates a huge global development industry to teach people in "developing states" how to run democratic-capitalist societies. So a new form of "control" called tutelage emerges. Another difference is that, once successfully developed, these new states are not destined to be incorporated into the American Union (as occurred in America's first empire). Instead, they are to be enmeshed, as client states, into the trade networks of the Pax Americana.

## 8.4.2 Universalizing "Democratic Capitalism"

Jefferson's "empire of freedom" idea was always grounded in an Enlightenment optimism and belief that Americans had found a "recipe" to make the world a better place. It is an optimism that powerfully informed twentieth-century Wilsonian thinking about building a Pax Americana and Roosevelt's thinking on internationalizing the New Deal. But an important consequence of transferring Jefferson's "empire of freedom" idea across to America's new informal empire is that Jeffersonian optimism tends to slide into a new kind of imperialism within which Americans increasingly look like they want to build a "universal" global civilization. Effectively, Jefferson's democratic ideals are transformed into something akin to a religion of "democracy" and "universal human rights." Within this religion, "American democracy" and "American values" slide into a pan-human universalism, which is intensely utopian. Hence, the Pax Americana, as globalization II, becomes ensnared in a new Jacobinism:

> It employs an idiom somewhat different from that of the earlier Jacobinism, and it incorporates various new ideological and other ingredients, but it is essentially continuous with the old urge to replace historically evolved societies with an order framed according to abstract, allegedly universal principles, notably that of equality. Like the old Jacobins, it does not oppose economic inequalities, but it scorns traditional religious, moral, and cultural preconceptions and social patterns...The new Jacobins...regard today's Western democracy as the result of great moral, social and political progress...they see it as an approximation of what universal principles require....[The new Jacobins] think globally. They put great stress on the international implications of their principles. The new Jacobinism is indistinguishable from democratism, the belief that democracy is the ultimate form of government and should be installed in all societies of the world. The new Jacobinism is the main ideological and

political force behind present efforts to turn democracy into a worldwide moral crusade. (Ryn 2004: 21)

The idea that Americans had found the "recipe," which if applied to the whole world would make it a better place, is grounded in an "American faith that it is a universal nation [which] implies that all humans are born American, and become anything else by accident – or error. According to this faith American values are, or will soon be, shared by all humankind" (Gray 2002: 132).

At heart is the belief "democratic capitalism" (in the form developed by Americans) is good for everybody (Gray 2002: 2). Universalizing democratic capitalism certainly would be good for America, because a fundamental feature of America's organization is that "American domestic vitality depends on continued expansion" (Latham 2000: 14). This is because America's economy has to keep growing, which means continually searching for new markets. All three of America's empires have been grounded in this need to find new markets in order to keep the economy growing. The post-1945 informal empire is simply the most recent means to feed this market-driven economy. Clearly America's informal empire would work best if lots of little American-type "democratic capitalist" states were cloned (or at least hybridized) across the globe. It would be better still if they were run by people who broadly shared American middle-class values. Hence, not surprisingly, the Pax Americana's "development industry" has not simply been intent on building infrastructures facilitating economic growth and trade; it has also been concerned with (Western) education and inculcating the sort of (Western secular materialist) values that drive American economic growth. This implies the simultaneous destruction of "old" values and the traditional cultures in which they were embedded. Thus has the Pax Americana been systematically destroying "traditional elites" since 1945 beginning in occupied Europe and Japan and then spreading to Africa and Asia.

### 8.4.3 A Wilsonian Informal Empire

The problem faced by an informal empire like the Pax Americana is that when its trade promotion agendas are impeded in the "developing states" by poor infrastructures, poor governance, and populations without the cultural capital required to run Western economies, it needs to employ *indirect* methods to try and rectify the situation. These indirect methods may not involve such drastic action as annexation (as used by formal empires), but they still involve intervention. And the preferred intervention of the Pax Americana has been to deploy specialists in Third World development – i.e., people working in the "development industry" (often linked to the West's "foreign aid industry"). The development industry has come to be inspired by a form of Jacobin thinking – i.e., a belief in middle-class-driven social engineering (of either a liberal or socialist variety). It is social engineering that intervenes into other people's lives to change them. Presumably those doing the intervention believe what they are doing is good – i.e., bringing about socioeconomic change. In many ways this makes the development industry a form of contemporary "missionary" activity in which (Western) "believers" work to "improve" non-Western societies (The "Western believers" can, of course, be Third World people who have been Westernized.). But today's missionaries do not promote Christianity. Instead, the development agenda has been driven by a secular vision of social engineering that grew up in the context of building a Pax Americana after 1945. This agenda has been closely associated with a strong belief in "education" as a key mechanism to roll out the skills (and attitudes) needed for "progress." Not surprisingly, we find the Wilsonian-liberal thinking of the Roosevelt team (outlined in Table 2) has strongly influenced the broad trajectory of the development industry's vision of how to "improve" the world. The above discursive framework, which has been widely propagated by Western intellectuals and journalists from the 1950s onward, is highly compatible with the economic needs of America's informal trading empire and with the "moral universe" of middle-class America. When development programs encoding these (Liberal) values are successfully rolled out,

**Table 2** Liberal thinking underpinning the post-1945 Western "development industry"

| Positive | Negative |
|---|---|
| Democracy<br>Religious freedom within a secular state/society (relativist pluralism)<br>Cosmopolitanism | Aristocratic elites and hierarchies/oligarchies/fascism/communism<br>Theocratized states/society<br>Ethnic nationalism |
| Anti-imperialism<br>Decolonization<br>Self-determination within nation states<br>Nation building | Imperialism<br>Colonialism<br>Confederalism/holism |
| Free trade/free movement of capital<br>Capitalism<br>Consumerism/materialism<br>Property rights<br>Liberalization<br>Government creates conditions for capitalist enterprise<br>Modernization<br>Development<br>Western education<br>Individualism | Anticompetitive behavior/restrictions on trade/restrictions on enterprise and capital/tariffs/protectionism/imperial preference<br>Government decisions impeding capitalist enterprise<br>Traditional society and tribalism<br>Collectivism |
| State formation around viable economic units (wherein different groups/nations/religions live together harmoniously)<br>Facilitating movement of labor:<br>Migration<br>Melting pot/assimilation<br>Multiculturalism<br>Cultural hybridization | Economic nationalism<br>Limiting movement of labor:<br>Segregation<br>Apartheid<br>Racism<br>Cultural autonomy |
| Peace (assured by multilateralism)<br>Universalized "human rights" | War (caused by imperialist and nationalist competition) |
| Progressive (teleology geared toward America as the "universal state") | Reactionary |

it produces populations sufficiently "modernized" (i.e., "Westernized" or culturally "hybridized") to run self-governing client states and/or run Western economies that are valuable to America's informal trading empire.

### 8.4.4 Continuities and Discontinuities Between American and British Models

Both the Pax Britannica and Pax Americana engaged in "development." And if we look at the development work done in the two empires, there are some striking similarities. Most obviously, both the British and Americans built global trading empires which served to integrate the world's economy and create an international division of labor (Fieldhouse 1999: 164). Effectively, the Pax Britannica and Pax Americana have jointly been the drivers of globalization. In the process, both Britain and America have spread capitalism, industrialization, intensive agriculture, free trade, liberalism, democracy, Anglo-culture, and the English language. But despite these similarities, there are some major differences between the British and American models of development. This is because the two models were driven by different assumptions and different social groups.

"Development" brought about by the British occurred when British merchants encountered weak states or peoples deemed unable to govern themselves well enough to be viable trading partners. Britain annexed such places and then dispatched colonial officials to these territories to "develop" satisfactory conditions for commerce and trade. Significantly, this British development was driven by an alliance of Britain's upper and middle classes. Middle-class "gentlemanly capitalists" built businesses and trading networks, while the upper classes went to the colonies to build the machinery of bureaucracy and law and order. Significantly, these two classes shared no ideological vision of social change. Only Britain's church missionaries were motivated by any ideological zeal to change people in the colonies; but they were never central to the "development" processes in Britain's empire. As a result, for most of the history of the British Empire, "development" did not occur according to any "ideological plan"; rather it happened organically as gentlemanly capitalists built their trade networks, British settlers built businesses and farms, and colonial officials established governance and infrastructures (like railways, roads, hospitals, and schools) to cope with demands being made on them. Further, Britain's middle class was driven by a desire to make money, not by any utopian social engineering vision; and Britain's governing upper class eschewed Jacobin social engineering projects (generally associated with middle-class utopians). So instead of trying to "change" or Westernize "natives," British colonial governance generally worked closely with traditional authorities. Consequently, British "development" was more respectful of indigenous traditions and culture than the Pax Americana approach to development, because Britons were not driven to "engineer" or change indigenous cultures so as to produce clones of the British middle class. On the other hand, Britons were "disrespectful" in so far as they made no apologies for intervening in places they deemed "primitive." And, unlike contemporary Americans, Britons made no pretense at political correctness when they perceived differences between cultures, races, and societies functioning at different levels. Further, British "development" involved no pretense at "consultation" and sometimes even bypassed the local "natives" by importing change/development agents from outside, e.g., British managers, or British settlers (some free and other compelled like convicts to Australia), or Indians to Fiji, Malaysia, Africa, and the Caribbean or Chinese to Malaysia.

The post-1945 Pax Americana approach to development has been far more coherently organized, planned, and professionalized. This is partly because Pax Americana "development" has been organized by a group of professional "developers" who share a relatively coherent "development ideology." This coherence is assisted by the fact those driving Pax Americana "development" tend to be drawn from a relatively narrow social strata – i.e., middle-class professionals educated to share a similar vision of the world (involving "democracy," "universal human rights," and diffusing modern skills, ideas, and technologies).

Although Americans like to believe their way of running the world is less brutal and interventionist than the methods used by the old European empires, Pax Americana "development" has ironically generally been far more culturally interventionist than British development was. This has occurred precisely because America has run an informal empire. Informal empires imply "indirect" governance, which means the Pax Americana (ideally) needs middle-class comprador partners in each independent state (Bacevich 2002: 41). This translates into a need to try and systematically build (Westernized) middle-class elites in the Third World through education and "nation building" and "development" exercises. This means changing indigenous people through ethnocentric (i.e., Westernizing) social change projects. Because of the need to build comprador partners, Pax Americana "development" has often involved Jacobin social engineering which has been culturally destructive of traditional societies in the Third World (Middle-class Jacobins also successfully unleashed the rhetoric of "democratic capitalism" against traditional elites of Europe.). Liberal-Jacobins also learned to successfully deploy the rhetoric "democracy" as a powerful agent for undermining traditional societies across the globe (Ryn 2004: 25). Once these traditional elites have been undermined, they are replaced with middle-class elites who then become local agents for secular nation building and comprador partners for the Pax Americana. In many cases these new middle-class compradors have not served the Pax American well because they become exploitative and repressive and/or incompetent and corrupt rulers. But because the Pax Americana is an informal empire, it lacks the sort of direct imperial mechanisms Britain possessed to correct such problems. Instead Washington imposes sanctions or launches "regime change" wars or has to operate anti-piracy operations off the coast of those places with poor or weak governance.

Another difference is that, whereas Pax Britannica "development" deliberately sets aside "native reserves" where traditional cultures could survive alongside the areas of "British development," Pax Americana development intrudes everywhere. This is hardly surprising given Pax Americana's liberal-Jacobins are driven by a compulsion to universalize their vision of a "better world." Because America is seen as a "universal state," Americans believe they are doing the world a service by spreading their "empire of liberty." The resultant Pax Americana's "development" unleashes a complex mix of middle-class cloning (i.e., comprador-building), the destruction of traditional elites (though promoting the ideologies of "democratization" and "egalitarianism"), an ever-growing network of trade relations, Westernization, Anglicization, the undermining of traditional cultures and indigenous languages, cultural homogenization (in some places), and cultural hybridization (in other places). This "development" has also unleashed resistance, e.g., Al-Qaeda.

## 8.5 The Mutating "Development Industry"

In the two decades after 1945, European empires were dismantled and replaced by the Pax Americana. It was Washington's intention to expand its informal empire over these areas, and from this arose the need to develop a policy framework for how to deal with these decolonized regions. Effectively, America discovered granting self-determination did not necessarily mean the production of functioning states with the capacity for effective self-governance or the capacity to manage their own economies. Instead, decolonization produced "underdeveloped" states that were not the valuable trading partners America had hoped for. If they were to be turned into valuable trading partners, intervention was required to build a core of (middle-class) people in each new state able to govern effectively and run the businesses, mines, and farms America wanted to trade with. The recognition that intervention was required gave rise to the development industry.

But in the 1950s–1960s, America was not going to explicitly frame discussions of "development" in terms of the obstacles that "underdevelopment" posed to building their informal empire, because:

- Americans had argued for the self-determination of these states precisely as a means to end empires.
- Americans had only recently been arguing empires represented an unjust system of European domination and exploitation.
- If America was seen to be behaving imperially, Third World states might align themselves with the Soviet bloc. In fact, the Third World became the key sphere for American-Soviet Cold War rivalry.

So instead, the issue of underdevelopment (and its solution) was reframed within a discourse of modernization. This made it sound as if "development" was driven by Third World needs rather than Pax Americana (trading) needs.

### 8.5.1 Development as Modernization

Modernization is liberalism's theory of change. It is premised upon a binary opposition between "traditional" and "modern" societies. To be "modern" is supposed to be better than "traditional." Within modernization discourse, America is the ultimate modern state – it is seen to have advanced furthest in a teleological chain leading from "backwardness" ("traditional") to "advanced" ("modern") (Latham 2000: 58–59). This picture of the world encodes similar assumptions to those that drove the British Empire – i.e., Britain was deemed a civilized and advanced society which, when it encountered backward and primitive societies, annexed them in order to develop and uplift them. This was often called the civilizing mission of empire. American modernization theory sanitized, disguised, and replaced this "civilizing mission" idea by deploying the language of positivist science (Engerman et al. 2003: 35). But ultimately, "modernization" is a language of (informal) imperialism serving to justify Americans telling the rest of the world how they ought to be organizing themselves. This new language of imperialism encoded Western superiority (Engerman et al. 2003: 38) and assumed there was only one pathway to modernity (Engerman et al. 2003: 39) – i.e., following the path America had already taken. The latter implied America was the benchmark to be reached or "universal state" to be cloned. And the cloning mechanism was to be found in the scientific method of "development." This method offered a blueprint to facilitate liberal-capitalist "takeoff" (Latham 2000: 210).

Modernization theory presented itself as benevolent and grounded in universally valid scientific laws (i.e., it claimed all societies passed through the same stages of development) (Latham 2000: 16). Effectively, this encoded into a "scientific theory" the idea of America's manifest destiny as a "universal state" worthy of being copied by all others. Modernization theory was American middle-class utopianism which postulated the "American way" as the endpoint of a progressive teleology. So just as the British had previously seen their empire as a progressive force in the world, now it was America's turn to be this progressive force – and to take up the role of being a "civilizer," "developer," and agent of "universal" values. But in the American version, social engineers sought to actively plan and implement social change so as to bring about their (utopian) vision for a better (universalized) world.

This notion of being a "developer" of the world was formulated at a particular moment in time – i.e., a post-World War II world where the only obstacle to implementing Wilson's dream was Soviet power. Modernization theory therefore encoded three American concerns (Latham 2000: 19):

- Transforming Third World decolonized states into reliable trading partners within America's global trading network
- Preventing the spread of communism to the Third World ("containment") by ensuring the new states were developed according to the principles of democratic capitalism
- A desire to spread American sociopolitical organization and values (i.e., the American "way of life")

So modernization was the American dream writ large. If the "American way" could be cloned across the world then, the liberal utopians would argue, one could eventually end up with global peace – this would be an American Peace (Pax

Americana) where there would no longer be any reason to fight wars because the whole world would agree on the rules governing the global trading network; America's vision of "human rights" would have been universalized; all self-determining states would be ruling elites governed by people educated to share American values and aspirations; and ethnic, cultural, and religious conflicts would be a thing of the past because of a process of globalized cultural homogenization/ hybridization (i.e., Americanization/secularization). This liberal-utopian vision assumes some kind of Francis Fukuyama-type "end of history," or "end of ideology" can be attained where wars end, because all countries have become the same and have McDonald's hamburgers! For the utopians, promoting modernization was the way to socially engineer this "better world" into existence. So modernization theory was not about analyzing the Third World, it was about what America was already like (Engerman: 57) and about the modernizers' desire to clone America globally. This did not mean "foreign-ness" would disappear, but it did mean foreigners would become more like Americans and that "deficient" (i.e., traditional) cultures would disappear (Latham 2000: 213).

The key drivers of this modernization vision were Walt Rostow, Lucian Pye, Daniel Lerner, Gabriel Almond, and Edward Shils (Samuel Huntington (*Political Order in Changing Societies*) produced a more sophisticated modernization theory in the late 1960s.). Some were Cold War warriors (Leys 1996: 10). The origins of this theory lay in American sociology, and so it encoded functionalism, the idea of linear evolutionary change, an optimistic belief in progress, and the assumption that universal models cold be built because all humans were essentially the same (irrespective of different historical, environmental, or cultural factors) (Latham 2000: 30–35). Rostow (1960) and Lerner (1958) laid the foundation for modernization theory by establishing the "traditional" versus "modern" dichotomy. Significantly Rostow's five-stage model of moving from traditional to modern ended with a mass consumer society – clearly an ideal situation for an American global trading empire. Pye was instrumental in shifting the focus toward what developers needed to do to bring about "political modernization." American-assisted nation building was deemed to be central to the process of modernizing the Third World.

Modernization theorists saw traditional culture as an obstacle to modernization and modern economic growth. Their fundamental concern was with how to remove the obstacles, so a Rostovian takeoff could be initiated. The aim was to develop "primitive" (traditional) people so they could "catch up" with and "imitate" the developed world (Servaes 2003: 3). Hence a core objective was diffusing "modern values" to elites in the Third World (Leys 1996: 10). The modernizers also put great emphasis on transferring capital and technology to the Third World and diffusing the cultural capital required to operate this technology and to run Western economies. In many ways modernization was about diffusing not only Western technology and skills but also diffusing Weber's Protestant work ethic (Latham 2000: 62). In this regard education and mass media were important agents for change. Ultimately the development industry sought to diffuse modernity in a variety of ways, including the building of (Western) schools in the Third World, encouraging Third World people to attend Western universities, dispatching agricultural extension officers, funding

NGOs as development agents, and sending Western "role models" into the Third World, such as the American Peace Corps (Latham 2000).

An important feature of modernization theory was its utopian belief in teleological progress. These modernizers held true to the spirit of American (Enlightenment) optimism and to the spirit of universalizing the Roosevelt team's New Deal – i.e., they believed the world was perfectible, and Americans had discovered the way to perfect it. A key feature of this belief was traditional societies were fated to vanish because, advanced/civilized societies (i.e., modern America) were "destined" (according to modernism's teleological logic) to replace traditional/'backward" societies. This ideological belief that traditional societies were destined to vanish because of the "intrinsic deficiencies" of "backward"/primitive cultures served to justify the modernizers' Jacobin interventionism (Latham 2000: 59) and served to hide their role as social engineers who were effectively destroying other peoples' cultures. Those inside the development industry with a missionary belief in what they were doing simply did not recognize the cultural arrogance that underpinned their modernist development schemes. So when, as social engineers, they designed "development" interventions to detach Third World people from traditional family and religious and cultural bonds (Latham 2000: 210), they believed they were doing a good thing by sifting these people toward the modern world. As Jacobins they showed no concern about the destruction of indigenous cultures and languages or the individual social dislocation and pain they caused. And because American modernizers saw no value in traditional society, they left no spaces (such as the British Empires' "native reserves") where people had the choice to opt out of "development." In the Pax Americana, traditional culture was fated to disappear because the liberal-Jacobin development industry was going to work hard to make sure it did disappear. The liberal development industry was extraordinarily disrespectful of others if they happened to be traditional people. These traditional people apparently had no right to survive in the face of the universalizing of American middle classism. So the outcome of applying Roosevelt's decolonization and "development" model was often a legacy of people who were both traumatized and not able to function in either traditional or Western societies. Many of these people now live marginal lives in the shanty towns that surround cities like Nairobi, Johannesburg, Karachi, Manila, and Jakarta.

### 8.5.2 Critics of Modernization

Although the modernization paradigm was dominant among Western academics from 1950 to the mid-1960s, it was not without its critics. On the right emerged opposition to Third World "Aid" on the grounds that aid undermined local entrepreneurship. On the left, dependency theorists argued Western development served to tie Third World countries into subservient relationships with First World capitalists which facilitated the exploitation of the Third World (Latham 2000: 5).

An American Marxist, Paul Baran (1957), initiated dependency theory, which then grew into an influential anti-modernization paradigm. Baran argued Third

World underdevelopment, and First World developments were interrelated processes. He argued the underdeveloped peripheries remained locked into socioeconomic relationships with the First World core because Western imperialism effectively continued after the end of the European colonial era. From this grew the idea that underdevelopment was caused by the decisions made by First World capitalists. Hence, Third World backwardness was deemed to be caused by Third World dependency on the First World – i.e., First World development/growth and affluence was seen to be the result of capitalist exploitation in which wealth was being transferred from the Third to First Worlds. The most sophisticated formulation of this argument was developed by a non-Marxist, Immanuel Wallerstein (1980), who discussed how the world had been enmeshed into a single (capitalist) economic system – a globalization process that had begun in the fifteenth century when European imperialism got underway. Wallerstein's world systems theory proposed equal development was impossible within this global capitalist system because the system encoded unequal exchange between the core and the periphery. Dependency theory became very influential in Latin America, where it was deployed by intellectuals like Andre Gunder Frank (1969).

The Latin American dependency theorists argued modernization taken to the Third World by the development industry was bad because it caused underdevelopment on the peripheries and generated growing inequalities in Third World countries. These dependency theorists argued if Third World countries wanted real development, they needed to cut themselves off from America's global trading network and achieve self-reliance by developing themselves. This led to the call for a New International Economic Order (NIEO) which, if implemented, would have seriously disrupted the Pax Americana's global trading network. Dependency theory and NIEO consequently became enmeshed in the Cold War competition for influence and control over the Third World. From the 1970s onward, the expanding influence of dependency theory across the Third World became a growing problem for the Pax Americana.

The Cold War struggle between modernization and dependency theorists hinged on two opposing explanations about the causes of underdevelopment. The modernization theorists saw the problem as being *internal* to these underdeveloped states – i.e., they believed the indigenous people of these states lacked the appropriate cultural capital, skills, and know-how to effectively manage their own political and economic systems. Among other things, these deficiencies meant they were unable to accumulate capital. This lack of capital and skills was deemed the key obstacles to Third World growth and development. The development and aid industries proposed intervening from the outside to rectify these deficiencies. The dependency theorists, on the other hand, believed the problem was not internal to these states; rather the problem was *external* – it was caused by a global capitalist system that exploited Third World states. The dependency theorists did much to popularize the idea across the Third World that Western imperialism was responsible for pauperizing Third World peoples by destroying their indigenous economies.

The idea that globalized capitalism was responsible for pauperization and underdevelopment served as a very useful ideology for bringing together Marxists and

Third World nationalists during the Cold War era (Smith 1981: 82). Politicians in Africa, Asia, and the Middle East looking for an excuse for the poor performance of their countries found dependency theory invaluable. This theory helped them to deflect attention away from their own poor performance and fed into a culture of victimhood. The latter was used to argue the West was morally obliged to provide aid to the Third World because the West was responsible for their poverty. Ironically this served to create another form of dependency in which many Third World people learned not to take responsibility for their own lives, but instead waited for outsiders to come and "fix" things for them and even to feed them.

Dependency theory may have been correct in pointing to the interconnectedness of the global economy brought about by the building of globalized trade networks by Britain and America. However, when dependency theorists grafted onto this the notion the West could be blamed for pauperizing the Third World, they were simply constructing an ideology useful for political mobilization. This ideology was only tenable if contradictory evidence was ignored:

- Labor rates were not necessarily worse in the Third World (Leys 1996: 10), especially when one factored in the often low labor productivity rates in parts of the Third World caused by cultural capital deficits.
- Poverty and underdevelopment existed in parts of world largely untouched by the West (e.g., Tibet, Afghanistan, and Ethiopia) (Engerman et al. 2003: 67). Pauperization theory did not explain this.
- The behavior of Third World politicians and leaders was ignored as an important factor responsible for creating negative conditions that undermined development and created poverty (Smith 1981: 80).
- Capitalist economies require certain conditions for success, such as the security provided by good governance, a legal system protecting property rights, a labor force with an appropriate work ethic and skills, transport and communication infrastructures, etc. Many Third World contexts have not provided these conditions. This was precisely one of the reasons Britain's empire annexed many regions – i.e., to establish the conditions required for economic development and trade.
- Western imperialism did carry out much successful development across the globe – i.e., building infrastructures and creating many viable new economies. Significantly, a number of the economies and infrastructures they built collapsed *after* decolonization (which caused an increase in poverty in places like Africa).

Dependency theory had always serviced the alliance between Third World nationalists and Marxists. So when the Soviet Union collapsed, dependency theory fell on hard times and the NIEO idea became untenable. Third World leaders found themselves confronting the harsh new reality that the "Cold War leverage" American-Soviet competition had provided them had now evaporated. Washington was no longer prepared to overlook the corruption, poor governance, and even tyranny practiced by many of their Third World middle-class compradors. Washington had previously ignored this out of a fear that if they complained, compradors would simply realign themselves with the Soviet Union. When the Cold War ended, all Third World states found that aid as a mechanism to "buy allegiance" ended, plus they had no choice but to now negotiate new relationships with the Pax Americana. And what America now demanded was competent compradors who could manage their states so as to provide reliable trading partners. As a result, the 1990s saw many regime changes across the Third World to try and improve Third World governance. The 1990s also saw modernization put firmly back onto Washington's agenda (which was coupled with the discourse of encouraging "good governance"). These late twentieth century Washington policies would have been instantly recognizable to nineteenth century Britons who argued Britain's empire needed to intervene in African and Asia to improve conditions for trade.

### 8.5.3 Modernization Revived

The collapse of the Soviet Union and the rise of postmodern theory made Marxism unfashionable in intellectual circles, and so dependency theory, which had been the only serious anti-modernization theory, found itself out of favor.

It was Francis Fukuyama (1989) who heralded the rehabilitation of modernization theory when he proclaimed liberalism triumphant. Fukuyama effectively proclaimed the idea of the USA as the "universal state" which the rest of the world should emulate. Significantly, he also saw the spread of Western technology as the mechanism that "guarantees an increasing homogenization of all human societies" (Fukuyama 1995: 16). Naturally the end goal of this homogenization was the cloning of America and the universalizing of liberalism. And so the old idea of development as modernization (and liberalization) – entailing the diffusion of Western technology and skills – was revived.

However, by the 1990s liberal-capitalism had itself evolved. And one of the changes this evolution had brought about was that capitalist businesses and the liberal state increasingly outsourced much of their work. This principle of outsourcing also came to underpin the development industry; and the bodies to which "development work" was increasingly outsourced were NGOs. In fact, not only did Washington's traditional development and aid bureaucracies turn to NGOs for the delivery of their projects, but the US State Department and US Military also came to see NGOs as important partners in the execution of American foreign policy (Quarto 2005: 1–2, 6). This was expressed in the 2001 *Joint Doctrine for Civil-Military Operations* which laid procedures for joint American military-NGO operations. This shift to working with NGOs was to impact on the modernization paradigm, because many of those working in the NGO sector were on the left of the political spectrum, which meant they were sympathetic to one of more of the following: green politics, a liberalized version of dependency theory (which believed Western greed was responsible for Third World poverty), and participatory democracy. From this arose a new conceptualization of Third World interventionism – as articulated by the Swedish Dag Hammarskjold Foundation and German Green

Movement – called "another development" (Servaes 2003: 11). This meant although modernization theory had been revived, it was rearticulated to make it palatable to NGO partners who would be implementing it. The result has been a revitalized modernization paradigm in which some of the errors of the past are acknowledged and efforts are made to correct these, by incorporating the perspectives of "another development" (Servaes 2003).

The concept of "another development" (as promoted by the Dag Hammarskjold Foundation plus the German Greens) rejected the modernization idea that there was only one pathway to development (Servaes and Malikhao 2003: 7–10 and 11). Instead "another development" (drawing on the ideas of Sartre, Marx, Freire, and Jayaweera) advocated a rejection of top-down modernization and cultural imposition. Grounded instead in Paulo Freire's participatory communication model, "another development" pushed for whole new approach to social change and development in the Third World. As consequence the post-1990 Pax Americana development industry learned to run two different development formats side by side. On the one hand the revitalized modernization paradigm pushed ahead with top-down technological transfer, and the cultural change (Westernization) model geared to producing more reliable/competent compradors. But on the other hand, the Pax Americana has also funded "another development." However, it should be noted that in the era of the revitalized modernization paradigm "another development" has tended to be looked upon as an "idealistic side show" (Servaes and Malikhao 2003: 7–31) to the main game – which is to use top-down modernization (to try and speedily build competent comprador elites for the Pax Americana) by using an extensive system of NGO development agencies.

The effect has been to turn the NGO sector into something akin to the Pax Americana's Colonial Office. America still needs to build middle-class elites (to function as its comprador partners). Since the 1990s many of the old incompetent and corrupt Third world elites have been removed – some peacefully (e.g., Uganda) and others violently (e.g., Iraq). This meant a major exercise in training up new middle-class elites. Within this process the NGO sector has become central, because many Third World-based NGOs have become vehicles for simultaneously grooming and training new elites and for destabilizing regimes Washington deems illiberal. Consequently, since the 1990s, NGOs have been playing an important role in facilitating "upward mobility" into the ranks of the Pax Americana's new comprador partners (Petras and Veltmeyer 2001: 129) and for mobilizing mass popular "democratic" uprisings (i.e., "liberalization campaigns") against regimes Washington does not like (Petras and Veltmeyer 2001: 130). These NGOs have also helped disseminate the discourse of human rights – which effectively universalizes American middle-class values and American-style democracy (Petras and Veltmeyer 2001: 131).

Interestingly, the old and revived modernization paradigms share much in common since both, after all, service the same American informal empire. But there are also differences caused by Washington having learned (in an era when outsourcing is fashionable) to even outsource elements of its informal empire's governance to the NGO sector. The result is that:

> NGOs foster a new type of cultural and economic colonialism – under the guise of a new internationalism. Hundreds of individuals sit in front of high-powered PCs exchanging manifestos, proposals and invitations to international conferences with each other. They then meet in well-furnished conference halls to discuss the latest struggles and offerings with their "social base" – the paid staff – who then pass on the proposals to the "masses" through flyers and "bulletins". When overseas funders show up, they are taken on "exposure tours" to showcase projects where the poor are helping themselves and to talk with successful micro-entrepreneurs (omitting the majority who fail the first year). The way this colonialism works is not difficult to decipher. Projects are designed based on the guidelines and priorities of the imperial centres and their institutions. They are then "sold" to the communities. Evaluations are done by and for the imperial institutions. Shifts of funding priorities or bad evaluations result in the dumping of groups, communities, farmers and co-operatives. Everybody is increasingly disciplined to comply with the donors' demands and their project evaluators. The NGO directors, as the new viceroys, supervise the proper use of funds and ensure conformity with the goals, values and ideology of the donors. (Petras and Veltmeyer 2001: 132–133)

So now we have an outsourced "modernization and development" industry which helps to make Washington look even more removed from its informal empire. Much of the NGO-driven development process now appears so removed from Washington that many development missionaries inside the NGO sector are unaware of their role in promoting the Pax Americana, their role in diffusing the skills and values required for cloning democratic capitalism, or their role in "stabilizing" the peripheries for the Pax Americana (through the delivery of health care, Aids prevention, refugee management, etc.). But perhaps this is not so surprising – after all, many nineteenth century Christian missionaries would not have thought of themselves as the agents of imperialism either.

## 8.6 And What of the Future?

Development was born from the needs of the Pax Americana to create reliable and competent compradors, plus viable trading partners in order that its post-1945 informal empire would function. This means any changes to the Pax Americana would presumably have impacts upon development-as-idea and development-as-industry. And changes are in the wind as heralded by the 2016 Trump and Brexit revolutions. Both of these signal a growing antipathy to ideas that lie at the very core of the Pax Americana-as-informal-empire, namely, globalization, global free trade, multilateralism, multiculturalism, and global mass migration. It is not inconceivable that the Trump era may signal the beginning of the end for the Pax Americana as conceptualized by the Roosevelt team and Woodrow Wilson. Hence we see in the Trump administration little enthusiasm for global free trade, multilateralism, "democratizing" the Third World, Westernizing Third World elite, or development and Aid. Essentially many in the Trump administration are not keen on America running an informal empire. This does not mean an end of the Pax Americana.

Rather it means a redesign of the Pax Americana. Among the changes we already see are a realignment from multilateral to bilateral foreign policy, aid budgets being cut and aid budgets merged with traditional foreign policy budgets, a move away from "informal empire" toward more direct interventions from both the USA and EU, a move away from "indirect" methods (e.g., outsourced NGOs) toward direct state actions, and a decline in tolerance for poor Third World governance. Simultaneously, the West appears to be conceptualizing the Fourth Industrial Revolution as offering an alternative to the old "global division of labor" – which adds impetus to the possibilities for reordering (not ending) the Pax Americana. The implications of this for "development" and "Western aid" are obvious. So the future may not look like the past in the development sector – because as the old "informal empire" methodology is abandoned (thanks to new global economic and power relationships), there will be declining need for (and support of) the "development" of Third World compradors/economies.

## References

Bacevich AJ (2002) American empire. Harvard University Press, Cambridge, MA
Baran P (1957) The political economy of growth. Monthly Review Press, New York
Cox R (1987) Production, power and world order. Columbia University Press, New York
Engerman DC et al (2003) Staging growth. University of Massachusetts Press, Amherst
Fanon F (1968) The wretched of the earth. Grove Press, New York
Fieldhouse DK (1999) The west and the third world. Blackwell, Oxford
Frank AG (1969) Capitalism and underdevelopment in Latin America. Monthly Review Press, New York
Fukuyama F (1989) The end of history? The National Interest 16(Summer)
Fukuyama F (1995) On the possibility of writing a universal history. In: Melzer AM et al (eds) History and the idea of progress. Cornell University Press, Ithaca
Gallagher J (1982) The decline, revival and fall of the British empire. Cambridge University Press, Cambridge, MA
Goldsworthy D (1971) Colonial issues in British politics 1945–1961. Clarendon Press, Oxford
Gray J (2002) False dawn. Granata, London
Herbst J (2000) States and power in Africa. Princeton University Press, Princeton
Latham ME (2000) Modernization as ideology. University of North Carolina Press, Chapel Hill
Lerner D (1958) The passing of traditional society. Free Press, New York
Leys C (1996) The rise & fall of development theory. James Currey, London
Louw PE (2010) Roots of the Pax Americana. Manchester University Press, Manchester
Lugard FD (1922) Dual mandate in British tropical Africa. W. Blackwood & Sons, Edinburgh
Mandelbaum M (2002) The ideas the conquered the world. Public Affairs, Oxford
Metz S (1984) American attitudes towards decolonization in Africa. Polit Sci Q 99(3), 515–533
Onuf PS (2000) Jefferson's empire. University of Virginia Press, Charlottesville
Owen N (1999) Critics of empire in Britain. In: Brown JM, Louis WR (eds) The Oxford history of the British empire, vol 4. Oxford University Press, Oxford
Petras J, Veltmeyer H (2001) Globalization unmasked. Fernwood Publishing, Halifax
Quarto FC (2005) US Military/NGO interface: a vital link to successful humanitarian intervention. Unpublished MA dissertation, US Army War College, Pennsylvania

Roosevelt FD (1944) Excerpts from the Press Conference for the Negro Newspaper Publishers Association, February 5. http://www.presidency.ucsb.edu/ws/?pid=16453
Rostow WW (1960) The stages of economic growth: a non-communist manifesto. Cambridge University Press, Cambridge, MA
Ryn CG (2004) America the virtuous. Transaction Publishers, New Brunswick
Servaes J (2003) Approaches to development. Unesco, Paris
Servaes J, Malikhao P (2003) Development communication approaches in international perspective. In: Servaes J (ed) Approaches to development. Unesco, Paris
Shepherd R (1994) Iain Macleod. Hutchinson, London
Smith T (1981) The pattern of imperialism. Cambridge University Press, Cambridge, MA
Underhill GRD (2000) Global money and the decline of state power. In: Lawton T et al (eds) Strange power. Ashgate, Aldershot
Wallerstein I (1980) The modern world system, 2 vols. Academic, New York, 1974

# Part III
# Normative Concepts

# Media and Participation

## Nico Carpentier

## Contents

9.1  Introduction .................................................................................. 196
9.2  Theoretical Approaches to Participation ................................................ 196
9.3  The Many Locations of Media Participation (Analysis) ............................... 199
    9.3.1  Audience Theory and the Active/Passive Dimension ........................... 199
    9.3.2  Marxist and Anarchist Media Studies and the Media Participation Debate ..... 200
    9.3.3  Deliberation and the Public Sphere ............................................. 201
    9.3.4  UNESCO, Communication Rights and Development Communication ......... 203
    9.3.5  Participation in Specific Media Technologies, Organizations, and Genres ...... 204
9.4  Conclusion .................................................................................. 211
References ........................................................................................ 212

### Abstract

The wide variety of meanings attributed to the concept of participation both enables and complicates its academic deployment. The first part of the chapter provides a roadmap through this diversity, by distinguishing between two main theoretical approaches – the sociological approach, which defines participation as taking part, and the political (studies) approach, which sees participation as sharing power. Grounded in the political approach, the second part of the chapter engages with a series of theoretical and research-based subfields within Media and Communication Studies. One overview, of the use of participation in audience studies, Marxist and anarchist media studies, deliberation and public sphere approaches, and development/international communication, demonstrates the

---

N. Carpentier (✉)
Uppsala University, Uppsala, Sweden

Vrije Universiteit Brussel (VUB), Brussels, Belgium

Charles University, Prague, Czech Republic
e-mail: nico.carpentier@im.uu.se

theoretical diversity in participation studies. A second overview, of the analyses of participatory practices in three locations – community and alternative media, television talk shows and reality TV, and online media/Internet studies, then shows the differences in participatory intensities and their normative evaluations.

> **Keywords**
>
> Participation · Power · Sharing power · Taking part · Audience · Marxism · Anarchism · Deliberation · Public sphere · Development communication · International communication · Community media · Alternative media · Television talk show · Reality TV · Online media

## 9.1 Introduction

Within Media and Communication Studies, participation re-established its presence from the turn of the century onward, mostly driven by the need to theorize user practices in relation to online media, and in particular Web 2.0 and social media. Even if participation appears to be a straightforward and easy-to-use concept, its political-ideological nature and its complex set of overlapping histories require a more careful unpacking of the concept.

In this chapter, the two main theoretical approaches – the sociological approach (participation-as-taking-part) and the political (studies) approach (participation-as-sharing-power) – will be mapped out, both in general and in relation to the media field. This will allow in showing the theoretical richness of participation studies, but also the choices that need to be made when engaging in the analysis of participatory practices. Grounded in the political approach, the second part of this chapter will then engage with a series of theoretical and research-based subfields – within Media and Communication Studies – in order to show the different articulations of participation in a series of theoretical subfields and the different participatory intensities and their different evaluations in three locations of media participatory practice.

## 9.2 Theoretical Approaches to Participation

This first part outlines the different theoretical approaches to participation and reuses text that has been published before in Carpentier (2016). The starting point of this overview is that the literature on participation, including media and participation, has produced many different positions (see, e.g., Jenkins and Carpentier (2013) and Allen et al. (2014) for two recent media-related debates). Arguably, two main approaches to participation can be distinguished in these debates: a sociological approach and a political (studies) approach (see also Lepik 2013). The sociological approach defines participation as taking part in particular social processes, a definition which casts a very wide net. In this approach, participation includes many (if not all) types of human interaction, in combination with interactions with texts and technologies. Power is not excluded from this approach, but remains one of the many secondary concepts to support it. One example of how participation is defined in this approach is Melucci's (1989: 174) definition, when he says that participation has a double meaning: "It means both taking part, that is, acting so as to promote the interests and the needs of an actor as well as belonging to a system, identifying with the 'general interests' of the community."

The sociological approach results, for instance, in labeling consumption as participatory, because consumers are taking part in a consumption culture and are exercising consumer choices (Lury 2011: 12). Also for doing sports, the label of participation is used, as exemplified by Delaney and Madigan's (2009) frequent use of the participation concept in their introduction into the sociology of sports. We can find a similar approach in what is labeled cultural participation, where participation is defined as individual art (or cultural) exposure, attendance, or access, in some cases complemented by individual art (or cultural) creation. As Vander Stichele and Laermans (2006: 48) describe it: "In principle, cultural participation behaviour encompasses both public and private receptive practices, as well as active and interactive forms of cultural participation." In practice, this implies that the concept of participation is used for attending a concert or visiting a museum.

Within media studies, the sociological approach can, for instance, be found in how Carey (2009: 15) defines the ritual model of communication in *Communication as Culture*, as the "representation of shared beliefs," where togetherness is created and maintained, without disregarding the many contending forces that characterize the social. For Carey, the ritual model of communication is explicitly linked to notions of "'sharing', 'participation', 'association', 'fellowship' and the 'possession of a common faith'" (2009: 15), where people are (made) part of a culture through their ritualistic participation in that very same culture. (Mass) media, such as newspapers (used by Carey as an example), play a crucial role by inviting readers to participate in a cultural configuration, interpelating them – to use an Althusserian concept – to become part of the society by offering them subject positions or, as Carey (2009: 21) puts it, social roles, with which they can identify (or dis-identify). This type of ritual participation (Interestingly, Carey (2009) does not use the concept of ritual participation in *Communication as Culture*. He does use "ritual of participation" (2009: 177), which refers to a very different process, namely, the emptying of the signifier participation as an elitist strategy. This use of the participation concept, mainly to be found in Chapter Seven of *Communication as Culture* ("The History of the Future," co-authored with John J. Quirk), is much more aligned with the political approach toward participation.) again defines participation as taking (and becoming) part, through a series of interactions, with – in Carey's case – media texts. Others have also used the ritual participation concept (and the sociological approach to participation it entails), for instance, in relationship to media (Real 1996; Dayan and Katz 2009: 120).

In contrast, the political approach produces a much more restrictive definition of participation, which refers to the equalization of power inequalities, in particular decision-making processes (see Carpentier 2011; Carpentier et al. 2014). Participation then becomes defined as the equalization of power relations between privileged and non-privileged actors in formal or informal decision-making processes. For instance, in the field of democratic theory, Pateman's (1970) *Participation and Democratic Theory* is highly instrumental in showing the significance of power when introducing the two definitions of partial and full participation (Pateman 1970: 70–71). Also in the field of urban planning, Arnstein (1969: 216) in her seminal article *A Ladder of Citizen Participation* links participation explicitly to power, saying "that citizen participation is a categorical term for citizen power."

The political approach also emphasizes that participation is an object of struggle and that different ideological projects (and their proponents) defend different participatory intensities (One complication is that the concept of participation itself is part of these power struggles, which renders it highly contingent. The signification of participation is part of a "politics of definition" (Fierlbeck 1998: 177), since its specific articulation shifts depending on the ideological framework that makes use of it.). More minimalist versions of participation tend to protect the power positions of privileged (elite) actors, to the detriment of non-privileged (non-elite) actors, without totally excluding the latter. In contrast, more maximalist versions of participation strive for a full equilibrium between all actors (which protects the non-privileged actors).

The more restrictive use of the notion of participation in the political approach necessitates a clearer demarcation of participation toward a series of related concepts that are, in the sociological approach, often used interchangeably. One key concept is engagement (Despite its importance, this will not be used in this chapter in order not to complicate things too much.), which Dahlgren (2013: 25) defines as the "subjective disposition that motivates [the] realization [of participation]," in order to distinguish it from participation. In earlier work, Dahlgren (2009) argues that the feeling of being invited, committed, and/or empowered and also the positive inclination toward the political (and the social) are crucial components of engagement. In his civic cultures circuit, Dahlgren also emphasizes (apart from more materialist elements like practices and spaces) the importance of knowledge, trust, identities, and values for (enhancing) engagement. Engagement is thus different from participation (in the political approach) as engagement refers to the creation, or existence, of a social connection of individuals or groups with a broader political community, which is aimed at protecting or improving it.

Other related, but still distinct, concepts are access and interaction. In earlier work, I have argued that access refers to the establishment of presence and interaction to the creation of socio-communicative relations. As a concept, access is very much part of everyday language, which makes clear definitions rather rare. At the same time, access – as a concept – is used in a wide variety of (academic) fields, which we can use to deepen our understanding of this concept. One area where access is often used is geography, when the access to specific spaces and places is thematized. More historical (spatial) analyses, for instance, deal with access to land and the enclosure of the common fields (Neeson 1996). The importance of presence for defining access can also be illustrated through a series of media studies examples: in the case of the digital divide discourse, the focus is, for instance, placed on the access to (online) media technologies, which in turn allows people to access media content. In both cases, access implies achieving presence (to technology or media content). Access also features in the more traditional media feedback discussions, where it has yet another meaning. Here, access implies gaining a presence within media organizations, which generates the opportunity for people to have their voices heard (in providing feedback).

A second concept that needs to be distinguished from participation is interaction. If we look at the work of Argentinean philosopher Bunge (1977: 259), we can find the treacherously simple and general definition of interaction "two different things x and y interact if each acts upon the other," combined with the following postulate: "Every thing acts on, and is acted upon by, other things." Interaction also has a long history in sociological theory, where it often refers to the establishment of socio-communicative relationships, as was already mentioned. An example can be found in Giddens's (2006: 1034) definition of social interaction in the glossary of *Sociology*, where he defines social interaction as "any form of social encounter between individuals." A more explicit foregrounding of the socio-communicative can be found in Sharma's (1996: 359) argument that the "two basic conditions of social interaction" are "social contact and communication." While the social dimension of the definition of interaction can be found in concepts like contact, encounter, and reciprocity (but also [social] regulation), the communicative dimension is referred to by concepts such as response, meaning, and communication itself.

## 9.3 The Many Locations of Media Participation (Analysis)

The analysis of media participation has a diverse theoretical background, combining – among others – media theory, political theory and philosophy, and development theory (This part is a shortened and updated version of Carpentier (2011).). The specificities of each of these theoretical fields have had their impact on the concept of participation and how it was deployed in research practice. Moreover, research into media participatory practices also brought in their specificities, given, for instance, the different participatory intensities that characterized these locations. This part of the chapter first discusses four more theoretical (sub)fields – audience theory, Marxist-anarchist media theory, public sphere theory, and development theory (combined with UNESCO debates) – making sure that also the frameworks that include more maximalist participatory approaches are represented. Then this part of the chapter turns the attention to three locations of media research which have prominently featured media participation: community and alternative media, television talk shows and reality TV, and online media.

### 9.3.1 Audience Theory and the Active/Passive Dimension

There are many approaches to structuring how the concept of audience is theorized and a "totalizing account [is] a logical impossibility" (Jenkins 1999). Nevertheless, Littlejohn's (1996: 310) *Theories of Human Communication* offers a good starting point when he writes that "disputes on the nature of the audience seem to involve two related dialectics. The first is a tension between the idea that the audience is a mass public versus the idea that it is a small community. The second is the tension between the idea that the audience is passive versus the belief that it is active."

In particular the first dimension – the active/passive dimension – is of importance here. The assumption of the human subject as an active carrier of meaning is already echoed in the development of Eco's (1968) aberrant decoding theory and Hall's 1973 encoding/decoding model (published in 1980). The concept of the active audience (see, e.g., Fiske 1987) emanated from these models. In addition, uses and gratifications theory of (among others) Katz et al. (1974) and deduced models, for example, Palmgreen and Rayburn's (1985) expectancy-value theory and Renckstorf et al.'s (1996) social action model, all rely to a large degree on the concept of the active audience (Livingstone 1998: 238).

But this "traditional" active/passive dimension often takes an idealist position by emphasizing the active role of the individual viewer in the processes of signification. This position risks reducing social activity to such processes of signification, excluding other – more materialist – forms of human practice. In other words, the active dimension hides another dimension, termed here the participation/interaction dimension. The interaction component of audience activity refers to the processes of signification and interpretation triggered by media consumption. Obviously, polysemic readings of media texts are an integrative part of this component. But also work on identity, where audiences engage with the media texts offered to them, is included in the interaction component of audience activity.

The participatory component of audience activity refers to two interrelated dimensions which can be termed participation in the media and through the media, inspired by Wasko and Mosco (1992: 7) who distinguish between democratization in and through the media. Participation *through* the media deals with the opportunities for mediated debate and for self-representation in the variety of public spaces that media sphere serves as a location where citizens can voice their opinions interacting with other voices. Participation *in* the media deals with participation in the production of media output (content-related participation) – captured by concepts such as user-generated content (UGC) and produsage (Bruns 2008) – and in media organizational decision-making (structural participation). These forms of media participation allow citizens to be active in one of the many (micro-)spheres relevant to daily life and to put into practice their right to communicate.

### 9.3.2 Marxist and Anarchist Media Studies and the Media Participation Debate

Marxist theory – in its broad sense – directly or indirectly has contributed to media studies in a wide variety of ways. Wayne's (2003) *Marxism and Media Studies* and Fuchs's (2016) *Reading Marx in the Information Age* are good examples of the direct application of the Marxist toolbox, but Marxist theory has also played a key role through its integration into the political economy of communication (Fuchs and Mosco 2016) and cultural media studies (Ogasawara 2017). One of the main concerns of political economy approaches is related to the colonization of public spaces, where the (growing) domination of corporate power in the communication industry is deemed problematic for media production, distribution, content, and reception. Participation in the media becomes blocked by the communication industry's market logics and focus on professional employment, but participation through the media is also hampered by the media's circulation of dominant ideologies that continue to serve the interests of the dominant class (Wayne 2003: 175). In cultural media studies, we find similar concerns, elaborated more through a mixture of (post)structuralist and (post)Marxist theory. Here, the focus is on the hegemonizing capacities of media and the (potential) diversity of audience interpretations. An early example appears in *Policing the Crisis: Mugging, the State, and Law and Order* (Hall et al. 1978), where Hall and his colleagues research the moral panics caused by the appearance of a "new" form of criminality (mugging) and the way that it supported a dominant societal (repressive) order.

Also from an anarchist theory perspective, there have been some substantial contributions to the media and participation debates owing to the fact that the focus of anarchist theory is not always on the state as such. The more Chomskian strands of anarchist theory have incorporated the vitriolic critiques of the mainstream media system, although even alternative media come in for some criticism, as Bradford's (1996: 263) analysis of pirate radio suggests. Some authors have managed to incorporate anarchist theory in subtler ways, with participation and self-management featuring prominently. Downing et al. (2001: 67 ff), in their book entitled *Radical Media*, distinguish two models for the organization of radical media: the Leninist model and the self-management model. Downing et al. (2001: 69) explicitly relate the latter – where "neither party, nor labor union, nor church, nor state, nor owner is in charge, but where the newspaper or radio stations runs itself" – to what they call a "socialist anarchist angle of vision." Another author that should be mentioned here is Hakim Bey – the pseudonym used by Peter Lamborn Wilson – who, in his essay *Temporary Autonomous Zone* (TAZ) (1985), reflects on the upsurge (and disappearance) of temporary anarchist freespaces. Distinguishing in this essay between the net and the web, he sees the net as the "totality of all information and communication transfer" (Bey 1985: 106), while the web is considered a counter-net that is situated within the net.

### 9.3.3 Deliberation and the Public Sphere

Communication plays a key role in the deliberative democratic model and in Habermas's model of the public sphere, described in Habermas's (1996: 360) definition of the public sphere as "a network for communicating information and points of view." It is no surprise that these models play a prominent role in theorizing the connection between participation and communication. First, participation in the public sphere is seen as an important component, since it relates to the basic assumptions that characterize the communicative action that takes place within the public sphere, and where "participants enter into interpersonal relationships by taking positions on mutual speech-act offers and assuming illocutionary obligations" (Habermas 1996: 361). But, in Habermas's two-track model of deliberative politics, there is also a strong emphasis on the connection of the public sphere to realities external to it and on participation through the public sphere.

However, we should not forget that Habermas's (1991) older work on the public sphere – *The Structural Transformation of the Public Sphere*, first published in German in 1962 – is a historical-sociological account of the emergence, transformation, and disintegration of the bourgeois public sphere. But later, Habermas (1996: 360–362) more exclusively emphasizes the formal-procedural nature of the public sphere (instead of conflating it with its historical origins), while maintaining a focus on communicative rationality and consensus in which the public sphere is seen as a social space generated through communicative action. The public sphere, positioned as part of the two-track model of deliberative politics, remains for Habermas a crucial site of participation and communication, which simultaneously "relieves the public of the burden of decision making; the postponed decisions are reserved for the institutionalized political process" (Habermas 1996: 362 – emphasis removed).

Even if Habermas currently remains the most significant theoretical voice in the study of the public sphere, a variety of public sphere theories and models have been developed (Dewey 1927; Arendt 1958; Negt and Kluge 1983; Fraser 1990; Hauser 1998) which have – together with Habermas's work – been used in a variety of ways in Media and Communication Studies. One important dimension is the more normative versus the neutralized-descriptive usages of the public sphere model, where the latter has – following Nieminen (2006) – two sub-approaches: a normative-prescriptive sub-approach and a historical-sociological sub-approach. In the normative-prescriptive sub-approach, the public sphere is a regulative idea, "an ideal which may never be fully realized but which can act as a normative framework for critical evaluation" (Nieminen 2006: 106). Equally normative, the historical-sociological sub-approach focuses on the "historical and sociological (pre)conditions of the phenomenon we call the [...] public sphere" (Nieminen 2006: 107). Characteristic of the normative approach – and its two sub-approaches – is that the public sphere is articulated with a concern for the democratic-participatory process, where participation, in the Habermasian version, is "governed by the norms of equality and symmetry; all have the same chances to initiate speech acts, to question, to interrogate, and open debate" (Benhabib 1994: 31). There are also neutralized-descriptive approaches to the public sphere, which do not use an explicit normative framework. In these approaches, the public sphere concept is still used to demarcate "a network for communicating information and points of view" (Habermas 1996: 360), but they do not attribute specific democratic characteristics to the public sphere, which makes them less relevant for participation studies.

### 9.3.4 UNESCO, Communication Rights and Development Communication

A fourth location where media participation is discussed intensively is located at the intersection of international and development communication. As Servaes (1999: 83) argues, several Latin American scholars, such as Beltran, Bordenave, and Martin-Barbero, in the 1970s discussed the role of participatory communication as a tool to create a more just world. Indicative of this is Bordenave's (1994) article *Participative Communication as a Part of Building the Participative Society*, in which he defines participatory communication as "that type of communication in which all the interlocutors are free and have equal access to the means to express their viewpoints, feelings and experiences" (Bordenave 1994: 43). This re-articulates communication as "a two way process, in which the partners – individual and collective – carry on a democratic and balanced dialogue" (MacBride Commission 1980: 172). Freire's (1992) theories, in particular, have had a considerable impact on this domain, as Thomas (1994: 51) remarks.

The debate on participation and development moved onto the global stage in the 1970s when the struggle over the New International Economic Order (NIEO) and New World Information and Communication Order (NWICO) began within UNESCO. For instance, the 1976 General Conference approved the *Recommendation on Participation by the People at Large in Cultural Life and their Contribution to it* (UNESCO 1977), which contained a section on audience participation that stated that:

> Member States or the appropriate authorities [should] promote the active participation of audiences by enabling them to have a voice in the selection and production of programmes, by fostering the creation of a permanent flow of ideas between the public, artists and producers and by encouraging the establishment of production centres for use by audiences at local and community levels. [...]

In December 1977, the 16-member International Commission for the Study of Communication Problems was established, headed by Sean MacBride. Its report (MacBride Commission 1980), *Many Voices, One World. Towards a New More Just and More Efficient World Information and Communication Order*, took a strong position on audience participation. The chapter on the Democratization of Communication describes the following four approaches to breaking down the barriers to the democratization of communication, with the objective to strengthen the so-called right to communicate:

> (a) broader popular access to the media and the overall communication system, through assertion of the right to reply and criticize, various forms of feedback, and regular contact between communicators and the public [...]; (b) participation of non-professionals in producing and broadcasting programmes, which enables them to make active use of information sources, and is also an outlet for individual skill and sometimes for artistic creativity; (c) the development of 'alternative' channels of communication, usually but not always on a local scale; (d) participation of the community and media users in management and decision-making (this is usually limited to local media). Self-management is the most

radical form of participation since it presupposes an active role for many individuals, not only in the programmes and news flow, but also in the decision-making process on general issues. (MacBride Commission 1980: 169)

The debates on participation and the right to communicate continued in the 1980s and featured in a number of general conferences, but the concept of the right to communicate (almost) received its coup de grace when the USA (1984) and the UK (1985) pulled out of UNESCO (for the first time) (Jacobson 1998: 398). During the 1990s, the right to communicate disappeared almost completely from UNESCO's agenda (and from the agendas of other international organizations), with the exception of forums such as the MacBride Round Tables (Hamelink 1997: 298). Only in 2003, in the slipstream of the UN WSIS (http://www.itu.int/wsis), was the debate on communication rights reinvigorated (For more recent publications on communication rights, see, e.g., Dakroury et al. (2009), Raboy and Shtern (2010), and Padovani and Calabrese (2014).), in part, thanks to initiatives such as the Communication Rights in the Information Society Campaign (CRIS) (http://www.crisinfo.org). However, in the final texts of the summit meetings, participation played only a minor role; it received minimalist significations and did not feature very often. More maximalist meanings of participation, linked to communication rights, were missing, which provoked rather skeptical evaluations: "A surge, at least in the short term, in the political will to incorporate such rights in a new international declaration is unlikely, regardless of the WSIS document's recommendations about the need to respect human rights" (Mansell and Nordenstreng 2006: 30–31). This does not mean that participation and communication rights were erased from the summit. The WSIS Civil Society Plenary (2003), which published an "alternative" declaration, uses the concepts of full participation and empowerment and elaborates on communication rights, providing shelter for the more maximalist articulations of participation.

### 9.3.5 Participation in Specific Media Technologies, Organizations, and Genres

In this last part, the attention is turned to three (academic) debates on participatory media practices (and to the articulations of participation they contain), since these debates have generated a considerable literature on media participation: community and alternative media, talk shows and reality TV, and online media. All three cases of media practices provide unique perspectives on the notion of (media) participation in different contexts, different periods, and different participatory intensities.

#### 9.3.5.1 Community and Alternative Media
The discussion of the role of participation in the NWICO debates contained a considerable number of references to community and alternative media, and it is difficult to ignore their key role in participatory theory and practice. At the same time, their diversity makes them difficult to capture: They can take many different organizational forms and use various technological platforms (print, radio, TV,

web-based, or mixed). As Atton (2015: 8 – emphasis in original) wrote in his *Introduction* to *The Routledge Companion to Alternative and Community Media*: "[...] the diversity of alternative media presented in this *Companion* shows how content will always be linked with social, cultural and political contexts."

Despite their differences, community and alternative media share a number of key characteristics, which distinguish them from other types of (mainstream) media organizations like public service or commercial media. Their close connection to civil society and their strong commitment to maximalist forms of participation and democracy, in both their internal decision-making process and their content production practices, are especially important distinguishing characteristics that establish community and alternative media as the third media type, distinct from public service/state and commercial media.

One way to capture their diversity and understand what unites them is to combine the four approaches that have been used in the literature for the study of community media (discussed in Carpentier et al. 2003; see also Bailey et al. 2007; Carpentier 2011). Support for this position can be found in Atton's (2002: 209) argument that "This encourages us to approach these media from the perspective of 'mixed radicalism', once again paying attention to hybridity rather than meeting consistent adherence to a 'pure,' fixed set of criteria [...]." Taken together, the four approaches to community and alternative media allow their complexity and rich diversity of to be unveiled, together with the role of participation:

- The community approach focuses on access by, and participation of, the community, the opportunity given to "ordinary people" to use media technologies to have their voices heard, and the empowerment of community members through valuing their skills and views.
- The alternative approach stresses that these media have alternative ways of organizing and using technologies, carry alternative discourses and representations, make use of alternative formats and genres, and remain independent from market and state.
- The civil society approach incorporates aspects of civil society theory to emphasize that citizens are being active in one of many (micro-) spheres relevant to everyday life, using media technologies to exert their rights to communicate.
- Finally, the rhizomatic approach uses Deleuze and Guattari's (1987) metaphor to focus on three aspects: community media's elusiveness, their interconnections (among each other and, mainly, with civil society), and the linkages with market and state. In this perspective, community media are seen to act as meeting points and catalysts for a variety of organizations and movements.

Participation features in all four approaches, as the community that is being served through the facilitation of its participation, as the provision of a maximalist participatory alternative to nonparticipatory (or minimalist participatory) mainstream media, as the democratic-participatory role of civil society, or as the participatory rhizome. Tabing's (2002: 9) definition of a community radio station as "[...]

one that is operated in the community, for the community, about the community and by the community" makes clear that participation in the community (and alternative) media organization is not only situated at the level of content production but is also related to management and ownership or to use Berrigan's (1979: 8) words, community media "[...] are the means of expression of the community, rather than for the community."

### 9.3.5.2 Television Talk Shows and Reality TV

Within mainstream media, a series of genres and formats have allowed for a certain degree of participation by ordinary people. It should be emphasized immediately that participation in this context is structurally limited, as mainstream media only rarely allow for structural participation (or participation within the media organization's decision-making structures themselves). Moreover, mainstream media have a variety of objectives, and the organization of societal participation and audience empowerment is not always part of their primary objectives (despite some authors protesting that it should – see Keane 1998). Nevertheless, mainstream media remain significant societal players that merit our attention, also because the achievements and failures of the participatory processes they organize can be very enriching to the debates on media participation. In this section I want to focus on two audiovisual genres – talk shows and reality TV – in the knowledge that other genres (in print and in audiovisual media) provide equally relevant (albeit slightly less well-discussed) examples.

The talk show genre does facilitate participation within the mainstream media again to a certain extent, but the existence of many subgenres complicates this discussion. As is often the case with media genres, this label clusters together a wide diversity of actual program formats. They have the common core element of a host talking to people in a studio setting (Leurdijk 1997: 149) and are aimed at a mixture of entertainment and information; as Hallin (1996: 253 – emphasis in original) puts it, "These new forms of media are often described as providing 'unmediated' communication. This is clearly not accurate: they are *shows*, often carefully scripted, each with its own logic of selection and emphasis." One subgenre, the audience participation talk show or audience discussion program, is of particular interest here. This subgenre finds its origins (partially) in what Munson (1993: 36–37) calls the "interactive talk radio" format, the call-in or the phone-in. These talk shows successfully transferred to television and became in the 1990s a popular subgenre (Iannelli 2016: 26ff). Despite their intra-genre differences, all these programs were based on the principle that "an active role is accorded to the studio audience which participates in a discussion about social, personal or political problems under the supervision of a presenter" (Leurdijk 1999: 37, my translation).

Audience discussion programs (and other talk show subgenres with participatory components) have provoked many and very different evaluations. In the academic literature on these talk shows, two main approaches can be distinguished: the emancipation and the manipulation approaches. The emancipation approach argues that these programs contribute to the democratization of the mass media (Hamo 2006: 428) since they provide access for ordinary people to public spaces and allow them to participate in the production of media content. Not surprisingly, the concept of the public sphere is often deployed here (e.g., Carpignano et al.

1990). Livingstone and Lunt (1996: 19) refer to the participants in audience discussion programs as citizen-viewers, who are "seen as participating, potentially at least, in democratic processes of the public sphere."

In contrast to the emancipation approach, several authors highlight the manipulative or pseudo-participatory nature of these programs. One line of argument focuses on the production context, which does not escape the processes of commodification. Their discursive diversity is seized on by some authors as a point of criticism, in which the absence of a rational discussion that results in a critical consensus (à la Habermas) leads to a risk of trivialization of the performed utterances (Priest 1995: 17). Livingstone and Lunt (1996: 175) argue: "It remains problematic that giving voice may not affect real decision making and power relations in society, but only give the illusion of participation." In addition, the power imbalances within these programs, with media professionals unavoidably playing a significant role in organizing the participation in a context that is "theirs" to control, are approached critically. For instance, Leurdijk (1997), White (1992), Tomasulo (1984), and Gruber (2004) describe how participants lose control over the narration of personal stories because editorial teams try to orchestrate, canalize, structure, and/or manage the debate.

The second genre, reality TV, became very popular in the 1990s and 2000s, but also has a long history going back to such programs as Candid Camera (1948). As a genre, it is based on the construction of both people and situations as real. In other words, ordinary people are featured prominently (although their presence in these programs is sometimes unplanned) and are placed in situations related to everyday life. The claim to reality, which is supporting the genre, is translated into a series of visual strategies, through the use, for instance, of fly-on-the-wall camera techniques, which, in turn, are supported by technological evolutions such as the lightweight camera and the possibilities for audiences to create their own content. Nevertheless, the genre spans a very large group of structurally very different programs, which has led some authors to call reality TV a trans-genre (see Van Bauwel and Carpentier 2010).

The combination of reality TV's reality claim, its focus on the ordinary and the everyday, and the management by media professionals who control many of reality TV's production aspects also makes it a highly relevant genre in relation to the discussion on media participation. Again, in this debate we can find approaches that focus on the concept of emancipation, while the more critical approaches point to intervention, manipulation, and pseudo-participation. First, reality TV provides ordinary people access to the TV sphere or to the machineries of mediation that render their existence, practices, and utterances visible to an outside world and (in some cases) allows them to acquire celebrity status (Biressi and Nunn 2005: 148). As Andrejevic (2004: 215) phrases it, "The promise of reality TV is not that of access to unmediated reality [...] so much as it is the promise of the access to the reality of mediation."

Of course, all is not well with the participatory process of reality TV programs, as pointed out in the manipulation approach. Even if some ordinary people are granted access to the TV sphere and the TV screen, the kinds of presences, practices, and discourses they are allowed to generate are questionable, and the levels of interaction and participation are often considered problematic. One of the harshest critics is Andrejevic (2004: 215), who claims that reality TV might result not in the demystification of TV but rather in the fetishization of TV. A crucial factor limiting the participatory intensity of reality TV is the specific position of its media professionals and the skewed power balance between them and the ordinary participants. This type of argument can be found in Turner's (2010: 46) critique on the democratainment concept, which, Turner says, "over-estimates the power available even to these newly empowered [...] citizens." Of course, the broad context of the commodified media sphere creates a context in which (some) ordinary people are transformed into what Rojek (2001) calls "celetoids," or people whose public careers cater to the interests of the media industry itself. In practice, this implies that media professionals exert strong levels of control over the participants and their actions. Participants are invited into these program contexts and then find themselves exposed to this heavy management, which is legitimized through the (psychological and legal) ownership of the program by the production team.

### 9.3.5.3 Online Media/Internet Studies

The arrival of another generation of so-called "new" media drastically affected the nature of the discussion on participation and the media. From the 1990s onward in particular – and in some cases earlier (for instance, Bey's *TAZ* [1985]) – the focus of theoreticians of participation and audience activity shifted toward online media. The development of the Internet, and especially the web – the focal point of this text – was to render most information available to all and to create a whole new world of communication, the promise of a structural increase in the level of (media) participation, within its slipstream, extending to the more maximalist versions of participation.

These academic debates on online media and participation contain a wide variety of articulations of the key concepts of access, interaction, and participation. Ordinary users are seen to be enabled (or empowered) to avoid the mediating role of the "old" media organizations and publish their material (almost) directly on the web, on self-managed websites and blogs, or on social media. These novel practices have affected discussions over access and participation in a fundamental way. In a first (pre-Web 2.0) phase, the two key signifiers of access and interaction gained dominance, although participation did not (completely) vanish from the theoretical scene. Later, the concept of participation made a remarkable comeback to reach a prominent position in the 2000s.

In the 1990s (and in some cases before then), the importance of access increased structurally, as techno-utopianism emphasized access for all, to all information, at all times (Negroponte 1995). This argument had an explicit political component since the increased potential of access to the public sphere was also emphasized. The potentially beneficial increase in information, which challenges the "existing political hierarchy's monopoly on powerful communications media" (Rheingold 1993: 14), the strengthening of social capital and civil society, and the opening up of a new public sphere, or a "global electronic agora" (Castells 2001: 138) began to take primacy. In turn, increased access was seen as affecting subject positions. To use Poster's (1997: 213) words, "the salient characteristic of Internet community is the diminution in prevailing hierarchies of race, class, age, status and especially gender." The critical backlash to these and other rather bold statements focused on the lack of access of some, foregrounding the notion of a digital divide, but still remaining within the realm of access. At the same time, in the pre-Web 2.0 phase of the Internet, the notion of interaction and the derived concept of interactivity played a significant role in discourses about online media, much more than participation did. For instance, in Rheingold's (1993) summary of new media consequences – supporting citizen activity in politics and power, increased interaction with diverse others, and a new vocabulary and form of communication – interaction is featured prominently.

The mainstream pre-Web 2.0 approach favored interaction over participation, implicitly reducing the intensity of participation, but the concept of participation still managed to maintain a presence in a number of academic subfields. The theoretical reflections on electronic (direct) democracy and online media especially offered a safe haven for participation. These elaborations partially continued the work of earlier participatory-democracy theorists such as Barber. In *Strong Democracy*, Barber (1984: 289) focuses mainly on "interactive video communications," but already is referring in a balanced way to the potential use of networked computers: "The wiring of homes for cable television across America [...], the availability of low-frequency and satellite transmissions in areas beyond regular transmission or cable, and the interactive possibilities of video, computers, and information retrieval systems open up a new mode of human communications that can be used either in civic and constructive or in manipulative and destructive ways" (Barber 1984: 274). In a later work (1998: 81), Barber refers more explicitly to the web: "the World Wide Web was, in its conception and compared to traditional broadcast media, a remarkably promising means for point-to-point lateral communication among citizens and for genuine interactivity (users not merely passively receiving information, but participating in retrieving and creating it)."

At the end of the 1990s, the situation changed, as Web 2.0 (Using Web 2.0 as a time-delineating concept has a problematic side. Nevertheless, at the beginning of the 2000s, Web 2.0 became a nodal point of online media discourses on democracy, which legitimizes its use. However, the label "Web 2.0 period" is used in a broad sense, referring to new media discourses on democracy that were articulated from the 1990s onward.) slowly came into existence, and the concepts of participation and democracy became more explicitly (re-)articulated within the realm of online media, allowing for more discursive space for the maximalist versions of participation. Arguably, four clusters can be distinguished in these debates:

- First, a series of e-concepts (such as e-governance, e-democracy, e-campaigning, e-canvassing, e-lobbying, e-consultation, and e-voting – see Remenyi and Wilson 2007; and see Pruulmann-Vengerfeldt (2007) for three Estonian examples) was used to point to the possibilities for increased participation in institutionalized

politics, but also to discuss the increased possibilities for political actors to reach out to the political community (see, e.g., Williamson et al. (2010) on the digital campaign).
- A second cluster of debates within the domain of participation through the media uses the concept of deliberation, as the deliberative turn also affects online media studies. Online media are here seen as potentially instrumental in providing a (series of) sites where deliberation can be organized. To use Gimmler's (2001: 30) words, "Here it is clear that, since it generates knowledge and functions as a medium of interaction, the internet can play a significant role in the deliberative public sphere." One strand of work is concerned with fragmentation and narcissism, where the Internet becomes (seen as) a series of echo chambers for the likeminded (Sunstein 2001). There are also more positive positions: Gimmler (2001: 31), for example, points to the positive contribution of the Internet to deliberative democracy by providing access to information and opportunities for interaction and by encouraging the exchange of services and information.
- Thirdly, alternative approaches to the public sphere as, for instance, captured by the concept of counterpublics, open up space to see online identity politics (and the politics of identity), culture jamming, cyborg politics, and pop protests (Iannelli 2016) as forms of participation of counterpublics in the cultural realm, while retaining the broad political dimension. Within this framework, online media become articulated as mobilization tools, assisting in political (in the broad sense) recruitment, organization, and campaigning, again contributing to participation *through* the Internet (Langman 2005; Dynel and Chovanec 2015; Iannelli 2016; Vromen 2016).
- In contrast to participation through the Internet (and ICTs), participation in the Internet focuses on the opportunities provided to non-media professionals to (co-) produce media content themselves and to (co-)organize the structures that allow for this media production. One entry into this debate, which also captures some of its complexities, are the already-mentioned concepts of user-generated content (UGC) and produsage (Bruns 2008), which again incorporate many different meanings and practices: Mainstream media organizations that have organized participation through online media often revert to the concept of citizen journalism to label the involvement of ordinary people in the media production process. However, the broad category of UGC also leaves room for a wide variety of non-mainstream practices and forms of eye-witnessing (Mortensen 2015) and online alternative journalism (see Atton and Hamilton 2008, for a broad approach to alternative journalism), which allow for more maximalist versions of participation (e.g., the Indymedia network [Kidd 2003]).

However, we should not lose sight of the impact of the contemporary media industries on the participatory process, where structural participation in the decision-making process of the involved companies, for instance, is excluded. Fuchs (2015: 102) refers to the "corporate colonization of social media," while Jenkins (2006: 175) argues that the corporate interpretation of participation still leans toward the more minimalist forms, while consumers strive for more maximalist versions (which, especially at the level of structural media participation, are often out of reach). Still, examples such as the online encyclopedia Wikipedia show that collaborative cultural production can still be grounded into a maximalist participatory model (O'Sullivan 2009: 186), even though, as always, not without problems.

## 9.4 Conclusion

We can find a wide diversity of articulations of participation in (and beyond) the field of Media and Communication Studies; this is a diversity that has only increased after Web 2.0 and social media studies provided a new impetus for the study of media participation. This diversity is a challenge for academic research, as it complicates dialogue and comparison, but if also reflects the political-ideological nature of participation, within both the field of academia and media. The concept, the definition, and the practice of participation all remain objects of political struggle, where some defend the more maximalist versions of participation, and call for a strengthening of the already existing pockets of maximalist media participation, as, for instance, can be found in the worlds of community and alternative media, in the alternative web, and in the more activist usages of media technologies. Others would consider the more minimalist participatory intensities as sufficient, applauding the increased visibility and activity levels of citizens without asking for "the impossible" (defined as the structural equalization of power relations in and beyond the media field). The high levels of interaction that mainstream media, including mainstream social media such as Facebook and Twitter, have to offer are then met with approval, without problematizing the low levels of structural participation.

Even if this political struggle remains unresolved and a quick resolution remains out of sight, this struggle also shows the coexistence of interactive, minimalist-participatory, and maximalist-participatory practices in the media field. Even if one could argue that the interactive and minimalist-participatory practices remain the dominant mode in the second decade of the twenty-first century, maximalist-participatory practices also continue to exist. What is also important here is that no particular media technology has become wholly minimalist- or maximalist-participatory. There are simply too many platforms, interfaces, genres, formats, organizations, usages, practices, and people involved to be able to have a linear connection between one particular media technology, or even one media organization, and one particular level of participatory intensities, whether it is minimalist or maximalist. This means that one media organization, whether it is a television broadcaster, a newspaper, or a social media organization, can have very different participatory intensities in its constitutive components (its editorial teams, its pages, its programs, its groups, etc.), which renders it often very difficult to make generalizing statements about particular media technologies or organizations, without studying these "lower" levels.

Moreover, the popularity of participation research has also shown that a series of research-related problems continue to surface. Arguably, these problems are located at three distinct, but still interrelated, levels. Firstly, there is still hardly a consensus on how participation should be theorized, or even defined. The resulting plurality of approaches toward participation can only be welcomed and embraced, but at the same time there is a need for clarity to ensure that academic dialogues can be organized and academics can build more on each other's work to better understand the role of participation in contemporary societies. Secondly, there is also considerable vagueness about how participation should be researched. All social practices are characterized by complexity, but, in the case of participatory processes, this complexity is further enhanced by the discursive and material struggles that are intimately connected with these participatory processes. Analytical models, which might support researchers better in tackling this complexity, remain rare. Thirdly, there is no sufficient debate on how participation should be evaluated. On some occasions, the impression might arise that any kind of social action can be labeled as participatory and then celebrated as part of the trajectory toward a democratic nirvana. There is a need to acknowledge the political-ideological nature of participation, which brings about normative discussions into the desirability of particular participatory intensities.

## References

Allen D, Bailey M, Carpentier N, Fenton N, Jenkins H, Lothian A, Qui JL, Schaefer MT, Srinivasan R (2014) Participations: dialogues on the participatory promise of contemporary culture and politics. Part 3: politics. Int J Commun 8: Forum 1129–1151. Accessed 27 Dec 2017. http://ijoc.org/index.php/ijoc/article/view/2787/1124

Andrejevic M (2004) Reality TV. The work of being watched. Rowman & Littlefield, Oxford

Arendt H (1958) The human condition. University of Chicago Press, Chicago

Arnstein SR (1969) A ladder of citizen participation. J Am Inst Plann 35(4):216–224

Atton C (2002) Alternative media. Sage, London

Atton C (2015) Introduction. Problems and positions in alternative and community media. In: Atton C (ed) The Routledge companion to alternative and community media. Routledge, London/New York, pp 1–18

Atton C, Hamilton JF (2008) Alternative journalism. Sage, London

Bailey O, Cammaerts B, Carpentier N (2007) Understanding alternative media. Open University Press, Milton Keynes

Barber B (1984) Strong democracy. Participatory politics for a new age. University of California Press, Berkeley

Barber B (1998) A place for us. How to make society civil and democracy strong. Hill and Wang, New York

Benhabib S (1994) Deliberative rationality and models of democratic legitimacy. Constellations 1(1):25–53

Berrigan FJ (1979) Community communications. The role of community media in development. UNESCO, Paris

Bey H (1985) The temporary autonomous zone, ontological anarchy, poetic terrorism. Autonomedia, Brooklyn

Biressi A, Nunn H (2005) Reality TV. Realism and revelation. Wallflower, London

Bordenave JD (1994) Participative communication as a part of building the participative society. In: White SA (ed) Participatory communication. Working for change and development. Sage, Beverly Hills, pp 35–48

Bradford G (1996) Media-capital's global village. In: Ehrlich HJ (ed) Reinventing anarchy, again. AK Press, Edinburgh, pp 258–271

Bruns A (2008) Blogs, Wikipedia, second life, and beyond – from production to produsage. Peter Lang, New York
Bunge MA (1977) Treatise on basic philosophy, vol 3: Ontology I: The furniture of the world. Springer, Berlin
Carey JW (2009) Communication as culture. Essays on media and society. Revised edition. Routledge, New York
Carpentier N (2011) Media and participation: a site of ideological-democratic struggle. Intellect, Bristol
Carpentier N (2016) Beyond the ladder of participation: an analytical toolkit for the critical analysis of participatory media processes. Javnost: The Public 23(1):70–88
Carpentier N, Lie R, Servaes J (2003) Community media-muting the democratic media discourse? Continuum 17(1):51–68
Carpentier N, Dahlgren P, Pasquali F (2014) The democratic (media) revolution: a parallel history of political and media participation. In: Carpentier N, Schrøder K, Hallett L (eds) Audience transformations: shifting audience positions in late modernity. Routledge, London, pp 123–141
Carpignano P, Anderson R, Aronowitz S, Difazio W (1990) Chatter in the age of electronic reproduction: talk show and the "public mind". Social Text 25/26:33–55
Castells M (2001) The internet galaxy. Reflections on the internet, business, and society. Oxford University Press, Oxford
Dahlgren P (2009) Media and political engagement. Cambridge University Press, Cambridge
Dahlgren P (2013) The political web. Palgrave Macmillan, Basingstoke
Dakroury A, Eid M, Kamalipour YR (2009) The right to communicate. Historical hopes, current debates and future premises. Kendall/Hunt Publishing, Dubuque
Dayan D, Katz E (2009) Media events: the live broadcasting of history. Harvard University Press, Cambridge, MA
Delaney T, Madigan T (2009) The sociology of sports: an introduction. McFarland, Jefferson
Deleuze G, Guattari F (1987) A thousand plateaus. Capitalism and schizophrenia. University of Minnesota Press, Minneapolis
Dewey J (1927) The public and its problems. Holt, New York
Downing J w F, Villarreal T, Gil G, Stein L (2001) Radical media. Rebellious communication and social movements. Sage, London
Dynel M, Chovanec J (eds) (2015) Participation in public and social media interactions. Amsterdam / Philadelphia: John Benjamins
Eco U (1968) La Struttura Assente. La Ricerca Semiotica e il Metodo Strutturale/The absent structure. Semiotic research and the structuralist method. Bompiani, Milano
Fierlbeck K (1998) Globalizing democracy. Power, legitimacy and the interpretation of democratic ideas. Manchester University Press, Manchester
Fiske J (1987) Television culture. Taylor & Francis, London
Fraser N (1990) Rethinking the public sphere. Social Text 25/26:56–80
Freire P (1992) Pedagogy of hope. Continuum, New York
Fuchs C (2015) Social media. A critical introduction. Sage, London
Fuchs C (2016) Reading Marx in the information age: a media and communication studies perspective on capital volume, vol 1. Routledge, New York
Fuchs C, Mosco V (eds) (2016) Marx and the political economy of the media. Brill, Leiden
Giddens A (2006) Sociology, 5th edn. Polity Press, Cambridge
Gimmler A (2001) Deliberative democracy, the public sphere and the internet. Philos Soc Critic 27(4):21–39
Gruber H (2004) The "conversation on Austria": a televised representation of Austria's internal condition after the National-Conservative "Wende". J Lang Polit 3(2):267–292
Habermas J (1991) The structural transformation of the public sphere. An inquiry into a category of bourgeois society. MIT Press, Cambridge
Habermas J (1996) Between facts and norms. In: Contributions to a discourse theory of law and democracy. MIT Press, Cambridge, MA
Hall S (1980) Encoding/decoding. In: Hall S (ed) Culture, media, language, working papers in cultural studies, 1972–79. Hutchinson, London, pp 128–138

Hall S, Critcher C, Jefferson T, Clarke J, Robert B (1978) Policing the crisis. Mugging, the state and law and order. Macmillan, London

Hallin DC (1996) Commercialism and professionalism in the American news media. In: Curran J, Gurevitch M (eds) Mass media and society. Arnold, London/New York/Sydney/Auckland, pp 243–264

Hamelink C (1997) The politics of world communication. Sage, London

Hamo M (2006) Caught between freedom and control: "ordinary" People's discursive positioning on an Israeli prime-time talk show. Discourse Soc 17(4):427–446

Hauser G (1998) Vernacular dialogue and the rhetoricality of public opinion. Commun Monogr 65(2):83–107

Iannelli L (2016) Hybrid politics: media and participation. Sage, London

Jacobson TL (1998) Discourse ethics and the right to communicate. Gazette 60(5):395–413

Jenkins H (1999) The work of theory in the age of digital transformation. http://web.mit.edu/~21fms/People/henry3/pub/digitaltheory.htm. Accessed 27 Dec 2017 (also published in Miller T, Stam R (eds) (1999) A companion to film theory. Blackwell, London, pp 234–261)

Jenkins H (2006) Convergence culture. Where old and new media collide. New York University Press, New York

Jenkins H, Carpentier N (2013) Theorizing participatory intensities: a conversation about participation and politics. Convergence 19(3):265–286

Katz E, Blumler J, Gurevitch M (1974) Utilisation of mass communication by the individual. In: Blumler J, Katz E (eds) The uses of mass communications. Current perspectives on gratifications research. Sage, Beverly Hills/London, pp 19–32

Keane J (1998) Democracy and civil society. University of Westminster Press, London

Kidd D (2003) Indymedia.org: a new communications commons. In: McCaughey M, Ayers MD (eds) Cyberactivism. Online activism in theory and practice. Routledge, New York, pp 47–70

Langman L (2005) From virtual public spheres to global justice: a critical theory of internetworked social movements. Sociol Theory 23(1):42–74

Lepik K (2013) Governmentality and cultural participation in Estonian public knowledge institutions, PhD thesis. University of Tartu Press, Tartu. Accessed 27 Dec 2017. http://dspace.utlib.ee/dspace/bitstream/handle/10062/32240/lepik_krista_2.pdf?sequence=4

Leurdijk A (1997) Common sense versus political discourse. Debating racism and multicultural society in Dutch talk shows. Eur J Commun 12(2):147–168

Littlejohn SW (1996) Theories of human communication. Wadsworth, Belmont

Livingstone S (1998) Relationships between media and audiences. In: Liebes T, Curran J (eds) Media, ritual and identity. Routledge, London, pp 237–255

Livingstone S, Lunt P (1996) Talk on television, audience participation and public debate. Routledge, London

Lury C (2011) Consumer culture. Polity, Cambridge

MacBride Commission [The International Commission for the Study of Communication Problems] (1980) Many voices, one world. Towards a new more just and more efficient world information and communication order. Report by the International Commission for the Study of Communication Problems. UNESCO/Kogan Page, Paris/London

Mansell R, Nordenstreng K (2006) Great media and communication debates: WSIS and the MacBride report. Inf Technol Int Dev 3(4):15–36

Melucci A (1989) Nomads of the present. Social movements and individual needs in contemporary society. Temple University Press, Philadelphia

Mortensen M (2015) Journalism and eyewitness: digital media, participation, and conflict. Routledge, New York

Munson W (1993) All talk. The talkshow in the media culture. Temple University Press, Philadelphia

Neeson JM (1996) Commoners: common right, enclosure and social change in England. Cambridge University Press, Cambridge, pp 1700–1820

Negroponte N (1995) Being digital. Knopf, New York

Negt O, Kluge A (1983) The Proletarian public sphere. In: Mattelart A, Siegelaub S (eds) Communication and class struggle. 2. Liberation, socialism. International General/IMMRC, New York/Bagnolet, pp 92–94

Nieminen H (2006) What do we mean by a European public sphere? In: Carpentier N, Pruulman-Vengerfeldt P, Nordenstreng K, Hartmann M, Vihalemm P, Cammaerts B (eds) Researching media, democracy and participation. Tartu University Press, Tartu, pp 105–119

O'Sullivan D (2009) Wikipedia. A new community of practice? Ashgate, Surrey

Ogasawara H (2017) Marx, and Marxism as method in Stuart Hall's thinking. Inter-Asia Cult Stud 18(2):178–187

Padovani C, Calabrese A (eds) (2014) Communication rights and social justice: historical accounts of transnational mobilizations. Palgrave Macmillan, Basingstoke

Palmgreen P, Rayburn JD II (1985) An expectancy-value approach to media gratifications. In: Rosengren KE, Palmgreen P, Wenner L (eds) Media gratification research. Sage, Beverly Hills/London, pp 61–73

Pateman C (1970) Participation and democratic theory. Cambridge University Press, Cambridge

Poster M (1997) Cyberdemocracy: internet and the public sphere. In: Porter D (ed) Internet culture. Routledge, London, pp 201–218

Priest PJ (1995) Public intimacies. Talk show participants and Tell-All TV. Hampton Press, Cresskill

Pruulmann-Vengerfeldt P (2007) Participating in a representative democracy. Three case studies of estonian participatory online initiatives. In: Carpentier N, Pruulmann-Vengerfeldt P, Nordenstreng K, Hartmann M, Vihalemm P, Cammaerts B, Nieminen H (eds) Media technologies and democracy in an enlarged Europe. The intellectual work of the 2007 European Media and Communication Doctoral Summer School. University of Tartu Press, Tartu, pp 171–185

Raboy M, Shtern J (eds) (2010) Media divides: communication rights and the right to communicate in Canada. UBC Press, Vancouver/Toronto

Real MR (1996) Exploring media culture: a guide. Sage, Thousand Oaks

Remenyi D, Wilson D (2007) E-democracy: An e too far? In: Griffin D, Trevorrow P, Halpin EF (eds) Developments in e-government. A critical analysis. IOS Press, Amsterdam, pp 89–100

Renckstorf K, McQuail D, Jankowski N (1996) Media use as social action. A European approach to audience studies. Libbey, London

Rheingold H (1993) The virtual community. Homesteading on the electronic frontier. Addison-Wesley, Reading

Rojek C (2001) Celebrity. Reaktion, London

Servaes J (1999) Communication for development. One world, multiple cultures. Hampton Press, Cresskill

Sharma RK (1996) Fundamentals of sociology. Atlantic Publishers & Dist, New Delhi

Sunstein C (2001) Echo chambers: Bush v. Gore, impeachment, and beyond. Princeton University Press, Princeton. http://press.princeton.edu/sunstein/echo.pdf. Accessed 27 Dec 2017

Tabing L (2002) How to do community radio. A primer for community radio operators. UNESCO, New Delhi

Thomas P (1994) Participatory development communication: philosophical premises. In: White SA (ed) Participatory communication. Working for change and development. Sage, Beverly Hills, pp 49–59

Tomasulo FP (1984) The spectator-in-the-tube: the rhetoric of Donahue. J Film Video 36(2):5–12

Turner G (2010) Ordinary people and the media. The demotic turn. Sage, London

UNESCO (1977) Records of the general conference, nineteenth session Nairobi, 26 Oct–30 Nov 1976, vol 1, Resolutions. UNESCO, Paris. http://unesdoc.unesco.org/images/0011/001140/114038e.pdf. Accessed 27 Dec 2017

Van Bauwel S, Carpentier N (eds) (2010) Trans-reality television. The transgression of reality, genre, politics, and audience. Lexington, Lexington

Vander Stichele A, Laermans R (2006) Cultural participation in Flanders. Testing the culture omnivore thesis with population data. Poetics 34:45–56

Vromen A (2016) Digital citizenship and political engagement: the challenge from online campaigning and advocacy organisations. Palgrave Macmillan, Basingstoke

Wasko J, Mosco V (eds) (1992) Democratic communications in the information age. Garamond Press/Ablex, Toronto/Norwood

Wayne M (2003) Marxism and media studies. Key concepts and contemporary trends. Pluto Press, London

White M (1992) Tele-advertising. Therapeutic Discourse in American Television, Chapel Hill, NC and London: University of North Carolina Press

Williamson A, Miller L, Fallon F (2010) Behind the digital campaign. An exploration of the use, impact and regulation of digital campaigning. The Hansard Society, London

WSIS Civil Society Plenary (2003) Shaping information societies for human needs. Civil Society Declaration to the World Summit on the Information Society, December 8. http://www.itu.int/wsis/docs/geneva/civil-society-declaration.pdf. Accessed 27 Dec 2017

# Empowerment as Development: An Outline of an Analytical Concept for the Study of ICTs in the Global South

10

Jakob Svensson

## Contents

| | | |
|---|---|---|
| 10.1 | Introduction ................................................................. | 218 |
| 10.2 | Intersectional Level: Empowerment and Axes of Oppression ......................... | 221 |
| 10.3 | Contextual Level: Empowerment, Opportunities, and Constraints in the Situation .... | 223 |
| 10.4 | Agency Level: Empowerment, Capabilities, and Critical Awareness .................. | 225 |
| 10.5 | Technological Level of Communication Platforms: Empowerment, ICTs, and Affordances ............................................................. | 228 |
| 10.6 | Concluding Remarks .......................................................... | 230 |
| References ............................................................................. | | 233 |

### Abstract

This chapter turns to the concept of "empowerment" as a result of disenchantment with the concept of "development" in the study of information and communication technologies (ICTs) and social change in the global South. It is a certainty that the proliferation of ICTs (mobile phones in particular) has opened up a range of possibilities and new avenues for individuals, aid agencies, and NGOs. However, overviews of communication supposedly for development reveal a field based on economic understandings of development biased toward techno-determinism. Moreover, these understandings lack sufficient critique and do not take larger contextual factors into account. Therefore, it is argued that empowerment is a better concept to draw upon in the critical study of ICTs and social change. However, empowerment is not an easy concept to define, and no analytical outline of the concept has been found in the existing body of literature. Addressing this lack, this chapter will trace the roots of empowerment in community psychology and in feminist and black power movements as well as

---

J. Svensson (✉)
School of Arts and Communication (K3), Malmö University, Malmö, Sweden
e-mail: jakob.svensson@mau.se

explore different understandings of the concept from various disciplines. From this overview, the chapter suggests that empowerment should be studied on a) an intersectional level, b) a contextual level, c) an agency level, and d) a technological level. It further argues that these four levels intersect and must be studied in tandem to understand whether processes of empowerment are taking place, and if so, in what ways? The chapter ends by shortly applying these levels to a study involving market women's use of mobile phones in Kampala.

**Keywords**

Affordances · Capability · Critical consciousness · Development · Empowerment · ICTs · Opportunity structures · Situatedness

## 10.1 Introduction

Interest in the concept of *empowerment* stems from a general disenchantment with the concept of *development* when studying mobile communication in so-called *developing* regions (see Svensson and Wamala-Larsson 2016; Wamala-Larsson and Svensson 2018). In these publications, the aim was to understand the role mobile phones play for market women in Kampala. These studies were influenced by the development paradigm in general and the burgeoning area of mobile communication for development in particular (see Svensson and Wamala-Larsson 2015 for an overview).

Development has its roots in *modernization* theory, which can be argued to reflect an underlying *West is best* worldview (see Servaes and Lie 2015). The theory assumes that so-called *modern* societies in the industrial globalized North are the model toward which more rural and traditional societies in the global South should develop (Nederveen Pieterse 2010: 21, 23). A similar top-down approach is also present in the theory of *diffusion of innovation* (DoI), in which expert knowledge and innovations (such as information and communication technologies, ICTs) should trickle down to traditional and often economically less wealthy societies (see Tufte 2017: 12). Similarly, the *bottom of the pyramid* theory explains that businesses, governments, and aid agencies should start thinking of poor people as consumers and should start developing regions as a potential market for investors (see Prahalad and Hart 2002). Market competition is expected to drive prices down and allow low-income earners from the bottom of the pyramid to get ahead and stay connected.

When approached in this manner, the critique of development should come as no surprise. Nederveen Pieterse (2010) argues that ICTs for development are primarily driven by market logic, market deepening (p. 166), and technological fetishism (p. 168) and serve as a rationale for trade and investment liberalization (p. 172). Hann and Hart (2011) similarly argue that the ultimate drive for development in postcolonial times is a world in which the rich, developed countries help poorer regions (often former colonies) in order to improve their *economic* prospects. However, while economic aspects are important for development, they are not the only aspects to be considered (see Servaes and Malikhao 2014: 171). Such critique is also reflected in postcolonial perspectives of Western development paradigms (see Tufte 2017: 1). Here, development is questioned and even considered hypocritical in times when debt repayments drain the income of countries in the global South and undermine these governments' ability to protect their citizens, while, at the same time, aid levels have shrunk to the point of being merely symbolic (Hann and Hart 2011: 116). Furthermore, conceiving of people in so-called developing countries as mere consumers, situated at the bottom of a postcolonial pyramid, and, as such, are only interesting for enterprises that aim to accumulate capital, is not only unattractive but also leads to questionable development interventions (see Dodson et al. 2013). The concept of development has thus been gradually replaced by social change, and the top-down models mentioned above are becoming replaced by more bottom-up approaches that promote participatory communication (see Servaes and Malikhao 2014; Servaes and Lie 2015). More evidence-based learning models have also been called for, such as Oxfam America LEAD's (Learning, Evaluation and Accountability Department) perspective on development as rights (see Van Hemelrijck 2013, who the chapter will return to).

It is apparent that development has not been used as a *critical* concept – understood here as the questioning and challenging of existing power relations. Empowerment, on the contrary, directly concerns unequal power relations (Sadan 2004; Kabeer 1999: 437). Here, power is often linked to Marxist theory and thus located in relations between those who control the means of production and the so-called market (Burton and Kagan 1996). The consequences of the market on the development of individuals, culture, and communication have been developed in critical theory (see Meyer-Emerick 2005: 542). Today, when the market is seen as the harbinger of all good things, there is a risk of identifying empowerment with consumer choice in a commodity market (Burton and Kagan 1996). Thus, to approach empowerment as a critical concept entails both the questioning of the logic of market capitalism and a focus on those without power, with the aim of improving their situation.

However, the concept of empowerment is not easy to draw upon when analyzing communication in the global South, as there is no shared definition of it (Choudhury 2009: 343). It is considered fuzzy (Kabeer 1999: 436) and elusive (Hill Collins 2000: 19). It is both under-defined and overused in academic, policy, and public discourses (Kleine 2013: 31). Moreover, it occurs on many different levels (individual, group, organization, and community [see Sadan 2004]) and covers and materializes through a variety of different processes (Drolet 2011: 633). Empowerment can be understood both as an outcome and a process (Sadan 2004), as a value orientation for implementing change, and as a theoretical model for understanding such implementation (Zimmerman 2000).

Etymologically, the concept comprises the root, *empower*, and the suffix *-ment*. Similarly, *empower* is composed by *em-* and *power*. Here, the prefix *em-* is probably used with the same meaning as *in-*. Thus, "empower" literally means "into/towards power" (Lincoln et al. 2002: 271–272). While it is the powerless who are the objects of empowerment, the agent of such change has varied in the history of the concept.

According to the Oxford English Dictionary, *empowerment* is defined as the granting/authorizing of power to an individual or group (by someone else), as oneself becomes stronger and more confident in controlling one's life and claiming one's rights, as well as someone else supporting others to discover and claim power. Today, many emphasize that those who live in poverty should be viewed as empowerment's primary agents (Van Hemelrijck 2013: 29).

In this chapter, empowerment is defined as those without power taking control over their life situation, destiny, and environment. As such, it is a normative concept (see Servaes and Lie 2015: 126). Empowerment is also a deeply political concept (Meyer-Emerick 2005: 543), as it conceptualizes social change in terms of the powerless gaining greater control (Sadan 2004) and challenging existing power relations (see also Kabeer 1999: 437). This resonates in Oxfam America's definition of empowerment as a "significant and sustainable change in power relations that enables excluded and marginalized people to realize their rights to access and manage the resources, services and knowledge they need" (Van Hemelrijck 2013: 31). It is important to point out that power is not a zero-sum game (Lincoln et al. 2002: 275), which means that empowering those without power does not necessarily mean taking power away from those already in charge of their life situation, destiny, and environment. This is nicely illustrated by Hall (1992, in Lincoln et al. 2002: 275) when discussing the empowerment of women: It is not the goal of women to take power away from men, but rather, the goal is to develop their own power.

A look at the history of the concept takes us back to the 17th century, when its use was first recorded. At that time, to empower meant to authorize/license someone to do something (Lincoln et al. 2002: 272). Contemporary academic use of the concept is accredited with having originated in American community psychology. It is claimed that Rappaport introduced the concept into social work and social psychology in the 1980s. He defines empowerment as the mechanism by which people, organizations, and communities gain mastery over their own lives (Rappaport 1984). Social psychologists identified the need for more interventionist approaches in order to move beyond mere descriptions of powerlessness to identify viable strategies of intervention for change to happen (Burton and Kagan 1996). Thus, empowerment came to underline intentional, ongoing processes in local communities, which involve mutual respect, critical reflection, and group participation through which people in distress, who lack the aptitude and an equal share of resources, can gain greater access to and control over those resources (Zimmerman 2000). Here, empowerment revolves around self-help, meaning the capacity of individuals, groups, and/or communities to take control of their circumstances, exercise power, and achieve their self-realized goals (Adams 2008: 6).

This history of empowerment in the field of social work and social psychology has made its mark, especially in the more capability-centered approach to which the chapter will return. For now, it suffices to point out that empowerment is used in many different fields. But as Lincoln and her co-authors (2002) emphasize, empowerment is contested and is used differently when addressing women, minority groups, education, community care, politics, or management theory. This ambiguity is problematic if it is aiming at employing empowerment as an analytical tool for studying and understanding communication and social change. Therefore, the purposes of this chapter are to develop empowerment as an analytical concept and to suggest a way of studying this.

The chapter will take as its point of departure feminist and black feminist theoretical concepts which have been influential in the development of the concept of empowerment. Indeed, perhaps empowerment is most commonly associated with feminism and gender equality. Feminists have emphasized the importance of an intersectional level of axes of oppression when studying and understanding empowerment. In addition, feminists emphasize that it is not only the intersectional level of larger power structures that is important to consider but also the contextual level, in terms of the particular situations women find themselves in; therefore, opportunity structures and constraints on the contextual level are important to include. However, it is not enough to study and understand empowerment by only attending to axes of oppression; contextual opportunity structures and constraints – the agency of those without power – must also be accounted for in terms of their capabilities and their awareness of the situation they find themselves in. Finally, the communication platforms used and their affordances also need to be attended to when studying communication and social change. These four levels intersect and feed off each other, but for the sake of analytical clarity, they will be individually outlined here. The chapter ends by applying these levels briefly in a study of mobile phones and market women in Kampala.

## 10.2 Intersectional Level: Empowerment and Axes of Oppression

Feminism has been present in development discourses from the 1980s in "women in development" approaches (later replaced by Gender and Development; see Servaes and Lie 2015: 126). From an empowerment perspective, feminists have been successful in connecting empowerment to challenging existing patriarchal power structures and to support women to take charge of – and transform – their life situations (Drolet 2011: 634, 639). Empowerment is often linked to women and girls, especially when connected to development such as in United Nations Millennium Development Goals (MDGs) and in the Sustainable Development Goals (SDGs) that replaced the MDGs in 2016. Here, it is claimed that (women's) empowerment is necessary to overcome obstacles associated with poverty and development. Given that women's empowerment concerns women challenging existing patriarchal norms to improve their own wellbeing (Bali Swain 2012: 59, 63), it directly points to patriarchal power structures as key obstacles (structures that are often entangled with traditional social beliefs and practices; see, e.g., Choudhury 2009: 342). But there are many more structures at play, such as capitalism (as previously discussed), sexism, ageism, and, especially, ethnicity.

The Civil Rights Movement in the USA has been important in the conceptualization of empowerment. One of the first articles written on the topic was titled *Toward Black Political Empowerment* (Conyers 1975), which served as inspiration for more articles that discuss empowering the black community. Feminism and the Civil Rights Movement intersect in black feminist thought (see Hill Collins 2000). Historically, black feminism grew out of a disenchantment with the sexism of the Civil Rights Movement and racism of the feminist movement and asserts that sexism, gender inequality, class oppression, and racism are bound together. It became apparent that black women were positioned within structures of power fundamentally differently than that of white women. In line with this, Hill Collins (2000: 18) introduced what she calls a *matrix of domination*. Through this concept, she explores how intersecting axes of oppression are organized, for example, through schools, housing, employment, et cetera (see Hill Collins 2000: 228). Along the same line, black feminism (and related concepts of intersectionality) contributes with an understanding that those without power are fundamentally different from each other because of their specific positions in the intersecting axes of oppression. As Hill Collins (2000) states, "Developing a Black feminist politics of empowerment requires specifying the domains of power that constrains Black women" (p. 19), which focuses on how "the matrix of domination is structured along certain axes – race, gender, class, sexuality and nation" (p. 288–289).

When empowering women, women themselves and their experiences should be included (Harding 1991: 152). Women's experiences result in diverse *standpoints*, which should be used as resources that can provide multiple vantage points (ibid.: 119). However, women's experiences are mostly absent because conceptual frameworks generated from within a patriarchal system fail to enable women to make sense of their experiences (Haraway 1991). Therefore, women's experiences from their own particular *standpoints* need to be listened to and considered not only in terms of collecting and sharing knowledge of oppression and domination but also in terms of enabling those without power to define their own realities (Hill Collins 2000: 274; see also Communication for Social Change, Tufte 2017: 13). This is about leaving behind "the status of objects to assume the status of historical subjects" (Freire 1996: 141). Partiality, not universality, is the condition of being heard (Hill Collins 2000: 270; see also Freire 1996: 90). This is not about emphasizing how different groups are affected differently along the different axes of oppression (e.g., white women from black women from black men) but rather about highlighting that black women's experiences serve as specific *standpoints* from which connections to other disempowered groups could be explored (Hill Collins 2000: 270). Each woman speaks from her own standpoint and contributes with her specific *situated knowledge* (a concept the chapter will return to). By acknowledging this, those without power can become better at considering others without power – and their standpoints – without relinquishing the uniqueness of themselves (Hill Collins 2000: 270). It is important to note here that axes of oppression and people's positions within them are not static but constantly changing.

The feminist discussion of axes of oppression – in particular how they intersect and lead to unequal power relations – tends to refer to women as a collective group. However, it is important for the argument in this chapter to move beyond the individual-collective dichotomy, as the individual cannot be fully separated from the collective and vice versa. Individual black women's experiences have fostered group commonalities that encourage the formation of a group-based collective standpoint (Hill Collins 2000: 24). Standpoint feminism helps us understand this, as it emphasizes the individual (woman) as imperative when understanding (women's) empowerment on a collective level (see Ristock and Pinnell 1996: 1). In other words, although women as a collective can act together to "displace oppressive power relations" (Riordan 2001: 282), change also has to happen at the individual level. The disempowered, who are forced to remain motionless on the outside, "can develop the 'inside' of a changed consciousness" (Hill Collins 2000: 118). What is highlighted here is that any analysis of women's empowerment on a collective level needs to take into account individual women's interpretations and experiences. Without doubt, the female collective is important for such individual experiences. Belonging to a group may lead to the creation of social capital and could provide support structures which are important in empowering the individual (Bali Swain 2012: 60; see also Tufte 2017: 2). After all, *the personal is political*, as the famous feminist slogan goes. Feminist scholars contribute with the idea that the individual and the collective are not in a dichotomized relationship to each other and that processes of empowerment move and morph between individuals, groups, communities, and collectives. Thus, any theory of empowerment needs to convincingly integrate levels of individuals, collectives, and societal structures (Sadan 2004: 137).

## 10.3 Contextual Level: Empowerment, Opportunities, and Constraints in the Situation

Connected to the previous discussion of standpoint feminism and the individual versus the collective is the idea of *situatedness*. Feminists have shown how knowledge and power inequalities are situated in the different contexts women find themselves in (Haraway 1991). As Hill Collins (2000) argues, "Race and gender may be analytically different, but in Black women's life they work together" (p. 269). People are beings in a situation, "rooted in temporal-spatial conditions which mark them and which they also mark" (Freire 1996: 90). But whereas the previous level dealt with larger intersecting axes of oppression on a macro level, situatedness puts an emphasis on the contextual level in which particular situations give rise to particular constraints as well as opportunities. What brings about empowerment among women in the globalized North is not necessarily the same in African countries, to make a crude comparison. For example, Kole (2001) underlines the significance of gender, geographical difference, context, and other micro aspects when examining gender empowerment. Highly aware of the different contexts in Africa, Kole realizes that disparities exist between and within the various women's groups; thus, different needs and problems are seen in different parts of the continent (Kole 2001: 160). Resources and limitations exist in different situations (contexts, cultures, and regions) and need to be taken into consideration, for example, when women without power struggle against patriarchal structures. Social change takes place in specific situations which both constrains and provides opportunities for change to happen. In other words, the specific situations that those without power find themselves in need to be foregrounded when studying empowerment (as also highlighted in Oxfam America's perspective; see Van Hemelrijck 2013: 29). These contextual levels also concern relational aspects and the social networks one is situated in (Van Hemelrijck 2013: 233; Servaes and Lie 2015: 126).

For an analytical study of the structures at the contextual level, Kitschelt's (1986) article on *opportunity structures* is helpful. In his study of multinational social movements that campaign for an antinuclear world, he identifies *opportunity structures* as comprised of a specific "configuration of resources, institutional arrangements and historical precedents for social mobilisation which facilitates the development of protest movement in some instances and constrain them in others" (p. 58). Opportunities do not determine whether social movements are successful or social change is possible, but should be understood as a factor that influences an actor's choice of strategy. While not fully determining, opportunities appear to be an important factor in understanding the actual possibility of those without power to take control over their life situation, destiny, and environment. The concept of opportunity structures underlines contextual factors when studying actors' possibilities for empowerment in a given situation. Here, opportunity structures are proposed as an analytical concept, not a strategic concept for a cost-benefit analysis for making rational choices when evaluating different courses of action (see, e.g., Cammaerts 2012: 118–119; Tufte 2017: 85–86).

Kitschelt claims that although unequal power relations are necessary conditions for social mobilization to arise, they are not enough to bring about social change. Actors also need to discern an opportunity to influence and induce social change in order to mobilize (Kitschelt 1986). Thus, a crucial dimension of opportunity structures is the perceived openness or closeness of institutions for change as well as the institutional capacity to advance change and effectively implement policies once they are decided. Institutions can range from being open and responsive in attempting to assimilate demands from actors seeking change to being closed and repressive. Political/social systems' institutional capacity and willingness to allow actors access to development processes also differ. In addition, Oxfam America highlights the role of institutions in its perspective (Van Hemelrijck 2013: 33). In other words, institutions need to be taken into account when studying and understanding empowerment.

While Kitschelt's main focus concerns the institutions in the context under study, it is important to underline that opportunities, as well as constraints, can be found on many other levels such as the larger sociocultural community and economic levels. Sen (1999: 38–39), for example, outlines what he labels "opportunity aspects" in terms of political freedoms, economic facilities, healthcare, education, trust, and social safety (the chapter will return to this). Furthermore, contextual structures are not the only important aspect for empowerment to happen. Kitschelt (1986) also highlights the actors themselves and their agency. Those without power need not only actual opportunities and institutional openings to become empowered but also the ability to perceive such opportunities in order to act upon them (as actors with agency). This leads to the next section, which turns to agency and the capabilities of those without power.

## 10.4 Agency Level: Empowerment, Capabilities, and Critical Awareness

To understanding empowerment, we also have to consider the agency of individuals and groups. People are the true agents of change (Tufte 2017: foreword viii) and "development is thoroughly dependent on the free agency of people" (Sen 1999: 4). Empowering the powerless to challenge power relations requires those without power having/sensing that they have the capability to do so – that they believe another world is possible and that this world is within reach.

Many scholars' definition of empowerment relates to abilities and capabilities (on the level of the individual as well as the collective). Adams (2008) defines *empowerment* in the field of social work as "the *capacity* of individuals, groups and/or communities to take control of their circumstances, exercise power and achieve their own goals" (p. 6, italics added). Bush and Folger (1994) similarly understand empowerment as a process that restores one's sense of self and value, at the same time, intensifying one's *ability* to address social problems. Bali Swain (2012: 60) argues that empowerment is an individual or a group's *capability* to choose and, from this choice, have the ability to alter their situation. Lennie and Tacchi (2013) approach empowerment as "recognizing your power to create/induce change" (p. 108). Van Hemelrijck (2013) understands empowerment as "the expansion of an individual or a group's *ability* to make transformative life choices" (p. 34, italics added). Servaes and Malikhao (2014) define empowerment as the ability of people to influence "the wider system and take control of their lives" (p. 175). Often cited here is Kabeer (1999), who defines empowerment as the process through "which those who have been denied the *ability* to make strategic life choices acquire such an *ability*" (p. 437, italics added).

Thus, it seems that empowerment concerns agencies, abilities, and choices. This focus resonates in Sen's (1999) *capability approach*. This approach has been widely used in studies of development (Nederveen Pieterse 2010) and has challenged the growth-oriented focus of development (Tufte 2017: 27). Sen (1999) theorizes development as a kind of freedom that lends itself toward the capacity of individuals to not only assess their situations but also have the capability to transform them. Within the area of mobile communication and development (M4D), Smith et al. (2011) have applied Sen's theory by emphasizing how mobile phones alter users' capabilities through increased access to timely and relevant information as well as through expanded possibilities for connectedness between people. Furthermore, Smith et al. (2011) argue that Sen's capability approach is useful for looking beyond traditional economic measures of development and, instead, consider empowerment (which coincides with the aim of this chapter). In other words, development and empowerment almost become the same thing when discussed through the lens of Sen's capability approach (see also Van Hemelrijck's 2013 discussion of development as rights and empowerment, p. 30). Nevertheless, Sen does not say much about empowerment in his book, nor does he define it.

While feminist scholarship claims that powerlessness (and thus also empowerment) is different in different intersecting axes of oppression and takes place on different levels (from the individual to the collective), Sen highlights that empowerment entails possessing certain capabilities in order to induce change, for example, resources, agencies, and skills (see Kabeer 1999). Capabilities can be located both in the individual and in the collective. This is important to underline because Sen has been criticized for placing too much focus on the individual at the expense of a structural understanding and thus reducing empowerment to individual capabilities. Feminist scholars have warned against solely focusing on the individual, arguing that this unwittingly advances a sense of entitlement rather than poses a challenge to patriarchal and capitalist structures (Riordan 2001: 282). Nevertheless, Sen (1999) does make it clear that what people can achieve "is influenced by economic opportunities, political liberties, social powers (...)" (p. 5). He also emphasizes "the role of social values and prevailing mores" (p. 9) and devotes an entire chapter to women's agency and social change (p. 189). Although Sen places agency and capabilities at the center of his analysis, he is not oblivious to the influence of power structures and contextual aspects on capabilities. He connects opportunities that people have to their personal and social circumstances (Sen 1999: 17). In other words, a proper understanding and analysis of empowerment needs to take both the levels of the individual and the collective/structural into account and acknowledge how the power to change one's life situation, environment, and destiny moves and morphs between these different levels.

A similar critique can be directed at Oxfam America's perspective on development as rights (and empowerment), as it tends to be too focused on the individual at the expense of larger structural aspects. Indeed, Oxfam America offers a sophisticated and holistic understanding of empowerment in terms of *agency, institutions*, and *societal relationships* (see Van Hemelrijck 2013: 33). Such an understanding is similar to the proposed model of this chapter, as institutions and societal relationships resonate on the contextual level and agency does so on the agency level. However, larger structures are lacking in their understanding and measure of empowerment. This is perhaps not surprising, as this rights-based approach is explicitly proposed as a consequence of a "disillusionment with the welfare state" (Offenheiser and Holcombe 2003: 270) and a desire to move human rights "beyond its state-centric paradigm" (ibid.: 274). In their overview of the rights-based approach, Cornwall and Nyamu-Musembi (2004: 1417) discusses this in relation to replacing a needs-based approach with a rights-based approach to development. Such a move places the individual at the center of the analysis. Basing their analysis on Sengupta's argument on the right to development as a human right, Offenheiser and Holcombe (2003) further argue that human rights are granted to the people by themselves (see p. 277), begging the question of who to hold accountable for the failure of delivering such rights and how. The individual is of course important to consider, as this section on the agency level of empowerment highlights. This should however not be done at the expense of a larger macrostructural analysis. It must be possible to both consider larger axes of oppression while, at the same time, acknowledging the role of individuals and their capabilities and right "to access and manage the resources, services and knowledge they need for strengthening their livelihood, improving their well-being" (Van Hemelrijck 2013: 33).

Acquiring capabilities is one thing, but becoming aware of that change is possible is another. Change requires choice – something which suggests that alternatives exist (Kabeer 1999: 437), or, at least, the possibility to imagine alternatives. This can be connected to Gramsci's idea of *hegemony* – the mechanisms through which some can exert domination without crude force and with some level of consent from those being dominated (Burton and Kagan 1996). To acquire the capability to change one's life situation, environment, and destiny, such consent needs to be challenged. The powerless need to understand why they are disempowered. In his well-known writings on pedagogy and emancipation of the poor, Freire (1996) elaborates on this through the concept of *conscientizaçao* – or critical consciousness, for lack of a better translation. He underlines the importance for the oppressed to become aware of their situation and argues that being passive means being complicit in their oppression and indirectly giving consent to the very mechanisms/structures/groups that oppress them. Critical consciousness entails acquiring an understanding of the sociocultural conditions that shape one's life (Sadan 2004: 82–83), which involves a deepening awareness of both power relations and the possibility of their transformation (Burton and Kagan 1996). Indeed, for Freire (1996), empowerment entails "being encouraged and equipped to know and respond" (p.12) to the concrete realities of one's world. Here, emotions are important to account for as well as the need to feel and be emotionally aware (see Tufte 2017: 87–88) – a need that perhaps can be conceptualized as *emotional conscientizaçao*.

It is possible to also relate critical consciousness to standpoint feminism's emphasis on enabling those without power to define their own realities on their own terms (Hill Collins 2000: 274). Black women's literature contains many examples of how individual black women become personally empowered by a change of consciousness (ibid.: 118). Independent self-definitions – defining their own situation and how they want to change it – empower those without power to bring about change (ibid.: 117).

Freire also connects critical consciousness to action. Social change will not happen by the mere understanding of one's oppression – it must be followed by a critical intervention (Freire 1996: 34). Similarly, Hill Collins (2000: 273) explains that the history of black American women's activism shows that becoming empowered requires more than changing the consciousness of individual black women – it requires transforming unjust social institutions. In turn, action needs what Freire (1996: 47) labels "reflective participation." How reflection can lead to action is where Sen and his focus on capabilities provides useful insights. And when Freire (1996) talks about the "awareness of the self" (p. 15), this could be understood as an awareness of one's capabilities. Critical consciousness thus entails conceiving of oneself as a capable actor of change in a particular situation and being aware of this situation's history and its intersecting relations of power on both micro- and macro levels. In other words, it is not enough to acquire the capabilities to change one's life situation, environment, and destiny; one must also recognize that one possesses these capabilities and believe that another world is possible and that this world is in reach. As Sadan (2004: 144) points out, perceived abilities are as important as actual abilities.

## 10.5 Technological Level of Communication Platforms: Empowerment, ICTs, and Affordances

Not only do we need to account for intersecting axes of oppression, the opportunities and constraints in a given situation, and actors' capabilities to change their life situation, destiny, and environment but also we need to recognize that the role of communication is important for social change to happen (Tufte 2017). In media and communication studies, changes in communication are often linked to societal changes at large. For example, it was argued that the advent of the printing press was tied to the rise of mass society and mass culture. Critical scholars (such as Horkheimer and the Frankfurt School) were early to relate communication practices to larger societal structures. Today, media scholars have made a similar argument, claiming that in tandem with the rise of more individualized and reflexive forms of communication – not the least through information and communication technologies (ICTs) – we are leaving mass society behind (see, e.g., Castells 1999). Communication has been linked to development through perspectives of the fields of communication for development and communication and social change (Tufte 2017). Indeed, political scientists, anthropologists, and sociologists increasingly recognize the importance of communication practices in everyday life for social and political change (ibid.: 80).

ICTs involve the electronic means of capturing, processing, storing, and communicating information. As a broad and catchy term, it has been endorsed in multiple positive contexts, including empowering marginalized and minority groups (Masika and Bailur 2015). The United Nations illustrates the potential for social and economic empowerment offered by ICTs and underlines the importance of women to make good use of this potential (UNDAW 2002: 3–4). However, access to ICTs is unequal. The so-called digital divide actually includes several gaps: the technological and content divide as well as the gender divide (UNDAW 2005: 1). People from the same social context may not have equal access to ICTs or similar abilities to deploy ICTs due to certain constraints such as poverty and illiteracy (ibid.).

Overviews of the study of ICTs in the global South reveal a field becoming biased toward techno-determinism and not sufficiently taking larger contextual and societal factors into account (Svensson and Wamala-Larsson 2015). Nevertheless, part of the techno-deterministic endurance could likely be explained by its appeal for funders (such as the World Bank; see Dodson et al. 2013: 19–20). If money is going to be invested in development measures, an optimistic and transformative view of what technology can achieve is important. However, there is the great risk of a backlash when studies show that up to 80 percent of ICT4D interventions fail (see Heeks 2002) – a stark reminder that the poor will not benefit from dumping hardware "and [the] hope [that] magic will happen" (Trucanero, in Dodson et al. 2013: 29).

Indeed, the role of ICTs in empowerment is a contested issue. This chapter started with a critique of development in general and ICT and development in particular. It is easy to fall into utopianism and determinism when talking about change in relation to communication technology. However, a critique of techno-determinism should not lead to an uncritical adoption of a sociocultural deterministic view of technology and its uses (i.e., that the context and sociocultural practices of technology determine everything). Communication platforms are not neutral technologies for people to use. They bring norms and values with them in the sense that there are certain preferred ways of interacting with the devices inscribed both in their design and how users use them.

Being mindful of the risk of different kinds of determinisms, the chapter turns to affordance theory (Gaver 1991). It is important to account for the properties of the particular communication platform used as well as the perception of these platforms by its users if one aims to understand and study the role of ICTs in empowerment in the global South. ICTs may, for example, afford new voices to be both raised and heard (Tufte 2017: 44), more dialogic and bottom-up communication, new forms of being political at the everyday level (ibid.: 65–66), greater possibilities for networking, and increasingly flat and flexible forms of social organization (ibid.: 83). It can be argued that such affordances exist independently of the users' perception of them, but it is not before communication platforms are perceived as tools that afford action for social change that they will be used as such. Consequently, most communication platforms appear to offer *perceptible affordances*, (i.e., information about the properties of the object which may be acted upon). Hence, communication platforms need to convey information about its affordances in a format that is compatible with the users' perceptual system. The actions communication platforms afford also need to be socioculturally relevant to the user. In line with this, Gibson (1977) argues that people do not interact with an object without first perceiving what it is good for.

Here, affordances can be connected to opportunities, meaning that ICT affordances could be labelled "communication opportunities" in a given situation or as "media opportunities" (see Tufte 2017: 136). As Cammaerts (2012) notes, some opportunities and constraints "are inherent to media itself" (p. 119). One can also argue that technology constraints change in terms of lack of access, know-how, and its affordance of surveillance (see Svensson and Wamala-Larsson 2016). This can, in turn, be connected to Sen's capability approach – having the capability/ability to use and understand the platforms at hand (i.e., media literacy). Thus, an analysis of empowerment should account for how communication platforms' affordances are interpreted and explored based on the users own defined needs from their standpoints and in their particular situations. By conceiving of communication platforms as affording particular practices while at the same time being open for users' needs and interpretations of such practices, it becomes possible to move beyond framings of technology as inherently good or bad in themselves, while avoiding resorting to the argument that they are neutral. Communication platforms dialectically intersect with society and interact with the users as they appropriate the technology to suit their particular life situation.

It is also important to underline that ICTs are only one part of an overall *communication ecology* of a given community (Treré and Mattoni 2015). Communication should be understood in a holistic and ecological manner (Tufte 2017: 21) and, therefore, not be studied in isolation. Understanding ICTs as part of a communication ecology will overcome the kind of communicative reductionism, one-platform bias, presentism, and fetishization of technological novelty that has been prevalent in studies of communication for social change (see Treré and Mattoni 2015). A communicative ecology lens also allows for a restoration of the communicative complexity in the study of actors struggling for empowerment and social change because it brings a holistic perspective beyond specific media instances (ibid.).

When studying communication platforms in the global South, deterministic stances should be discarded, which an affordance lens makes possible. Nevertheless, it is important to understand how affordances interact with both the sociocultural/political context/situation and the capabilities of individuals and groups to make use of communication platforms in empowering ways. To summarize, ICTs, or any communication platform for that manner, should not be studied in isolation. A communication platform itself is situated within a larger media and communication ecology in the particular context it is used. To understand the potentially empowering role of communication, an ecological perspective allows for a deeper and more context-aware analysis.

## 10.6 Concluding Remarks

Hill Collins (2000: 274) argues that we cannot develop an approach to empowerment without understanding how power is organized and how it operates. This chapter proposes that power is located in and, thus, operates from the following:

1. How those who are powerless are situated in intersecting structures of oppression (such as ethnicity, gender, class, and sexuality).
2. The opportunities and constraints of the social and political systems/institutions of the particular situations that those without power find themselves in.
3. The agency of individuals as well as collectives (and their interrelations).
4. Communication platforms and how they are perceived and used.

To summarize, empowerment is a tool for understanding how those who are powerless are situated in the different intersecting axes of oppression, how they have different opportunities and constraints in the various situations they find themselves in, their agency, and the role of communication platforms affordances. This tool is in line with the critique of development at the beginning of the chapter. A focus on empowering the disempowered through acquiring a critical consciousness and recognizing their capabilities avoids the kind of determinisms that have been prevalent and criticized in the field. Rather, this chapter proposed to study and understand empowerment on four levels – the intersectional, the contextual, the agency, and the technological. It is important to stress that relations, capabilities, situations, and intersections are constantly changing and not static; hence, these levels need to be regularly reviewed. As Van Hemelrijck (2013: 34) emphasizes, persistent and collaborative efforts are needed to realize empowerment and social change.

Development problems are complex (Servaes and Lie 2015: 141). Oppression and powerlessness are also complex (Hill Collins 2000: 289); therefore, the study of empowerment and resistance must also be complex (ibid.). One way to summarize empowerment and its attached levels is to connect questions to each of them. This model of empowerment can also be applied to an analysis of ICTs in the global South.

**Intersectional level:** This level concerns the roots of powerlessness in terms of intersecting structures of power such as ethnicity, gender, class, and sexuality. How are the powerless under study situated in the larger structures of power? Such a description should not be made *von oben*, but rather involve the standpoints of those under study in order for them to define their own realities on their own terms. Here, it is also important to avoid the individual versus collective dichotomy, as it is more appropriate to explain how these dimensions interrelate instead.

**Contextual level:** This level concerns the particular context of those without power and how their actions are situated in this context. How do their contexts provide both opportunities and constraints (in terms of institutions, local community, local culture, economic and family arrangements, et cetera)? Which configuration of resources, institutional arrangements, and historical precedents exist in the contexts of those under study? How open or closed to change are important institutions, authorities, or organizations? What is the capacity of these institutions, authorities, or organizations to advance change and effectively implement policies once they are decided?

**Agency level**: This level concerns the capabilities of individuals and groups under study and their ability to make choices and thus control their life situation, destiny, and environment. What capabilities, resources, agencies, skills, choices, and achievements can be discerned? How critically conscious are those under study of their situation? Do they have possibility/ capability to imagine that change is possible? And does such critical consciousness lead to action and, thus, change?

**Technological level:** This level concerns the affordances of the communication platforms used by those under study. What access to communication platforms do they have? What is their communication/media literacy? How do they use these platforms? What affordances do they perceive these platforms to have? Do they find new affordances over time? And how are these communication platforms situated in larger media/communication ecologies, and how does this influence their use?

The social world is complex, holistic, and dynamic in a manner which is difficult to outline in a linear text such as the format of this chapter. Also apparent in the chapter is that these levels intersect, merge into, and influence each other. It is easier to separate these levels in writing than in an empirical reality. Therefore, the chapter will conclude by applying these levels to the situation that market women in Kampala find themselves in (see Svensson and Wamala-Larsson 2016; Wamala-Larsson and Svensson 2018 for details on the study and how it was carried out).

The study shows that, while mobile phones helped these women stay in control of their small businesses and juggle both family life and coordinate sales goods delivery, larger patriarchal structures and the lack of state welfare held them back. While the mobile phone was used for increasing their customer base and making transactions through mobile payments, the same device was used by their male partners for surveillance – to control them. Moreover, just when the women began to make money on their own– which, to an extent, led to greater independence and the ability to take control of their own life situation – their male partners, and the fathers of their children, reneged on their financial obligations. The lack of a welfare state in Uganda to provide basic services such as free education intersects with patriarchal notions that women should be in charge of the household and raise children. The fact that these women made money on their own was used as an excuse for their male partners to renege on their parental obligations.

On the contextual level, the market women's stories showed that women in Uganda (and Africa in general) are those who historically are supposed to trade in street markets. This opened up opportunities for these women to earn a small living. Mobile communication in general and mobile money, in particular, have been helpful here. With mobile money, the market women do not need to handle large amounts of cash. Mobile money has also brought banking to segments of the population that previously did not use banks. Via the mobile phone, these women could also keep a repository of customers and their preferences in order to contact them in the future when their stock was renewed. At the same time, the local authority (KCCA – Kampala Capital City Authority) challenged the street markets and tried to push the women out of the open-air street markets in favor of more expensive shopping grounds in a general attempt to beautify the city. The street markets had been burned down many times. Also, stories from the women suggested that mobile service providers, through mandatory SIM card registration, collaborated with KCCA in order to keep track of them. Once again, the drawback of mobile communication in terms of empowerment seems to be its affordance of surveillance – from male partners as well as the local authorities.

When asked questions relating to their empowerment from the agency level, the market women responded that they did not sense they could do anything about their life situation, destiny, and environment. For example, they did not feel they had the capabilities to fight KCCA in order to keep their vending spots in the Kampala street markets. Nevertheless, it cannot be claimed that the women lacked critical awareness. Rather, they did not foresee any possibilities/opportunities for action. Concerning other issues, they did act; for example, the women organized saving circles among themselves to make sure they would all have money when school fees were due. They were also very skillful in dealing with money transactions, in particular mobile money, and could quickly secure mobile money transactions during times of hectic trading with multiple traders.

The women claim that mobile phones also make it easier to keep a repository of customers to contact when goods came in or refer customers to their colleagues (often in the same saving circles). Some of their customers prefer not to go into these maze-likes markets. By sending pictures over their mobile phone and by offering mobile money transactions, the women could secure a deal without meeting in person. The market women would then send the goods by bus to a point where the customer could then collect them. This is about affordances of mobile communication and mobile money and the ability to imagine other possibilities over time.

Developing intricate communication systems (without airtime being lost) via so-called beeping (i.e., lost calls) is another example of this. From an ecology perspective, mobile communication is clearly connected to the important traditional oral culture of Uganda. Mobile communication does not require texting, but rather talking, and thus suits the communication ecology of Uganda.

These short glimpses into an empirical case have hopefully shown that the suggested framework can provide a critical analysis of ICTs in the global South which neither resort to techno-determinism nor view technology as neutral. This analysis also shows how empowerment is a complex concept and how a group – in this case Kampala market women – can both be empowered and disempowered at the same time, depending on the level and situation. This hopefully shows how empowerment can be used as an analytical concept in a defined and structured manner.

**Acknowledgments** The argument developed in this chapter was influenced by collaborations, discussions and conference contributions together with Dr Caroline Wamala-Larsson (SPIDER, Karlstad and Stockholm University), Dr Cecilia Strand (Uppsala University) and YuQin Xi, a master's student at Uppsala University who is doing a research internship with Dr Wamala Larsson and the author.

## References

Adams R (2008) Empowerment, participation and social work. Palgrave Macmillan, New York
Bali Swain R (2012) The microfinance impact. Routledge, London
Burton M, Kagan C (1996) Rethinking empowerment: shared action against powerlessness. In: Parker I, Spears R (eds) Psychology and society: radical theory and practice. Pluto Press, London, pp 197–208
Bush RAB, Folger JP (1994) The promise of mediation: responding to conflict through empowerment and recognition. Jossey-Bass, San Francisco
Cammaerts B (2012) Protest logics and the mediation opportunity structure. Eur J Commun 27(2):117–134
Castells M (1999) The information age. Blackwell, London
Choudhury N (2009) The question of empowerment: women's perspective on their internet use. Gend Technol Dev 13(3):341–363
Conyers J (1975) Towards black political empowerment. Can system be transformed. Black Scholar 7:2
Cornwall A, Nyamu-Musembi C (2004) Putting the 'rights-based approach' to development into perspective. Third World Q 25(8):1415–1437
Dodson LL, Sterling SR, Bennett JK (2013) Considering failure: eight years of ITID research. Inf Technol Int Dev 9(2):19–34
Drolet J (2011) Women, micro credit and empowerment in Cairo, Egypt. Int Soc Work 54(5):629–645
Freire P (1996) Pedagogy of the oppressed. Penguin Books, London (first published 1970)
Gaver WW (1991) Technology affordances. Proceedings of the SIGCHI conference on human factors in computing systems, ACM, pp 79–84
Gibson J (1977) The theory of affordances. In: Shaw R, Bransford J (eds) Perceiving, acting, and knowing. Lawrence Erlbaum, New York, chapter 3
Hann C, Hart K (2011) Economic anthropology. History, ethnography, critique. Polity Press, Cambridge

Haraway D (1991) Simians, cyborgs, and women: the reinvention of nature. Routledge, New York
Harding S (1991) Whose science? Whose knowledges? Thinking from women's lives. Cornell University Press, Ithaca
Heeks R (2002) Information systems and developing countries. Failure, success, and local improvisations. Inf Soc 18:101–112
Hill Collins P (2000) Black feminist thought: knowledge, consciousness, and the politics of empowerment, 2nd edn. Routledge, London
Kabeer N (1999) Resources agency, achievements: reflections on the measurements of women empowerment. Dev Chang 30:435–464
Kitschelt A (1986) Political opportunity structures and political protest: anti-nuclear movements in four democracies. Br J Polit Sci 16:57–85
Kleine D (2013) Technologies of Choice? ICTs development and the capability approach. MIT Press, Cambridge
Kole SE (2001) Appropriate theorizing about African women and the internet. Int Fem J Polit 3(2):155–179
Leenie J, Tacchi J (2013) Evaluating communication for development. A framework for social change. Routledge, London
Lincoln ND, Travers C, Ackers P, Wilkinson A (2002) The meaning of empowerment: the interdisciplinary etymology of a new management concept. Int J Manag Rev 4(3):271–290
Masika R, Bailur S (2015) Negotiating women's agency through ICTs: a comparative study of Uganda and India. Gend Technol Dev 19(1):43–69
Meyer-Emerick N (2005) Critical social science and conflict transformation. Opportunities for citizen governance. Int J Organ Theory Behav 8(4):541–558
Nederveen Pieterse J (2010) Development theory, 2nd edn. Sage, London
Offenheiser RC, Holcombe SH (2003) Challenges and opportunities in implementing a rights-based approach to development: an Oxfam America perspective. Nonprofit Volunt Sect Q 32(2):268–301
Prahalad CK, Hart SL (2002) The fortune at the bottom of the pyramid. Strategy Bus 26. https://www.strategy-business.com/article/11518?gko=9a4ba. Accessed 31 Dec 2017
Rappaport J (1984) Studies in empowerment: introduction to the issue. Prev Hum Serv 3:1–7
Riordan E (2001) Commodified agents and empowered girls: consuming and producing feminism. J Commun Inq 35(3):279–297
Ristock JL, Pinnell J (1996) Community research as empowerment: feminist links, postmodern interruptions. Oxford University Press, Oxford
Sadan E (2004) Empowerment as community practice. http://www.mpow.org/elisheva_sadan_empowerment.pdf. Accessed 15 Dec 2017
Sen A (1999) Development as freedom. Oxford University Press, New York
Servaes J, Lie R (2015) New challenges for communication for sustainable development and social change: a review essay. J Multicult Discourses 10(1):124–148
Servaes J, Malikhao P (2014) The role and place of communication for sustainable social change (CSSC). Int Soc Sci J 65:171–183
Smith M, Spence R, Rashid A (2011) Mobile phones and expanding human capabilities. Inf Technol Int Dev 7(3):77–88
Svensson J, Wamala-Larsson C (2015) Approaches to development in M4D studies: an overview of major approaches. In: Kumar V, Svensson J (eds) Promoting social change through information technology. IGI Global, Hershey, pp 26–48
Svensson J, Wamala-Larsson C (2016) Situated empowerment. Mobile phones practices among market women in Kampala. Mobile Media Commun 4(2):205–220
Treré E, Mattoni A (2015) Media ecologies and protest movements: main perspectives and key lessons. Inf Commun Soc 19(3):290–306
Tufte T (2017) Communication and social change. A citizen perspective. Polity Press, Cambridge
United Nation Division for the Advancement of Women (UNDAW) (2005) Gender equality and empowerment of women through ICT. Department of Economic and Social Affairs, New York

United Nations Division for the Advancement of Women (UNDAW) (2002) Information and communication technologies and their impact on and use as an instrument for the advancement and empowerment of women. Report of the expert group meeting, Seoul

Van Hemelrijck A (2013) Powerful beyond measure. Measuring complex systemic change in collaborative setting. In: Servaes J (ed) Sustainability, participation and culture in communication: theory and praxis. Intellect-University of Chicago Press, Bristol/Chicago, pp 25–58

Wamala-Larsson C, Svensson J (2018) Mobile phones and the transformation of the informal economy. Stories from market women in Kampala. J East Afr Stud. http://www.tandfonline.com/doi/full/10.1080/17531055.2018.1436247

Zimmerman MA (2000) Empowerment theory: psychological, organizational and community levels of analysis. In: Rappaport J, Seidman E (eds) Handbook of community psychology. Springer, New York, pp 43–63

# The Theory of Digital Citizenship

## 11

Toks Dele Oyedemi

## Contents

| | | |
|---|---|---|
| 11.1 | Introduction: Rethinking Citizenship in the Digital Age | 238 |
| | 11.1.1 Theoretical Approaches to Citizenship and Implications for Internet Access | 239 |
| 11.2 | The Theory of Digital Citizenship | 241 |
| | 11.2.1 Participation and Digital Citizenship | 246 |
| | 11.2.2 Cultural Citizenship and Digital Citizenship | 248 |
| | 11.2.3 The Elements of the Theory of Digital Citizenship | 250 |
| 11.3 | Conclusion | 253 |
| 11.4 | Cross-References | 253 |
| References | | 253 |

### Abstract

Increasingly new digital technologies continue to shape the ideal of participation in relation to citizenship in society. The Internet is not only a tool for engendering development and social change, but it provides spaces where this change occurs. Citizens need information for effective participation in the political, economic, and social spheres and for their ability to foster social change in these spheres. The Internet as a medium and a tool provides this information, and as such, it becomes relevant for the realization of citizenship in society. As the Internet continues to evolve technologically and culturally, concern about access to it still remains relevant to scholarly and policy discourses. The concerns about unequal access to the Internet have a historical trajectory that has evolved into many concepts, approaches, and metaphors. Examples abound: digital divide, digital inequality, digital entitlement, virtual inequality, and many others. The concept of

---

T. D. Oyedemi (✉)
Communication and Media Studies, University of Limpopo, Sovenga, South Africa
e-mail: toyedemi@gmail.com

© Springer Nature Singapore Pte Ltd. 2020
J. Servaes (ed.), *Handbook of Communication for Development and Social Change*,
https://doi.org/10.1007/978-981-15-2014-3_124

digital citizenship is not new; it has been applied in the European and American contexts to draw attention to the importance of the Internet to citizens' ability to participate in society. However, it remains a strong symbolic term without practical implementable framework. Through critical theoretical engagements with the concepts of citizenship, various theoretical articulations of rights, and a critical analysis of the discourse of participation, this chapter proposes a theory of digital citizenship with attendant elements that form its framework. This framework provides a guide for the pursuit of cultural, economic, and social change required for active citizenry aided by technology access. It highlights the relevance of the Internet together with other socioeconomic and political resources in shaping today's ideal of citizenship.

**Keywords**

Internet · Citizenship · Participation · Human rights · Digital citizen

## 11.1 Introduction: Rethinking Citizenship in the Digital Age

This chapter conceptualizes digital citizenship in relation to access to the Internet. Although the concept has been applied to studies of access and use of the Internet, it remains a strong symbolic term without practical implementable framework. There is a need for a critical attempt to expand this concept from its current descriptive and symbolic connotations to a theoretical framework for studying access to the Internet. Citizens need information for effective participation in the political, economic, and social spheres in society. The Internet is a medium and a tool that provides this information, and as such, it becomes relevant that citizens have access to this technology. Digital citizenship describes the ability to address the inequalities that exist in access and use of the Internet. For Mossberger et al. (2008), digital citizenship is a concept that describes the ability of an individual to participate in a society online. D'Haenens et al. (2008) apply the term to describe how citizens use the Internet for certain types of context in engaging civic and social activities, and Servaes (2003) applies the concept in tackling the rapid revolution in communication technologies and how lack of access could create or expand the notion of information inequalities.

These applications of the concept imply that digital citizens are those who use the Internet effectively on a regular, preferably daily, basis. The argument is that the concept of digital citizenship is a move away from the narrow concepts of "haves" and "have-nots" that describe the regime of digital divide, where research on digital technologies is engaged in terms of physical intermittent access to the Internet and, at times, skills to use this technology. As Mossberger et al. (2008) note, daily Internet use in the concept of digital citizenship implies sufficient technical skills for effective use along with frequent means of access. They note further that the concept of digital citizenship highlights the importance of the Internet in enhancing citizen's participation in society on three aspects: (1) the inclusion in prevailing forms of communication through regular and effective use of the Internet, (2) the impact of Internet use on the ability to participate as democratic citizens, and (3) the effects of the Internet on the quality of opportunity for citizens in the marketplace and in the economic domain generally. Hence, digital citizens are those who use digital technologies frequently for fulfilling social activities, for sourcing information to fulfill civic duty, and for financial and economic activities. But to further engage this concept, a theoretical analysis of citizenship is required in order to justify its application to technology and society.

## 11.1.1 Theoretical Approaches to Citizenship and Implications for Internet Access

The concept of citizenship in relation to the discourse around the Internet predictably elicits simple, but critical, questions about the relevance of the political concept of citizenship to Internet access and use. It raises question about what it means to invoke citizenship in envisioning universal access to the Internet. The rights attached to citizenship make it an attractive concept for Internet access; it locates access within the framework of rights, social justice, and equity. Janoski (1998) provides an analysis of the political theory of citizenship. He defines citizenship as the "passive and active membership of individuals in a nation-state with certain universalistic rights and obligations at a specific level of equality" (p. 9). This definition is relevant, for it touches on the core issues essential to the discourse here, and thus warrants further explanation. Firstly, as he explains, this definition identifies "membership" as one of the characteristics of citizenship. Citizenship bestows membership in a nation-state to the individual or groups of individuals irrespective of ethnicity, race, gender, class, or disability. In a situation that this categorization becomes a tool to deprive the individual effective participation in a society, the arguments are often made of deprivation of citizenship or relegation of one's citizenship. This argument has supported calls for policy of inclusion for stigmatized groups based on the aforementioned categories.

Secondly, citizenship involves "passive and active rights," passive rights of existence and active rights that include present and future capacities to participate and influence politics and economy in a democratic polity. Thirdly, Janoski (1998) notes that citizenship rights are universalistic rights enacted into law and implemented for all citizens. Lastly, citizenship is a statement of equality. Consequently, these concepts of membership, inclusion, rights, and equality become relevant in the way we reframe access to the Internet.

There are different approaches to the political theory of citizenship. The liberal concept of citizenship is based on the idea of individual freedom and liberty (Schuck 2002; Janoski 1998). Influential proponents of liberalism, such as
John Locke and John Stuart Mill, expounded individual rights, freedom, and liberty as cornerstones of citizenship. The liberty in the liberal notion is the claim of equal opportunity for participation in a society, a claim that the state, according to Berlin (1958), should act affirmatively to create or secure those substantive entitlements,

such as income, health care, and education that individuals need in order to lead the dignified, independent lives essential to their freedom (Schuck 2002).

In this notion, policy on access to education is seen as attempt to provide individuals with the capacity to participate and pursue individual freedom – socially, economically, and politically. Likewise with regard to the Internet, the benefits that the Internet provides make access an important issue in order for citizens to gain equal opportunity to information that make them participate effectively in society. Also included in the liberal idea of individual equality and participation is the notion of free market, which explicitly implies individuals' equal participation in the market and implicitly affirms that competition is fair. The assertion of individualism borders and conflicts with egalitarianism, and as a result, it draws many criticisms. For instance, liberal individualism presupposes equality; however, due to capitalism and other social formations, inequality is always present in individual's ability to access utility that makes citizenship attainable. Technological innovations, such as the Internet, always generate new forms of inequality. Consequently, economic and social opportunities in this tradition may justify expanding individual access to the Internet, beyond the market efficiency arguments and rhetoric of effective competition.

Communitarianism is another approach to citizenship. It puts strong emphasis on the collective as in a community or a nation. The goal in this tradition is toward achieving effective and just society, which is built through mutual support and group action, not atomistic choice and individual liberty of the liberal tradition. With regard to citizenship, the ultimate goal is to "build a strong community based on *common* identity, mutuality, autonomy, participation, and integration" (Janoski 1998, p. 19).

Republican citizenship or republicanism is similar to communitarianism. Dagger (2002) argues that the key ideas of republican citizenship are the concepts of community and the common good, as against the liberal idea of individualism. He notes that citizens in the republican ideology are members of the public, who must be prepared to overcome personal inclinations and set aside their private interests, when necessary, to do what is best for the public as a whole. Citizenship in this tradition requires commitment to common good, and public interest, and places the interest of the community ahead of personal interest. For example, the notion of public education is seen as enhancing public and common good rather than the liberal version of education serving to promote individual equality of opportunity (Oyedemi 2015). The notion of public interest in communication policy is another example. Specifically, individual access to the Internet may be argued as enhancing the common good of the community or the nation as a whole. In essence, this ideology implies that personal welfare is dependent on the general welfare in order to enhance the well-being of a community (Bellah et al. 1985).

In studying the American polity and politics, Roger Smith (1993) refers to ascriptive hierarchy to engage the "inegalitarian" ideologies that defined the political status of racial and ethnic minorities and women through most of US history. This ascriptive hierarchy tradition describes the exclusion of large segment of the population from full citizenship based on ascriptive traits such as race, gender, or ethnicity. Ascriptive hierarchy describes the inequality in a society based on these social categorizations. In a similar note, Janoski (1998) discusses expansive democracy theory as the approach that deals with the expansion of rights, mostly of individual and organizational rights of people who have been discriminated against including most class, gender, and ethnic groups. The deprivation of millions of native Africans citizenship in South Africa during apartheid regime stands as strong instance of ascriptive hierarchy, and the expansive democracy theory explains the post-apartheid expansion of rights to those previously disenfranchised. These two theories reflect the different shapes of social inequalities evident in contemporary society. These theories highlight the enduring inequalities which are visible in the current context of society online, and research studies have shown that income, education, age, race, gender, and ethnicity all have impacts on access to the Internet.

T. H Marshall (1950, 1964, 2009) provides a seminal analysis of citizenship with a tripartite framework of civil, political, and social rights. Civil rights include the rights necessary for individual freedom and the liberties of the citizen. This includes rights enshrined in freedoms of speech, thought, and faith. This also includes the right to justice and the right to own property. The right to participate in the political and democratic systems is encapsulated in "political rights"; this right involves participation based on membership in political authority and participation in the electoral process. Social rights include a broad range of rights, from rights enshrining basic economic welfare and security to rights to share and participate in social heritage and "to live the life of a civilized being according to the standards prevailing in the society" (Marshall 2009, p. 149). Internet access is important in the realization of freedoms and participation enshrined in civil rights, political rights, and social rights of citizens (Oyedemi 2015).

Irrespective of these disparate theorizations of citizenship, a universal agreement is derived from a belief that citizenship is a concept that encapsulates several basic rights. These rights can be categorized as legal rights, political rights, social rights, and participation rights; see Table 1.

In this typology, access to the Internet is important for sourcing the information that makes all these rights possible. The emphasis on rights leads us to the issue of social justice and human rights, and it strengthens the application of citizenship to Internet access.

The notion of equality as a component in framing human rights is particularly relevant to a theorization of citizenship. The Universal Declaration of Human Rights (1948) is one of the main sources of the conceptual understanding of human rights and the framing of citizenship. The rights included in the declaration are critical in the formulation of the contemporary notion of citizenship.

## 11.2 The Theory of Digital Citizenship

Bearing in mind the rights embedded in citizenship, digital citizenship describes the universal availability of the Internet, the importance of every citizen to gain access to this technology for effective participation in society. A digital citizen is someone with *regular* and *flexible* access to the Internet, the *skills* to apply this technology,

**Table 1** Four types of citizenship rights

| Legal rights | Political rights | Social rights | Participation rights |
|---|---|---|---|
| A. Procedural rights<br>1. Access to courts and counsel<br>2. Rights to contract<br>3. Equal treatment under the law<br>4. Rights of aliens to immigrate and citizens to emigrate | A. Personal rights<br>1. Enfranchisement of the poor, gender groups, ethnic/racial groups, age categories, and immigrants<br>2. Rights to run and hold office<br>3. Rights to form and join a political party | A. Enabling and preventive rights<br>1. Health services<br>2. Family allowances<br>3. Personal and family counseling<br>4. Physical rehabilitation | A. Labor market intervention rights<br>1. Labor market information programs<br>2. Job placement programs<br>3. Job creation services |
| B. Expressive rights<br>1. Freedom of speech<br>2. Freedom of religion<br>3. Choice of friends, companions, and associates<br>4. Rights to privacy | B. Organizational rights<br>1. Political lobbying<br>2. Political fund raising<br>3. Legislative and administrative consultation<br>4. Political bargaining | B. Opportunity rights<br>1. Pre-primary education<br>2. Primary and secondary education<br>3. Vocational education<br>5. Educational assistance for special groups | B. Firms and bureaucracy rights<br>1. Job security rights<br>2. Workers' councils or grievance procedure rights<br>3. Client participation in bureaucracy or self-administration<br>4. Collective bargaining rights |
| C. Bodily control rights<br>1. Freedom from assault and unsafe environment<br>2. Medical and sexual control over body | C. Naturalization rights<br>1. Right to naturalize upon residency<br>2. Right to information on naturalization process<br>3. Refugee rights | C. Distributive rights<br>1. Old-age pensions<br>2. Public assistance<br>3. Unemployment compensation | C. Capital Control rights<br>1. Codetermination rights<br>2. Wage earner and union investment funds<br>3. Capital escape laws<br>5. Regional investment and equalization programs |
| D. Property and service rights<br>1. Hold and dispose of property and services<br>2. Choice of residence | D. Oppositional rights<br>1. Minority rights to equal and fair treatment<br>2. Political information and inquiry rights<br>3. Social movement and protest rights | D. Compensatory rights<br>1. Work injury insurance<br>2. War injury pension<br>3. War equalization | |

(continued)

**Table 1** (continued)

| Legal rights | Political rights | Social rights | Participation rights |
|---|---|---|---|
| 3. Choice of occupation | | 4. Rights infringement compensation | |
| E. Organizational rights<br>  1. Employee organizing<br>  2. Corporate organizing<br>  3. Political party organizing | | | |

Source: Janoski (1998).

and a *regular use* of the Internet for participation and functioning in all spheres of the society. Digital citizenship has been applied by scholars to address the impact of technology on current society: Servaes (2003) in the debate around the European Information Society applies the term to describe the need to understand the role of ICTs in relation to people's ability to participate in society.

D'Haenens et al. (2008) also use digital citizenship in describing how ethnic minority groups in the Netherlands use ICTs for civic activities and participation. Mossberger et al. (2008) define digital citizenship as the ability to participate in society online, and digital citizens are those who use the Internet regularly and effectively. The work of Amartya Sen, on inequality, capability, and functioning, becomes relevant in this context. Drawing from Sen's (1992) capability approach, Mossberger et al. (2008, p. 2) view digital citizenship as representing capability, belonging, and the potential for political and economic engagement in society. Garnham (1999) also draws on Sen's capability approach to argue for access to communication technologies for citizens; he asserts communication technology as a citizen's entitlement in achieving the notion of "informed citizenship."

The application of the term, digital citizenship, is relevant in highlighting the importance of the Internet for citizens in enhancing their rights and responsibilities and being able to participate in all spheres of society. However, one shortcoming of the application of this term is the lack of a theoretical framework to apply to the concept. Consequently, the term risks being perceived like many other metaphors and concepts used to describe issue of access to ICTs, which are strong in invoking emotions, but lacks a theoretical framework. Besides employing concepts such as equality, participation and citizen's rights, digital citizenship invokes emotions, but lacks a framework.

To buttress this criticism, let us engage Mossberger et al.'s (2008) use of the term. They apply the term extensively in arguing for participation in society through Internet use. For them, digital citizenship is mostly achieved through frequent use of the Internet, specifically daily use. They argue that "the frequency of use, especially daily use, more accurately measures digital citizenship than home access or simply having used the Internet at some point. Those who have Internet connections at home may still lack the ability to find and evaluate information online" (p. 10). Based on this, they define digital citizens as "those who use the Internet daily" (p. 120) and that the frequency of use is a better measure of capacity for digital citizenship. The core issues of digital citizenship, as discussed by Mossberger et al., are the application of citizenship values; the opportunities that the Internet offers for social, economic, and political participation; and a daily use of the Internet. They paid limited emphasis on skills as essential component of being a digital citizen, assuming that daily use presupposes "sufficient technical competence and information literacy for effective use along with some regular means of access" (p. 1).

Undoubtedly, Mossberger et al.'s discussion of digital citizenship provides a useful compass for the application of the term; especially the focus on political idea of citizenship values, issues of inequalities, and the daily use of the Internet to participate in society are all relevant to the theoretical framework developed here. However, a theorization of digital citizenship needs to extend beyond these issues. The relative widespread of the Internet in the United States may have led to a large focus on Internet use, but in many other countries, especially the developing ones, other issues of physical access, skills, and effective policies are still central to how digital inequalities are engaged. While home access does not automatically connote use, as they argue, and the frequency of use is a better measure, it is also critical to acknowledge that home connection provides huge advantages, presupposes regular use, and allows for flexibility of use without the constraints of public access or access at work place.

The definition of digital citizen as daily Internet users is limited and needs to be expanded. Someone that receives and sends e-mail on a daily basis cannot be referred to as a digital citizen; the assumption that daily use presumes information literacy skills is a narrow approach to digital skills. There is a need to expand the scope of digital skills beyond information searching capability to include the ability to create content. This includes creating relevant information, posting multimedia content, and many more. Consequently digital citizen is defined here, not merely as daily Internet user but as *someone with regular and flexible access (implicitly individual and household access) to the Internet, the skills to apply this technology, and a regular use of the Internet for participation and functioning in all spheres of the society.*

Flexibility of access offers the ability to use the Internet anytime one wishes without any structural or institutional hindrances. Individual and household types of Internet access offer the highest flexibility of access and are crucial in attaining digital citizenship, contrary to the constraints often confronted in public access places or workplaces. Individual access includes, but not limited to, broadband access on mobile devices. Essentially, there is a need to further develop the concept of digital citizenship and propose elements that may comprise a theory of digital citizenship. This leads to two critical questions: Can a theory of digital citizenship be developed? And how might this be done? A key contribution of this chapter is the attempt to move the concept of digital citizenship from being a descriptive terminology to a theory that can easily be applied to issue of access to communication technologies in different geographical contexts.

The developments in Internet technologies and the resultant changes occurring in human interactions and sociability call for theories to explain the role of ICTs in today's cultural and economic practices. It requires that we explain the dynamic nature of the Internet and its effect on human interaction beyond the technology deterministic theoretical approaches. Scholars have theoretically engaged the impacts of the Internet on various aspects of cultures, such as identity (Turkle 1995, 2005) and community (Willson 2006).

Burnett and Marshall (2003) engage a compendium of theories in discussing "web theory." Rather than identifying a universal and possibly a unified theory of the web, they apply various forms of theoretical approaches in engaging the web as a technology, as networks and information, as communication, as platform of identity configurations, and as political economic domain of economy, policy, and regulation. Webster's (1995) analysis of information society is an early and seminal attempt to theorize the concept in explaining the impact of digital technologies on society. He draws on a structural-functional theoretical approach to develop a theory of an information society. He engages the information society from technological, spatial, economic, occupational, and cultural perspectives.

A theory of digital citizenship is identified as a critical theory because it attempts to address issues of inequality. It is also an empirical theory, as Williams (2003) notes, empirical theory seeks to build a model or develop a conceptual framework to explain the phenomena it is trying to understand. Digital citizenship addresses how personal access to the Internet allows people to function in every sector of society. It is largely used to tackle social inequalities in society as they shape access to the Internet and how this access can conversely help address certain layers of social inequalities.

Digital citizenship focuses on individual Internet access, daily Internet use, digital skills, and the citizen's right to gain access to this resource for effective participation in society. This approach is contrary to the concept of "digital divide" and its policy implementations that largely focus on physical access that is often publicly available.

The rationale for developing a theory of digital citizenship is based on the fact that many concepts, approaches, and metaphors have been developed to tackle the issue of access to the Internet. Examples abound: digital divide (NTIA 1995; Servon 2002), digital inequality (Dimaggio and Hargittai 2001), digital entitlement (Mansell 2002), virtual inequality (Mossberger et al. 2003), etc. As Kuhn (1996) notes, when theories proliferate, it leads to a crisis, and the crisis leads to theoretical revolution where a new theory develops to engage the theoretical crisis. While many theorists believe in the same ideology of providing access to the unconnected, it has been engaged and studied differently in multiple of ways. The theoretical landscape calls for a new approach that adopts a universal concept. The universal concepts of human rights and social justice are critical to the understanding of digital citizenship. Other rationalizations of the concept are as follow:

1. It moves the discussion away from mere access to the issue of rights. Having connection to the Internet should be a right for a functional participation in society.

2. It touches on citizenship rights, social justice, entitlement, and equality of opportunities.
3. It moves discussion away from bridging the digital divide to locating access to the Internet as a precondition for full citizenship in the digital age.
4. It recognizes that the Internet is crucial for an effective democracy and citizenship participation. Just as the news media have been critical in fostering an informed citizen, which is crucial for the enrichment of democracy, the Internet makes many social, civic, economic, and political activities of citizens possible.

Research and opinions about digital citizenship are largely focused on technological and social inequalities that hamper full citizenship participation in the current society that relies on new communication technologies as the predominant platforms for most social and cultural activities and interactions.

## 11.2.1 Participation and Digital Citizenship

The concept of participation is essential in theorizing digital citizenship. From its seminal use in democratic theory, participation has materialized as a philosophical framework that has been applied to studies in many fields, such as politics, spatial planning, development and social change, arts and museums, and communications (Carpentier 2011). The concept of participation is critical in understanding the role of the individual in connection with the community and the society as a whole. It is an important concept in the study of democracy and democratic theory (Pateman 1970; Held 2006), has been used to study social inclusion and citizenship (Mertens and Servaes 2011), and juxtaposed against concept of alienation (Mejos 2007).

In democratic theory, Carpentier (2011) identifies a minimalist and maximalist views of participation. Minimalist participation reduces the political role of citizenry to participation in the election process resulting in representation and delegation of power. Maximalist democratic participation is more substantial, is multidirectional, and is not limited to mere election of representatives. Similarly, Thomas (1994) identifies micro- and macro-participation, where macro-participation reflects participation in the national political spheres of a country and micro-participation is about the local spheres of school, family, workplace, church, and community. The concept of participation is a critical aspect of the study and process of communication and media through theories of audience participation and communication rights (Carpentier 2011). The concept is a prominent framework in the theories, process, and study of communication for development and social change (Servaes 1999; Servaes and Malikhao 2008; Huesca 2008); here the emphasis is on the equality of participants in decision-making and their empowerment to be actively involved in shaping phenomenon that affects their being, individuality, community, and cultural identities.

With the current regime of Internet technologies and web 2.0 and beyond, the concept of participation becomes strongly articulated into the discourses of new media and citizenship (Chadwick 2006; Jenkins 2008; Kann et al. 2007). Firstly, the analysis is on the participatory culture that new media and the Internet engender through collaborative authoring and the interactivity aspect of social media. This culture is important for citizenship values of openness and democracy; as Jenkins (2008) notes, this participatory culture contrasts with older ideas of a passive public and inert media spectatorship, and it creates a "current moment of media change [which] is affirming the right of everyday people to actively contribute to their culture" (p. 136). Secondly, new media and the Internet allow citizens to participate in many aspects of society, for example, as Carpentier (2011) observes, it leads to clusters of debate on citizens' participation in institutionalized politics, through many e-concepts (such as e-government, e-democracy, e-campaigning, etc.).

The concept of participation as applied here takes the maximalist approach; it refers to the ability of citizen to gain access to resources that facilitate their inclusion and decision-making process in all spheres of society. It involves the ability to participate in the economic, political, cultural, and social spheres of society as they relate to everyday practices. Specifically, it highlights how the Internet becomes additional resource for the individual to participate effectively toward full citizenship in society. Mertens and Servaes (2011) note citizenship is beyond the legal and political ideologies; citizenship should also be a social citizenship, which implies social participation.

Mertens and Servaes (2011) engage the concept of participation as a core philosophy of social inclusion and relate this to how information and communication technologies are resources in achieving social inclusion. They identify two types of participation that are relevant to the current discourse on digital citizenship. The first type of participation is called *structural participation*, this involves having a good economic position to facilitate participating in society structurally, and the dimensions of this include being employed and unemployed. In addressing this structural participation, they acknowledge that some aspects of ICT policy promote skills for the unemployed. Another dimension of structural participation is the quality and level of the labor position that people hold, as in the case that jobs with a higher level of ICT use are structurally better jobs. A third dimension of structural participation is education, and educational qualification can have correlation with one's economic position. With reference to ICT, they argue that policy activities oriented toward ICT education can be seen as processes that promote structural participation.

A second type of participation is *sociocultural participation*. This refers to the contacts people have in their networks with family and friends, their social contact in organizations, and their activities and influence in politics and political organizations (Mertens and Servaes 2011). These dimensions of sociocultural participation can equally be influenced by technology and ICT policy. Access to the Internet can foster the ability of citizens to participate structurally, socially, economically, politically, and culturally in society and foster digital citizenship. In essence, the concept of digital citizenship addresses the ability of the individual to utilize the Internet as a resource in participating in every sphere of society. The overall assertion is that citizenship is hampered if some people are unable to participate in society based on their lack of access to a relevant medium of information and communication.

## 11.2.2 Cultural Citizenship and Digital Citizenship

The discussions of citizenship, participation, and social inclusion highlight the notion that citizenship is more than a legal and political concept that describes the rights and civic obligations of an individual. Citizenship is equally social and cultural (Delanty 2003; Stevenson 2001). Culture, broadly defined, reflects every facet of human life – "the way we do things, the way we represent the world, the way we construct notions of self and other" (Ommundsen 2010, p. 4). In a narrow sense, culture may be seen as aesthetic practices as in high and low (or popular) culture or a signifier of group or communal identity (ibid.). Culture reflects the individual in relation to everyday practices in which individuals engage and participate to enrich and give meanings to their lives (Melleuish 2010). As Raymond Williams (1989, p. 4) argues, culture can be seen in two ways: as a whole way of life, which is the common meaning, and as arts and learning, which is the special process of discovery and creative effort. Culture defines the self and identity – hence the struggles for cultural rights and equality. It involves the daily practices of consumption, the engagement with arts and education, and their material institutionalizations. Culture is not bounded in space; it involves the realm of postmodern globalized world of disembodied practices. Critically, culture can be studied as a marker of power, class, inclusion, and exclusion in society (Bourdieu 1984, 1986, 1989).

To have access to cultural citizenship is to be able to participate and make interventions into the public sphere at the local, national, and global levels (Stevenson 2001). The themes of participation and equality are equally prominent in understanding cultural citizenship. Cultural citizenship can be studied from many perspectives: engaging the term empirically, one needs to focus on issues of participation and the ability of individuals to function within a cultural group of which one claims membership. In this sense, it invokes issue of human and cultural rights. Cultural citizenship also invokes the issues of liberty, cultural identity, and empowerment as citizens negotiate multicultural space in the discourses of identity and nationhood (Rosaldo 1994.).

Delanty (2003) engages cultural citizenship as a learning process about some important aspects of citizenship, which include language, cultural models, narratives, and the discourses that people use to make sense of their society, understand their place in it, and create courses of action. Citizenship is thus connected to culture in a cognitive relationship by which "learning processes in the domain of citizenship are transferred to the cultural dimension of society" (p. 604). Delanty notes further, "the power to name, create meaning, construct personal biographies and narratives by gaining control over the flow of information, goods and cultural processes is an important dimension of citizenship as an active process" (p. 602).

The control of information flow and cultural process by citizens makes cultural citizenship relevant to cultural technologies, such as mass media and the Internet. The role of communication technologies in shaping cultural citizenship is critical. On one hand, global communication in the epoch of rapid globalization and hypercommodification of media operations has led to global culture and global idea of citizenship. On the other hand, access to these technologies allows citizens to participate in everyday cultural practices, define their own cultural space, and access information that enhance the knowledge necessary for a sense of identity and involvement in social sphere of a local community – a process that leads to social inclusion in society. It is in this notion that cultural citizenship becomes relevant to the discourse of digital citizenship. Digital citizenship is an aspect of cultural citizenship; it draws on the political rights of citizenship to argue for participation in society through the use of a cultural resource (the Internet). It focuses on empowering the individual to gain control of information flow and cultural processes that define their participation as citizens in a society.

Ommundsen (2010) notes that the values associated with cultural citizenship are equality, respect, recognition, identity, belonging, agency, and empowerment. All these values "are in most cases regarded as the preserve of pluralist societies, in which the measure of cultural citizenship is the extent to which cultural rights and *resources* are available to citizens regardless of background" (p. 270, italics added).

Applying the concept of cultural citizenship to the study of the Internet follows in a tradition of invoking citizenship to social and cultural issues in the scholarship of media, communication, and cultural studies. For instance, Toby Miller (2006) applies the concept to the study of the television culture and critique of the media. He identifies three key types of citizenship: political citizenship (the right to reside and vote), the economic citizenship (the right to work and prosper), and cultural citizenship (the right to know). Miller notes that classical political theory of citizenship accorded the representation to the citizen through the state. The modern notion of citizenship is an economic addendum that the state promised a minimal standard of living. The postmodern notion is a cultural idea of citizenship, which guarantees access to the technologies of communication (p. 35). He applies the term cultural citizenship to address the political economy of communication – commodification by corporation and identity politics in the media and how these impact on what sort of information is available to the citizen and, most importantly, what is kept off the media. Consequently, he addresses the implication of this on the notion of citizenship.

Joke Hermes (2006) engages cultural citizenship, from a cultural studies perspective, as the process of bonding and community building and contemplation on that bonding, with regard to the production, consumption, and reflection of cultural text – in media and other cultural platforms. It is how we use media text and everyday culture to generate and reflect on identities embedded in different communities of any kind, and it is manifested by the range of emotions and appropriations that we connect with the use of popular media. Hermes argues that cultural citizenship comes with the rights to belong to a community and to offer one's views and preferences and the responsibilities to respect other members of the communities whose opinion and taste may be different. Hermes (2006) also locates citizenship to the Internet by drawing on its political and the cultural theorizations. Hermes' theorization and Miller's (2006) postmodernist view of citizenship as cultural with access to communication technologies provide additional rationalization for the theory of digital citizenship. Digital citizenship is a form of cultural citizenship that draws extensively on the rights provided by political theory of citizenship to offer theorization for access to the Internet and the appropriation and use of this technology for participation in society.

## 11.2.3 The Elements of the Theory of Digital Citizenship

Recalling the definition of digital citizenship as a *citizenry with the fulfilled rights to regular and flexible access to the Internet – implicitly individual and household forms of access, the skills to apply the Internet, and a regular use of the Internet for participation in all spheres of society* – it is essential to enumerate and discuss what digital citizenship entails. The following elements are proposed and discussed here as components of the conceptualization of digital citizenship:

1. Citizenship rights and the attendant issues of human rights, social justice, equality, and participation
2. Access to communication technologies (specifically here, the Internet)
3. Skills to apply these technologies
4. The regular/daily use of these technologies
5. The policies that make all the above possible

1. Citizenship Rights
   The political ideology and framing of citizenship have been elaborately engaged at the onset of this chapter. Citizenship rights stand as the overarching principle in achieving digital citizenship. Arguments may be made that the application of basic human and citizen rights to communicate and participate in the social, cultural, and political milieus of a society is what is missing in previous theorization of access to technologies. The concept of human rights based on the principles of fairness, equality, and social justice and the political concept of citizenship are important in the theorization of digital citizenship. Because all the rights of citizenship – individual, group, legal, social, political, economic, and participation rights (Marshall 1964; Janoski 1998) – can be advanced by access to the prevailing information and communication technologies, lack of access to these information and technologies thereof compromises these rights. Irrespective of how citizenship is theorized, either as a liberal ideology of the pursuit of individual freedom and liberty or republican and communitarian ideologies of community integration and public good, the Internet as a medium of information and communication has become a dominant factor of citizenship and participation in society. Hence, the argument here is that access to a prevailing information technology (specifically the Internet) should be recognized as a citizen's right.

   Oyedemi (2015) has argued that Internet access is a citizen's rights by drawing on Marshall's (2009) citizenship framework. Civil rights, with regard to possession of resources, are about the legal capacity to strive for things an individual would like to possess but not assured of their possession (Marshall 2009). A property right is a right to acquire it, if one can, and to protect it, but it is not a right to possess property. Based on this rationalization, critics of Internet access as a human right tend to argue for Internet access as a civil right (Cerf 2012). But contrary to affordability and ability notions of civil rights, social rights tend to be absolute rights of which their attainment is not limited by the economic value of the individual claimant. As a result, the state often draws on the dictates and expectations of social rights to make civil rights feasible for the citizens. So as Oyedemi (2015) argues, while civil rights offer the right to acquire a property if you can, and not the right to possess a property, states draw on social rights to provide housing for the poor. Consequently, Internet access as a right can be framed as civil rights, to be acquired, if one can, but the state should draw on social rights to assist those who cannot.

2. Access to Technology
   This is the next element of a theory of digital citizenship. With the role that broadband Internet is playing in every area of life, access should be seen as a citizenship right. If the rights to communicate are enshrined in the rights of citizenship, access to the prevailing form of communication is a human right. Access to the Internet in a regime of digital citizenship should be addressed mostly from an individual, household, and home access rather than the emphasis placed on public access that defined the implementation of "digital divide" policies. However, this does not imply a total abandonment of public access policy implemented in public education institutions, public libraries, and community centers; this policy should be maintained. With home-based access and personal access come issues of affordability and pricing – which government intervention in terms of policy should address. Public access comes with some constraints to individual flexibility in accessing the Internet. People have to compete with other users, often with limited time to use the Internet, and there is shame or embarrassment for users lacking adequate skills to use computers in public view of other people (cf Kvasny 2006). Hence, the argument here is that individual – including mobile access – and household forms of Internet access are essential for the fulfillment of digital citizenship.

3. Digital Skills
   Skills to use technology to search and understand the information that the Internet offers are important. However, in this approach, skills also mean the ability to create different forms of content, including multimedia content. Many social software and applications allow individuals to create wikis, blogs, post videos, and audio contents without learning codes or using complicated web language or algorithm. The bourgeoning development of social media has made content creation less complicated. The ability to create digital content is essential for effective digital citizenship. While the ability to source social, political, cultural, and economic information is necessary, the ability to form, create, and share opinion is relevant for effective participation as a citizen.

4. Technology Use
   This refers to the use of the Internet for social, political, and economic activities on a daily basis. The focus is not intermittent use but daily use to communicate, to be informed, to transact business and personal economic activities, and to be able to engage many aspects of daily life that the Internet has made easy.

5. Policy
   This is a very critical element of the theory of digital citizenship. It is the element that makes all the other elements possible. There should be effective policies in many areas in order to achieve digital citizenship. The overarching thrust of

policies should be citizen's rights. That is policy should be rights-based rather than market-based. Specifically, technology policy should address the issues of access, skills, and use as they relate to all sectors of society. The overall policy framework should include policies such as:

- Political and legislative policy that states ICT access as citizens' right. Although many national laws and constitutions recognize the freedom of expression and freedom of information, in most cases the rights to the technology that makes access to information possible are not stated. For instance, as noted in Article 19 of the Universal Declaration of Human Rights, everyone has the right to freedom "to seek, receive and impart information and ideas through any media and regardless of frontiers." While this declaration acknowledges the rights of access to media of information regardless of frontiers, there is a need to recognize the need for state governments and related agencies to recognize the rights to access the communication and information media that make the rights to "seek, receive and impart information and ideas" possible. Many national constitutional frameworks also adopt this idea of freedom to seek and impart information. This makes it imperative for government to come up with appropriate policies that recognize the rights for all citizens to gain access to the prevailing technology of information, especially in this digital age. There are few examples of countries such as France, Greece, and Finland that have recognized Internet access as human rights and have adopted this as law.
- Education policy that states access to basic education is universally accepted as a right of every child. Education in general provides the resource for every individual to participate in society and "level the playing field" for economic opportunities, likewise is access to the Internet. In this regard:
  (i) Internet access must be available in all public schools both in rural and urban regions of a nation.
  (ii) Internet skills, and digital literacy programs must form part of educational curriculum. The benefit of this is that young school children will learn the basic tools necessary for educational, social, and economic participations in the digital age. Most importantly, they will be able to create online content in their own languages.
- Technology policy that focuses on improving the use of appropriate technology, for example, broadband policy that explores the different approaches and technologies to make broadband available to all citizens.
- Economic policy:
  (i) Policies that foster a market that enhances affordability of Internet access and oversee the market by enhancing competition and monitoring of pricing strategies; policies that address the market operations of telecom sector to avoid abuse of corporate power.
  (ii) Public-private partnerships in building Internet infrastructure, such as fiber optic cables, wireless technologies, and other broadband transmission technologies.
- Social policy that encourages the cultural adoption of the Internet in all social sectors from health services to criminal justice systems. This type of policy will foster effective use of e-government platform for civic activities,

telemedicine for health activities, or for patients to be able to schedule appointment online in non-emergency cases. This will foster the diffusion of a culture of technology use among the general population. This form of policy guides all state institutions and departments to create online sites which citizens can engage with for public service delivery. Social policy also addresses the placing of public interest and public good in access policies.

## 11.3 Conclusion

The concept of digital citizenship has been applied in the European and American contexts as exemplified by Servaes (2003); D'Haenens et al. (2008); and Mossberger et al. (2008), but the application of the term has been mostly figurative expression. To a large extent, Mossberger et al.'s analysis of digital citizenship offers a useful compass for the application of the term in this chapter. Particularly their focus on political idea of citizenship values, issues of inequalities, and the use of the Internet to participate in society are all relevant to the theoretical framework developed here. However, there are few limitations to their approach, especially when applied to a different context outside of the United States.

Drawing on theoretical traditions and approaches to citizenship and the discourses of rights, participation, and culture in theorization of citizenship, access to technologies of information and communication, specifically the Internet, is acknowledged as essential. A definition of digital citizenship is offered with attendant elements that form its framework. This framework provides a guide for the pursuit of cultural and social developments and active citizenry aided by technology access, among other socioeconomic and political resources that shape today's ideal of citizenship.

## 11.4 Cross-References

▶ A Threefold Approach for Enabling Social Change: Communication as Context for Interaction, Uneven Development, and Recognition
▶ Capacity Building and People's Participation in e-Governance: Challenges and Prospects for Digital India
▶ Empowerment as Development: An Outline of an Analytical Concept for the Study of ICTs in the Global South
▶ Media and Participation
▶ Political Economy of ICT4D and Africa

## References

Bellah R, Madsen R, Sullivan W, Swidler A, Tipton S (1985) Habits of the heart: individualism and commitment in American life. Perennial Library, New York
Berlin I (1958) Two concept of liberty. Clarendon press, Oxford
Bourdieu P (1984) Distinction: a social critique of the judgment of taste. Routledge, London

Bourdieu P (1986) Forms of capital. In: Richardson J (ed) Handbook of theory and research for the sociology of education. Greenwood Press, New York
Bourdieu P (1989) Social space and symbolic power. Soc Theor 7(1):14–25
Burnett R, Marshall PD (2003) Web theory: an introduction. Routledge, New York
Carpentier N (2011) Media and participation: a site of ideological-democratic struggle. Intellect, Bristol
Cerf VG (2012) Internet access not a human right. The New York Times, January 4, 2012. Accessed 4 June 2012. http://www.nytimes.com/2012/01/05/opinion/internet-access-is-not-ahuman-right.html?_r.0
Chadwick A (2006) Internet politics: states, citizens and new communication technologies. Oxford University Press, New York
Dagger R (2002) Republican citizenship. In: Turner B, Isin E (eds) Handbook of citizenship studies. Sage, London
Delanty G (2003) Citizenship as a learning process: disciplinary citizenship versus cultural citizenship. Int J Lifelong Educ 22(6):597–605
D'Haenens L, Koeman J, Saeys F (2008) Digital citizenship among ethnic minority youths in the Netherlands and Flanders. New Media Soc 9(2):278–299
Dimaggio, P & Hargittai, E. (2001). From the 'Digital Divide' to 'Digital Inequality': studying internet use as penetration increases. Retrieved 07 Dec 2009 from www.webuse.umd.edu/webshop/resources/Dimaggio_Digital_Divide.pdf
Garnham N (1999) Amartya Sen's "capabilities" approach to the evaluation of welfare: its application to communication. In: Calabrese A, Burgelman J (eds) Communication, citizenship and social policy: rethinking the limits of welfare state. Rowan & Littlefield, New York
Held D (2006) Models of democracy. Polity, Cambridge, UK
Hermes J (2006) Citizenship in the age of the internet. Eur J Commun 21(3):295–309
Huesca R (2008) Tracing the history of participatory communication approaches to development: a critical appraisal. In: Servaes J (ed) Communication for development and social change. Sage, Thousand Oaks
Janoski T (1998) Citizenship and civil society. Cambridge University Press, New York
Jenkins H (2008) Convergence culture: where old and new media collide. New York University Press, New York
Kann M, Berry J, Gant C, Zager P (2007) The internet and youth participation. First Monday 12(8):1
Kuhn T (1996) The structure of scientific revolution. University of Chicago Press, Chicago
Kvasny L (2006) The cultural (Re)production of digital inequality, Information, Communication and Society, 9(2):160–181
Mansell R (2002) From digital divides to digital entitlements in knowledge societies. Curr Sociol 50:407–426
Marshall TH (1950) Citizenship and social class. Cambridge University Press, London
Marshall TH (1964) Class, citizenship and social development. University of Chicago Press, Chicago
Marshall TH (2009) Citizenship and social class. In: Manza J, Sauder M (eds) Inequality and society: social science perspectives on social stratification. W.W. Norton and Co., New York, pp 148–154
Mejos DEA (2007) Against alienation: Karol Wojtyla's theory of participation. Kritike 1(1):71–85
Melleuish G (2010) Cultural citizenship: a world historical perspectives. In: Ommundsen W, Leach M, Vandenberg A (eds) Cultural citizenship and the challenges of globalization. Hampton Press, Cresskill
Mertens S, Servaes J (2011) ICT policies on structural and socio-cultural participation in Brussels. In: Adomi EE (ed) Handbook of research on information communication technology policy: trends, issues and advancements. IGI Global, New York
Miller T (2006) Cultural citizenship: cosmopolitanism, consumerism, and television in a neoliberal age. Temple University Press, Philadelphia, PA

Mossberger K, Tolbert CJ, Stansbury M (2003) Virtual inequality. Georgetown University Press, Washington DC

Mossberger K, Tolbert CJ, McNeal RS (2008) Digital citizenship: the internet, society and participation. MIT Press, Cambridge, MA

National Telecommunications and Information Administration (1995) Falling through the net: a survey of the 'have nots' in rural and urban America. US Department of Commerce, Washington, DC

Ommundsen W (2010) In: Ommundsen W, Leach M, Vandenberg A (eds). Cultural citizenship and the challenges of globalizationProductive crisis: cultural citizenship in social theory. Hampton Press, Cresskill

Oyedemi T (2015) Internet access as citizen's right (?): citizenship in the digital age. Citizen Stud 19(3-4):450–464

Pateman C (1970) Participation and democratic theory. Cambridge University Press, New York

Rosaldo R (1994) Cultural citizenship and educational democracy. Cult Anthropol 9(3):402–411

Schuck P (2002) Liberal citizenship. In: Turner B, Isin E (eds) Handbook of citizenship studies. Sage, London

Sen A (1992) Inequality reexamined. Harvard University Press, Cambridge

Servaes J (1999) Communication for development: one world, multiple cultures. Hampton, Cresskill

Servaes J (2003) Digital citizenship and information inequalities: challenges for the future. In: Servaes J (ed) The European information society: a reality check. Intellect books, Bristol

Servaes J, Malikhao P (2008) Development communication approaches in an international perspectives. In: Servaes J (ed) Communication for development and social change. Sage, Thousands Oak

Servon LJ (2002) Bridging the digital divide. Blackwell, Malden

Smith RM (1993) Beyond Tocqueville, Myrdal, and Hartz: the multiple traditions in America. Am Polit Sci Rev 87(3):549–566

Stevenson N (ed) (2001) Culture and citizenship. Sage, London

Thomas P (1994) Participatory development communication: philosophical premises. In: White SA, Nair KS, Ascroft J (eds) Participatory communication: working for change and development. Sage, New Delhi

Turkle S (1995) Life on the screen: Identity in the age of the Internet. Simon & Schuster, New York, NY

Turkle S (2005) The second self: computers and human spirit. MIT Press, Cambridge, MA

United Nations (1948) Universal Declaration of Human Rights, 1948. Retrieved August 1, 2018 from http://www.un.org/en/universal-declaration-human-rights/

Webster F (1995) The theories of information society. Routledge, New York

Williams R (1989) Resources of hope. Verso, London

Williams K (2003) Understanding media theory. Arnold, London

Willson M (2006) Technically Together: Rethinking Community Within Techno-Society. Peter Lang, New York

# Co-creative Leadership and Self-Organization: Inclusive Leadership of Development Action

## 12

Alvito de Souza and Hans Begeer

## Contents

| | | |
|---|---|---|
| 12.1 | Introduction | 258 |
| 12.2 | An Emerging Leadership Culture Shift | 261 |
| | 12.2.1 Co-creation in Social Action | 262 |
| 12.3 | Co-creative Leadership Concept | 262 |
| | 12.3.1 Co-creation and Self-Management | 263 |
| | 12.3.2 Principles and Prerequisites | 263 |
| | 12.3.3 Levels of Self-organization | 264 |
| | 12.3.4 Five Learning Phases | 265 |
| | 12.3.5 Attitude and Culture | 267 |
| 12.4 | Conclusion: Not Top-Down, Not Bottom-Up, but 360° Co-creation | 267 |
| References | | 268 |

### Abstract

The quest for genuine local participation and ownership in development initiatives has perplexed and bedevilled development discourse for over 40 years. Practical application, despite some genuinely sincere efforts, has remained relatively elusive. New developments in communication technology are often cited as the solution to attain this elusive beneficiary-led development paradigm shift. A lot has been written about communication for development (C4D) and in recent decades information and communication technology for development (ICT4D), yet we still struggle to attain an authentic paradigm shift and genuine local participation or beneficiary-led development praxis. This chapter seeks to shift

A. de Souza (✉)
Co-Creative Communication, Brussels, Belgium
e-mail: alvito.desouza@gmail.com

H. Begeer
BMC Consultancy, Brussels, Belgium
e-mail: hans@bmc-consultancy.be

the focus from the trendy (not so) new communication technologies and focus instead on human beings and the way we organize ourselves. We argue that emerging trends in organizational development toward, flatter (less hierarchical) organizations, co-creation and self-organization may indicate the way forward to authentic (beneficiary-led) development practice. Rather than the trendy bottom-up paradigm, we advocate an all-round and do-it-yourself leadership structure. C4D and ICT4D remain crucial as tools for communication to facilitate and ease information flow and critical debate, but it is self-organization, trust building, and co-creative process that enable authentic equality and dignity in decision making and design innovation for development.

**Keywords**

Co-creation · Co-creative leadership · Self-organization · Social change

## 12.1 Introduction

Wisdom is like a baobab tree; no one-individual can embrace it.
(African Proverb attributed to the Akan and Ewe peoples of present-day Ghana)

In their aptly entitled book *Moving Targets: Mapping the paths between communication, technology and social change,* editors Servaes and Liu (2007) showcase multidimensional communication models that focus on local communities in the earnest quest for authentic communication for development (C4D). Scanning best practices in the use of communication and technologies for social change, the authors in the book advocate the recognition of context-specific cultural identity and the multidimensionality of communication for social change. There is general acceptance of the need for participatory processes to empower local communities, yet, as Flint and Meyer zu Natrup contend, the provision of development remains an "expert-led enterprise" (Flint and Meyer zu Natrup 2014). Harrison laments the mind gap between policy rhetoric and practice in the relationship between experts and communities at the receiving end of social programs (Harrison 2014).

So if there is acceptance, and even promotion of the "bottom-up" concepts of local ownership, participation, and stakeholder accountability, how is it that we are still struggling with donor-set agendas and expert-led development agency?

With the rapid expansion of the aid industry in recent years, there has been significant push for the professionalization of the development sector in terms of project management and administration. This has been accompanied by an explosion of innovative communication technologies that are increasingly affordable and in widespread use for development agency. Flint and Meyer zu Natrup suggest that the push to professionalize the development sector has, in a sense, eroded emphasis on ownership, participation, and partnership with local beneficiaries in favor of an expert-led culture. They further contend that, while exciting and clearly expanding the scope and speed of communication, recent advances in information and communication technologies for development (ICT4D) have, at times, only served to accentuate the pendulum swing toward expert-led "professionalization" that alienates local communities (Flint and Meyer zu Natrup 2014). Harrison notes that the "professionalization" pressures have led to an expert-led "superman" approach to practice with accountability reduced to "bean counting" of outcomes (Harrison 2014).

Clearly there is a need for transformation of the work culture in development agency. In principle there is consensus on facets of organizational approaches that hold out promise of an authentic local community-led culture of development agency.

Servaes and Malikhao reiterate the imperative of grassroots participation. The starting point of any communication for development and social change must be the context and specificity of the local community (Moving Targets Ch2). This is the nonnegotiable point of departure. Rather than donor-set agendas, expert-led strategies, or need assessments, it is first of all for the communities to deliberate and define what they *WANT* for themselves. Local communities are, in the end, the recipients and underwriters of any social change that is produced. Development agency and social change must begin with the local community.

Local communities, however, are not homogeneous in themselves. They are a network of diverse perspectives in constant interaction with each other. Neither are they isolated entities sealed off from the rest of the world. All communities are subject to global, national, and other stimuli that kindle and influence change. Recognizing the complexity of ad intra and ad extra interactions, Servaes and Malikhao assert a concept of development that accentuates cultural identity and multidimensionality (Moving Targets Ch2). Social change agency has to embrace the complexity and fluidity of context-specific community identities, including all necessary stakeholders.

Again, cultural identities are not static constants. Identities are created through a nexus of interactions that continually interrogate individual and collective representations, persistently influenced by external realities. In the closing chapter in *Moving Targets*, Claude-Yves Charron suggests a concept of community beyond the *geographic cognate* (Moving Targets Ch13). A social change project engages a broad sense of community where the various stakeholders participate in the project with the baggage of their particular identity standpoints. In a sense they form a living complex organism of diverse voices, socially organized in search of agreed action. The project, in itself, assumes a work culture identity that is a construct of multiple identities in a dynamic, organic, and continuously transforming social interaction. The multidimensional and complex communication within such organization implies sharing, comparing, and finding out about viewpoints toward the creation of common purpose. Social change is not about reproducing models developed by outside experts; "it is taking the internal dynamics of the community itself, with both its internal and external pressures, with its network of interdependencies" (Moving Targets Ch13). Social change and development is beyond expert-led instruction and model reproductions. It is about here and now, not there and then!

In a social change project, multiple perspectives interact, individuals (women, men, girls, and boys), local community leaders, farmers and traders, political leaders and public servants, donor agents, and NGO representatives. They form organic project or program structures that are essentially social processes. They are complex processes that are "in essence, the nurturing of knowledge aimed at creating a consensus for action that takes into account the interests, needs and capacities of all concerned" (Moving Targets Ch2). The social change project brings together a diversity of interests, needs, capacities, and levels of power interrelating in the quest for agreed action. It implies that people are capable of agreeing to disagree. Methodologies like Future Search are fitting this purpose.

The process of consensus creation for action is a discussion process (sharing, comparing, worrying through, and finding out), where existing power relations need to be laid bare and deconstructed to enable the construction of common ground purpose. All cultural identities involved in the project must be able to engage the creative process with equal voice. It is "not the technology, not the medium nor the message, but the concept of 'voice': voice in the identification of problems and the search for solutions by and for the community's own stakeholders... a multiplicity of voices in constant negotiation with each other" (Claude-Yves Charron – Moving Targets p.253). Social change can only be authentically appropriated through a process of reciprocal listening, learning, and discussing common purpose for action. Authentic ownership of the common action cannot be bestowed by the more powerful and dominant voices; it must be appropriated (taken as their own) by all the separate stakeholder voices through their genuine engagement in the collective exercise of creating the agenda, work process, and desired outcomes.

Once power relations have been laid bare and deconstructed, it is vital that all the diverse perspectives are fully expressed, listened to, and debated in the process of creating the consensus goals to be achieved. In their book *Future Search, Getting the Whole system in Room for Vision, Commitment and Action,* Weisbord and Janoff use the analogy of the ancient Indian parable of the blind men and the elephant (Weisbord and Janoff 2010, p.49) to illustrate the importance of examining all perspectives in search of common ground action. In *Moving Targets*, Charron uses the analogy of Akira Kurosawa's 1950 film *Rashomon* to illustrate the same point. The standpoint of all stakeholders voiced, debated, and pulled together leads to knowledge aggregation and common ground creation.

The question is how can diverse perspectives be harnessed to bring about sustainable social change? Flint and Meyer zu Natrup propose the concept of beneficiary-led aid (BLA) to harness the collective knowledge of beneficiaries by way of crowdsourcing (Flint and Meyer zu Natrup 2014). Pulling away from expert-led interventions, they draw inspiration from Surowiecki's *Wisdom of the Crowds*, "...under the right circumstances, groups are remarkably intelligent, and are often smarter than the smartest people in them. Groups do not need to be dominated by exceptionally intelligent people in order to be smart. Even if most of the people within a group are not especially well-informed or rational, it can still reach a collectively wise decision" (Surowiecki 2005: xiii–xiv).

They further suggest a functional reform where beneficiaries design their own development agenda with agencies and NGO experts assuming the role of facilitators (not designers) of the project (Flint and Meyer zu Natrup 2014). The concept of facilitation is beautifully captured in Weisbord and Janoff's (2007) book, fittingly entitled *Don't Just Do Something, Stand There.* Facilitation is to let go of all impulse to control the content, direction, and outcome of a project. It is, rather, focusing only on the process itself, enabling the co-creative space of communication and trusting all the rest to the community.

Here we propose the concept of co-creative leadership and self-organization, drawn from the work of Hans Begeer and inspired by the works of Marvin Weisbord and Sandra Janoff and the experience of a learning community that Hans set up.

## 12.2 An Emerging Leadership Culture Shift

Despite the continued dominance of an aggressive, immediate-profit-margin-focused business culture today, there is a perceptible emerging desire for more meaningful corporate practice. A "market" is developing of enterprises seeking to fulfill existential meaning beyond the precept of maximizing profit. Profit making is still at the helm of enterprise, but there is an intensifying thirst for meaning that transcends profit margins. Research shows that it is possible to work on both profit and meaning successfully! (Begeer 2018).

In our quick-paced and shifting global culture, enterprises (private, public, and civil society) find themselves increasingly reliant on the ingenuity of their human resources to remain competitive and even to simply survive. This emerging need is providing a new significance to the concept of return on investment (sustainable livelihood in enterprise), a meaning that necessarily focuses on the human person(s).

Only a work force that is inspired, enthusiastic, and motivated can meet the challenges and opportunities emerging from the pace of change today. For this to happen, each individual and the organization as a whole need to find authentic and profound sense in their work, their relationships, and their lives. Organizations seek to achieve meaningful livelihoods, as institutions and for each individual within their structures.

There is also increasing recognition that top-down management structures cannot meet this need. Developing and maintaining a sharp, quick-paced, and innovative work force requires a radical overhaul of the command and control culture and concept of leadership. A new kind of management leadership is emerging. One that is inclusive, where the creative energies of human beings are harnessed and where open systems develop a work force/management relationship in which all are inspired to be CO-CREATIVE LEADERS.

The concept of co-creative leadership and collective management is not really new, Charron alludes to the Japanese management by consultation culture of nemawashi (Moving Targets p.254), and other examples can be found in our human history. What is emergent is the renewed enthusiasm for more horizontal, human-centered organizational management process today. See also *Productive Workplaces: Dignity, Community and Meaning in the 21st Century* (Weisbord 2012).

Frederic Laloux's *Reinventing Organizations* has generated increasing fellowship in search of evolutionary organizations, horizontal management practices that focus on people and enterprises that are "soulful and purposeful" (Laloux 2014). The Corporate Rebels (https://corporate-rebels.com/) are a small group of former corporate employees who now travel the globe to gather stories of inspiring organizations that have successfully created more engaging and meaningful workplaces and unleashed the potential of their employees. Begeer has created a learning platform on self-organization, where people from different organizations meet to exchange best practices on self-organization. In his new book (2018), he presents 16 cases from this platform.

There are numerous organizational development (OD) tools and methods that pursue this swelling demand for innovative and inspired organizations, yet there is NO ONE magic tool or method. Each venture (or social change project) must find its own "way," using the insights and knowledge presented in the various methodologies, but ultimately designing its own pathway, drawing from its specific history, culture, and especially its people. Begeer warns, however, that the conditions for co-creation and self-organizations should be right, and attention to these conditions should not be underestimated (2018).

### 12.2.1 Co-creation in Social Action

The use of co-creative leadership and self-organization in development and social change enterprise is also not new. Outlining his memoirs of development work in his self-published book *My Development Experiences in Asia, Africa, and the Americas*, Christopher Roesel recounts the frustrations he faced with donor-set agendas and expert-led processes. His discovery of *Future Search*, and its transformative power in application, is inspiring, although the reversion to donor agendas, in some cases, is regretful (Roesel 2016).

Mark Harrison aligns co-creation with the principles of social action in *Social Action – co-creating social change: A Companion for Practitioners.* He also redefines the role of the development-expert away from directive leadership to that of facilitator who is "now responsible for the process, for asking questions, for enabling the group to set and realise its own goals. In traditional practice the practitioner is seen as the agent of change and community members/service users are, more or less, passive participants. In Social Action the facilitator is the catalyst for change and the agents of change are the community members/service users" (Harrison p.13).

## 12.3 Co-creative Leadership Concept

Co-creative leadership can be described as an inclusive leadership attitude that harnesses the creative capacities of individuals within organizations (viewed as living complex systems) toward collectively agreed goals while synchronously allowing for the divergence and ingenuity of individual viewpoints to consistently challenge and reinvigorate these same goals.

It is an approach that perceives leadership as collective, harvesting divergent (even conflicting) perspectives within an enterprise to creatively chart common action pathways in constantly changing environments. It enables the ingenuity of self-motivated individuals, who source their energy from the diversity of perspectives. It is a leadership attitude that inspires entrepreneurial flexibility in work environments and gives existential meaning (sense of collective purpose and motivation) to an enterprise. Co-creative Leadership allows individuals to collectively inspire each other to personally steer enduring processes of resourcefulness to meet opportunities and challenges through self-leadership. It fashions organizational values of trust, transparency, and accountability that nurture talent and craft work systems that are durable. But the individual stakeholders need to pull their acts together through flexible self-steering teams!

### 12.3.1 Co-creation and Self-Management

Co-creative leadership sounds promising and attractive but, for some, maybe somewhat mythical or unrealistic. Beyond the hype of inspiring concepts is co-creation credibly feasible? Inspired, by the work of Marvin Weisbord and Sandra Janoff in the *Future Search Network* (http://futuresearch.net/), and his own consulting experience, Hans Begeer coauthored the book *Co-creation... 13 Myths Debunked*. Co-creation, he demonstrates, is practicable, but it requires meticulous preparation and planning and rests on setting the groundwork based on some fundamental principles.

### 12.3.2 Principles and Prerequisites

| Six principles of co-creation | Conceptual framework |
|---|---|
| Get the whole system in the room | System perspective on organizing |
| Explore the whole elephant before fixing any parts | |
| Control what you can, let go of what you cannot control | Self-determination theory |
| Let people be responsible | |
| Search for common ground to base action upon | Open dialogue |
| Use differences in opinion to allow fresh and new ideas | |

Getting the whole system in the room reflects Servaes and Milakhao's concern for ensuring that the needs, capacities, interests, and knowledge of *all* concerned with the social change project participate in creating consensus action. Weisbord and Janoff use the AREIN criteria to ensure that all voices in the system are included in a Future Search meeting.

A-Those who have *Authority* to act on their own (make decisions).
R-Those who have *Resources* in terms of time, money, access, and influence.
E-Those who have *Expertise*, social, economic, technical, etc. in the topic.
I-Those who have *Information* that others need.
N-Those who have *Need*, which are people who will be affected by the outcome.

Explore the whole elephant before fixing the parts refers back to the parable of the blind men and the elephant (or Kurosawa's Rashomon). The point is to ensure that all perspectives on the subject are listened to, discussed, and seen for what they are before diving into structures and work processes. Openly identifying alternative perceptions and deconstructing power relations enables constructive listening, debating, and learning.

The third principle of co-creation is learning to control what you can and let go of what you cannot control. The axiom points to two fundamental attitudes behind co-creative leadership. It underlies the need for meticulous planning and preparation of the co-creative space and process, but it also highlights the need to trust the *Wisdom of the Crowds* and let go of control of content and objectives. This is sometimes the hard part, trusting that the collective intelligence of the community will lead to the most beneficial outcomes. Arriving at this level of trust often entails a steep culture shift, which should not be undervalued.

Letting people be responsible is about actively entrusting self-organization and self-management. It is crucial for any co-creative project. When conducting Future Search meetings, Weisbord and Janoff purposely use self-managing groups where "everybody shares information, interprets it and decides on action steps" allowing "every person to share leadership" (Weisbord and Weisbord and Janoff 2010).

By listening to distinctive opinions and expressing their own points of view, individual identities, together, draw out their disparities and synergies. The separate interpretations can then be drawn together to search for common ground. The enterprise, or project, becomes a social organization where common ground purpose for collective action is discussed and discovered through difference.

Differences of opinion, even conflicting opinions (as distinct from interests!), are the kernel of innovation and creativity; "it is not a desire for harmony or constant agreement that underlies co-creative leadership. On the contrary, differences are recognised as being something positive, since within them lies the potential for change" (Schieffer 2006, p.12). Managed constructively, differences lead to new perspectives and growth. There are many ways of mitigating difference and conflict, but essential to all is a clear distinction between opinion and interest. The discussion of common ground purpose goes a long way to mitigating the distinction. A transparent identification and deconstruction of power relations, along with enthusiastically appropriated discovery of common ground purpose, allows for a clear and open differentiation of opinion from interest by sharing, comparing, and finding out.

### 12.3.3 Levels of Self-organization

Begeer delineates three system levels to self-organization. The overarching level is where the entire social organization allows for maximum autonomous growth and horizontal functioning. Within the overarching project are self-steering teams, flexible working groups with enough space to manage their own workflows and processes. They are able to set their own goals and define the members of their community and their individual roles. Central to the entire process is self-leadership which Begeer describes as "the individual will and capacity to question and develop oneself" (Begeer 2018). It implies the desire to grow through giving and receiving critical feedback, inquisitive exploration, and the ability to correct and innovate.

The decision and desire to allow self-organization for a project must be explicit, and the facilitation of the process requires meticulous preparation and attention. Usually this is done by a so-called Planning Group, which, in itself, includes many different stakeholders and is self-steering. There is no need for a formal leader as group members will shift roles. It does, however, require professional facilitation, the art of which can be learnt.

Over a 2-year period, Begeer and his team helped a Brussels public service intuition for unemployment learn to work more horizontally and apply co-creation. The organization structure was typical bureaucratic with many different hierarchical levels and subdepartments all functioning from their islands. A tasting workshop co-creation was organized with 30 participants. Through "facilitation" guidance, 20 internal facilitators were trained. The mixed groups of facilitators were coached in organizing large-scale interventions, like World Cafés in which up to 400 participants worked together to develop actions for improvement. The participants included external shareholders and, especially, their clients (unemployed people). This resulted in a first step of cultural change: more cooperation, transparency, and self-initiative. Although the hierarchical structures still exist, people now work more horizontally and apply the six principles of co-creation where ever they can (Begeer 2018).

### 12.3.4 Five Learning Phases

Begeer proposes five learning phases on the path to self-organization. He expressly notes, however, that the path is a dynamic depiction to assist praxis and NOT a miracle prototype toolkit for application.

Step 1 Mobilize – This is a required phase in every change project – particular to self-organization is the fact that people often cannot imagine what it feels like or conceive its practical application. Therefore it is important to offer the group **tangible** experiences of self-organization to enable them to perceive and internalize its real possibilities, for example, with some short "taster" workshops, where they learn by doing.

Step 2 Collective Ambition – Ideally this is a phase in which all stakeholders are involved to develop **together** a shared vision with a collective ambition toward self-organization. Large-Scale Intervention methods like Future Search or Appreciative Inquiry can be applied here. It is imperative that the community engaged in the social change project have a shared vision of how they want to work together.

Step 3 Application and Support – The kernel of the social change project is application of the common ground action(s) agreed upon through self-organization. How do we operate here? Implementation requires a conducive environment for change. This is principally about **culture and values** and to a lesser degree about structure. Without shared values like **trust, transparency, and support**, self-organization cannot function. Working on this cultural change and the development of shared values should precede work on organizational structures. This often takes a long time and asks for repeating actions (routines) of explicating beliefs and values which drive our behavior.

Step 4 Analysis and Anchoring – This is the phase in which the **structure** of the work processes is analyzed. **Together** stakeholders discuss work streams, processes, and roles needed **to make it work**. It is a moment of action, action review, and (if necessary) path reorientation, visualizing the ideal situation to compare with the present reality. The GAP analysis results in choices for different ways of self-organizing to achieve the shared outcome. What do we want to achieve? The shared values described in step 3 play an important role in how this is done.

Step 5 Learning to Learn – Essential to a co-creative process is the development of the collective capacity to learn. As Begeer points out, people learn, not organizations, however organizations can create the conditions that stimulate learning. A **safe space** needs to be developed where people are at ease to learn to develop themselves, to ask for feedback, and to provide feedback, inspiring a learning culture where people together **continuously learn to learn**, without ever finishing.

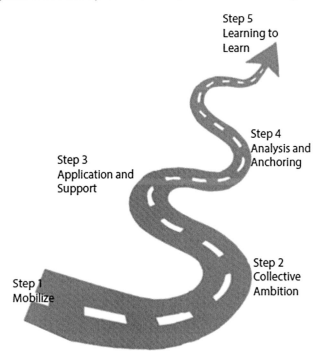

The two concepts of co-creation and self-organization are intimately intertwined feeding off each other to build flexible and innovative organizations with meaningful outcomes. Co-creation enables all stakeholders to engage in designing a social change project; create, together, the work processes required; and apportion individual roles in line with abilities. All appropriate the common purpose and goals through self-leadership and manage the project more horizontally. Self-organized teams commit to action, evaluate progress, and estimate capacity development needs toward self-organization.

### 12.3.5 Attitude and Culture

The co-creative leadership is an attitude that implies acceptance of a radical shift in working culture. It goes beyond technologies, mediums, and even message content. It also goes beyond project management methodologies and structures. Co-creative leadership is a process of continuous learning and growing. It will have its high points and moments of reckoning, yet its main objective is that all stakeholders grow in their co-creative leadership and self-leadership abilities.

This implies intentionally deconstructing power relations and cultivating open dialogue for active listening and learning. It entails nourishing a culture of authentic trust through transparency in communication with reciprocal accountability for action. It requires a working environment that is safe space for critical feedback and co-creative dialogue. It entails commitment and intentionality of purpose. It also necessitates meticulous preparation and planning in facilitation, as we mentioned earlier by a Planning Group. Ultimately, it is an attitude and a cultural mind-set shift that needs to be nurtured and maintained.

Co-creative leadership and self-organization does not happen just because it is desired. Maintaining the co-creative space requires painstaking planning and attentive oversight from committed facilitators, possibly from a select team specifically trained and dedicated to sustaining the co-creative space and supporting self-organization (see Begeer 2018).

## 12.4 Conclusion: Not Top-Down, Not Bottom-Up, but 360° Co-creation

Clearly the starting point of all social change agencies must be the specific context of the local community. Before any examination of a needs analysis, a social change project should start with an intentional dialogue on what the local community wants and a transparent discovery of common purpose, through a Large-Scale Intervention method (Begeer and Vanleke 2016).

Co-creation must be intentional, meticulously prepared, and facilitation planned to create a working culture of transparency and trust. Openly deconstructing power relations, the perspectives of all stakeholders should be engaged on an equal basis and from the competencies of each participant and in search of common ground. "It is not about finding the single correct solution to a problem. Indeed, it is even assumed that this cannot be found at all since it does not exist. It is much more about delivering a solution that members of an organisation can collectively agree upon, following a thorough examination of the various perspectives" (Schieffer 2006, pp. 608–609).

Co-creation is about facilitating the processes and environment for transparent communication. "We understand leadership as the continuous formation of creative and communicative contexts that facilitate a cooperative process for developing solutions for the organisation as a whole. We call this understanding of leadership co-creative leadership" (Schieffer 2006, p.615).

Co-creative leadership and self-organization goes hand in hand. We distinguish three levels: self-leadership, self-steering teams, and self-organization (Begeer 2018). They are founded on developing a culture based on trust. Without values like trust, support, and transparency, self-organization cannot function. A common identity and working culture based on these values must be developed through intentionality and persistent facilitation. "Supported by modern leadership methods (dialogue, World Café, and semantic mapping, etc.), co-creative leadership allows for the initiation of continual solution-finding processes and the decisive strengthening and maintenance of the capacity for action and success of the organisation as a whole. On the basis of co-creative leadership, organisations can develop a shared identity that serves as a guide for action and allows for the focussed action of the whole organisation" (Schieffer 2006, p.622).

Co-creative leadership and self-organization authentically engages community voice in the co-creation of *their* social change. Local communities are not on the periphery of social change design; they are at the very core of co-creative leadership of the change. Ultimately co-creative leadership is not about top-down control; neither is it really about bottom-up direction, but rather 360° inclusive leadership for development agency.

## References

Begeer H, Vanleke L (2016) Co-creation... 13 myths debunked. Lannoo Campus, Leuven
Begeer H (2018) Doe-het-zelf leiders. De hiërarchie voorbij, Praktijkboek voor zelforganisatie. Lannoo Campus, Leuven
Flint A, Meyer Zu Natrup C (2014) Ownership and participation: towards a development paradigm based on beneficiary-led aid. J Dev Soc 3(3):273–295
Harrison, Mark (2014) Social action – co-creating social change; A companion for practitioners. Social Action Solutions. www.socialaction.info
Laloux F (2014) Reinventing organizations. Nelson Parker, Brussels
Roesel CJ (2016) My development experiences in Asia, Africa and the Americas. Self-published on Amazon
Schieffer A (2006) Co-creative leadership: an integrated approach towards transformational leadership. Transit Stud Rev 13(3):607–623
Servaes J, Liu S (2007) Moving targets: mapping the paths between communication, technology and social change in communities. Southbound, Penang
Surowiecki J (2005) The wisdom of the crowds. Anchor, New York
Weisbord M (2012) Productive workplaces revisited. Dignity, meaning and community in the 21th century, 3rd edn. Jossey-Bass, San Francisco
Weisbord M, Janoff S (2007) Don't just do something, stand there, ten principles for leading meetings that matter. Berrett-Koehler Publishers, San Francisco
Weisbord M, Janoff S (2010) Future search – getting the whole system in the room for vision, commitment, and action, 3rd edn. Berret-Koehler Publishers, San Francisco

# Communication for Development and Social Change Through Creativity

**13**

Arpan Yagnik

## Contents

| | | |
|---|---|---|
| 13.1 | Tilling | 271 |
| 13.2 | Creativity | 271 |
| 13.3 | Neglect of Creativity | 272 |
| 13.4 | Communication for Development and Social Change Through Creativity | 276 |
| 13.5 | Positioning of Creativity Enhancement | 277 |
| 13.6 | Assumptions | 278 |
| 13.7 | Creativity Enhancement Intervention | 279 |
| | 13.7.1 Finding Facts | 279 |
| | 13.7.2 Naming Names | 280 |
| | 13.7.3 Finding Similarities Between Dissimilars | 280 |
| | 13.7.4 Creating New Meanings for Existing Phrases | 281 |
| 13.8 | Adaptation of Creative Aerobics in Development and Social Change | 281 |
| 13.9 | Benefits of Creativity Enhancement Intervention | 282 |
| 13.10 | Summary | 284 |
| References | | 284 |

## Abstract

As we celebrate the 80th year of the successful Marshall plan, it is unfortunate that practitioners and scholars have struggled, since: to achieve the same level of success in development and social change efforts. This indicates a need to seriously assess the field and even introduce new and unique approaches to boost the efficacy of such efforts. But there is a general neglect of creativity in the field of communication for development and social change. Creativity is a powerful force with tremendous potential to enhance efficacy of development and social change efforts. A gardening analogy is apt to elaborate on the value and role of creativity in communication for development and social change. Imagine you have a piece of

---

A. Yagnik (✉)
Department of Communication, Penn State, Erie, Erie, PA, USA
e-mail: arpanyagnik@gmail.com

© Springer Nature Singapore Pte Ltd. 2020
J. Servaes (ed.), *Handbook of Communication for Development and Social Change*,
https://doi.org/10.1007/978-981-15-2014-3_102

fertile soil where you are trying to grow a garden. You have good seeds, you sow them at the right depth, you water them, and you ensure that they receive appropriate balance between shade and sunlight. Despite everything, only half the seeds germinate leaving the idea of a flourishing garden biting the dust. This is the state of communication for development and social change. However, the one thing missing here was tilling. Tilling allows movement (upward and downward) and breathing, enabling and empowering a seed to transform into an independent plant. The element of tilling in gardening is what creativity is to communication for development and social change. Before the initiation of communication for development effort, planners and executioners should ensure the tilling of the soil, which translates to increasing creativity index of the target community or individuals because a creative individual or a creative community is relatively more open to the existence and acceptance of alternate or new ideas and behaviors. Creativity can substantially aid development and social change efforts because of its organic fit with the values of dialogue, participation, empowerment, social justice, and equality.

**Keywords**
Creative aerobics · Creativity enhancement · Creativity enhancement intervention · Millennium development goals · Social change

Creativity is a beautiful enigma.

A quick look at the history of communication for development and social change reveals that communication for development and social change efforts have been a mixed bag of successes and failures. After the success of Marshall Plan as the development efforts moved towards Africa, Asia, and Latin America practitioners and scholars have struggled with discovering solutions for the intractable problems faced by the underdeveloped communities. Fortunately, under the enlightened guidance of stellar communication for development and social change scholars (Melkote, Rogers, Mefalopulos, Servaes, Jacobson, and Sen to name a few), the field has cruised past beyond the top-down, outsider, methodological, nonparticipatory, and etic or emic cultural biases.

Despite the advances in the body of knowledge, meager success of millennium development goals and lack of substantial progress in sustainable development goals is not a matter of pride. Even today, allotment of resources by transnational agencies are done to fulfill fundamental life necessities. This indicates a need to assess and introduce universal elements in the communication for development and social change efforts with potential to increase efficacy and thereby success of development efforts.

Creativity is a universal element, which is an integral value that is appreciated and accepted in every cohort, culture, and community. Capitalizing on this universal element of creativity, and adding it to the equation of communication for development and social change to argue in favor of the limitless possibilities in "Communication for development and social change through creativity" is the goal of this chapter.

The conceptualization of the strategic role and value of creativity and creativity enhancement in communication for development and social change is best explained using a gardening analogy. Imagine that you have a piece of fertile soil where you are trying to grow a garden. You have good seeds, you sow them at the right depth, you water them, and you ensure that they receive appropriate balance between shade and sunlight. Despite doing everything as specified with the right intention, only half the seeds germinate leaving your idea of a flourishing garden biting the dust. This is the state of communication for development and social change. Despite good intention, expending millions of dollars on acquiring the best intellect, and implementing plans according to the specifications, the hope of development and social change ends up biting the dust.

Gardening is a heuristic activity. It is a great teacher, if you are willing to learn and change. So based on what you read in the previous paragraph, here is a question for you: Was there something missing in the gardening analogy? Was there something that could have made a difference in the final outcome, if it were included in the overall gardening activity? Let us see if you can get the right answer before it is revealed...

## 13.1 Tilling

If you got it right before reading it, then pat your back!

The one thing missing in the gardening analogy was tilling. Tilling of the soil enables the seed to generate movement (upward and downward). It allows for more space to navigate and grow. It also facilitates breathing. Keeping other variables such as sunlight, shade, water, and depth same, tilling the soil enables a seed to transform into a blooming plant. The element of tilling in gardening is what creativity is to communication for development and social change.

Theoretically, before the initiation of communication campaign, planners must ensure tilling of the soil. In other words, efforts in enhancing creativity levels of the target community or individuals should be taken before the initiation of a communication campaign. A creative individual is more likely to be open to the existence and acceptance of alternate or new ideas and behaviors. This will increase the probability of success of the communication campaign for development and social change.

A shift in the framework and models of communication for development and social change are proposed here. The contribution of this chapter is at a theoretical and conceptual level. It proposes communication for development and social change through creativity. This chapter suggests a new theoretical approach to communication for development and social change. It proposes a new flow of activities, which will theoretically increase the effectiveness of communication campaigns. Creativity enhancement of the communities can substantially aid development and social change efforts because of its organic fit with the values of dialogue, participation, empowerment, social justice, and equality. The development and subsequent publication of this approach also invites scholars and practitioners to examine it in the lab and the field, and evaluate it for its value.

## 13.2 Creativity

Creativity is a beautiful enigma. Scholars and practitioners have been working tirelessly for decades to unravel creativity and develop a definition of creativity. Runco (2014) provides an in-depth understanding of what creativity is and what it is not. He explains how different elements such as intelligence, imagination, originality, invention, discovery, innovation, logic, serendipity, and flexibility come together in various proportions to constitute creativity but not all the time. Scholars have made great advances in clarifying what creativity is not. But the one definition that everyone can unanimously agree upon as the definition of creativity is still a goal that is further away. Most definitions are incomplete and do not encompass the true vastness of creativity. They are myopic and suffer from one or the other bias such as the product bias or the art bias. Along with the ongoing disagreement about the definition and scope of creativity, interestingly there is a striking undercurrent of agreement among scholars and practitioners regarding the merit, value and utility of creativity.

Among the definitions of creativity out there, the one definition that encompasses the essence of creativity is the one provided in the FAQ section of the MacArthur Foundation website.

> Creativity comprises the drive and ability to make something new or to connect the seemingly unconnected in significant ways so as to enrich our understanding of ourselves, our communities, the world, and the universe that we inhabit. Creativity can take many forms: asking questions that open onto fields of enquiry as yet unexplored; developing innovative solutions to perplexing problems; inventing novel methods, tools, or art forms; fusing ideas from different disciplines into wholly new constructions; producing works that broaden horizons of the imagination. (MacArthur Foundation Webpage)

The real strength of creativity is in being a force that produces works that broaden the horizons of imagination. To begin with, it is difficult to believe that imagination has a horizon. Imagination in itself is a seamless and limitless notion. Now come to think of it that creativity is a force that has the ability to broaden the horizons of something that is already seamless. This force has got to be tremendously powerful.

Thus, being inspired by creativity, and to exemplify how creativity produces thoughts and works that have the ability to broaden the horizons of imagination, creativity will be introduced to communication for development and social change to make it more effective. The work in this chapter is not a claim equivalent to inventing an engine. It is more as the one that introduces a new catalyst (diesel instead of coal) to optimize the performance of the engine and making it ore effective so that carriages of development and social change may move faster and smoother.

## 13.3 Neglect of Creativity

Creativity has aided humanity in progressing from hunter-gatherers to modern civilizations. At every major crossroads of the 10,000 years old history of mankind, creativity, through its manifestations and avatars, has unfailingly stood by mankind and aided it in successfully navigating the crises of the times to emerge alive and well. Despite repeatedly withstanding the tests in the fires of time and crises, creativity is categorically and systematically neglected.

This neglect stems from its abundant and unceasing supply. Due to its abundant supply, its demand has gone down considerably. What has increased in return is this odd and unnatural significance of specificity. This demand has been fueled by the geometric progression of reliance on machines. Machines rely on specificity. Machines cannot handle ambiguity and therefore we traded estimation for precision. Machines require specific commands to perform optimally. One thing that makes us human, in my humble opinion, is the ability to create and apply context to ambiguous situations. Machines are not yet entirely capable of doing this. They are baffled by it and the ambiguity is then further referred to as a bug that needs to be exterminated or fixed. In favoring specificity and machines, we may have sacrificed creativity to the extent that individuals risk being referred to as deviants at the slightest expression or demonstration of creative tendencies. Travel and time-keeping were some of the early contenders of specificity and precision. As the technology moved from analog to digital, we too moved from fluidity to rigidity.

The neglect towards creativity, unfortunately, starts in schools. There is enough evidence about this and the most watched TED Talk by Sir Ken Robinson and his books discuss in detail the role of schools in killing creativity. As students move from one school year to another, their natural creative instincts are repressed. Every kid who has gone through the standardized funnel of education has had his/her creativity diminished. The reason behind this is simple. It is easier to teach a standard curriculum to many than to teach multiple curriculums to different students.

Another example of neglect of creativity is a quick search of all the courses offered in The Pennsylvania State University. The Pennsylvania State University is listed among the top 100 universities in the world. It is one of the few great public universities. The university charter promotes creativity and the faculties are evaluated based on their creative accomplishments. However, upon doing a quick search of the course catalog, only four courses were found that listed creativity in its course title or description. There are many other universities that share similar neglect. There are a few concentrated pockets such as San Francisco area where there is lesser neglect towards creativity. But such communities are rare. The creative class is concentrated in the developed nations and communities leaving the developing and underdeveloped communities and nations practically devoid of the creative class (refer to Fig. 1).

Creativity, in general, despite its utility is neglected. The negligence of creativity in communication for development and social change efforts sadly too comes as no surprise. This negligence is demonstrated in the limited number of approaches that practitioners and scholars have discovered and relied on in over half a century.

Apart from a genuine need, it is mainly training, traditions, and tribalism that have afforded the field of communication for development and social change its relevance and applicability. However, these very trainings, traditions, and tribalism have also acted as concealed impediments to the true manifestation of the efficacy and results of communication for developments and social change efforts. New or unique approaches, ideas, individuals, or organizations face obstinate resistance rendering them toothless. This chapter will highlight creativity and argue in favor of its value and utility in increasing the success of communication for development and social change efforts.

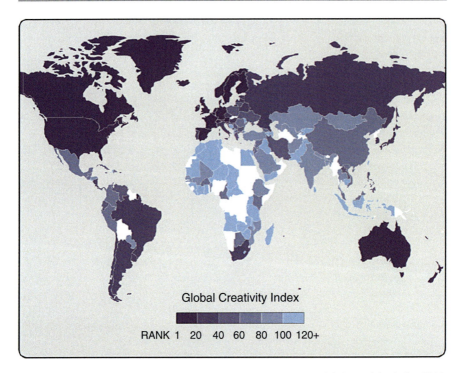

**Fig. 1** Global creativity index. (Source: Florida et al. (2015). The global creativity index 2015. Martin Prosperity Institute, University of Toronto)

My major argument in favor of negligence of creativity in communication for development and social change comes also from the lack of efforts to address or increase the creativity index in Africa, South Asia, and parts of Central and Southern America. This disparity is evident from their low ranking on the global creativity index (refer Fig. 1). "The Global Creativity Index is a broad-based measure for advanced economic growth and sustainable prosperity based on the 3Ts of economic development – talent, technology, and tolerance." (Florida et al. 2015, p. 6). Western Europe, Scandinavian countries, North America, and Australia are leading in the global creativity index with Australia at number one. This report brings to light a possible relation between higher creativity index and more development. These essentially become the variables worth testing with creativity index as an independent variable and development/well-being/prosperity as dependent variable. In my review of literature, this relationship has not been thoroughly examined within the context of communication for development and social change.

The second argument stems from the skewed spread of creative class (See Fig. 2). The proliferation of creative class is again restricted to first world nations or the developed nations or the Global North. "The creative class includes workers in science and technology and engineering; arts, culture, entertainment, and the media; business and management; and education, healthcare, and law." (Florida et al. 2015,

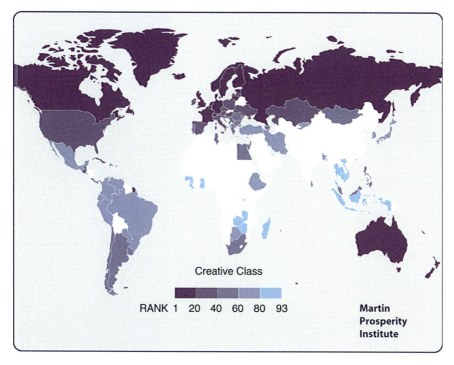

**Fig. 2** The global creative class map. (Source: Florida et al. (2015). The global creativity index 2015. Martin Prosperity Institute, University of Toronto)

p. 14). Based on the global creativity index-2015, it is clear that developing and underdeveloped nations have terribly low or nonpresent percentage of creative class (refer Fig. 2).

Majority of Africa, South Asia, and parts of Central and Southern America are devoid of the creative class and low in creativity index. Not surprisingly, these also happen to be the regions where the communication for development and social change efforts are focused. It might be theoretically plausible that if creativity index in these regions could be increased it would most likely lead to betterment of well-being in these regions. If the betterment of well-being does not come about on its own, even then the effectiveness of communication for development and social change campaigns would substantially increase if the creativity index were higher.

A creative individual, a creative citizen, a creative professional, or a creative leader is immensely sought after by our society at large. The evidence of the above statement is given below:

- An individual with creative abilities expresses more confidence and is seen as a possessor of a unique trait (Bungay and Vella-Burrows 2013; Garcia-Ros et al. 2012; Kienitz et al. 2014).

- Individuals with creative abilities are more desirable to potential employers (Jin Nam et al. 2009; Pace and Brannick 2010).
- Contemporary American culture emphasizes on creativity and an individual's creative abilities (Kern 2010). Several reasons for this emphasize come up in the research.
- In daily life, creative thinking is related to flexibility and adaptation (Csikszentmihalyi 1996; Reiter-Palmon et al. 1998).
- Creative thinking is also a crucial element discovered among successful entrepreneurs (Amabile 1997; Kern 2010).
- Additionally, research conducted by several scholars (Coholic et al. 2012; Greene et al. 2012; Lynch et al. 2013; Metzl 2009) reflected that individuals with higher creative capacity show greater resilience to trauma and tragedy (e.g., Hurricane Katarina and holocaust survivors) indicating enhanced psychological well-being among creative folks.

A creative individual can be a major force and cheerleader of positive change. However, this has not been thought about deeply. Scholars have continually strived to come up with new and creative ways to deliver the message, produce the message, involving the participant, solving local problems locally, etc. But they have not entirely focused on the tilling of the minds or enhancing creativity.

Ideas are the new currency. Creativity is at the heart of every idea. Thankfully, creativity is not limited like other factors of production that diminish or deplete. On the contrary, it is "infinitely renewable and continually replenished and deepened" (Florida et al. 2015; Romer 1986). This basic difference between creativity and other factors of production has worked both in favor and not when it comes to comprehending the utility of creativity.

A creative individual is resourceful and is open to new ideas and suggestions. And this is the important quality needed in individuals and communities spread across the spectrum of development and social change. Openness to listening, comprehending, and actually considering others perspective or others idea requires an open and empathic attitude.

It is safe to argue that creative individuals have this ability and if the creativity quotient could be increased prior to launching a communication for development and social change campaign then the likelihood of success of a campaign increases many fold. Outright rejection of development and social change ideas and approaches by the impacted community is a major obstruction. An individual or a community with higher creativity index will at the least consider it and see it as one more approach to the common well-being, irrespective of whether it is from inside or outside, and not reject it outright.

## 13.4 Communication for Development and Social Change Through Creativity

Currently, the communication campaign is launched after conducting a needs assessment and the effect is evaluated. This can be expressed in a formulaic fashion as below.

$$CM \; -i------I-> D\&SC$$

where,

CM = Communication Campaign
D = Development
SC = Social Change
-i――I-> = Impact/change on the Individual/Community

In the new approach, the only substantial change is the addition of creativity enhancement training in the mix. Creativity enhancement training is included along with the communication campaign. This, however, does not clarify whether the training has to happen before or after. That elaboration on the strategic positioning of the creativity enhancement training is provided in the next section.

$$CM + C \; -i-----I-> D\&SC$$

where,

CM = Communication Campaign
D = Development
SC = Social Change
-i――I-> = Impact/change on the Individual/Community
C = Creativity

## 13.5 Positioning of Creativity Enhancement

Communication for development and social change is a process that utilizes communication, as a harbinger for impacting the change. Since it is a process, it is systematic and therefore it can be explicated in a step-by-step format. The positioning of creativity enhancement as one of the steps has to be strategic. Before elaborating on the positioning of creativity enhancement, let me briefly share with you the steps in the communication for development and social change process.

1. An assessment is conducted to fully comprehend the need.
2. Based on the findings, a strategy is determined, and a plan is designed.
3. Next is the implementation of the plan according to the predetermined timeline.
4. The execution is evaluated and the change is monitored.

Based on the explanation above, communication for development and social change systematically follows steps more or less in an order. To fit in the order and for it to have an impact, creativity enhancement has to be positioned strategically. For that, it is pertinent to go back to the original gardening analogy. You till the ground before you plant the seed, which means that creativity enhancement sessions or workshops ought to happen before the actual communication campaign is

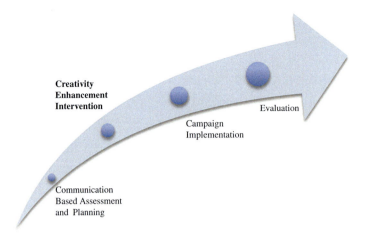

**Fig. 3** Creativity enhancement intervention positioning

implemented. While the results of the need assessment are being analyzed and a strategy is being determined, creativity enhancement should be rolled out. Figure 3 illustrates the positioning of creativity enhancement intervention.

The creativity enhancement intervention has to be positioned precisely so that when the communication campaign is rolled out, the campaign benefits from the enhanced creativity.

## 13.6 Assumptions

- This concept assumes that the initiation of a campaign or the idea of social change required in a community is etic. In other words, an external observer instead of an internal player or a local resident initiates the development campaigns and efforts. This is not to say that creativity and creativity enhancement does not have a role to play in the case when an internal player initiates a campaign. Explication of what role creativity and creativity enhancement can play in those circumstances will be dealt at another time. As of now, it is outside the scope of this work.
- This concept assumes that creativity and creativity enhancement discussed here are used as tools, before the actual campaign starts, to prepare the community members for the incoming change. The creativity that is used in designing the campaign is not where this concept is trying to make a contribution. That creativity has always existed, to what extent might be a matter of debate.
- This concept also assumes that the author does not have a preference or a bias for one communication for development and social change approach over another. The author looks at all approaches and models equally.

## 13.7 Creativity Enhancement Intervention

Creativity is a skill. Just like every other skill, creativity can also be taught and can be enhanced with practice.

The million-dollar question, now, is how to enhance creativity. There are numerous strategies to enhance or generate creativity. Psychologists have studied creativity now for over many decades and the area of creativity, which is not restricted by product bias or art bias has been studied extensively. As a result of the practice and research, many creativity enhancement strategies have been developed such as Brainstorming, Design thinking, Seven Hats, Creative Aerobics, and others.

Creativity enhancement intervention is to be conducted by introducing people to a creativity enhancement strategy. Based on the evaluation of the advantages and disadvantages of the different creativity enhancement strategies, the creativity enhancement strategy that is optimal for this intervention is creative aerobics.

Creative Aerobics is a set of four specific, interrelated word-driven exercises that alternate left- and right-brained thinking (George and Yagnik 2017). Creative Aerobics is a process-driven system instead of a solution driven system. Using it, one can generate lots of rich information quickly. The four exercises are:

1. Finding facts
2. Naming names
3. Finding similarities between dissimilars
4. New meaning for existing phrases

Creative Aerobics is a relaxing process that rids an individual of any anxiety or panic. Creative Aerobics has delivered results in boardroom and classroom for over two decades. Creative Aerobics also aligns well with the notions of participation, empowerment, social justice, and equity. Creative Aerobics does not require any special qualification or training to master. It is strongly rooted in one's own daily life and experiences.

Typically, due to the standardized schooling and parenting, we are trained to utilize our left-brain more and therefore our ability to use the right side of the brain is diminished. Creative Aerobics allows one to access solutions and ideas that were previously available but not accessible. Creative Aerobics allows you to explore your what it isn't fearlessly and be tolerant of ambiguity. The four exercises are explained in detail below.

### 13.7.1 Finding Facts

Finding Facts is Creative Aerobics 1 (CA1). This is the simplest of all the four exercises. In CA1, all you are required to do is to find facts about a product or an object and list them. List as many facts as you can. The only caveat here is to try and list as many lesser-known facts as possible. The lesser-known the facts, the better it is. To make this easy task easier, you can apply one of the three tactics to come up with facts:

- Research: A participant can perform secondary research, wherein s/he can do a quick Google search and list facts from there. A participant can also use the more traditional route of going to a library to find relevant information.
- Observation: A participant can use her/is observation and other senses to come up with facts that are seen or felt.
- Experience: The participant can list facts from her/is own past experiences with that object. This is accomplished by asking ordinary and out of ordinary questions to oneself.

### 13.7.2 Naming Names

Naming names is your Creative Aerobics 2 (CA2). "CA2 asks for new what it isn't names for existing object" (George and Yagnik 2017). Here, the learner is tasked to come up with new what it isn't (WII) names. This is the first step towards actively involving the right side of your brain, while being introduced to the key concepts of "what it is" and "what it isn't." What it is and what it isn't are layered concepts. This characteristic makes their comprehension impactful. What it is serves as an explanation of the current state of existence. WII is the imagination or elaboration of what one's state of existence could be. The only caveat here is that the new WII names or scenarios cannot be synonymous or antonyms and in the case of a product have to be a noun.

CA2 requires you to come up with new WII names. A trainer, in the development and social change context, after teaching the technique and strategies to come up with new names can ask the participants to come up with new WII state of existence. These new WII states of existence can be helpful in realizing and understanding the following:

- What state does the individual and his/her community live?
- What state does the individual and his/her community not live in?
- What state could the individual and his/her community live in?

These distinctions through creativity enhancement and imagination may enable individual to work towards while being accepting of the communication campaigns, development, and social change. There are five strategies to come up with new WII names. The first strategy will be especially a good fit for the development and social change context.

### 13.7.3 Finding Similarities Between Dissimilars

One of the major things that our brain is very good at, due to the training it receives, is the ability to distinguish between objects. It is relatively normal to figure out differences as opposed to similarities. Neither is difficult, it is just that one of it is seldom applied and therefore feels more unnatural.

Creative Aerobic 3, otherwise known as finding similarity between dissimilars, makes us utilize our right side of the brain actively. You can take any two dissimilar objects and still find similarities between them. For example, take orange and clock, they are both round, they both have partitions, they both get eaten up, they both can be opened, and more. Keep in mind, no matter how dissimilar the objects are, you will always be able to find similarities between them when you allow the right side of the brain to inspire your thoughts. Let us take another example, sugar and shoe. They both start with letter "S," they are both used to cover things, they can both be bought, and using them both is comforting. In my class, I have an open challenge for my students. If they are able to come up with any two objects that do not have any similarity between them, then they can get a $100. I have never had to give my $100 away.

This can be utilized in the communication for development and social change context where they can find similarities between being scenarios of being underdeveloped and being developed. It is guaranteed that they will be able to find similarities between the two. These similarities can be built upon as foundation to mitigate the issue and accentuate the well-being.

### 13.7.4 Creating New Meanings for Existing Phrases

CA-4 is creating new meanings for existing phrases. This CA exercise teaches you to utilize the right and left brain in tandem. Let me start by giving an example of CA-4.

I woke up today and finished all my cereal. Who am I?

A cereal killer.

You see how the phrase serial killer has been given a new meaning by capitalizing on the inherent ambiguity that is made available to us. In CA-4, we discover colloquially popular phrases and give them new meaning. For example, venetian blind: a blind person from Venice; Fishing pole: a Polish person who is fishing or a voting poll among fishes.

CA-4 can be used in the development and social change context to give new meanings to existing situations and scenarios. These new meanings can help people in realizing whether a situation is good or bad, positive or negative, right or wrong, happy or unhappy. Creativity here has no limits, so you never know what realizations may emerge from this.

From a training perspective, this is the exercise that usually leads to the "Aha!" moment. This powerful realization makes it easy for the communication campaign to succeed because by then an individual is cognitively involved with the issue.

## 13.8 Adaptation of Creative Aerobics in Development and Social Change

Creative Aerobics, due to its ease and effectiveness, can be used for creativity enhancement of an individual or a large number of individuals constituting a community. Once a participant is trained in creative aerobics, then s/he can be asked to utilize CA in a particular context. Let us take the issue of higher female illiteracy due to menstruation. Millions of girls drop out of school when they hit menarche. This is due to several reasons but a few are lack of toilets, ridicule and shaming, marriageable because they can now reproduce, and negligence of the importance of education for women independence and empowerment.

If the Water, Sanitation, and Hygiene (WASH) section in UNICEF were to launch a communication campaign for development and social change targeted at retaining more female students after menarche, then a few actions will need to be taken. Individuals and a community will need to go through knowledge, attitude, and behavior changes regarding menstruation. They will also need to be made cognizant of the long-term implications of female illiteracy. To prepare an individual and a community for comprehending the communication that is going to be directed at them or for realizing the problem and coming up with a solution, healthy and high levels of creativity and imagination are required. Without one's ability to imagine the consequences that are far away in the future her/is ability to receive and understand the communication efforts are limited. This is where creativity enhancement training comes in. It will be a two-step process:

1. Engage members of the community in the creativity enhancement training.
    (a) When they join, a creativity test should be administered to determine the level of creativity. This will serve as a pretest.
    (b) Then the individuals should be introduced to CA, as an ideation system aimed at aiding them in enhancing their creativity.
    (c) This introduction should not be in the context of menstruation and female illiteracy.
    (d) The duration of the sessions can vary from 2 to 4 h. This can be done in one day or can be spread depending on the circumstances.
    (e) Once they are done with the training, another creativity test should be administered. This will serve as a posttest.
    (f) The results will present with two options:
        (i) No significant change: Repeat the CA training.
        (ii) Significant change: Move onto the next step.
   Based on my professional experience, I can say with almost certain confidence that by now participants will be thrilled at the discovery of this new potential and power.
2. Invite them for an advanced CA session for community development.
    (a) When they join, inform them about the issue.
    (b) Task them with performing CA exercises about a given issue.
    (c) Document the knowledge, perception and solutions that emerge from it.
    (d) Based on the nature of the communication campaign and the level of participation utilize the knowledge emerged from the CA sessions.

## 13.9   Benefits of Creativity Enhancement Intervention

The benefits of a creativity enhancement intervention are multipronged.

- The important benefit is most likely to be seen in the reduced resistance towards the communication campaign. This can also be looked at increased acceptance or speedy acceptance of the communication campaign.

- The second aspect is that creative aerobics is a great motivator. It motivates people by giving them confidence in themselves and their ability. This is very important for a participatory approach because when you involve the people in a dialogue or an exchange you want them to participate, you want them to contribute, you want them to act like a stakeholder, you want them to come up with local and more sustainable solutions or ideas to increase sustainability. CA enables them to connect the seemingly disconnected and to give new meanings to existing situations. This is empowerment at an individual level, which can easily transpire into empowerment at a local or a community level.
- It fits in well with the existing frameworks and models of communication for development and social change. Given below is the modified version of the multitrack approach (Mefalopulos 2008) including creativity enhancement intervention.

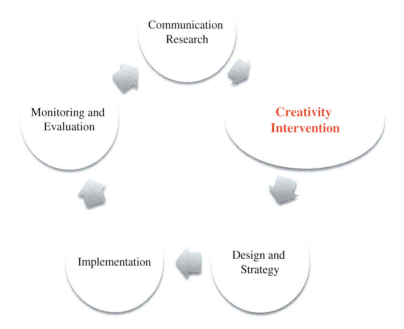

- This advancement in theory may propel testing of this approach through research or practice and lead to further refinement of this approach or development of new directions for the field of communication for development and social change.
- By exploring and examining this, there is nothing to lose. Creativity enhancement will help individuals and communities in dealing other issues in their life not related to development and social change. It may give them a new life perspective, a new career, confidence, joy, and maybe even a way to cope with stress and calamity.

## 13.10 Summary

Communication for development and social change through creativity is an attempt at broadening the horizons of imagination in regards with approaching development and social change. This approach has not been tested yet but the hope is that considerable amount of time will be spent in testing this by scholars and practitioners. One of the immediate next steps is be to create a creativity enhancement training manual that can be utilized by trainers to train individuals in Creative Aerobics.

Enhanced creativity index of people also has other advantages such as boosting self-confidence, increasing flexibility, and ability to deal with calamity. The creativity enhancement intervention can also be a great fun way to connect with individuals and communities and build a rapport with them. This will also help them open up and share candidly during participatory or dialogic meetings. If nothing else, it will surely bring these communities and nations higher up in the creativity index ranking.

As mentioned earlier, if ideas are the new currency then clearly creativity is the mint where these ideas are generated. As much as it is important that the neglect of this universal force comes to an end, it is also important for creativity to be actively included and utilized in development and social change. Incorporating creativity into the equation is bound to have an effect. Remembering the gardening analogy, the effect of creativity enhancement training is likely to be positive social change and individual empowerment.

## References

Amabile TM (1997) Entrepreneurial creativity through motivational synergy. J Creat Behav 31(1):18–26

Bungay H, Vella-Burrows T (2013) The effects of participating in creative activities on the health and well-being of children and young people: a rapid review of the literature. Perspect Public Health 133(1):44–52

Coholic D, Eys M, Lougheed S (2012) Investigating the effectiveness of an arts-based and mindfulness-based group program for the improvement of resilience in children in need. J Child Fam Stud 21(5):833–844

Csikszentmihalyi M (1996) Creativity: flow and the psychology of discovery and invention. Harper Collins Publications, New York

Florida R, Mellander C, King K (2015) The global creativity index 2015. Martin Prosperity Institute, University of Toronto, Toronto

Garcia-Ros R, Talaya I, Perez-Gonzalez F (2012) The process of identifying gifted children in elementary education: teachers' evaluations of creativity. Sch Psychol Int 33(6):661–672

George L, Yagnik A (2017) Creative aerobics: fueling imagination in the 21st century. Sage Publishers, New Delhi

Greene RR, Hantman S, Sharabi A, Cohen H (2012) Holocaust survivors: three waves of resilience research. J Evid Based Soc Work 9(5):481–497

Jin Nam C, Anderson TA, Veillette A (2009) Contextual inhibitors of employee creativity in organizations: the insulating role of creative ability. Group Org Manag 34(3):330–357

Kern F (2010) What chief executives really want. Bloomberg Businessweek. Retrieved from http://www.businessweek.com/innovate/content/may2010/id20100517190221.htm

Kienitz E, Quintin E, Saggar M, Bott NT, Royalty A, Hong DW, Liu N, Chien Y, Hawthorne G, Reiss AL (2014) Targeted intervention to increase creative capacity and performance: a randomized controlled pilot study. Think Skills Creat 13:57–66

Lynch M, Sloane G, Sinclair C, Bassett R (2013) Resilience and art in chronic pain. Arts Health 5(1):51–67

Mefalopulos P (2008) Development communication sourcebook: broadening the boundaries of communication. The World Bank, Washington, DC

Metzl ES (2009) The role of creative thinking in resilience after hurricane Katrina. Psychol Aesthet Creat Arts 3(2):112–123

Pace VL, Brannick MT (2010) Improving prediction of work performance through frame-of-reference consistency: empirical evidence using openness to experience. Int J Sel Assess 18(2):230–235

Reiter-Palmon R, Mumford MD, Threlfall KV (1998) Solving everyday problems creatively: the role of problem construction and personality type. Creat Res J 11(3):187–197

Romer P (1986) Increasing returns and long-run growth. J Polit Econ 90:1002–1037

Runco M (2014) Creativity theories and themes: research, development, and practice, 2nd edn. Elseiver, London

Sternberg R (2018) A triangular theory of creativity. Psychol Aesthet Creat Arts 12(1):50–67

# The Relevance of Habermasian Theory for Development and Participatory Communication

## 14

Thomas Jacobson

## Contents

| | | |
|---|---|---|
| 14.1 | Introduction | 288 |
| 14.2 | Part I: Modernization Theory | 289 |
| 14.3 | Part II: The Theory of Communicative Action | 291 |
| | 14.3.1 Cultural Value Sphere: Science, Technology, and the Objective World | 295 |
| | 14.3.2 Cultural Value Sphere: Law, Morality, and the Social World | 296 |
| | 14.3.3 Cultural Value Sphere: Art, Literature, and the Person | 297 |
| | 14.3.4 Criticism | 299 |
| 14.4 | Part III: Communicative Action and Development | 301 |
| | 14.4.1 Cultural Value Sphere: Science, Technology, and the Objective World | 302 |
| | 14.4.2 Cultural Value Sphere: Law, Morality, and the Social World | 303 |
| | 14.4.3 Cultural Value Sphere: Art, Literature, and the Person | 304 |
| 14.5 | Research Opportunities | 304 |
| 14.6 | Conclusion | 306 |
| References | | 307 |

### Abstract

This chapter examines Jurgen Habermas's theory of communicative action as a theory communication for development and social change. In the main his theory has focused on analysis of industrial or postindustrial societies, but it is also relevant to the study of development and social change. This relevance extends beyond questions related to the public sphere. Habermas's work involves a general theory of social evolution. This he shares with early modernizationists along with his defense of at least some of the enlightenment traditions they shared. At the same time, the theory of communicative action has deep and

T. Jacobson (✉)
Department of Media Studies and Production, Lew Klein College of Media and Communication, Temple University, Philadelphia, PA, USA
e-mail: tlj@temple.edu

© Springer Nature Singapore Pte Ltd. 2020
J. Servaes (ed.), *Handbook of Communication for Development and Social Change*,
https://doi.org/10.1007/978-981-15-2014-3_13

systematic differences from modernization theory. Its critical theory shares with the many critics of modernization theory a deep apprehension regarding the negative effects of scientism and uncontrolled market rationality. It is fundamentally critical of the functionalist reason underlying early modernization research. Its theory of the lifeworld provides an elaborate framework for analyzing cultural change. It offers a sophisticated defense of universalizable human rights that eschews the idea of a priori universal rights. In sum, the theory defines development and social change as a process of increasing dialogical, and participatory, communication in all sectors of society. The chapter reviews modernization theory and its weaknesses, outlines the theory of communicative action, and then considers Habermas's theory as an approach to the study of development and social change. A concluding chapter suggests research opportunities using the communicative action framework.

**Keywords**

Habermas · Modernization · Communicative action · Development

## 14.1   Introduction

The relevance of Jurgen Habermas's work to studies of development and social change has been noted in a number of respects within the literature on communication. Chiefly these apply his theory of the public sphere to democratic and participatory processes in the global south. While these studies are productive, the relevance of Habermas's theory to development studies is broader than this. His theory of communicative action as a whole embodies a systematic theory of social change that is historical and evolutionary (Habermas 1984, 1987). It addresses the challenges of cultural change, institutional growth, economic fairness, and other matters of importance to development studies, and it does so within a unified theoretical framework.

The core of his theory is relevant to studies of development communication not only because of its broad scope. It is also relevant because the theory of communicative action identifies the chief driving force of development, in relation to economics, law, democratic politics, socialization, and more, as a participatory form of communication. For Habermas, development and social change is essentially a process in which communication, in the form of communicative action, plays an increasingly central role in all social subsystems across modern historical time.

In the paper that follows, the theory of communicative action is treated as a theory of development and social change, focusing on its analysis of communicatively driven social evolution. The paper begins in Part I with a sketch of modernization theory. This material is well known, but a selective review will help to make clear the relevance of Habermas's theory to a wide range of social and analytic subjects, as well as differences between his work and modernization theory. The theory of communicative action itself is presented in Part II. A number of criticisms are reviewed, some general and others pertinent specifically to development studies. Afterward follows a comparison of modernization theory and the theory of communicative action, in Part III, identifying ways in which the two theories are similar and more importantly ways in which they are different.

Parts II and III are organized in a tripartite fashion. Each is divided into three parts by identifying what Habermas calls, following Max Weber, three cultural value spheres (Habermas 1984, pp. 143–273): science and technology, law and morality, and art and literature. A concluding section offers suggestions for research.

## 14.2 Part I: Modernization Theory

Modernization theory was based in considerable part on the work of founding sociologists such as Emile Durkheim, Ferdinand Tonnies, and Wax Weber (Black 1966). Perhaps the most influential of the founding sociologists was Weber. He more than anyone is identified with the idea that modernity proceeds through the institutionalization of rational ways of thinking, resulting in social specialization and bureaucratization. He characterized progress as evidence of a "great rational prophesy."

The 1950s saw these classical characterizations of the birth of modern society clothed in the biologically based metaphor of evolution. The evolution of social stratification was viewed naturalistically in terms of biological differentiation. Naturalistic as it was, this evolution was considered inevitable. As Marion Levy explained, "The patterns of the relatively modernized societies, once developed, have shown a universal tendency to penetrate any social context whose participants have come in contact with them..." (Levy 1967, p. 190). The result of this penetration was global cultural assimilation. Modernization was expected to reduce the variety of local cultures expressed across societies and move them toward a common vision. "As time goes on, they and we will increasingly resemble one another ... because the patterns of modernization are such that the more highly modernized societies become, the more they resemble one another" (Levy 1967, p. 207).

Within this teleological model of change in societies as a whole, individual studies addressed subsystems within societies. The most widely influential of these studies early on was Walter Rostow's study of economic development, including his proposal of a series of five stages through which traditional societies must move in order to achieve self-sustaining economic "takeoff" (Rostow 1960). Closely associated with this economic growth, as a necessary input to production, was an increased incorporation of technology and scientific rationality. Following Weber, and speaking of modernization as an emerging universal culture, Lucien Pye explains, "It is based on a scientific and rational outlook and the application in all phases of life of ever higher levels of technology" (Pye 1963, p. 19).

Evolution in economy and the technification of culture required concomitant evolution in social structure. Most generally, modern economic institutions were characterized by the differentiation and specialization of social roles (Levy 1967). Because self-sustaining growth requires market economies and these in turn require democracy, modernization theory also included in its studies of differentiation the emergence of politically specialized democratic institutions. Almond and Verba viewed democratization as a necessary component of modernization and saw it as moving from traditional authoritarianism toward a world culture of democratic participation, what they called "the participation explosion" (Almond and Verba 1963, p. 4).

To occupy roles in these new economic and social institutions, a new kind of individual person was required, one that was less bound to tradition and the stasis of traditional society. In order for individuals to escape rural torpor, move to urban areas, and become involved in a modern workforce, Daniel Lerner argued that traditional mentalities had to change. Individuals had to learn how to ". . . see oneself in the other fellow's situation" in order to be able to envision new and desirable ways of life (Lerner 1958, p. 50). Only then could one be expected to want to "achieve" – an attitudinal condition studied by David McClelland (1961). Given importance assigned to economic growth, rational thought, and individual achievement, the modern individual came to appear in the form of a modern factory worker (Inkeles and Smith 1974, p. 19).

Communication processes were focal in a considerable amount of research. Lerner's singular contribution was to study relations among literacy, media participation, and political participation. He was among the earliest to recognize that for economic and social institutions to change, change was required in individual knowledge, attitudes, and aspirations. Representing deliberations of the Committee on Comparative Politics, Lucien Pye was another early exponent of a communication-based approach. "A scanning of any list of the most elementary problems common to the new states readily suggests the conclusion that the basic processes of political modernization and national development can be advantageously conceived of as problems in communication" (Pye 1963, p. 8).

In general, Lerner's and Pye's approaches to conceptualizing communication corresponded with others in the transmission school of communication research. Harold Lasswell's definition of communication held that it was, "Who says what to whom through what channel with what effect" (Lasswell 1948). This expressed normal science in communication at the time. The media centrism of this approach to the study of communication was evidenced in the title of Wilbur Schramm's major work on development communication, *Mass Media and National Development* (1964), as well as in Everett Rogers' widely influential research on the diffusion of innovations (Rogers 1963).

The modernization process turned out to be more complicated and more difficult than assumed in early theoretical statements. Economically, Rostow's anticipated takeoff into sustained growth turned out for the majority of developing nations to be a takeoff into sustained debt (Korten 1995). Socially, newly modern institutions were often troubled by authoritarianism, corruption, and exploitation. At the level of individual experience, Lerner noted as early as 1976 that the "tide of rising expectations" he had foreseen was in many places turning into a "tide of rising frustrations" (Lerner and Schramm 1976).

Studies of the demise of modernization theory have included a wide range of theoretical perspectives. Dependency, neo-dependency, and then world systems theory analyzed the political economy of global economic structures (Amin 1997;

Chew and Denemark 1996; Frank 1984; Wallerstein 1979). Postmodernism and cultural studies have addressed cultural power (Escobar 1995; Spivak 1987; Chambers and Curti 1996). Rural participation studies have highlighted the need for local self-determination (Cernea 1991; Brohman 1996).

The theoretical variety found in the development literature can be found in communication research as well. Studies of political economy have explored distortions in information flows resulting from corporate power (Schiller 1976). Awareness of the importance of culture is reflected in calls for multicultural approaches to development policy (Servaes 1998). Participatory communication emphasizes dialogic processes in grassroots and community organizations (White et al. 1994; Servaes et al. 1996). Neo-modernizationist research addresses information campaigns and program evaluations without reference to the broad theoretical claims made earlier (Hornik 1988).

This brief account of modernization theory provides only a sketch of theoretical trends and is not intended to represent the full variety of modernization theory or its criticism. It allows one observation and sets the stage for one claim.

The observation concerns the breadth of social processes that are inevitably involved in development and social change and the relative paucity of contemporary theories addressing this breadth. Development studies over recent years have placed a worthy emphasis on participatory processes and on a lack of participation in much development work. However, this emphasis on participation focuses on local social change processes and to some extent on political participation. The understanding of large-scale social change requires, in addition, the analysis of large-scale institutions, including institutions responsible for economics, law, change in social norms, federated political systems, global institutions, and more. The claim is that the study of communication for development and social change must address all these.

Very few theories address the full breadth of these issues in a systematic way. Amartya Sen's capabilities approach is one of the few. Habermas's is another. Although it is not primarily oriented to development studies, Habermas's theoretical framework addresses a wide range of the social processes required for development. It does so systematically and critically.

## 14.3 Part II: The Theory of Communicative Action

Major differences between the theory of communicative action and modernization theory can be seen in their different appropriations of Weber's work. Noting its 1950s origins, Habermas explains that modernization theory had taken up Max Weber's program but associated it with social-scientific functionalism. "The theory of modernization performs two abstractions on Weber's concept of 'modernity.' It dissociates 'modernity' from its modern European origins and stylizes it into a spatio-temporally neutral model for processes of social development in general" (Habermas 1995, p. 2).

With this observation Habermas begins by criticizing both Weber's account of modernity and its functionalist employment by modernization theorists. At the same time, he wants to retain some elements of Weber's approach and of the enlightenment legacy generally. White summarizes Habermas's position. "One of the underlying goals of Habermas's work has been to construct a middle position between an uncritical, endorsement of modernity and a 'totalized critique' of it" (White 1987, p. 128). This middle position rests on an analysis of Weber's historical view of societal rationalization and the position that Weber's treatment of societal rationalization relied on a one-sided conceptualization of reason generally.

It is in relation to this *critique of reason* that Habermas offers a communicative theory of reason upon which is built a new treatment of societal rationalization. A review of Habermas's treatment of social evolution therefore requires an understanding of his theory of communicative reason.

The theory assumes a certain linguisticality of experience (Habermas 1984, pp. 94–95). Language is more than a tool. It embodies beliefs and preunderstandings below the level of conscious reflection before it is used in any tool-like respect. Nevertheless, against this background of preunderstandings, language is used to communicate in more or less deliberate ways. And this use rests on three reciprocal assumptions (see Fig. 1). Any communicator assumes of his or her interlocutor that they are (1) speaking the truth, (2) acting appropriately in light of social norms when speaking, and (3) speaking truthfully or sincerely (Habermas 1984, p. 99).

These "validity claims" of truth, rightness, and truthfulness, respectively, are each treated as necessary. They do not always obtain insofar as any of the claims can be false during an actual, given interaction, and yet their existence as behavioral operating assumptions is considered necessary for any kind of communication at all, even deceitful communication.

If an implicit validity claim is challenged explicitly, then discussion can be attempted in order to resolve the disagreement. The relevant claim can be said to have been "thematized." Discussion, or discourse, may then follow in order either to justify or to disprove the claim. Discussion oriented toward understanding is characterized by the willingness of participants to meet certain speech conditions. They must allow a symmetric distribution of opportunities to contribute to discussion. They must be willing to entertain any proposition raised during discussion. And they must be willing to offer justifications for their own claims when asked to do so.

Within the framework of these validity expectations, communicative action is treated as interaction involving the willingness to justify validity claims, or as "action oriented toward understanding." When action is oriented toward success rather than understanding, or when the speech is manipulative, then it is treated not as communicative but as strategic action (see Fig. 2).

> I shall speak of *communicative* action whenever the actions of the agents involved are coordinated not through egocentric calculations of success but through acts of reaching understanding. (Habermas 1984, pp. 285–286) [original emphasis]

This formulation of communicative action provides the basis for Habermas's account of reason. Enlightenment thought generally holds reason to be a property of individual consciousness, characterized by its dispassionate and instrumental grasp

| Conceptual category | Construct |
|---|---|
| Validity claims | Truth (accuracy) |
| | Appropriateness |
| | Truthfulness (sincerity) |
| | Comprehensibility |
| Speech conditions | Symmetric distribution of opportunities to contribute |
| | Ability to raise any proposition |
| | Full treatment of propositions raised |

**Fig. 1** Empirical concepts – validity claims and speech conditions

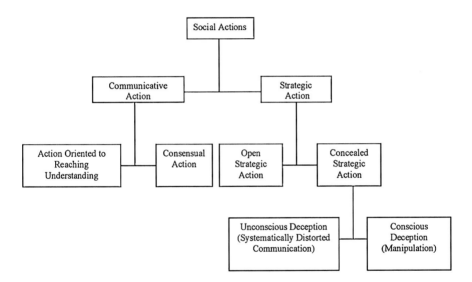

**Fig. 2** Kinds of social action. (Modified from Habermas 1979, p. 209)

of truths about an external world. This treatment is assumed among both the classical sociologists, including Weber, and later modernizationists. A communicative approach, alternatively, treats reason not as the individual's instrumental grasp of the external world but as a process through which individuals interact to coordinate action with regard to the world.

Action coordination takes place with regard to different kinds of affairs, as indicated by the kinds of validity claims involved. And it requires different forms of discussion, or discourse, appropriate for each. Claims pertaining to truths about the world must be adjudicated using theoretical discourse, ultimately including the tools of observation, testing, and so on or reference to them. It should be noted that

| Validity Claims | Cultural Value Spheres | World Relations |
|---|---|---|
| Truth | Science and Technology | Objective World |
| Rightness | Law and Morality | Social World |
| Truthfulness (sincerity) | Art and Literature | Subjective World |

**Fig. 3** Validity claims, world relations, and cultural value spheres

*truths* about the *objective* world are not considered to be absolute, but rather are defined in terms of a pragmatist epistemology (Habermas 2003). Claims pertaining to agreements within the social world must be handled with normative discourse of an ethical, moral, or political nature. Claims pertaining to the subjective world of individual sincerity can only be addressed using expressive discourse and, in the end, can only be justified through consistent performance of expressed positions (Habermas 1984, pp. 19–22).

Although they are usually implicit, or unconscious, these claims play an essential and pragmatic role in all social interaction. Assumptions regarding truth make it possible for individuals to interact in relation to the objective world. Assumptions regarding rightness make it possible for individuals to interact in relation to the social world. And those regarding truthfulness make it possible to interact in relation to subjective experience. These "world relations" enable individuals to coordinate their actions with regard to these different spheres of existence. And they do so through the option of giving good reasons for claims, in anticipation of the success of the better argument, regardless of the kind of discourse employed (see Fig. 3).

Given the different forms of discourse appropriate for each, then, these world relations require different forms of rationality. Reason is not a unitary skill, but is multiple. Or, in specific, it is tripartite. And reason is not only cognitive in an instrumental or scientific sense. It is also employed cognitively in matters related to social and subjective affairs, assuming communicative forms appropriate to each.

Returning to the question of the rationalization of society, the theory of communicative reason provides a foundation upon which social rationalization need not be understood as bureaucratization through the application of scientific reasoning and technology. More broadly, social rationalization can be seen as tripartite. The emergence of modernity comprises the application of reason within three different "cultural value spheres:" science and technology, law and morality, and art and literature. "With science and technology, with autonomous art and the values of expressive self-presentation, with universal legal and moral representations, there emerges a differentiation of *three value spheres*..." (Habermas 1984, pp. 163–164) [original emphasis]. Each cultural value sphere performs characteristic functions and accumulates knowledge and tradition. Habermas treats this in terms of the "differentiation of the structural components of the lifeworld."

## 14.3.1 Cultural Value Sphere: Science, Technology, and the Objective World

In modern scientific cultures, the economic sphere became more autonomous as instrumental rationality produced tools suitable for facilitating increases in productive capacity. Government became more autonomous as the state took on its own management, freed from the meddling of external social forces such as the aristocracy and the church. For the cultural activities of the arts, accepted artistic practices and autonomous standards of aesthetic appreciation evolved. In each case, one sphere might be informed or affected by the others, but the most effective mode of discourse for each is that in which its characteristic form of communicative rationality holds sway. Each operates "according to its own logic."

More generally, Habermas speaks of the rationalization of the lifeworld. And he emphasizes the manner in which the broader conceptualization of reason employed makes possible a more adequate grasp of the historical evolution of the legal, moral, and aesthetic characteristics of modern society, in addition to those associated with science and technology.

> The release of a potential for reason embedded in communicative action is a world-historical process; in the modern period, it leads to a rationalisation of life-worlds, to the differentiation of their symbolic structures, which is expressed above all in the increasing reflexivity of cultural traditions, in processes of individuation, in the generalisation of values, in the increasing prevalence of more abstract and more universal norms, and so on. (Habermas 1992a, p. 180)

This triadic approach to social rationalization in modernity differentiates the theory of communicative action from Weber's work, from modernization theory, and from previous approaches to modernity generally. And it is the chief means by which Habermas has attempted to develop his critical theory of society. For it provides an account of society in terms of which modernity's chief problem can be analyzed. This problem is theorized as the "colonization of the inner lifeworld."

The lifeworld as a whole includes activities dedicated to the transmission of cultural norms, to the integration of social institutions, and to the formation of identity. These activities are values-oriented, and their discussion requires communicative action utilizing all forms of discourse. Colonization occurs when instrumental reason is inappropriately used in place of more appropriate discourse in the attempt to fulfill such lifeworld functions. It occurs when, for example, science and/or bureaucracy is used in place of family interaction to make determinations on matters of child rearing. It occurs when market values come to determine the direction of aesthetic and artistic trends. And it occurs when bottom-line reasoning moves from the business to the editorial side of pressrooms.

The colonization thesis identifies the negative aspects of instrumental rationality, of social differentiation, and of confusing complexity in such a manner that they can be treated not as incidental misfortunes of modernity but as systematic expressions of unbalanced development. Colonization is the result of a "selective pattern of rationalization," in which normative and aesthetic spheres of society have not been sufficiently worked up and have been distorted by outsized growth in the importance accorded to technical and administrative reason (Habermas 1984, p. 240).

## 14.3.2 Cultural Value Sphere: Law, Morality, and the Social World

The view that spheres of society become increasingly differentiated while interacting with and mutually affecting one another predicates, or grounds, Habermas's classification of historical stages into three periods. These include pre-conventional, conventional, and post-conventional stages of social evolution. Through this classification of stages, a framework for the analysis of social evolution is created that he treats as a "... *directional variation of lifeworld structures* ..." (Habermas 1987, p. 145) [original emphasis].

In pre-conventional societies, the organization principle is reflected in kinship systems. In these, natural, social, religious, and artistic norms are interwoven within mythic systems. Conventional societies have class stratification as their organizing principle, where the production and distribution of wealth shifts away from regulation by family lineage and into the hands of centralized authority. The justification of this situation takes place through "dogmatized knowledge," in which the authority of king, pharaoh, or Caesar is proclaimed. Post-conventional society begins to emerge with what we know as the rise of modernity. The force of dogma wanes. The basis of legitimation for social practices is raised for examination free from otherworldly justifications.

It is only in the post-conventional stage of social evolution that communicative action becomes more fully institutionalized. For in this stage it is necessary. The complexities of later mercantile and capitalist economies required relative independence among the value spheres. In each sphere, new social arrangements can only be legitimately ordered through communicative action among affected social members without undue interference from the other spheres. Insofar as such new arrangements required modification of traditional practices, it became necessary that discourse be undertaken in the manner of a "reflective appropriation of tradition." This reflective appropriation is a collective social process in which choices are made about desirable forms of life, i.e., what to abandon through change and what to retain. In democratic societies, many such decisions as are of public concern are intended to be expressed politically. Here, input into governance requires the formation of public opinion. To differentiate it from the mere answering of opinion polls, Habermas treats this opinion formation process as being highly discursive in nature, referring to it as "collective will formation." Thus, communicative action emerges socially in the form of a public sphere (Habermas 1989).

To be clear, processes of reflective appropriation take place within public spheres both at the general level of a political public sphere and in more specialized spheres of science, law, and the arts, as well as within various social and cultural groups. In all cases, the discursive nature of the process is essential. In the political public sphere, discursive will formation is required to legitimize newly emerging political arrangements. Freeflowing discussion is required, taking into account the full range of possible concerns. Unless these concerns are all addressed, along with the interests of those who express them, then any decision will be considered illegitimate by those excluded. This makes modern life highly dependent on communication. Increasing freedom from the unreflective practices of tradition increases the social need for "interpretive accomplishments." Finally, if these accomplishments are to become socially embodied in a fair and lasting way they must be validated within the institutions of law and justice.

### 14.3.3 Cultural Value Sphere: Art, Literature, and the Person

The freeing up of communicative practice that results from structural differentiation takes place not only at the societal and institutional levels. It also affects processes of identity formation and individual socialization.

> Under the functional aspect of mutual understanding, communicative action serves to transmit and renew cultural knowledge; under the aspect of coordinating action, it serves social integration and the establishment of solidarity; finally, under the aspect of socialization, communicative action serves the formation of personal identities. (Habermas 1984, p. 137)

Communicative action can be seen as operating both downward and upward, relating micro- and macro-level social processes. Microlevel, or individual behavior, is analyzed in terms of George Herbert Mead's symbolic interactionism (Morris 1962). For the requirements of communicative understanding are ultimately defined not in terms of sentential or grammatical sufficiency but rather in terms of the ability to coordinate social action.

Following Mead, the self is constructed insofar as it sees itself reflected in the behavior of others and in turn reacts to what it has seen. This is a process in which role taking is learned, in the ability to assume and to consider the perspective of the other. The emerging ability of an individual to coordinate action for the self and with the other comprises the development of ego identity. Through the course of maturation, in this sense, the ability to consider the perspectives of others is widened, resulting in a "decentered" ego.

As the process of identity formation takes place, so too takes place the process of socialization. Insofar as maturation takes place within the lifeworld of a given linguistic community, the interactions through which identity is formed reproduce among other things the sanctions and prohibitions of the community's values. Thus it is that the self becomes imbued with the particular values of a community at the same time as it learns who it is and how to behave (See Habermas 1987, p. 142).

A common attribute of both the individual and the social expression of values is moral/ethical maturation. Following Mead, the self is highly dependent early on. But eventually the self learns to operate in relative independence. Through stages of cognitive development, the self is increasingly differentiated off from significant others as it gains its own identity. And, as Habermas argues using Lawrence Kohlberg's (1984) work, this is associated with the development of moral reasoning.

First the self understands only the imperative of submission due to dependency. Then it learns the nature of norm conformative group expectations. Eventually the self learns to evaluate group expectations autonomously. And thus, the ego becomes socialized within a community's lifeworld but at the same time becomes relatively independent in terms of its ability to assume ethical and moral positions. This includes the expectation that others will behave the same and thus allow such autonomy universally. This final state is that of an ethically mature individual.

The theory of communicative action provides here the anchor for a discourse theory of ethics that emphasizes the universality of discourse assumptions but denies the existence of universal substantive rights. The group expectations comprising a community's values may extend globally, and may thus embody positions on questions such as human rights. These rights are not universally valid *a priori* in the sense of natural law, but they may always be universalizable in the sense that they can be valued globally (Habermas 1990, 1993).

Finally, it is in the value sphere of personality and subjectivity that aesthetic expression finds its home. As social conditions change rapidly, the individual experience of such change can find its most singular expression through art, as long as norm conformity is not absolutely required. Fears, promises and interpretations of change can be explored and shared. Along with the rise of governments whose authority was legitimated independently from religious institutions, during the seventeenth and eighteenth centuries, the world of art and aesthetics also began to evolve its own, autonomous domain. Thus, modernity has been characterized, in part, by the appearance of artistic expression freed from domination by outside concerns.

On one hand, the differentiation of artistic practices away from those of the church and state represents a gain in freedom for the individual artist. On the other hand, this differentiation expresses itself in benefits at the social level as well. Art becomes a means for collective social reflection on evolving social conditions, through aesthetic criticism and public appreciation. According to Habermas's analysis, in premodern eras values changed in an unconscious, naturalistic manner. Alternatively, in modernity, values can be held up for conscious, public evaluation. Values become "discursively fluidified." This makes individual aesthetic experience available for circulation through the spheres of law and morality and science and technology.

> To the extent that art and literature have become differentiated into a sphere with a logic of its own, and in this sense have become autonomous, a tradition of literary and art criticism has been established which labours to reintegrate the innovative aesthetic experiences, at first "mute," into ordinary language, and thus into the communicative practice of everyday life. (Habermas 1992a, pp. 168–169)

All of this illustrates the sense in which communicative action is not simply an abstract, formal process. It is not a rarefied form of interaction between hypothetical individuals, but is rather a form of behavior that is presumed in daily interaction between parents and children, teacher and pupil, performer and audience, and among scientists, artists, and members of all groups who cooperate in the normal production and reproduction of society. And it emphasizes the manner in which social values are not mere attitudes or opinions. Social values are reflections of social structure, deeply built into identities formed during early socialization though subject to change through maturation. They are also the subject of aesthetic reflection, subject to public expression in a manner that both represents the social status quo and also facilitates social change.

### 14.3.4 Criticism

Of the many debates in which Habermas has been engaged while elaborating this theory, a small number must be addressed in the review undertaken here. I will briefly mention one standard objection related to the charge of idealism and then turn to somewhat lengthier treatments of objections particularly relevant to development debates.

One of the most common criticisms of the theory of communicative action is that it is idealistic and that it suggests people always try to understand one another. Both Foucault's analysis of power/knowledge and Lyotard's treatment of *agonistics* argue that power is inextricably bound up with interaction and that mutual understanding is seldom if ever at the heart of intentions (Foucault 1980; Lyotard 1984). From Habermas's perspective, this is an over reading of the theory, which aims to understand the basis upon which rests the possibility of communication of any kind. As noted, the theory identifies a variety of action forms, including as strategic action, some of which are agonistic. As Bernstein explains, the theory of communicative action is intended not to dispute the prevalence of power or difference, but rather to outline a program of research into difference without, "... ignoring or repressing the otherness of the Other" (Bernstein 1992, p. 313).

Other charges against Habermas's system are reflected in debates regarding the nature of national development processes. One is the charge that the theory of communicative action is ethnocentric. On this account, the ideal speech situation treats as universal values that in truth are only Western. Richard Rorty, for one, holds that Habermas's project falls prey to the same foundationalism characteristic of Western philosophy generally (Rorty 1985). The idea of reason alone, no matter how communicative, is enough to reveal the cultural particularity of Habermas's system.

This is one of the most difficult charges for Habermas's system to answer. The problem is indeed recognized. Speaking with regard to his discourse ethics, Habermas says that any system must do more than just "...reflect the prejudices of a contemporary adult white male Central European of bourgeois education" (Habermas 1990, p. 197). Following the simple fact of recognition, his counterargument emphasizes the idea that the content of any discourse relies on the background of preunderstandings and values embodied in lifeworlds. Any challenge of a validity claim, and any attempt to redeem that claim, takes place within the lifeworld of actual participants. And the content of lifeworlds is not universal but is rather particular. Habermas differentiates the universality of the *kinds* of claims that are assumed in communication from the content of communication. "... the validity claimed for propositions and norms transcends spaces and times, but in each actual case the claim is raised here and now, in a specific context, and accepted or rejected with real implications for social interaction" (Habermas 1992b, p. 139).

The content of communication embodies particular cultural values that play a fundamental role in communicative exchange. This is a central tenet of the theory, and therefore the charge of ethnocentrism must focus on the presumption of validity claims rather than on cultural content. To this extent, the theory answers its critics. But the idea that validity claims are universally presumed among communicators is an ambitious one nevertheless.

Another charge related to development debates concerns ethnocentrism and universalism in a different way. This charge holds that structural differentiation of the lifeworld can no more be universal than anything else and that all social structures are equally deserving of respect, differentiated or otherwise. Habermas argues that differentiated social and cognitive structures are the hallmark of modernity's universalism. He also maintains that structural requirements do not determine cultural content and traditions that a society may wish to retain within such differentiated structures can be freely chosen. Therefore, the theory is culturally relativistic. However, it is clear that differentiation is not entirely independent of content. Tradition must be abandoned wherever it blocks differentiation. And this raises the problem of ethnocentrism in another light.

Consider theocracy. The theory of communicative action cannot count as modern the undue intrusion of religious reasoning into governance. Modernity requires the separation of church and state through a constitutional form that guarantees freedom of religious choice. This requirement in effect determines that religiously based states cannot be fully modern.

> The universalist position does not have to deny the pluralism and the incompatibility of historical versions of "civilized humanity"; but it regards this multiplicity of forms of life as limited to cultural contents, and it asserts that every culture must share certain formal properties of the modern understanding of the world, if it is at all to attain a certain degree of "conscious awareness." (Habermas 1984, p. 180)

Yet another place where the theory's universalism may be found objectionable, on related grounds, is in its treatment of historical stages. Habermas formalizes his conceptualization by drawing from the developmental psychology of Jean Piaget (1970). He does not intend to suggest that individual psychological processes can be used to explain social processes. But he does want to borrow, at an abstract level, Piaget's model of developmental stages. In Piaget's model, each stage of an individual's cognitive development is able to process more complexity than its previous stage, and thus some stages presuppose others (Habermas 1984, pp. 66–74). Similarly, Habermas argues that history's logic is developmental. Social stages presuppose certain other stages. But this logic does not require that development be inevitable, as it generally was in modernization theory. Progress gained can be lost. It might be, "...eroded – even finally crushed" (Habermas 1992a, p. 196).

Neither is progress treated very hopefully. The consumer society envisioned by modernization theory seemed to promise much, a land of plenty, the good life. Alternatively, Habermas sees modernity simply as the enhanced capacity for responding to survival challenges. Quality of life, as inferior or superior, is not at issue. "For example, medieval societies, notwithstanding higher degrees of oppression than modern or primitive societies, could in certain circumstances be clinically healthier than either of them" (Habermas 1992a, pp. 204–205). Nor, finally, and to make the point clear, does progress imply happiness. "Emancipation – if we are to give this word a meaning that is foolproof against all misunderstanding – makes humanity more independent, but not automatically happier" (Habermas 1994, p. 107).

## 14.4 Part III: Communicative Action and Development

These reviews of modernization theory and the theory of communicative action provide the basis for an instructive comparison of the two. Similarities are evident at the most general level. Both seek to explain the evolution and characteristics of modern nations within a historically globalizing framework. They both try to relate individual and social processes, and both employ a notion of progress. Both address processes of change in tradition, social norms, and political systems. However, there are also significant differences between them at the general level.

Modernization theory was bipolar, involving two opposing stages, i.e., tradition and modernity, while the theory of communication action sees the emergence of modernity through three stages, including pre-conventional, conventional, and post-conventional stages. Modernization was in the main naturalistic, referring to the theory of biological evolution for its model of growth through differentiation, growth which was treated as being inevitable, while the theory of communicative action does not. Habermas consciously eschews the biological metaphor for social change while embracing a form of historicism (Habermas 1979, pp. 167–178).

Modernization theory began as a study of economic growth and later recognized the interdependence of economic, social, and cultural institutions. Nevertheless, modernization theory remained an economically centered theory, not only because economic growth was the major concern of involved governmental and financial institutions but also because the theory never fully abandoned its economistic origins. The theory of communicative action, alternatively, begins with a model of social evolution having three distinct and essential social spheres. The importance of economic growth is not denied, but it is removed from the central role it previously occupied. Evolution in the normative and subjective spheres is seen as being as important as evolution in economics.

In modernization theory, communication's role is to facilitate social evolution through attitude change. As students of Weber's thought, Lerner, Pye, and others shared the assumption that modernity was enabled primarily through the application of instrumental reason. This approach to rationality fits well with a transmission theory of communication. But it is fundamentally different from a linguistically oriented approach to rationality emphasizing at the outset that communication and reason are intersubjective. Only with an intersubjective, or linguistic, starting point can attitudes, motivations, and identity be adequately explained in relation to social and cultural institutions.

The theory of communicative action does see communication as serving the spread of rationality. However, it refers not to instrumental rationality but rather to a tripartite scheme of communicative reason. This tripartite scheme is understood using a theoretical reconstruction of the conditions necessary for any kind of communication and includes the presumption of an inescapable and largely unconscious orientation toward mutual understanding. Because this orientation is always already assumed by all, it constitutes a steady pressure for its own historical embodiment in social institutions, in each of the lifeworld spheres. Communication is thus elevated from its role as an assistant to economic growth through the creation of scientifically rational attitudes. It becomes instead the main explanatory construct for social evolution in all three value spheres.

Given an intersubjective account of reason, a new approach to problems in each of the three cultural value spheres is possible. We can now turn to the differences between modernization theory and the theory of communicative action in relation to each.

### 14.4.1 Cultural Value Sphere: Science, Technology, and the Objective World

Modernization theory treated the emergence of a modern global culture as a consequence of the spread of rationality. In the marriage between technology and economy, reason was seen as leading the way to a future of material well-being. By treating individuals largely as rational consumers and economy as a matter of rational markets, social processes were flattened. Other social elements were analyzed insofar as they supported this flattened account. On one hand, market dysfunctions were almost invisible. On the other, traditional cultures were relegated to the dustbin of history. The theory of communicative action redresses the focus on economy and instrumental rationality with its triadic treatment of society, enabling criticism both of market dysfunctions and excessive reliance on such a narrow conception of rationality. And it does so in a manner similar to critiques of modernization theory.

While the theory of communication action does not address economics in the manner of dependency theory, its thesis on lifeworld colonization does provide its own justification for critical economic analysis under the rubric of the "suppression of generalizable interests." Habermas has reworked dialectical materialism in order to take into account methods employed by the modern state to diffuse contradictions, methods not foreseen by Marx (Habermas 1979, pp. 130–177).

For Habermas, the mechanism of social control is not ideology, but systematically distorted communication, and the result is not false consciousness, but fragmented, colonized world views. Instrumental reason operates in such a manner as to justify the emphasis on scientific economic expertise, which in turn mystifies power relations through the use of specialist jargon and non-discursive information diffusion. The apparent mystifications of market behavior represent one example. The rush of diffusion expertise witnessed in the South exemplifies another.

### 14.4.2 Cultural Value Sphere: Law, Morality, and the Social World

Modernization theory's account of rationality rendered it unable to accord culture sufficient importance. Because modernization theory assumed that traditional values would simply disintegrate, its account of social progress could assume that values from the West would fill the cultural void left by this disintegration. However, once it is recognized that tradition cannot be simply abandoned, then a process in terms of which culture is carried forward and hybridized must be specified. Here the theory of communicative action uses the lifeworld as the background of tradition, and change becomes a matter of "selective appropriation of tradition." As a social process this can be done accidentally and unconsciously, or it can be done reflectively through recognized institutions and social practices. Elements of tradition and change can be carried forward intermixed. Insofar as this reflection requires institutionalization, it is likely to arise not only in political discourse but also in law and the arts.

The matter of public discourse receives very little treatment in modernization theory. In Lerner's work, media and political participation require media consumption along with the formation and expression of public opinion. But the formation of opinion is treated largely as a private matter, as an acquired skill of individuals, and public expression is treated as little more than voting and answering opinion polls. Public discourse and the mechanisms that facilitate discourse are never mentioned.

In his analysis of the public sphere, Habermas argues that commercial values have severely jeopardized the ability of mass media to facilitate collective will formation and public opinion. This "transformation of the public sphere" had already taken place in the North well before the time of political decolonization in the South. Therefore, the media institutions implanted in the South during and after the colonial period had already in a sense been lifeworld colonized. Thus, in the case of the South, lifeworld colonization didn't grow from inside as it did in the North. But Northern lifeworld colonization processes can reasonably be seen as having taken place in the South as well, imported as a form of neocolonialism.

The process of cultural change is much more complex than was suggested by the attitude change models employed in modernization theory, and if it is to be democratic, it requires discursive processes involving the public at large, taking place in all lifeworld spheres.

### 14.4.3 Cultural Value Sphere: Art, Literature, and the Person

Modernization theory's account of the person was individualistic. Drawing lightly on Freud, and more heavily on the Protestant Ethic, modernization theory produced a theory of the person that was anemic. Insofar as it was removed from its local and cultural context, the ego ostensibly became susceptible to Western influence. Attitudes, it was assumed, could readily be changed and modernized. Given the assumption that tradition could be abandoned wholesale, possessive individualism could be expected to immediately take on an attractive hue (Macpherson 1962).

As it turns out, wholesale abandonment of tradition was never a realistic option. Frantz Fanon (1967) analyzed the identity choices facing local blacks in colonized North Africa. Particularly insofar as race and class were concerned, healthy models of identity were unavailable. And thus, using Northern models, there emerged indigenous classes in which presentation of self became tantamount in Fanon's eyes to self-betrayal. On another continent, Paulo Freire's teachings on teaching illustrated the submersion of the self that results from systematic oppression (Freire 1968, 1973). And novelists from Chinua Achebe to Ruth Prawer Jhabvala to Salman Rushdie have explored the cultural gaps and disjunctions individuals must somehow transverse while making their way among cultures.

Compared to present need, the theory of communicative action's treatment of personality is thin too. It does not explore these problems from the subject's viewpoint in empirical detail. But it does provide a framework within which such viewpoints can be explored. The fact that Third World populations were driven off their farms and into factories is not news, but analysis of lifeworlds in the global South provides some leverage in understanding social problems. Social policy must monitor the health of lifeworlds and minister to them through a variety of mechanisms including cultural policy. Habermas points out:

> ... Central concepts such as the happiness, dignity, integrity of the person are now changing as if before our very eyes. Diffuse experiences, which crystallize out under transformed life circumstances produced by changes in social structure, find their illuminating, suggestive, visible expression through cultural productivity. (Habermas 1992a, p. 169)

Habermas speaks here of conditions of the North. But in the South too, culture is changing as if before our eyes. Cultural policies must be used to carry a heavy burden in attempts to facilitate reflective appropriation of tradition in ways that facilitate health through the expression of subjective experience.

## 14.5 Research Opportunities

Debates over the nature of development and social change have over recent decades problematized the manner in which modernization theory analyzed social processes. The theory of communicative action offers an alternative approach for the study of many of these.

One line of inquiry could examine the theory's claim regarding the universality of lifeworld differentiation. If cultural value spheres must differentiate in modern societies, then science, government, and religion are expected to stay in their own backyards. Consider the possible impact of religion on governance. Does democracy require that the separation between religion and the state take the specific form that it has in Europe? Or, is it possible that a dialog between religious values and government might take various forms? Habermas seems to be suggesting as much in his recent thinking on religion (Habermas 2008, 2010).

Another line of inquiry could address the universalistic claim found in discourse ethics. Habermas's intent is to provide a minimal theory of ethics, one that makes the thinnest claims necessary to support a universalist account of justice, a "moral standpoint." He is well aware of the problems attending this project but has not yet addressed them in a multicultural setting, preferring so far to address them abstractly, within the framework of his own theory and that of philosophical ethics. The development context provides a setting that is particularly suitable for such multicultural inquiry.

The universalist element offers something akin to the right to communicate (Fisher and Harms 1983). The relativist element of discourse ethics is found in the idea that validity claims are necessarily embodied in statements whose content comes from the lifeworlds within which these statements are made and thus are culturally and historically specific. Given this account, substantive rights principles can be universal only insofar as they come to be generally recognized through adoption, as what Habermas calls "generalizable" rights or interests. Research could be undertaken to explore how rights become generalized and how particular cultural forms articulate with new rights as they emerge historically, including communication rights (Jacobson 1998).

The theory of communicative action of course provides an approach to evaluating the health of public discourse. Libertarian press theory places its guarantee of democracy in private ownership of the press (Siebert et al. 1963). Habermas's treatment of the public sphere, alternatively, places it in collective will formation. His recent work explores collective will formation in the context of industrialized nations, emphasizing the reliance of such will formation on discursive processes anchored in public lifeworlds (Habermas 1996). In the Southern context, emphasis on lifeworld integrity would suggest that local cultural forms should be integrated with modern press institutions. Development journalism (Shah 1996) and Third World applications of the more recently evolved "public" journalism could be examined in this light (Merritt 1995). Traditional forms of collective communication expressed in theater, music, and storytelling can be employed in the discussion of community issues (Brokensha and Werner 1980).

In the sphere of personality, the theory of communicative action suggests a number of research possibilities. Cultural hybridity resulting from globalization of cultural products is plainly important, but its effects are difficult to evaluate. Individual and social adoption of globally circulating aesthetics can be judged to be either positive expressions of free choice or negative expressions of market domination. But it is more difficult to explain why some are adopted and others are not.

The theory of communicative action provides its own approach for the study of such processes. On one hand, where trends represent the market-driven circumvention of public will, these trends could be considered to be a form of lifeworld colonization. On the other hand, where such trends represent public will regarding normative and aesthetic matters, they must be accepted as reflective appropriation.

## 14.6 Conclusion

The vision of modernization shared among students of national development in the postwar period has long faded in scholarly research. But current theories do not offer a systematic analysis incorporating what has been learned across fields of application and levels of analysis. The theory of communicative action addresses this situation with a critical theory of society. The terms of progress have been changed from that offered by modernization theory. Modernization is seen not as economic development driven by instrumental rationality. Instead, modernization is treated as lifeworld rationalization in a manner that accords equal importance to in the spheres of economy and culture, social equity, and personal emancipation. The terms for criticism of unbalanced development have changed as well. Rather than class-based analysis, as in dependency theory, a theory of lifeworld colonization is highlighted that identifies dysfunctions affecting all classes of society, in terms of distortions of communicative practice. Cultural critique is retained in a way that relies on insights shared with postmodernism, but holds on to the ideas of justice and progress nevertheless.

These analyses, of progress, of unbalanced development, and of culture, are all based on the role Habermas assigns to communication, or communicative action, in social change. In his approach to development, advances in social institutions, whether economic, political, or cultural, increasingly occur through processes of collective reflection.

The outcome of research into his universalist claims will determine the uses to which Habermas's theory can be put in the long run. He hopes to blunt charges that the staged theory of history is eurocentric by holding onto the relatively modest claim that stages represent neither inevitable change nor happiness, but only enhanced survival capability. He hopes to blunt charges against ethical universalism by holding onto the relatively modest claim that there are no *a priori* universal rights, but only the presumption of reciprocity underlying speech that explains the intuitive appeal of justice and so on.

If Habermas's formulations are to be taken as he intends them, then his critical theory of society provides the basis for a theory of justice that is universal, a theory of politics that is participatory, and a theory of the self that foregrounds personal emancipation. If these universalistic formulations are denied, then the theory nevertheless provides an elaborate framework for analysis of these subjects and their related institutions simply as products of the evolution of Western-styled societies. The structures and values they embody would then be taken only to represent options from which Western nations are choosing and others can choose to adopt, and hybridize, or not. In either case, the framework provides an intricate approach to studying long-standing problems in global development: at once critical, hermeneutic, and modernist.

## References

Almond G, Verba S (1963) The civic culture: political attitudes and democracy in five nations. Princeton University Press, Princeton
Amin S (1997) Capitalism in the age of globalization: the management of contemporary society. Zed Books, London
Bernstein RJ (1992) The new constellation: the ethical-political horizons of modernity/postmodernity. The MIT Press, Cambridge, MA
Black CE (1966) The dynamics of modernization; a study in comparative history. Harper & Row, New York
Brohman J (1996) Popular development: rethinking the theory & practice of development. Blackwell Publishers, Cambridge, MA
Brokensha DM, Werner O (eds) (1980) Indigenous knowledge systems and development. University Press of America, Washington, DC
Cernea M (1991) Putting people first: sociological variables in rural development. Oxford University Press, New York. Revised Second Edition
Chambers I, Curti L (1996) The post-colonial question: common skies, divided horizons. Routledge, London
Chew SC, Denemark RA (1996) The underdevelopment of development: essays in honor of Andre Gunder frank. Sage, London
Escobar A (1995) Encountering development: the making and unmaking of the third world. Princeton University Press, Princeton
Fanon F (1967) Black skin, white masks. Grove Press, New York
Fisher D, Harms LS (eds) (1983) The right to communicate: a new human right. Booble Press, Dublin
Foucault M (1980) In: Gordon C (ed) Power/knowledge: selected interviews and other writings 1972–1977. Pantheon, New York
Frank AG (1984) Critique and anti-critique: Essays on dependence and reformism. McMillan, New York
Freire P (1968) Pedagogy of the oppressed. Seabury Press, New York
Freire P (1973) Education for critical consciousness. Continuum Press, New York
Habermas J (1979) Communication and the evolution of society. Beacon Press, Boston
Habermas J (1984) The theory of communicative action volume 1: reason and the rationalization of society. Beacon Press, Boston
Habermas J (1987) The theory of communicative action, volume 2: a critique of functionalist reason. Beacon Press, Boston
Habermas J (1989) The structural transformation of the public sphere: an inquiry into a category of bourgeois society. The MIT Press, Cambridge
Habermas J (1990) Moral consciousness and communicative action. The MIT Press, Cambridge
Habermas J (1992a) Autonomy and solidarity. Verso, London
Habermas J (1992b) Postmetaphysical thinking: philosophical essays. The MIT Press, Cambridge, MA
Habermas J (1993) Justification and application: remarks on discourse ethics. The MIT Press, Cambridge, MA
Habermas J (1994) Past as future. University of Nebraska Press, Lincoln
Habermas J (1995) The philosophical discourse of modernity: twelve lectures. The MIT Press, Cambridge

Habermas J (1996) Between facts and norms: contributions to a discourse theory of law and democracy. The MIT Press, Cambridge, MA
Habermas J (2003) Truth and justification. The MIT Press, Cambridge
Habermas J (2008) Between naturalism and religion: philosophical essays. Polity Press, Cambridge
Habermas J (2010) An awareness of what is missing: faith and reason in a post-secular age. The MIT Press, Cambridge
Hornik R (1988) Development communication. Longman, White Plains
Inkeles A, Smith DH (1974) Becoming modern: individual change in six developing countries. Harvard University Press, Cambridge, MA
Jacobson T (1998) Discourse ethics and the right to communicate. Gazette 60(5):395–413
Kohlberg L (1984) Essays in moral development: the psychology of moral development. Harper & Row, New York
Korten DC (1995) When corporations rule the world. Kumarian Press, West Hartford
Lasswell H (1948) The structure and function of communication in society. In: Bryson L (ed) The communication of ideas. Harper & Brothers, New York, pp 37–51
Lerner D (1958) The passing of traditional society. The Free Press, Glencoe
Lerner D, Schramm W (1976) Communication and change: the last ten years and the next. University of Hawaii Press, Honolulu
Levy MJ (1967) Modernization and the structure of societies; a setting for international affairs. Princeton University Press, Princeton
Lyotard J (1984) The postmodern condition. University of Minnesota Press, Minneapolis
Macpherson CB (1962) The political theory of possessive individualism: Hobbes to Locke. Clarendon Press, Oxford
McClelland D (1961) The achieving society. Van Nostrand, Princeton
Morris C (ed) (1962) Mind, self & society. University of Chicago Press, Chicago
Merritt D (1995) Public journalism and public life: why telling the news is not enough. Erlbaum, Hillsdale
Piaget J (1970) Genetic epistemology. Columbia University Press, New York
Pye LW (ed) (1963) Communications and political development. Princeton University Press, Princeton
Rogers EM (1963) Diffusion of innovations. The Free Press, Glencoe
Rorty R (1985) Solidarity or objectivity? In: Rajchman J, West C (eds) Post-analytic philosophy. Columbia University Press, New York, pp 3–19
Rostow WW (1960) The stages of economic growth: a non-communist manifesto. Cambridge University Press, Cambridge
Schiller HI (1976) Communication and cultural domination. International Arts and Sciences Press, New York
Schramm W (1964) Mass media and national development: the role of information in the developing countries. Stanford University Press, Stanford
Servaes J (1998) One world, multiple cultures: a new paradigm on communication for development. Hampton Press, Creskill
Servaes J, Jacobson T, White S (1996) Participatory communication and social change. Sage, New Delhi
Shah H (1996) Modernization, marginalization, and emancipation: toward a normative model of journalism and national development. Communication Theory 6:143–159
Siebert FS, Peterson T, Schramm W (1963) Four theories of the press; the authoritarian, libertarian, social responsibility, and soviet communist concepts of what the press should be and do. University of Illinois Press, Urbana
Spivak G (1987) In other worlds: essays in cultural politics. Methuen, New York
Wallerstein I (1979) The capitalist world-economy. Cambridge University Press, Cambridge
White S (1987) Toward a critical political science. In: Ball E (ed) Idioms of inquiry: critique and renewal in political science. State University of New York Press, Albany, pp 113–136
White S, Nair KS, Ascroft J (1994) Participatory communication: working for change and development. Sage, New Delhi

# The Importance of Paulo Freire to Communication for Development and Social Change

## 15

Ana Fernández-Aballí Altamirano

## Contents

| | | |
|---|---|---|
| 15.1 | Introduction | 310 |
| 15.2 | Basic References About Paulo Freire's Life and Work | 311 |
| 15.3 | Contributions of Paulo Freire to Praxis in Communication and Education | 312 |
| | 15.3.1 Pedagogy of the Oppressed and the Breaking of the Subject-Object Approach in Social Sciences | 312 |
| | 15.3.2 The Paulo Freire Literacy Method, Popular Education, and Critical Pedagogy | 312 |
| | 15.3.3 Participatory Action Research and Participatory Development Communication | 313 |
| 15.4 | Freirean Epistemology | 314 |
| | 15.4.1 Basic References About Freirean Epistemology | 314 |
| | 15.4.2 Basic Concepts of Freirean Epistemology | 315 |
| | 15.4.3 Cognitive-Emotional Cycle of Freirean Epistemology | 317 |
| | 15.4.4 Levels of Conscientization: Cultural, Dialogic, and Power Structure Awareness | 320 |
| 15.5 | Implications of Freirean Epistemology to Communication for Development and Social Change | 322 |
| 15.6 | Conclusion | 323 |
| 15.7 | Cross-References | 325 |
| References | | 326 |

### Abstract

Paulo Freire (1921–1997) was the father of critical pedagogy, popular education, and participatory action research (PAR), as well as a basic reference of liberation theology and of Latin American critical theory and a precursor of the epistemologies of the south. His main work, *Pedagogy of the Oppressed*, marked a before

A. Fernández-Aballí Altamirano (✉)
Department of Communication, Universitat Pompeu Fabra, Barcelona, Spain
e-mail: ana@laxixateatre.org; ana.fernandezaballi@upf.edu

© Springer Nature Singapore Pte Ltd. 2020
J. Servaes (ed.), *Handbook of Communication for Development and Social Change*,
https://doi.org/10.1007/978-981-15-2014-3_76

and after in the fields of education, research, and communication, initially in Latin America and later on spreading to other regions both North and South. Particularly in the case of development communication and communication for social change, Freire's work had a definitive impact. As these disciplines were configuring into fields of thought and practice during the second half of the twentieth century, these currents were both fed and challenged by Freire's understanding of praxis. Freirean-based participatory action research became the basis of what we know today as participatory development communication (PDC), which ended up becoming, toward the end of the 1980s, a common ground between development communication and communication for social change. However, as PDC evolved, its practice was still – and is still – challenged by Freirean concepts. Currently, the field of Communication for Development and Social Change needs to continue reaching out to Freire's teachings, particularly to his epistemological and ontological framework based on his *Pedagogy of the Oppressed*.

**Keywords**

Paulo Freire · Communication for social change · Communication for development · Epistemology · Awareness · Pedagogy of the oppressed · Participatory action research · Participatory development communication

## 15.1 Introduction

*"Paulo Freire: 'He who in the middle of the night caused the sun to crack.'"*
Carlos Alberto Torres in *Paulo Freire: Uma biobibliografia* (1996).

This chapter examines the relevance of Paulo Freire's work to the field of Communication for Development and Social Change. Freire raises questions regarding "who," "why," and "how" agents are involved in communicative and social processes from a focus of power use and abuse and provides the framework to engage with social transformation through awareness of power, cultural, and dialogic structures. Dehumanizing others, through the abuse of power, which is depriving any person of the ability or space to problem-pose, can only serve the oppressor, regardless of how progressive, good-willed, or apparently emancipatory certain proposals may seem. To identify power abuse in communication for social change, it is always useful to go back to Freire and particularly to his epistemological and ontological proposals.

Communication for Development and Social Change is constantly confronted with the role which oppressed groups should have in their own liberation to avoid what Freire calls cultural invasion. Understanding communicative interactions based on cultural, dialogic, and power structure awareness – conscientização – does not always find its way through the implementation of these projects and programs. In 1992, Freire publishes *Pedagogy of Hope*, where he revisits his *Pedagogy of the Oppressed*, and three decades later still reinforces and calls upon the need for critical radicalism as a way to fight oppression from whomever and however it may come. A year before his death, the Paulo Freire Institute in the Brazilian city of São Paulo edited a biobibliographical volume of Freire's life and work. This compilation shows the extent to which Freire's both actions and thoughts reached out to change the lives of many, particularly in Latin America, and to challenge abuse of power anywhere. Currently, the field of Communication for Development and Social Change needs to continue reaching out to Freire's teachings, particularly to his epistemological and ontological framework which assures all liberation processes are of – and never for – the oppressed.

The following chapter provides a brief overview of the basic references about Freire's life and work, his contributions to praxis in the fields of communication and education, and how *Pedagogy of the Oppressed* has influenced social science, popular education, critical pedagogy, participatory action research, and participatory development communication. Furthermore, the chapter explores Freirean epistemology and how it generates awareness on cultural, dialogic, and power structures. Toward the end of the chapter, the reader will find a set of premises on the implications of Freirean proposals applied to the field of Communication for Development and Social Change.

## 15.2 Basic References About Paulo Freire's Life and Work

Paulo Freire (1921–1997) was a Brazilian thinker and educator born in Recife. He was incarcerated and exiled during the military dictatorship in Brazil in 1964. Freire goes on to live in Chile, the United States, and Europe until he returns to Brazil in 1980. His pedagogic, academic, and political trajectory is still relevant today. Freire is the father of critical pedagogy, popular education, and participatory action research, as well as a main figure in theology of liberation and epistemologies of the South.

Freire's work has had a significant influence in the development of critical thought disciplines. To name some of the most relevant in-depth studies of Freire's life and work are the texts by Carlos Alberto Torres *Estudios Freireanos* (1995), *La Praxis Educativa y la Educación Cultural Liberadora de Paulo Freire* (2005), and *First Freire: Early Writings in Social Justice Education* (2014). As a basic text to study Freire's work is *Paulo Freire: Uma Biobibliografia* (1996) written by Moacir Gadotti et al. Also of interest is the article collection titled *Paulo Freire: Contribuciones a la educación,* published by CLACSO in 2008.

Some authors who have built theoretical corpus which advance Freire's work are Orlando Fals Borda, who can be studied in detail in his anthology *Ciencia, compromiso y cambio social* (2014), Miguel Escobar's erotic pedagogy described in his book *Pedagogía Erótica: Paulo Freire y el EZLN* (2012), as well as the union between Pedagogy of the Oppressed and Theatre of the Oppressed developed by Tânia Baraúna and Tomás Motos in their book *De Freire a Boal* (2009).

## 15.3 Contributions of Paulo Freire to Praxis in Communication and Education

Although a complete and thorough overview of Freire's work is detailed in *Paulo Freire: Uma Biobibliografia* (1996), it is important to highlight specific contributions which were particularly relevant to the field of Communication for Development and Social Change.

### 15.3.1 Pedagogy of the Oppressed and the Breaking of the Subject-Object Approach in Social Sciences

*Pedagogy of the Oppressed* was first published in Portuguese in 1968. The text presented a system to anthropologically reinterpret the relations between philosophy, education, and politics (Gadotti et al. 1996). Half a century later, this text continues to generate debate all around the world about the nature, significance, and importance of education as a form of cultural policy (Gadotti et al. 1996) and is currently considered one of the founding texts of critical pedagogy.

The text is composed of four chapters. The first chapter defines the concepts and relations between oppressors and oppressed. The second chapter explains Freire's models of "banking education" versus liberating or problem-posing education as means to liberate oneself and the world in communion with others. The third chapter focuses on describing the basic epistemological and methodological elements of research and knowledge creation from the perspective of oppressed groups. The fourth chapter details different characteristics regarding dialogical action which are key to understanding how to achieve liberation through solidarity.

The main implication of this text is that Freire provides a theoretical and practical framework to break the subject-object relationship in educational, communicative, and research contexts. Orlando Fals Borda (2014) compares the significance of Freire's subject-subject approach in social sciences to what nuclear energy has meant for physics, in that it provides a new framework to understand and transform reality which is transversal through disciplinary and geographical frontiers.

Freire opens the way to a reconfiguration of science, politics, history, and social communication understood from the perspective of the oppressed. It is the oppressed – "the wretched of the Earth" – in communion, dialogue, and solidarity with each other, who are able to most accurately analyze, know, and change reality, since it is them who carry, see, and suffer the burden of the structural inequalities and injustices which conform the world (Fanon 1963; Freire 1975).

### 15.3.2 The Paulo Freire Literacy Method, Popular Education, and Critical Pedagogy

The Paulo Freire Literacy Method, extensively used in adult literacy programs throughout Latin America and beyond, is more than just learning to read and write. Reading and writing are instruments to the method, where the main objective is that the person perceives him or herself as a maker of culture. In order to achieve this, Freire proposed a methodology where (1) the literacy group, conformed in what Freire denominates "circles of culture," identifies about 17 words – generative words – which have great phonemic richness and are relevant and commonly used in the community; (2) using these words, the participants explore their syllables and phonemic families, forming new words and learning the technicalities of reading and writing as they gain awareness of the significance of the word in naming and shaping their reality; and (3) the literacy becomes the stepping stone where individuals and groups become able to critically name, assess, and transform their surroundings (Gadotti et al. 1996). Once groups have learned to read and write, and most importantly to understand themselves as cultural makers, these processes can enter more complex stages of social transformation conforming into the field of popular education.

Popular education began in Latin America, namely, in Brazil, and spread rapidly to other territories, greatly due to the enthusiastic welcome of Freire's ideas and methods on education by progressive activists, educators, politicians, and religious leaders. Although this enthusiasm was abruptly shut down by the 1964 Brazilian coup – and the wave of bloody repression that was shaking the entire Latin American territory – the seed of resistance and radical liberation among different base organizations had already been planted. With the influence of Marxism, Bourdieu, Passeron, Illich, and Vasconi, the questioning of the school system was radicalized, the political character of the educational practices was made evident, and alternative pedagogical proposals were generated (Torres 2007). Popular education as a movement is still one of the most relevant theoretical and methodological fields in communication for social change, counting with worldwide networks, organizations, and grassroots experiences.

Both the Freirean literacy method and his approach to popular education were crucial to the field of critical pedagogy. However, in Freire's approach, critical pedagogy is not only centered in the development of critical thinking skills, rather it is necessarily linked to the social transformation of the self and of the community (Chege 2009). Freirean critical pedagogy goes beyond the "critical" to emphasize the political nature of education, inviting a reflection on power structures within our teaching-learning contexts to transform education into a "pragmatic and political art" (Arnett 2002, p. 490). Critical thinking in itself is not enough to construct solidarity and communal subjectivities, since it can be instrumentalized following the rational self into a neoliberal understanding of what critical thinking is and should be.

### 15.3.3 Participatory Action Research and Participatory Development Communication

Participatory action research (PAR), based on Freire's approach to dialogical action and later systematized by numerous authors and practitioners (Alberich Nistal 2008; Ortiz and Borjas 2008; Villasante 2010; Fals Borda 2014), is a reflection-action spiral in which all persons involved in the research have the ability to propose and redefine the research itself. Participatory development communication (PDC) is a PAR-based practice within the participatory paradigm of communication for development. PDC focuses on the use of communication strategies and tools to facilitate and support the dynamics of local development through processes that actively involve the community – citizens, authorities, external agencies, researchers, technical teams, and associations, among others. PDC is a systemic and planned exchange process that is based on two pillars. On the one hand, it is based on the participatory processes of horizontal communication between agents through PAR methodology and on the other hand on the use of communication media and interpersonal communication as tools to support development initiatives identified in the participatory processes (Servaes 2003; Bessette 2004; Mari Sáez 2010). PAR, PDC, and the relationship between them have been studied thoroughly in the discipline (Alejandro and Vidal 2004; Bessette 2004; Alberich Nistal 2008; Tufte and Mefalopulos 2009; Barranquero 2014) both theoretically and methodologically. This chapter will focus concretely on the Freirean epistemological underpinnings of these fields.

## 15.4 Freirean Epistemology

### 15.4.1 Basic References About Freirean Epistemology

Freirean epistemology has already been defined as a field of study (Morrow and Torres 2012). In *Uma Biobibliografia*, Jacinto Ordóñez Peñalonzo specifically mentions Freirean epistemological proposal as one of the most significant contributions of his work (Gadotti et al. 1996). He establishes a difference between Freirean epistemology and other approaches in the field in that according to Freire any person can generate knowledge as long as that person is aware of the interrelation between three elements: the knower, the known object, and the method which relates them both. For Freire, in order to gain awareness of these three elements, the knowledge creator must begin with a process of critical pedagogy – a dialogue of teaching-learning based on cultural, dialogic, and power structure awareness – and end up with a theory to understand and explain reality (Gadotti et al. 1996). This process which links critical pedagogy to knowledge creation is made up of four complementary dimensions: the logical dimension (the method), the historical dimension (the relationship between knowledge and context), the gnoseological dimension (the purpose of knowledge), and the dialogic dimension (the communication of knowledge). As a result, all knowledge can be studied from its pedagogic, humanistic, scientific, and political implications (Freire 1994; Gadotti et al. 1996).

In the text *Reading Freire and Habermas: Critical Pedagogy and Transformative Social Change* (Morrow and Torres 2012), Freirean epistemology is described as a distinctive form of critical theory based on "a critical realist ontology that rejects essentialism; a constructivist and pragmatist conception of epistemology that is antipositivist but not antiscientific or purely subjectivist; and a pluralist conception of methodology within which participatory dialogue provides a regulative ideal, but is not an exclusive model of social inquiry" (Morrow and Torres 2012).

Orlando Fals Borda refers to Freirean epistemology as extensive, critical, and holistic (Fals Borda 2014). This is further developed by Boaventura de Sousa Santos (2010, 2011, pp. 16–17) in his concepts of epistemologies of the South where "[a] The understanding of the world is more extense the the Western understanding of the world...[b] The diversity of the world is infinite...and [c] This great diversity of the world, which must and should be activated, as well as transformed theoretically and practically in plural ways, cannot be monopolized by a single general theory." Following these premises, de Sousa Santos (2010, p. 50) introduces the concept of *ecologies of knowledge* understood as "a general epistemology of the impossibility of a general epistemology."

### 15.4.2 Basic Concepts of Freirean Epistemology

One of the fundamental contributions of Paulo Freire to communication for social change is his systematization of the interrelation between "oppressor" and "oppressed" groups and the implications of these power relationships in the generation of knowledge and subsequent social processes (message creation, channels, educational actions, programs, etc.). In this sense, the first important question to address regarding Freirean epistemology derives from the notion that Freire's work is OF the oppressed and not FOR the oppressed. Freire (Freire 1975) frames this basic condition in his most known work, *Pedagogy of the Oppressed*, with significant implications for the configuration of how communicative processes can be understood within the field of communication for development and social change.

Freire defines three categories: the radical man/woman, the oppressed, and the oppressor. The definitions that he provides of these three concepts are the following:

- The radical man/woman is one who is committed to the liberation of all men/women and is not trapped in "circles of security" which he/she also entraps reality. On the contrary, he/she inserts him/herself in this reality in order to know it better and hence transform it better (Freire 1975).
- The oppressor refers to those who exploit, abuse, and exercise violence due to a given power and/or privilege (Freire 1975).
- The oppressed are those who have been dehumanized by the violence of the oppressor and who nevertheless still hold within themselves the idea that *to be* means to *look alike* the oppressor (Freire 1975).

The Freirean proposal is directed toward the liberation of the oppressed man/woman, so that he/she can become a radical man/woman. Graphically, this distinction between OF and FOR the oppressed can be expressed according to Diagram 1.

This matter, which is far from trivial, determines the following concepts and premises which are the basis underlying all Freirean communication:

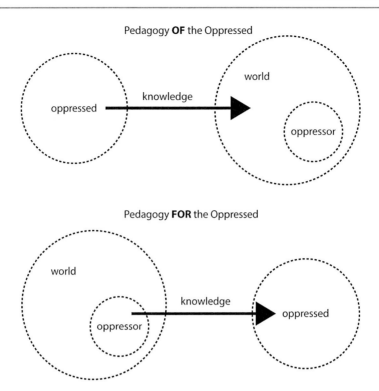

**Diagram 1 Pedagogy OF and not FOR the oppressed.** Diagram shows graphically the difference in knowledge flows depending on whether it is generated by the oppressed or by the oppressor group

- We call **oppression** a relationship in which a group uses its power to pressure, coerce, or prevent another group from doing something, which eventually perpetuates or increases the inequality of power between both groups.
- We call **liberation** the process by which the oppressed person or group undertakes a process of reflection-action (**awareness** through **praxis**) which eliminates the situation of oppression (Baraúna and Motos 2009).
- The oppressed group creates and instrumentalizes knowledge (**critical pedagogy**) to liberate itself from the oppressor group.
- The oppressor group and the oppressed group are mutually exclusive.
- The oppressor group cannot liberate the oppressed group.
- The oppressed group liberates itself if and only if it liberates the oppressor group.
- The oppressed group can be an oppressor of itself (**adherence**).
- Different groups oppressed by the same oppressor can liberate themselves and simultaneously liberate each other (**solidarity**).

These premises cannot be applied unambiguously since each context requires its own articulations and definitions. In any case, what can be considered objectively from the Freirean point of view is the overcoming of the situation of oppression

(Freire 1975), which results in the process of liberation/transformation. The Pedagogy of the Oppressed, as epistemology, can only be used rigorously when those who, deprived of privileges or having renounced them for their own gain out of solidarity and with no other intention than to put an end to injustice, undertake the task of transforming themselves to transform the world. This does not mean that in order to make rigorous use of Freire's epistemology, one must necessarily be part of the oppressed group, but one must be in solidarity, that is to say, one must have identified – or be in the process of identifying – his/her own oppressions to be able to undertake a process of union in solidarity. What is relevant is the act of reasoning about how, why, for what, and who benefits from a particular communicative action or program. This requires an awareness of one's own culture, ways of communicating, and the structures of power to which one is subjected to (both as a person who exercises power and a person over whom power is exercised). This union based on awareness and solidarity is what Freire calls *unity in diversity*, which is opposite of *cultural invasion*. That said, being aware of oneself implies trust in one's own abilities as well as in the abilities of others to overcome oppression. Solidarity implies making oneself vulnerable together with the people with whom we dialogue. In order to embark on a path of transformation in the Freirean sense, we must start from a shared vulnerability. In this sense, vulnerability does not mean being a victim nor is it a condition to be eradicated; nor does it eliminate the ability to act. On the contrary, vulnerability is an essential condition for dialogue and action (hence for knowledge creation) based on love, trust, hope, faith, humility, and indignation. Therefore, we could say that Freirean epistemology is based on an ontology of vulnerability. By definition, Freirean-based knowledge must always spur from those without voice and without power. This is the core identity of Freirean communication.

### 15.4.3 Cognitive-Emotional Cycle of Freirean Epistemology

Freirean epistemology, which is the basic underpinning of participatory action research, can be structured based on a cognitive-emotional cycle (Freire 1975, 1984, 1991, 1994, 2002a, b, 2007, 2012; Freire and Faundez 2010). The epistemological framework that Freire proposes was influenced by a diverse range of authors that marked his path (Freire 2002a): Marx, Lukács, Fromm, Gramsci, Fanon, Memmi, Sartre, Kosik, Agnes Heller, Merleau Ponty, Simone Weil, Arendt, Marcuse, Latin American popular groups, the Cuban revolution, Camilo Torres, Che Guevara, Mao Tse-tung, and student movements, among others. We also find in Freire's works countless anecdotes that refer to peasants, youth, colleagues, relatives, and friends who he acknowledges as co-creators of many of his concepts. Therefore, to speak of a Freirean epistemology is also to speak of an epistemology of collective creation.

Freire speaks of an epistemology composed of two moments (Freire 1994, p. 137): a first moment of *awareness* – or conscientization – and a second moment of *distancing*. These two moments are not exclusive; instead they are interrelated.

This dual process of recognition of reality allows for the identification of automated habits that make it impossible for the mind to work epistemologically and disable the curiosity to look for "the reason why things are" (Freire 1994). The process of *awareness/distancing* uses *coding* and *decoding* (see Diagram 2) to name, identify, and research collectively about the *generative themes* (Freire 1975). During this process, epistemological curiosity is developed (Freire 2004; Freire and Faundez 2010) in order to then change the beginning position of the group and to understand the reasons behind that change (Freire 1984). This change or transformation that begins with reflection-action is knows as *praxis*, which must always be analyzed from four complementary dimensions: the pedagogical, the scientific, the human, and the political one (Freire 1994). The relationship between these concepts is expressed graphically in Diagram 2.

The generative concepts of praxis proposed by Freire are framed in a cognitive-emotional circular process with two juxtaposed dimensions as shown in Diagram 2.

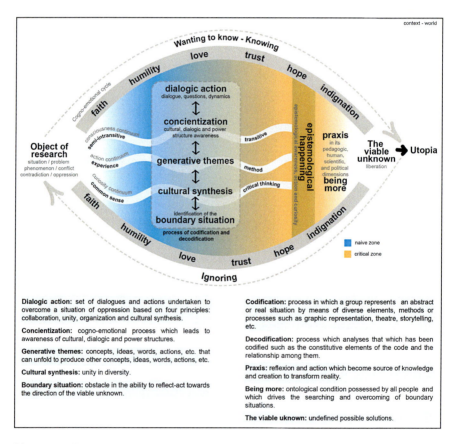

**Diagram 2** Synthesis of the Freirean epistemology. Diagram shows the cycle of knowledge creation applied to any communicative process based on Freirean epistemology (Fernández-Aballí Altamirano 2016)

The first dimension refers to the ontology of Freirean theory that is hope, faith, humility, love, trust, and indignation (Freire 2002a). Defining the Freirean ontological basis in terms of a relationship of emotions generates an immediate conflict with the claimed objectivity and neutrality of hegemonic knowledge – even that of critical schools of thought (Freire 1984); but with the emotionality as the basis of knowledge, Freire invites to consider ourselves as emotional beings. Negating the emotional condition of humans dehumanizes and distorts the way in which reality is learned, taught, and transformed. From a Freirean perspective, rigor is not found in the denial of the human emotional condition but in its rescue, along with consistency, method, and, above all, listening (Freire 1994).

The second dimension of the cognitive-emotional circular process is the ignoring-wanting to know-knowing-ignoring cycle. Referring to these concepts, Freire writes (Freire 2002b) that knowledge does not have a permanent character but is rather a continuous and contextualized process. Knowledge and ignorance are both relative and interrelated. Once knowledge is achieved, it automatically becomes a form of ignorance. There is no such thing as absolute knowledge or absolute ignorance, since knowledge is always relative and relational. The entire cognitive-emotional process defined by Freire occurs simultaneously between what we have called the *naive zone* and the *critical zone*. Table 1 characterizes these two cognitive areas.

Some characteristics of the naive zone are (Freire 2002b) that when one's mind is in this frame, it reveals a simplicity in naming the causality among the different elements and agents of a system and/or problem, reaching conclusions which are rash, superficial, and with a tendency to idealize the past. In the naive zone, it is easy to fall into generalizations of behavior including the underestimation of those who have less access to power, and it considers reality to be static. In the critical zone, there is a longing for a deeper analysis of the problematic under consideration, with an understanding of the limitations and lack of tools to reach this analysis. In this cognitive area, there is a recognition of the changing character of reality, but the knowledge of how that reality is conformed is linked through causality, verified, and always open for revision. This area requires the elimination of prejudice and status conforming positions and does not reject "the old" just for being "old" or accept "the new" just for being "new" but rather evaluates all relevant options and chooses according to their validity in relation to the context and the problem.

The transitive quality of consciousness – the meta-skill to become self-aware of one's own ability to change – allows for fluctuation from naive consciousness to

**Table 1** Characterization of the naive zone and the critical zone. Characterizes the different elements of cognitive-emotional continuum present in Freirean epistemology

| Element of the cognitive-emotional continuum | Naive zone | Critical zone |
| --- | --- | --- |
| Conscientization (awareness) | Intransitive and semi-transitive conscience | Transitive conscience |
| Action | Action based on experience | Action based on method |
| Curiosity | Curiosity based on common sense | Curiosity based on critical thought |

critical consciousness (Freire 2002b). Likewise, dialogical action allows movement from experience to method – which is nothing more than systematized experience – and from a curiosity based on common sense to a curiosity based on critical thinking. That is, dialogic action makes it possible to reach the *epistemological happening* in which consciousness, action, and curiosity are aligned in the critical area. The epistemological happening allows us to overcome the *limit situation* (oppression) through praxis making the *viable unknown* possible and hence moving closer to utopia (Escobar Guerrero 2012).

It is important to emphasize that for the Freirean epistemological process, the naive area is as relevant as the critical area. Common sense and experience are of equal importance as critical thinking and method in order for the process of transformation to happen. These processes are circular, dynamic, and contextualized. That is, knowledge based on Freirean epistemology is of a phenomenological nature (Torres 2006; Key 2009).

To recapitulate the basic characteristics of Freirean epistemology:

- It is of the oppressed and not for the oppressed.
- It is based on an ontology of vulnerability.
- It is formed from collective knowledge.
- It is circular, dynamic, and contextualized.
- It is phenomenological.

### 15.4.4 Levels of Conscientization: Cultural, Dialogic, and Power Structure Awareness

Cultural awareness, dialogic awareness, and awareness about power structures are inherent to Freirean epistemology – and thus in participatory action research – during the various stages of communication for social change. We can graphically express the interrelation among these three levels of conscientization as shown in Diagram 3.

**Diagram 3** Levels of conscientization. Diagram shows the different levels in the process of conscientization or awareness, where conscientization is achieved at the intersection of the three levels

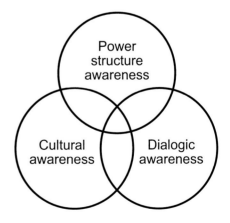

**Cultural awareness** responds to the recognition of one's own cultural heritage (Freire 1975). Freire writes (Freire 1994):

> We cannot doubt the power of our inherited culture, of how it conforms us and hinders us. But the fact of being programmed, conditioned and aware of this conditioning, and not determined, is what makes possible to overcome the strength of our cultural inheritance. The transformation of our material world, of our material structures, where a critical pedagogic effort is simultaneously added, is the way to – the never mechanical – overcoming of this inheritance.

Cultural awareness relies on the extent to which one has explored one's own cultural framework (Cohen-Emerique 1999): language structure, will and participation frameworks, prejudices and stereotypes regarding different social groups, concept and organization of work, concept and organization of leisure time, concept/representation of time, concept/representation of space, sense of ownership/property, concept of development/progress, use habits of ICT, needs, motivations, ideals, local referents, nonlocal referents, religion, family composition, predominant educational system, verbal language, nonverbal language, beauty ideals, family/local economy, morphological cultural practices (traditions, festivities, folklore, music, food, etc.), life/death concept, personal ideology/local politics, organization of daily life, imaginary of alterity, and forms of consumption, among other cultural and communication dynamics.

**Dialogic awareness** involves questioning how communication happens in terms of occupation of space, themes, times, etc. What words does one use? How is one's discourse composed? How does one exercise power through discourse? How does one listen? Freire's conception of dialogic awareness is based on his concept of liberating pedagogy where (Freire 1975):

- All agents have the ability to teach and learn.
- All agents have some kind of knowledge.
- All agents are subjects of the process and never objects of the process.
- All agents have an equal right to speak and to be heard.
- All agents have the same right to propose problems, options, contents, and solutions.

To address the question of **power structure awareness**, it is useful to complement Freire's conceptualizations with Arnold and Amy Mindell's proposals of rank and privilege (Mindell 2002, 2008). According to their theories framed within the field of Process Work, power does not only function intersectionally adhering to criteria of gender, ethnicity, social class, and functional diversity, among other social configurations but also is actually a much more complex structure.

Amy and Arnold Mindell explain power structures referring to the anthropological and psychological concept of rank and define it in five categories: social rank, structural rank, contextual rank, psychological rank, and spiritual rank.

- Social rank is defined by our class, gender, age, appearance, ethnicity, etc.
- Structural rank is given by our position in an institution or a specific field.
- Contextual rank is defined by the imminent context of space-time in which the interaction occurs.
- Psychological rank depends on our level of awareness regarding our own experience and abilities.
- Spiritual rank is given by inner peace and the notion of being part of something that transcends the individual, such as faith or the sense of belonging to a collective or a community.

One's rank with respect to each of these categories provides the individual with certain privileges that are not chosen but are rather assigned socially. However, oppressed groups are not merely tied down to their low social rank to access and use power but can actually become aware and gain contextual, structural, psychological, and spiritual rank through Freirean communicative processes in order to resist and overcome oppression.

## 15.5 Implications of Freirean Epistemology to Communication for Development and Social Change

The application of Freirean epistemological and ontological premises to the current fields in Communication for Development and Social Change, particularly in participatory development communication, has practical implications in the configuration of projects and programs.

A project or program is Freirean-based when its definition and cycle comply with the following principles (Fernández-Aballí 2014):

- **All agents are subjects** and never objects of a project or program.
- Questioning **why, for what, for whom, with whom, and who** defined a particular problem addressed by a communication project or program is essential at the beginning of any transformation process (Hernández 2010).
- Development as a goal is not taken for granted. Communication for Development and Social Change should be **based on awareness raising** in the Freirean sense for all agents involved (project promoters included). Development is not necessarily a desirable outcome, nor does it have to be part of a community's world view (Tortosa 2011).
- Goals and objectives do not necessarily determine the final outcomes of a communication project. Project **cycles are open** to continuous change in all of its phases as agents involved learn and gain awareness of their own problems, desires, alternatives, and solutions.
- Methodologies based on **participatory tools and methods** are used in the planning, implementation, and evaluation of the projects or programs.
- Projects are framed in processes of **reflection about the conditions derived from the existing hegemonic context** in global and local relations among agents, both

at the micro-level and at the macro-level (economic, educational, racial, gender, cultural, environmental, etc.).
- Projects are **thought and validated collectively**. This means that roles and responsibilities are agreed upon and clarified, and time is assigned for participatory brainstorming and proposal validation during project implementation.
- **Learning processes are explicit** for all agents involved (community participants, project promoters, project leaders, etc.). This requires assigning specific time allocations through project planning to address such learning processes where all agents must speak dialogically in the Freirean sense about the impact of the process to their own understanding of self and of the context.
- **New media and ICTs are facilitators in participatory processes** of awareness raising and are always used inclusively and with consideration of access and resources available by the community and agents involved in the project.

Regarding Freirean-based actions and content creation within communication projects and programs (Fernández-Aballí 2016):

- The **complexity and diversity** behind any given problem and context are taken into account and made explicit.
- **Historical and background information** about a context (not only statistics) is provided using that context's point of view in order to build critical spectators within and outside a community.
- **Local efforts** are recognized and reinforced explicitly to facilitate network creation, collaborative learning, and building up knowledge about a specific problem.
- **Historical memory, conflicts, and inequalities are acknowledged** to open up spaces for recognition of oppressed groups and facilitate healing and reconciliation within or among communities.
- **Power structures are pinpointed** as concretely as possible to allow for the identification of where and what type of actions are necessary for the enactment of change and to avoid diffusion of the responsibility of power abuse.
- All content created within a project or program is verified to **avoid symbolic violence** through the identification of discursive practices that might reinforce negative prejudices and stereotypes against a community or which in any form negates the capacity of agency of any group.

## 15.6 Conclusion

Paulo Freire, Brazilian educator and thinker of the twentieth century, evokes faith, humility, love, trust, hope, and critical thinking in communication to dialogue from the consciousness and to fluctuate in the teaching-learning continuum. This invitation is in itself a way of knowing, understanding, and naming the world, and Freire provides a broad framework of ideas, concepts, and practices that generate a knowledge-creating space for transformation: an epistemological frame.

All this results in a synthesis of the Freirean epistemology posed as a cognitive-emotional cycle that is characterized as being: (a) of the oppressed and not for the oppressed; (b) based on an ontology of vulnerability; (c) formed from collective knowledge; (d) carried out through circular, dynamic, and contextualized processes; and (e) of a phenomenological nature. This cognitive-emotional cycle transits toward utopia understood as a world where "no one names on behalf of others" (Freire 1975), that is, it allows for *another possible world* to come to exist.

Freire writes (Freire 2012):

> ...to accept the dream of a better world and to adhere to it is to accept entering the process of creating it, a process of struggle deeply anchored in ethics. A fight against any type of violence, against violence exerted on the life of trees, rivers, fish, mountains, cities, on the physical footprints of cultural and historical memories; against the violence exerted on the weak, on the defenseless, on the outraged minorities; against the violence suffered by the discriminated against, no matter the reason for the discrimination. A fight against the impunity that at this moment foments between us the crime, the abuse, the contempt towards the weakest, the ostensible contempt towards the life.

For Miguel Escobar Guerrero (Escobar Guerrero 2012), Freire's *Pedagogy of the Oppressed* is the text that *pronounces-announces* the basis and cognitive paradigm change required for the dream of *another possible world*.

This chapter provides the basic ideas behind Freirean knowledge generation and the implications this approach has to the field of Communication for Development and Social Change. Regarding project and program definition and implementation, the shift to Freirean-based paradigm requires that all agents become subjects of the process; that critical question-posing is introduced from the beginning of the proposal; that projects are based on awareness raising of the self and the context about cultural, dialogic, and power structures; that project cycles are participatory and open; that learning process is explicit; and that the use of new media and ICTs facilitates – rather than impose – a path based on collective consensus. As important as project/program, definition is the content that is created within program framework. Freirean-based communicative content must contemplate the complexity and diversity of social contexts; provide historical and background information; recognize and reinforce local efforts; acknowledge historical memory, conflict, and inequalities; pinpoint power structures; and avoid symbolic violence.

Despite Freire's influence in the field, dehumanization of agents in project definition, implementation, and content creation is still a challenge (Easterly 2006; Tortosa 2011; Fernández-Aballí 2016). Peter McLaren (1999) writes in memory of Freire:

> Freire's life vehemently unveils the imprints of a life lived within the margins of power and prestige. Because his work was centered around the issue of social and political change, Freire has always been considered controversial, especially by educational establishments in Europe and North America. While he is recognized as one of the most significant

philosophers of liberation and a pioneer in critical literacy and critical pedagogy, his work continues to be taken up mostly by educators working outside of the educational mainstream. The marginal status of Freire's followers is undoubtedly due to the fact that Freire firmly believed educational change must be accompanied by significant changes in the social and political structure in which education takes place. It is a position most educators would find politically untenable or hopelessly utopian. It is certainly a position that threatens the interests of those who are already well served by the dominant culture.

For progressive, radical, and critical thought currents, revision of their role in maintaining hegemonic systems is not only urgent but essential in generating emancipatory knowledge and change. This revision, which can be based partly on Freire's chapter of cultural invasion in the text *Pedagogy of the Oppressed,* must become a never-ending task due to the complexity and responsibility of the processes which compose the field of Communication for Development and Social Change. Freire's meta-approach opens a discussion on legitimacy and entitlement as part of the communicative process: Who is entitled to name truth? Whose truth is legitimate? How is this truth constructed, shared, taught, or imposed? For Freire, truth lies in how questions are produced, not in the answer; the answer is merely instrumental for the process of question-posing. How are progressive, radical, and critical thought currents, particularly in Communication for Development and Social Change, generating question-posing debates that break away from cultural invasion?

Nevertheless, the questioning and critical nature of Freire's epistemology forces a cautious approach to concepts to avoid a sense of exclusive ownership of them, since it is the words, their meanings, and their existence in terms of dialogue that define the transforming reflection-action, or praxis. Freire's invitation is to think from the word of "other possible worlds." Thinking-saying-acting in communion and solidarity – in unity in diversity – is urgent in the Freirean sense, since, as José Luis Sampedro described, *another world* is not only possible but certain.

## 15.7 Cross-References

- ▶ A Threefold Approach for Enabling Social Change: Communication as Context for Interaction, Uneven Development, and Recognition
- ▶ Development Communication in Latin America
- ▶ Fostering Social Change in Peru Through Communication: The Case of the Manuani Miners Association
- ▶ Key Concepts, Disciplines, and Fields in Communication for Development and Social Change
- ▶ Multidimensional Model for Change: Understanding Multiple Realities to Plan and Promote Social and Behavior Change
- ▶ Participatory Mapping

# References

Alberich Nistal T (2008) IAP, Mapas y Redes Sociales: desde la investigación a la intervención social. Cuad Cimas 8:1–32

Alejandro M, Vidal JR (2004) Comunicación y Educación Popular. Selección de Lecturas, FEPAD, Cen. Editorial Caminos, La Habana

Arnett RC (2002) Paulo Freire's revolutionary pedagogy: from a story-centered to a narrative-centered communication ethic. Qualitative Inquiry 8(4):489–510

Baraúna T, Motos T (2009) De Freire a Boal. Ñaque Editora, Ciudad Real

Barranquero A (2014) La formación en comunicación/educación para el cambio social en la universidad española. Rutas para un diálogo interdisciplinar. Cuadernosinfo 35:83–102

Bessette G (2004) Involving the community: a guide to participatory development communication. Southbound & International Development Research Centre, Ottawa

Chege M (2009) Literacy and hegemony: critical pedagogy Vis-a-vis contending paradigms. International Journal of Teaching and Learning in Higher Education 21(2):228–238

Cohen-Emerique M (1999) Análisis de incidentes críticos: un modelo para la comunicación intercultural. Rev Antípodes 145:465–480

De Sousa Santos B (2010) Descolonizar el saber, reinventar el poder. Ediciones Trilce, Montevideo

De Sousa Santos B (2011) Introducción: las epistemologías del sur. In: Formas-Otras. Saber, nombrar, narrar, hacer. CIDOB, Barcelona, pp 9–22

Easterly WR (2006) The White Man's burden: why the West's efforts to aid the rest have done so much ill and so little good. Oxford University Press, Oxford

Escobar Guerrero M (2012) Pedagogía erótica: Paulo Freire y el EZLN, México DF: El autor

Fals Borda O (2014) Ciencia, compromiso y cambio social. El Colectivo – Lanzas y Letras – Extensión Libros, Montevideo

Fanon F (1963) The Wretched of the Earth. Grove Weidenfeld, New York

Fernández-Aballí A (2014) En busca de la horizontalidad: variables clave y convergencia metodológicas en el proyecto "Art d Kambi". Una propuesta para la creación de proyectos de comunicación participativa glocal. IC – Rev Científica Inf y Comun 11:15–34

Fernández-Aballí A (2016) Advocacy for whom? Influence for what? Abuse of discursive power in international NGO online campaigns: the case of amnesty international. Am Behav Sci 60:360–377. https://doi.org/10.1177/0002764215613407

Fernández-Aballí Altamirano A (2016) Where is Paulo Freire? Int Commun Gaz 78:677–683. https://doi.org/10.1177/1748048516655722

Freire P (1975) Pedagogía del oprimido. Siglo XXI, Madrid

Freire P (1984) ¿Extensión o comunicación? La concientización en el medio rural. Siglo XXI, Montevideo

Freire P (1991) La importancia de leer y el proceso de liberación. México DF: Siglo XXI

Freire P (1994) Cartas a quien pretende enseñar. Siglo XXI, Buenos Aires

Freire P (2002a) Pedagogía de la esperanza. Un reencuentro con la pedagogía del oprimido. Siglo XXI, Buenos Aires

Freire P (2002b) Educación y cambio. Galerna-Búsqueda de Ayllu, Buenos Aires

Freire P (2004) Pedagogía de la autonomía. Saberes necesarios para la práctica educativa. Paz e Terra SA, Sao Paulo

Freire P (2007) La educación como práctica de la libertad. Siglo XXI, Mexico D.F.

Freire P (2012) Pedagogía de la indignación: cartas pedagógicas en un mundo revuelto. Siglo XXI, Buenos Aires

Freire P, Faundez A (2010) Por una pedagogía de la pregunta. Ediciones del CREC, Xátiva

Gadotti M, Araújo Freire AM, Antunes Ciseski A et al (eds) (1996) Paulo Freire: Uma biobibliografia. Cortez Editora, Sao Paulo

Hernández L (2010) Antes de empezar con metodologías participativas. Cuad Cimas 10:1–31

Key R (2009) El conocimiento académico, científico y crítico en el pensamiento educativo de Paulo Freire. Sapiens 10:261–276

Mari Sáez VM (2010) El enfoque de la comunicación participativa para el desarrollo y su puesta en práctica en los medios comunitarios. Razón y Palabra

McLaren P (1999) Research news and comment: a pedagogy of possibility: reflecting upon Paulo Freire's politics of education: in memory of Paulo Freire. Educ Res 28:49–56. https://doi.org/10.3102/0013189X028002049

Mindell A (2002) The deep democracy of open forums. Hamptons Roads, Charlottesville

Mindell A (2008) Bringing deep democracy to life: an awareness paradigm for deepening political dialogue, personal relationships, and community interactions. Psychother Polit Int 6:212–225

Morrow RA, Torres CA (2012) Reading Freire and Habermas: critical pedagogy and transformative social change. Teachers College, New York

Ortiz M, Borjas B (2008) La Investigación Acción Participativa: aporte de Fals Borda a la educación popular. Espac Abierto 17:615–627

Servaes J (ed) (2003) Approaches to development communication. UNESCO, Paris

Torres CA (1995) Estudios freireanos. Buenos Aires: Libros del Quirquincho

Torres CA (2005) La praxis educativa y la educación cultural liberadora de Paulo Freire, Xátiva: Diálogos

Torres CA (2006) Lectura crítica de Paulo Freire Materiales para una crítica de la pedagogía liberalizadora de Paulo Freire. Institut Paulo Freire de Espanya y Crec, Xátiva

Torres CA (2014) First Freire: early writings in social justice education. Teachers College, New York

Torres CA (2007) Paulo Freire y la educacion popular. Rev EAD – Educ adultos y Desarro 69:1–10. Retrieved from https://www.dvv-international.de/es/educacion-de-adultos-y-desarrollo/ediciones/ead-692007/el-decimo-aniversario-de-la-muerte-de-paulo-freire/paulo-freire-y-la-educacion-popular/

Tortosa JM (2011) Mal desarrollo y mal vivir: pobreza y violencia a escala mundial. Producciones Digitales Abya-Yala, Quito

Tufte T, Mefalopulos P (2009) Participatory communication: a practical guide. The International Bank for Reconstruction and Development/The World Bank, Washington, DC

Villasante TR (2010) Reflexividades socio-práxicas: esquemas metodológicos participativos. Cuad Cimas

# De-westernizing Alternative Media Studies: Latin American Versus Anglo-Saxon Approaches from a Comparative Communication Research Perspective

## 16

Alejandro Barranquero

## Contents

| | | |
|---|---|---|
| 16.1 | Introduction | 330 |
| 16.2 | Debates and Dilemmas in Alternative Communication Research | 330 |
| 16.3 | Latin American Alternative Communications During the Twentieth Century | 331 |
| 16.4 | An Overview of Alternative Media Research and Experiences in the Northern Countries | 334 |
| 16.5 | Potentialities and Limits of the New Literature on Digital Alternative Media | 336 |
| 16.6 | A Claim to De-westernize Alternative andCommunity Communication | 337 |
| References | | 338 |

### Abstract

This chapter reviews and compares the historical grounds of alternative media in Latin America and in the Anglo-Saxon world. A number of theoretical works and case studies have been summarized in order to present a comparison between both academic communities from the perspective of communication for development and social change. What is the origin and evolution of theory and practice within both contexts? Are there differences between them? What are the basic features of Latin American approaches to citizen communication? The answers to these questions will be critically developed in order to shed light on a number of issues regarding communication processes made by, from, and for the people. Furthermore, we will try to build bridges between both academic communities from a comparative media perspective.

A. Barranquero (✉)
Universidad Carlos III de Madrid, Madrid, Spain
e-mail: abarranq@hum.uc3m.es

© Springer Nature Singapore Pte Ltd. 2020
J. Servaes (ed.), *Handbook of Communication for Development and Social Change*,
https://doi.org/10.1007/978-981-15-2014-3_69

Keywords

Community communication · Alternative media · Communication for social change · Comparative studies · Communication theory

## 16.1 Introduction

In recent years, academic and activist literature in communication for development and social change is expanding and gaining in complexity and diversity (Tufte 2017). Furthermore, the last two decades have witnessed a growing interest in the specific field of community media, in special from 2001, when three relevant publications laid solid foundations for further research (Atton 2001; Downing 2001; Rodríguez 2001). On the other hand, the concern for media and communication repertoires has gained importance in the field of social movement studies, where many academics have started to understand research from the perspective of a deeper collaboration with activists (Hinz and Milan 2010: 842).

This chapter reviews and compares the historical grounds of alternative media theories and practices both in Latin America and in the Anglo-Saxon context. We will first approach the field of community communication and its main academic debates. Then, we will compare knowledge and practices that emerged in both contexts, underlining the main continuities and differences and showing up a few limitations in recent literature on digital activism. Finally, a de-westernizing perspective will be adopted in order to encourage for the progressive incorporation of Latin American and Southern approaches in the Anglo-Saxon literature.

## 16.2 Debates and Dilemmas in Alternative Communication Research

Communication studies have historically centered their attention either in public or in private commercial media. Therefore, research on alternative media has been frequently dismissed as ephemeral and irrelevant (Downing 2010: 12), becoming thus "a blind spot in media historiography" (Howley 2010: 4). This neglect can be traced back to the very early origins of media studies in the USA and Europe, which systematically undervalued the transformative role of media and communication processes made by, from, and for the people. This disregard is also evident in the field of social movements' studies which has habitually focused either on the media framing of social movements or on their tactics they use to draw the attention of mainstream media in hope of winning adepts and political success.

Despite this oblivion, alternative media have historically operated at the margins of the dominant media system, and this represents "a dizzying variety of formats and experiences, far greater than mainstream commercial, public, or state media" (Downing 2010: xxv). To interpret this complexity, community media research is strictly interwoven with practice, and alternative communication is, in fact, not a model

(Gumucio 2011: 36), but the result of reflection about situated and local practices. This is perceivable in the multiple definitions generated along history to name the field: radical media (Downing 2001), popular media (Kaplún 1985), citizen media (Rodríguez 2001), community media (Rennie 2006), social movements' media (Downing 2010), etc.

Although creative and nuanced, the burgeoning of labels has also become a matter of controversies and even pit the different experiences against each other, especially at the moment of building citizen coalitions to advocate for the right to communicate. Besides, the Northern and Southern academic communities have developed separate bodies of knowledge which have rarely interacted, although dialogues have started to be undertaken in recent times. This is the case of a few ambitious compendia in the area, such as "Making waves" (Gumucio 2001), a systematization of 50 experiences along the world, and the more theoretical "Anthology" (Gumucio and Tufte 2006) and "Handbook" (Wilkins et al. 2014) of communication for social change. These efforts have been also taken on in the specific field of alternative and civic media (Atton 2015; Gordon and Mihalidis 2016), in particular after the publication of the first *Encyclopedia of Social Movement Media*, which included 250 essays on diverse media experiences over the planet (Downing 2010).

The introductory chapters of all the former anthologies demonstrate that there is an emerging interest to advance toward comparative communication research, which is a way to compare macro-units of knowledge (regions, language areas, social milieus, etc.) beyond the traditional borders of the nation-states (Esser and Hanitzsch 2012: 5). Nevertheless, cross-territorial comparisons have been far more recurrent in the field of communication for development (e.g., Servaes 1999; Manyozo 2012; Melkote and Steeves 2001) than in the embedded area ofcommunity communication, with a few recent exceptions (Cammaerts 2009; Hamilton and Atton 2001; Harlow and Harp 2013). The following lines attempt to initiate this dialogue between Latin American and Anglo-Saxon traditions in the field ofalternative media.

## 16.3 Latin American Alternative Communications During the Twentieth Century

In Latin America, the early precedents of alternative communication can be traced back to Pre-Columbian times (Beltrán et al. 2008). Nevertheless, these forms gained consistency during the anti-colonial struggles of the twenty-first century, which typically took the form of partisan press and seditious satirical posters (Vinelli 2010: 28). Later on, during the second half of the twentieth century, many grassroots media projects were conceived as an education and liberation tools (Freire 1970) against the historical dependency of local oligarchies and US imperialism. The pioneering initiatives were basically fueled by two major agents: the reformist sectors of the Catholic Church – also called liberation theology – and workers' unions and associations in both urban and rural areas (Vinelli 2010: 28). Theology of liberation was behind many radio schools along the region. These experiences combined radio broadcasts, workshops, and other teaching materials with the aim of promoting distant and face-to-face education. On the other hand, the Bolivian miners' radio stations – such as *Radio La Voz del Minero* (1947), *Radio Sucre* (1947), and others – were integrally conceived, financed, and managed by the miners themselves as a way to challenge the mining oligarchies and offer an autonomous programming in local languages (O'Connor 2004).

Beside these landmarks, there is also a vast amount of pioneering initiatives which combined education and political claims. Many of them were influenced by the revolutionary and reformist cycles of the second half of the twentieth century, exemplified by the Cuban revolution (1959) and Chilean Salvador Allende's reforms (1970–1973). Among these projects, we can quote the experience of militant and guerrilla radios in Central America during the 1980s, which broadcasted from hidden environments not to be discovered by the governments. Other projects took the form of video and cinema, such as the so-called third cinema and many other community video projects (Gumucio 2014). In particular, third cinema aimed at representing groups which have been traditionally marginalized in the mainstream media (popular classes, indigenous people, etc.), and opposed the traditional patterns of Hollywood in order to make emerge the voice of the third world.

Despite their differences, many of the forerunning experiences aimed at gaining both cultural recognition and material values such as the improvement of labor conditions, housing, and education (Inglehart 1990). Within a context dominated by private broadcasters and weak public media systems, Latin American community media accomplished the public service functions neglected by the states, in particular by providing citizen access and participation in the media (Madriz 1988). In other cases, they helped to raise the voice of dissident groups that defied dictatorships or corrupted governments, although these kinds of experiences had to work underground or in the limits of censorship. The following table provides an overview of a number of projects developed in Latin America along the second half of the twentieth century (Table 1).

From the early 1970s, alternative media inspired a vast tradition of essays and empirical works. This task was led by a generation of thinkers who combined reflection and activism (Díaz Bordenave 1976; Kaplún 1985; Prieto 1980) and proposed to reconsider communication theory by approaching the concepts of dialogue (Freire 1970) and horizontal communication (Beltrán 1979). These studies were early qualified as "liberation communicology" since they aimed at building an autonomous communication science for Latin America, adapted to the necessities of the region and explicitly committed to social change (Beltrán 1974). Compared to the more empirical and theoretical academic traditions of, respectively, the USA and Europe, the importance of praxis – or theories emanated for and from practice – was at the core of manyalternative media approaches (Freire 1970). This idea also shaped the features of the so-called Latin American (Critical) Communication School. This denomination was used by the Brazilian pioneer José Marques de Melo (2009) to highlight that the region had contributed to a synthesis of both US administrative and European critical theories from the perspective of hybridization, attention to local problems, and search for social change.

**Table 1** Paradigmatic community media experiences in the twentieth century

| Radio | Video and cinema | Press and news agencies | Other cultural expressions |
|---|---|---|---|
| The Bolivian miners' radio station (1950s–1980s) | Community cinema, participatory video, and the so-called third cinema (*Tercer Cine*) (1970s onward) | Brazilian *nanica* press (*imprensa nanica*) (1970s) | Distant literacy projects based on the use of radio schools and radio forums (e.g., *Radio Sutatenza*, 1947–1989 in Colombia) |
| Militant radios in Central America: *Radio Rebelde* in the Cuban revolution (from 1958), *Radio Venceremos* & *Radio Farabundo Martí*, Salvadorian guerrilla radios (1980s) | Manuel Calvelo's Massive Audiovisual Pedagogy (*Pedagogía Masiva Audiovisual*) addressed to rural environments (1970s–1990s) | The industrial zone newspapers in Santiago de Chile, during Salvador Allende's government (1970–1973) | Grassroots edu-communication projects inspired by Paulo Freire & Orlando Fals Borda's participatory-action research |
| Community radios in impoverished urban environments such as *Radio Favela* in Brazil (1980s–...) | Indigenous video in Bolivia, Ecuador, Peru, Guatemala, México, Colombia, Brazil, etc. (e.g., *CRIC*, *Tejido de Comunicación*, *CLACPI*, *Ojo de Agua*, *Video nas Aldeias*, etc.) | The Nicaraguan political press along the Sandinist revolution (1980s) | Edu-tainment programs and methods developed by Mexican Miguel Sabido (e.g., soap operas for social change) and by Uruguayan Mario Kaplún (e.g., cassette forum) |
| Educational (catholic) radios integrated in networks such as the Latin American Association for Radiophonic Education (ALER) | Brazil's Worker's TV (*TV dos Trabalhadores*) and participatory video experiences in the Southern Cone (Argentina, Chile) before and after dictatorships | Rodolfo Walsh's *Agencia Clandestina de Noticias*-ANCLA (1976–1977) and *Agencia Latinoamericana de Información* (ALAI) (1977–...) | Augusto Boal's Theater of the Oppressed in Brazil |

Own elaboration based on Beltrán (2005), Vinelli (2010), Gumucio (2001), and Downing (2010)

Research in alternative media peaked during the 1980s. At this period, many scholars shifted their attention to small-scale media projects (Reyes Matta 1983; Simpson 1986), considering the difficulties to implement the recommendations of the UNESCO McBride Report (1980), which encouraged the states to promote communication policies in order to control media monopolies (Madriz 1988). Nevertheless, during the 1980s, the debate ended up trapped into a too simplistic position that observed alternative media as spaces of alleged purity and goodness at the margins of mainstream media (Huesca and Dervin 1994). In fact, many scholars at the period were concerned with how to construct alternatives from small media projects, neglecting structural constraints (such as regulation and policies) and sharing a certain "small is beautiful" mentality.

The solution to this theoretical impasse was partly proposed by Spanish-Colombian Jesús Martín Barbero (1987) who suggested to transit from "media" to "mediations" as a way to explore the multiple and reciprocal "interpenetrations" between the mainstream and popular cultures. In other words, Martín Barbero stated that popular cultures are influenced by mass media and "a major reason for the success of commercially produced mass culture" is the cooptation of "numerous elements of popular culture expressions" (Downing 2001: 4). His work was very influential, and it even inspired the widespread concept of "citizen media," by Colombian scholar Clemencia Rodríguez (2001). Her notion invited alternativist thinkers and practitioners to detach from the too simplistic and essentialist positions we have already described. Instead, she claimed that citizen media are perfectly able to be elitist, racist, or misogynist and they are not directly connected to democratization and social change. To face this complexity, Rodríguez invited scholars to concentrate not just on the realm of alternative contents or media outlets themselves but rather on the cultural processes that trigger when local communities appropriate information technologies to "create one's own images of self and environment" (Rodríguez 2001: 3).

## 16.4 An Overview of Alternative Media Research and Experiences in the Northern Countries

In the European context, a number of scholars have studied the long history of pre-modern popular communication forms, marked by oral or handwritten songs and ballads, almanacs, and other cultural expressions such as the satiric theater and the carnival (Bakhtin 1968; Burke 1978). Along the sixteenth and seventeenth centuries, the Reformation in Germany and English Civil War (1642–1651) demonstrated the potentials of the printing press to disseminate oppositional ideas, especially "in the language of the ordinary people which could take on the weight of a material force for change" (Conboy 2002: 28). The Enlightenment ideals were also widely spread thanks to a vast web of underground printers and publishers, who disseminated clandestine books, pamphlets, and seditious papers and set the grounds of the French and American Revolutions (Hesse 2007: 374). The concept of "public sphere," originally proposed by Jürgen Habermas (1962/1989), shed light on the reading of newspapers and public discussions as the origin of a bourgeois public sphere during the Enlightenment. Nevertheless, his approach became a matter of controversy because it undervalued other historical dissident forms by marginalized groups such as women (Fraser 1992) and the working classes (Kluge and Negt 1993; Thompson 1963/1980).

In the first half of the twentieth century, a few seminal works contributed to the understanding of the immigrant and popular press in US urban environments (Park 1922; Janowitz 1952/1967), while in Europe, Bertold Brecht had started to describe the emancipatory potentials of the first radio transmissions (Brecht 1927/2006). However, academic interest in alternative media was scarce until the 1970s. Within a context of youth upheavals (e.g., May 68) and post-material values – feminism, pacifism, etc.

(Inglehart 1990) – this decade witnessed an explosion of critical art ensembles, culture jamming actions, and underground publishing. The 1970s is also the period when the first free radios started to operate in France, Italy, and the UK. The latter radios claimed for a citizen space in the radio-electric spectrum, and they were normally driven by cultural and political groups aiming at different vindications. Paradigmatic examples are British pirate radios – *Radio Caroline, Radio City, Radio Veronica*, etc. – and, in special, the free stations influenced by the May 68 spirit: Italian *Radio Alice* and *Radio Popolare* and French *Radio Verte*, among others.

The 1960s and 1970s were also prolific in counterculture and underground media experiences in the USA. Magazines, movies, arts, and other popular expressions gave voice to the demands of youngsters, feminists, pacifist, and environmentalists. They also helped to canalize the claims for cultural rights by the emergent Black and Latino movements (Hamilton and Atton 2001). Free and low-power radio stations also mushroomed in different regions of the USA, with pioneering landmarks such as *Pacifica Radio, Dale City Television, WTRA radio, Paper Tiger*, and the larger *Prometheus Radio Project* network. Accompanying these developments, the first systematizations of experiences were published in both sides of the Atlantic during the 1980s (Downing 1984; Lewis and Booth 1989; Siegelaub and Mattelart 1983; Soley and Nichols 1986) and, in special, along the 1990s, when institutions such as AMARC and UNESCO commissioned a few large-scale studies (Berrigan 1977; Díaz Bordenave 1977; Girard 1992; Lewis 1984, 1993). According to this overview, the following table summarizes the main differences between Latin America and Anglo-Saxon approaches to community media (Table 2).

**Table 2** Academic traditions in alternative media along the twentieth century

|  | USA and Western Europe | Latin America |
| --- | --- | --- |
| **Birth of the field** | Anecdotic academic studies since the 1920s (Park 1922; Brecht 1927) | Precursors in the 1970s (Freire 1970) |
| **Institutionalization** | 2001 onward | Mid-1980s |
| **Promoters and values of the first alternative media experiences** | Social movements guided by post-material values: critical art collectives, youngsters, university students (May 68), women, immigrants, environmentalists, etc. | Civic organizations aiming at material values (unions, theology of liberation) and progressive presence of post-material demands (e.g., indigenous and feminist media) |
| **Demands** | Counterculture movements, fights to recognize identity and cultural rights and struggles to construct a community media sector beyond public and commercial media | Educate and empower marginalized groups; fight against oligarchies and colonial powers; community projects to alleviate the absence or the weak presence of public media |
| **Most popular concepts** | Alternative, community, citizen, radical | Alternative, popular, horizontal, participatory, media for development and for social change |

Source: Own elaboration

Along the twenty-first century, community communication has exponentially expanded and gradually institutionalized as a research area (Coyer et al. 2007; Downing 2010; Rodríguez 2009). This development is closely connected with the ever-growing academic interest in the potential of ICTs for activism in globalized societies. In fact, the anti-globalization movement from 1999 (i.e., *Indymedia*) and the 2011 protest cycles – Spanish 15-M, Occupy, Arab Spring, etc. – demonstrated how digital media and social networks can be tactically used to expand protests and gain new participants. Furthermore, community media started to associate and construct large-scale networks from the end of the twenty century, and organizations such as the Association of Community Radio Broadcasters (AMARC) have extensively contributed to strengthen the sector and lobby for favorable policies in international and national regulations.

## 16.5 Potentialities and Limits of the New Literature on Digital Alternative Media

From the 2000s, the unstoppable digital revolution has shaken the debates about the potentialities of ICTs to shape global transnational social movements. The uses and appropriations of Internet platforms by civil organizations have risen academic interest, especially from the birth of the so-called Web 2.0: blogs, social networks, wikis, etc. Within this new scenario, digital media have been interpreted as platforms for the birth of "participatory cultures" (Jenkins 2006), "mass-self communication" spaces for horizontal interchange of ideas (Castells 2012), and enhancers of new collective actors: "smart mobs" (Rheingold 2003) and "collective intelligence" (Lévy 1997). This emergence has led Jankowski (2006) to suggest the existence of a third wave of reflections in the field of alternative media, after a first and second waves, which respectively, focused on the press and on audiovisual media: radio, video, etc.

However, literature in digital resistance presents a few weaknesses that limit its potential to understand the new scenario. First, many studies show a too decontextualized, ahistorical, and presentist bias, since they are usually disconnected from historical reflections about non-digital media (Rennie 2006). Second, recent research in the Anglo-Saxon context surpasses in volume any other geographic community, but this literature is very Western in focus, and it systematically ignores the theoretical advancements of Latin America and the Global South. For example, there is a powerful research line on the so-called participatory cultures that emerge around the Internet (Jenkins 2006), but this literature neglects the prolific reflection about participatory communication developed in Latin America from the 1970s (Freire 1970/2000; Beltrán 1974; Díaz Bordenave 1976).

Third, the oblivion of historical research and Southern perspectives may have driven into an excess of technological determinism in recent alternative media studies, especially when they analyzed the role of the social networks within the new cycle of protests starting in 2011: Spanish *indignados* movement, Occupy Wall Street, revolts in Iran and in the Arab world, etc. Within this context, a few studies (such as Castells 2012; Juris 2012; Shirky 2008) partly revitalized the "modernizing" idea of media as "magic multipliers" of development (Lerner 1958). Luckily, this trend has started to be criticized by a new generation of scholars who warn about the rampant techno-fascination in corporate, political, and media discourses (e.g., Fuchs 2014; Mosco 2004; Gerbaudo 2012; Morozov 2012).

Contemporary research in social movements and alternative media is currently insisting on the concepts of "media practices" and "mediatization" (Cammaerts et al. 2013) that help to "shift attention from the specific categories of media texts, outlets and technologies to what social movement actors do with the media at large, in order to grasp activist groups' agency in relation to media flows" (Couldry 2006: 27). Nevertheless, it is important to warn that the already studied concepts of praxis, mediations and citizen media, were deeply elaborated by Latin American scholars, although just a limited number of Northern authors acknowledge this heritage (Couldry 2013; Mattoni and Treré 2014).

## 16.6 A Claim to De-westernize Alternative and Community Communication

In the last years, postcolonial and decolonial perspectives from critical scholarship, indigenous people, and social movements – feminism, environmentalism, etc. – are vindicating the need to articulate new epistemologies from the side of exclusion and from the voice of the invisible (Santos 2010). This calls have been also heard in the field of communication, where different scholars are claiming to de-westernize (Curran and Park 2000), decolonize (Dutta 2015; Torrico 2015), and internationalize media studies (Simonson and Park 2016), beyond the "self-absorption and parochialism of much Western media literature" (Curran and Park 2000: 3).

In the last years, recent anthologies (e.g., Downing 2010) have started to build bridges between different alternative media traditions. Furthermore, these works offer a fruitful path to compare the different media cultures that have developed in large areas beyond the boundaries of the nation-states (Esser and Hanitzsch 2012). In fact, despite the multiple differences, Latin America can be partly understood as an intellectual unit, at least until the end of the twentieth century (Waisbord 2014). As we exposed above, the region developed a set of fruitful concepts and theorizations that helped to understand alternative media, even before the Anglo-Saxon academia started to systematically research on the topic, and setting the grounds of the participatory approach to communication for social change (Gumucio and Tufte 2006).

Lastly, this chapter has tried to summarize a long period of critical research and practices. Although limited and incomplete, we hope that this summary has helped to understand alternative media from situated local perspectives and historical contexts. In fact, community media can have different origins and motivations – from political demands to cultural and educational claims. But, as we have demonstrated, many of them tend to expand under contexts of crisis, when mainstream media neglect their commitment with citizenship and public service.

# References

The original dates of each publication are indicated before the consulted version
Atton C (2001) Alternative media. Sage, London
Atton C (ed) (2015) The Routledge companion to alternative and community media. Routledge, New York
Bailey OG, Cammaerts B, Carpentier N (2008) Understanding alternative media. Open University, Maidenhead
Bakhtin M (1968) Rabelais and his world. MIT, Cambridge, MA
Beltrán LR (1974) Communication research in Latin America: the blindfolded inquiry. In: Scientific conference on the contribution of the mass media to development of consciousness in a changing world, Leipzig
Beltrán LR (1979) Farewell to Aristotle horizontal communication. International Commission for the Study of Communication Problems, vol 48. UNESCO, Paris
Beltrán LR (2005) La comunicación para el desarrollo en Latinoamérica. Un recuento de medio siglo. In III Congreso Panamericano de la Comunicación. Buenos Aires. 12–16 June
Beltrán LR et al (2008) La comunicación antes de Colón. Tipos y formas en Mesoamérica y los Andes. CIBEC, La Paz
Berrigan FJ (ed) (1977) Access: some Western models of community media. UNESCO, Paris
Brecht B (1927/2006) The radio as an apparatus of communication. In: Gumucio A, Tufte T (eds) Communication for social change anthology: historical and contemporary readings. Communication for Social Change Consortium, South Orange, pp 2–5
Burke P (1978) Popular culture in early modern Europe. Wildwod House, Aldershop
Cammaerts B (2009) Community radio in the West. A legacy of struggle for survival in a State and capitalist controlled media environment. Int Commun Gaz 71(8):635–654
Cammaerts B, Mattoni A, McCurdy P (eds) (2013) Mediation and protest movement. Intellect, Bristol
Castells M (2012) Networks of outrage and hope: social movements in the Internet Age. Polity Press, Cambridge, UK
Conboy M (2002) The press & popular culture. Sage, London
Couldry N (2006) Listening beyond the echoes: media, ethics, and agency in an uncertain world. Paradigm, London
Couldry N (2013) Media, society, world: social theory and digital media practice. Polity Press, London
Coyer K, Dowmunt T, Fountain A (eds) (2007) The alternative media handbook. Routledge, London
Curran J, Park M (eds) (2000) De-westernizing media studies. Routledge, London
Díaz Bordenave J (1976) Communication of agricultural innovations in Latin America: the need for new models. Commun Res 3(2):135–154
Díaz Bordenave J (1977) Communication and rural development. UNESCO, Paris
Downing JD (2001) Radical media: rebellious communication and social movements. Sage, Thousand Oaks
Downing JD (ed) (2010) Encyclopedia of social movement media. Sage, Thousand Oaks
Dutta MJ (2015) Decolonizing communication for social change: A culture-centered approach. Commun Theory 25:123–143. https://doi.org/10.1111/comt.12067
Esser F, Hanitzsch T (2012) On the why and how of comparative inquiry in communication studies. In: Esser F, Hanitzsch T (eds) Handbook of comparative communication research. Routledge, London, pp 3–22
Fraser N (1992) Rethinking the public sphere: A Contribution to the critique of actually existing democracy (pp. 109–142). In: C. Cailhoun (Ed.), Habermas and the public sphere. Cambridge, MS: MIT
Freire P (1970/2000) Pedagogy of the oppressed. Continuum, New York
Fuchs C (2014) Social media. A critical introduction. Sage, London
Gerbaudo P (2012) Tweets and the streets. Social media and contemporary activism. Pluto, London

Girard B (ed) (1992/2001) A passion for radio. Black Rose Books, Montreal
Gordon E, Mihalidis P (eds) (2016) Civic media. Technology, design, practice. MIT, Cambridge, MA/London
Graziano M (1980) Para una definición alternativa de la comunicación. Anuario ININCO. Investigaciones de la Comunicación 1(1):71–82
Gumucio A (2001) Making waves. Participatory communication for social change. Rockefeller Foundation, New York
Gumucio A (2011) Comunicación para el cambio social: clave del desarrollo participativo. Signo y pensamiento 30(58):26–39
Gumucio A (2014) El cine comunitario en América Latina y el Caribe. Friedrich Ebert Stiftung, Bogotá
Gumucio A, Tufte T (eds) (2006) Communication for social change anthology: historical and contemporary readings. Communication for Social Change Consortium
Habermas J (1962/1989) The structural transformation of the public sphere: an inquiry into a category of bourgeois society. The MIT Press, Cambridge, MA
Hamilton J, Atton C (2001) Theorizing Anglo-American alternative media: toward a contextual history and analysis of US and UK scholarship. Media History 7(2):119–135. https://doi.org/10.1080/13688800120092200
Harlow S, Harp D (2013) Alternative media in a digital era: comparing news and information use among activists in the United States and Latin America. Commun Soc 26(4):25–51
Hesse C (2007) Print culture in the Enlightenment. In: Fitzpatrick M et al (eds) The enlightenment world. Routledge, London, pp 366–380
Hinz A, Milan S (2010) Social science is police science: researching grass-roots activism. International Journal of Communication 4:837–844
Howley K (2010) Introduction. In: Howley K (ed) Understanding community media. Sage, Thousand Oaks, pp 1–14
Huesca R, Dervin B (1994) Theory and practice in Latin American alternative communication research. J Commun 44(4):53–74
Inglehart R (1977/1990) The silent revolution: changing values and political styles among Western publics. Princeton University, Princeton
Jankowski N (2006) Creating community with media: history, theories and scientific investigations. In: Lievrouw LA, Livingstone S (eds) The handbook of new media. Updated student edition. Sage, London, pp 55–74
Janowitz M (1952/1967) The community press in an urban setting. The social elements of urbanism. University of Chicago, Chicago
Jenkins H (2006) Convergence culture. Where old and new media collide. New York University Press, New York
Juris J (2012) Reflections on #occupy everywhere: social media, public space, and emerging logics of aggregation. Am Ethnol 39(2):259–279
Kaplún M (1985) El comunicador popular. CIESPAL, Quito
Kluge A, Negt O (1972/1993) Public sphere and experience: toward an analysis of the bourgeois and proletarian public sphere. University of Minnesota, Minneapolis
Lerner D (1958) The passing of traditional society: modernizing the middle east. The Free Press, Glencoe
Lewis P (ed) (1984) Media for people in cities: a study of community media in the urban context. UNESCO, Paris
Lewis PM, Booth J (1989) The invisible medium: public, commercial and community radio. Macmillan, London
Lewis P (ed) (1993) Alternative media: linking global and local. UNESCO, Paris
Lévy P (1997) Collective intelligence: mankind's emerging world in cyberspace. Plenum, New York
Madriz MF (1988) De los puntos marginales a los mapas nocturnos. Anuario ININCO Investigaciones de la Comunicación 1(1):81–108
Manyozo L (2012) Media, communication and development. Three approaches. Sage, London

Marques de Melo J (2009) Pensamiento comunicacional latinoamericano. Entre el saber y el poder. Comunicación Social, Sevilla

Martín Barbero J (1987) De los medios a las mediaciones. Comunicación, cultura y hegemonía. Gili, Barcelona

Mattoni A, Treré E (2014) Media practices, mediation processes, and mediatization in the study of social movements. Commun Theory 24:252–271

Melkote S, Steeves L (2001) Communication for development in the third world: theory and Practice. Sage, New Delhi

Morozov E (2012) The dark side of internet control. The Net Delusion. Public Affairs, New York

Mosco V (2004) The digital sublime: myth, power, and cyberspace. MIT, Cambridge, MA

O'Connor A (ed) (2004) Community radio in Bolivia. The miners' radio stations. The Edwin Mellen, Lewiston

Park R (1922) The immigrant press and its control. Harper & Brothers, New York

Prieto D (1980) Discurso autoritario y comunicación alternativa. Edicol, México

Rennie E (2006) Community media: a global introduction. Rowman & Littlefield, Lanham

Reyes Matta F (Comp) (1983) Comunicación alternativa y búsquedas democráticas. México: ILET/Friedrich Ebert Stiftung

Rheingold H (2003) Mobs: the next social revolution. Basic Books, New York

Rodríguez C (2001) Fissures in the mediascape. An international study of citizen's media. Hampton, Cresskill

Rodríguez C (2009) De medios alternativos a medios ciudadanos: trayectoria teórica de un término. Folios 21–22:13–25

Rodríguez C, Ferron B, Shamas K (2014) Four challenges in the field of alternative, radical and citizens' media research. Media Cult Soc 36(2):150–166

Ryan C et al (2013) Walk, talk, fax or tweet: reconstructing media-movement interactions through group history telling. In: Cammaerts B, Mattoni A, McCurdy P (eds) Mediation and protest movement. Intellect, Bristol, pp 133–158

Santos B d S (ed) (2010) Voices of the world. Verso, London

Servaes J (1999) Communication for development. One world, multiple cultures. Hampton, Cresskill

Shirky C (2008) Here comes everybody. Penguin, London

Siegelaub S, Mattelart A (eds) (1983) Communication and class struggle. 2 Vols. International General, New York

Simonson P, Park DW (eds) (2016) The international history of communication study. Routledge, London

Simpson M (Comp) (1986) Comunicación alternativa y cambio social. México: Premià

Soley LC, Nichols JC (1986) Clandestine radio broadcasting: a study of revolutionary and counterrevolutionary electronic communication. Praeger, Westport

Thompson EP (1963/1980) The making of the English working class. Penguin, Toronto

Torrico E (2015) La comunicación 'occidental'. Oficios Terrestres, vol 32, pp 3–23

Tufte T (2017) Communication and social change: a citizen perspective. Polity Press, Cambridge

Vinelli N (2010) Alternative media heritage in Latin America. In: Downing JD (ed) Encyclopedia of social movement media. Sage, Thousand Oaks, pp 27–30

Waisbord S (2014) United and fragmented: communication and media studies in Latin America. Journal of Latin American Communication Research 4(1):23

Wilkins KG, Tufte T, Obregon R (eds) (2014) The handbook of development communication and social change. Wiley, Malden

# Part IV

# Context for Communication Activities for Development and Social Change

# 17 A Threefold Approach for Enabling Social Change: Communication as Context for Interaction, Uneven Development, and Recognition

Gloria Gómez Diago

## Contents

| | | |
|---|---|---|
| 17.1 | Introduction | 344 |
| 17.2 | A Proposed Paradigm: Communication as a Context for Interaction | 345 |
| | 17.2.1 Communication as a Dependent Variable | 348 |
| | 17.2.2 Communication as Space | 348 |
| 17.3 | Two Concepts: Uneven Development and Recognition | 349 |
| | 17.3.1 Uneven Development | 350 |
| | 17.3.2 Recognition | 351 |
| 17.4 | Conclusion | 353 |
| 17.5 | Cross-References | 354 |
| References | | 354 |

### Abstract

In a context where research aimed to intervene on an increasingly unequal society is very necessary, we are bringing a theoretical framework intended to enable social change. The frame presented here is conceived for assessing environments communicatively, and it is articulated through a proposed paradigm (communication as a context for interaction) and two concepts: uneven development and recognition.

### Keywords

Epistemology · Communication research · Methodology · Recognition · Social change · Uneven development

G. G. Diago (✉)
Department of Communication Sciences and Sociology, Rey Juan Carlos University, Fuenlabrada, Madrid, Spain
e-mail: gloria.gomez.diago@urjc.es

© Springer Nature Singapore Pte Ltd. 2020
J. Servaes (ed.), *Handbook of Communication for Development and Social Change*,
https://doi.org/10.1007/978-981-15-2014-3_7

## 17.1 Introduction

Social and financial crisis, specially the high levels of corruption that stifle institutions and an increasing number of unemployed and of badly employed people (people who have precarious jobs, who receive salaries that do not allow them to survive), has generated and is generating an ongoing discontent and despair on citizens who are encouraged to click and to participate into online polls, but without having opportunities to experiment a real engagement with matters of their interest, with issues that have a direct influence on their lives, provoking that words such as participation, development, or empowerment are invoking empty signifiers (Ninam 2015). Notwithstanding the radical transformations on different dimensions provoked by the extension in the access and in the use of new technologies and by the global interconnection of computers, mobile telephones, and other devices, in order to be autonomous in our judgements and choices, free access to information and transparency as conditions for individual and collective autonomy is still needed (Papadimitropoulos 2017). Access to information and transparency are mainstream requirements for people to progress depending on their effort and their capacities. *Access to information* and *transparency* are core elements for allowing people to accomplish a social change to live worthily. Social change must be understood as the possibility for people and for groups of people to improve their life conditions. Social change must be approached as a criteria and as an objective which has utterance in the possibilities that people has for at least, having a salary that permits them to afford the minimum requirements that a human being needs: a roof, food, heat, light, and water. Further, social change must be granted by public institutions because it is the skyline that motivates people to continue in the effort. Without possibilities for social change, society will be more unequal and corrupt.

In a context of profound social crisis and ongoing social differences and in a context where political parties situated within extreme right are conquering the helm of the governments, communication but also communication research and, specially, academic communication research can do much for humankind.

Communication research is mainly focused on studying issues related with mass media. In this line, according to several investigations (Cáceres and Caffarel 1992; Martínez-Nicolás and Saperas 2011, 2016; Caffarel et al. 2017), for 25 years, Spanish communication research has been focused on "mass media contents." We contend that studying media contents is still needed, specially the study of the impact and the routines of production of media contents because these are lines of inquiry which can provide clues for reshaping journalism in a moment where, as Mosco (2018) states, Facebook, with its 2.3 billion users, was in 2018 the largest distributor of news and information in history. However, apart from working within the cited lines of inquiry more focused on studying contents and uses of mass media and social media contents, it is possible for communication scholars to analyze environments communicatively, assessing whether they meet essential requirements for facilitating relationships among citizens based on equality and respect. This type of analysis is especially urgent in public contexts, spaces, and institutions funded by public money, but we can also apply it to enterprises because they are expected to comply criteria related with equal treatment to persons. For assessing environments from a communicative perspective, it is needed to transcend the consideration of communication as a linear process, approach from which aroused most of communication research, generating frameworks intended to identify the effects of media but also frames focused on relationships between communication and development, line of inquiry which generated *diffusion and participatory models (*Morris 2005), and also *modernization and dependency paradigms* (Tehranian 1991).

In this chapter we are proposing a framework for conducting research intended on generating environments that enable social change. We based our proposal on a paradigm that approaches communication as a context for interaction and on two concepts: uneven development and recognition. As we already expressed elsewhere (Gómez Diago 2017), the idea is to conduct research aimed to analyze whether citizens can participate in contexts exercising their rights and their duties while enjoying opportunities for progress. The frame here proposed seeks to contribute to fulfil the need of a type of research attained to society and to humankind while being of help for building a critical approach claimed three decades ago by most authors of the articles included in two internationally recognized volumes specialized in meta-research in communication and published in 1983 and 1993, respectively: the volumes of the *Journal of Communication Ferment in the Field*, 33(3), and *The Future of the Field-Between Fragmentation and Cohesion*, 43(3). As we concluded in the dissertation entitled *For Communication Research. 400 ideas from the last three decades and a proposal to update the interaction paradigm* (Gómez-Diago, 2016a), most of the authors of these flagship volumes consider that research must solve social needs. More specifically, Melody and Mansell (1983); Mosco (1983); Grandi (1983); Mattelart (1983); Davis and Jasinski (1993); Tunstall (1983); Curry Jansen (1983); Schiller (1983); Stevenson (1983); or Monahan and Collins-Jarvis (1993) pointed out the importance of conducting research to understand and to solve social problems, claiming the need of attaining investigations to real world, thus to professionals, to associations of workers, to organizations of women, etc.

To combat inequalities and to contribute to generate a society where citizens can progress according to their efforts, we need theoretical frames intended on assessing environments from a communicative perspective so it will be possible to study how interaction takes place between actors that shape the context under analysis while assessing whether relationships among them base on transparent procedures. Below we are introducing our theoretical framework composed by a paradigm (communication as a context for interaction) and two concepts (uneven development and recognition). Firstly, we will be introducing the paradigm proposed (Fig. 1).

## 17.2 A Proposed Paradigm: Communication as a Context for Interaction

Kuhn (1962) understands theories as social constructs that carry historical traces of its creators, of the time, and of the place where were created. A paradigm embraces a theory as well as a number of assumptions that are taken for granted and influence

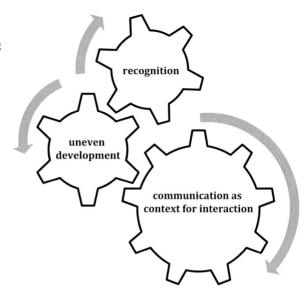

**Fig. 1** A theoretical framework for conducting research intended on enabling a social change: communication as a context for interaction and uneven development and recognition (Gómez Diago 2017)

the research process and the formulation of questions, defining the methods used, influencing the selection of criteria for deciding what is accepted as data while shaping the standards for assessing the legitimacy of the claims made. Differences between research traditions depend on economic, social, and political objectives (Melody and Mansell 1983). Technologies, their uses, and their regulation also condition paradigms. In this line, Chaytor (in Meyrowitz 2009) argues that mechanization of writing created a new space of authorship, intellectual property, reforming the literary style, boosting the birth of nationalist sentiment, and modifying the interaction of words and thought.

A *paradigm shift* happens when investigators locate anomalies that they cannot explain by using the accepted paradigm (Kuhn 1962). Three decades ago, Rogers and Chaffe (1983) in the article they prepared for the volume of the *Journal of Communication*, "Ferment in the Field" and that is configured as a dialogue between both scholars, claimed that "source" and "receiver" cannot be distinguished in an interactive communication system, and they wondered whether a term like "participant" could adequately identify an individual member of a computer-mediated system or if the word "communication" modifies its meaning in the context of these new technologies. Rogers and Chaffee (1983) situate interactive communication as a historic turning point with respect to the one direction transmission model embedded on television, radio, cinema, press, and mass media, and the researcher considers that these transformations that unsettle society in several labels should motivate scholars to abandon his long dedication to the linear effects and move toward convergence patterns. The key element is, according to Rogers, the massively apparition of interactive communication and the role of computer as a type of communication channel and not only as a tool for obtaining data.

Thirty-five years after, the Internet has modified many practices related to how citizens use and access information, as well as how they communicate, motivating, according to McQuail (2013), the need of new paradigms to overcome a linear conception of communication that does not reflect anymore how communication takes place. The loss of power of mass media as content filters and the possibility that political actors have now for addressing citizens, without depending so much on the media, has debilitated the concept of mediatization (Schulz 2004), whereas the idea of interaction has been reinstated as useful device to deepen how contexts are communicatively built. Three decades ago, Melody and Mansell (1983) demanded to assess research in light of its ability to promote beneficial change, warning of the danger of focusing only on discrete static relations among individuals or atomistic organizational units.

In a moment where research is needed that intervene on society and where new technologies have dramatically transformed communication, we are proposing to approach communication as a context for interaction. We will be introducing the paradigm that approaches communication as a context for interaction by outlining its epistemological basis. Instead of starting from a pre-defined concept of communication, this frame puts attention on how communicative elements give utterance to an environment. Considering communication as a context for interaction entails starting from the idea that communication cannot be an ultimate goal, but a context. From this lens, we approach communication as dependent variable and as space (Gómez-Diago 2015, 2016a, b, 2017), making possible to assess issues such as the capability of an environment to promote equality between the actors that shape it; its ability to generate transparency; its capability to enable, generate, and promote exchange of ideas; its ability to motivate participation or to boost and encourage collective creation; etc. Besides understanding communication as a tool necessary for other scientific fields, in words of Shepherd (1989), it is mainstream for communication studies to analyze the ways whereby manifestations of existence (individuals and or societies) are communicationally constructed. Rothenbuhler (1993) emphasizes that, as well as political analysis or psychological analysis of the "communicating" problem is possible, it is possible to perform communicative analyses of psychology and politics. From an epistemological perspective, approaching communication as a context for interaction implies to understand communication as a mechanism that shapes social manifestations simultaneously at the individual and at the collective level, working as an articulating axis of societies. This approach overcomes dichotomies based on a more or less deterministic conception of the nature of man, such as *superstructure/ structure* (Marx), *systems/worlds of life* (Habermas), or *social actors/agents* (Giddens), frames that approach communication as a bridge between two worlds, between two "realities." Considering communication as a context for interaction implies adopting a holistic view to identify the actors and elements that intervene and shape what we are studying. From this angle, we do not worry much about analyzing communication, but about studying a context or situation from a communicative perspective. Below, we will be explaining two heuristics that we can use to apply the paradigm proposed: understanding communication as dependent variable and understanding communication as space.

### 17.2.1 Communication as a Dependent Variable

Most communication research followed a component-based approach where a source, message, channel, and receiver are variables, being modified to determine the consequences of this manipulation on the communication effects. Changes in the source, message, channel, or receiver are the independent variables, and the resulting communication effects constitute the dependent variables, largely provided by scholars from other disciplines such as political science, marketing, psychology, etc. As Krippendorff (1993) states, *message-oriented research* was conducted despite messages were not so obvious in these media. Within this context Lasswell (1948) proposed the formula "Who says what, in what channel, to whom, with what effect?"; Berelson and Lazarsfeld (1948) presented content analysis as an objective, systematic, and quantitative description of the manifest content of communication; and in 1948 Shannon and Weaver published the mathematical theory. None of these approaches regards at human participants as capable of constructing their own meanings, negotiating relationships with themselves or reflecting on their own realities. Rogers and Kincaid (1981) cite, as example of this type of research focused on effects, an experiment on the sources' credibility carried by Hovland et al. on which researchers provided participants leaflets with equal content but with different types of sources. The aim of the research was to see whether participants had different perceptions depending on the sources cited but participants could not discuss among them, so the variable credibility was not allowed to vary in effects through their interaction.

While communication research has mainly approached communication as a factor that influences other issues, communicational behavior as a dependent variable has been less studied. By asking on what forms of communicational behavior individuals depend, we can focus on communicative behavior while building a coherent discipline. One area that shifted toward communicative explanations is the subfield of organizational communication. Scholars within this area consider communication as the underlying constitutive force behind all organizational activities, structures, and processes (Koschmann 2010). Foucault et al. (2014) bring also a holistic communicative angle from which to investigate when highlighting the fact that when two individuals communicate, their communication models will depend on their characteristics as individuals at that moment but also on the history of their communication models and on the communication models of the members of the group. Approaching *communication as a dependent variable* is an appropriate starting point to carry research from a communicative perspective because we can focus on how different elements and actors generate different types of relations. Below, we are explaining what we do mean with the other heuristic proposed consisting on understanding communication as space.

### 17.2.2 Communication as Space

Recent developments in postmodern geography highlight the advantages of placing space as a construct of multiple and heterogeneous sociomaterial interrelationships that coexist and are affected. Approaching communication as space allows to focus in the capacity of communication for generating contexts oriented to different aims. This capacity of communication for giving utterance to environments has been multiplied with the spread in the use of new technologies of communication, especially of the Internet and of the possibilities brought by wireless connection because communication is no longer dependent on geographic locations. In this line, similar to the proposal of McLuhan, who identified that a global village emerged from mass media, Sassen (2012) posits that the circuits of interaction that formed through ICT-mediated exchange generate a new "layer" in the social order, something similar to a microglobal community. Contending that it is not a microglobal community but a lot of communities that have surged on cyberspace, we propose considering *communication as space*. From this perspective, we understand that communication functions as an environment while it gives form to it. Our understanding of communication as space resembles how Bakhtin approaches the concept of landscape. According to the scholar (in Folch 1990), space goes beyond the visual criteria that the geographer handles, turning into a continuous historical development that holds and destabilizes the "natural harmony" of a given region through an ongoing interaction between meanings created by a conversation.

Researchers underline the need to understand communication as both an object (communicational event) and a lens or model (communicative explanation) of research (Vázquez and Cooren 2013). In this line, Friedland (2001) proposes to study *communication ecologies* to analyze how communities are communicatively integrated. The researcher refers to "communication ecologies" as the range of communication activities linked to networks of individuals, groups, and institutions in a domain of a specific community. Friedland (2001) considers that it is no longer possible to separate the social structure from communication, if ever it was possible and seeks to develop a perspective inspired by the one that was according to him the first serious attempt to fuse structure and communication: the Chicago School of Sociology. Similarly, Meyrowitz (1985) contends that what determines the nature of interaction is not the physical context but the information flow models, and he emphasizes that by changing the limits of social situations, electronic media not only give us faster access to events and behaviors but also provide new events and new behaviors. The researcher proposes to approach situations as information systems to overcome the distinction made between studies on face-to-face interaction and mediated communication studies, and he also proposes to see media as certain types of social environments that include or exclude, unite, or divide people in particular ways. In line with this last idea, we are introducing two concepts to apply the proposed paradigm consisting on considering communication as a context for interaction: uneven development and recognition.

## 17.3 Two Concepts: Uneven Development and Recognition

We chose these two concepts for building our theoretical frame because they are core concepts in what respect to humanity. Uneven development and recognition are in constant interplay; they are simultaneously cause and consequence. Uneven development is, most of the time, a consequence of a lack of recognition suffered by an individual or by an institution, and when there are a person or a group of persons that suffer uneven development, there are less opportunities for them to be recognized and, consequently, to have an opportunity to progress. Below we are introducing what we mean by uneven development and by recognition.

### 17.3.1 Uneven Development

Until 1970, the indices to assess development focused on quantitative measures that did not bother to perform comparative analyses. Poverty was associated with underdevelopment, and to be developed, countries should resemble developed countries in a context in which developed countries controlled the rules of the game. From this paradigm, the causes of underdevelopment are in the underdeveloped nations instead of being external to them or being produced by an interrelation of internal and external causes (Rogers 1976). Usually the line of inquiry *Communication for Development and Social Change (CDCS)* is cited as mainstream in attaching communication and social change, but this line of research encounters barriers when conducting empirical investigations because both terms (development and social change) convey different meanings depending on context, and therefore, applying an approach by using those concepts will mean to bring conceived ideas from one context to another making it difficult to deepen the features of the environments under analysis. Development is often associated with an interest on translating practices of capitalist environments than on achieving a comfortable situation for the majority of the population. In his line, Rist (2010) considers development as a toxic concept born without a little attempt to define itself while proposed as an unquestioned assumption. According to the researcher, the height of absurdity was reached when the Brundtland Commission (1987) created the concept "sustainable development" to reconcile the contradictory requirements to be met in order to protect the environment from pollution, deforestation, and greenhouse while ensuring that the economic growth was still considered a condition for general well-being. Rist contends that the essence of development is the general transformation and destruction of the natural environment and of social relations while increasing production of commodities (goods and services) geared, by means of market exchange.

Because of the problems that motivate the use of the concept development, we propose to utilize the concept of *uneven development*, understood as a situation when development occurs unequally. The idea of uneven development is born within a comparative perspective that we can apply to assess countries and locations but also to assess situations of citizens. Harris (2007) points out that uneven development occurs by persistent economic differences in levels and rates across many different countries during a long time. Avdicos (2007) explains that from the 1980s and onward, Marxism had brought an epistemological and methodological lens for looking at uneven development from an angle that sees evolution of local trajectories through the spatial power relations in localities. According to Bieler and Morton

(2014), Trotsky introduced the notion of uneven and combined development in his book *Results and Prospects* (1960), when analyzing the situation of disadvantage of Russia within the world economy. The researcher concluded that capitalist expansion is a result of the pressures of capitalist development on less developed countries. Van der Linden (2007) highlights one observation of Elster, who placed the theory of uneven development of Trotsky between two points of view from which history of world can be studied: (1) as the rise and decline of nations (Thorstein Veblen, Mancur Olson) and (2) as the rise and decline of institutions (Marx, Douglass North). We could add one third level of analysis which we are interested on covering from the frame here proposed: (3) the rise and decline of the lives of citizens.

We can apply the concept of uneven development to evaluate how citizens live. Are citizens enjoying a progress in their lives in line with their effort and their work? Is taking place an uneven development? If this happens, why is it happening? In what contexts? Which are the manners in which citizens can progress? Are there criteria that allow people to progress? What criteria manage public institutions and private enterprises to integrate and to promote workers? Are similar criteria being used around the world? Related to the criteria used or not used for assuring the development of people, we propose the second concept on which we build our approach: recognition.

## 17.3.2 Recognition

We base on the theory of recognition of Honneth (1995) to build our framework for the importance of this concept to articulate a society where citizens can achieve what is worth achieving. Recognition is the baseline of communication and the most important criteria because without it, actors who give form to a context are in unequal position to contribute and, hence, to provide ideas and insights, to intervene, and to develop their interests and objectives. A society where citizens are recognized according to their efforts is a healthy society. Citizenry has the right to enhance their social conditions because of their education, their effort, and their work. When this does not happen, citizens feel as strangers and discontent appears. According to Honneth (1995), every subject is given the chance to experience oneself to be recognized, in light of one's own accomplishments and abilities, as valuable for society. As the researcher acknowledges, this definition lacks specifics regarding the type of *rights* associated to individuals and regarding the mode of legitimating whereby rights are generated within a society. It is mainstream to identify the criteria that orient the social esteem because abilities and achievements are judged intersubjectively. Fuchs (2016) highlights that Honneth reformulates Hegel and Mead's approaches by identifying three modes of recognition at society: (1) emotional support provided by the family and friends, (2) cognitive respect supported by legal rights, and (3) social esteem given by solidarity communities of value. The three forms of recognition are social conditions whereby persons can develop a positive attitude toward themselves because it is through the cumulative acquisition of basic self-confidence, of *self-respect*, and of *self-esteem* – provided by the experience of those three forms of recognition that a person can come to see himself or herself, as both an autonomous and an individuated being. Fuchs (2016) identifies forms of alienation/reification in society in three realms: economic, political, and cultural. Depending on the realm we are talking about, citizens can suffer work dissatisfaction, political dissatisfaction, and cultural discontent. We are including a table where the forms of *alienation* proposed by Fuchs are described by differentiating their consequences on four realms: subject (experiments, emotions, attitudes), intersubjectivity (social agency and interaction), object (structures and products), and struggles (Table 1).

The situations of alienation identified by Fuchs (2016) are related to recognition, to uneven development, and to communication; thus, the contexts that give utterance to the three realms proposed by the researcher (economic, political, and cultural) can be evaluated from a communicative perspective intended on identifying how the actors involved interact among them and on studying how they interact with information or resources available. Fuchs's approach is articulated around a conception of society as something opposed to the individual. From a *communicative perspective*, we can identify elements that produce and perpetuate alienation in the different contexts within they perform their lives. This perspective will allow us to identify which factors are those that contribute to generate more or less alienation, and it will enable us to detect elements to combat it.

**Table 1** Forms of reification/alienation in society's three realms (Fuchs 2016)

| Forms of alienation/ reification | Subject (experiences, emotions, attitudes) | Intersubjectivity (social agency and interaction) | Object (structures, products) | Struggles |
|---|---|---|---|---|
| Economic reification | Work dissatisfaction | Lack of control/ alienation of labor power: exploitation | Lack of control/ alienation of the means of production and output: property lessens | Individual: anti-capitalism Social: unionization, class struggles |
| Political reification | Political dissatisfaction | Lack of control/ alienation of political power: disempowerment and exclusion | Lack of control/ alienation of decisions: centralization of power | Individual: politicization Social: social movements, protests, parties, revolutions |
| Cultural reification | Cultural discontent | Lack of control/ alienation of influential communication: insignificance of voice, disrespect, misrecognition | Lack of control/ alienation of public ideas, meaning and valued: centralization of information | Individual: cultural literacy Social: struggles for recognition |

## 17.4 Conclusion

The field of communication should aspire to become what Compte wanted for sociology, the queen of the social and behavioral sciences. It fulfils one of the conditions, the object of study, because communication and information play a key role not only in all issues of life but also in the central theories and models of all relevant disciplines (Beniger 1993). The objective is not only to critically understand communication today, which is an important aim, but also to assess environments from a communication perspective. For doing this, it is needed to reverse the widespread tendency that says that researchers do not use a theoretical reasoning in Media and Communication Studies, lacking epistemological validity beyond bringing information for control purposes (Erdogan 2012). We articulate our framework in a time where communication research must renew its perspectives and methods to intervene on society to combat inequalities and to contribute to generate environments where countries and citizens are enabled to progress according to their efforts. It is mainstream to develop approaches to evaluate contexts from a communicative perspective oriented to assess interaction between the actors that shape environments and to evaluate how relationships among these actors are built. It is mainstream to put attention on how recognition is granted and to evaluate whether there are elements, rules, and procedures intended on combating uneven development. In contexts where information is denied or where information is used to benefit few persons to the detriment of others, recognition is directly bounded with power, occasioning uneven development and impeding social change for most of citizens. The grid brought here is dynamic and flexible for its capacity to integrate combined perspectives, so apart from the two proposed here (uneven development and recognition), it is possible to add more conceptual devices to assess environments communicatively. One possible concept to be added is citizen participation. Most of the definitions available of citizen participation are articulated through indicators usually referred to a context of collective action and of necessary protest (demonstrations, associations), to actions related to politicians (affiliations, communication with politicians, etc.), or to the responses of the citizens to requests of collaboration from public administrations. We contend that when studying citizen participation, it is very important to take into account whether citizenry is reporting situations of abuse suffered in the contexts where they perform their lives and also to assess the nature of these complaints. Inquiring about these issues will allow us to deepen on citizen participation. It is necessary to put attention on complaints made by citizenry in the labor context and to the denunciations regarding the use of services. It is needed to investigate what types of complaints are being made and on what issues they are centered, as well as to identify the reasons why these complaints are not being made, if the case. In this way, we will know what citizens can do and what they cannot do and what they are doing. We will be in conditions to answer questions such as the following: Are there means for citizens to communicate with public institutions? For what? In which situations? It is needed to identify the existing shortcomings in a society where the economic and social gap between citizens is getting bigger. To investigate the genesis, characteristics and impact of citizen participation are a means to identify ways to delve into relationships between society, public institutions, media, political parties, governments, and policies, and it is also a way to identify interests, needs, and opinions of citizens. Social relationships always occur between citizens who have different assigned functions in social life, depending on the position they perform. It is essential to regulate relationships between citizens through transparent procedures and criteria, and it is necessary to spread the opportunities, the advantages, and the possibilities of development for the citizens. Further, when there is a lack of procedures and means to denounce, to progress, and to achieve, a social change is very difficult. Communication research can have a function in achieving healthy environments where social change is possible. The framework here proposed is a step in that endeavor.

## 17.5 Cross-References

▶ Capacity Building and People's Participation in e-Governance: Challenges and Prospects for Digital India
▶ Protest as Communication for Development and Social Change in South Africa

## References

Beniger JR (1993) Communication – embrace the subject, not the Field. J Commun 43:18–25. https://doi.org/10.1111/j.1460-2466.1993.tb01272.x

Berelson B, Lazarsfeld PF (1948) The analysis of communication content. University of Chicago Press, Chicago

Bieler A, Morton D (2014) Uneven development, unequal exchange and free trade: what implications for labour?. Retrieved from http://adamdavidmorton.com/2014/02/uneven-development-unequal-exchange-and-free-trade-what-implications-for-labour/

Cáceres MD, Caffarel C (1992) La Investigación sobre Comunicación en España. Un Balance Cualitativo. Telos 32:109–124. Retrieved from http://www.quadernsdigitals.net/datos/hemeroteca/r_32/nr_447/a_6136/6136.pdf

Caffarel Serra C, Mohedano O, Moya G (2017) Investigación en comunicación en la universidad española en el período 2007–2014. El profesional de la información 26(2). Retrieved from https://www.researchgate.net/publication/315639940_Investigacion_en_Comunicacion_en_la_universidad_espanola_en_el_periodo_2007-2014

Curry Jansen S (1983) Power and knowledge: toward a new critical synthesis. J Commun. Ferment in the Field 33(3):342–354. https://doi.org/10.1111/j.1460-2466.1983.tb02434.x

Davis DK, Jasinski J (1993) Beyond the culture wars: an agenda for research on communication and culture. J Commun 43(3):141–149. https://doi.org/10.1111/j.1460-2466.1993.tb01286.x. The Future of the Field Between Fragmentation and Cohesion

Erdogan I (2012) Missing Marx: the place of Marx in current communication research and the place of communication in Marx's work. Triple C Cogn Commun Co-oper 10(2):349–391. http://www.triple-c.at/index.php/tripleC/article/view/423

Folch-Serra M (1990) Place, voice, space: Mikhail Bakhtin's dialogical landscape. Environ Plaining Soc Space 8:255–254. https://doi.org/10.1068/d080255

Foucault B, Vashevko A, Bennet N, Contractor N (2014) Dynamic models of communication in an online friendship network. Commun Methods Meas 8(4):223–243. Retrieved from https://www.researchgate.net/publication/273959298_Dynamic_Models_of_Communication_in_an_Online_Friendship_Network

Friedland LA (2001) Communication, community, and democracy. Commun Res 28(4):358–391. https://doi.org/10.1177/009365001028004002. Retrieved from https://goo.gl/AW57aX

Fuchs C (2016) Critical theory of communication. University of Westminster Press, London. https://doi.org/10.16997/book1.a. License: CC-BY-NC-ND 4.0. Retrieved from http://fuchs.uti.at/books/critical-theory-of-communication/

Gómez Diago G (2015) Communication in crowdfunding online platforms. In: Zagalo N, Branco P (eds) Creative technologies: create and engage using art and play. Springer, London. Retrieved from https://www.researchgate.net/publication/284044910_Communication_in_Crowdfunding_Online_Platforms

Gómez Diago G (2016a) The role of shared emotions in the construction of the Cyberculture. From cultural industries to cultural actions. The case of crowdfunding. In: Tettegah S (ed) Emotions, technology and social media. Elsevier, London/San Diego/Cambridge/Oxford, pp 49–62. Retrieved from https://www.researchgate.net/publication/303791737_The_Role_of_Shared_Emotions_in_the_Construction_of_the_Cyberculture_From_Cultural_Industries_to_Cultural_Actions

Gómez Diago G (2016b) Dissertation: for communication research. 400 ideas from the last three decades and a proposal to update the interaction paradigm. Dissertation defended on 22 Jan 2016. Awarded with the extraordinary award of doctorate (Sept 2014–Sept 2016) Rey Juan Carlos University. Supervisor: Professor Dr. Manuel Martínez-Nicolás. Abstract. Retrieved from https://www.researchgate.net/publication/318969195_FOR_COMMUNICATION_RESEARCH_400_ideas_from_the_last_three_decades_and_a_proposal_to_update_the_interaction_paradigm

Gómez Diago G (2017) The role of research in communication before citizen participation. Proposal to investigate understanding communication as a context for interaction. In: Herrero J, Mateos C (eds) From the verb to the bit, Cuadernos Artesanos de Comunicación, 2nd edn, pp 1879–1899. https://doi.org/10.4185/cac116edicion2. Retrieved from https://www.researchgate.net/publication/315671849_El_papel_de_la_investigacion_en_comunicacion_ante_la_participacion_ciudadana_Propuesta_para_investigar_entendiendo_la_comunicacion_como_contexto_para_la_interaccion

Grandi R (1983) The limitations of the sociological approach: alternatives from Italian communications research. J Commun 33(3):53–58. https://doi.org/10.1111/j.1460-2466.1983.tb02406.x. Fement in the Field

Harris (2007) Uneven Development, The new Palgrave dictionary of economics, 2nd edn. Macmillan, London

Honneth A (1995) The struggle for recognition. The moral grammar of social conflicts (Transl.: By Joel Anderson). The MIT Press, Cambridge

Koschmann M (2010) Communication as a distinct mode of explanation makes a difference. Commun Monogr 77(4):431–434. https://doi.org/10.1080/03637751.2010.523593. http://koschmann.webstarts.com/uploads/Koschmann__2010__comm_as_distinct_mode_of_explanation.pdf

Krippendorff K (1993) The past of Communication's hoped-for future. J Commun 43(3):34–44. https://doi.org/10.1111/j.1460-2466.1993.tb01274.x

Kuhn TS (1962) The structure of scientific revolutions. In: International encyclopedia of unified science, vol 2(2). University of Chicago, Chicago, pp 1–210. http://projektintegracija.pravo.hr/_download/repository/Kuhn_Structure_of_Scientific_Revolutions.pdf

Lasswell HD (1948) The analysis of political behaviour, an empirical approach. Oxford University Press, New York

Martínez-Nicolás M, Saperas E (2011) Communication research in Spain, 1998–2007. An analysis of articles published in Spanish communication journals, en Revista Latina de Comunicación Social 66: 101–129. Retrieved from http://www.revistalatinacs.org/11/art/926_Vicalvaro/05_NicolasEN.html

Martínez-Nicolás M, Saperas E (2016) Research focus and methodological features in the recent Spanish communication studies (2008–2014). Rev Lat Comun Soc 71:1365–1384. Retrieved from http://www.revistalatinacs.org/071/paper/1150/70en.html

Mattelart A (1983) Technology, culture, and communication: research and policy priorities in France. J Commun 33(3):59–73

McQuail D (2013) Communication research paradigms reflections on paradigm change in communication theory and research. Int J Commun 7:216–229. http://ijoc.org/ojs/index.php/ijoc/article/viewFile/1961/85

Melody WH, Mansell RE (1983) The debate over critical vs administrative research: circularity or challenge. J Commun 33:103–116. https://doi.org/10.1111/j.1460-2466.1983.tb02412.x

Meyrowitz J (1985) No sense of place. Amazon Kindle edn 2010. Oxford University Press, New York/Oxford

Meyrowitz J (2009) Medium theory. An alternative to the dominant paradigm of media effects. In: Nabin R, Oliver MB (eds) The Sage handbook of media processes and effects. https://www.academia.edu/12688710/_Medium_Theory_An_Alternative_to_the_Dominant_Paradigm_of_Media_Effects

Monahan JL, Collins-Jarvis L (1993) The hierarchy of institutional values in the communication discipline. J Commun 43(3):150–157

Morris N (2005) The diffusion and participatory models: a comparative analysis. In: Hermer O, Tufte T (eds) Media & glocal change rethinking communication for development. Publicaciones Cooperativas, Buenos Aires. CLACSO. Retrieved efrom https://www.researchgate.net/publication/227975430_A_Compative_Analysis_of_the_Diffusion_and_Participatory_Models_in_Development_Communication

Mosco V (1983) Critical research and the role of labour. J Commun Fement Field 33(3):237–348. https://doi.org/10.1111/j.1460-2466.1983.tb02424.x

Mosco V (2018) Social media versus journalism and democracy. Journalism 20(1):181–184

Ninam Thomas P (2015) Communication for social change, making theory count. Nordicom Rev 36(Special Issue):71–78. Available in: http://www.nordicom.gu.se/en/node/35941

Papadimitropoulos V (2017) The politics of the commons: reform or revolt? tripleC 15(2):563–581. Retrieved from http://www.triple-c.at/index.php/tripleC/article/view/852/1019

Rist G (2010) Development as a buzzword (19–28). In: Cornwall A, Eade D (eds) Deconstructing development discourse. Buzzwords and Fuzzwords. Practical Action Publishing in association with Oxfam GB, Bourton on Dunsmore, Rugby, Warwickshire. Retrieved from https://www.guystanding.com/files/documents/Deconstructing-development-buzzwords.pdf

Rogers EM (1976) Communication and development. The passing of the dominant paradigm. Commun Res 3(2)

Rogers EM, Chaffee S (1983) Communication as an academic discipline: a dialogue. J Commun 33:18–30. https://doi.org/10.1111/j.1460-2466.1983.tb02402.x

Rogers EM, Kincaid DL (1981) Communication networks: toward a new paradigm for research. Free Press, New York, p 386

Rothenbuhler EW (1993) Argument for a Durkheimian theory of the communicative. J Commun 43(3):158–163

Sassen S (2012) When the center no longer holds: cities as frontier zones. J Cities. https://doi.org/10.1016/j.cities.2012.05.007

Schiller HI (1983) Critical research in the information age. J Commun 33(3):249–257

Schulz W (2004) Reconstructing mediatization as an analytical concept. Eur J Commun 19(1):87–101

Shepherd GJ (1989) Building a discipline of communication. J Commun 43(3):83–91

Stevenson ARL (1983) A critical look at critical analysis. J Commun Ferment Field 33(3):262–269. https://doi.org/10.1111/j.1460-2466.1983.tb02427.x

Tehranian M (1991) Communication and theories of social change: a communitarian perspective. Asian J Commun 2(1):1–30. https://doi.org/10.1080/01292989109359538

Tunstall J (1983) The trouble with U.S. communication research. J Commun,. Ferment in the Field 33(3):92–95. https://doi.org/10.1111/j.1460-2466.1983.tb02410.x

Van der Linden M (2007) The 'law' of uneven and combined development: some underdeveloped thoughts. Hist Mater 15:145–165. Retrieved from https://socialhistory.org/sites/default/files/docs/vanderlinden.doc

Vásquez C, Cooren F (2013) Spacing practices: the communicative configuration of organizing through space-times. Commun Theory 23:25–47. https://doi.org/10.1111/comt.12003

World Commission on Environment and Development (1987) Brundtland report. https://www.are.admin.ch/are/en/home/sustainable-development/international-cooperation/2030agenda/un-_-milestones-in-sustainable-development/1987–brundtland-report.html

# Shifting Global Patterns: Transformation of Indigenous Nongovernmental Organizations in Global Society

## 18

Junis J. Warren

## Contents

18.1 Introduction ................................................................... 360
    18.1.1 One Theoretical Framework ................................................. 361
    18.1.2 A Second Theoretical Framework .......................................... 361
    18.1.3 A Third Theoretical Framework ............................................. 362
18.2 Strategies of Guatemalan Nongovernmental Organizations .......................... 363
    18.2.1 Employing a Different Mode of Thinking ................................... 363
    18.2.2 IL Mondo Immaginare's Shifting Paradigm ................................ 364
    18.2.3 Transnational Alliances with NGOs and Other Institutions ................... 365
18.3 GNGOs Across Local, National, and International Borders ........................... 366
18.4 Integrating Technology and Communications Network ............................. 369
18.5 Impact for Sustainable Development Policy ......................................... 371
18.6 Conclusion ................................................................... 373
Appendix 1: Survey of Guatemalan Nongovernmental Organizations (GNGOs) ............. 374
References ......................................................................... 376

### Abstract

Conventional nongovernmental organization (NGO) theory describes NGOs' function within predetermined neoliberal frameworks – migration, social movements, and developmental sustainability. Yet other organizational theories establish that NGOs can, in fact, be placed in particular frameworks or models. Therefore, attributing this transformation in global society to an adjustment to changes made by other types of NGOs pursuit in expanding social and equal justice to local communities. These theoretical assumptions describe NGOs role as either tied to international industries or controlled by multinational corporations and states. However, emerging case study analysis of NGOs in Latin

---

J. J. Warren (✉)
York College, City University of New York, New York, USA
e-mail: juniswarren@gmail.com; warrenjj@outlook.com

America (Guatemala) and the Caribbean (Barbados) shows that local NGOs are redefining strategies and methodologies through a network of transnational indigenous nongovernmental organizations (TINGOs). (Warren J (2016) Case Study Analysis: "Integration Programs for Deportees in Latin America and the Caribbean." Global Stud J 9(2), Common Ground Publishing: Oneglobalization. Retrieved from: https://www.Oneglobalization.com. Manuscript Proposal: IL Mondo Immaginare: A New Framework for Assessing Transnational Indigenous Nongovernmental Organizations (TINGOs) (2017) is a grant-funded research project by The Gaston Institute, Andres Torres Series, University of Massachusetts, Boston. Retrieved from https://www.umb.edu/gastoninstitute) This chapter is a result of a grant-funded research project for the Gaston Institute, *Andres Torres Paper Series*, which analyzes and discusses four factors of GNGOs: (1) methods and strategies; (2) organizational structure across local, national, and international borders; (3) technology and communications network; and (4) impact on sustainable development policy.

**Keywords**

Guatemalan Diaspora Community (GDC) · Guatemalan nongovernmental organizations (GNGO) · IL Mondo Immaginare (IMI) · Multistranded approach · Nongovernmental organization (NGO) · Shifting global patterns · Sustainable development policy · Transnational indigenous nongovernmental organization (TINGO)

## 18.1 Introduction

The epistemological basis of this study conducted with NGOs in Guatemala, Honduras, and the USA explains a paradigm shift between traditional and transnational indigenous nongovernmental organization (TINGOs). Although much of the existing theoretical frameworks' attention has been paid to the traditional roles of NGOs and their role in the UN, Fernando and Heston (1997) write that "despite the lack of consensus about the meaning of the term 'NGO,' a large literature has been produced on NGOs and many claims have been made concerning their role as if there were a true and authentic NGO consistent over time and context" (p. 10). With the globalization of human rights and the role of the UN, not only has there been a transformation of nongovernmental organizations but also, as this study confirms, a realignment among local indigenous NGOs has occurred. Verifiable evidence demonstrates that these organizations that are venturing outside the normal modes of operating in the global community are restructured the distribution of power by INGOs. In this vain, Falk's (1993) "globalization-from below" discusses how a global civil society linked by transnationals social forces (environment, human rights), hostile to human community based on diverse cultures seeking an end to poverty, oppression, humiliation and collective violence will be linked through a "multi-stranded opposition" (p. 39). Falk's theory is mirrored with the kind of "bottom-up" method employed by NGOs assessed in this study. Moreover, it precisely describes the structure and functions of transnational indigenous nongovernmental organizations (TINGOs).

Therefore, this chapter builds on the study on Guatemalan Nongovernmental Organizations (GNGO) and uses some of these frameworks. However, it explores the function and new configuration of NGOs in the global community and provides an understanding of the linking of NGOs across borders and how they operate outside the institutional structures of the UN and the state. Edelman (2001) likened these "diverse sectors organizing across borders are explicitly directed against the elite and corporate-led globalization from above" (p. 304). Moreover, the literature shows that new groups are creating new organizational structures to continue addressing local issues in the global community. And just as Falk (1993) described assessment of NGOs, this study shows that TINGOs are not relying on traditional institutions but instead effecting changes in a variety of ways, including collectivism – the bottom-up and linking across geographically borders to find solutions to global problems.

### 18.1.1 One Theoretical Framework

The post-development alternative practice declares, Nederveen Pieterse (1996), is concerned more with participatory and people-centered development and redefining the goals of development with newly liberated nation-states. Post-development theory criticizes postcolonialism and developmental theorists like Arturo Escobar (1995) and Gustavo Esteva's (1998) who challenged traditional concepts of global development. For them, any understanding of development begins with colonial ideas of western dogma regarding the lack of development in the south. For Escobar (1995), development theories should focus on the intellectual and local agency's assertion into global development from two perspectives: local communities are encouraged to address their own problems because of similar resources and power to the west and, second, local organizations should criticize global economic and political misrepresentations that limit local communities' ability to develop independently (p. 3). During this period, different ideologies about postcolonial determination advocated for Third World strategies on local indigenous economic development, away from western concepts of development.

### 18.1.2 A Second Theoretical Framework

The scaling-up paradigm according to Uvin et al. (2000) is another way of describing current behavior of NGOs in the international community and their structure (p. 1409). The scaling-up theoretical framework categorizes NGOs as "thinking large and acting small," a new way of exerting or flexing legitimacy and influence in the global community. Scaling-up is also about the increasing way in which local community-based indigenous NGOs have selected to participate in the international arena on issues directly affecting their lives. *Shifting Global Patterns: Transformation of Indigenous Nongovernmental Organizations in Global Society* draws on the work of Uvin et al. (2000), Falk (1993), Edelman (2001), Tvedt (2002), Escobar (1995), and Esteva (1998) to make the argument that transnational indigenous NGOs have contributed to a paradigm shift on global international organizations and how they are connected, interact, and organized. Furthermore, if, as indicated by Uvin et al., "scaling-up is about capacity and not size," then it is about TINGOs' shared practical tools and techniques for building influence and legitimacy within the global sphere. Empirical evidence illustrated in the literature coupled with the information gathered during the assessment of NGOs confirms that some of these approaches have been refined by TINGOs to help them navigate the increasing complexities that come with challenging the state to adopt or improve policies for ingenious populations throughout the Third World.

### 18.1.3 A Third Theoretical Framework

Sociopolitical development framework theorist describes the process by which NGOs acquire the capacity for action in the political and social systems necessary to interpret and resist oppression. This perspective contends that resistance becomes a factor to the interests and incentives facing organizations to gain knowledge on how politics, policies, and promotion of development hinder a community. In this vein, Tvedt (2002) suggested "new theoretical approaches are needed to encourage additional comparative research on the role of evolving NGOs as new actors in the new international social system" (p. 363). Generally, the literature shows that sociopolitical development theorists reason that organizations should be familiar with how formal political structures, rules of law, and elections impact their communities. Indigenous NGOs should understand relationships between formal structures and third sectors to identify power dynamics and recognize how resources are distributed and contested and how policy choices are determined (Mitlin et al. 2007; Atack 1999).

This approach is consistent with building on new theoretical approaches to further comprehend how NGOs and other indigenous organizations are focusing on new forms of economic and sociopolitical development going forward. Research and observation of IMI (IL Mondo Immaginare) show that this GNGO has morphed into a multipurpose and multidimensional organization. In essence, the assessment of TINGOs further illustrates that IMI's connection and collaboration with other NGOs across borders in a hierarchal structure demonstrate that indigenous NGOs have adopted Edelman's description of Falk's (1993) theory of these types of NGOs as employing a multi-stranded approach to sociopolitical development. This approach differs from other studies on NGOs and proves that these organizations actually impact the global addenda and, in fact, are offering solutions from a bottom-up approach that exemplifies Folk's theory of a multi-stranded approach to global solutions for indigenous populations. Notwithstanding, the increases in membership of states in the UN or that non-state actors have become worthwhile agents on the global stage, there are still some significant challenges emerging from the institutionalization of the UN and state's NGOs. Theoretical frameworks for conducting critical social research of GNGOs show that there is a shift in paradigm away from traditional NGOs that were either part of the state or receive funding from state institutions and the organizations that are hammering their basic human rights.

## 18.2 Strategies of Guatemalan Nongovernmental Organizations

### 18.2.1 Employing a Different Mode of Thinking

The philosophical underpinning of the architect of IMI, Senior Edgar Guillermo Castillo Barillas' worldview, "is rooted in new ways of thinking and in political theorist such as Arendt's (1961) nature of power and man's need to learn to 'how' to think about traditions of freedom, education, authority, culture, history and politics." Other theoretical influences include Huntington and Fukuyama (1996) who argued that "the differences between societies are not political distinctions but the degree of governments in society and that deficient states lack consensus, community, legitimacy, organizations and effectiveness" (p. 2). According to Senior Castillo Barillas, factors like these are the fundamentals of building organizations that are part of the community. For example, Taleb's (2007) theory focuses on the occurrence of highly improbable events and how society reacts to them, while de Bono's (1967) lateral thinking is on breaking habitual patterns of linear thought. Thinking in terms of dimensions of civil society and social movements, Senior Edgar Castillo Barillas discussed the need for civil society and third sector NGOs to change direction as de Bono suggested with regard to looking at trade, the economic sectors, technology innovations, political institutions, and the state in different ways versus the old in breaking barriers of structural oppression. These frameworks inspired his "out of the box" perspective about indigenous NGOs in a global society. And the challenges faced by civil society and social movements in Latin America had a profound impact and helped shape the core values and focus of IMI.

The name for IMI was adopted from Einstein's idea that "imagination is more important than knowledge because imagination embraces the entire world." A former career diplomat and NGO director in Honduras confirmed during an interview that the theoretical underpinning about revisiting Central America's engagement in the third sector, guiding economic principles and the core values of Senior Castillo Barillas' idea of self-sustainability, is essential for nongovernmental organization's ability to deliver services to the Guatemalan Diaspora Community (GDC). This thinking provides innovative frameworks for addressing sociopolitical issues in Guatemala and the rest of the Third World. Similarly, Mr. Pantis, Director of an NGO in Honduras, indicated that "Honduras and Guatemala have identical problems in human rights and women's rights; therefore, we are sharing ideas where we can continue to build alliances between the two organizations in reinvesting in Central America in tackling migration and immigration issues." Since its creation in 2007, IMI's guiding principle has been "service to others," rooted in a philosophy that economic sustainable is necessary for destabilizing structural constraints that create poverty.

The inspiration of IMI's model is an entrepreneurial social responsibility concept, which, according to Senior Guillermo Edgar Castillo Barillas, became an important concern in creating a deep consciousness among alliances that would build a better world through continuous investments in educational institutions and opportunities for economic investments in local businesses to change the minds and lives of young people in Guatemala and Central America. Economic structure along with technology systems brings communication and collaborative networks in as many spaces as possible, "making the invisible visible." Some of the ideas and approaches promoted by IMI are justice, rule of law, and education as a way not to migrate and to generate opportunities in the third sector for the betterment of the GDC.

### 18.2.2 IL Mondo Immaginare's Shifting Paradigm

IMI's global shift began with the President of IMI, Eduardo Alvarez Castillo, Jr., due to financial irregularities in the financial sector of Guatemala and a series of events which changed the direction of the organization to a global focus, thereby building a collaborative relationship with NGOs in Guatemala, Honduras, and the USA between 2012 and 2015 and volunteering and working with different institutions from the international sector, private sector, government sector, and law enforcement in Guatemala, Honduras, El Salvador, and the USA. According to Eduardo Alvarez Castillo, "Traveling through the northern triangle of Central America teaching women and entrepreneurs how to open markets and develop communities and to be self-sustaining, help IMI to generate a new focus and direction on a global scale." In addition, document reviews showed that IMI travel to Mexico lending the organization's knowledge of immigration and expertise on Tijuana and conducting trainings and workshops on domestic violence with women's organizations helps propelled this GNGO to international status, which included addressing issues about lack of justice, human trafficking for sexual exploitation, and organized crime in the rural areas of Latin America with other NGOs concerned about the same issues.

In assessing the relationship between IMI's global and migration policy for the Guatemalan Diaspora Community (GDC) and the sociopolitical relationships of IMI's connections and collaborations with other GNGOs along with other non-governmental organizations (NGOs) from a holistic perspective, empirical evidence emerged that demonstrated how a local NGO gained a more global focus across borders. Over a 2-year period, the assessment of IMI's alliance presented a deeper understanding of its practices in Guatemala, Honduras, and the USA (Connecticut, New York, Florida, Kentucky, Maryland, Rhode Island and Virginia,). A single-case approach was applied in this assessment and a thorough investigation of a sociopolitical phenomenon that steadily suggests that INGOs are becoming the lifelines for the Guatemalan communities on a global scale.

This ethnographic qualitative approach documented and described a network of TINGOs through a comprehensive case study that investigated the sociopolitical relationships between IMI and NGOs by interpreting the functions, nature, and strength of these alliances to further understand how IMI's connections and collaborations translate into migration policy changes affecting the GDC (see Appendix 1: Survey of GNGOs). Semistructured interviews conducted by directors of GNGOs and representatives from government, technological, and media agencies having

**Table 1** IMI and GNGOs areas of concern and decentralized alliance network. (Source: Author)

| Sustainable Local Community Development preventing migration through an economic sustainability development Agenda among the GNGOs. | E-Government - Treasure and Transparency linked to the environment, energy and transportation and housing. | |
|---|---|---|
| Entrepreneurs & Small Business Franchises to the Empower Women & Households. | Migrant/Deportee Prevention of Crisis with Sustainable Recovery for the Region | Technology and Communications - Reach GDC through Social Media networks |

connections to IMI were used. Questionnaires were also used with the same participants as part of a larger assessment process to comprehend and quantify the alliance networks' sociopolitical involvement with IMI, as the two most important sources of data (see Appendix 1: Survey of Guatemalan NGOs). In addition, the review of UN policies around human rights of indigenous populations and the growing importance of NGO's role within the global community along with documentation and records of services provided a better picture of the Guatemalan Diaspora Community and the GNGOs that provide services to communities across national and international borders (see Table 1).

Table 1 shows how IMI's multi-stranded method of building social capital among NGOs and the areas of concern among a decentralized alliance network helped IMI transformed into a global entity. Senior Guillermo Barillas Castillo referenced that "it is an operation goal of IMI," but Eduardo Alvarez Castillo, Jr., asserted that "the organization has assumed a more resourceful and active role in linking with other NGOs in the third sector across borders." Building on the organizations' shift, he continued, "I kept knocking on doors," and in 2015 began working with President Jimmy Morales' presidential campaign traveling to states in the USA, where IMI had established alliances with NGOs and the Guatemalan community to have a meaningful impact on policy initiatives. This was an opportunity to address migrant and immigration issues directly with a presidential candidate who presented himself and his future administration as the only option against organized crime, corruption, and impunity. Unfortunately, the lack of commitment to migrant and returning nationals' concerns by the state while remaining faithful to IMI's convictions of "learning to service" did not produce any meaningful policy initiatives from the state.

### 18.2.3 Transnational Alliances with NGOs and Other Institutions

Another area that helped shaped IMI's shift toward a more active global role has been its association with NGOs in Guatemala, Honduras, El Salvador, and the USA. In building new frontiers of opportunities, IMI combined a variety of factors: development centers, migrant cooperation, hometown associations, entrepreneurs, philanthropy, church, and volunteers. To accomplish this task, IMI envisions that 20% would come from "faith, perseverance, vision, trust, and loyalty among alliances but that 80% of the solutions would come from prosperity, change, understanding and community philanthropy from outside the network." The synergy to infinity indicated Mr. Eduardo Alvarez Castillo, Jr., is "that all members are important and it is the mission of third sector NGOs to help people." Results show that NGOs' areas of concern overlap and that organizations are amalgamating expertise, knowledge, and skills in building transnational third sector solutions for migrants, deportees, and returning nationals. One central theme that emerged was sustainable solutions for migrants in the Central America region and solutions that would stop migration by creating opportunities domestically. As such, Table 1 further summarizes IMI's new frontiers of opportunity within a decentralized network of NGOs in two or more geographical units of services to the GDC with 22% of GNGOs in Guatemala, 17% US based, and 39% in Guatemala and the USA. There are some NGOs within the first frame of networks that are in close alliance with IMI, but this structural design also contains NGOs in subnetworks and micronetworks each cumulating into one unit (see Appendix 1: Survey of GNGOs).

## 18.3 GNGOs Across Local, National, and International Borders

Surveys and interviews of 22 GNGOs show that the structure and background of IMI's network alliance are composed of GNGOs operating in Guatemala, Honduras, and the USA. GNGOs have male and female directors attempting to change and influence policy in Central America, particularly the Northern Triangle around migration and deportation issues. Epistemological basis of this research is divided into four distinct parts. Part I establishes the GNGO's background, where they operate, the populations they serve, and the services provided. Part II reflects past collaboration among IMI network alliances. Part III shows collaboration among IMI and GNGOs. Lastly, Part IV illustrates the effectiveness of IMI and GNGOs impact on policy changes in the Northern Triangle. When directors were asked: Are you currently collaborating with IMI? Their responses are mirrored in Chart 1, 78% of GNGOs are currently collaborating with IMI, and 22% indicated that they are not in current collaboration with the organization. This shows that most of those surveyed are presently collaborating with IMI on projects or in some capacity whether in Guatemala, Honduras, or the USA.

Likewise, when respondents were asked how satisfied they were in collaborating with IMI, 57% of GNGOs indicated that they are very satisfied, 36% are somewhat satisfied, and the remaining 7% are neutral. Chart 2 demonstrates the level of commitment and cooperation GNGOs have expressed in partnering with IMI and that GNGOs are receptive to new ideas and sharing resources and that the partnership among GNGOs are overall committed to improving conditions within the GDC community. Similarly, interviews with respondents stated GNGOs were able to gain a broader understanding of the issues affecting the GDC. Overall, the GNGOs

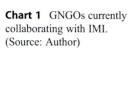

**Chart 1** GNGOs currently collaborating with IMI. (Source: Author)

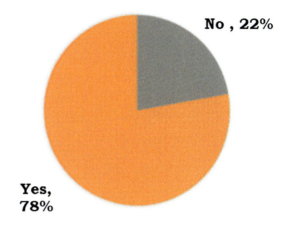

**Chart 2** GNGOs overall satisfaction collaborating with IMI. (Source: Author)

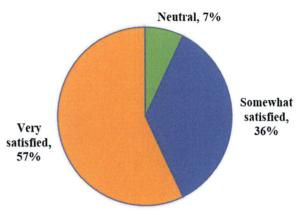

gained a better understanding of the needs affecting the GDC through the decision-making process of the network alliance.

The study also discloses that 32% of GNGOs operate in Guatemala and the USA presently; 23% operate in Guatemala only and another 23% in the USA only; an additional 23%, although not all GNGOs, operate as other types of NGOs across borders. Nine percent operate in Honduras and Guatemala only and 5% operate in Honduras, Guatemala, and the USA (see Appendix 1: Survey of GNGOs). Additionally, the involvement of GNGOs with various services provided to the GDC shows that 50% offer basic social services such as food, shelter, and healthcare, 41% provide immigration services such as citizenship assistance and other issues surrounding migration, 36% provide education services, another 36% address miscellaneous community need, and 9% provide housing in the USA and Guatemala.

Furthermore, the general view among GNGOs shows that 71% (Chart 3) would be somewhat likely to recommend this type of arrangement to other nongovernmental organizations, while 29% indicated they would be very likely to

**Chart 3** Recommendation of these types of arrangements. (Source: Author)

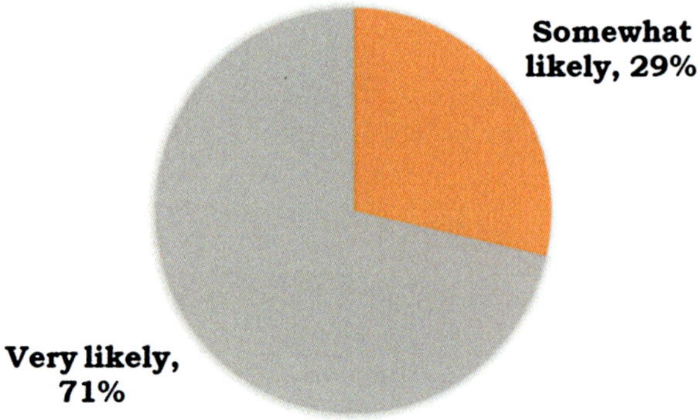

Somewhat likely, 29%

Very likely, 71%

**Chart 4** Improvement services of these types of collaborations. (Source: Author)

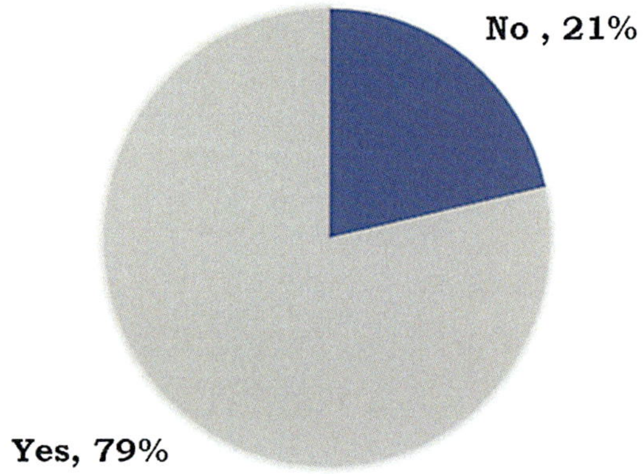

No, 21%

Yes, 79%

recommend connecting and collaborating with other organizations. To add, the data results also show other important aspects of GNGO's past collaboration with IMI: (a) 39% have been working with IMI for a period of 1 to 3 years; 28% collaborated for a period of 4 to 7 years, 22% for a period of 8 to 10 years, and 11% for at least 10 years or more. (b) Respondents expressed why they have collaborated with IMI, and 72% said they partnered with IMI to handle common illegal infractions (deportation, migrant, and other immigration issues), 33% in employment verification, another 33% on other related immigration issues, and 11% on detention matters with an additional 11% on worksite enforcement issues in the USA (see Appendix 1: Survey of GNGOs).

Similarly, Chart 4 shows that 79% of GNGOs would like to improve the services of these types of connection and collaborations with IMI, nongovernmental organizations, and other institutions; 21% of GNGOs expressed no interest in improving the services or collaborations with other NGOs; 50% are more likely to have collaborated with IMI on a weekly basis rather than once year and 39% on a monthly basis, and the remaining 11% indicated that they have collaborated once a year.

## 18.4 Integrating Technology and Communications Network

Fernando and Heston (1997) contend, "that contemporary NGOs occupy a place between a readily changing political economy at both the national and the international level" (p. 9). As such, the GNGOs alliance network makes them a new kind of NGO, playing significant "bottom-up" sociopolitical roles between states and the global community. Through various connections and collaborations with IMI, GNGOs have not only experienced stability but have also strengthened their communications globally by integrating technology and participating in the communication network.

This is exemplified in Fig. 1, with 86% of GNGOs providing services and sharing resources with the GDC; 32% served the private sector, and another 32% provided for the public sector; 18% serviced international clients, while another 18% helps agency cliental from GNGOs. This demonstrates that GNGOs are reaching a diverse group, and this was evidenced with the adjustment made to the questionnaire showing the expansion of services provided in multiple places, and the types of services provided the GDC often overlapped.

Integrating technology and communications is further illustrated in Fig. 2 documenting GNGOs ability to provide services to locations with very little resources and only the availability of social media and a strong community in the USA. There are 36% of GNGOs operating in Guatemala and 32% in the USA; 18% provide services in both locations, with another 9% in Guatemala and Honduras, and the remaining 5% provide services in Honduras. This confirms that GNGOs are not only knowledgeable about what service are needed but also where they are needed (see Appendix 1: Survey of GNGOs).

It is not surprising in Fig. 3 when respondents were asked what other types of services they would be interested in collaborating on the majority, 68% responded with educational services; 55% want micro-business opportunities; 45% felt they needed skills and training and 43% employment, and 32% want more focus on immigration issues, while 18% did not show a preference. Furthermore, when GNGOs were asked what are some of the advantages when combining forces with

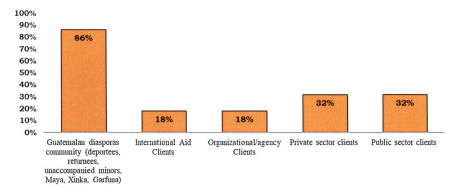

**Fig. 1** Types of populations served by GNGOs. (Source: Author)

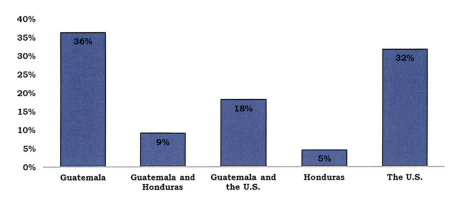

**Fig. 2** Locations of GNGOs. (Source: Author)

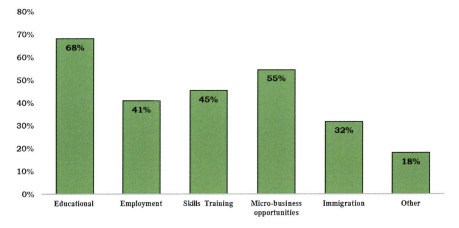

**Fig. 3** Other types of collaborations GNGOs want. (Source: Author)

IMI and other alliance network members, 79% responded that it is a good idea for unifying policy initiatives under one umbrella; 36% believes it strengthens effective and efficient program delivery; another 36% feel it improves communication between NGOs and agency, and an additional 36% expressed enhancement of professional development (see Appendix 1: Survey of GNGOs).

It is clear that this structure has contributed to the following: (1) With limited resources, IMI and GNGOs have carved out other vehicles to reach a greater portion of the GDG through the linkage of organizations that operate in multigeopolitical areas in Central America and the USA. (2) Majority of GNGOs feel positive about connections, collaborations, and exchanges with IMI and are eager to extend involvement with other NGOs in Central America and the USA. (3) Extended periods of integration across borders have allowed GNGOs to link and extend service delivery in multi-areas, in three geopolitical locations (Guatemala, Guatemala/Honduras, and Guatemala/USA). Moreover, 55% are providing immigration services and 41% immigration and constitutional services; 27% of affiliations are connected only on entrepreneurship for the GDC, with another 18% focused on constitutional rights and 14% on foreign and central affairs (see Appendix 1: Survey of GNGOs). GNGOs' structured and overall purpose leads to the realization that these organizations exist with the certainty that they can make a difference in ending migration through socioeconomic development and at the same time address human rights in Guatemala and neighboring Central American countries.

## 18.5 Impact for Sustainable Development Policy

IMI and GNGOs have adopted and integrated existing frameworks and models of socioeconomic development for sustainability and economic growth in Guatemala and Central America within a predetermined set of goals, strategies, and a philosophy grounded in third sector community development theory. Central to this theory are connection, collaboration, technology, and communication between IMI and GNGOs, which transpired on two different levels. On one level, IMI's efforts in working with GNGOs, NGOs, and institutions became evident by the implementation of a transnational strategy integrating local-national GDC policy agenda migration globally. On the other level, partnering with USNGO, the Children's Aid Guatemala (CAG), a medical healthcare facility, put IMI in contact with different groups domestically and with foreign personnel. According to Judy Freeman-Schwank, "four things are impacting Guatemala: child starvation, droughts, hydroelectric (mining is polluting the water of indigenous populations) and foreign companies' lack of social responsibility in giveback to the country." With the lack of corporation's social reasonability in Central America, GNGOs often respond with volunteer help within the community when displaced populations need for services increased. It is within these categories that IMI and Children Aid of Guatemala (CAG) began working to deliver services to remote areas of Guatemala and the transnational relationship developed between the two organizations and others. Significantly, the adoptability and implementation of GNGOs into an integrative best practice framework with a global strategy with little resources on projects with a cross-border emphasis (Table 2).

Using an integrative framework with other international NGOs allows IMI to continue working on building connections and forging collaborations with emerging opportunities for broader global involvement and building association or affiliation with organizations/institutions (religious, academic, and government) that GNGOs have expressed interests (see Table 2). These actions further demonstrate and support the discoveries in this study by displaying a shifting view of

Table 2  IMI integrative best practice model. (Source: Author)

| GNGO | Integration of services across borders | Emerging opportunities |
|---|---|---|
| IMI | Macro model of immigration and deportation into socioeconomic development model: Migration – Health – Education training – Small business development | Growing numbers of migrants and deportees flooding Guatemala and central America. GNGOs integrate ideas and goals |
| US-based PPLC | Children's aid Guatemala a nonprofit NGO linked with IMI to deliver healthcare to remote area in Guatemala • Created three different NGOs from 1985 providing 3000 women and children from 14 countries medical care • Women's health and domestic violence | Defends immigration from the northern triangle • Infrastructure exist to establish and link services with other institutions and NGOs on immigration • Create programs the community needs in ESL, immigration, education, and employment training • Facility can also serve as a site for graduate internship on migration, community development for students in urban or Latin American studies programs |
| Northern-triangle NGO | Work with private/public/religious sectors to deliver services to GDC residents, migrants, deportees, and African refugees | Build upon existing structures to pursue further inquiry into regional phenomenon. Support and develop further studies of migration |

indigenous nongovernmental organizations that warrants further exploration and investment. For example, the geographical scope of this network reaches across seven states within the USA and connects with other NGOs in Guatemala and Honduras. And as such, civil society and social movements' scholars are encouraging additional theoretical understandings of this social phenomenon in the global community.

Some scholars agreed that research on indigenous NGOs should expand to include empirical data on what impact these organizations make. Fisher (1997) is correct when he explains that "changes in studies illuminate understandings of trans local flows of ideas, knowledge, funding and people; shed light on changing relationships among citizenry, associations and the state" (p. 439). This perspective was reinforced by Senior Guillermo Castillo in explaining that "IMI's three-dimensional diagram of commercial division, social organizations and a thinking model of organizations were used to create IMI. It was designed to show people that they have shortages and constraints." In addition, Ana Judith Ramirez Garcia of the Dioceses Fronterizas in Esquipulas explains a three-dimensional diagram as "integration structures with solutions and resources that are urgently needed to assist the GDC, but also to help with an increasing flow of other displaced populations of Africans who travel from Brazil – into Venezuela – to Guatemala,

Mexico and on the U.S. in recent years." Moreover, 79% of GNGOs wants to improve services with IMI, other NGOs, and institutions in the USA. Some directors voiced a need for the USA to understand the problems of immigration and to support programs with local indigenous NGOs. Furthermore, 78% of GNGOs are presently in partnership with IMI, while document reviews and interviews confirmed that although GNGOs are not assisted by government or corporations, they managed to adapt to existing political structures. In addition, the data shows that GNGOs provide services to 31.8% of the public sector clients and 36% have little resources and a staff of volunteers.

A majority of GNGOs are positive about the impact on policy changes with 71% of them believe that connection and collaboration with IMI have been effective in policy changes on immigration and deportation. Forty-three percent said that changes also happened in education, health, and welfare issues; 21% saw changes in family and child custody issues; 14% said changes occurred in criminal issues, while an additional 14% responded that GNGOs are ineffective in producing policy changes. Furthermore, the impact and sustainability from GNGOs validate the following: IMI has been overall effective in influencing policy changes for the GDC while others believe it to be somewhat effective; the level of commitment and cooperation of the GNGOs shows that 79% are more receptive to new ideas and sharing resources; 43% voiced that the partnership among GNGOs is, overall, committed to improving conditions within the GDC community; similarly, 43% indicated GNGOs were able to gain a broader understanding of the issues affecting the Guatemalan community; and the remaining 43% gained a better understanding of the needs affecting the GDC in the USA specifically through the decision-making process of the network alliance.

## 18.6 Conclusion

In sum, assessment of IMI and 22 GNGOs provided a better understanding of how and what makes indigenous nongovernmental organizations transnational. Empirical analysis of these types of organizations shows three significant things: First, IMI and GNGOs share and are guided by a set of beliefs grounded in third sector socioeconomic development as the vehicle for change in Guatemala, the USA, and the global GDC. Second, there is verifiable evidence of their connection and collaboration on issues such as immigration services, housing and healthcare issues, and deportation. Lastly, these types of organizations have overlapping ideas and pooled resources to achieve common interests across geographical locations which resulted in changing policy for the GDC.

# Appendix 1: Survey of Guatemalan Nongovernmental Organizations (GNGOs)

| Gender | Guatemala | | Guatemala And Honduras | | Guatemala and Honduras The U.S. | | The U.S. | | Comparative | |
|---|---|---|---|---|---|---|---|---|---|---|
| Female | 3 | 37% | 1 | 50% | 1 | 25% | 0 | 0% | 2 | 29 | 7 | 32% |
| Male | 5 | 63% | 1 | 50% | 3 | 75% | 1 | 100% | 5 | 71% | 15 | 68% |
| Total | 8 | 100% | 2 | 100% | 4 | 100% | 1 | 100% | 7 | 100% | 22 | 100% |
| **What is your title?** | | | | | | | | | | | | |
| CEO, President | 1 | 13% | 0 | 0% | 0 | 0% | 0 | 0% | 2 | 29% | 3 | 14% |
| Director, Manager | 4 | 50% | 2 | 100% | 2 | 50% | 0 | 0% | 3 | 42% | 11 | 50% |
| Executive Director | 0 | 0% | 0 | 0% | 1 | 25% | 1 | 100% | 0 | 0% | 2 | 9% |
| Some Other | 3 | 37% | 0 | 0% | 1 | 25% | 0 | 0% | 2 | 29% | 6 | 27% |
| Total | 8 | 100% | 2 | 100% | 4 | 100% | 1 | 100% | 7 | 100% | 22 | 100% |
| **How long has your organization existed?** | | | | | | | | | | | | |
| 13 years or more | 2 | 25% | 1 | 50% | 3 | 75% | 1 | 100% | 4 | 57% | 11 | 50% |
| 2-4 years | 2 | 25% | 0 | 0% | 0 | 0% | 0 | 0% | 1 | 14% | 3 | 14% |
| 5-8 years | 2 | 25% | 0 | 0% | 1 | 25% | 0 | 0% | 0 | 0% | 3 | 14% |
| 6 months – 1 year | 1 | 12.5% | 1 | 50% | 0 | 0% | 0 | 0% | 2 | 29% | 4 | 18% |
| 9-12 years | 1 | 12.5% | 0 | 0% | 0 | 0% | 0 | 0% | 0 | 0% | 1 | 4% |
| Total | 8 | 100% | 2 | 100% | 4 | 100% | 1 | 100% | 7 | 100% | 22 | 100% |
| **Does your NGO receive assistance or aid from government/agency or international NGO?** | | | | | | | | | | | | |
| No | 6 | 75% | 2 | 100% | 4 | 100% | 1 | 100% | 7 | 100% | 20 | 91% |
| Yes | 2 | 25% | 0 | 0% | 0 | 0% | 0 | 0% | 0 | 0% | 2 | 9% |
| Total | 8 | 100% | 2 | 100% | 4 | 100% | 1 | 100% | 7 | 100% | 22 | 100% |

| GNGOs Services to Guatemalan Diaspora Community (GDC) | Guatemala | | Guatemala & Honduras | | Guatemala & The U.S. | | Honduras | | The U.S. | | Comparative | |
|---|---|---|---|---|---|---|---|---|---|---|---|---|
| Less than 25 | 0 | 0% | 0 | 0% | 0 | 0% | 0 | 0% | 4 | 57% | 4 | 18% |
| 25-50 GDC clients | 0 | 0% | 0 | 0% | 0 | 0% | 0 | 0% | 1 | 14% | 1 | 5% |
| 50-100 GDC clients | 2 | 25% | 0 | 0% | 0 | 0% | 0 | 0% | 1 | 14% | 3 | 14% |
| 101-200 GDC clients | 1 | 12.5% | 0 | 0% | 0 | 0% | 0 | 0% | 0 | 0% | 1 | 5% |
| 201-500 GDC clients | 1 | 12.5% | 1 | 50% | 0 | 0% | 0 | 0% | 0 | 0% | 2 | 9% |
| 500 or more | 3 | 38% | 1 | 50% | 2 | 50% | 1 | 100% | 1 | 14% | 8 | 36% |
| Other | 1 | 12.5% | 0 | 0% | 2 | 50% | 0 | 0% | 0 | 0% | 3 | 14% |
| Total | 8 | 100% | 2 | 100% | 4 | 100% | 1 | 100% | 7 | 100% | 22 | 100% |
| **Length of Time GNGOs Worked with IMI** | | | | | | | | | | | | |
| 1-3 years | 2 | 33% | 1 | 50% | 1 | 33% | 0 | 0% | 3 | 50% | 7 | 39% |
| 4-7 years | 2 | 33% | 0 | 0% | 1 | 33% | 0 | 0% | 2 | 33% | 5 | 28% |
| 8-10 years | 1 | 17% | 1 | 50% | 1 | 33% | 1 | 100% | 0 | 0% | 4 | 22% |
| 10 or more years | 1 | 17% | 0 | 0% | 0 | 0% | 0 | 0% | 1 | 17% | 2 | 11% |
| Total | 6 | 100% | 2 | 100% | 3 | 100% | 1 | 100% | 6 | 100% | 18 | 100% |
| **Where GNGOs Collaborate With IMI** | | | | | | | | | | | | |
| Guatemala | 3 | 50% | 1 | 50% | 0 | 0% | 0 | 0% | 0 | 0% | 4 | 22% |
| Guatemala & Honduras | 0 | 0% | 0 | 0% | 0 | 0% | 1 | 100% | 0 | 0% | 1 | 6% |
| Guatemala & U.S. | 3 | 50% | 0 | 0% | 2 | 67% | 0 | 0% | 2 | 33% | 7 | 39% |
| Guatemala, Honduras & U.S. | 0 | 0% | 1 | 50% | 0 | 0% | 0 | 0% | 2 | 33% | 3 | 17% |
| U.S. Alone | 0 | 0% | 0 | 0% | 1 | 33% | 0 | 0% | 2 | 33% | 3 | 17% |
| Total | 6 | 100% | 2 | 100% | 3 | 100% | 1 | 100% | 6 | 100% | 18 | 100% |

| | Guatemala | | Guatemala & Honduras | | Guatemala & The U.S. | | Honduras | | The U.S. | | Comparative | |
|---|---|---|---|---|---|---|---|---|---|---|---|---|
| **Frequency of GNGOs Collaboration with IMI** | | | | | | | | | | | | |
| Weekly | 2 | 33% | 1 | 50% | 2 | 67% | 0 | 0% | 4 | 66% | 9 | 50% |
| Monthly | 3 | 50% | 1 | 50% | 1 | 33% | 1 | 100% | 1 | 17% | 7 | 39% |
| Once a Year | 1 | 17% | 0 | 0% | 0 | 0% | 0 | 0% | 1 | 17% | 2 | 11% |
| **Total** | **6** | **100%** | **2** | **100%** | **3** | **100%** | **1** | **100%** | **6** | **100%** | **18** | **100%** |
| **GNGOs Current Collaboration with IMI** | | | | | | | | | | | | |
| No | 2 | 33% | 0 | 0% | 0 | 0% | 0 | 0% | 2 | 33% | 4 | 22% |
| Yes | 4 | 67% | 2 | 100% | 3 | 100% | 1 | 100% | 4 | 67% | 14 | 78% |
| **Total** | **6** | **100%** | **2** | **100%** | **3** | **100%** | **1** | **100%** | **6** | **100%** | **18** | **100%** |
| **Length GNGOs Participation with IMI** | | | | | | | | | | | | |
| 1 week-3 months | 1 | 25% | 1 | 50% | 0 | 0% | 0 | 0% | 0 | 0% | 2 | 14% |
| 6 months -1 year | 0 | 0% | 0 | 0% | 0 | 0% | 0 | 0% | 1 | 25% | 1 | 7% |
| 1 year | 0 | 0% | 1 | 50% | 0 | 0% | 0 | 0% | 0 | 0% | 1 | 7% |
| 1 - 2 years | 0 | 0% | 0 | 0% | 2 | 67% | 0 | 0% | 2 | 50% | 4 | 29% |
| 3 years or more | 3 | 75% | 0 | 0% | 1 | 33% | 1 | 100% | 1 | 25% | 6 | 43% |
| **Total** | **4** | **100%** | **2** | **100%** | **3** | **100%** | **1** | **100%** | **4** | **100%** | **14** | **100%** |
| **Frequency GNGOs Meet/Communicate with IMI** | | | | | | | | | | | | |
| Every week | 1 | 25% | 1 | 50% | 2 | 67% | 0 | 0% | 2 | 50% | 6 | 43% |
| Every month | 1 | 25% | 1 | 50% | 1 | 33% | 1 | 100% | 2 | 50% | 6 | 43% |
| Every 2-3 months | 1 | 25% | 0 | 0% | 0 | 0% | 0 | 0% | 0 | 0% | 1 | 7% |
| Once/Twice a year | 1 | 25% | 0 | 0% | 0 | 0% | 0 | 0% | 0 | 0% | 1 | 7% |
| **Total** | **4** | **100%** | **2** | **100%** | **3** | **100%** | **1** | **100%** | **4** | **100%** | **14** | **100%** |
| **GNGOs Overall Satisfaction with IMI** | | | | | | | | | | | | |
| Neutral | 1 | 25% | 0 | 0% | 0 | 0% | 0 | 0% | 0 | 0% | 1 | 7% |
| Somewhat | 3 | 75% | 2 | 100% | 0 | 0% | 0 | 0% | 0 | 0% | 5 | 36% |
| Very | 0 | 0% | 0 | 0% | 3 | 100% | 1 | 100% | 4 | 100% | 8 | 57% |
| **Total** | **4** | **100%** | **2** | **100%** | **3** | **100%** | **1** | **100%** | **4** | **100%** | **14** | **100%** |
| **GNGOs Effectiveness of Changes** | | | | | | | | | | | | |
| Little difference | 0 | 0% | 0 | 0% | 0 | 0% | 1 | 100% | 0 | 0% | 1 | 8% |
| Somewhat Effective | 1 | 50% | 2 | 100% | 1 | 33% | 0 | 0% | 2 | 50% | 6 | 50% |
| Very effective | 1 | 50% | 0 | 0% | 2 | 67% | 0 | 0% | 2 | 50% | 5 | 42% |
| **Total** | **2** | **100%** | **2** | **100%** | **3** | **100%** | **1** | **100%** | **4** | **33%** | **12** | **100%** |
| **GNGOs Recommendation of these types of Collaboration** | | | | | | | | | | | | |
| Somewhat Likely | 2 | 50% | 1 | 50% | 1 | 33% | 0 | 0% | 0 | 0% | 4 | 29% |
| Very Likely | 2 | 50% | 1 | 50% | 2 | 67% | 1 | 100% | 4 | 100% | 10 | 71% |
| | **4** | **100%** | **2** | **100%** | **3** | **100%** | **1** | **100%** | **4** | **100%** | **14** | **100%** |
| **Suggestions for Improving these types of Collaborations** | | | | | | | | | | | | |
| No | 0 | 0% | 1 | 50% | 1 | 33% | 0 | 0% | 1 | 25% | 3 | 21% |
| Yes | 4 | 100% | 1 | 50% | 2 | 67% | 1 | 100% | 3 | 75% | 11 | 79% |
| **Total** | **4** | **100%** | **2** | **100%** | **3** | **100%** | **1** | **100%** | **4** | **100%** | **14** | **100%** |

| | Guatemala | | Guatemala & Honduras | | Guatemala & The U.S. | | Honduras | | The U.S. | | Comparative | |
|---|---|---|---|---|---|---|---|---|---|---|---|---|
| **Assistance of Policy Initiatives for the GDC** | | | | | | | | | | | | |
| Don't Know/Not Sure | 1 | 12% | 0 | 0% | 0 | 0% | 0 | 0% | 0 | 0% | 1 | 4% |
| No | 5 | 63% | 1 | 50% | 2 | 50% | 1 | 100% | 5 | 71% | 14 | 64% |
| Yes | 2 | 25% | 1 | 50% | 2 | 50% | 0 | 0% | 2 | 29% | 7 | 32% |
| **Total** | **8** | **100%** | **2** | **100%** | **4** | **100%** | **1** | **100%** | **7** | **100%** | **22** | **100%** |
| **GNGO Interested in Partnering with other NGOs** | | | | | | | | | | | | |
| Don't Know/Not Sure | 0 | 0% | 1 | 50% | 1 | 25% | 0 | 0% | 0 | 0% | 2 | 9% |
| No | 0 | 0% | 0 | 0% | 0 | 0% | 0 | 0% | 2 | 29% | 2 | 9% |
| Yes | 8 | 100% | 1 | 50% | 3 | 75% | 1 | 100% | 5 | 71% | 18 | 82% |
| **Total** | **8** | **100%** | **2** | **100%** | **4** | **100%** | **1** | **100%** | **7** | **100%** | **22** | **100%** |
| **GNGOs Seeks Collaboration Align with U.S. Engagement in Guatemala** | | | | | | | | | | | | |
| Need More Information | 2 | 25% | 1 | 50% | 0 | 0% | 0 | 0% | 1 | 14% | 4 | 18% |
| Yes | 6 | 75% | 1 | 50% | 4 | 100% | 1 | 100% | 6 | 86% | 18 | 82% |
| **Total** | **8** | **100%** | **2** | **100%** | **4** | **100%** | **1** | **100%** | **7** | **100%** | **22** | **100%** |
| **GNGOs seeks Future Collaborations** | | | | | | | | | | | | |
| Need More Information | 0 | 0% | 1 | 50% | 0 | 0% | 0 | 0% | 1 | 14% | 2 | 9% |
| Yes | 8 | 100% | 1 | 50% | 4 | 100% | 1 | 100% | 6 | 86% | 20 | 91% |
| **Total** | **8** | **100%** | **2** | **100%** | **4** | **100%** | **1** | **100%** | **7** | **100%** | **22** | **100%** |

# References

## Books

Arendt H (1961) Between past, present and future. Viking Press, New York. Retrieved from: http://www.hannaharendtcenter.org. Ask, Sten and Mark-Jungkvist, Anna (2006)

Escobar A (1995) Encountering development: the making and unmaking of the third world. Princeton University Press, Princeton/Oxford. Retrieved from: http://www.jstor.org/stable/j.ctt7rtgw

Falk R (1993) The making of global citizenship. In: Childs JB, Brecher J, Cutler J (eds) Global visions: beyond the New York order. South End, Boston, pp 39–50. Retrieved from: http://sk.sagepub.com/books/the-condition-of-citizenship/n10.xml

Taleb NN (2007, 2010) The black swan: the impact of the highly improbable. Published in USA by Random House Trade Paperback. Retrieved from: https://www.amazon.com/Black-Swan-Improbable-Robustness-Fragility/dp/081297381X

## Interviews for Research Study

Alvarez Castillo E (Personal communication, January 17, 2017), is the co- director of IL Mondo Immaginare (IMI) is a nongovernmental organization operating in Guatemala, Honduras and the United States

Barillas Castillo E (Personal communications, February 17, 2017), is the entrepreneur and founder of IL Mondo Immaginare (MI) is a nongovernmental organization operating in Guatemala, Honduras and El Salvador

Delmar Vibizo Pantis J (Personal conversation January 17th, 2017 via skype in Honduras)

Freeman-Schwank J (Personal Communications, January 18th, 2017), is the Director of Children's Aid Guatemala and Law Firm in Kentucky, USA

## Journal

Atack I (1999) Four criteria of development NGO legitimacy. World Dev 27(5):855–864. Retrieved from: https://ideas.repec.org/a/eee/wdevel/v27y1999i5p855-864.html

Edelman M (2001) Social movements: changing paradigms and forms of politics. Ann Rev Anthropol 30:285–317. Retrieved from: http://www.annualreviews.org

Esteva G (1998) "Beyond development, what?", with M.S. Prakash. Dev Pract 8(3). Retrieved from: https://newint.org/blog/editors/2008/07/24/gustavo-esteva-the-man-who-invented-no-corn-no-country

Fernando JL, Heston AW (1997) The role of NGOs: charity and empowerment: introduction: NGOs between states, markets, and civil society. Am Acad Polit Soc Sci. 554 Annals 8. Retrieved from: http://journals.sagepub.com/doi/pdf/10.1177/0002716297554001001

Fisher WF (1997) Doing good? The politics and Antipolitics of NGO practices. Ann Rev Anthropol 26:439–464. Department of Anthropology, Harvard University, Cambridge, MA. Retrieved from: http://www.annualreviews.org

Huntington S, Fukuyama F (1996) Political order in changing societies. Yale University Press, New Haven/London. Retrieved from http://www.jstor.org/stable/j.ctt1cc2m34

Mitlin D, Hickey S, Bebbington A (2007) Reclaiming development: NGOs and the challenge of alternatives. World Dev 35(10):169–1720. Retrieved from: http://citeseerx.ist.psu.edu/viewdoc/download?doi=10.1.1.517.5100&rep=rep1&type=pdf

Nederveen Pieterse, J (1996). My paradigm or yours: alternative development, post- development, reflexive development. ISS Work Pap Ser/Gen Stud 229:1–36. Eramus University, Rotterdam. Retrieved from: http://repub.eur.ril/pub/18959

Tvedt T (2002) Development NGOs: actors in a Global Civil Society or in a New International Social System? Volunt Int J Volunt Nonprofit Org 13(4):363–375. Global Civil Society, Springer. Retrieved from: http://www.jstor.org.stable27927806

Uvin P, Jain PS, Brown DL (2000) Think large and act small: toward a new paradigm for NGO scaling up. World Dev 28(8):1409–1419. Retrieved from: http://www.elsevier.com/locate/worlddevelopment

## On-Line Article

de Bono E (1967) Lateral thinking. Retrieved from: http://www.creativethinkingwith.com/Edward de-Bono-html

Einstein A. Quotable Quotes. Retrieved from: http://www.goodreads.com/quotes/556030-imagination-is-more-important-than-knowlege-for-knowledge-is-limited

# Women's Empowerment in Digital Media: A Communication Paradigm

## 19

Xiao Han

## Contents

| | | |
|---|---|---|
| 19.1 | Introduction | 380 |
| 19.2 | Power as a Theoretical Basis: Uncovering the Communicative Character of Women's Empowerment | 381 |
| 19.3 | Rethinking Women's Empowerment as a Communicative Process and the Role of Digital Communication | 384 |
| | 19.3.1 Developing Women's Consciousness Through Communicative Action of Voice | 384 |
| | 19.3.2 Constructing a Public Identity: Emergence of Social Empowerment | 385 |
| | 19.3.3 New Dynamics of "Power with" for the Organization of Women's Collective Action | 386 |
| 19.4 | Digital Communication and the Emergence of a Political Process of Women's Empowerment? | 388 |
| | 19.4.1 Providing Organizational Information for Identity Projection | 388 |
| | 19.4.2 Disseminating Information to Raise Awareness of Gender Issues | 389 |
| | 19.4.3 Creating a Sense of the Collective Through the Internet | 390 |
| | 19.4.4 A New Space for Political Campaigns Within an Authoritarian Context? | 391 |
| | 19.4.5 Developing Interorganizational Relations | 391 |
| 19.5 | Conclusion | 391 |
| References | | 392 |

### Abstract

The study of women's empowerment, inspired by various women's movements from the late 1960s onward, encompasses a wide range of concepts, approaches, and practices in different disciplines. Moreover, while the Internet has experienced dramatic technological development since the late 1980s, fierce debates about the empowering potential of the Internet for women's liberation have also

X. Han (✉)
Mobile Internet and Social Media Centre, Communication University of China, Beijing, China
e-mail: han.xiao@cuc.edu.cn

© Springer Nature Singapore Pte Ltd. 2020
J. Servaes (ed.), *Handbook of Communication for Development and Social Change*,
https://doi.org/10.1007/978-981-15-2014-3_79

raged. In particular, feminist theorists have grasped this opportunity to enquire whether women can become empowered by the Internet. However, the existing feminist research lacks systematic theoretical frameworks that would help us investigate what roles digital media play in the process of women's empowerment. This chapter aims to address this gap by drawing on the literature from digital media studies to investigate the communicative affordances of online platforms for the development of women's empowerment. To approach this problem, there is a need to place communication at the center of the issue in order to conceptualize the open-ended process of women's empowerment. This chapter starts by acknowledging that women's empowerment is a multifaceted concept and drawing upon the feminist interpretation of power – the core concept of empowerment – to anchor the proposed argument. In regard to digital media, the existing literature well documents that women have had a great opportunity to capture various forms of horizontal, participatory communication to exercise power in their living circumstances. This sets the scene for creating possible theoretical links between digital media and women's empowerment. The role of Chinese women's groups in the process of women's empowerment will be presented in this chapter to demonstrate the explanatory value of the framework.

**Keywords**

China's Communist Party (CCP) · Communicative action of voice · Empowerment · Feminist · Social networking sites (SNSs) · Women's empowerment

## 19.1 Introduction

The study of women's empowerment, inspired by various women's movements from the late 1960s onward, encompasses a wide range of concepts, approaches, and practices in different disciplines, including anthropology, sociology, development, and feminism. However, since the 1990s, the term "empowerment" has been transformed from a political concept into a *buzzword*, as major international development agencies, from a top-down perspective, have institutionalized it as mainstream policy or agenda while neglecting its original meaning and conceptual value (Batliwala 2007).

Indeed, within the academic field, many scholars insist that we need to treat women's empowerment as a concept and explore its theoretical underpinnings (Batliwala 2007; Kabeer 2010). In particular, "women's empowerment" has become a pressing issue in contemporary feminist scholarship, at least since the second wave of women's movements. This body of literature, not least from the strand of feminist political theory, has developed various approaches to uncover the theoretical underpinnings of women's empowerment by focusing on process rather than finished product (Allen 1998; Carr 2003; Collins 2000; Young 1990, 1994; Yuval-Davis 1994). Furthermore, some feminist political theorists have cast light on the constitutive role of *communication* in the process of empowerment of women where "dialogue" (Mansbridge 1993; Young 1994; Yuval-Davis 1994), "solidarity" (Allen 1998; Yuval-Davis 1994), "recognition" (Fraser 2000), or "participation" (Fraser 2009; Phillips 1992) is considered paramount. A stepping stone in this line of enquiry is a group of theorists from the feminist development communication who advance the view that *empowerment is fundamentally a communicative process* (Papa et al. 2000).

As the discussion about women's empowerment intensifies, the Internet is simultaneously emerging, developing, and expanding around the world. Of particular note is an intense questioning of whether the new communication infrastructures afforded by the Internet may offer empowering potential for women's liberation. One of the most recent strands is that, following the studies of the relationship between social media and protest in the current context, many feminist writers have marked out communicative characteristics of the Internet to renew feminist politics (Baer 2016; Fotopoulou 2016). An emerging strand in this field devotes attention to participation and investigates how women participate in digital spaces and how they challenge power structures through online participatory practices (e.g., Hasinoff 2014; Portwood-Stacer 2014). Within such discourses, participation does not simply involve online activities such as logging, clicking, writing, commenting, or networking – which is not necessarily empowering. Rather, it involves women's unique participatory practices – in terms of "a bodily reaction to flows of information" and "a positioning of subjectivity in an experiential relationship" (Levina 2014, p. 279) – which has a necessary relationship with empowerment.

However, there seems to be no analytical framework that would help us link the communicative characteristics of the Internet with the understandings of women's empowerment as a communicative process. To approach this problem, there is a need to place communication at the center of the issue in order to conceptualize the open-ended process of women's empowerment. This chapter starts by acknowledging that women's empowerment is a multifaceted concept and drawing upon the feminist interpretation of power – the core concept of empowerment – to anchor the proposed argument. In regard to digital media, the existing literature well documents that women have had a great opportunity to capture various forms of horizontal, participatory communication to exercise power in their living circumstances. This sets the scene for creating possible theoretical links between digital media and women's empowerment. The role of Chinese women's groups in the process of women's empowerment will be presented in this chapter to demonstrate the explanatory value of the framework.

## 19.2 Power as a Theoretical Basis: Uncovering the Communicative Character of Women's Empowerment

As the root concept, it is no surprise that the existing theoretical discussions of women's em*power*ment have attached great importance to the notion of *power*. As Collins (2000) suggests, "how does one develop a politics of empowerment without understanding how power is organised and operates?" (p. 292). In this sense, the feminist critique of power offers a useful starting point for developing a thorough understanding of women's empowerment.

Amy Allen's (1998) synthesized triple models of power represent a valuable addition to this strand as they provide a multidimensional view of power that helps to illuminate the dynamic process of women's empowerment. The power models outlined by Amy Allen are analytically integrative, but each model entails a different focus.

The first model, "*power over*," offers a useful entry point in recognizing "what women's empowerment is fundamentally about." From the feminist perspective, Allen's (1998, p. 38) "power over" goes as follows:

> Not only is male power over women exercised by men who do not deliberately intend to do so, [but it also] is exercised by men who deliberately do so. This is because, whatever their intentions are, they still act within a set of cultural, institutional, and structural relations of power that work to women's disadvantage.

Underlying this interpretation of "power over" is the recognition of the existence of patriarchy and the subsequent urge to end men's domination over women as the starting point for women's empowerment. In other words, it is the exercise of "power over" that motivates women to initiate their empowerment process. This is because social systems are, overall, structured around relatively stable patterns in which rules, norms, beliefs, and practices are defined by the dominant group of men, which leads to patriarchy being a very significant order of the social reality (Fraser 2009). In such a male-dominated world, not only are women excluded from political participation but their experiences, values, and interests are also neglected (Mansbridge 1993; Phillips 1992).

However, this "power over" model only provides a one-dimensional view of women's empowerment, as it primarily carries masculine connotations of domination and control (Allen 2005). In fact, many feminists, albeit from different theoretical stances, have noticed this problem and argued for an alternative conception of power to complement "power over," i.e., the second power model of "power to." Compared with "power over," "power to" focuses on the positive elements of power, instead of seeing power as a negative force in association with repression, prohibition, or coercion. In this respect, power refers to a capacity that drives women as individual actors to achieve their goals and ends but not necessarily at the expense of others (Allen 1998, 2005). Thus, women's empowerment is closely linked to the realization of women's potential to take actions to attain a series of ends and goals in their lives.

Nevertheless, it is important to bear in mind that "power to" needs to be considered as a response to "power over." In other words, "power to" is not located within a vacuum, wherein women only place their personal experiences at the center of an appeal for freedom, but instead is conceived as resistance to "power over." Women's resistance, therefore, lies in their ability to "attain an end or series of ends that serve to subvert domination" (Allen 1998, p. 35). In this formulation, "power to" and "power over" models are in a dialectical relationship. That said, women can be described as being "both dominated and empowered at the same time and in the context of the same norm, institution, or practice" (Allen 1998, p. 31).

Women's empowerment is also more than just an individual or personal matter, as collective power is much greater than isolated efforts of individual women, even if they have benefited from "power to" to bring about changes to their subordination (Collins 2000; Phillips 1992; Young 1990, 1994; Yuval-Davis 1994). In this respect, feminist researchers have adopted a new approach – the third model of "power with" – to shed light on the collective aspect of power. As such, "power with" suggests a necessity for the formation of a collectivity in the empowerment process. In this way, "power with" can be understood as "the ability of a collectivity to act together for the shared common purpose of overturning a system of domination" (Allen 1998, p. 36). Within this perspective, women's empowerment needs to be organized around common goals to tackle the social problems associated with gender-related inequalities.

Again, similar to "power to," "power with" is also motivated by resistance to "power over." A typical example in this sense is the women's movement that challenges men's position of dominance by bringing women into the decision-making process (Young 1990; Yuval-Davis 1994). As Allen (1998) notes, to challenge systematic relations of domination and oppression, "some sort of collective action or struggle seems necessary" (p. 164). Moreover, it should be noted that the essence of "power with" involves people participating in decision-making based on the "working together" model rather than in a top-down manner to achieve social change. For feminists, this is "a form of coalition politics" (Yuval-Davis 1994), which involves women's negotiation, joint effort, collaborations, and mutual support to pursue their common goals.

The feminist conceptions of power introduced above serve as a particularly useful starting point in setting the scene for a clear theorization of empowerment, as mentioned already – power is the root concept of em*power*ment. Power, based on Amy Allen's synthesis of key feminist ideas that underpin the concept of power, is thereby understood as a dynamic process that involves the development of women's capacities to exercise "power to" and "power with" to make significant changes to the ways in which men exert "power over" women. These three power relations are not isolated from each other; rather, they need to be seen as the coproduct of questioning of and reaction against patriarchal structure by women.

In fact, the aforementioned power models have indicated the communicative character of women's empowerment that entails a process of change. That said, in the process of challenging gender-related structural constraints, it is through communication that female actors recognize their subordinate status, raise consciousness, express their standpoint, make decisions, participate in actions, build solidarity, and organize collective action. Thus, underlying the process of women's empowerment is, indeed, a communicative process whereby individual and collective actors produce generative senses of "power to" and "power with" to change gender inequalities where men exert "power over" women.

Furthermore, the extensive use of digital media for contemporary feminist politics has brought communication into a central place, leading to communication as an important entry point to explore the relationship between women's empowerment and the Internet. A prevailing tendency in this regard is to suite communication capacities of the Internet to the communicative practices that underline the process of women's empowerment.

## 19.3 Rethinking Women's Empowerment as a Communicative Process and the Role of Digital Communication

In this section, three specific aspects will explain how the communicative action shapes the empowerment process: development of women's consciousness, construction of a public identity, and organization of collective action, as well as the role of digital communication in these processes.

### 19.3.1 Developing Women's Consciousness Through Communicative Action of Voice

The above discussion on women's empowerment as a process seems to underline the importance of *consciousness-raising* as the point of departure. A possible explanation in this regard might relate to the fact that the concept of empowerment has roots in Paulo Freire's theory of "conscientization" in the field of popular education. In regard to women, conscientization suggests that "women can connect their experiences of oppression with those of other women and thereby see the political dimensions of their personal problems" (Carr 2003, p. 15).

Following this logic, *communicative action of voice* would play an important role in the process of "conscientization" for female actors. Through the lens of women's empowerment, voice refers to "the different ways in which women might seek, individually or collectively, to bring about desired forms of change in their lives and relationships" (Kabeer 2010, p. 107). Women's capacity for voice is thus closely related to the communication processes through which women can narrate their personal stories and express their concerns.

In this respect, the Internet is claimed to offer a combination of communicative elements (e.g., profiles, posts, comments, bookmarks, and hyperlinks) that allow people to become storytellers in their own right. This activity has been recently referred to as *digital storytelling* (see Lundby 2008). In light of this, we can see that digital storytelling provides a means to distribute more widely the capacity to reflect on both one's thoughts and oneself (Couldry 2008). This also suggests that the self can be realized through the process of digital storytelling.

One key aspect of digital storytelling associated with women is the portrayal of their intimate stories in real life. The emergence of hashtag feminism in relation to Twitter actions of #PussyRiot, #YesAllWomen, and #Slutwalk (see Baer 2016; Fotopoulou 2016) can be put forward here as one of the most recognizable examples to illustrate how digital media have helped women become a social, and potentially political, agent through the posting of their intimate and sexual stories. At this moment, following the observation of Brickell (2012), it is "a matter of empowered becoming" (p. 30).

This also leads to the Internet's potential for *self-expression*. In this sense, self-expression involves the process through which women have their views heard and achieve personal change. As Papacharissi (2010) notes, "self-expression values are connected to the desire to control one's empowerment, a strong desire for autonomy, and the need to question authority" (p. 134). These values are aligned with, as Stavrositu and Shyam Sundar (2012) argue, "ultimately a deep sense of empowerment" (p. 370). A growing body of evidence has also suggested that the expressive affordances of digital media can be adopted by women as a mediated activity aimed at raising awareness, demonstrating personal beliefs and values, and challenging feminine stereotypes (Antunovic and Hardin 2013; Sinanan et al. 2014).

The notion of voice further links well with the concept of *dialogue*. By drawing on various perspectives from communication studies, Ganesh and Zoller (2012) identify that dialogue can be viewed as "a special form of mutual relationship building and collaboration" (p. 70). This is also alluded to in studies of development communication, suggesting that dialogue is "the route to self-reflection, self-knowledge, and liberation from disempowering beliefs" and "the route to mutual learning, acceptance of diversity, trust, and understanding" (Papa et al. 2000, p. 94). Moreover, if we examine dialogue from feminist perspectives, this form of communication would eschew "persuasion, which is associated with patriarchal attempts at domination and control" (Ganesh and Stohl 2013, p. 70). A similar point is used by Yuval-Davis (1994) to place dialogue as central in the building process for women's empowerment. For Yuval-Davis, women, especially those who are powerless, would engage in dialogue with each other to interpret and understand their subordination status and to see the possibility of creating successful alliances to bring about social change.

Underlying this process is a necessity of a more participatory, dialogical, non-directive, and horizontal communication space in which all "powerless" women in a group could have a conversation with each other to promote a collective sense of consciousness-raising. In this vein, the Internet is claimed to offer the potential for participants to seek to engage in a constructive dialogue, which in turn facilitates the building of mutual relationships, allies, and collaborations. This is particularly evident in the use of "comments." In her research on the ways in which commentary occurs between Swedish female top bloggers and readers, Lövheim (2013) finds that the Internet helps women to build self-centered social relations that serve as a possible generator of a sense of empowerment.

### 19.3.2 Constructing a Public Identity: Emergence of Social Empowerment

The foregoing analysis draws our attention to the communication processes with which women become aware of their subordinated position and thus form a collective consciousness, which deepens their sense of empowerment. In this respect, it is important to note that these processes occur in the context of social relations. Thus, the processes of women's empowerment also involve the construction of public identities through which both individual and group actors can produce a desired impact on others (see Buckingham 2008; Jenkins 2008). This is very relevant to the ways in which women produce their subjective self through social practices within a cultural framework. It is worth noting here that while cultural norms modify women's social interactions, they do not determine them. Instead, individual women do not merely draw from but transcend gendered roles and norms to manage the self-impressions given to other participants in the interaction, which contributes to the development of a creative and reflexive identity (Elliott 2014).

In regard to online space, the open architecture of the Internet is claimed to provide powerful new communication platforms for the projection of personal self-disclosure. For women, the Internet would be seen alternatively as a manipulated resource that allows them to think in new ways about their identities. There is some early work investigating the ways in which women can use the text-based online environment as a playground to experiment with different gender identities regardless of the existence of their physical bodies (e.g., Danet 1998; Nakamura 1999; Turkle 1995). Current work has, in contrast, focused on the embodiment of women's everyday experience as an online performance of their gender or sexuality (e.g., van Doorn et al. 2007, 2008). More recent studies in this field present a commonly adopted approach to the ways in which women can perform their gendered and sexualized identities in online contexts, either constrained or empowered by the rules of online platforms and the opportunities they offer, and in offline contexts (Dobson 2013; Paechter 2013).

In other words, within online environments, people, especially those who are marginalized, have come to realize the importance of online self-presentation, as they have a great opportunity to become visible to others. For these people, "the performance of identity and practicing code-switching are a matter of survival and a way of life" (Mann 2014, p. 293). We can advance these important points by noting that the Internet constitutes a new source of power for women to exert control over their identities, which comes close to realizing the Internet's communication capacity for women's empowerment.

Furthermore, many scholars have noted the capacity of digital media for the promotion of the self and the management of personal reputation and their potential for public presence and popularity (e.g., van Dijck 2013). This point has also been supported by some empirical evidence in relation to women, whereby self-promotion is a crucial aspect for female actors to meet their social needs and interact with the outer world (e.g., Carah and Dobson 2016; Marwick 2015). In turn, these experiences of integration into public space and the satisfaction of social needs pertain to women's feelings of competence, confidence, and assertiveness, which subsequently translate into a sense of empowerment.

### 19.3.3 New Dynamics of "Power with" for the Organization of Women's Collective Action

The possible mechanisms underlying the relationship between the Internet and women's empowerment discussed so far outline an analytical framework that helps us understand the "power to" sense of empowerment through digital communication. That said, the communicative affordances of the Internet confer a high degree of autonomy to women who now can engage in an array of activities to express their concerns and perform identities. To be fair, this individual facet can only present a partial picture of women's empowerment, as becoming empowered requires more than an enhancement in a sense of agency. This subsection goes beyond women's experiences of personal empowerment and speaks to the political-collective possibilities brought about by the communicative potential of the Internet for the goal of collective empowerment. Writings on the role of digital communication in collective action can be helpful here, as they can turn our attention to the communicative mechanisms that underlie the development of "working together" – an important means by which to profoundly impact the system of male domination.

It has been widely documented that collective action in the new communication environment is characterized by diversity, decentralization, fluidity, and open communication (Fenton 2016; Papacharissi 2015; Tufte 2017). This synergy also heralds a new world of feminist movements and activism – a salient aspect of "power with," as the Internet makes it easier for women, and young women in particular, to be actively engaged in micro-politics and less formal activities concerning lifestyle issues (Baer 2016; Fotopoulou 2016). All these cases suggest that the emergence of digital media has opened up a discursive avenue for women through which to arrive at common understandings of their problems, which in turn might help forge alliances, coalitions, and networks based on the common causes resulting from dialogue and knowledge exchange occurring in online spaces.

In digital environments, perhaps the most striking aspect of the change has been the emergence of heterogeneous forms of organizations that facilitate women's collaborative work, bonding, and coalition. At the outset, the Internet helps women's organizations share knowledge in relation to feminism and gender with individuals, which fundamentally reflects their capability to manage problem perceptions and their ways to advance a political agenda for women (see Foot and Schneider 2006). In this respect, women's organizations would play a critical role in constructing issues, leading to powerful sources of identity for a movement's own constituency (see Ganesh and Stohl 2013).

Subsequently, organizations become important actors to persuade potential participants to join in, make an effort, remain, and manage and coordinate their contributions and resources (Bimber et al. 2012). In this way, women's organizations can use the Internet to facilitate the process of interaction with individuals, to consolidate the sense of being in a collective. This is because, to challenge a male-dominated structure, not only do women's organizations have standpoints but involved participants have their own perspectives as well. Thus, one process vital to an organization is the learning gained from its members' perceptions, opinions, and viewpoints for joint agenda setting, so that both organizations and individuals have a clearer idea about what they are going to achieve (Papa et al. 2000). As such, through the use of the Internet, women's organizations can actively engage in the production and maintenance of values, causes, and cultural trends in association with gender issues to guide their participants as one overall unity.

In addition, the Internet allows participants to take part in virtual, ephemeral actions, such as online petitions, memes, or practices of political satire (Earl et al. 2014). These online tactics, according to Earl et al. (2014), are particularly suitable for authoritarian contexts, as they are "harder for authorities to control and less costly for protesters than street protests" (p. 2). Following this logic, the Internet enables women's organizations to mobilize participants into flash mobs to express their concerns, which is also an efficient way to escape the control of government censorship.

Finally, the connective affordances of the Internet allow women's groups to form interorganizational coalitions and alliances to pursue their political and social aims (Fotopoulou 2016). In this process, organizational actors can form two types of networks: "a hyperlink network" and "a frame network" (Ackland and O'Neil 2011). In the former, women's groups can simply use hyperlinks to connect with each other without exchange of ideas, resources, and information. In this sense, hyperlinks are just a public strategy to signal who is fighting the same cause. In contrast, in the latter, it is the "mutual use of a particular frame component" (Ackland and O'Neil 2011, p. 180) that links organizations into a network.

## 19.4 Digital Communication and the Emergence of a Political Process of Women's Empowerment?

Then, how can we demonstrate the explanatory value of the proposed framework to study the relationship between digital media and women's empowerment? In doing so, women's groups in the context of China will be used as examples to illustrate how they use the Internet to persuade participants to join in, negotiate with them, construct common issues, and collaborate with allies to exercise collective power.

The women's groups were chosen for their common goal to promote gender equality or change the status quo of women in Chinese society but also for their differences in causes, organizational resources, and campaign tactics. In addition, they were selected in light of four organizational types based on their relationship with the government and the organizational structure: "government agencies," "civic associations," "university-based centers," and "online communities." These organizational types serve as a way to shape female group actors' political purposes and priorities in relation to a breadth of Chinese women's issues, including gender equality, domestic violence, sexual abuse and harassment, law reforms, women's development, homosexual rights, poverty of rural women, and equal pay.

### 19.4.1 Providing Organizational Information for Identity Projection

From the outset, women's groups showed their desire to catch up with technological development for fear of being left out, as they were already in a periphery position in the Chinese society. In their eyes, showing their online presence seemed to be the "only choice" to draw public attention. Through the use of online platforms, these studied organizational actors had opportunities to present information about themselves and emphasize their identities. However, given the different organizational types involved, the underlying purposes in this regard were varied. Represented as the mouthpiece for China's Communist Party (CCP), the "government agencies" predominantly used the Internet as a new marketing tool in public relations exercise. The aim of their online content was to present their work accomplishments, "singing the praises of the government." Hence, it is no surprise that engaging or exchanging ideas with the audience was not the prioritized focus of their use of the Internet.

This ran in sharp contrast to "civic associations": although this type of actor also relied heavily on the Internet to present information about the group to the general public, their underlying intention was to "make their voices heard." However, because of government control, they did not have sufficient money to develop their online platforms into a resourceful space. At times, several groups needed to censor their online content, as their organizational goals were to promote law reforms – a key target of online surveillance in China. Worse still, "civic associations" had to compete with big commercial sites to be heard, which was also the case for "online communities." Thus, even though "civic associations" and "online communities" showed their willingness to draw public support, a lack of resources and marginalized position constrained them.

### 19.4.2 Disseminating Information to Raise Awareness of Gender Issues

At the same time, for women's groups, online platforms were used to shape audiences' perceptions of gender ideas, beliefs, and values and served as an important means of generating women's awareness about their subordinate or even oppressed status. Women's groups frequently updated the postings in retweeted news, commentaries, and articles to enhance the visibility of various women's issues to their visitors. Among these issues, the most important was to educate the public about Chinese gender equality, which had been established as a basic state policy since 1995.

However, different types of organizational actors approached this issue from different perspectives. For "government agencies," the focus was mainly on policy aspects to introduce the government's efforts to protect women's rights and interests, such as having the same retirement age for men and women, women's health, and women's participation in decision-making. However, again, this can be interpreted as an alternative strategy to present the CCP's achievement in the promotion of gender equality in China.

In the case of "civic associations," more light was shed on specific issues and on addressing different groups of Chinese women. These included fighting domestic violence, rural women's rights, sexual harassment in the workplace, equal pay, women's development, and stereotyped gender representations in the media. A possible reason for this may be that all these issues have been closely related to women's lives and circumstances since the economic reform era in the late 1980s and are written into the national legislative framework. Thus, although "civic associations" published information on a wide range of women's issues on their online platforms, they still followed structural rules to help women gain greater control over their own lives. In other words, the views, ideas, and beliefs they spread through the Internet were within the governmental framework but did not challenge or change the existing structure.

In terms of "university-based centers" and "online communities," they paid more attention to reshaping the discourse on gender. For instance, the group actors "university-based centers" used their online platforms as a public lecture space to spread proper knowledge on feminism. Apart from feminist rights, "online communities" also mobilized participants to debate issues of gay rights and sexual rights but only attracted a small group of followers.

### 19.4.3 Creating a Sense of the Collective Through the Internet

Furthermore, women's groups also used online platforms to facilitate communication with individual supporters, but there was an absence of a collective sense created between them. On the surface, the Internet allowed for greater ease of affiliation between women's groups and individuals. While all groups welcomed individuals to directly email them, a limited number provided the general public with access to register as "official members." This might be the result of the way in which interested people registered. To become part of the "inner ring" of certain groups, people needed to offer detailed personal information, which in turn put their privacy at risk. In this sense, the majority of women's groups provided much easier access to encourage people to join them, such as becoming email subscribers or followers of their social media accounts.

In addition to using online platforms as a way for like-minded people to form an aggregate, many of my studied women's groups also realized the potential of the Internet to construct interaction with people to reinforce a sense of the collective. However, the results show that these groups were reluctant to develop interactive features on their social networking sites (SNSs), even though they knew that they ought to do so. The main reason for this was the *public character* of these sites, in which readers' comments were visible to all, including the Internet police. The situation was worsened if some commentators posted messages against the government – which would put the group at risk of being blocked, as the group's SNS coordinators could not control what they could say. This contextual factor hindered many groups from fully developing the interactive features on their SNS. Furthermore, a lack of resources was another reason for the groups' reluctance to increase the degree of online interactivity. In addition, although social media could facilitate communication with audiences in a more efficient manner, some groups expressed their concerns over the tension between speed and organizational credibility.

However, this does not mean that the Internet did not play a role in developing strong ties between women's groups and their supporters. Instead, the empirical evidence shows that a few groups turned to more private communication platforms (e.g., QQ discussion group) to maintain participants' engagement through which mutual interactions emerged between them, something that was, at times, accompanied by face-to-face communication.

### 19.4.4 A New Space for Political Campaigns Within an Authoritarian Context?

In regard to campaigning, it is clear that women's groups, aside from "government agencies," included limited conservative online actions on their online platforms to mobilize individuals to take part in their activities. Among the few conservative features, "plans of offline activities" was the most relevant, particularly for "civic associations." The reason this is called "conservative" is more evident in this respect, as the activities presented online were mostly lectures, seminars, photography/drawing contexts, community services, or training activities – none were related to "protests."

Thus, in the online public space, it was impossible for women's groups to show "protests" or "radical" elements. However, it is worth keeping in mind that all of the information about activities published online were about plans but not actions – these instead occurred in the offline realm. From this perspective, no one could promise that the content of offline actions would not be related to "radical" elements, even if the activities or events organized by women's groups were in conservative form. However, online censorship did not diminish the need for women's groups to organize ephemeral campaigns (e.g., e-petitions) to express their concerns. Sometimes, some "civic associations" used their online platforms as broadcasting platforms to disseminate information about their offline lobbying activities, aiming to present their efforts to change the structural constraints on women's status.

### 19.4.5 Developing Interorganizational Relations

Finally, although the Internet relaxed the constraints of geographical distance to facilitate connection building among organizations who fight for a common cause, the women's groups I studied were not obsessed with using their online platforms as a base for alliance-building. Instead, most of them heavily relied on face-to-face contact to build, maintain, or reinforce interorganizational relations. Still, it is worth noting here that the data revealed that hyperlinks did indeed provide new opportunities for women's groups to connect with each other and to form an information-sharing circle among themselves, an activity that could form a virtual interorganizational relationship. Nonetheless, the fact remains that such collectivity, regardless of "the virtual" or "the real," was placed at the periphery of Chinese society. Hence, it was still a long process for them to claim a legitimate political voice and bring about social change.

## 19.5 Conclusion

Women's groups, albeit with different positions and prioritized goals, are under no illusion that the Internet harbors the source of success for women's empowerment. For one thing, the Internet has not been used as an effective means to interact with prospective individual participants or to help develop the sense of a collective. The fact remains women's groups have predominantly treated their online platforms as a cost-efficient tool to enhance group recognition. In contrast, instead of the Internet, women's groups have primarily relied on face-to-face communication to form interorganizational alliances. Thus, in the present study, the potential for the Internet to form "coalition politics" (Yuval-Davis 1994) to bring about changes in existing structures of power is rather limited.

Therefore, the Internet is claimed to provide potential, opportunity, and new practices for women, but it falls short of determining the emergence of women's empowerment. We also should keep in mind that the research on the digital media and women's empowerment is connected indelibly to the context. For example, in China, the promise of empowerment is mainly undermined by state control. Contextual factors prevent women's groups from fully realizing the possibilities of the Internet for increasing their organizational visibility, promoting public awareness about gender issues, building a sense of the collective, campaigning, and networking. In this respect, it is also important to note that exploring the relationship between women's empowerment and digital media has generated a complex picture, rich in detail but difficult to capture entirely.

## References

Ackland R, O'Neil M (2011) Online collective identity: the case of the environmental movement. Soc Networks 33:177–190

Allen A (1998) Rethinking power. Hypatia 13(1):21–40. https://doi.org/10.1111/j.1527-2001.1998.tb01350.x

Allen A (2005) Feminist perspectives on power. In: The Stanford encyclopedia of philosophy. Metaphysics Research Lab, Stanford University. https://plato.stanford.edu/archives/fall2016/entries/feminist-power/

Antunovic D, Hardin M (2013) Women bloggers: identity and the conceptualization of sports. New Media Soc 15(8):1374–1392. https://doi.org/10.1177/1461444812472323

Baer H (2016) Redoing feminism: digital activism, body politics, and neoliberalism. Fem Media Stud 16(1):17–34. https://doi.org/10.1080/14680777.2015.1093070

Batliwala S (2007) Taking the power out of empowerment – an experiential account. Dev Pract 17(4–5):557–565. https://doi.org/10.1080/09614520701469559

Bimber B, Flanagin A, Stohl C (2012) Collective action in organizations: interaction and engagement in an era of technological change. Cambridge University Press, New York

Brickell C (2012) Sexuality, power and the sociology of the Internet. Curr Sociol 60(1):28–44. https://doi.org/10.1177/0011392111426646

Buckingham D (2008) Introducing identity. In: Buckingham D (ed) Youth, identity, and digital media. The MIT Press, Cambridge/Malden, pp 1–24

Carah N, Dobson A (2016) Algorithmic hotness: young women's "promotion" and "reconnaissance" work via social media body images. Soc Media Soc 2(4):2056305116672885. https://doi.org/10.1177/2056305116672885

Carr ES (2003) Rethinking empowerment theory using a feminist lens: the importance of process. Affilia 18(1):8–20. https://doi.org/10.1177/0886109902239092

Collins PH (2000) Black feminist thought: knowledge, consciousness, and the politics of empowerment. Routledge, New York/London

Couldry N (2008) Mediatization or mediation? Alternative understandings of the emergent space of digital storytelling. New Media Soc 10(3):373–391. https://doi.org/10.1177/1461444808089414

Danet B (1998) Text as mask: gender, play, and performance on the Internet. In: Jones S (ed) Cybersociety 2.0: revisiting computer-mediated community and technology. SAGE, Thousand Oaks, pp 129–158

Dobson AS (2013) Performative shamelessness on young women's social network sites: shielding the self and resisting gender melancholia. Fem Psychol. https://doi.org/10.1177/0959353513510651

Earl J, Hunt J, Garrett RK, Aysenur D (2014) New technologies and social movements. In: della Porta D, Diani M (eds) The Oxford handbook of social movements. Oxford University Press, Oxford

Elliott A (2014) Concepts of the self, 3rd edn. Polity, Cambridge/Malden

Fenton N (2016) Digital political radical. Polity, Cambridge/Malden

Foot K, Schneider SM (2006) Web campaigning. The MIT Press, Cambridge, MA/London

Fotopoulou A (2016) Feminist activism and digital networks: between empowerment and vulnerability. Palgrave Macmillan, Basingstoke

Fraser N (2000) Rethinking recognition. New Left Rev 3:107–120

Fraser N (2009) Feminism, capitalism and the cunning of history. New Left Rev 56:97–117

Ganesh S, Stohl C (2013) Community organizing, social movements and collective action. Paper presented at the annual meeting of the International Communication Association, London

Ganesh S, Zoller HM (2012) Dialogue, activism, and democratic social change. Commun Theory 22(1):66–91. https://doi.org/10.1111/j.1468-2885.2011.01396.x

Hasinoff AA (2014) Contradictions of participation: critical feminist interventions in new media studies. Commun Crit/Cult Stud 11(3):270–272. https://doi.org/10.1080/14791420.2014.926242

Jenkins R (2008) Social identity, 3rd edn. Routledge, London/New York

Kabeer N (2010) Women's empowerment, development interventions and the management of information flows. IDS Bull 41(6):105–113. https://doi.org/10.1111/j.1759-5436.2010.00188.x

Levina M (2014) From feminism without bodies, to bleeding bodies in virtual spaces. Commun Crit/Cult Stud 11(3):278–281. https://doi.org/10.1080/14791420.2014.926243

Lövheim M (2013) Negotiating empathic communication: Swedish female top-bloggers and their readers. Fem Media Stud 13(4):613–628

Lundby K (2008) Introduction: digital storytelling, mediatized stories. In: Lundby K (ed) Digital storytelling, mediatized stories: self-representations in new media. Peter Lang, New York, pp 1–17

Mann LK (2014) What can feminism learn from new media? Commun Crit/Cult Stud 11(3):293–297. https://doi.org/10.1080/14791420.2014.926244

Mansbridge J (1993) Feminism and democratic community. Nomos 35:339–395

Marwick AE (2015) Instafame: luxury selfies in the attention economy. Publ Cult 27(1 (75)):137–160. https://doi.org/10.1215/08992363-2798379

Nakamura L (1999) Cyberspace: race, ethnicity, and identity on the Internet. Routledge, New York

Paechter C (2013) Young women online: collaboratively constructing identities. Pedagogy Cult Soc 21(1):111–127. https://doi.org/10.1080/14681366.2012.748684

Papa MJ, Singhal A, Ghanekar DV, Papa WH (2000) Organizing for social change through cooperative action: the [dis]empowering dimensions of women's communication. Commun Theory 10(1):90–123

Papacharissi Z (2010) A private sphere: democracy in a digital age. Polity, Cambridge/Malden

Papacharissi Z (2015) Affective publics: sentiment, technology, and politics. Oxford University Press, New York

Phillips A (1992) Must feminists give up on liberal democracy? Polit Stud 40(1_Suppl):68–82. https://doi.org/10.1111/j.1467-9248.1992.tb01813.x

Portwood-Stacer L (2014) Feminism and participation: a complicated relationship. Commun Crit/Cult Stud 11(3):298–300. https://doi.org/10.1080/14791420.2014.926246

Sinanan J, Graham C, Zhong Jie K (2014) Crafted assemblage: young women's 'lifestyle' blogs, consumerism and citizenship in Singapore. Vis Stud 29(2):201–213

Stavrositu C, Shyam Sundar S (2012) Does blogging empower women? Exploring the role of agency and community. J Comput-Mediat Commun 17(4):369–386. https://doi.org/10.1111/j.1083-6101.2012.01587.x

Tufte T (2017) Communication and social change: a citizen perspective. Polity, Cambridge/Malden

Turkle S (1995) Life on the screen: identity in the age of the Internet. Simon & Schuster Paperbacks, New York

van Dijck J (2013) 'You have one identity': performing the self on Facebook and LinkedIn. Media Cult Soc 35(2):199–215. https://doi.org/10.1177/0163443712468605

van Doorn N, van Zoonen L, Wyatt S (2007) Writing from experience: presentations of gender identity on weblogs. Eur J Womens Stud 14(2):143–158. https://doi.org/10.1177/1350506807075819

van Doorn N, Wyatt S, van Zoonen L (2008) A body of text: revisiting textual performances of gender and sexuality on the Internet. Fem Media Stud 8(4):357–374. https://doi.org/10.1080/14680770802420287

Young IM (1990) Justice and the politics of difference. Princeton University Press, Princeton

Young IM (1994) Punishment, treatment, empowerment: three approaches to policy for pregnant addicts. Fem Stud 20(1):33–57

Yuval-Davis N (1994) Women, ethnicity and empowerment. Fem Psychol 4(1):179–197

# Development Communication and the Development Trap

## 20

Cees J. Hamelink

## Contents

| | | |
|---|---|---|
| 20.1 | Development Communication and Development | 396 |
| 20.2 | Development as Delivery | 398 |
| 20.3 | Development Trap | 400 |
| 20.4 | From Envelopment to Development | 402 |
| 20.5 | Development Communication Agencies as Convivial Institutions | 404 |
| References | | 405 |

### Abstract

The basic argument of this chapter is that development communication as "part of overall development" (UNGA 1962) can only contribute to "positive social change" if the field manages to escape from the "development trap." A constitutive element of this powerful trap is the colonial conception of development as external intervention in a process of linear transfer. This process contributes more to "envelopment" than to genuine development as mobilization of inherent potentialities. For a productive way, forward development communication needs to be based upon the grounds of evolution theory and postcolonial cosmopolitanism. This requires rethinking development as an evolutionary adaptive process, reconceptualizing communication as a complex adaptive system, and challenging agencies operating in the field of development communication to transform into convivial institutions.

---

The colonizers came to wither the flowers. To nourish their own flower, they harmed and sucked the flower of others. Chilam Balam, Maya prophet, 1500

C. J. Hamelink (✉)
University of Amsterdam, Amsterdam, The Netherlands
e-mail: cjhamelink@gmail.com

© Springer Nature Singapore Pte Ltd. 2020
J. Servaes (ed.), *Handbook of Communication for Development and Social Change*,
https://doi.org/10.1007/978-981-15-2014-3_30

**Keywords**

Convivial institutions · Darwinian evolutionary theory · Digital solidarity fund · Envelopment · MDGs · Sustainable Development Goals

## 20.1 Development Communication and Development

Development communication is the container descriptor of projects, strategies, and policies that use human communication – in a multitude of formats – to achieve positive social change. (In most definitions of development communication, there is a strong link with the overall field of international development. In most definitions of development communication, one finds a strong link with overall international development. According to the World Bank, development communication is the "integration of strategic communication in development projects" based on a clear understanding of indigenous realities. And UNICEF views it as "...a two-way process for sharing ideas and knowledge using a range of communication tools and approaches that empower individuals and communities to take actions to improve their lives." In UNDP documents, the term "development support communication" was used to refer to the role of communication in promoting UN agricultural and development programs.) Nora C. Quebral – sometimes referred to as the mother of development communication – defined the field as "the art and science of human communication applied to the speedy transformation of a country and the mass of its people from poverty to a dynamic state of economic growth that makes possible greater social equality and the larger fulfillment of the human potential" (2001). The field of development communication finds its origin in the concern of UN member states to "develop media of information in underdeveloped countries" (United Nations General Assembly, -UNGA- 1957]. In 1962 the UNGA stated that "development of communication media was part of overall development." From here, an impressive array of communication projects came to be executed, funded, and evaluated by a growing community of international states and nongovernmental institutions. Development communication projects have focused on such issues as mass poverty, malnourishment, women's empowerment, educational capacity building, agricultural extension, mortality of women and their babies, and illiteracy. Its agenda coincides pretty much with the global MDGs (The eight Millennium Development Goals (MDGs) were established following the Millennium Summit of the United Nations in 2000 and the adoption of the United Nations Millennium Declaration. All 191 United Nations member states at that time, and at least 22 international organizations, committed themselves to help achieve the following goals. For the post-2015 period, the UN member states adopted in September 2015 "Transforming our World: The 2030 Agenda for Sustainable Development. The Agenda 2030 establishes 17 Sustainable Development Goals to be achieved over the coming 15 years." According to former UN Secretary-General Ban Ki-moon "The seventeen Sustainable Development Goals are our shared vision of humanity and a social contract between the world's leaders and the people. They are a to-do list for people and planet, and a blueprint for success." Early 2016 the SDGs went into effect.):

1. Eradicate extreme poverty and hunger.
2. Achieve universal primary education.
3. Promote gender equality and the empowerment of women.
4. Reduce child and maternal mortality.
5. Improve maternal healthcare.
6. Combat HIV and AIDS, malaria, and other major diseases.
7. Ensure environmental sustainability.
8. Develop global partnership for development NOTE ON MDG.

The MDGs represent positive social change, and the field of development communication is by and large fueled by the need to contribute to positive social change.

The irony of the field is that development communication refers – in the colonial perception of the world – to social change to be achieved in the global South. There are few if any communication projects that are initiated by consultants from the South to achieve social change in countries of the global North. And yet there are numerous countries like the member states of the EU or the USA that would benefit from positive social change to deal with such burning issues as rising inequality, drug abuse, organized crime, racism, populism, refugee policies, banking crises, dwindling space for civil action, or urban violence. In all these issues, failing communication (fake news, commercial censorship, political propaganda, or the lack of social dialogue) plays a crucial role. However, if one surveys the field, then judging by numbers of projects and volumes of financial resources development communication was and remains Northern business and very much part of international aid programs that have been described as "the circus that never leaves town."

The targets of the MDGs, for example, practically all focus on changes in the global South and ignore that the problems of the South may only be solved in case the global North implements radical changes in its trade policies, military projects, and environmental politics and faces the reality of the colonial mind which is still prevailing in North-South relations. The key aspects of colonial rule continue to shape global power relations. Most colonizers never really gave up, and many of the colonized became accomplices in continued external domination. More often than not the Western image of how humanity should live continues to be the standard. Needed is – according to Dutta – a "postcolonial approach to communicative processes" that examines the colonial discourses that circulate and reify the material inequities across the globe (2011, 5).

The global contemporary development effort is best represented by the MDGs that were mentioned above. However, these laudable were not achieved – as hoped by 2015 – and were relabeled Sustainable Development Goals for All. There are evidently – as always – exceptions where on a small scale through the application of local knowledge by local communities, remarkable changes have been realized. What has been called the "world's largest promise" is for its realization obviously dependent upon communication in the broad sense of information provision, conscientization, stimulation, and conversation. To achieve the positive social change that the MDGs and the new SDGs strive toward communication is a crucial tool (In the context of the MDGs or SDGs, communication is rarely mentioned except for the references to advanced information and communication technologies. The emphasis on the need to use the ICTs is remarkable in the light of the absence of concrete evidence that these technologies have substantially contributed to the changes their protagonists promised.), but with the exception of community radio projects (Good illustrations are found in Jallov 2012.) across the globe, the dominant media continue to function as conduits for cultural invasion and globalization as neoliberal market integration.

## 20.2 Development as Delivery

In the debates on the global digital divide (during the United Nations World Summit On The Information Society (2003 and 2005)), it was not critically questioned whether rich-poor divides can at all be resolved within the framework of the prevailing development paradigm. In this paradigm development is conceived of as a state of affairs which exists in society A and, unfortunately, not in society B. Therefore, through some project of intervention in society B, resources have to be transferred from A to B. Development is thus a relationship between interventionists and subjects of intervention. The interventionists transfer such resources as information, ICT, and knowledge as inputs that will lead to development as output. In this approach, development is "the delivery of resources" (Kaplan 1999, 5–7). This delivery process is geared toward the integration of its recipients into a global marketplace. There is no space for a different conceptualization of development as a process of empowerment that intends "to enable people to participate in the governance of their own lives" (Kaplan 1999, 19). In the WSIS discourse there was a strong tendency to consider the global digital disparity as a problem in its own right. This divide was not primarily seen as a dimension of the overall global "development divide." Since this bigger problem was not seriously addressed, a romantic fallacy prevailed which suggested that resolving information and communication problems, and bridging knowledge gaps or addressing inequalities in access to technologies, can contribute to the solution of the world's most urgent and explosive socioeconomic inequities. This "ghettoization" of communication ignores the insight that the solution of the "global development divide" has little to do with information, communication, or information and communication technology. This divide is a matter of political will which is lacking in a majority of nation-states. Instead of the strong political commitment that is needed, the WSIS discourse focused on the possibility of a global "Digital Solidarity Fund." This is an almost scandalous proposition in view of the fact that since the 1970s all the efforts to develop and sustain such funds for communication development, telecom infrastructures, or technological self-reliance have failed because of the lack of political will. The WTO ministerial meeting in Cancún (September 2003) demonstrated once again that not all stakeholders are equally intent on solving rich-poor divides. As Walden Bello commented, "Not even the most optimistic developing country came to Cancún expecting some concessions from the big rich countries in the interest of development" (Bello 2003, 16). Fortunately, the poor countries understood that the rich countries (particularly the USA and the EU countries) intended to impose yet another set of demands on them that would be very detrimental to their societies and their people. In this sense the Cancún meeting was a great success. That same sense of alertness did not inspire the poor country representatives at the December 2003 WSIS.

A difficult problem is that if indeed greater global equality in access to information could be achieved, this would not guarantee an improvement in the quality of people's lives. "Even when these disparities are recognised and new organisational models such as telecentres are proposed, the policy emphasis is frequently biased towards improving access to networks rather than towards content creation and the social processes whereby digital content can be converted into socially or economically useful knowledge" (Mansell and Wehn 1998, 8). Including people in the provision of basic public services does not create egalitarian societies. The existing social inequality means that people benefit from these services in highly inegalitarian ways. Actually, the growing literacy in many societies did not bring about more egalitarian social relationships. It certainly did have some empowering effect, but did not significantly alter power relations. Catching up with those who have the distinct social advantage is not a realistic option. They too use the new developments, such as ICTs, and at a minimum the gap remains and might even increase. It is a common experience with most technologies that the powerful players know best how to appropriate and control new technological developments and use them to their advantage. In the process they tend to further increase their advantage. The conventional approach to development may correct social disadvantages through the equal treatment of unequal but does not structurally change relations of power. Providing equal opportunities to unequal parties often functions in the interest of the more powerful.

The conventional discourse has no strong interest in the cosmopolitan ideals of communal responsibility and collective welfare. A cosmopolitan discourse accepts reciprocal obligations among the members of society. The essential issue of cosmopolitanism is the conversation with the other. The question however is on whose terms is this global conversation conducted. Often the engagement with the other – despite assurances of participation, deliberation, and empowerment, managed by a dominant, missionary culture – thus remains a colonial adventure. Establishing a complex adaptive framework for development communication in the twenty-first century requires a postcolonial [institutional] mindset that accepts reciprocal obligations in development and communication. It needs a global community that realizes that the destinies of the powerful and the powerless are intertwined.

In his final report on the MDGs, the UN Secretary-General (in 2010) did not address the question of the adequacy and effectiveness of the conceptual framework within which the goals are embedded. The report stated that the goals are achievable and that the shortcomings in progress can be fully explained by a lack of political will, insufficient funds, a lack of focus and sense of responsibility, and insufficient interest in sustainable development. However, this is only part of the explanation. More than anything else, the "development trap" stood in the way of achieving positive social change.

## 20.3   Development Trap

The development goals as formulated by the international community are laudable moral aspirations that are possibly achievable, and communication (in different formats from information provision to social dialogue) is undoubtedly a crucial factor in realizing them. However, the "world's largest promise" cannot become reality as long as overall development (of which development communication is part) is captured by a colonial conceptualization of development as an interventionist, linear process. Too many development communication projects were and are based on the colonial mental model that assumes that one can develop others and that social change can be achieved through external intervention. However, development is always from within. It is an inherent process and "In that sense, you cannot grow potatoes. Potatoes grow themselves" (Sankatsing argues, 2016, 34). Genuine development is a process of natural evolution that has been largely obstructed by development projects and policies. As Sankatsing proposes, "Development is the mobilization of inherent potentialities in iterative response to challenges posed by nature, habitat and history to realize a sustainable project with an internal locus of command" (2016, 35). Against this what has been called development by international agencies and development scholars is better described as "envelopment." As Sankatsing defines it "Envelopment is the paternalistic, disempowering control of an entity by an external locus of command at the expense of its internal life process and ongoing evolution" (2016, 38). Contrary to a process of unfolding inherent human potentialities development as envelopment became a "unidirectional process of transformation by incorporating the other into an alien destiny" (2016, 38). Envelopment is disruptive "since it prevents a community from responding in a natural way to contextual conditions and environmental challenges" (2016, 39). Many development communication projects – as part of overall development projects – fell in the trap where development was really envelopment and where positive social change in the evolutionary sense was seriously obstructed. It is tempting to counter this reasoning by pointing to successful "empowerment through communication" projects, but also here the focus is usually more on the empowerment of the other and not on the self-empowerment of the other. Empowerment projects often take place in asymmetrical relations and hamper "a process in which people liberate themselves from all those forces that prevent them from controlling decisions affecting their lives" (Hamelink 1994, 142).

As part of overall development, development communication can only escape the development trap if the field is reconceptualized as an evolutionary, complex adaptive process and implicitly accepts nonlinearity in thinking about communication. This means that development communication should be based on the solid scientific basis of evolutionary thinking. The most general observation in Darwinian biology is that species (and their behaviors) evolve over time through successful adaptation to their environment. The key to biological evolution is the finding of solutions to adaptive problems.

It seems sensible to argue that a similar process occurs in forms of nonbiological evolution, such as cultural and psychological evolution. In these forms, human beings find nongenetic solutions to adaptive problems. One such adaptation is human communication: an evolutionary response to problems in our environment. Communication is essentially linked to human evolution: evolutionary theory. The origins of human communication can only be meaningfully approached from evolutionary theory. Although "evolutionary communication is a powerful theoretical perspective that applies to all forms of biological and social interaction" (Lull and Neiva 2012, 16) and evolutionary theory "is one of the most powerful scientific explanations ever put forward" (ibid., 16), there is no impact on communication theory. The social sciences have marginalized biology and particularly Darwinian evolutionary theory from its efforts to understand the world. Therefore, Lull and Neiva argue "for a major infusion of Darwinian thinking into communication theory..." (ibid., 16).

Communication is a crucial instrument in both our biological and cultural evolution, and yet the social sciences and the humanities have been unable to provide a solid understanding of how culture evolves. The thinking has been too much focused on the individual and not on the group. Cultural evolution and with it the evolution of human communication would be best studied at the level of the group since human life is characterized by living in large groups (nations, corporations, tribes). The focus of much research has been the individual, whereas human communicative behavior can be best understood if studied as behavior in social groups (see Richerson and Boyd 2006 on "population thinking"). The basic Darwinian algorithm for successful adaptation is based upon variation, selection, and replication. In the domain of human communication this can be applied as follows. Communicative behavior evolves through variation. A great variety of modalities of communication evolve because of the need to adapt to different and changing environments. This evolution is both nonintentional and intentional and limited by both genotypical and historical factors. Much of it proceeds (as in the evolution of knowledge in science) by trial and error. In the evolutionary process, the best adaptive solutions are kept. Communication forms that optimally serve human survival and reproduction be retained. Those forms of attention and memory that are designed to notice, store, and retrieve information inputs and that are useful to solving adaptive problems will further evolve. Inadequate communicative solutions will disappear. The most adequate adaptations will be transmitted to future generations.

The field of development communication equals the complexity of a tropical rainforest in which everything is related to everything else (interdependence), where small events may have big and unpredictable effects (nonlinearity) and where one cannot make reliable forecasts as flows of ideas and opinion may unexpectedly and rapidly change (uncertainty). A complex system is a collection (universe) of interacting agents that compete for a scarcity of essential resources (like space in traffic or oxygen + glucose in cancer cells). Characteristic of such systems is a mixture of order and disorder. Without a central controller, the emerging disorder (the traffic jam) may resolve as if nothing had happened and without external interference. The traffic jam appears for no reason and then disappears again (Johnson 2010, 19). Development communication is both orderly and chaotic. The conduct of its agents is dynamic. To understand this complexity, the prevailing scientific preference for determinism, reductionism, and quantification is not helpful to understand nonlinear processes in which unpredictable and surprising phenomena emerge without a central controller (Hamelink 2015, 46).

From the study of evolutionary biology, we learn that human communication was always driven by the instinct to cooperate. The species homo understood early on that their communities would benefit from cooperative communication. Communication made the kind of coordination that hunting required possible and facilitated the organization of complex societies. There is a good deal of evidence to safely suggest that the origin of human communication lies in the instinct to cooperate. Through cooperative communication, humans designed adequate adaptive systems that secured their survival and reproductive capacity. Human communication is a key player in the biological and cultural evolution of humanity (Tomasello 2008). Human cooperative communication emerged as a collaborative activity that was needed for an effective adaptation to rapidly changing environments. Understanding development and by implication development communication as a complex adaptive system helps to see how these processes have to be – almost chaotically diverse, innovative and the dynamic so they can cater for the constant changes in human communicative behavior and the nonlinear and probabilistic nature of human communication.

## 20.4 From Envelopment to Development

Leading the road toward development as the liberation of inherent potentialities is an evolutionary understanding of development as adaptation to external realities. If the field could be reconceptualized, this would also imply a redefinition of the agencies (statal and nongovernmental) that design and implement strategies for the application of communication tools to development.

As development communication is practically always an organizational effort, it is important to address the quality of institutional formats that prevail in these organizations.

Large protocol-driven bureaucracies (such as UN agencies and the larger INGOs) are well equipped for envelopment but not equally well for development. As Ramalingam argues that what is most needed is the creative and innovative transformation of how aid works. Foreign aid today is dominated by linear, mechanistic ideas that emerged from early twentieth-century industry and is ill-suited to the world we face today. The problems and systems aid agencies deal with on a daily basis have more in common with ecosystems than machines: they are interconnected, diverse, and dynamic; they cannot be just simply re-engineered or fixed. A crucial element in institutional transformation will have to be the way in which policies, strategies, and projects will be evaluated in terms of their contribution to positive social change. Today, the key agencies have not developed reliable instruments for assessing effects of communicative interventions. The tendency to quantification and thus simplification makes it impossible to seriously monitor complex processes such as development communication. The agencies and their result-based management strategies and evaluation instruments focused on the measurement of quantity, and "...despite technically sophisticated efforts to develop tools, techniques and frameworks, the majority of accountability solutions and practices retain an upward focus to donors" (Ramalingam 2015, 120).

Institutions evolved as adaptive responses to the need to cooperate (in order to make communal living possible), to the need for collective action (for survival purposes; in hunting big game and distributing it; defense against predators or con-specific bullies), and to the need of social behavior. Pro-social behavior stands for cooperation, altruism, generosity, trust, and trustworthiness. Antisocial behavior stands for competition, selfishness, distrust, and unreliability. Although institutions such as the law may be intended to prevent antisocial behavior, they may end up engaging in antisocial behavior. The direction that institutionalization has taken in modern societies – in such fields as education, healthcare, or law – tends to produce unsettling antisocial behavior. In development at large and development communication, the institutional factor is important: effective and safe institutions may lead to cooperation, generosity, altruism, trust, and trustworthiness, and poorly functioning, corrupt institutions may lead to the opposite (Jordan et al. 2015, 95).

Human beings are institutionalizing animals. "The role of institutions in human cooperation has received much less attention among experimentalists using economic games than direct and indirect reciprocity. But several recent studies have begun to demonstrate the power that institutional incentives have over human cooperative behavior in the lab" (Ibid., 91). Humans introduced institutions (like law, market, education, social communication, religion, healthcare, politics, family, money, and arts) as abstract notions that were operationally translated into organizations such as courts, media, cultural industries, schools, hospitals, political parties, the family, the banks, or trade bodies. These organizations came to be governed by hierarchical divisions of roles, by policies, codes of conduct, rules, directives, best practices, and feedback processes. Through these organizations we meet the satisfaction of basic needs. Not even our closest associates such as chimpanzees or bonobos design – for the satisfaction of their alimentary needs – agrobusiness conglomerates or mega meat processing organizations. We institutionalize the satisfaction of our needs that are increasingly characterized by organizational criteria such as (applied by the fast food industry) efficiency (we love our clients but they should not hang on too long), predictability (we hate uncertainty: the same French fries across the globe), calculation (the biggest hamburger for the lowest price), and rationalization (we mechanize the frying of burgers and the service to our clients). In addition, modern organizations feature hierarchization (although we know that in hospitals with strict hierarchies more patients die and that airlines with strict hierarchies have more crashes), professionalization (we ask people to outsource their own capabilities to those who know better), and protocolization. The latter is arguably the most essential feature of current institutionalization processes. Experience is replaced by protocols. In educational institutions, the competent teacher is no longer the person with a great experience in teaching but someone who follows the teaching instructions through which his/her teaching activities are monitored and evaluated. In healthcare organizations, the competent physicians are the ones that rely on evidence-based practices and not on experiential knowledge. The road to reliable diagnosis is blocked by protocols designed by outside interim managers. The core of the protocols is the rationalization of accountability. Desire for rational justification stands in the way of creativity and innovation in organizations; in education, it leads to uninspiring and boring teachers; in healthcare institutions, it kills people. The rules of the book are taking the place of the capacity for intelligent assessment based upon experience.

Rules and protocols originate from decisions: often incorrect decisions and often badly informed decisions. The rules once laid down in protocols are no longer open to intelligent human assessment and lead to wasting time in hospitals and universities on filling out forms and ticking off boxes whereas that time should have been used for teaching and caring. Evidently personal assessment –however competent – can be abused and should be subject to some sort of check but not without a reasonable balance between freedom and control. We have to try to undermine the tyranny of playing by the book and open up organizations for the use of intelligent capacities at assessment based upon experience.

## 20.5 Development Communication Agencies as Convivial Institutions

The concept "convivial" denotes the combination of cheerfulness with cooperative helpfulness. The link is important as we know from a range of experiments in social psychology that people who are cheerful tend to be more helpful and cooperative; they evaluate themselves and others more positively and think more creatively. This is precisely what organizations that operate in the field of development and development communication need. Their major challenge is that they are often driven by the idea that bigger is better and try to reach economies of scale through expansion, take-overs, and mergers. However, the scale at which humans can conduct meaningful conversations about cooperation is limited. Technology allows us to communicate with many more people, but our cognitive abilities do not. The suggestion of unlimited social networks is deceptive. Whatever the technology facilitates, our mind will prefer a limited number of people in our networks. Convivial organizations are nonhierarchical places where humans learn, develop, and produce through creative storytelling. People are not seen as resources or as raw processable materials. The convivial organization will dismantle its office of human resources. People are seen as creative artists. The convivial organization will establish an office of creative production. In the convivial organization, transformation is fun. It is realized that when we want people to change types of behavior or mindsets, then we should be able to demonstrate to them that the alternative conduct or mentality is fun. People are willing to do all kinds of (new) things if they are pleasurable. Pleasure is an underrated motivating force in processes of change!

The conclusion is that the "world's greatest promise" can be met if humanity institutionalizes the best moral ideas it has created (such as the global development goals) into organizations in which participants cooperate with cheerfulness because they found the common challenge of using the basis of human existence: cooperative communication to release human potentialities.

## References

Bello W (2003) The meaning of Cancún. South Lett 39:16–18
Dutta M (2011) Communicating social change. Routledge, London
Hamelink CJ (1994) Trends in world communication. Southbound, Penang
Hamelink C (2015) Global communication. Sage, London
Jallov B (2012) Empowerment radio. Empowerhouse, Gudhjem
Johnson N (2010) Simply complexity. Oneworld Publications, Oxford
Jordan J, Peysakhovich A, Rand DG (2015) Why we cooperate. In: Decety J, Wheatley T (eds) The moral brain: multidisciplinary perspectives. MIT Press, Cambridge
Kaplan A (1999) The development capacity. UN Non-Governmental Liaison Service, Geneva
Lull J, Neiva E (2012) The language of life. How communication drives human evolution. Prometheus Books, New York
Mansell R, Wehn U (1998) Knowledge societies: information technology for sustainable development. United Nations Publications, New York
Quebral NC (2001) Development communication in a borderless world. In: Paper presented at the national conference-workshop on the undergraduate development communication curriculum, "New dimensions, bold decisions". Continuing Education Center, UP Los Baños: Department of Science Communication, College of Development Communication, University of the Philippines Los Baños. pp 15–28
Ramalingam B (2015) Aid on the edge of chaos. Oxford University Press, Oxford
Richerson PJ, Boyd R (2006) Not by genes alone. The University of Chicago Press, Chicago
Sankatsing G (2016) Quest to rescue our future. Rescue Our Future Foundation, Amsterdam
Tomasello M (2008) Origins of human communication. The MIT Press, Cambridge, MA
UNGA (1962) Resolution adopted by the United Nations General Assembly during its 17th session. New York, United Nations

# The Soft Power of Development: Aid and Assistance as Public Diplomacy Activities

**21**

Colin Alexander

## Contents

| | | |
|---|---|---|
| 21.1 | Introduction | 408 |
| 21.2 | Morality and Foreign Policy | 410 |
| 21.3 | Public Diplomacy, Soft Power, and Foreign Aid | 414 |
| 21.4 | The Domestic Audience | 416 |
| 21.5 | The Governments of Recipient States | 417 |
| 21.6 | Other International Audiences | 418 |
| 21.7 | Recipients of Aid and Assistance | 419 |
| 21.8 | Conclusion | 419 |
| References | | 420 |

### Abstract

Following a political communications framework can provide useful critical understanding of the international philanthropic industries beyond the more traditional approaches of political economy, anthropology, and postcolonial studies. To this end, this chapter frames foreign aid and development assistance through theories of soft power, arguing that these activities are acts of public diplomacy and thereby conducive to the source government's power accumulation motive. This is open to some contest across the literature as research framed under international political economy or social anthropology often assumes that international power redistribution is the primary motive. Analysis of these programs under the soft power framework allows for the discussion of the multitude of audiences that the activities engage with beyond the direct recipients of assistance as part of the power accumulation precedent. The chapter will hereby discuss the role of morality and compassion within the policy-making process,

C. Alexander (✉)
School of Arts and Humanities, Nottingham Trent University, Nottingham, UK
e-mail: colin.alexander@ntu.ac.uk

© Springer Nature Singapore Pte Ltd. 2020
J. Servaes (ed.), *Handbook of Communication for Development and Social Change*,
https://doi.org/10.1007/978-981-15-2014-3_74

which leads to the question of whether we should really be considering whether most aid and development is in fact meant to work rather than the more popular query of why so much of it does not work.

**Keywords**

Department for International Development (DfID) · Foreign aid and development assistance · Neoliberalism · Morality and foreign policy · Public diplomacy · Soft power

## 21.1 Introduction

The propaganda that surrounds foreign aid describes its altruistic and compassionate motives. However, far from any empathetic or developmental intent, foreign aid is primarily motivated by the consolidation of the power status of the source as usurper and the recipient as usurped. To this end, an understanding of foreign aid and development assistance through the prism of soft power – the term coined by Joseph Nye (see 2004) at the end of the Cold War to describe when power in international politics is achieved through attraction rather than coercion or persuasion – is useful to wider understanding of the grand purpose of these industries. At the crux of this paper then is an attempt to provide an explanation for why the so-called foreign *aid* and *development assistance* does not work. If, by "work," it is meant that programs facilitated by international governmental actors reduce extreme poverty, relieve destitution, improve educational attainment, and deliver vital skills training while at the same time respecting aspects of culture, human movement, and the environment, then foreign aid and development assistance has been a catastrophic failure as it has made no contribution to a more egalitarian and fairer world that purports to be its motivation. To this end, the question may be better phrased as "is foreign aid meant to work?" because its perpetuation seems to be based on it satisfying alternative power ambitions to those that it claims.

The nineteenth century German philosopher Arthur Schopenhauer (1995) argued that compassion should be the only motivation if an action is to be considered as philanthropy. As such, the foreign aid industries communicate from a disingenuous platform as they primarily exist to fulfil the power ambitions of the source or self-styled "donor" state. We understand this because, as Arturo Escobar (1995: 21) explains in jest, the poverty of certain parts of the world appears to have been "discovered" after World War II when the narrative surrounding the planet's poorest people became one of deliverance from their destitution. It was as though the former colonial exploiters had suddenly discovered their moral compass. To this end, the postwar growth of the foreign aid industries was part of attempts to ensure political and economic continuity in the postcolonial era and was only interested in poverty alleviation to the extent that it benefited the major powers as they continued their exploitation of these territories.

The key argument made by this chapter is that communications analysis of foreign aid can provide an alternative framework to those more commonly used from political economy, anthropology, or postcolonial studies and that this may assist with the repositioning of wider questions concerning development studies generally. As such, this chapter will discuss foreign aid as acts of public diplomacy: a distinct political communications terminology referring to attempts made by governments to engage with foreign publics as befitting strategic goals. The intent of these activities is normally the generation of soft power for, or the bestowment of attractiveness upon, the source of these communications. To be clear therefore, the mechanisms of foreign aid tacitly communicate messages of virtuosity regarding the source to numerous domestic and international audiences. Thus, while public diplomacy is predominantly considered to have international priorities, it can also be linked to governmental priorities concerning its domestic public. Beyond this, foreign aid under a public diplomacy framework also forms part of governmental attempts to demonstrate congruence with prevailing international ideological norms and deontological guidelines regarding the responsible behavior of wealth states. In today's case this involves the propagation that supposed international philanthropy can be an adequate counterweight to the acceleration in global inequality that has been the result of neoliberalism. However, much like other activities that are propagated as philanthropy under this ideology, foreign aid and development assistance have little intent to offer social justice and should always be considered part of the exploitation rather than any offset against it.

Herein lies one of the fundamental limitations of democracy: that it cannot deliver meaningful change, partly because group consensus almost always produces an egotistical outcome (see Niebuhr 1932). For example, democracy is ill-equipped to reverse or stifle the effects of climate change because its institutions refuse to enforce the necessary restrictions on the consumptions of the electorate lest the leadership become unpopular. Democracy involves the art of dream selling and the limitation of toil, and this rarely involves a utopia of human reintegration with the natural world. Thus, political leaders who fail to adequately nourish the collective ego of the electorate normally have short-lived political careers. Consequently, governments maintain the fiction that foreign aid is motivated by philanthropic rather than strategic and economic interests because this assists with the nourishment of a domestic collective ego that the country and its citizens make a positive contribution to the planet beyond consumption of its resources. It then gains uncritical support from many echelons of the public who misguidedly support foreign aid as a form of state charity because this feeds the individual ego's desire to perceive oneself as virtuous within the prevailing ideology. However, this is a subconscious self-deception that has been incentivized by ideological hoaxing seeking to create feelings of worthiness within the consumer that lead to optimal consumption. As such, worthiness is normally maintained in one of two ways under these conditions: either through forms of collective or individual voluntary penance that are paid at the point of transaction to permit a guilt-free continuation of a supposedly virtuous existence of inconsequential behavior or in the form of foreign aid, wherein a government claims that taxation monies are assisting the world's poor who suffer misfortune rather than the effects of an organized system of denial. These propagandas essentially seek to maintain capitalism's profit-making imperative by detaching the consumer experience from the reality of its implications on the world's poor, and indeed the environment, and keep hidden that a reduction in consumption would assist both the balance of humans with the planetary ecosystem while also improving the relations of humans with other humans.

Much of the philanthropic industries then exist in a doublespeak world reminiscent of George Orwell's classic dystopian novel *1984*. Orwell (1949) provides the adage of the Ministry of Peace, which makes war; the Ministry of Truth, which lies; the Ministry of Love, which tortures; and the Ministry of Plenty, which starves. Orwell argued that these acts of doublespeak are essentially attempts by the state to misconstrue the consciousness of the people. To this end, much like our Ministries of Defence are more concerned with outward aggression, it can be argued that the use of words like "aid," "development," "assistance," and the notion of "donors" and "donations" are Orwellian-style propaganda. The industry communicates in a philanthropic language using terms like "support," "dignity," "compassion," "capacity," and the "alleviation of hardship." However, perhaps rather than being called the *United States Agency for International Development* (USAID) or the British *Department for International Development* (DFID), it would be more accurate to call them Agencies or Departments for *International Diminution*.

The remainder of this chapter has been split into three subsections. The first provides a wider academic discussion of the compatibility of morality with the foreign policy process. This is important because it helps to frame the extent to which foreign aid misrepresents itself as an act of state philanthropy. The chapter will then provide a specific discussion of soft power and public diplomacy within the context of development studies. Finally, the chapter will provide some concluding remarks concerning foreign aid and development and wider positions regarding prevailing ideology and the propaganda that justifies itself.

## 21.2 Morality and Foreign Policy

The eighteenth-century philosopher Jean-Jacques Rousseau wrote in *The Social Contract* that:

> Since no man has a natural authority over his fellow, and force creates no right, we must conclude that conventions form the basis of all legitimate authority among men. (Rousseau 1993: 185)

These "conventions" may pose as moral authority. However, they are upheld for the primary purpose of the maintenance of power structures by those who propagate their value. Transferred to international politics then, it can be argued that no state has entitlement to its international power. As a result, states strategically communicate with domestic and international audiences in an attempt at orchestrating value judgments about themselves that serve their power ambitions. To this end, public diplomacy activities form part of the alibi that states propagate in an attempt at convincing others of their power ambitions despite a lack of legal or moral entitlement therewith. Indeed, if power was guaranteed, then there would be little need for these strategic communications. As such, public diplomacy is often the ground on which governments propagate an ethical compliance of sorts with prevailing international norms. This is notwithstanding the authenticity of these communications as it is power rather than compliance that is their primary objective.

On the role of morality in foreign policy then, Nicholas Spykman has argued that:

> The statesman who conducts foreign policy can concern himself with values of justice, fairness, and tolerance only to the extent that they contribute to or do not interfere with the power objective. They can be used instrumentally as moral justification for the power quest, but they must be discarded the moment their application brings weakness. The search for power is not made for the achievement of moral values; moral values are used to facilitate the attainment of power. (Spykman 1942: 18)

Henceforth, the role of the state in international affairs is always, as Spykman says, to "facilitate the attainment of power," with the harboring of any contrary opinions only resulting in eventual disappointment. Accordingly, the moral accompaniment to power ambitions is delivered explicitly and implicitly through communications activities like public diplomacy, which includes, but is not limited to, international broadcasting, cultural diplomacy, education and knowledge exchange, endorsements of other governments, state visits, gestures of goodwill, media and public relations, and foreign aid and development assistance. However, all of these activities have the potential to become redundant if they are deemed to be unimportant or even detrimental to the power quest. This includes all foreign aid. Indeed, foreign aid's usefulness to the power equation is its association with the philanthropy that forms a key part of the prevailing neoliberal ideology's moral justification of itself. In short, neoliberalism's exploitation is propagated as being acceptable as long as a remedy for the suffering that it induces is also provided in some measure. However, this industry, indeed most supposedly charitable acts under neoliberalism, is motivated not by collective compassion but by ego. In short, many of those engaged in these industries are at least partly motivated by the self-confirmation of virtuosity that such a career brings. Additionally, positive external acknowledgment may also be sought for their actions, and offence may be taken if the person being assisted does not bestow an expected level of gratitude. Alternatively, recognition is sought for charitable acts as part of the formation of public persona, now most commonly pursued on social media platforms.

Therefore, neoliberalism has conveniently created a multibillion dollar foreign aid and development industry ready to dispense the advocated prescription and provide some outlet for the ego that capitalism prioritizes over compassion. In many ways, therefore, foreign aid resembles a global form of the medical illness Munchausen's syndrome by proxy where sickness is induced in a person by a caregiver in order that attention be drawn to themselves. This rather uncomfortable analogy is at the crux of the conceptualization of foreign aid as public diplomacy that we will return to later in the chapter.

The writings of the Italian philosopher and political prisoner Antonio Gramsci (1891–1937) have been used in many contexts related to the study of power and communications. Gramsci wrote the following on the uses of morality and culture by the powerful in their pursuit of authority:

> In my opinion, the most reasonable and concrete thing that can be said about the ethical and cultural state is this: every state is ethical in as much as one of its most important functions is to raise the great mass of the population to a particular cultural and moral level, a level (or type) which corresponds to the needs of the productive forces of development, and hence to the interests of the ruling classes. The school as a positive educative function, and the courts as a repressive and negative educative function, are the most important state activities in this sense: but, in reality, a multitude of other so-called private initiatives and activities tend to the same end – initiatives and activities which form the apparatus of the political and cultural hegemony of the ruling classes. (Gramsci, cited in Forgacs 1988: 234)

Thus, for Gramsci, whatever that "particular cultural and moral level" ought to be for those within and beyond the powerful state was decided by whatever the forces of production require it to be. To this end, the landscape of foreign aid at a given time can be linked to the requirements of the means of production through its delivery of a cultural or moral authority that befits the priorities of the prevailing world system and manufactures a sense of power by attraction under the ideological guise. To this end, the key to the maintenance of hegemonic power was through what Gramsci called the "guardians" of any society. Most likely inspired by Plato's *The Republic*, Gramsci's guardians essentially amounted to intellectuals whose moral leadership is admired by society and who gain credibility and authority through the perception that they act as intermediaries between the powerful and the suppressed. Think: celebrity endorsement of charities. Gramsci concluded that this is a fabrication as the intellectuals seek the maintenance of at least a semblance of the power status quo. Indeed, they would not be conferred as guardians if they sought anything more radical or revolutionary.

This position is articulated well by the American ethicist Reinhold Niebuhr who wrote that:

> No society has ever achieved peace without incorporating injustice into its harmony. Those who would eliminate the injustice are therefore always placed at the moral disadvantage of imperilling its peace. [...] This passion for peace need not always be consciously dishonest. Since those who hold special privileges in society are naturally inclined to regard their privileges as their rights and to be unmindful of the effects of inequality upon the underprivileged [...]. (Niebuhr 1932: 78)

Thus, those who are held to be guardians of a community will not be advocates of any anarchy, false or otherwise. Niebuhr continues:

> The moral attitudes of dominant and privileged groups are characterized by universal self-deception and hypocrisy. The unconscious and conscious identification of their special interests with general interests and universal values, which we have noted in analyzing national attitudes, is equally obvious in the attitude of classes. (Niebuhr 1932: 72)

This argument concerning the association between privileged class interests and supposed "universal values" helps to explain the educational narrative of foreign aid and development assistance, which preaches that if only the world's poor could become more efficient and learn from the developed nations then they would somehow "catch up." However, this is propagated as part of the creation of attraction that the source seeks for itself as part of its soft power ambitions.

The generation of attraction leads to the question of access to, and control of, cultural spaces and fora, in short, whose voices are heard, who is listened to, and who is marginalized, vilified, or even absent altogether. The notion of cultural space, and its control, was discussed at length by the French philosopher Henri Lefebvre during the late twentieth century. Lefebvre's (1991) explained how the narratives of cultural space embody future notions of utopia of the prevailing ideology at a given time. These narratives are the result of "knowledge" that is propagated as objective but which is framed by the means of production. Lefebvre (1991: 9) wrote that "[t]his ideology carries no flag, and for those who accept the practice of which [...] it is indistinguishable from knowledge." It is also indistinguishable partly because of the individual's egotistical desire to believe in their own virtue and the validity of their life choices. This denial or repression is lived at such a level that various subconscious defense mechanisms are engaged so as not to encounter the discomfort brought by the realization of contradiction. This critical path is also no doubt discouraged by the mass distractions that all ideologies employ to prevent the potential for cognitive dissonance within their advocates.

In terms of foreign aid then, Foucault (1991) discussed the notion of the "humanitarian mode of power" in which a subversive authority portrays themselves as the provider of solutions to prevalent social issues while simultaneously utilizing natural resources and human labor to meet their own ends. Foucault discussed this within the context of the prison system and the notion of rehabilitation. However, the argument can also be used to explain foreign aid narratives. During the colonial period, for example, the notion of "civilizing" the "uncivilized" to Western cultural standards became the popular moral justification that accompanied economic exploitation, and this has continued in the postcolonial period in practices of foreign aid and development assistance within the "catch up" mentality. Foucault's work on the dynamics of discourse and power in the representation of social reality also revealed the mechanisms by which a certain order of discourse produces permissible modes of being and thinking while placing other beyond the pale. This creates senses of what is to be revered and what is unattractive. Thus, the argument that foreign aid represents state philanthropy, rather than an essential part of the power accumulation strategy that dominates policy objectives, likely sits so safely within the perimeters of prevailing discourses that the majority of people consider it to be no argument at all.

The great literary figure Oscar Wilde (1900: 2) once said that "[...] the worst slave-owners were those who were kind to their slaves, and so prevented the horror of the system being realised by those who suffered from it." To this end, therefore, in many ways foreign aid is an unsatisfactory stopgap that prevents its recipients from receiving social justice and perpetuates, even justifies, their exploitation. Some of those engaged in the industry may mean well, particularly those on the ground performing its functions. However, they are ultimately misguided or self-deceiving because, far from any moral catalyst, foreign aid symbolizes much of the alibi that this deeply unfair world system uses to preserve its legitimacy.

## 21.3 Public Diplomacy, Soft Power, and Foreign Aid

Public diplomacy can be defined as the governmental act of attempting to communicate with foreign publics to assist the power ambitions of the source of these communications. These ambitions are always strategic and economic, such is the overlap between the motives, and involve efforts at engineering cognizant or incognizant positive value judgments about the source and its domestic and international actions and intentions, in the hope that these transfer into social, economic, and political gains. Thus, public diplomacy works well when it balances tendencies toward sharing, equality, understanding, and connectedness with other humans while also stimulating the ego which often prevents the enjoyment of a collective compassionate experience to any great extent. This represents public diplomacy's exploitation of human dialectic for its own ends. As such, public diplomacy's engagement with human tendencies toward compassion for others is one of the main reasons for the positivity and sometimes romanticism that can be found in much of its academic literature. However, its stimulation and incentivization by the powerful for accumulative and conservative ends, much like foreign aid, ultimately result in the industry existing on a fallacy of compassionate authenticity.

The terms public diplomacy and soft power are frequently found together in academic literature with the rationale for this usually that public diplomacy is concerned with the generation of soft power for the source of the communications. Consequently, an actor cannot "do" soft power, so to speak, as soft power can only be bestowed upon an actor by another. Furthermore, soft power can be bestowed upon a benign actor who makes little purposeful or proactive attempt to improve their soft power status but simply acts in a way deemed to be attractive. Thus, soft power status can be achieved without engaging in public diplomacy activities. Finally, while it can be acknowledged that public diplomacy normally seeks to make the source more attractive, this is not a precondition. Indeed, public diplomacy activities have been known to present the source as less attractive to some groups as strategic objectives determine, for example, when seeking to deter some economic migrants or refugees from travelling to the source.

Public diplomacy narratives and activities often epitomize the cultural and national identity of their source. To this end, public diplomacy often uses the knowledge capital of the source's domestic public as part of the persona that it attempts to create in the international sphere. Thus, domestic publics can perform an informal ambassadorial role when public diplomacy activities become an international outlet for the skills of that group. This is particularly prominent when public diplomacy concerns foreign aid, development assistance, and some forms of cultural diplomacy, which often seek to harness the talents, knowledge, and emotions of the public. What is more, public diplomacy often seeks to massage the collective ego of the domestic public by linking to nationalist desires to believe in the positivity of one's own country's role in the international system and one's individual contribution to it. It can also assist with community building at home.

This domestic function of public diplomacy can be overlooked such is the focus of the literature on international communications. However, most governments considered it important that a country has a clear sense of its past, present, and future and incentivizes propaganda to induce this. To this end, most countries are founded on myths concerning the shared history of all echelons of society. These are propagandas designed to manufacture the obedience of those experiencing a power deficit and may involve the propagation of a glorious past involving the defeat of barbarous enemies who posed existential threats, a tragic past involving gallant defeat to external forces displaying immense brutality, or a past from which history has been reckoned with or learned from and from which the domestic public can now provide assistance to others around the world as befits public diplomacy priorities.

As this author has argued in other publications (see Alexander 2014a, b, 2015), foreign aid and development assistance fit well within the remit of public diplomacy and soft power because it seeks to improve the attractiveness of the source within the minds of target domestic and international audiences, such is its positive propagation under prevailing ideology. This chapter has already discussed the notion of morality and foreign policy, and this has provided a platform from which specific details of the audiences that foreign aid seeks to engage with can now be understood. This section will now discuss four specific audiences in turn. However, the political elite's consistent unwillingness to acknowledge that foreign aid and development assistance are linked to power ambitions results in the following discussion being largely based on critical analysis alone. Indeed, in many years of interviewing politicians, civil servants, and even volunteers involved in foreign aid from a variety of countries, this author has only rarely heard an interviewee admit the power motive of the industry.

One example of such a narrative came from Yen Ming-hong, who, when interviewed in 2013, was the Deputy Director for Technical Cooperation at the Taiwan International Cooperation and Development Fund (ICDF). He stated the following about his government's primary concern over the health of diplomatic relationships as motive for providing foreign aid. For Yen, Taiwan's foreign aid had previously amounted to an intergovernmental bribe that befitted Taiwan's international priorities at the time.

> In the 1960s the reason why we were involved in international aid was totally a diplomatic concern. As you know, we are no longer a member of the United Nations, and during that time the Taiwan government tried to keep our allied countries in Africa. So what they came up with was to send out a lot of technical missions to stay there and to provide agricultural assistance in the hope that we can have their diplomatic support. In that time, we were not focused on the results of the projects. We did not care about the results. We just wanted the support of the local governments. (Yen 2013)

Nevertheless, Yen was adamant that Taiwan's contemporary intent was more philanthropic. This was despite a volume of evidence to the contrary (see Alexander 2014a, 2015).

## 21.4 The Domestic Audience

Let us begin this subsection by reviewing transcripts from the leaders' debates in the lead up to the 2017 UK general election. These narratives do much to reveal how political parties in wealthy states present the notion of foreign aid to their domestic publics.

> Transcript of ITV Leaders Debate 18 May 2017
> Paul Nuttall (PN) (Leader of UK Independence Party): [...] *We also would put £1.4 billion a year into social care. And how would we pay for this? Well we would take that money directly from a foreign aid budget that is costing us around £13 billion every single year. We would cut that back to 0.2% of GDP, which is exactly the same as it was in the United States under Barack Obama.*
> — *(later in the debate)* —
> PN: *We take that money from the foreign aid budget which is costing us £30 million every single day [...]*
> Leanne Wood (LW) (Leader of Plaid Cymru): *Taking it away from refugees then, yeah?*
> PN: *[...] Hang on, hang on. If we reduce foreign aid to 0.2% that is, as I said earlier, the same as the United States under Obama, the point is [...]*
> LW: *You're stopping refugees coming here, preventing them.*
> PN: *[...] this is about low hanging fruit, there's no need to put up taxation, we can simply take it from the foreign aid budget [...]*
> Nicola Sturgeon (Leader of the Scottish National Party): *Take it from the poorest people in the world?*
> PN: *[...] supporting British people first. Charity begins at home.*
> (ITV 2017)
> Transcript of BBC Leaders Debate 31 May 2017
> PN: *[...] we would slash the foreign aid budget that is costing the British people £30 million a day.*
> [Incoherent commotion]
> Caroline Lucas (Joint Leader of the Green Party): *Shame on you!*
> (BBC 2017)

These short transcripts present the political propaganda that foreign aid amounts to a form of state charity. Wood, Sturgeon, and Lucas, whose parties all occupy the center or center-left ground in UK politics, do not question the philanthropic guise of foreign aid that the more right-wing Nuttall emphasizes and are appalled, genuinely or not, by his apparent lack of compassion. The discussion within these debates then, and one that is replicated in political debates in wealthy countries around the world, surrounds how much international state charity should be provided, not whether foreign aid amounts to a form of philanthropy at all. These narratives form a key part of the public discourse that conceives of foreign aid as an act of kindness on the part of government with the primary topic of debate how generous the government ought to be on the electorate's behalf. This nourishes the notion of a domestic collective ego concerning a country and its citizens making a positive contribution to the planet beyond the consumption of its resources but which is, in reality, motivated by the manufacture of the consent of the domestic public to be ruled.

## 21.5 The Governments of Recipient States

The second audience of foreign aid that can be discussed is the governments and powerful people within recipient states. In the postcolonial era, these individuals and institutions have been responsible for ratifying the contracts that the governments of wealthy states and international corporations have made to rent and purchase land for a variety of purposes within these countries. This quality land then becomes unavailable for use by the publics of recipient states, squeezing them either into city slums or onto land that is susceptible to climatic variation and/or not suited to the agriculture that it is then asked to produce. This system produces a ready supply of destitute communities primed for "saving" by foreign aid and development assistance while also allowing wealthy governments to propagate their moral responsibility to the world's poor and providing an international outlet for philanthropic, but ultimately misguided, members of their domestic publics. To this end, the next time an appeal from a charity or aid agency appears on the television stating that someone must walk many kilometers from their home to find fresh water, perhaps it should be asked why they live so far from fresh water as this has been the basic premise of all human habitation since the birth of organized society.

Although somewhat limited in its breadth and analysis, Saskia Sassen's (2014) account of the *expulsions* of people from land that their families have farmed for generations is a powerful argument about the responsibility for the plight of the world's poor ultimately lying with the rich. Sassen also notes that the twenty-first century has seen an intensification of land acquisition with two significant factors being the main contributors to this. The first is that the rise in food prices has made land an even more desirable investment by international buyers, even for speculative reasons; and the second is the growing demand for industrial crops, notably palm for biofuels, and for food crops, the latter particularly coming from the rapidly developing states of East Asia and the Middle East. The intensive industrialization of these countries has resulted in additional demands for the products that tend to follow economic growth, with greater meat consumption one area of note.

What Sassen does not discuss, however, is the process by which the governments of wealthy states and international corporations engineer elections around the world in an attempt at ensuring that the candidates who will provide a preferential treatment of sorts are elected to office. To this end, elections in countries where a significant percentage of land is used for export purposes tend to be swamped with clandestine foreign monies, which by and large control the menu of political choice for the voters. Indeed, even moderately socialist or protectionist groups struggle to start a campaign such is their lack of access to the capital needed to compete in a national election resulting in some turning to violence. This is before consideration is given to the communications professionals who join the teams of preferred candidates to structure the delivery of campaign messages and researchers who "muckrake" candidates less favorable to the priorities of international interests, attempting to find any kind of underhand dealing or personal demon. As such, foreign aid in this context is used as an attractive political incentive committed by wealthy states and international corporations to their preferred candidates in recipient states as part of an exchange package that essentially limits the role of the government of poorer states to the welfare of its citizens and largely thwarts any movement toward them achieving social justice within the international system. The governments of these states, and those close to power, thus form one of the key audiences of foreign aid under soft power analysis.

## 21.6   Other International Audiences

This subsection is formed of a variety of groups and clusters who may have contact with, or become aware of, a government's contribution to the foreign aid industries and who may then make positive value judgments about that contribution and its source. It includes international political, economic, and societal groups that a source government conceives of and wants to appeal to as part of their broader strategic international ambitions. The most important of these is the United Nations as this is the primary fora for the debate of global governmental legislation and the international platform for the articulation of the prevailing values of the world system and the policies and positions that follow from them. Other international political groups that source governments regularly attempt to attract include the World Health Organization, the International Monetary Fund, the World Bank, the European Union, and a variety of regional military alliances.

At the crux of this process of manufacturing attractiveness through foreign aid is a desire to demonstrate conformity with perceived international norms of responsible or dutiful behavior as have been determined to befit the power status of the source. This process does not necessarily exist on a platform of morality, but moral obligation may form part of the propaganda here. Perhaps the most prominent articulation of this ambition comes from separatist regional governments of wealthy states, who, in a bid to demonstrate congruence with prevailing values and consensus on global issues, engage in small amounts of foreign aid as part of what is known as "protodiplomacy." By engaging in protodiplomacy, these regional governments are implicitly confirming that in the event of their independence from the parent state they would become consistent, regular, and reliable members of the so-called international community as their power status would determine. A strong recent example of this has been the Scottish government, which has used foreign aid and development assistance as part of its pro-independence international communications strategy and charm offensive toward major international governmental organizations (see Alexander 2014b). However, this is a regular activity of all state governments engaged in aid provision as they all articulate their contribution to the industry as part of their political communications toward a host of international audiences beyond the immediate recipients of foreign aid. Indeed, some governments may have reservations about the manner in which aid operates, concerns over corruption, for example, but will nevertheless continue to engage in the practice in some form as it assists their power ambitions and prevents any negative attention being drawn to themselves lest they be labelled as some kind of foreign aid "scrooge" by those they seek to influence.

## 21.7 Recipients of Aid and Assistance

The final audience that requires discussion is those in receipt of an aid package or development program. Often the receipt of aid or assistance is recognition of some form of capital or usefulness within the priorities of the world system. Beyond this, those facing a nadir of destitution, with seemingly minimal capital, may become recipients because of the so-called CNN effect, where the international mainstream news media shine a light on suffering and, if certain criteria are met, successfully pressurize governments to act (see Robinson 2002). Linking more to the domestic audience within donor states, this may also present as a form of the so-called poverty porn, wherein media viewers watch images of suffering as a form of fetish that they subconsciously do not want to see eradicated for it excites them, makes them feel good about themselves, or presents them with a guilt narrative that can be conveniently alleviated by charity donation or endorsement of government programs (see Zizek 2009). The notions of *jouissance* and *schadenfreude* are noteworthy here. Finally, the foreign aid industries require a regular supply of "success" stories as part of their propaganda that it can make a positive impact on reducing global poverty and capacity building.

## 21.8 Conclusion

Foreign aid and development assistance are very motivated by the source's power imperative. To this end, these industries ought to be thought of as public diplomacy outputs aimed at the generation of power for the source government in the form of attractiveness rather than power as a result of coercion or persuasion. Foreign aid's lack of regular inclusion within the portfolio of ministers assigned to public diplomacy is synonymous with the fictional demarcation of government departments and a likely attempt by government to prevent the growth of a sense of hypocrisy within their own publics over the international role of those who govern them. This is primarily because public diplomacy tends to be thought of as a communications arm of regular state foreign policy rather than any form of philanthropy. As such, while ministries of foreign affairs are concerned with the consolidation or accumulation of the power status of the source, departments of international development operate separately under the pretense of being motivated by power dispersal. They are not, however, and this demarcation represents part of the Orwellian misconstruction of government.

To conclude then, the evidence presented here strongly suggests that foreign aid is conducted for purposes beyond that of philanthropy despite much of its propaganda narrative indicating philanthropic intentions. One of the most significant factors behind this is because the modern corporate state is increasingly run like a business that is power and capital hungry. To this end, governments employ spin doctors, undertake branding initiatives, perform "reputation management," and are increasingly subservient to corporate industry rather than the people to which they remain accountable (at least within democracies). Why then should departments of international development be any less "spun," so to speak? What is more, calls for transparency mean the publication of accounts, much like a public limited company, and subsequent scrutiny and justification of expenditure by the domestic public and media. Therefore, it follows that the funding of foreign aid must be in line with the economic and political interests of the source, if the government is to avoid accusations of financial mismanagement or squandering of public funds. Indeed, public diplomacy has had a long and arduous battle with government finance because its results are essentially unquantifiable such is the solipsistic nature of its activities. Thus, while outright plunder cannot occur as it did in colonial times, the ideologically constructed "benefit" that foreign aid has for any recipient remains utterly subordinate to the source's self-interests.

## References

Alexander C (2015) The communication of international development assistance: the case of the Taiwan international cooperation and development fund (ICDF). J Glob Gover 21(1)
Alexander C (2014a) China and Taiwan in Central America: engaging foreign publics in diplomacy. Palgrave Macmillan, New York
Alexander C (2014b) The public diplomacy of sub-state actors: the case of the Scottish government with Malawi. Palgrave J Public Dipl Place Brand 10(1)
BBC (2017) 2017 UK general election political leaders debate: 31 May 2017. Available from world wide web: [https://www.youtube.com/watch?v=xmopAnM_exA]. Accessed 18 July 2017
Escobar A (1995) Encountering development: the making and unmaking of the third world. Princeton University Press, Princeton
Forgacs D (ed) (1988) A Gramsci Reader. Lawrence and Wishart, London
Foucault M (1991) Discipline and punish: the birth of the prison. Penguin, London
ITV (2017) 2017 UK general election political leaders debate: 18 May 2017. Available from world wide web: [https://www.youtube.com/watch?v=wEgi50lP8co]. Accessed 18 July 2017
Lefebvre H (1991) The production of space. Blackwell, Oxford
Niebuhr R (1932) Moral man and immoral society: a study in ethics and politics. Charles Scriber's Sons, New York
Nye J (2004) Soft power: the means to success in world politics. Public Affairs, New York
Orwell G (1949) 1984. London: Penguin
Robinson P (2002) The CNN effect: the myth of news, foreign policy and intervention. Routledge, London
Rousseau JJ (1993) The social contract and the discourses. David Campbell Publishers, London
Sassen S (2014) Expulsions: brutality and complexity in the global economy. Harvard University Press, Cambridge, MA
Schopenhauer A (1995) On the basis of morality. Hackett Publishing International, Indianapolis
Spykman N (1942) America's strategy in world politics: the United States and the balance of power. Harcourt, Brace and Company, New York
Wilde O (1900) The soul of man under socialism. Free River Community, London
Yen M-H (2013) [Interview] Headquarters of the Taiwan ICDF, Taipei, Taiwan, with C. Alexander. 10th May 2013
Zizek S (2009) First as tragedy, then as farce. Verso, London

# Asian Contributions to Communication for Development and Social Change

## 22

Cleofe S. Torres and Linje Manyozo

## Contents

| | | |
|---|---|---|
| 22.1 | Introduction | 422 |
| 22.2 | Development Communication in Asia: Overview | 423 |
| 22.3 | Asian Contributions to Development Communication | 424 |
| 22.4 | Schools of Thought | 425 |
| | 22.4.1 Participatory Planning | 425 |
| | 22.4.2 Rural Communication | 427 |
| | 22.4.3 Learning as Capacitation | 431 |
| | 22.4.4 Professional Degrees in Devcom | 434 |
| | 22.4.5 Mainstreaming of Devcom in Civil Service | 435 |
| | 22.4.6 ICT for Development | 436 |
| | 22.4.7 Evaluation Framework for Capacity Development | 437 |
| 22.5 | Conclusions and Challenges | 439 |
| References | | 440 |

### Abstract

The debates on the role of media and communications in promoting sustainable development acknowledge the multiple experiments that were undertaken from the 1940s and 1950s as being key to laying the foundations of the field we now call development communication or communication for development. Yet, as has been emphasized over the years, it is the agriculture communication and extension experiments at the University of the Philippines, Los Baños, that would

---

C. S. Torres
College of Development Communication, University of the Philippines Los Baños, Laguna, Philippines
e-mail: cleofe.torres@yahoo.com

L. Manyozo (✉)
School of Media and Communication, RMIT University, Melbourne, Australia
e-mail: linje.manyozo@rmit.edu.au

© Springer Nature Singapore Pte Ltd. 2020
J. Servaes (ed.), *Handbook of Communication for Development and Social Change*,
https://doi.org/10.1007/978-981-15-2014-3_22

herald the pioneering contestation of development thinking using media and communication in Asia. Considered together with the Asian Tigers' approaches toward radical economic development, the Asia's schools, institutions, organizations, and governments demonstrated that there was a different way of thinking about and doing development beyond what the western modernist theories and perspectives were proffering.

This chapter is a celebration of the Asian contributions to development communication, which was a term employed by the pioneers of the field. Over the years, other terms have come into use, including communication for development and social change. While the emphasis has been placed in the fact that these were pioneering contributions, this chapter examines the specific attributes of such a contribution and how, considered in retrospective, students of society will come to appreciate the fundamental aspects of theory and practice today that trace their origins to these experiments. The chapter highlights seven of these contributions in order to demonstrate the fundamental concerns that drive the agenda, processes, and partnerships in the field. The chapter acknowledges that the field has moved on; there is a growing danger of a growing disregard for history, as if the field just emerged without a historical context. Yet it is this very historicity, in Asia, as much as in Latin America and Africa, that continues to shape the practice on the ground, even in the face of the historical revisionism of modernist development approaches mostly being promoted by international development institutions.

**Keywords**

Development communication · Participation · Development thinking · Knowledge management · Capacity building · ICT for development

## 22.1 Introduction

The chapter aims to celebrate the contributions from Asia to the theory and practice of development communication. There are many reasons for this. First, there is an increased western footnoting of the history of the field, in which the pioneering contributions from the Asian region are being footnoted and written out of global scholarship in development communication. Such an aversion is very much orientalist (Said 1978), with the consequence that the liberatory and postcolonial ideals that shaped the emergence of the field have been submerged under the hysteria of instrumentalist participatory communication or social and behavior change communication. It seems to us, there is a deliberate proposition suggesting that either the field emerged from the west or it just emerged from without. The release of various books, handbooks, or articles on the field by major western publishers is a case in example. Second, there is an increased emphasis in a media-centric discourse that either celebrates media and development or social and behavior change communication. This is not problematic in itself. Yet, it seems as if development communication is just about changing people's behaviors, as reflected in the preeminence of social and behavior change models and theories. Third, there is a need to remind current students of society, of the specific attributes of the pioneering contributions from Asia that shaped the theory and practice of development communication, and continue to do so in much of the global south. This is the main agendum of this chapter.

The chapter acknowledges the simultaneous pioneering contributions from Latin America, observing that the emphasis on liberation and deconstruction of inequality provided a postcolonial prose and discourse for rejecting the modernist development, which continues to undergo what Mansell (1982) describes as "superficial revisionism." Particularly for this chapter, the discussion demonstrates that for Asia, there were seven areas of focus that would shape the emergence of the field of development communication. Consideration of these areas allows for the appreciation of the fact that, at its center, development communication remains as alternative theory of development in practice that rejects and undermines the modernist, capital-intensive, or technology-driven development models that emerge from the metropoles of western capitals and think tanks. These seven contributions were (1) participatory planning (Bessette 2004; Quebral 2012), (2) rural communication (Sainath 2016; Loo 2009; Verghese 1976), (3) learning as capacitation (Cadiz and Dagli 2010); (4) devcom degree training (Quebral 1988, 2012), (5) mainstreaming development communication, (6) ICT (information, communication technologies) for development (Agrawal 2006; Kumar 1981; Mansell 2017), and (7) evaluation frameworks (Campilan et al. 2009).

## 22.2 Development Communication in Asia: Overview

Development communication or, as widely accepted now, communication for development has been described as a phenomenon more akin to the Third World (Quebral 1976/2012) or the developing countries. As a planned social intervention, Quebral categorically stated that its allegiance is to the poor, the powerless, and the disadvantaged in developing societies, many of whom are found in Asia. The pro-poor emphasis is also evident in the Latin American scholarship and experiments (Freire 1970; Gutierrez 1971/1988). In fact, development communication actually grew in response to the pressing problems of underdevelopment, namely, poverty, unemployment, high population growth, inequality, environmental degradation, malnutrition, and even ethnic conflicts (Ongkiko and Flor 2003; Quebral 1988, 2012).

Hence, the Asian version of development communication thinking and practice is closely tied up with Quebral's (1976/2012, 9) definition of the concept as "the art and science of human communication linked to transitioning of communities from poverty in all its forms to a dynamic, overall growth that fosters equity and the unfolding of individual potential." This is an important definition for students of society to ponder, perhaps in more ways than one. In referring to the "art," Quebral recognized that the creative industries are critical in providing pedagogy of the oppressed. In the Philippines and Asia as a whole, music, folk media, theater, and indigenous communications, groups of men and women were already creating productions and performances that aimed to empower people. Also critically, is the fact that development communication is a collective endeavor, whose process of empowerment is not scripted off site. For Quebral, art was always about two things, co-design and learning. Likewise, the incorporation of science was in a way prophetic. Current debates in the field emphasize the evidence-based, method-driven, and theory-informed discussions, focus being on investment thinking. In fact one of the most tenuous debates today concerns the contestation between learning-based and evidence-based approaches and perspectives (Lennie and Tacchi 2013).

Alongside the pioneering experiments at the University of the Philippines, Los Baños, where Nora Quebral and colleagues would introduce development communication degree training programs in the 1960s (within the College of Agriculture) and the 1970s (within a new College of Development Communication), were other schools and institutes within the region. Over and above the numerous experiments that are too many to accommodate in this discussion, a couple stand out for their overall contribution to the field. This chapter explores the key ones.

## 22.3 Asian Contributions to Development Communication

In Asia, development communication struggled to be recognized as a legitimate field of study and practice in the early years. Quebral (1976/2012) discussed the suspicions that stakeholders had with this concept. For governments, they were concerned with the focus on empowering people. For other development organizations, the focus on promoting the country's goals appeared as a form of legitimation of authoritarian governments. But now, it is acknowledged as a discipline in its own right. It has established its own body of knowledge, built up its theories, and accumulated good practices in various development themes and settings.

Among other regions in the world, Asia has been the setting of many development struggles. It is, thus, expected that development communication will find it as a good breeding ground in shaping the philosophy and theory that drive its actual practice on ground. To acknowledge categorically what those from Asia have contributed to its growth and to its evolving challenges is but a fitting tribute. The topics discussed in succeeding paragraphs may not be that comprehensive; they are, however, indications of how vibrant the field has remained despite that "the old guards have left and institutional support for the field petered out" (Quebral 2002).

Contributions of Asian scholars, practitioners, and community partners to development communication or devcom, as mentioned earlier, include participatory planning, rural communication, learning as capacitation, devcom degree training, mainstreaming devcom, ICT for development, and evaluation frameworks. They are not meant to be mutually exclusive as some features of one tend to be also present in others.

## 22.4 Schools of Thought

Manyozo (2012) introduces the six schools of thought that he argues have shaped the emergence and current approaches to the theory and practice of development communication, namely, Bretton Woods, Latin American, African, Indian, Los Baños, and the Communication for Social Change Schools. Devcom literature will be incomplete without acknowledging that two of its schools of thought originated from Asia: the Indian School and the Los Baños School. Though influenced by different circumstances and development challenges, both produced systematic, theory-based, and method-informed development communication practice through time (Manyozo 2012). In its emergence, the Indian School focused more on broader social issues such as family planning, dynasty, gender equality, and social harmony. Today, it is characterized more by the use of folk media, indigenous communication, entertainment-education, development broadcasting, and use of information and communication technologies (ICTs) for development.

The Los Baños School, on the other hand, grew out of its earlier agricultural extension function, it being situated in an agricultural university. Hence, it focuses more on new farm practices and technologies yielded by agricultural research. In terms of media, Los Baños School in early years thrived on the use of broadcast-based distance learning, development broadcasting, community newspapers, and others that fall under the category of community media. The growth of both schools can be attributed also to the assistance, financial and otherwise, extended by universities from the West. What has not been explored well in development communication scholarship is the Chinese approaches and practices in development communication, which seem to have a deeper history than many anticipate. Over the years and until today, both schools are engaged in developing and testing theories on how people-oriented communication may well serve the goals and facilitate the process of human development. So what were the key contributions from Asia to development communication?

### 22.4.1 Participatory Planning

Asia's landmark contribution to devcom is undisputedly in the area of participatory planning. Though devcom evolved parallel with other western communication models, its thrust has always been co-design – the engagement of relevant stakeholders in the development process itself. As elucidated by Quebral (2002), the devcom philosophy includes that "the people be participants in their own development."

While many development workers would easily buy-in the noble idea of participatory planning, there was no clear method, model, nor guide readily available that would show the way. This is especially so in Asia where many countries are still living with the remnants of their colonization days, thus their tendency to cling to the modernization and diffusion mindset. The early experiments in integrated pest management would emphasize this notion of participatory planning. Ironically, in some of these experiments, the funding and support has come from western partners and stakeholders.

Giving flesh to the concept of participatory planning would also occur around 2001–2002 when the International Development Research Centre (IDRC) of Canada piloted the *Isang Bagsak* (IB) program in Cambodia, Vietnam, and Uganda. The learning program was undertaken by teams of Asian researchers and development practitioners. The program would field-test and implement a carefully designed communication approach anchored on participation called participatory development communication or PDC (Bessette 2003). It was applied in the implementation of community-based natural resource management projects (CBNRM) in selected rural communities of the three pilot countries.

PDC refers to "a new form of dialogue that aims to engage local communities in the process of charting their own future" (Kavanagh et al. 2006). It employs clearly planned communication activities to facilitate community participation in a development initiative (Bessette 2004). Community participation in turn means facilitating the active involvement of different community groups including the development and research agents working with the community and decision-makers (Bessette 2004). It is planned to facilitate dialogue among stakeholders to achieve an agreed upon goal. It puts premium on learner participation and ownership of the learning process.

As a methodology, PDC follows a logical process composed of ten clear steps (Fig. 1): (1) establishing relationship with local community and understanding local setting, (2) involving local community in identification of problem and solutions, (3) identifying relevant stakeholders, (4) identifying communication objectives and activities based on needs, (5) identifying appropriate communication tools, (6) preparing and pretesting communication materials, (7) facilitating building of partnerships, (8) producing an implementation plan, (9) planning the monitoring

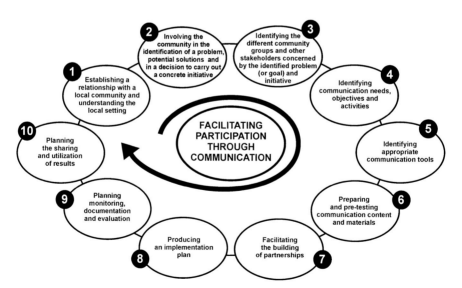

**Fig. 1** Participatory development communication process (Bessette 2003)

and evaluation, and (10) planning the sharing and utilization of results. All throughout the cycle, communication facilitates the participation of the community stakeholders. On top of this, the process also builds mutual understanding and collaboration which favorably push the development process forward.

One year of pilot testing (2001–2002) led to the refining of steps and activities of PDC using the peculiarities of the Southeast Asian (SEA) setting. The process was again worked out in the field using selected SEA countries in 2003–2005; this time adding the Philippines to Cambodia and Vietnam. This leveling up of experience contributed to the firming up of the steps in the PDC process and the wider application of PDC to other NRM settings.

Evidence generated about *Isang Bagsak* indicate that the participatory nature of appraisal, planning, implementation, and evaluation as well as engagement in constant dialogues and innovative communication methods helped raise awareness about existing NRM problems. This in turn enabled community members to agree and work on their own solutions. Bottom-up communication provided local government and researchers the window to actively engage with their respective communities and later on secure commitment from them (Cadiz and the College of Development Communication IBSEA Program 2004).

Non-Asian scholars might have started the operationalization of PDC, but it has been enriched together with their Asian counterparts. In no certain measure was it nurtured and brought to fruition in the Asian setting. This peculiar context in the region and the deep involvement of Asian players in the process provided the needed flavor and medium where PDC was able to mold its authentic form.

### 22.4.2 Rural Communication

Writing in the 1970s, Quebral (1976/2012, 4) observed that "when we speak of development today, we are mainly concerned with the majority of people in the developing societies, most of whom live in the countryside." This is a fundamental observation. Yes over the years, the attention has shifted from rural to other economic and social issues that exist in urban economies as well, but the rural was always the starting point. For Quebral, mastering development in the rural space would provide the blueprint for dealing with urban development questions, after all, they were both products of marginalization, exclusion, and inequality. It therefore makes sense that much of the pioneering experiments in Asia aimed to address problems of rural communication as well. Two of these problems stood out. Problems existing as a result of inadequate communication systems and literacy in the rural area and the problem of communication between policy makers and rural audiences. In emphasizing the significance of the rural polity, Quebral (1976/2012, 11, 14) argue that:

> One problem in the developing country is to get the majority into the dialogue. Where the old and new media are concentrated in cities, the rural populace is virtually voiceless. What is passed of as public opinion is urban opinion. How to give voice to the voiceless is a concern

of the development communicator. [...] When we speak of the poor and the disadvantaged, are they still in the rural areas? They are the families of small farmers, labourers and fishermen most of whom are still in the rural areas though many of them have migrated to urban slums. They are also the indigenous people who live apart from the main population and are economically and socially deprived.

To deal with these challenges, three strategies of development communication emerged, both of which aimed to bring the rural population into the decision-making space. First were the development journalism experiments in India, the Philippines, and the whole region. As has been noted before (Agrawal 2006; Kumar 1981; Manyozo 2012), India's approaches to development communication were rooted in its pre-colonial rich traditions of folk media and indigenous knowledge communication; the post-independent entertainment-education experiments; and of course, the growth of indigenous language press. Important to mention is that India has relied on its academic institutions to expound development communication thinking and practice (Manyozo 2012). Notable among the academic centers were the University of Poona, the Centre for the Study of Developing Societies, the Christian Institute for the Study of Religion and Society, and the University of Kerala. A major experiment in development journalism would emerge with the *Hindustan Times* in the 1960s, when George Verghese introduced a column on rural issues, "Our Village *Chhatera*," based upon a village located about 25 miles from Delhi (Verghese 1976, 2009). This initiative enabled the *Hindustan Times* newspaper to "win the affection of a village community and encouraged it to grow" (Verghese 1976, 3). As explored by Eric Loo (2009) in great detail, alongside this experiment, were other similar experiments in development journalism focusing on rural populations that would take hold in the whole Asia.

Second, were community communication experiments. Community communication has long been part of development communication literature, but it is noticeably not widely spoken of in the Asian setting, though its practice is pervasive. Community media such as community radio or newspapers are usually associated with it. The strongest distinction of this domain is its being interactive and participatory. It may or may not be mediated (Berrigan 1979). Even if it is mediated, the medium is only a means to facilitate the social interaction in small groups. If it is not mediated, it uses interpersonal or indigenous means for communicating like folk songs and theater. It is associated with grassroots and, hence, is biased toward stories, orality, proverbs (Manyozo 2012), and other folk media.

Historically, Asian countries would have closer affinity with community communication given their cultural tradition for being communal rather than individualistic. Farmers, women, or youth usually decide and behave on what the community thinks, feels, and believes in despite the irrationality and inconvenience this may create. Despite the advent of ICTs, many of the intended stakeholders of devcom are located in areas with very limited access to electricity, virtual connectivity, and other modern communication infrastructure. As such, community communication as an area of practice is highly relevant in development communication.

As experienced in many rural areas of Asia, community communication facilitates community decision-making, problem-solving, conflict resolution, interest articulation, advocacy, and social mobilization (Ongkiko and Flor 2003). In the 1960s and 1970s, development programs in the Philippines particularly those associated with population experimented on the use of folk media to promote family planning methods among the rural poor (Flor 2004). Employed were puppet shows, shadow plays, rural theater, folk songs, *balagtasan* (poetic debate), and *balitaw* (an extemporaneous exchange of love verses between a man and a woman). These helped popularize the message as the medium enabled it to cater both to the heart and head.

In later years, the traditional media were tapped to boost environmental conservation campaigns. Legends and myths about mountains, rivers, and other natural landscapes became popular as a way of explaining how man can harm the spirits that reside in nature and how the latter can take their vengeance and punish man (Oroza 2008). Though folklore may run against the grain of science, they definitely were effective to some extent. Fear of nature's wrath worked to minimize man's abuse of nature as commonly shared in informal and media discussions and documented in some academic studies.

In one particular case, community communication facilitated the coffee farmers' adaptation to climate variability in Cavite, Philippines (Ilagan 2013). As part of their regular routine, four to five farmers gathered together after visiting their farms to talk about what was happening with their coffee plants. They discussed as peers or coequals about how climate has become so erratic and highly variable such that their coffee plants also became unproductive. In the course of their gatherings, they shared and studied deeper the information gathered from extension workers, traders, and cooperative officials about their problem. As they experimented on the technological advice they obtained from various sources, they themselves sorted out what worked.

As a result, the coffee farmers adapted their farming practices to climate variability by introducing another crop for intercropping, adjusting their time of fertilizer application and harvesting, more frequent weeding, regular pruning, covering of harvested coffee beans, and stocking their harvest before selling. All these they have learned from intensive discussions of the realities they were facing and the price they have to pay for changing some of their practices. In other words, the adaptation practices were not something merely "extended" to them (as in typical agricultural extension work) but out of their own self-discoveries of what will work for them at certain cost. These adaptations manifest their potentials that were being unfolded as they went through the community communication experience.

Third were the considerations for inclusive Rural Communication Services (RCS). RCS refers to the provision of need-based, interactive, ICT-enabled, participatory communication processes (FAO 2014). It is delivered regularly to the rural population to perform three major functions: (1) facilitate equitable access to knowledge and information, (2) enable social inclusion in decision-making, and (3) build stronger links between rural institutions and local communities.

For the first function, it strategically employs interpersonal and mediated communication (community media, ICTs) to enable stakeholders to access information and knowledge for their unmet needs. For the second, it purposively selects participatory methods to engage stakeholders in various stages of development work, i.e., planning, implementation, monitoring, and evaluation. For the third, it provides platforms through which institutions and members of local communities can be linked together and work in partnership in meeting their common needs.

While many might think that RCS is just a new name for agricultural extension, it is quite different in that it does not subscribe to the one-way information service flow. It does not focus only on the farmers but tries to bring into partnership the research and extension services, local institutions, NGOs, academe, media, and private sector so that the needed communication services are orchestrated well to serve the needy population (Torres et al. 2015).

In 2011–2013, the Agriculture Information Service (AIS), Ministry of Agriculture in Bangladesh, ventured into the establishment of RCS primarily through rural radio and mobile phones in Patuakhali, Barguna district, a distant coastal village in southern part of the country. The demand for agriculture information was high in this area due to the effects of climate change on their already marginal and saline farmlands. Radio and mobile phones were noted in the baseline study as the most available and highly accessed media in the area.

Following the basic project planning cycle, various stakeholders were identified, information/communication needs assessed, RCS strategy crafted, action plans and budget laid down, and mechanisms for participatory monitoring and evaluation completed. All these involved the participation of relevant stakeholders through a series of meetings. Likewise, sustainability plan was put in place as early as the conception of the project. With radio as the locomotive medium, rural radio planning and programming were done together with the needed capacitation for various stakeholders.

After 2 years, a rural communication unit (RCU) was set up at AIS, manned by personnel redeployed from existing AIS units. RCU coordinates with the different stakeholders for their contributions to RCS: research institutions for the needed technologies, extension service for field demonstrations and training, academe for baseline and other needed studies, community media for local awareness raising, farmer groups for organizing listenership groups, youth for their performance numbers in community assemblies, and women in gathering local news. Everybody helped in monitoring radio listenership.

RCU coordinates with the rural radio station, called Krishi Radio, set up in the pilot area. Stakeholders listened to radio programs mostly through mobile phones. Program content during the pilot focused on the introduction of the saline-tolerant rice variety (STRV). Radio and mobile phones were gradually complemented by computer-based services provided by AIS in the established ICT centers in villages.

When FAO funding ended, the project was continued by IFAD in 2015–2017 in cooperation with the Local Government Engineering Department. This time, three more existing rural radio stations were tapped in Radio Nalta, Radio Lokobetar, and Radio Sundarban, all located in poverty-stricken coastal areas in the south. Program content was expanded to include climate-resilient technologies, aside from STRV, that address water siltation, soil erosion, deforestation, and others as expressed by members of local communities.

Despite some constraints, the initiative brought about appreciable outcomes: improved institutional and local capacities in the design and implementation of communication strategies; initial establishment of an RCU unit to respond to information priorities demanded by farmers; strengthened linkages among institutions in agricultural extension, research, information services, as well as NGOs, local government, and farmer organizations; effective use of rural radio talk shows for explaining demanded technologies; proactive crafting of sustainability efforts; and drawing out of lessons learned that drive iterative planning.

The four rural radio stations have formed an alliance called "Voice of the Coastal People" and have organized a number of listener groups whose members also form part of the radio program production and hosting. There have been 100 listener clubs established, with 15–20 members per listener club, or a total of 15,000 listeners for all the four rural radio stations. Being part of the communities, these volunteers know exactly what information would cater best to their needs and how these may be delivered in the most comprehensible manner.

### 22.4.3 Learning as Capacitation

For devcom to create a dent on poverty and an appreciable impact on development, it needs a critical mass of practitioners who would be able to operationalize consistently its philosophy, theories, methods, tools, and techniques. While there are tertiary level schools that produces professionals, the need for a sizeable number to help multiply the efforts in a shorter time can only be augmented by nonformal training courses.

A number of devcom trainings must have been undertaken in various parts of Asia by more than a hundred devcom training providers spread out all over the region. As documented and captured in CCComdev Web Portal (cccomdev.org/index), these providers include universities, development organizations, NGOs, media organizations, and some civil society organizations. The full profile, contact details, and description of courses offered by nonformal education providers in Asia can readily be accessed from the portal using the training map and under training opportunities.

One regional program can be cited to have contributed to the systematic capacitation of devcom practitioners, development workers, and other development actors in the region: the Adaptive Learning and Linkages in Community-Based Natural Resource Management or ALL in CBNRM. This program evolved out of IB and was implemented in 2006–2008. Its thematic focus broadened from PDC to a wider array of participatory development approaches. It may be considered significant in terms of the breadth of geographical coverage, program objectives and content, number of institutional learning facilitators, number of learning groups, and combination of learning modalities.

ALL in CBNRM is an adaptive learning network and a collaborative program for CBNRM covering six countries in SEA: Cambodia, Indonesia, Lao, the Philippines, Thailand, and Vietnam. Adaptive learning implies customized learning that should enable people to cope with their changing environment. It is based on the premise that learning from experience empowers people to respond more effectively to new uncertainties, enabling them to change old ways of doing things and allowing them to make better decisions (Cadiz and Dagli 2010). In all these, communication plays a central role.

As a capacity building program, ALL in CBNRM was designed for CBNRM researchers and practitioners in SEA working in different ecosystems such as forestry, coastal, and wetlands. It aims to create a community of participatory-oriented NRM researchers and practitioners, a wider knowledge base and network, and develop synergies among the community or network members. Here, the community covered a regional level but was more virtual, mediated by an online medium, the Internet.

Content-wise, it adopted the perspectives and methods of participatory research and development including social and gender analysis, adaptive and social learning, participatory rural appraisal, participatory action research, and participatory monitoring and evaluation (Cadiz and Dagli 2010). The program was facilitated by a regional partnership of recognized institutional players in rural development: the Regional Community Forestry Training Center for Asia and the Pacific (RECOFTC), International Potato Center-Users' Perspectives with Agricultural Research and Development (CIP-UPWARD), Community-Based Natural Resources Management Learning Center (CBNRM LC), International Institute of Rural Reconstruction (IIRR), and CDC- UPLB. Using adaptive learning, they worked with nine learning groups composed of CBNRM researchers, development workers, and local community members engaged in NRM.

The program involved the participation of nine learning groups from the academe, NGOs, government agencies, and people's organizations engaged in CBNRM. Each of the group was in itself a case in search for more solid, dynamic, and ultimately, useful and meaningful learning processes (Campilan et al. 2009). Learning within and across groups took place through a combination of face-to-face discussions, regional workshops, field mentoring, technical back stopping, and regional online discussions.

Based on program evaluation conducted in a participatory manner, ALL in CBNRM accomplished several outcomes (CDC-UPLB 2008). In terms of capacity development, improved capacity of the individual led to improved capacity of the group, which in turn led to improved capacity of the organization. Subsequently, this led to development outcomes at the community level (Fig. 2). More specifically, the change in individual knowledge, attitude, and practices led to strengthened teamwork and improved group processes. This then led to improved organizational performance vis-a-vis their mandates. Consequently, it led to improved communities' capacities, their livelihoods, and to conserved natural resource base.

Other reported outcomes were manifested by strengthened collaboration and partnerships, heightened trust building, enhanced women participation and

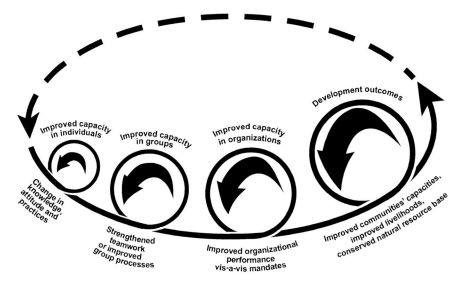

**Fig. 2** Capacity changes and development outcomes

development, stronger community-based organizations, improved livelihood, and more active community participation in agricultural technology development (CDC-UPLB 2008).

The program also contributed improvements in policy, governance, and conservation of the NRM sites. Moreover, it enabled the facilitators and learning groups to produce knowledge and learning resources. In a way, all these paved the way for the mainstreaming of adaptive learning and networking in CBNRM at local and regional level (CDC-UPLB 2008). It appears that adaptive learning strategies are effective way to respond to, cope with, and reshape people's new reality – although gradually and not without challenges (Campilan et al. 2009).

### 22.4.3.1 Knowledge Exchange as Capacitation

As early as 2010, an initiative to link devcom practitioners across various countries in the regions, and among regions in the world to create a space for knowledge sharing in devcom, has been put in place. Called the Collaborative Change Communication (CCComdev), it is being spearheaded by FAO in partnership with academia, community media organizations, farmer organizations, and field projects. Content covers the themes on agriculture, forestry, natural resource management, food security, climate change, and resilience, among others.

CCComdev primarily uses a global web portal for worldwide access as well as regional hubs for Asia (ComdevAsia.org), Africa (Yen Kasa Africa), and Latin America (Onda Rural). It aims to create a global community of practice, strengthen partnerships, and promote knowledge sharing among a variety of rural actors including farmer organizations, community media, rural institutions, universities, research organizations, and development and communication practitioners.

The ComdevAsia.org is managed by AMARC (World Association of Rural Radio Broadcasters) Asia Pacific office. As a regional platform, it documents experiences and promotes knowledge sharing and joint initiatives in communication applied to agriculture and rural development. It supports communication, development, and community media practitioners based in Asia Pacific in the areas of sharing experiences and knowledge resources on communication for rural development, strengthening capacities and facilitating fruitful partnerships at regional and country level, and promoting a regional agenda and advocating for devcom in the rural sector. It is still a work in progress.

The online platform features good practices and lessons learned on devcom as applied to rural development challenges in Asia and the Pacific. It regularly posts news and updates keeping an eye on what's going on in the region. It also offers multimedia resources, useful links, and virtual community where registered members can connect to participate in thematic group discussions and online exchanges with colleagues in the region and even worldwide.

A notable activity of the platform was the conduct of regional virtual consultation last 2014 on three concerns affecting family farming: contribution of community media and ICTs to family farming; trends, challenges, and priorities for rural communication services; and a common agenda to promote south-south collaboration in devcom. Participants included those from the farmers' associations, media organizations, devcom practitioners, development organizations, civil society organizations, and academicians. Results have been collected to constitute a compendium that could help guide policy and program direction for devcom in family farming.

### 22.4.4 Professional Degrees in Devcom

Quebral (2002), the devcom torch bearer herself, acknowledged that the Los Baños School does not have the monopoly nor the ownership of intellectual property rights to devcom study and teaching. But it is to her credit, and the devcom faculty she worked with, that the first and the only three-tiered formal degree program in devcom can be found only in Asia – the Philippines that is. Based at the College of Development Communication (CDC), University of the Philippines, Los Baños (UPLB), the course offerings in devcom were started in the 1970s for the baccalaureate (BS) and masters (MS) and in the 1980s for the doctoral (PhD) degree. It also offers the Master of Professional Studies in devcom jointly with the UP Open University. Through the years, the BS Devcom curriculum has been adopted as a regular program offering in nine other universities all over the country.

The undergraduate curriculum at CDC-UPLB is unique as it does not only cover core courses in communication and related social sciences. It also includes elective technical courses readily available in other colleges in the same university such as agriculture, environment, nutrition, human ecology, and forestry, among others. This is to equip the students with technical grounding on the context when they eventually get employed in either the government or private sector.

The MS Devcom program prepares the student to be grounded both in devcom theory and practice and enable them to apply these as they serve at mid-level career positions. The PhD Devcom aims to produce competent planners and leaders of institutions engaged in communication designed primarily to assist development work.

Outside the country, the CDC has helped institutions to establish postgraduate devcom programs perceived from local perspectives and contexts. A case in example being CDC's collaboration with the Kasetsart University in Thailand, helping the latter develop a postgraduate program in devcom, Manyozo (2012) observes that Kasetsart's devcom courses, such as broadcasting for development, writing for development, or communication systems management, reflect the influence of the agricultural origins of the Los Baños development communication. Similar programs have emerged in India (Govind Ballabh Pant University of Agriculture and Technology), Europe (University of Reading), and Africa (Universities of Malawi, Mozambique, and Zambia). It is important to note that some of these programs do not necessarily offer subjects directly dealing with development, poverty reduction, economics, or sociology, but their philosophical orientation leans toward poverty reduction as well as the farming and rural communities since the "sheer number and its central role in the economy give the rural family the edge" when it comes to development communication training (Quebral, 1976/2012, 5).

### 22.4.5 Mainstreaming of Devcom in Civil Service

Mainstreaming devcom has always been one of the big challenges for the practice. Initially and while there are project funds, support to devcom is steadfast. Once project ends, devcom ends with it. The experience of Sarawak, Malaysia, in mainstreaming devcom in state civil service is an exception. Sarawak is a rich state, but at some point its officials realized the value of devcom in bringing about the needed social transformations among its populace.

No less than the former Chief Minister Tun Pehin Sri Abdul Taib Mahmud acknowledged the value and contributions of devcom in state governance. He is among the noted leaders in Asia who applied devcom principles in his work during his term. In addition, the current State Secretary made this pronouncement: "Development communication is people-centric and that should be the core of any development project that intends to bring about change among its intended stakeholders," (Sarawak Tribune, June 23, 2015). He knew that in any major development plan, there is an urgent need to listen to the people's fears as well as aspirations. Support from the people of Sarawak is crucial for the success and smooth implementation of the development program.

In 2015, the state was about to roll out its Sarawak Corridor of Renewal Energy (SCORE) program, a master plan for its socioeconomic development until 2030. Aware of people's lack of awareness and hesitant support to SCORE, the State Secretary decisively chose to bring in devcom to the state civil service.

Immediately, he called for sensitization and orientation seminar about devcom among other state officials and heads of key agencies to insure that everybody will be on the same page about devcom. A few months later, all heads of key government offices representing various sectors (energy, forestry, agriculture, health, education, lands management, indigenous peoples, etc.) were gathered in a 4-day devcom strategic planning course. Not satisfied with the short training, the head of the state through the Sarawak Development Institute (SDI) arranged for an 18-week mentoring program for officers of major government agencies and statutory bodies. He likewise promised to send the five outstanding performers to be sent to graduate programs in devcom abroad under full scholarship.

Recognizing further that continuous partnership with a devcom institution would be key to strengthening his devcom pillar in governance, SDI sought for a memorandum of understanding with a devcom service provider in the Philippines. Though the document took time to get approved, it took only a few weeks for another major decision to be made: a Development Communication Centre for the Sarawak state has to be established. It is to serve as a research and training center in devcom for civil servants in support of the government development programs.

In December 2017, the first devcom training under the DevCom Center's programs was carried out. The set of officers earlier trained, and a devcom student, about to complete her PhD in devcom, took over the training with invited devcom resource persons. Sarawak is now on its way in making devcom a mainstay of their state development programs and operations. More programs are being developed by the Center to cater to Sarawak's needs.

### 22.4.6 ICT for Development

Discourses of ICT for development have their roots in some of the experiments carried out in the region. The emphasis then was not just about the penetration of technology in the rural areas. It was about ethical inclusion of socially marginalized groups who were being digitally marginalized as well. As such, from a political economy perspective, ICT for development has always been about breaking up of knowledge monopolies, so as to ensure the access and participation of citizens in the knowledge society (Mansell 2017). This was the case with the landmark SITE experiment in India. This experiment emerged in light of the questions of broadcasting and how it could be employed to serve the ethnically and culturally diverse post-independent India. Even though one concern in India always revolved around heavy state intervention in media and communication policy, the focus on employing local rural radios in rural and regional areas was always critical (Kumar 1981). Over the years, various policies and committees have highlighted the need for establishing local radio stations within various districts, to provide educational broadcasting for both formal and nonformal instruction (Agrawal 1981; Kumar 1981).

It would make sense, therefore, that in the 1950s and in collaboration with UNESCO, India introduced a carefully designed network of rural radio forums, known as the *Charcha Mandals* (Kumar 1981; Masani 1976). These forums provided rural development information especially on agriculture, which was transformed into horizontal communications through participatory discussions (Agrawal 2006, 1981). Binod Agrawal (2006, 5) observes the positive impact these listening forums were having, noting that the "summative impact evaluation indicated positive outcome of radio rural forum. Impressive knowledge gains as a result of radio listening were reported across illiterates and literates, agriculturists and non-agriculturists, village leaders and others." By 1960, the Indian government scaled out the project nationally by ensuring that these forums were part of the country's development plans (Manyozo 2012).

Probably building on the rural farm forum project, the Indian government would introduce two rural television projects; the Satellite Instructional Television Experiment (SITE) which would be followed up by Kheda Communication Project (Agrawal 1981, 2006; Kumar 1981). Binod Agrawal (2006, 6) notes, regarding SITE:

> The one-year experiment aimed to provide direct broadcasting of instructional and educational television in 2400 villages in the states of Andhra Pradesh, Bihar, Karnataka, Madhya Pradesh, Orissa and Rajasthan. Over 500 conventional television sets spread over 335 villages in Kheda district, Gujarat was also part of SITE. Satellite technologists had called SITE as leapfrogging from bullock cart stage to satellite communication, which did not discriminate between rural poor and urban rich for information and communication.

Even the educational model might have seemed very top-down and driven by technology, a major aspect of India's approach to ICT for development was the emphasis in the sociocultural context in which technology was made to operate. It was not just about technology. Important to mention is that all these ICT for development experiments were taking place alongside other national and regional development communication experiments.

### 22.4.7 Evaluation Framework for Capacity Development

As capacity development becomes a main staple in devcom interventions, the need for it to be part of evaluating the results, outcomes, and impacts has become an imperative. But beyond the typical changes in knowledge, attitudes, and practices that pervade devcom evaluation, two considerations arise: how can capacity development be effectively and meaningfully monitored and evaluated? What approaches may be used to systematically assess both the process and outcomes of capacity development?

This concern was at the heart of the "Evaluating Capacity Development" or ECD initiative launched in 2006 by CIP-UPWARD. It brought together nine Asian partner organizations working in ALL in CBNRM into an informal network. The network then embarked on the following tasks: develop and pilot methods for evaluating capacity development processes and outcomes, promote the effective use of evaluation by organizations engaged in capacity development efforts, and facilitate wider learning and use of evaluation in capacity development (Vernooy et al. 2009). These organizations represented the academe, regional network, and community-based organizations from China, Mongolia, the Philippines, and Vietnam. Developed and tried was the so-called 5Cs evaluation framework for capacity development: context, content, capacity, the capacitated, and capacity development process (Campilan et al. 2009).

***Context*** refers to existing historical, sociopolitical, and environmental conditions and realities in the area where capacity building takes place. It includes local needs and interests, previous efforts, resources at hand, and policy situation. ***Content*** has to do with the themes and topics covered by the learning strategies. ECD, being attached to ALL in CBNRM as the capacity development program, focused on the earlier ten themes of PDC (Fig. 1). ***Capacity*** refers to the practices and other competencies that need to be improved among individuals, learning groups, and organizations where they belong. Under ECD, CBNRM-focused capacities were as follows: identifying and addressing rural development issues coherently, dynamically, from multiple perspectives and grounded reality; working with other stakeholders and bridging worlds of knowledge, expertise, points of view, aspirations, and interests; conducting participatory action, research addressing key challenges, or exploring new opportunities; designing and supporting viable CBNRM strategies in local context; as well as designing and using effective participatory monitoring and evaluation strategies that reinforce learning.

Beyond individual capacities and onto organizational capacities, these were defined as the ability of the organization to "be" (i.e., to maintain its specific identity, values, and mission); its ability to "do," that is, to perform functions and achieve stakeholder satisfaction; and its ability to "relate," that is, to manage external interactions while retaining autonomy (Campilan et al. 2009).

***Capacitated*** refers to the learning participants – they could be researchers and practitioners as individual members, their learning groups as a collective entity, and the CBNRM organizations as the formal unit where the individuals and groups belong. They have been clearly identified at the onset of the initiative. ***Capacity development process*** refers to the learning strategies employed. The ECD initiative exemplified an egalitarian, partner-driven, open-learning approach and a responsive support system to empower all partners (Campilan et al. 2009). It employed a number of clearly defined learning strategies all throughout the ECD timeline. These included introductory workshop, online team discussions for the PDC themes, face-to-face regional forums, midterm workshop, online team evaluation and planning, cross-visits, and writeshop.

The experience enabled the participants to draw from their individual and collective experiences a basis for building up a practice-informed theory of ECD using the 5Cs. As an insight derived from the experience, it is recommended though that an adaptive mode of learning be used to continuously refine the concept and methods as better understanding of the context and the capacitated are acquired in the process.

## 22.5 Conclusions and Challenges

The chapter acknowledges that various practices and experiments in communication for development did emerge in different parts of the world, and in fact, for much of the global South, these would precede western modernity and development thinking. The Asian contributions have become widely celebrated as being the foundation stone of the field that is also known as communication for development and social change. The reason is that unlike other schools of thought that emerged in 1950s/1960s, the Asian contributions were supported by a comprehensive package of theory and scholarly experiments. These institutions have, over the years, comprised the Universities of the Philippines Los Baños, Kasetsart, Hyderabad, and regional research centers such as Southeast Asian Regional Center for Graduate Study and Research in Agriculture (SEARCA). It is out of these experiments and scholarly endeavors that the field of development communication would emerge as a distinct discipline of study.

It can thus be observed that the Asian contributions to the growth of development communication revolve more around practice and development of methodological models and techniques. More efforts on theorizing are needed to allow for a more robust growth of the discipline in the region.

Suffice to highlight however that these contributions have also been facilitated by technical and funding assistance from UN agencies and international development funding agencies. Though outsiders, the enabling environment they provided has favored the Asians to bring development communication practice to what it is today.

This discussion has thus attempted to spell out the seven areas in which Asia made original and substantial contributions to the field. As has been explored in the chapter, these contributions comprise the following aspects: (1) participatory planning, (2) rural communication, (3) learning as capacitation; (4) devcom degree training, (5) mainstreaming devcom, (6) ICT for development, and (7) evaluation framework for capacity development. Central to these contributions was the idea that a well-skilled graduate with the relevant tools for living with people could bring people and relevant stakeholders to co-create development interventions that would benefit families and communities in ways that they could explain and experience at that local level. And it is the cumulative contributions of these local-level interventions that would bring about social change. In this case, social change was not considered a social force that sweeps through a society, a community – rather social change is a confluence, a sum of gradual changes taking place in various local communities. This is the essence of the Asian contributions – and it explains why, then as well as now, a development communicator should first and foremost master the skills and tools for community engagement.

Yet with increasing influence of donor funding and multiple players, standardization of capacity building initiatives becomes a problem. There are increasing cases where capacity building through training has become arbitrary, rather than selective. Some important but simple questions may need to be raised to better rationalize the proliferation of development communication trainings – what knowledge, attitudes, and skills are needed to accomplish the task? Is training in development communication the best option to pursue the task? Nevertheless, some concept building initiatives are still relatively new in Asia and in the formative stage like ECD. More systematization is needed so that a practice-driven theory in evaluating capacity development in development communication may be enhanced to stand a firm ground.

The chapter recommends that a strong political will from a development communication champion can be a game changer. The afore-discussed Sarawak experience shows how a typically top-down state governance can introduce radical change if there is a devcom champion around. How do we look for or build devcom champions? Mainstreaming development communication is one such approach. There are now few successes in the area of mainstreaming devcom in development work. But on the whole and despite years of field work, capacitation, technical assistance, and evidences gathered, devcom in Asia is still considered a suspicious or unwelcome guest in many development programs. Development planners see it as something they can do without; and for project implementers, it is just a burdensome add-on to their undertakings. While some windows of opportunities are slowly opening up, the scale of appreciation for devcom remains insignificant. This is borne out by the simple fact that every time we do development communication in Asia, the first thing that will be asked of us is "what is devcom?"

## References

Agrawal B (1981) SITE social evaluation: results, experiences and implications. Space Applications Centre, ISRO, Ahmedabad

Agrawal B (2006) Communication technology and rural development in India: promises and performances. Indian Media Stud J 1(1):1–9

Berrigan FJ (1979) Community communications: the role of community media in development. UNESCO, Paris

Bessette G (2003) Isang Bagsak: a capacity building and networking program in participatory development communication. IDRC, Ottawa

Bessette G (2004) Involving the community: a guide to participatory development communication. International Development Research Centre/Southbound, Penang

Cadiz MCH, Dagli WB (2010) Adaptive learning: from Isang Bagsak to the ALL in CBNRM. In: Vernooy R (ed) Collaborative learning in practice: examples from natural resources management in Asia. Cambridge University Press India Pvt Ltd/International Development Research Centre, New Delhi/Ottawa, pp 55–93

Cadiz MCH, the College of Development Communication IBSEA Program (2004) Building capacities and networking in participatory development communication in natural resource management: the Isang Bagsak Southeast Asia experience. Philipp J Dev Commun 1(1):136–174

Campilan D, Bertuso A, Nelles W, Vernooy R (eds) (2009) Using evaluation for capacity development; community-based natural resource management in Asia. CIP-UPWARD, Los Baños/Laguna

College of Development Communication-University of the Philippines Los Baños (2008) Adaptive learning and linkages in community-based natural resource management. Program Terminal Report. Los Baños, Laguna

FAO (2014) Rural communication services for family farming: contributions, evidences and perspectives. Results of the forum on communication for development and community media for family farming. Food and Agriculture Organization, Rome

Flor AG (2004) Environmental communication. University of the Philippines Open University, Quezon City

Freire P (1970) Pedagogy of the oppressed. Continuum, New York and London

Gutierrez G (1971/1988) A theology of liberation: history, politics, and salvation (trans: Inda C and Eagleson J). Orbis Books, Maryknoll

Ilagan BJP (2013) Community communication and coffee farmers' adaptation to climate variability in Amadeo, Cavite, Philippines. Unpublished PhD dissertation. College of Development Communication, University of the Philippines Los Baños

Kavanagh P, Bessette G, MacKay B (2006) Learning in the field: Isang Bagsak helps people chart their own future. International Development Research Centre. Retrieved from https://idl-bnc.idrc.ca/dspace/handle/10625/34218

Kumar K (1981) Mass communication in India. Jaico Publishing, New Delhi/Bangalore

Lennie J and JoTacchi (2013). Evaluating communication for development: A framework for social change, Routledge and Earthscan, London and New York

Loo E (2009) Best practices of journalism in Asia. Konrad-Adenauer-Stiftung, Singapore. Available: http://www.kas.de/wf/doc/kas_18665-544-2-30.pdf

Mansell R (1982) The new dominant paradigm in communication: transformation versus adaptation. Can J Commun 8(3):42–60

Mansell R (2017) The mediation of hope: communication technologies and inequality in perspective. Int J Commun 11:4285–4304

Manyozo L (2012) Media, communication and development: three approaches. Sage, London/New Delhi

Masani M (1976) Broadcasting and the people. National Book Trust, New Delhi

Ongkiko IVC, Flor AG (2003) Introduction to development communication. SEAMEO Regional Centre for Graduate Study and Research in Agriculture and University of the Philippines Open University, Los Baños

Oroza A (2008) Indigenous culture as medium for communicating natural resource management practices among *Hanunuo-Mangyans* in Occidental Mindoro, Philippines. Unpublished Master of Science thesis. College of Development Communication, University of the Philippines Los Baños

Quebral N (1976/2012) Development communication primer. Southbound, Penang/Malaysia

Quebral N (1988) Development communication. College of Development Communication, Los Baños, Philippines

Quebral NC (2002) Reflections on development communication, 25 years after. College of Development Communication/University of the Philippines Los Baños, Laguna/Los Baños

Quebral NC (2012) Development communication primer. Southbound, Penang

Said E (1978) Orientalism. Random House, New York

Sainath P (2016) Rural reporting: opening spaces for the people's voices. Media Asia 43(3–4):127–137

Torres CS, Francisco RP, Tirol MSC, Catapang KSG, Tallafer LL, Velasco MTH (2015) The College of Development Communication extension programs in community communication: towards a new domain in development communication. Philipp J Dev Commun 7:146–186

Verghese G (1976) Project Chhatera: an experiment in development journalism. Occasional paper number 4. Asian Mass Communication Research and Information Centre, Singapore

Verghese G (2009) Reclaim public service values of journalism. In Eric Loo (Ed) Best practices of journalism in Asia. Konrad-Adenauer-Stiftung, Singapore. 48–52. Available: http://www.kas.de/wf/doc/kas_18665-544-2-30.pdf

### 23.1.1 A Long Way to Reach the Utopia of a Democratic Communication

*Farewell to Aristotle: Horizontal Communication (1979)*, a representative paper written by Luis Ramiro Beltran, Bolivian communicator, set the beginning of democratic communication. Because of communication for development studies, a radical shift from the traditional perception of the process of social communication to its one-way-oriented model started in Latin America. The linear structure – source, encoder, message, channel, decoder, receiver, and effect and its main purpose of persuasion – shifted to a relationship model or a reciprocal paradigm. The information structure under discussion allowed the use of notions, such as bidirectional relation, dialogic process, horizontal contact between equal parts through process and similar horizontal new elements. Later on, Beltrán used other scholars' previous work, such as Frank Gerace and Juan Díaz Bordenave's. These authors had previously attempted to look into the human communication experience as a bilateral sharing dynamic. Beltrán, therefore, devised a definition of the nature of the human communication process based on three basic elements: access, dialogue, and participation (Beltrán 1979). Both Bordenave, known historically and a pioneer of this kind of dynamic as the "returned communication" concept (Bordenave 1962), and Gerace came out before using the concept of "horizontal communication" for the first time (Gerace 1973), but Beltrán used these perceptions structuring a concept about the nature of the human communication process as a democratic experience rooted in three basic elements: access, dialogue, and participation (Beltrán 1979).

The concept of human communication process by Beltrán also recaptured the regional political debate of the 1980s which stated deep criticism toward the social field and its relation to power, in which practitioners and philosophers in the education and communication fields, among others, Antonio Pasquali (1970) and Paulo Freire (1970, 1973), took part. They brought back the notion of communication as a mental cooperation process and a shared dialogue. This led to Beltrán's core definition, "Communication is the process of democratic social interaction based upon exchange of symbols by which humans voluntarily share experiences under conditions of free and egalitarian access, dialogue and participation. Everyone has the right to communicate in order to satisfy communication needs by enjoying communication resources. Humans communicate to carry out multiple purposes. Exerting influence on others' behavior is not the main one" (Beltrán 1979: 16).

Beltrán introduced a horizontal or bidirectional approach challenging the unidirectional perceptions of the human communication process and the persuasion control-oriented models. However, loyally he considers as promoters of this notion in Latin America intellectuals experienced in the reality of this region; he refers: "Frank Gerace, an American living in Bolivia, was the first to suggest the notion of horizontal communication. Paraguayan specialist in educational communication Juan Diaz Bordenave (1962) soon engaged in creative reflection on the matter, which made him a chief proponent of theoretical bases for the democratization of communication for development" (Beltrán 2008: 253).

Until now Latin America and its local communication experiences have kept a tradition that characterizes the core thought in communication that was used in the 1980s which expressed the idea of social communication as a human-centered experience rather than an informative act. In this region, social communication practice maintains a close relation to communal styles of living not only due to a theoretical drive but also due to its particular social, cultural, and political context. Cultural roots of indigenous societies, whether Andean or Amazonian, connected to their sharing practices and community life and its relation to nature, which are all a characteristic of precolonial and indigenous societies, are closely related to today's communication practices, for instance, the oral tradition, the connection among their context, nature, and the spiritual aspect which provide people with content structure and message direction as well as knowledge transmission and preservation very different from sense diffusion oriented toward individualism or self-reference.

Beltrán explains Latin American scholars' position and perception by stating, "Latin Americans have been the earliest opponents of the traditional concept of communication derived from unilinear Aristotelian thinking, which prevailed across the world unchallenged until the late 1960s. It was them who, digging beyond the apparent simplicity of the paradigm, discovered its undemocratic implications. And, as a consequence, they were also among the first ones to propose new views on communication, new models to reshape it according to genuine democracy. This intellectual innovation not only had broad acceptance in the region, but it was also eventually acknowledged, and even adopted by some of the most prestigious U.S. theoreticians in this field, such as Schramm and Rogers, and even Lasswell himself" (Beltrán 1993:32).

Everett M. Rogers recognized the impact of Beltran in his own perception and new definition of development by saying, "The definition of development I proposed was very much influenced by a Latin American communication scholar, Luis Ramiro Beltran. If anyone ought to be given credit for this definition, he should be the one. And this new definition dealt not just with economic development but also with social development: Increases in literacy levels, more years of education, lower infant death rates, greater life expectancy, fuller equality and more press freedom. This is still the main definition of development" (Rogers 1994: 3).

The Latin American Critical Communication Approach named by the Latin American School of Communication, and historically proposed by José Marques de Melo (Marques Ferrari and Antoniacci Tuzzo 2013) in the 1980s, combined the intellectual production of social communication researchers and practitioners, concentrating their critical approach on central factors, such as challenging the concept of development, the concept of communication, the blindness of communication studies related to the social structure, foreign theoretical and methodological influence into the practice of communication research, and the need to put a human center approach into development studies.

Despite criticism to the western-oriented theory in the field of communication, Uranga acknowledges that "We will not fall, however, into the simplicity of assigning the full weight of our ills to the decisions that came from abroad. None of the above would have happened without the open and active complicity of many political and military leaders and, most importantly, individuals and economic groups within our countries. Many of them obtained, at least momentarily, huge profits and profits that were supported by the misfortune of many of their compatriots. Today, without totally neglecting any of the previous items, problematizing the relationship between communication and development requires putting the most relevant data of this social scenario on the table. This should also be done in the democratic system, citizenship, modes of participation, and the public space where all this is debated and introduced symbolically by communication. By choosing this path, we are clear that we will leave other traditional reflections on the contribution that communication has made to the development of other initiatives which are also valuable and can be the subject of other considerations. We start off with a concept of development that focuses on the human being, on his personal and cultural best, and seeks to harmonize all material and symbolic resources based on a social construction that contributes to a dignified life of all the citizens. Thus, understanding development from all aspects is tied to culture and communication processes. Communication and communicators are required to make fundamental contributions to the construction of development, but they acquire very different characteristics in each period" (Uranga 2005:77).

### 23.1.1.1 Theoretical Disagreement and Need for Social Change

Social communication in Latin America has been historically guided by its critical nature. Peirano refers to this characteristic pointing out, "In order to understand and study communication in Latin America it has always been said that we should understand and analytically process all several theoretical traditions that came to the region trying to explain communication advance worldwide" (Peirano 2017:11).

In addition, Peirano states that the ones interested in "thinking communication" have worked processing and stating that the worldwide knowledge that was inherited was not enough. It is a fact that the most important researchers in social sciences and the ones concerned about communication issues in Latin America approached the field demonstrating strong criticism toward social knowledge that was delivered along the countries of this region in the midst of the twentieth century, period in which the first communication programs arose showing a clear influence by US theories (Peirano 2017).

According to Peirano, this regional quality shown in communication studies has been a permanent proof of the close relation between communication and politics. In Latin America, supporting social change was related to questioning the status quo and confronting the establishment. This critical approach also influenced communication for development theories that started in this region. The first theory challenged the spread of innovations articulated by Everett Rogers, considering that the conception of development was oriented toward the modernization of the subcultures of the Third World. The use of the means of communication was meant to deliver information to promote the adoption of practices mainly in the fields of agriculture and health. Latin American practitioners opposed to straightforward models of communication linked their criticism to revealing ideological or non-contextual foreign approaches.

A trend in communication for development, introduced and spread in Latin America in the 1960s, was challenged by the recognition of practices in "alternative communication" (Reyes Matta 1983; Atwood and McAnany 1986). The forms of this alternative communication reached global goals in communication. Atwood and McAnany state, "Hence, the goals of alternative communication endeavors include the creation of a New International Information Order (NIIO) as well as democratic and popular communication systems" (Atwood and McAnany 1986:19).

### 23.1.1.2 The Nature of This Utopia

According to Beltrán, communication for development was introduced as a theory in Latin America by the end of the 1950s and early 1960s; however, this approach started years earlier through some practice (Beltrán 1993). This took place a decade before any kind of theory was developed, specifically by foreign universities mainly in the USA. It is recognized that the communication for development approach began thanks to two phenomena in Latin America. The first phenomenon has to do with the existence, nature, and style of broadcasting events called worker's radio that belonged to a network of radio stations, named the Bolivian Miner's radio. This experience is the genesis of the practice of communication for development because of its collective broadcasting style of property, use of local native languages, close contact with and participation of local listeners, production of dramas based on regional tales and daily life stories, and self-financed nature of their activities.

The second phenomenon, in Latin American history of communication for development, took place in Colombia by the end of the 1940s simultaneously. This was an experience of a radio for peasants called "Radio Sutatenza," a Catholic Church radio station located in the rural valley of Tenza in Colombia and directed by José Joaquín Salcedo, a priest. The original approach that was introduced in 1948 meant to use radiobroadcasting to promote educational content and even literacy for its large indigenous and peasant listeners. Radio Sutatenza contributed to the origin of a long-lasting tradition called educational radio experience in Latin America. Despite the closing of ACPO (Acción Cultural Popular), the leading program that was generated by Sutatenza's experience, its original effort using radio to promote social inclusion of rural population, as a model, was imitated in Bolivia, Peru, and Ecuador.

In this region, there has been a widespread belief regarding leading an economic expansion as a goal of progress since the 1950s. Then, a critique from the field of economy gradually came out at the very center of the Economic Commission for Latin America (ECLA), or CEPAL as it is known in Latin American countries, the most relevant center in this field located in Chile. Doubts regarding the idea of concentration of benefits from technological progress and development conditions began. An intellectual movement located in Portuguese, Mexico, Buenos Aires, and Caracas also expressed a need to understand development in light of wealth inequalities and opportunities that derive from the capitalist expansion and later on from the strengthening of imperialisms as it was mentioned by Fernando Henrique Cardoso and Enzo Faletto in their classic analysis of *Dependency and Development in Latin America* (1969).

A theory about understanding dependency conditions and its close relation to development possibilities was presented as a counter-critique, so the nature of unequal conditions for growing Latin American societies asked for another kind of development. This is because the imbalance affected most of the rest of the Third World. Beltrán refers to the emergence of a proposal of another kind of development that was presented in the United Nations General Assembly in 1975 by a group sponsored by the Dag Hammarskjold Foundation, in which Juan Somavía, a Latin American economist, played an important role. This model proposed a development based on the satisfaction of needs of the majorities, on endogenous and self-reliant approaches, and on harmony with the environment (Beltrán 1993:13).

More importantly, Cardoso and Faletto's ideas helped to change the linear models of development introducing a dual approach. As they mentioned, "our approach is both structural and historical. It not only emphasizes the structural conditioning of social life, but also the historical transformations of structures due to conflict, social movements, and class struggles. Thus, our methodology is historical-structural (. . .) In other words, our approach should bring both aspects of social structures: a mechanism of self-perpetuation and change prospect to the fore front" (Cardoso and Faletto, 1979:10–11).

Cicilia Krohling states, "a theory of dependency is somehow a theory of participatory development. A theory of modernization, as was delivered by Walt Rostow, considered development as a linear process, a point reached through a modernization strategy. A theory of dependency means to cut the vision of modernization. The pioneers of this approach, such as, Fernando Henrique Cardoso, criticized the unequal relation between central countries and peripheric countries, because the first ones looked for benefits and commerce been benefited by the peripheric nations. The central countries did not have the perception that development was further than economic growth and industrialization, then a theory of a participatory development was growing based on an active community participation, because development was a process built in common and directed to solve real conditions and concrete needs of a nation. In this sense Jan Servaes says that there is not only a universal model of development, furthermore, development is a dialectic process, integral and multidimensional, because this might be different for one nation in comparison to another" (Krohling 1997: 2).

Servaes and Malikhao in relation to the dependency theory that grew in Latin America conclude, "As a result of a general intellectual 'revolution' that took place in the mid-sixties, this Euro-or ethno-centric perspective on development was put into discussion by social scientists in Latin America, and gave birth to the theory of dependency and underdevelopment. The approach from the dependency was part of a general structural re-orientation in the social sciences. The 'dependentistas' at first, they concentrated on the effects of dependence in the countries peripheral, but implicitly in their analyzes was the idea that development and underdevelopment they should be understood in a global context (Chew and Denemark, 1996). The paradigm of dependency played an important role in the movement the New World Economic Order (NWEO) and the New World Order of Information and Communication (NOMIC) from the end of the 1960s to the first years of ninety. At that time, new states in Africa, Asia and the success of the Socialist movements in

Cuba, China, Chile and other countries provided the objectives of political, economic and cultural self-determination within the community of the nations. These new countries shared the idea of becoming independent of the superpowers and moved to form the Non-Aligned States. The movement Non-aligned defined development as a political battle" (Servaes and Malikhao 2012:12).

The dependency approach helped to make a clear difference between development and underdevelopment conditions. The analysis went further from the economic dependency to the cultural dependency. In this field found a relation between mass communication as a key tool for cultural domination. Some followers of the dependency theory conclude "...in short, that underdevelopment is not due to the survival of archaic institutions and the existence of capital shortage in regions that have remained isolated from the stream of world history. On the contrary, underdevelopment was and still is generated by the very same historical process which also generated economic development: the development of capitalism itself" (Cockcroft et al. 1972:9).

### 23.1.1.3 Understanding C4D Path in Latin America Before Communication for Social Change Started

Beltrán presented a three stage model of the communication for development evolution in Latin America in 1993. This trend started by the end of the 1950s and lasted until the end of the 1980s. This description includes three major conceptualizations that have prevailed in Latin America in reference to understanding the relationship between social communication and social development: "development communication," "development support communication," and "alternative communication for democratic development" (Beltrán 1993).

The latest period, alternative communication for democratic development, somehow is still alive leading some specific practices of C4D. It was characterized by "...expanding and balancing people's access and participation in the communication process, at both mass media and interpersonal grassroot levels. Development should secure, in addition to material gains, social justice, freedom for all, and the majority rule" (Beltrán 1993).

This perception was linked to Paulo Freire's ideal of social change in Latin America that started with the promotion of a radical shift in the institutional educational practices considered oppressive by its authoritarian and instructional style and which have impacted on the communication field claiming a change in perceptions and practices from extension structures to dialogic and democratic contacts among people. Freire's democratic utopia in communication found close relation to rejecting the sender/receiver schema that operates under a persuasive paradigm changing it with an alternative communication, participatory, democratic, popular, and horizontal experience bringing back the meaning of human communication process. Freire said, "What is utopian is not unattainable. It is not idealism. It is a dialectic process of denouncing and announcing; denouncing a dehumanizing structure and announcing a humanizing structure" (Beltrán 1979:1).

In Latin America, there is still a debate between searching objectives for social change through communication processes (utopia) and advocating technological use of the mass media expecting to reach large numbers of people to influence their behavior and attitude (information). These views arose and clearly clashed in the 1980s although the grassroots practices outside the theoretical framework showed their own perspectives and goals linked to producing social change.

### 23.1.2 Communication for Development in Latin America: A Trend of the Past 20 Years

Beltrán's three moments of the existence of communication for development in Latin America identified the timeline of this experience starting in the end of the 1950s and going along toward the 1980s (Beltrán 1993). Years later, Beltrán enlarged his analysis that took 10 years through his study *Communication for Development in Latin America: An Account of Fifty Years* (Beltrán 2005). This study extended the scope until the 1990s presenting a review of some of the advances and failures of the trend related to reaching the utopia of communication and development in Latin America under the principles of access, dialogue, and participation.

An attempt to draw the scenario from that path to the following years might be an almost impossible task due to its complexity and broadness. However, it might be useful to keep using Beltrán's three stages of communication for development to organize an approximate review so that it is necessary to take a look into the Latin American context in which the communication and development is located.

#### 23.1.2.1 Political Context in Latin America and the Growth of C4D After the 1990s

The political scenario of Latin America in the 1990s showed the consolidation of democratic structures of government that came back to the region as a common aspect by the end of the 1970s. Cayuela describes the political context of Latin America stating, "Throughout 1978 and 1979, a series of political events occurred that seemed a presage of the progressive democratization of Latin America, a continent traditionally ruled by military dictatorships. Presidential elections were held in Bolivia, Brazil, Colombia, Costa Rica, Ecuador, Guatemala, Paraguay, the Dominican Republic, and Venezuela. Other forms of popular vote also took place: parliamentary elections in Brazil, parliamentary and municipal elections in El Salvador and Guatemala, and intended elections for the formation of a constituent assembly in Peru" (Cayuela 1980: 29).

Despite the existence of democratic structures in this region, there were not only differences among countries and their realities but also perceptions and practices regarding the same democracy. Uranga expresses that democracies that came after strong dictatorships in Latin America can be known as minimal democracies because of factors such as election of authorities through voting, alternation in power, constitutional legality, and the fact that rights such as association and information might be far from a true democratic sense. This also turned into an obstacle for ideals or proposals on national projects, political sovereignty, or economical independency due to the inheritance of neoliberal governments that dominated the economy and policies in the 1980s leaving an effect of a greater concentration of political and economic power associated with the loss of prestige of political parties for new governments (Uranga 2005).

An undermined democracy also leads to an increase in inequalities and avoids equal opportunities for a good living among population, especially among historically neglected sectors of society. This context demanded communication for development or communication for social change as a necessary means to defend human rights and also to produce political structures that can represent the interest of the powerless. One of the tasks of democracy, within this context, is to allow living in a reality that is essentially diverse. Therefore, according to Uranga, communication for development needs to promote democratic communication by working on otherness and its complexity in order to avoid the traditional existence of unilateral powers (Uranga 2005).

Under actions and persistence to pursue an alternative communication in Latin America, debates on practice and concepts had been guided, since the 1990s, by civil movements and their expectation for bringing back and expanding the notion of the right to communicate that was proposed at UNESCO's debates and the MacBride's report long ago (MacBride 1980).

Throughout the 1990s, this critical approach of communication has shown its presence in the communication for development field in Latin America, mainly oriented to discuss the neoliberal economy models. This is because the political environment stated norms that affected people's interest in their access to telecommunication, radio frequencies, and the concentration of means of communication among private and even public networks.

Following up Beltrán's third stage of development for communication in Latin America, alternative communication for democratic development, three attempts to describe a new scenario after the 1990s were made. The first one refers to development-communication or communication for development coined by Adalid Contreras Baspineiro (2000). Contreras refers to this new moment as, "Development as a conscious process design built by people. It starts with a horizon that is constructed, everyday, from a substantial, contradictory field with cultural conflicts which is permanently reconstructed under continuous tension" (Contreras 2000:21). This scenario shows new routes that link communication and development. Contreras explains that communication is then enriched by studies on cultural consumption, drawing a paradigm that puts value on mediation and meaning shift regarding the appropriateness of messages due to cultural complexity.

The second one, within the evolution of communication for development, was proposed by Alfonso Gumucio Dagron who called the period after the 1990s as communication for social change (2007). Gumucio believes "The contribution of Latin America that has pioneered communication for development" is significant, both in generating concrete experiences and also critical thinking. We have developed critical thinking on communication for social change as well as for development, from a participatory perspective, which is one of our remarkable advantages. We have developed important experiences at a local, national, and regional level, and "we have proved that citizen and community communication that is based on dialogue and participation ensures sustainable and appropriate social development and social change" (Gumucio 2007:1).

Gumucio explains, "From communication to development, communication for social change has inherited a concern for culture and community traditions, respect for local knowledge, horizontal dialogue between development experts, and development subjects. While communication for development became an institutional model, and to some extent vertical, applicable and replicable, as shown in experiences supported by FAO, communication for social change does not intend to define neither the means, the messages, nor the techniques before hand because it is considered that it is from the process itself, inserted in the community universe, that proposals for action must emerge" (Gumucio 2003:22).

The concept of communication for social chance has main premises to be characterized. Gumucio takes into account seven aspects as strengths for CFSC: "a. the sustainability of social change is safer when affected individuals and communities embrace the process and the communication contents as theirs; b. the CFSC, which is horizontal and empowers community feelings, should spread the voice of the poorest, and have at its core local contents and the view of appropriation of the communicational process; c. communities must be agents of their own change and managers of their own communication; d. instead of the emphasis on persuasion and the transmission of information and knowledge from outside; e. the CFSC promotes dialogue, debate and negotiation from within the community; and. the results of the CFSC process must go beyond individual behaviors, and take into account social norms, current policies, culture and the context of development; f. the CFSC means dialogue and participation, with the purpose of strengthening the cultural identity, the trust, the commitment, the appropriation of the word and community strengthening; g. the CFSC rejects the linear model of transmission of information from a sender center to a receiver. It promotes a cyclic process of interaction between community shared knowledge and its collective action" (Gumucio 2003:23).

Gumucio says, "The central concept of communication for social change has been presented as a dialogic process and debate which is based on tolerance, respect, equity, social justice, and active participation of all" (Gumucio 2003:22).

The third one, view that came into Latin America after the 1990s, is Communication from Diversity coined by José Luis Aguirre Alvis, which might still be current. It explains that today new constants converge which have to do with the emergence of debates on decoloniality, complexity in the social sciences scenario, inclusive citizenship, as well as the observation of the same technological environment of telecommunications that unlike the previous belief that discourses and media would concentrate on a few hands can be seen as spaces of opportunity and inclusion of greater diversity. At the moment, communication makes its intercultural dimension visible, and perhaps because of this, it recovers its essential meaning: communication is only given. It is possible a goal thanks to the encounter of one ego with another one. They function in a constant discovery without ignoring the existence of tension as well as the possibility of experiencing the same dialogue as a transitory condition. This tension is also the product of being left or not occupied by an alter ego or another. It can also be considered that the communicative space itself is supposed to be given through the encounter between diverse communities because only then this difference, which creates the experiential, existential, and symbolic load, is capable of generating the construction of a space for dialogue in which one willingly participates without a desire of invading the Other. Let's imagine that the Other takes part in the construction of the meaning dynamics, contributing to the encounter and construction of this sense considering diversity and divergence which gives the communicative experience the demand of being willing to listen, being welcoming and generous, and accepting that communication is the means to learn who one is as well as who the other one is as a result of the interaction between them (Aguirre 2011:6).

### 23.1.2.2 Current Practice of Building Alternative Communication for Development

Latin America has been the ground of experiences that nurtured practice of alternative communication for democratic development in the past 20 years. This trend can be summarized on five lines of evidence that can describe part of the contemporary C4D context in this region: (1) the use of radio as a central promoter of communication for development; (2) the introduction of the right to communicate as a central part for a communication for social change; (3) the presence of diverse social sectors using communication to fight for rights; (4) knowledge production from bringing back the concept of C4D in Latin America, communication for Living Well, and a decolonial approach; and (5) training context, building capacity in C4D, and knowledge production.

## 23.2 The Use of the Radio as a Central Promoter of Communication for Development

Radio in Latin America is strongly and historically related to experiences of alternativity, ranging from alternative property, diverse actors and their needs, diverse bands of broadcasting, diverse power, and technology to a direct presence of listeners and community radio, which are part of a tradition which had been rooted in almost 70 years. Community radio, as it is known today, has gone through a long previous process using radio stations in the rural area and native speakers in urban radio stations. This had been somehow similar to practices of popular communication. Radio is still the means that creates proximity with its listeners and has more chances than others to establish participatory relations. Community radio, in several countries and regions in Latin America, created varied and rich diversity of experiences. This might have caused that many studies related to C4D in Latin America have radio as a central point. A tradition of community radio, which started in the 1940s with the genesis of the Miner's radio, is still going on with different conditions regarding the legal framework. After a long fighting in some countries, community radiobroadcasting is recognized legally. Radio in Latin America, especially in the rural and native areas, keeps the oral aspect of its local cultures which can fit contextual communicative systems as its most important advantage. Due to an unfair lack of acknowledgment of the existence and role of community radio in the 1990s, this condition has changed in some countries, so that it has become a legal practice that arouse in the middle of traditional commercial radiobroadcasting and government radiobroadcasting. Community radiobroadcasting is a third path and no longer an invisible experience of using a means of communication in order to reach social change. Not only segments of rural population at international borders and isolated regions but also cases of community radio within urban areas are a spread growing experience. This is the case of school radio that was spread in Colombian projects which is linking the use of radio and rebuilding a culture of peace.

However, despite the power it has as alternative media, community radio has found opposition in some countries where these cases are rejected by government or commercial interests. Regarding changes in relative regulations to allow the existence of community radiobroadcasting, Herrera explains, "Regarding secondary bills, the Organic Bill of Communication of Ecuador passed in 2013 proposes the democratization of the radio spectrum by allocating 34% of radio and television frequencies for community media, 33% for public media and 33% for commercial media. Allocating at least a third of the spectrum seeks that popular organizations, indigenous peoples and social movements (youth, environmentalists, feminists, etc.) could create means to freely and fully exercise their right to communication, producing and transmitting their own content. Something similar happened to audiovisual media bills of Argentina and Uruguay – passed in 2009 and 2014- which were discussed by the entire population, approved by a large majority in national congresses and endorsed by constitutional courts. However, the Argentine bill was changed by executive decree in 2016 generating a strong rejection from community media, social organizations, universities and even the office of the Rapporteur for Freedom of Expression of the Inter-American Commission on Human Rights (CIDH) which urged Mauricio Macri's administration to pass a new law in accordance with international standards of freedom of expression (Busso 2016)" (Herrera 2017:7).

Regarding Bolivia, its constitution asserts that the "State shall support the creation of communitarian means of communication with equal conditions and opportunities in Art. 107" (Bolivia 2008). Telecommunications Bill 164 of 2011, also acknowledges a new division of the spectrum allowing the existence of community radiobroadcasting, Article 10. Distribution of frequencies for broadcasting. I. Distribution of all frequency band channels for the service of modulated frequency broadcasting and analog television nationally where there is availability will be subject to the following:

1. "State, up to thirty-three percent.
2. Commercial, up to thirty-three percent.
3. Community social, up to seventeen percent.
4. Peasant indigenous peoples, intercultural and indigenous communities as well as African-Bolivian communities, up to seventeen percent" (Bolivia 2011:15).

In contrast to both cases, this scenario is unequal and complex for the rest of Latin American countries. Throughout the 162nd period of public hearings of the Inter-American Commission on Human Rights (IACHR) held in Buenos Aires, Argentina (May 2017), Javier García a representative of Observacom stated, "Out of the thirteen countries studied in Latin America, only four of them, Argentina, Uruguay, Bolivia, and Colombia, have a regulation that is in agreement with Inter-American standards developed by the Rapporteurship (Special Rapporteurship for Freedom of Expression) and IACHR. Then, the other nine countries recognize community media but with discriminatory rules of higher or lower severity."

At the same IACHR hearings, it was concluded that "In Guatemala, Chile, Brazil and Peru, communities that operate their radio stations and then arrange their regulation with the government, are criminally penalized. In Peru and Guatemala, advocates of these radios are punished with the crime of 'aggravated theft' or 'theft of frequency,' which the AMARC representative explained has no relation to radio broadcasting. Legally, you cannot steal a frequency because the theft is either of energy or heat, and a radio frequency does not adhere to either situation. This has been explained several times by the ITU (International Telecommunications Union, of the United Nations), Loreti explained." In addition, García from Observacom said that "Legal recognition of community broadcasting is not enough. Rules of recognition of this media contain discriminatory conditions that prevent their development." He added that "restrictions on the exercise of freedom of expression continue. Several Central American countries continue to deny this right to indigenous communities and indigenous peoples to have their own media outlets." This is the case of Guatemala. García said that another difficulty faced by community radio stations in many cases, according to Loreti, is their definition as rural radios. "Radio stations should be recognized with a community of interests that are not limited to a specific geographic location." Loreti explains that this is happening in Peru and Paraguay, where the definition of "community" unequivocally mandates that they should be rural or remote radios. One of the examples mentioned by this researcher was Brazil "whose legislation only allows community radios to have a kilometer, a little more than half a mile, of coverage in the radio spectrum."

To the field of communication for development, community radiobroadcasting in Latin America represents the ground in which the right to community is under fight. New community sectors, innovative approaches for using radiobroadcasting potential, and even finding convergence with new technologies to offer opportunities to deliver messages through online radio are arising in several countries of the region. Another case is the recent opening of online radio that has also been more affordable for social sectors that were neglected access to radiobroadcasting frequencies. This digital shift is not totally present in Latin America, but it is expected to do so in the upcoming years despite the fact that technology and access to this new system might be slow and somehow costly for listeners and viewers. However, this expansion of channels might be an opportunity to redistribute the access to sources that were scarce in the analogical system.

## 23.3 The Introduction of the Right to Communicate as a Central Goal of Communicational Changes

There are evident changes in the legal framework for communication and information of countries that consider the practice of communication as a human right. Regarding this, Latin America is the only region in the world that can show cases in which the right to communicate has scaled to reach the constitutional level. This happens in two countries: Bolivia and Ecuador.

The right to communicate was initially established by Jean D'Arcy in 1969. It expressed a dream in which one day, all states might recognize a new right, the right to communicate (J. D'Arcy Revue de l'UER, November 1969), and this seems to be the case of Latin America.

Herrera explains that Latin America reports important progress in the recognition of the right to communicate in regulation frameworks and its implementation through public policies aimed to building new media models that ensure content diversity and plurality of voice plurality in the public debate. In Bolivia and Ecuador, the right to communicate is recognized in their Constitutions. "In Argentina and Uruguay, bills were passed to democratize the spectrum of audiovisual media, and in El Salvador, there were legal reforms that at least recognize community media and democratize access to media and radio electric spectrum" (Herrera 2017:5).

This new legal framework creates conditions in favor of a communication for development and social change. According to Herrera, the Constitution of Ecuador, approved in 2008, establishes a section on Communication and Information in the Chapter of "Rights of Good Living." Article 16 states that all individuals, independently or collectively, have the right to:

- "Free, intercultural, inclusive, diverse and participatory communication, in all areas of social interaction, by any means and form, in its own language and with its own symbols.
- Universal access to information and communication technologies.
- The creation of social communication media, and equal access to radio frequencies for management of public, private and community radio and television stations, and free bands for wireless networks functioning.
- Access and use of all forms of visual, auditory, sensory and other communication that allow the inclusion of people with disabilities.
- Incorporating participation spaces established in the Constitution in the field of communication" (Herrera 2017: 5).

Regarding Bolivia's case, the Constitution of the Plurinational State of 2008 introduces a specific chapter for social communication and in two articles (106 and 107) states "I. The State ensures the right to communication and the right to information (Article 106). II. The State ensures Bolivians the right to freedom of expression, opinion and information, to amend and reply, and the right to freely publish ideas by whatever means of publication, without prior censorship. Freedom of expression III. The State ensures freedom of expression and the right to communication and information to workers of the press. IV. The conscience clause on workers' information is recognized. Article 107. I. Public means of communication must contribute to the promotion of ethical, moral and civic-minded values of different cultures of this country with the creation and publication of multi-lingual educational programs and in an alternative language for the disabled. II. Information and opinions issued by public means of communication must respect principles of truth and responsibility. These principles shall be put into practice through rules on ethics and self-regulation of journalists' organizations and means of communication and their law. III. Public means of communication shall not form, either directly or indirectly, monopolies or oligopolies. IV. The State shall support the creation of communitarian means of communication with equal conditions and opportunities" (Bolivia 2008: 42–43).

The right to communicate is also promoted by the civil segment and academic organizations in Latin America. This is the case of the CRIS (Communication Rights in the Information Society) movement that was articulated to represent civil interest into WSIS debates (2005 and 2006) even if the goals of this organization could not greatly affect international WSIS (World Summit of Information Society) debates at the local level and that experience was later focused to promote the introduction of the article of right to communicate throughout the Constituency process (2005) in Bolivia allowing its introduction. Besides this, the ecumenical movement of the World Association for Christian Communication (WACC) in Latin America is also a promoter of this debate and supports experiences on the ground regarding right to communicate even expanding the concepts to protect alternative practices such as the right to communicate of people with disabilities or traditional indigenous practices of communication seen in several Latin American countries.

## 23.4 Existence of Diverse Social Sectors Fighting for the Right to Communicate

Indigenous people, youth, children, women, rural women, people with disabilities, and migrants can be seen as part of the promoters of this media and strategies of communication that are spread in Latin America. These civil society movements have a common goal which is to develop capacities in communication or using the media to promote their visibility and to express their demands particularly in cultural aspects, reach security and peace, and being acknowledged as citizens with equal rights.

This kind of practice occurs in several countries of this region, and some of the recent cases represent the struggles for preserving their security, cultural survival, and use of native languages. As a short reference, a few cases of C4D in Argentina, Guatemala, and Colombia can be mentioned. According to a comparison of experiences presented by WACC (www.waccglobal.org), those cases are as follows.

In Argentina, MOCASE, an organization that belongs to the National Movement of Indigenous Peasants, located in Santiago del Estero in northern Argentina works to protect land rights of indigenous peasants and promote solidarity, healthy food, agroecology, development, justice, and social change. The indigenous peasant people in the area are constantly threatened to be evicted from their land and have limited access to means of communication. The mass media in the area replicates discrimination and marginalization felt by the rural indigenous people. MOCASE felt that it is essential to construct their own means of communication that can help communities to generate and distribute their own information that reflects their life and struggles and help to build consensus as they move forward. This project established a community radio station and provided technical and academic training in radio production to local youth. The radio station allows youth to articulate and share views on the problems they face and protect their indigenous identity and lifestyle. It also reinforces the links between rural people to advocacy of nature and food sovereignty.

Another case of C4D in Argentina is oriented to recuperating indigenous languages. This experience has been developed by Fundación ASOCIANA that works with a native organization called Lhaka Honhat located in Pilcomayo, Salta province. This project works with four native communities, Wichí, Chorote, Chulupí, and Toba, in the Pilcomayo area in order to defend their territory by developing training in communication within the communities so that they can make their voice be heard toward the whole Argentine society by demanding respect for their culture and practices.

In Guatemala, Central America, there is a set of C4D cases, and some of them are taking advantage of radiobroadcasting. One of these ones is the experience of The Voice of the Hills supported by Asociación Estoreña para el Desarrollo Integral (AEPDI). El Estor is a municipality where 80 percent of its population speaks Q'eqchi and considers itself to be part of the Mayan nation. The area receives little attention from the state in terms of development, and there is a marked lack of media content that reflects the community's priorities. Over the past several years, AEPDI, one of the most active community organizations in the area, has used commercial media – and paid high fees – to promote indigenous rights and demand the right to land and territory for their communities. In light of this, AEPDI has been attempting to establish a community-owned radio station for years hoping that the station would become a means to inform, educate, and entertain the community. This project consists of establishing a community radio station in El Estor. This entails the purchase and equipment installation and training for community reporters and editors. The establishment of the station will be the first step toward enabling people in El Estor to fully exercise their communication rights. This radio station also seeks to become a vehicle to strengthen indigenous governance systems and community organizing efforts.

"Freedom of Expression for the Q'eqchi' People," a project that aims to strengthen Radio Nimlajacoc in Guatemala, is another example. Nimlajacoc is a geographically isolated community in Guatemala. Nimlajacoc was impacted by civil war during the 1980s as many families were murdered, kidnapped, or displaced. After the conflict, local indigenous authorities established a community radio station as a way to facilitate communication among survivors, as well as to help local communities to get organized. Today, the Nimlajacoc community is facing uncertainty over a possible approval of a local hydroelectric project that would threaten natural resources and violate their right to Free, Prior and Informed Consent. The community radio station plays a critical role in spreading information and reinforcing bonds of solidarity and collaboration.

A third case, in Guatemala, is Radio Ixchel, a community radio station serving the community of Sumpango Sacatepéquez located far away from Guatemala City. The station works to meet the communication needs of community members including the need for locally relevant information, availability of culturally relevant and sensitive content, and preservation of the Kaqchikel language. In recent years, the Sumpango Sacatepéquez community has seen a decline of the use of indigenous languages and a diminished appreciation and understanding of indigenous world views, especially among youth. This is the result of, among other things, close proximity to the capital city. Through volunteers and community reporters' training, Radio Ixchel helps to reinvigorate social change objectives. This project focuses on creating new participation spaces for children and youth within Radio Ixchel.

A fourth case is a Training Building Project for community communicators at Radio Xilotepek, a Maya Poqomam community radio that serves in San Luis Jilotepeque community, department of Jalapa, in southeastern Guatemala. This community faces issues brought by the growth of the mining industry in this region. Most community members disagreed with the growth of extractive industry. This project seeks to strengthen Radio Xilotepek's ability to educate community members about their rights, especially in relation to the preservation of natural resources.

A fifth case in Guatemala is Radio Sinakan, a Kaqchikel community radio located in Chimaltenango, Guatemala. It plays a key role in the preservation of the Kaqchikel language as its use is declining due to the community's proximity to Guatemala City. This project trains volunteers so they get involved as community reporters and media editors able to make significant contributions in terms of media content that draws on rights frameworks. In the long-term, this project will help this region in Guatemala to preserve and reinvigorate indigenous identity and culture.

The project entitled Youth Building Peace, located in Colombia, sponsored by Comunicarte Group is our last example. This project seeks to engage youth and adolescents, living in Arauquita village, department of Arauca, in the Colombian Orinoco region, in media-based initiatives to build peace for their communities. Building on the momentum of peace agreements between the state and the FARC reached in Colombia in 2016, this project seeks to enable youth in Arauquita to use media to protect themselves from other armed actors in their communities that might fill the void left by the FARC, as well as to promote their participation and engagement in local peace initiatives. Special attention is given to providing youth with the opportunity to use media to tell their own stories, share their dreams and aspirations, and advocate for peaceful conflict resolution in their schools and neighborhoods. Project activities seek to foster active citizenship expressed through media and communication.

## 23.5 Knowledge Production from Returning to C4D in Latin America, Emerging Communication for Living Well, and a Decolonial Approach

Knowledge production in the field of C4D in Latin America has been showing reference materials in planning, theoretical proposals, and a return to a critical thinking tradition, systematizing experiences, and the alternative media approach. From an extensive list of publications, articles, and documents, we have selected some of them to give you a glance at this rich area of thought and action.

Regarding C4D's planning stage, the following is a glance at literature written in Spanish: *Strategic Communication-Communication for Innovation* by Sandra Massoni (2011) and *Towards a General Theory of Strategy-The Paradigm Shift in Human Behavior, Society and Institutions* by Rafael Alberto Pérez and Sandra Massoni (2009). An updated reflection on the C4D field is also proposed in *Grays of Extension, Communication and Development* by Ricardo Thornton and Gustavo Cimadevilla (2008). An extended review of the C4D field rooted in new visions of the South is discussed in tree volumes of *Communication, Technology and Development* compiled by Gustavo Cimadevilla, Debates and perspectives from the South (2002); Current debates (2004a), and Debates of a new century (2006). Gustavo Aprea also compiled a set of documents about debates in communication for development in *Problems of Communication and Development* (2004), and Gustavo Cimadevilla presents deep critics toward the practice of communication and development in *Domains Criticism of Interventionist Reason, Communication and Sustainable Development* (2004b). And the Latin American Institute of Communication for Development (ILCD) located in Paraguay designed an updated approach to C4D through an inclusive perspective in *From Communication. Bet for Inclusive Development* (2011).

Thought development in the C4D field also has been innovated by the emergence and new reflections connecting concepts of communication for Living Well and the theoretical approach of communication under a decolonization model.

The relation between communication and the paradigm called Living Well, which emerged in Latin America, came from the political arena and was related to the discourse of Latin American governments aligned to radical social changes. The concepts of Living Well are used depending on the country where the debate is set. The representative of this new approach is Adalid Contreras with *Sentipensamientos, from Communication-Development to a Communication for Living Well* (2014) written in Spanish; *The Limit is Infinity-Relations between Integration and Communication* (2015) also in Spanish; *The Word that Walks. Popular Communication for Living Well* (2016) in Spanish; and *Jiwasa: Participatory Communication for Coexistence* (2017). Regarding radio and development practice, the Latin American Association of Radiophonic Education (ALER) contributed by setting a framework for a practice of right to communicate and living well with *Sowings of Good living. Between utopias and Possible Dilemmas* (2016) written in Spanish.

Erick Torrico is a representative of communication from the paradigm of decolonization with *Towards Decolonial Communication* (2016a) and

*Communication Thought from Latin America; 1960–2009* (2016b), both written in Spanish. Another broad document in the same line is *Communication and Decolonization. Horizon under Construction* by Erick Torrico and others (2018) written in Spanish.

In Latin America, the intellectual production in C4D has also been oriented to bring back a tradition in critical thinking, systematizing experiences, and alternative media approaches. This is a glance of noteworthy literature such as *Another compass. Innovations in communication and development* by Rosa María Alfaro (2006). Compilation made by the Latin American and Caribbean Catholic Organization of Communication (OCLACC) and Universidad Técnica Particular de Loja *Communication, citizenship and values. Reinventing concepts and strategies* (2008). Additionally, *Communication, education and social movements in Latin America* by César Bolaño and others (2010) is a compilation that links communication for development and social movements in Latin America. Identifying experiences of C4D in the Colombian Andes, we can find a large map of cases in *Experiences of communication and development on the environment. Case studies and life stories in the Andean region of Colombia* by Herrera Huérfano and Eliana del Rosario (2011). A compilation written by José Miguel Pereira and Amparo Cadavid that presents a broad range of topics related to C4D is *Communication, development and social change. Interrelations between communication, citizen movements and media* (2011). From a perspective of the right to communicate and indigenous movements, we have *Right to communication; reality and challenges in Latin America. Indigenous peoples and public broadcasting policies* by Rosa Elena Sudario (2013). Another extended compilation of documents of the current debate of C4D in Latin America is *Thinking from experience. Participatory communication in social change* by Amparo Cadavid and Alfonso Gumucio (2014). Another compilation of cases of C4D and environment is *Emergency of the territory and local communication. Experiences of communication and development on the environment in Colombia* by Eliana Herrera Huérfano and others (2014).

In Brazil, we can also find work on C4D written by Cecilia M. Krohling from Universidade Metodista de Portuguese. She is one of the most prolific writers in this field. That is the case of her work *Communication for Development, Communication for Social Transformation* (2014) written in Portuguese.

*Communication, Citizenship and Democracy: for a Fulfilled and Plenty Life* by SIGNIS ALC (2018) written in Spanish is a document that recovers papers of the 5th Latin American and Caribbean Congress of Communication (V COMLAC) that took place in Paraguay and which brings back Juan Diaz Bordenave's ideas. He was the first theoretician of C4D and rural communication in that country.

Academic production on C4D reveals important additional sources in Latin America. One of them is Asociación Latinoamericana de Investigadores de la Comunicación (ALAIC), the largest organization of scholars of communication research in the region which started in 1978. Events promoted by ALAIC encourage the knowledge production under specific areas of interest. This is the case of Work Groups (Grupos de Trabajo – GTs) introduced in its 2nd Congress that took place in Guadalajara, México, in 1994. A GT of communication for social change is part of it. This GT is one of the most active groups among 20 others under ALAIC (www.alaic.org) that submit advances in the theory and practice of C4D in this region.

Additionally, *Anthology of Communication for Social Change: Historical and Contemporary Readings* by Alfonso Gumucio and Thomas Tufte (2008), an extensive and complete recompilation of main international documents that follow the trend of C4D, printed in Bolivia, should be mentioned due to its relevance for the study of C4D in Latin America. This anthology was originally written in English and later on was expanded and published in Spanish thanks to the support of Communication for Social Change Consortium (CFSC).

## 23.6 State of Training and Capacity Building in C4D

The field of communication for development is nurtured by training projects as well as university programs specifically oriented to this field in Latin America. Pontificia Universidad Católica del Perú has provided a venue to study communication for development which is offered as a concentration at Facultad de Ciencias y Artes de la Comunicación. This program's description is the following: "The concentration on communication for development promotes the analysis and management of communication strategies in order to generate or improve processes of interpersonal, group, and mass communication for social development. It seeks to prepare students to do research, design, manage and implement strategies, actions, and messages in communication projects and organizations, emphasizing current issues for the improvement of quality of life, such as health, education, citizenship and human rights, gender, institutional development, productivity and environment."

Gumucio (2002) states that, in Latin America, there are important academic efforts that favor C4D and comments on experiences of Universidad Nacional de Tucumán in Argentina, Universidade Metodista de Portuguese in Brazil, Universidad de Lima in Peru and Universidad NUR in Santa Cruz, and the Universidad Andina Simón Bolívar in La Paz, in Bolivia. All these cases are mentioned due to the existence of graduate programs in C4D training.

One of the most visible projects of training to develop capacities in C4D is Onda Rural "a regional initiative that was born to strengthen the exchange and collaboration between stakeholders interested in the participatory use of community media and ICT for family farming, resilience and sustainable rural development in Latin America. Onda Rural starts a new stage, promoted by the International Center for Advanced Communication Studies in Latin America (CIESPAL) and the United Nations Organization for Food and Agriculture (FAO), with the collaboration of the Specialized Meeting on Family Farming (REAF MERCOSUR), the World Association of Community Radio Broadcasters (AMARC) and the Latin American Association of Radiophonic Education (ALER)" (Onda Rural, www.ondarural.org).

Onda Rural offers a distance training course in Communication for Rural Development. According to its promoters, "This course provides a global view of the Communication for Development (CpD) approach, which seeks to address the information and knowledge needs of rural actors as well as to enable their participation in development initiatives. This course aims to guide users in the design and implementation of communication strategies for agricultural and rural development initiatives by combining participatory methods with processes, means, and communication tools, more appropriate to a specific context, which vary from community media to Information and Communication Technologies (ICT). This course mainly targets facilitators and field agents, communicators and development professionals who need to improve their skills to formulate and implement participatory rural communication strategies" (Onda Rural, www.ondarural.org).

Universidad Andina Simón Bolívar, UASB, created in 1985 as an academic mean of Comunidad Andina de Naciones, opened its studies in communication and journalism in 1996. After this, in 1998, the first Master's degree in Communication and Development was opened. Today this experience has changed its venue to become a virtual program; therefore, it is the first case of this concentration in the Andean region, and right now it's offering this program for the tenth time under the name of Strategic Communication (www.uasb.edu.bo).

Finally, an online tool which promotes debate and practice on C4D in this region is the Communication Initiative. Its proposal reaches Latin America and Caribbean regions in order to build communication for development practices and studies. This online resource is activated and permanently active on www.comminit.org.

## 23.7 Conclusions

Communication for development in Latin America has a long tradition of practical and theoretical history. This field was not a result of academic proposals, but it came as a practical result of the specific conditions of social and technological inequality and the need to protect the existence and cultural richness of local societies in the region. This approach was historically characterized in the region by its criticism and opposing concepts or models of unidirectional and persuasive structure. The theoretical introduction of C4D to Latin America presents different stages, and each one brought a specific view on communication as a human process and also a conceptualization of social development. Today, the ground of C4D has a close connection to social demands and people's movements like the right to communicate, language, and local systems of communication preservation, land security, and also social and cultural vindication. Several diverse community actors participate in these movements.

The soul of this utopia set by the most distinguished specialist in communication for development and C4D practices in Latin America, Luis Ramiro Beltrán, is persistent. These concepts have been expanded with the incorporation of different actors, means and resources, objectives, and demands showing that the practice of communication for development in Latin American acquires sense and an ongoing need supported by integral political, ethical, and aesthetic dimensions to envision more democratic and inclusive societies.

The means able to keep a tradition in C4D in Latin America is still marked by radiobroadcasting and community radio using native languages, new technological applications, and local innovations more widely. It's also marked by the voices of diverse social sectors which gradually recognize that communication is a skill on its own that empowers their presence in order to reach conditions for equal participation. Analyzing practice and the utopia of a democratic and inclusive society goes through empowering means to use communication processes which follow already old elements based on access, dialogue, and participation. These can lead to self-empowerment and to the possibility to reach social justice in societies structurally diverse and culturally plural. Communication is a means rooted in self-expression which might prompt effective social change.

A regional thought about communication and communication for development is providing new opportunities for reflection because of an increase of university programs so that the body of knowledge in this field can expand. Additionally, new approaches that bring back a critical debate on communication and social change are emerging. This is a new paradigm of Living Well. This also happens thanks to an understanding of how conditions rearrange communication processes as well as development under a decolonial approach. Latin America might have the opportunity to offer new dimensions to the communication for development field in the near future. They could avoid the use of the same view, but due to administrative or technical tendencies, as development and interdevelopment is a result of structural and historical forces within the construction of social systems, it could also be neglected once again.

## References

Aguirre Alvis JL (2011) Enfoques teóricos para una comunicación orientada al desarrollo y retos actuales para una comunicación y desarrollo desde la diversidad. Simposio Internacional de Interculturalidad y Educación Superior: Desafíos de la diversidad para un cambio educativo. La Paz

ALER (Asociación Latinoamericana de Educación Radiofónica) (2016) Siembras del buen vivir. Entre utopías y dilemas posibles. Quito

Alfaro Moreno RM (2006) Otra brújula. Innovaciones en comunicación y desarrollo, Lima, Calandria

Aprea G (compilador) (2004) Problemas de comunicación y desarrollo. Prometeo libros. Buenos Aires, Argentina: Universidad Nacional General Sarmiento

Atwood R, McAnany EG (1986) Communication and Latin American society. Trends in critical research, 1960–1985. The University of Wisconsin Press, Madison

LR Beltrán Salmón (1979) Farewell to Aristotle: horizontal communication. International Commission for the Study of Communication Problems. UNESCO. No. 48. Paris

Beltrán Salmón LR (1993) Communication for development in Latin America: a forty years appraisal. IV Roundtable on Development Communication, Lima

Beltrán Salmón LR (2005) La comunicación para el desarrollo en Latinoamérica: un recuento de medio siglo. III Congreso Panamericano de la Comunicación, Buenos Aires

Beltrán Salmón LR (2008) Development communication: Latin America. The international encyclopedia of communication. Volume III, Oxford

Bolaños C. et al. (organizadores) (2010) Comunicación, educación y movimientos sociales en América Latina. Brasilia, Brasil

Bolivia (2008) Nueva Constitución Política del Estado

Bolivia (2011) Law of telecommunications. Law No 164

Cadavid BA, Gumucio Dagron A (editores) (2014) Pensar desde la experiencia. Comunicación participativa en el cambio social. Bogotá, Colombia: UNIMINUTO

Cardoso FH, Falletto E (1979) Dependencia y desarrollo en América Latina. Mexico D.F.: Siglo XXI

Cayuela J (1980) El retorno a la democracia en América Latina. ¿Mito o realidad? Nueva Sociedad. No. 46 enero-febrero, pp. 29-38

Cimadevilla G (compilador) (2002) Comunicación, tecnología y desarrollo. Discusiones y perspectivas desde el sur. Río Cuarto, Argentina: Universidad Nacional de Río Cuarto

Cimadevilla G (compilador) (2004a) Comunicación, tecnología y desarrollo. Debates actuales. Río Cuarto, Argentina: Universidad Nacional de Río Cuarto

Cimadevillla G (2004b) Dominios. Crítica a la razón intervencionista, la comunicación y el desarrollo sustentable. Prometeo Libros. Buenos Aires

Cimadevilla G (compilador) (2006) Comunicación, tecnología y desarrollo. Discusiones del siglo nuevo. Vol. 3. Río Cuarto, Argentina: Universidad Nacional de Río Cuarto

Cockcroft JD, Gunder Frank A, Johnson DL (1972) Dependence and underdevelopment. Latin America's polityical economy, New York

Contreras Baspineiro A (2000) Imágenes e imaginarios de la comunicación-desarrollo. CIESPAL, Quito

Contreras Baspineiro A (2014) Sentipensamientos. De la comunicación-desarrollo a la comunicación para vivir bien. Ediciones La Tierra. Quito, Ecuador: Universidad Andina Simón Bolívar

Contreras Baspineiro A (2015) El límite es el infinito. Relaciones entre integración y comunicación. Ediciones CIESPAL, Quito

Contreras Baspineiro A (2016) La palabra que camina. Comunicación popular para el Vivir. Bien/ Buen Vivir. Ediciones CIESPAL, Quito

Contreras Baspineiro A (2017) Jiwasa. Comunicación participativa para la convivencia. FES Comunicación, Bogotá

D'Arcy J (1969) Revue de l'UER, noviembre. France, Paris

Díaz Bordenave J (1962) Latinoamérica necesita revolucionar sus comunicaciones. Revista Combate. Noviembre y diciembre. No. 25. San José, Costa Rica

Freire P (1970) Pedagogía del oprimido. Siglo Veintiuno Editores, México

Freire P (1973) Extensión o comunicación? La concientización en el medio rural. Siglo Veintiuno Editores, México

García J (2017) Hearings of the Inter-American Commission on Human Rights (IACHR) (https://knightcenter.utexas.edu/blog/00-18467-community-radio-stations-latin-america-discriminated-against-law-and-its-advocates-fac)

Gerace F (1973) La comunicación horizontal: cambio de estructura y movilización social. Studium, Lima

Gumucio Dagron A (2002) El cuarto mosquetero: La comunicación para el cambio social. IV Congreso de la Asociación Latinoamericana de Investigadores de la Comunicación. ALAIC, Santa Cruz

Gumucio Dagron A (2003) Comunicación para el cambio social: clave del desarrollo participativo

Gumucio Dagron A (2007) Three challenges of communication for social change

Gumucio Dagron A, Tufte T (compiladores) (2008) Antología de comunicación para el cambio social. Lecturas históricas y contemporáneas. La Paz, CFSC

Herrera H, del Rosario E (2011) Experiencias de comunicación y desarrrollo sobre medio ambiente. Estudios de caso e historias de vida en la región Andina de Colombia. UNIMINUTO, Bogotá

Herrera Lemus L (2017) Derecho a la comunicación y modelos de medios democráticos para el Buen Vivir. https://www.alainet.org/es/articulo/182790

Herrera H, E del Rosario et al. (editores) (2014) Emergencia del territorio y comunicación local. Experiencias de comunicación y desarrollo sobre medios ambiente en Colombia. Barranquilla, Universidad del Norte

ILCD (Instituto Latinoamericano de Comunicación para el Desarrollo) (2011) Desde la comunicación, apuesta por un desarrollo inclusivo. SICOM, Asunción

Krohling PC (1997) Escola Latino-Americana de Comunicação: Contribucoes de Luis Ramiro Beltrán. I Ciclo de Estudos sobre a Escola Latino-Americana de Ciencias da Comunicação. Sao Bernardo do Campo, São Paulo

Krohling PC (2014) Comunicação para o desenvolvimento, comunicação para a transformação social. In: Aristides Monteiro Neto. (Org.). Sociedade, política e desenvolvimento livro 2. 1ed. Brasília: IPEA, 2014, v. 2, p. 161-195

MacBride S (1980) Un solo mundo voces múltiples. Comunicación e información en nuestro tiempo. México: Fondo Cultura Económica

Marques Ferrari L, Antoniacci Tuzzo S (2013) José Marques de Melo and Latina American school of communication: work, thought and history. Comun Inf 16(1):98–112

Massoni S (2011) Comunicación estratégica. Comunicación para la innovación. Homo Sapiens. Santa Fe

OCLACC (Organización Católica Latinoamericana y Caribeña de Comunicación) (2008) Comunicación, ciudadanía y valores. Re-invetando conceptos y estrategias. Quito, Ecuador: Universidad Técnica Particular de Loja

Pasquali A (1970) Comprender la comunicación. Venezuela: Monte Ávila Editores

Peirano FL (2017) Entrar y salir (por el espejo) de los estudios de la comunicación. Balance temático de lo hecho y lo mucho por hacer. In. Comunicación y cambio. Carla Colona Guadalupe, Juan Jorge Vergara (compilators). Lima: FCE, Maestría en comunicacicones de la PUCP

Pereira JM, Contreras Baspineiro A (editores) (2011) Comunicación, desarrollo y cambio social. Interrelaciones entre comunicación, movimientos ciudadanos y medios. Pontificia Universidad Javeriana, Bogotá

Perez RA, Massoni S (2009) Hacia una teoría general de la estrategia. El cambio de paradigma en el comportamiento humano, la sociedad y las instituciones. Ariel. Barcelona, España

Reyes Matta F (1983) Comunicación alternativa y búsquedas democráticas. Instituto Latinoamericano de Estudios Transnacionales, México

Rogers EM (1994) The history of development communication. Comm Dev News, Newsletter of Ohio University's Communication and Development Studies Program. Vol. 5. No. 1

Servaes J, Malikhao P (2012) Communication and sustainable development. Selected papers from the 9th UN roundtable on communication for development. FAO, Rome

Sudario Manrique RE (2013) derecho a la comunicación realidad y desafíos en América Latina. Pueblos indígenas y políticas públicas de radiodifusión. Perú: SERVINDI

Thornton RD, Cimadevilla G (editores) (2008) Grises de la extensión, la comunicación y el desarrollo. INTA

Torrico EV (2016a) Hacia la comunicación decolonial. Universidad Andina Simón Bolívar, Sucre

Torrico EV (2016b) La comunicación pensada desde América Latina (1960–2009). Salamanca, España

Torrico EV et al. (organizadores) (2018) Comunicación y decolonialidad. Horizonte en construcción. La Paz

Uranga W (2005) Desarrollo, ciudadanía, democracia: aportes desde la comunicación. Anuario UNESCO/Metodista deComunicacaoRegional. Ano 9. n. 9.75–90, jan/dez. Brazil

WACC (World Association for Christian Communication). www.waccglobal.org

## Additional References

ALAIC. Asociación Latinoamericana de Investigadores de la Comunicación. www.alaic.org

Onda Rural. www.ondarural.org

Pontificia Universidad Católica del Perú. Facultad de Ciencias y Artes de la Comunicación. Program of Communication for Development http://facultad.pucp.edu.pe/comunicaciones/carreras/comunicacion-para-el-desarrollo/

# Development Communication in South Africa

## 24

Tanja Bosch

## Contents

| | | |
|---|---|---|
| 24.1 | Introduction | 470 |
| 24.2 | ICTs for Development in South Africa | 471 |
| 24.3 | Health Communication | 472 |
| 24.4 | Community Radio | 473 |
| 24.5 | Community Television | 475 |
| 24.6 | Social Movements and Nanomedia Strategies | 476 |
| 24.7 | Conclusion | 477 |
| References | | 478 |

### Abstract

Despite a peaceful transition to democracy in 1994, South Africa is still considered to be a transitional democracy, due to high levels of social equality which has resulted in a rise in social protests. This chapter provides a brief overview of development communication strategies in South Africa within this context, focusing on the following key areas of media for social change: ICTs for development, health communication, and community radio and community television initiatives. In addition, a brief overview of the communication strategies of civil society activist movements is provided.

### Keywords

Development communication · Community radio · Community television · Health communication · ICT4D · South Africa

---

T. Bosch (✉)
Centre for Film and Media Studies, Cape Town, South Africa
e-mail: tanja.bosch@uct.ac.za

## 24.1 Introduction

South Africa experienced a peaceful transition to democracy in 1994, with the country's first democratic elections and the end of the system of racial segregation known as apartheid. Despite a shift in the political landscape, a number of inequalities persist to present day, as a result of the legacy of apartheid political and socioeconomic policy. South Africa is one of the most economically unequal societies in the world, with 10% of the population owning 90–95% of the wealth and high unemployment levels (Von Fintel 2017). South Africa has characteristics of both a developing and an advanced economy, with access to technology, a strong private sector and fiscal resources, etc., though half of the population live below the poverty line (Gillwald et al. 2012). The South African economy has been impacted by international and domestic factors such as low economic growth, continuing high unemployment levels, lower commodity and higher consumer prices, lower investment levels, and greater household dependency on credit, with 55.5% of South Africans living in poverty (www.statssa.gov.za).

South Africa is thus considered to be a young, transitional democracy. Since the country's first democratic elections in 1994, there has been a steady rise in community protests, as a response to the inadequate provision of services such as water, sanitation, housing, etc. Growing citizen frustration with the new government's failure to share the dividends of democracy and high levels of government corruption are most often listed as reasons for these sometimes violent protests, described by Alexander (2010) as a "rebellion of the poor." A range of social movements have also emerged in the late 1990s, with civil society groups confronting "questions of social exclusion in terms of gender, sexuality, education, labor status, access to land, housing and services, poverty and citizenship: issues which sit at the intersection of recognition and redistribution" (Ballard et al. 2005: 615).

The role of the media and communications sector has been highlighted as a central tool in facilitating the participation of citizens in development activities.

This chapter provides a brief overview of development communication strategies in South Africa and focuses on the following key areas: ICTs for development, health communication, and community radio and community television initiatives. The South African government has implemented a number of projects around the country, with the aim of strengthening development communication and empowering previously disadvantaged communities. In particular, multipurpose community centers have been set up and identified as spaces through which the government can dialogue with citizens and address socioeconomic inequalities. The Government Communication and Information System (GCIS) was tasked by cabinet "to provide development communication and information to the public, to ensure that they become active participants in changing their lives for the better." (www.thusong.gov.za/documents/policy_legal/gdc.htm) Concepts such as community-based or people-centered development, citizen participation, and public-private partnerships are found in many documents of the postapartheid state, with calls for community involvement across a range of sectors (Emmett 2000).

## 24.2 ICTs for Development in South Africa

There is a relatively sophisticated ICT sector in South Africa after telecommunications reforms since the mid-1990s, but as Gillwald et al. (2012) argue, growth in the sector has not been accompanied by the primary policy objective of affordable access for all. In South Africa, the government has demonstrated the political will to drive programs aimed at improving the lives of the poor, through various policy initiatives. The Comtask Report, for example, mandated the government to establish Thusong Service Centres (multipurpose centers) to implement development communication (Naidoo and Fourie 2013).

The growth of the Internet and related technologies and platforms has created new possibilities for development in Africa. In the field of education, for example, ICTs have been listed as playing a number of roles, including providing a catalyst for rethinking teaching practice, developing the kind of citizens required in an information society, improving educational outcomes, e.g., pass rates, and improving the quality of teaching and learning (Jaffer et al. 2007).

There have been several initiatives related to the use of ICTs in education in South Africa, as one strategy for addressing teaching and learning challenges. Jaffer et al. (2007) explored the role of educational technology in higher education, with specific reference to the pressure on the sector to provide diverse graduates who can meet the social transformation and skill needs of the country. Specific challenges in the South Africa context include students' lack of academic preparedness, large and multilingual classes, and a range of other issues related to the previously unequal and racially divided educational system. The Fees Must Fall student protests highlighted these ongoing divisions in the higher education sector, with national protests highlighting various issues related to uneven access (Bosch 2016). Jaffer et al. (2007) list several ways in which ICTs can be used to respond to the range of educational challenges. These include using interactive spreadsheets to develop mathematical literacy skills, providing online feedback, using videos to simulate tasks such as video editing, and thus contributing to student learning experiences, as well as curriculum design.

The South African ICT sector has demonstrated dynamic growth, particularly with respect to growth of the mobile sector, though this growth has not met the national objectives of affordable access to the full range of communication services, and access to fixed broadband, for example, remains low in comparison to other lower-middle-income countries (Gillwald et al. 2012). The growth of mobile telephony has been a big factor with claims of penetration rates of over 100% and the availability of smart devices at lower costs, together with reductions in the cost of services (Gillwald et al. 2012). As Aker and Mbiti (2010) have shown, mobile phones or m-development can facilitate economic benefits by improving access to and use of information, creating jobs and income-generating opportunities, and being a vehicle for the delivery of financial, agricultural, health, and educational services.

There have been a number of projects set up by government which experience "sustainability failure" which means that they succeed at first but are then abandoned shortly thereafter. One example is the setup of a set of touch screen kiosks for remote rural communities in the North-West province, but the lack of recent and local content and lack of interactivity led to disuse, and the kiosks were later removed (Heeks 2002). Naidoo and Fourie (2013) argue that government communication often initiate projects which engage in partnerships with communities, but these are linear – communities do not have equal levels of influence and decision-making authority, preventing them from actively sustaining development efforts.

## 24.3 Health Communication

ICTs such as e-health, telemedicine, etc. are often considered ways to bridge the digital divide between rural and urban healthcare centers and to resolve challenges in the rural health sector (Ruxwana et al. 2010). The potential for mHealth, i.e., the use of mobile phone technology to improve health services is growing in South Africa as a result of the widespread availability of mobile phones, together with the low levels of literacy required to use them for this purpose (Leon et al. 2012).

However, in South Africa there is uneven access with some hospitals experiencing a limited number of computers, telephone services, and telemedicine equipment, while others had a wide range of these plus videoconferencing services. A study by Ruxwana et al. (2010) showed that while ICTs have the potential to improve the lives of people in rural communities, an Eastern Cape e-health project had limited success due to structural barriers such as a lack of computer equipment, skills, and poor or no internet connection.

While the international trend has been toward using a range of ICTs in health, cellphones have emerged as the most popular and cost-effective in Africa, with cellphone penetration on the continent increasing rapidly since the privatization of telephone monopolies in the mid-1990s (Bosch 2009).

There have thus been a range of mobile telephone-based provider-patient intiatives in South Africa, such as the Cell-Life Project which allows counselors to use mobiles to monitor the treatment and health status of HIV patients. Similarly, a company called On Cue uses a system to send SMS messages to patients reminding them to take their TB medication (Lucas 2008). Other projects include SIMPill which also helps patients to remember to take their medication, SMS helplines such as the Teen SMS Helpline of the South African Depression and Anxiety Group. Several other projects have been run, using mobile phones to improve patient referrals between local clinics and district hospitals and to facilitate communication between rural hospital doctors, e.g., photographing test results and sending them for instant analysis (Bosch 2009). A rural wireless tele-health system has been active in two remote rural areas of the Eastern Cape since 2004. There have also been initiatives focused on Mobile Deaf Communication and Mobile Finance.

## 24.4 Community Radio

Unlike many other developing countries, South Africa has a constitutionally protected public broadcaster, and community, and private free-to-air broadcasters; and radio and television remain primary modes of information and education (Gillwald et al. 2012). The growth of community radio and community television projects have created opportunities for the use of media in development communication; and South Africa is one of the only countries which makes legislative provision for community radio stations. Community broadcasting is defined in Section 1 of the Convergence Bill as a broadcasting service that is fully controlled by a nonprofit entity and is carried on for nonprofit purposes; serves a particular community; encourages members of the community served by it, or persons associated with or promoting the interests of such a community, to participate in the selection and provision of programs to be broadcast in the course of such broadcasting service; and may be funded by donations, grants, sponsorships or advertising, or membership fees or by any combination of the aforementioned.

The nationwide nonracial elections of 1994 marked only the start of the restructuring of the broadcast sector (Barnett 1999), and community broadcasting was acknowledged for the first time and defined as "initiated and controlled by members of a community of interest, or a geographic community, to express their concerns, needs or aspirations without interference, subject to the regulation of the Independent Broadcast Authority—IBA" (Duncan and Seleoane 1998). Hence, community-run, nonprofit broadcasting has only been possible in South Africa since 1994, when the IBA was established in terms of the 1993 *IBA Act*. The IBA accordingly granted the country's first set of community radio licenses in 1994–1995, which needed to be renewed every 12 months. In addition, the IBA granted 1-month "special-event" radio broadcast licenses to radio groups awaiting the outcome of longer-term license applications.

Globally, community radio stations give citizens a platform to create alternate public spheres for the promotion of local culture and language, as well as a space for dialogue and deliberation, and alternative political information in the face of concentrated media ownership. Radio is still the cheapest and most widespread medium, and in Africa it is particularly effective in reaching rural and illiterate constituencies. Community radio stations can provide local communities with development-related information, as well as a platform to engage in dialogue on a range of developmental issues. When community radio emerged and established itself in South Africa in the mid-to-late 1990s, it occupied a position that challenged the state-owned and commercial media that had existed until then (Hadland et al. 2006: 37). In allowing citizens to play a role in the production of community media, the medium fulfilled an important empowerment need and was seen to represent a radical shift in the country's media industry (Bosch 2011).

The emergence of participatory community radio in South Africa in the post-apartheid era represented a key shift in the country's media landscape. Until then media had been primarily owned and operated by the state and served as a propaganda tool. The opening up of the airwaves represented a major shift, and over 100 community radio stations around the country were granted broadcast licenses by the newly formed regulatory authority (Bosch 2006). These stations were tasked to "empower and educate the community," though there has not been much policy guidance on exactly what this entails in practical terms. There are several types of community radio stations in South Africa – geographic stations broadcast to specific areas, campus radio stations are located at universities, and community of interest stations cater to specific groups and include the religious Christian and Muslim stations. The key features of community radio are as follows:

1. Access, i.e., they provide access to all members of the target community and act as a public forum, giving everyone a voice
2. Participation, i.e., they encourage participation of all in management, planning, and production.
3. Training is provided in content and radio production as well as physical maintenance of equipment, giving community members the ability to operate their own station and empowering them with technical skills.
4. Nonprofit, i.e., they may receive funding, but profit is plowed back into the community. In South Africa community radio stations are nonprofit Section 21 companies with Boards of Directors.
5. Community owned, i.e., the highest decision-making body is usually an annual general meeting which is open to all, and leadership structures are often horizontal.
6. Volunteer driven and mostly run by volunteers who sometimes receive only a small stipend to cover, e.g., transportation costs.
7. Local, i.e., content is predominantly local and relevant to the specific community served by the station (Tucker 2013).

The notion of community participation is premised on the assumption that there exists a "community" which can participate in development projects, though in South Africa the concept of community has become associated with other referents such as race or class (Emmett 2000). The audiences of community radio stations in South Africa were originally conceptualized as formerly disadvantaged, predominantly black communities; and community radio was a symbolic development, granting access to and demystifying media production processes for a group of people who had until then only had access to the state-owned propaganda media system.

However in more recent times, communities are seldom homogenous, which often presents a challenge for community radio stations. In addition, South African community radio stations face the primary challenges of financial and social sustainability. While many stations were originally funded by internationally based donors, this funding dwindled in the postapartheid era. This, together with the fact that stations target audiences that are not perceived as high priority consumers by advertisers, has resulted in many stations struggling financially and often relying on government funding to remain operational. Social sustainability refers to the notion that stations are strongly rooted in communities who feel a strong sense of affinity or sense of ownership, which may result in financial contributions from local businesses, but this has been difficult for stations that attract socioeconomically and racially diverse individuals in urban areas, making it difficult to foster a sense of shared group identity. Stations rely on unpaid volunteers who rarely have broadcasting experience, resulting in a perception that community radio stations are "unprofessional." The primary goal of community radio stations is their development orientation and focus on encouraging ordinary citizens to because media producers (and not just media consumers), which makes it difficult for stations to compete with the more "professional" commercial stations that are their primary competitors (Bosch 2011).

Despite various challenges, community radio remains a key tool for development communication in South Africa. Audience numbers for community radio stations continue to grow, demonstrating the popularity and need for these alternative platforms in an increasingly centralized media space. While community radio stations in South Africa are themselves diverse, there are many success stories of stations who play a key role in terms of development communication. Radio Khwezi.

## 24.5 Community Television

During apartheid, South Africa's community media sector served as "a tool to counter state propaganda, inform, mobilise and educate the masses about their rights and to facilitate the building of strong community organisations" (Berger 1996: 4). Hadland, Aldridge, and Ogada (2006: 2) argue that the extent to which community media can play a role in participatory democracy is questionable without the presence of community TV, especially within the context of democracy in South Africa. While there was a strong postapartheid policy imperative in the case of community radio, the community TV sector remained without a position paper. It was only in August 2003 when ICASA released its *Local TV* (LTV) *Discussion Paper* that attention was turned to creating an enabling environment for community TV. It soon became clear that the struggle for public access to community TV still had a long way to go. For a long time, it appeared as if the political will to make community TV a reality was lacking, as the IBA and ICASA "sent mixed messages to community TV stakeholders eager to get on air" (Greenway 2005: 48). Finally, ICASA's position paper on *Community Television Broadcasting Services* was released on 30 November 2004, with the paper calling for, among other things:

- Four-year renewable full-time licenses for nonprofit, South African-controlled local TV groups able to prove local support.
- An amendment of the community TV special-event license rules to allow for the awarding of special-event licenses of as long as 1 year in duration (instead of the current 1-month limit).

- Financing of the stations through adverts, sponsorships, and donors, provided the stations remain nonprofit (i.e., don't have shareholders, don't pay out dividends).
- A 55-percent South African local content quota for community TV broadcasters.
- License applications from community TV groups to begin sometime in the 2005–2006 financial year.
- No public or private/commercial community TV stations.

On distinguishing between the two types of community television, i.e., geographic and "community of interest," ICASA has made it clear that the former will be favored. In the case of community radio, majority of the stations have adopted geographically defined interests (Armstrong 2005: 24). Today, there are five geographically defined television stations on air: *Cape Town TV* in Cape Town, *Soweto TV* in Johannesburg, *Tshwane TV* in Pretoria, *Bay TV* in Port Elizabeth, and *1KZN TV* in Richards Bay, and one community of interest, *TBN Africa*, which is also the oldest licensed community TV channel in the country. ICASA has also issued "test-licenses" to other stations such as *Fresh TV* and other community TV projects, including *Ekurhuleni TV* and *Alex TV*. However, until the launch of DTT, there is no new spectrum available for any other community TV stations (Thomas and Mavhungu 2013).

## 24.6 Social Movements and Nanomedia Strategies

Social movements in South Africa, and around the world, depend on the public discourse and to broaden the scope of their campaigns (Gamson and Wolfsfeld 1993), though they often compete for media attention with many other issues and organizations, alongside the imperative of news values. Biases in coverage of social movement activities have been well documented, with limited media attention even though media discourse is a crucial source of strategic information for activists, upon which they base decisions and provide information for their future strategies. (Koopmans 2004). As a result of skewed media attention, social movement activists have tended to use smaller-scale media to publicize their struggles, including demonstrations, dress, slogans, murals, songs, radio, dance, poetry, and political theatre. Dawson (2012) has highlighted these strategies as examples of nanomedia, arguing that they play a key role in mobilization processes. The term nanomedia was coined by Mojca Pajnick and John Downing (2008), to refer to small-scale media, which includes alternative media such as community radio or video but also refers to "popular song, dance, street theatre, graffiti, murals, dress along with print, broadcasting, film and the internet" (Downing 2010: 2). The term "media" is used broadly to include the symbolic, ritual, and performative modes of communication. Further examples of nanomedia include flyers, jokes and songs, revolutionary pamphlets, dance performances, street theatre, and even the diapers worn by the Mothers of the Plaza de Mayo who protested the Argentinian military dictatorship in the 1970s (Downing 2010). In South Africa, the most well-known example of nanomedia is the toyi-toyi street dance, which is often used in present-day protest marches and demonstrations.

South African social movements engage in a wide repertoire of strategies for internal mobilization, as well as for advocacy particularly in efforts to draw the state's attention to their problems. Civil society organizations and protest movements frequently employ traditional and "new" methodologies to achieve maximum impact. Members of these organizations cite pamphlets and t-shirts as critical mobilization tools, while murals and graffiti are also to raise awareness about social issues. In addition, music and songs, together with protest marches, are commonly used to mobilize support, together with the tactic of physically occupying spaces which is commonly used.

In addition to nanomedia strategies, digital media is increasingly used for mobilization. Social media, Facebook and Twitter in particular, is used by activists in parallel with traditional media, to raise awareness about their campaigns and social issues. Twitter, for example, is used by activists to disseminate general information and organization events and news and is also seen as a useful mechanism to communicate with journalists in order to get mainstream media coverage. Facebook is often used by social activists for lateral linkages, to connect with similar movements locally and internationally. Social media was primarily seen as important for raising awareness and for networking with activist groups nationally and even internationally. While social media platforms such as Twitter and Facebook are widely used, they complement existing traditional and nanomedia strategies and are not the primary tool for civic engagement. However, activists do social media as a tool used to enhance public engagement, coordinate offline action, and used for instant dissemination and exchange of information, in particular using it as a "choreography of assembly" (Gerbaudo 2012) to coordinate offline activities. But rather than being the primary method of communication and mobilization, digital and social media serve as complementary tools for South African activists (Bosch et al. 2018).

## 24.7 Conclusion

A range of government initiatives in the field of development communication exists in South Africa, but official government strategies often demonstrate a linear information-dissemination process which does not align with the basic principles of development communication, or Freirian notions of communication as participatory, dialogic, and reciprocal, because "communities are consulted but not fully involved" (Naidoo and Fourie 2013: 105). Moreover, in these projects, the feedback from communities is often not well structured, and monitoring and evaluation of projects are poor on nonexistent. However, in the South African context, participatory community media fills this gap. "Community media are strategic interventions into contemporary media culture committed to the democratization of media structures, forms, and practices" (Howley 2005: 2). Community media activists argue that the role of community media has risen to the fore as an invaluable tool to empower previously disadvantaged citizens (Opuku-Mensah et al. 1998). A range of community radio and community television projects fill the gap and provide local communities with access to the media and a space for debate and dialogue in local languages.

With respect to ICTs, policy exists to facilitate the introduction of ICTs in communication strategies, but this is often not prioritized by government. Most people do not have broadband internet access – and the digital divide within the country limits the possibilities for the use of ICTs for social change. However, the growth of the mobile internet and increased access to feature phones and smartphones with internet capability represents an opportunity for e-democracy and m-health projects; and social media is increasingly being used in the campaigns of social activists and civil society organizations.

## References

Aker J, Mbiti I (2010) Mobile phones and economic development in Africa. J Econ Perspect 24(3):207–232

Alexander P (2010) Rebellion of the poor: South Africa's service delivery protests–a preliminary analysis. Rev Afr Polit Econ 37(123):25–40

Armstrong M (2005) Public service broadcasting. Fiscal Studies 26(3):281–299

Ballard R, Habib A, Valodia I, Zuern E (2005) Globalization, marginalization and contemporary social movements in South Africa. Afr Aff 104(417):615–634

Barnett C (1999) The limits of media democratization in South Africa: politics, privatization and regulation. Media Cult Soc 21:649–671

Bosch T (2006) Radio as protest: the history of bush radio. J Radio Stud 13(2):249–265

Bosch T (2009) Cellphones for health in South Africa. In Lagerwerf L, Boer H, & Wasserman H (eds) Health communication in southern Africa: engaging with social and cultural diversity, Rozenberg Publishers, 11:279–300

Bosch T (2011) Community radio continues to provide an alternative. In: Brger G (ed) Media in Africa: twenty years after the Windhoek declaration on press freedom, Media Institute of South Africa (Misa), Cape Town

Bosch T (2016) Twitter and participatory citizenship: #FeesMustFall in South Africa. In: Mutsvairo B (ed) Digital activism in the social media era: critical reflections on emerging trends in sub-Saharan Africa. Palgrave Macmillan, United Kingdom

Bosch T, Wasserman H, Chuma W (2018) South African activists' Use of Nanomedia and Digital Media in Democratization Conflicts. International Journal of Communication, 12:18

Dawson M (2012) Protest, performance and politics: the use of "nano-media" in social movement activism in South Africa. Res Drama Educa J Appl Theatr Perform 17(3):321–345

Downing J (2010) Nanomedia: community media, network media, social movement media: why do they matter? And what's in a name? Text prepared for the conference "Mitjans comunitaris, moviments socials i xarxes", organised by the Unesco chair in communication InCom-UAB in collaboration with Cidob (Barcelona Center for International Affairs), Cidob, Barcelona. 15 Mar 2010. Available online at http://www.portalcomunicacion.com/catunesco/download/2010_Downing_Nanomedia_english.pdf. Accessed 3 May 2016

Duncan J, Seleoane M (1998) Media and democracy in South Africa. HSRC Press & Freedom of Expression Institute, Pretoria

Emmett T (2000) Beyond community participation? Alternative routes to civil engagement and development in South Africa. Dev South Afr 17(4):501–518

Gamson WA, Wolfsfeld G (1993) Movements and media as interacting systems. Ann Am Acad Pol Soc Sci 528:114–125

Gerbaudo P (2012) Tweets and the streets: social media and contemporary activism. Pluto Press, New York

Gillwald A, Moyo M, Stork C (2012) Understanding what is happening in ICT in South Africa: a supply and demand side analysis of the ICT sector. Evidence for ICT policy action. Policy paper 7. Available online at https://www.researchictafrica.net/publications/Evidence_for_ICT_Policy_Action/Policy_Paper_7_-_Understanding_what_is_happening_in_ICT_in_South_Africa.pdf. Retrieved 15 Jan 2018

Hadland A, Aldridge M, Ogada J (2006) Re-visioning television: policy, strategy and models for the sustainable development of community television in South Africa. HSRC Press, Cape Town

Heeks R (2002) Failure, success and improvisation of information systems projects in developing countries. In: Development informatics working paper series no 11/2002. Institute for Development Policy and Management, Manchester

Howley K (2005) Community media. University of Cambridge, Cambridge

Jaffer S, Ng'ambi D, Czerniewicz L (2007) The role of ICTs in higher education in South Africa: one strategy for addressing teaching and learning challenges. Int J Educ Dev using ICT 3(4). Available online at http://ijedict.dec.uwi.edu/viewarticle.php?id=421. Retrieved 11 Jan 2018

Koopmans R (2004) Movements and media: selection processes and evolutionary dynamics in the public sphere. Theory Socy 33:367–391

Leon N, Schneider H, Daviaud E (2012) Applying a framework for assessing the health system challenges to scaling up mHealth in South Africa. BMC Med Inform Decis Mak 12:123

Lucas H (2008) Information and communications technology for future health systems in developing countries. Soc Sci Med 66(10):2122–2132

Naidoo L, Fourie L (2013) Participatory anchored development in South Africa as evaluated at Thusong service centres. Communitas 18:95–115

Opuku-Mensah A, Kanaimba E, Mosaka N, Sitoe E, Lush D, Kaitira K, Golding-Duffy J, Vilakazi A, Banda F, Manyarara J (1998) Up in the air? The state of broadcasting in southern Africa. Panos Southern Africa, Lusaka

Pajnick M, Downing J (2008) Introduction: The challenges of nano-media. In M. Pajnick & J. Downing (eds) Alternative media and the politics of resistance: Perspectives and challenges, (pp. 7–16). Ljubljana, Slovenia: Peace Institute

Ruxwana N, Herselman M, Conradie D (2010) ICT applications as e-health solutions in rural healthcare in the eastern Cape Province of South Africa. Health Inf Manag J 39(1):1833–3583

Thomas H, Mavhungu J (2013) What is happening about community telly? The Media Online. Available at http://themediaonline.co.za/2013/10/what-is-happening-about-community-telly/. Accessed 2 Aug 2018

Tucker E (2013) Community radio in political theory and development practice. J Dev Commun Stud 2(2/3):2305–7432

Von Fintel D (2017) Unemployment, poverty and inequality: South Africa's triple scourge". Research on Socioeconomic policy and Department of Economics, Stellenbosch University. Institute of Labor Economics, Bonn

# Development Communication as Development Aid for Post-Conflict Societies

## 25

### Kefa Hamidi

## Contents

| | | |
|---|---|---|
| 25.1 | Development Communication as Education for Post-conflict Societies | 482 |
| 25.2 | Literature Review: Mass Communication and Development Countries | 483 |
| 25.3 | Development, Journalism, and Education | 487 |
| 25.4 | Journalism and Competence | 488 |
| 25.5 | Social Framework in Afghanistan | 489 |
| 25.6 | Media System and Journalist Education | 490 |
| 25.7 | Mediation as a (New) Journalistic Competence | 492 |
| | 25.7.1 Journalist Should Mediate Between State and the People | 492 |
| | 25.7.2 Urban and Countryside | 493 |
| | 25.7.3 State and Opponents | 493 |
| | 25.7.4 Social Consensus Between Ethnic Groups | 493 |
| | 25.7.5 Conflict Resolution | 493 |
| | 25.7.6 Constructive Journalism as an Agent for Change | 494 |
| | 25.7.7 Education and Enlightenment | 494 |
| 25.8 | Conclusion | 494 |
| 25.9 | Future Discussions | 496 |
| References | | 497 |

### Abstract

For a couple of years now, a discussion has been going on about the role that journalists should play in so-called fragile states. According to the following understanding, the journalist should not only report on the most up-to-date, neutral, and objective events but also "educate," "mediate," "convince," and thus adopt a "mediation function." Although this function is often seen as the

---

K. Hamidi (✉)
Institute communication and media Studies, Leipzig University, Leipzig, Germany
e-mail: kefa.hamidi@uni-leipzig.de

© Springer Nature Singapore Pte Ltd. 2020
J. Servaes (ed.), *Handbook of Communication for Development and Social Change*,
https://doi.org/10.1007/978-981-15-2014-3_101

core responsibility of the media in the developing countries, so far hardly any differentiation has been made as to what exactly this "concept" means for journalism education and how this could be integrated into the models of journalism education. Here, this gap is to be closed by developing a competence model for academic journalism education for a fragile state, using Afghanistan as a case study. This article presents the partial results of the project "Professionalization of Academic Journalist Education in Afghanistan."

**Keywords**

Community journalism · Conflict resolution · Constructive journalism · Journalism education · Organizational competence · Participatory action research · Professional competence · Social orientation

## 25.1 Development Communication as Education for Post-conflict Societies

Approach, aim, and purpose of the project "Professionalization of Academic Journalist Education in Afghanistan" funded by the Volkswagen Foundation was to develop a theoretical competence model for the reform of academic journalism training and initiate a communication studies program in Afghanistan based on the current state of research on the one hand and a discussion and negotiation of possible solutions and problems with the recipients on the other hand. They were specifically asked for their ideas and how the reform shall proceed. Thus, 40 interviews with Afghan experts of media practice, media policy, media studies, and higher education policy were conducted to determine which journalistic skills should be taught to students in academic journalism education in a fragile state such as Afghanistan. For the project, the premise was to derive the model for an academic journalist education not only from the scientific theory or from Western academic models and to introduce it from a "foreign" perspective. A sustainable reform of academic journalist education will only succeed if this reform takes into account the social and cultural value orientations, the structural conditions, as well as the interests, expectations, and opportunities of the actors involved in this reform process. An essential prerequisite for success, therefore, is to seek from the experts involved in this reform, on the one hand, opinions and assessments of the theoretical and practical training objects and, on the other hand, experience and recommendations for the process steps to ask their coordination and sequence, which are necessary at different levels, to successfully develop and test the model. To achieve this goal, the "participatory action research" (PAR) strategy was applied. The approach of participatory action research differs in some key points from other empirical social science methods. Characteristic of participatory action research is not a concrete method, but the underlying understanding of how and by whom research is to be conducted. "Participation" and "action" are here the key aspects. The term "participatory" means that persons in the field are not only included as research objects in the research, that is, as researched on which data are collected, but actively participate as "co-researchers" (Wöhrer and Arztmann 2017, p. 29). It is generally requested that persons, groups, and institutions are involved who are affected by the research topic and the expected results. For this study, 40 interviews were conducted with stakeholders from academic journalism training over a period of time from mid-February to the end of May 2017 via Skype. Each interview lasted for about 2–3 h and was in either Pashto or Dari. They were recorded, transcribed, and translated into English. This sample included (1) 20 university and college teachers, (2) 5 representatives of regional and national higher education policy, (3) 10 heads of regional media organizations (press, radio, public relations, news agency, online services), and (4) 5 executive boards of regional and national professional associations of journalism. The second important term in the PAR, "action," means that a change in the sense of problem-solving is sought. The PAR research process often starts with a problem that causes "pausing" or "stopping" and reflection begins. This reflection leads to a research question for which answers are sought. In order to be able to answer, data is needed that is collected, reflected, and analyzed. This brings new options and solutions. The purpose of this study, with the help of the interviewed experts, was to first assess the current state of academic journalism education in Afghanistan (problem) and identify the needs and expectations for improving (approach to) journalism education in Afghanistan both inside and outside state universities. Accordingly, the survey instrument was divided into three parts, each with a detailed explanation and a guide with 35–55 questions. The first part of the survey asked about the general function of the media and journalism in Afghanistan. In the second part, we asked the co-researchers which journalistic skills (theoretical and practical skills) in academic journalism education should be offered to students. Specifically, we asked the co-researchers referring to Weischenberg (1998) and others the items that belong to the theoretical skills in journalism education and asked them if they should also be offered training in journalism in Afghanistan and what additional skills should be taught. The third part of the survey then dealt with the implementation of solutions and suggestions for improving journalistic training. In this article only the partial results of the second part (theoretical competencies) are presented. The answers were evaluated by (MAXQDA) software and subjected to a cluster analysis.

## 25.2 Literature Review: Mass Communication and Development Countries

Hallin and Mancini's *Comparing Media Systems*, published in 2004, has become a classic in comparative media studies. These four comparative dimensions are not enough to summarize the media systems beyond Western countries (Shen 2012) or fragile state (Schulze and Hamidi 2016). The reason for this is that comparative cross-national or intercultural research too often starts from an "implicit Western bias" and "models and frameworks developed in a Western context are used as templates for evaluation and comparison" and "implicitly disregards, apart from the proverbial lip service, the importance of cultural dimensions in media systems and societies" (Servaes 2015). For some years now, there has been a discussion about how media should be organized (and described) in so-called developing countries and what functions they should include. According to this understanding, mass communication should not only report on the most up-to-date, neutral, and objective events, but primarily journalism should assume a kind of "mediation function" and not only inform, but "educate," "mediate," and "convince" the population. The common denominator of these approaches could be summarized under the following normative view: The mass media should contribute to *national development* (Kunczik 1985, 1988) and *social harmony* (Massey and Chang 2002; see also Bandyopadhyay 1988; Menon 1996). The central assumption of the *first conception* is that mass media (newspapers, radio, television, and ICT) have not only programmatically but have an extraordinarily important role as facilitators of development policy programs for the formation and consolidation of social change. Therefore, various international studies show that the professional attitudes of journalists are very different. Christians et al. (2009) argue that journalism tend to fulfill four core-normative roles, namely, a monitorial, a facilitative, a radical, and a collaborative role, and do point to different ways in which journalism is positioned on society: the *monitorial role* refers to the classic liberal role of a neutral and objective media watching the authorities; the *facilitative role* refers to the need for more independence from power structures as this refers to the media's role to provide a platform to citizens (ordinary citizens); the *radical role* is fulfilled by oppositional forces that challenge those in positions of power; and the *collaborative role* is taken up by those media and journalists who operate and act unequivocally to protect and safeguard the interests of those in power. Especially the last two rolls fit in this context. Hanitzsch et al. (2011, p. 275) refers to three central areas in which journalism cultures materialize the perception of journalism's institutional roles, epistemologies, and ethical ideologies. He mentions these three domains together constitute "the basic elements of difference between journalism's cultures and the domain of institutional roles refers to the normative and actual functions of journalism in society." His key empirical findings on role perceptions (Journalistic role perceptions denominate the journalists' self-image of their social roles and functions (Hanitzsch et al. 2011, p. 281). They can be defined as "generalized expectations which journalists believe exist in society and among different stakeholders, which they see as normatively acceptable, and which influence their behavior on the job" (Weischenberg 1995, 1998)) from a comparative survey of the "role perceptions" of 1800 journalists from 18 countries show that "that journalists across the globe pay high regard to the normative ideals of detachment, providing political information, and acting as a watchdog of the government." But the journalists from non-Western contexts tend to be "more interventionist in their role perceptions (...) for particular values, ideas and social change" (Hanitzsch et al. 2011, p. 273). The journalists in developing countries point that they pursue an "active, interventionist role" (Hanitzsch et al. 2011, p. 273) and to understanding "social change" as an elementary goal (see also Hamidi 2016; Ramaprasad 2004, 2006, 2007). Such a function, which can be placed in the context of the idea of "development journalism" (Hanitzsch et al. 2011, p. 273), is much more endorsed among journalists in developing societies and transitional contexts. In another paper Hanitzsch et al. (2011, p. 278) extracted the four global professional milieus (These professional milieus (re)constitute themselves on the ground of "shared assumptions and beliefs" that provide the cultural framework for journalistic practice (Bourdieu 1998, p. 47 zit n. Hanitzsch et al. 2011, p. 274)): the "populist disseminator," "detached watchdog," "critical change agent," and the "opportunist facilitator." He points out that the "detached watchdog milieu" clearly dominates the journalistic field in most Western countries, while the milieu of the "opportunist facilitator" reigns supreme in several "developing, transitional, and authoritarian contexts." The main characteristic of journalists in milieu of opportunist facilitator group he describes "their relatively strong opportunist view of journalism's role in society, namely as constructive partners of the government in the process of economic development and political transformation (286). So many scientists and practitioners in developing countries argue for a "mass media development model" that should or should be in line with key concerns in their countries, particularly "nation-building" (see McQuail 2000; Jayaweera 1987). McQuail (1983, p. 95) speaks in his approach of "a specific media model for the developing world" as an important common feature, considering "the acceptance of economic development per se" and the "emergence of a nation as the ultimate goal." The mass media can play a crucial role in social progress, as any transitional society will encounter new attitudes, a new mindset, and a new value system (Namra 2004, p. 17). Basically, it is assumed that mass media are able to influence the social development process by reporting on development programs (ibid., Grossenbacher 1988), because only mass media might reach the vast rural population and give them a voice in debates affecting their lives. In this context, Hedebro (1982, p. 16) attributes to the mass media the potential for conveying the meaning of "nationhood" (sense of nation-ness). He justifies this by arguing that "many developing countries are mixtures of different cultures, languages, political systems and religious beliefs. This is regarded as a serious obstacle to social change on the national level." For Lozare (2015) the key roles of communication in the "building of a nation" foster "meaningful dialogue among different sectors of society," "nurture a shared vision for the country's future," and "harness non-material and material resources to realize the national shared vision." In particular, reporting on national projects that could "constructively" contribute to the development and improvement of living standards is desirable (Kunczik 1992). Of course, one cannot attribute omnipotence to the media, as modernization theorists have thought. The media system should be considered as reflectors or indicators of progress, freedom, state of development, and modernization of a country rather than determinants. They can be a factor among others that can bring national development and harmony (ibid.). The *second view*, that the media and journalists have a "social responsibility" and are therefore responsible for the preservation of "social harmony," is represented in the context of so-called Asian values. The "classic models" to describe the media systems are based on a "too restricted (Western) description of concepts like "freedom," "democracy," "objectivity," and so on, which allow little or no generalizations" (Servaes 2015), for example, to explain journalism in the context of Asian values. In this context of "Asian values," journalists and the media are also expected to maintain "social harmony" through "sensitive reporting" and to be constantly aware of the effects that the impact of coverage may have on society (Bayuni 1996; Xu 1998). Therefore, Gunaratne (2000, p. 15) states that the concept of "social responsibility" is understood differently in Asian societies than in Western countries. While the Western approach is defined by the individual's political, social, and economic freedom, the Asian approach refers to "collective social security and economic prosperity." This means that mass media have to generate more than anything else "nation-building," "national consciousness" and "unity," and the "encouragement of co-operation and peaceful co-existence between diverse and sometimes hostile communities" (Massay and Chang 2002, p. 991). These *two views are interlinked with the definition of development communication*, which is mainly used by researchers from countries in global south, emphasizing the aspect of "societal consensus" enabled by "development communication" (Quebral 2012). Nora Cruz Quebral (2012, p. 14) from the Philippines, an important representative of this view of development communication, explains in detail what the usual tasks of communication media in developing countries are. She states four basic items: (a) circulate knowledge that will inform people significant events, opportunities, dangers, and changes in their community, the country, the region, and the world; (b) create and maintain a base of consensus that is or community life may be discussed to achieve a better life needed for the stability of a state; (c) provide a forum where issues affecting the national; and (d) teach those ideas, skills, and values that people need. First, Quebral (2006) sees "development communication" as "an art and science of human communication apply to the speedy transformation of a country and mass of its people from poverty to a dynamic state and the larger fulfillment of the human potential." Second, she reacts to the question "how can communication media help create and maintain consensus?" and argues "[l]ike a marriage, a nation is founded on a bedrock of common experiences and shared values. Conflicts among groups in a nation cannot be helped, but with a wide enough base of agreement, the answers are sought not in secession but in accommodation and compromise. It is communication through the mainstream and new media that can project this national identity to a people and that can demonstrate to them that a united nation can be fashioned out of diverse cultures." Third, she describes what "providing a forum" means. Discussion through communication media is one of the basic features of a democracy. There must be a marketplace where ideas and opinions on public issues may be heard, answered, or exchanged. How to give voice to the voiceless is a concern of the development communicator, as well as reporting events as they happen, which certainly is not a new task for communication media. Fourth, she considers the question why communication media should assume the role of a teacher and argues that for a country to develop, its citizens must be exposed to progressive ideas, skills, and accompanying values. Formal education that happens in schools cannot do this yeoman task alone. It must be supplemented and reinforced by other social institutions, not the least of which is a country's communication system. Therefore Quebral (2012, p. 7) defines "development communication" as it "circulates useful information and knowledge; provides a forum where problems and issues may be aired, teaches needed ideas, skills, and values; and creates a base of consensus that stabilizes the state."

## 25.3 Development, Journalism, and Education

The challenge to develop curricula and do research based on the realities in developing countries and steeped in contextualized theory is one that is acknowledged by several journalism educators (see Mensing 2010). Apart from creating the appropriate political and economic environments for an independent media system, it is crucial to educate journalists to the highest ethical and professional standards possible (Servaes 2015). Professional education of journalists should be based on local or national parameters (ibid.). Journalism that ignores a community dimension can end up being used as a tool to divide people by interests, rather than build relationships critical for functioning communities (Mensing 2010).

The journalism education in developing countries must contend with defining a new academic identity for itself, extricating itself from dependency on Western-oriented models of journalism education and training (ibid.). Many feel that the Western or "libertarian" model of journalism had not contributed fully toward national progress because of its emphasis on the entertainment function of the mass media and its treatment of information as a "saleable market commodity" (see Ali 2011). Western news values of timeliness, prominence, proximity, conflict, and the bizarre exclude the ordinary people in the news unless they are involved in accidents, violence, or catastrophes (ibid.). For very different reasons, journalists in developing countries are dissatisfied with the Western model of journalism and have been trying to develop models they feel would serve their needs better. Development journalism is stemming from the dissatisfaction with the Western news values that do not serve the cause of national development (Ali 2011; Yin 2008). Development journalism is an important part of development communication that can be conducted by professional or lay journalists. In the first case, one speaks of "constructive journalism" and in the second case of citizen journalism. Accordingly, community journalism is often associated with this type of journalism. In all cases, the aim is to report on answers to social problems and to shed light on the effectiveness of approaches to solutions so that the recipients of such reporting can learn from them or be inspired by them. The concept of development journalism has been developed and tailored to the needs of countries in the Global South, especially in Asia since the 1960s in contrast to Western watchdog journalism, but is currently also being adapted in Western countries. According to Xu (2009, p. 358), the main components are, among other things, not only to focus on current affairs, but long-term development processes; to work independently of the government and criticize the government constructively, but to place the journalistic focus on news about economic and social development and to constructively engage in nation-building with the government; and to empower "normal" people to improve their lives and those of their communities. Here communication not only means the transfer of information but also includes the participation in the society and in the community (Namra 2004) and focus on the educational function of the news, stories about social needs, self-help projects, and obstacles to development (Gunaratne 1996). Each journalist role (creator of forum and consensus, teaches or advocate) could be associated with a specific set of competencies (and skills and values), which the journalist (fragile states) of the future needs to acquire in order to perform adequately (see also Servaes 2015).

## 25.4 Journalism and Competence

The job description of journalists has changed over the past decades toward more professionalism. Nevertheless, there is still no uniform or legally prescribed way in which future journalists can gain their journalistic competence. Weischenberg (1995, p. 492) defines journalistic competence as "specific knowledge, values, norms and standards of conduct of the profession acquired through systematic training." Competence therefore includes knowledge, values, and behaviors. If skills are to be acquired during training, knowledge, values, and behavioral patterns must be systematized. Within the project "Journalism and Competence," Weischenberg (1990, p. 21f) developed together with Sigrid Schneider and Lutz Michel an analytical grid of journalistic competence. Despite some shortcomings, this competence model became an important basis for later competence and curriculum discussions in Germany (Nowak 2007, p. 82). He divides journalistic competence into professional competence (Fachkompetenz), special competence (Sachkompetenz), presentation competence (Vermittlungskompetenz), as well as social orientation (Soziale Orientierung). According to Weischenberg (1990, p. 45), "professional competence" includes research, selection, editing, and organizational techniques, often referred to as "craft," and specialist media knowledge of media studies, media economics, politics, law, history, and technology. This media knowledge should provide in the mediation process the theoretical basis for the fields of competence "functional awareness," "reflectivity," and "autonomy awareness." In terms of special competence, Weischenberg (1990, p. 34) attaches great importance to the fact that the requirements in this area are constantly increasing. On one hand, he demands the study of a special subject in order to acquire departmental knowledge. On the other hand, aspiring journalists should acquire orientation knowledge in order to classify their knowledge in social, political, and economic contexts. The presentation competence includes a general articulation ability and knowledge of the forms of presentation in order to present the contents in a recipient-oriented manner. According to Weischenberg, social orientation differentiates the university training program significantly from other offers. It is mainly about "thinking about journalistic actions." Functional awareness, the first component of "social orientation," is designed to help journalists use their influence responsibly while assuming their role as critics and controllers of politics and society. Not all assignments in Weischenberg's competence model are easily comprehensible; one or the other element could be supplemented (Nowak 2007, p. 82). The rapidly increasing mechanization of the journalistic working environment – not only of the means of work but also of the sources of information – therefore demands a high level of technical skills. Klaus Meier (2011, p. 223; see also Hanitzsch 1999, p. 119) adds another area to the four areas of competence proposed by Weischenberg: technical competence and composition competence. On one hand, knowledge of commonly used tools such as the Internet or basic computer software is meant; on the other hand, the journalist needs in particular ready-to-use knowledge in dealing with special technologies. In particular, this area of expertise includes cross-media competences. The organizational and conceptual competence strengthens the journalist's ability to innovate when he knows how media are used and with which concepts the target groups are optimally attained through quality journalism (Meier 2011, p. 223). On one hand, the subjects of organizational competence are specialist knowledge of editorial organization and quality management and, on the other hand, specific expertise on the current economic structures and fundamentals of journalism in society.

## 25.5 Social Framework in Afghanistan

The Islamic Republic of Afghanistan is a landlocked country in southern Asia with roughly 29 million inhabitants. The population is made up of more than 30 ethnic groups for whom their tribes and clans are far more important than their shared national identity (CIA 2015). Pashtun (ca. 40–50%), Tadzhik (ca. 20–35%), Hazara (ca. 7–20%), and Uzbek (ca. 8–15%) form the four largest ethnic groups (see Schetter 2009, pp. 123–133). Economically, Afghanistan is one of the poorest nations on earth predominantly practicing agriculture. Afghanistan is classified as a fragile state (one of roughly 46 fragile states worldwide), since Afghanistan still struggles after three decades of civil war with serious internal conflicts and problems (Deane 2013, p. 15; Barker 2008; Page and Siddiqi 2012, p. 4). Widely known is further the term "fragmented state," because those states are characterized by religious, political, ethnic, and other rifts (Putzel and Van der Zwan 2005, p. 41). No unified definition of fragile statehood exists so far, but as a rule, fragile states can be recognized when presenting with a "lack of security, capacity and legitimacy" (ibid.). An important solution for a successful state formation is the "increase in similarities between social groups" seen as the "key to success" (Putzel and Van der Zwan 2005, p. 39). Collier (2010, p. 9) also speaks of a "shared identity" and argues that the development of this identity is essential to the long-term stability of a state. Similarly, Hippler (2004, pp. 5f) sees "integrative ideology" as a very important component of successful state formation. This could arise from the "integration of a society" from the previously only "loosely connected groups." The basic point here is that "the communication patterns between the social groups are condensing so much that communication does not take place mainly within the groups." This implies a significant amount of internal social communication, which must be promoted by mass and communication media (ibid.). Kaltenborn-Stachau (2008, p. 21) describes the necessity of a "national dialogue process" in a "media public sphere" in which these similarities can emerge as the "heart" of any state formation. Key actors in media development cooperation, such as the USA and bilateral organizations such as the World Bank, recommend the liberalization and privatization of media in transition countries, including fragile states (Putzel and Van der Zwan 2005, p. 59). "Unregulated media revolution" is causing "fragmentation of the media sector" (Deane 2013, p. 7) as many minorities have their own community and local group-based media. The inevitable consequence of this expansion is the progressive fragmentation of target groups and a consequent strengthening of the "ethnicization of politics." The trend makes it increasingly difficult to mobilize larger groups of people for a task or a common goal, which is central to state-building (see more Deane 2013).

## 25.6 Media System and Journalist Education

The media system in Afghanistan has been through an extraordinary dynamic development since 2001, initiated by the political liberation and propelled by the commitment of international stakeholders, nongovernmental organizations (NGOs), and the International Security Assistance Force (ISAF). Afghanistan has a dual media system as to state and economically privately organized press, broadcasting, and online providers. Within the scope of this system, around 800 periodicals and more than 75 televisions, as well as 175 radio broadcasting providers, were registered at the end of 2015 next to state media institutions (Altai 2015, p. 118). In sum, it can be observed that the Afghan media system currently experiences the difficult transition from dependency on international subsidy to self-sufficient financial viability and future; yet in a fragile state, this is an enormous challenge (Hamidi 2016). The so-called media boom caused a rapid expansion and differentiation of the job market in the journalistic field. As a result of the media boom, the number of journalists increased significantly. It is assumed that between 8000 and 10,000 people in the country are somehow engaged in journalistic work. Through the expansion and differentiation, many unqualified workers got into the Afghan media industry. Journalism as a profession suffers from loss of credibility in the population and thus loses partially its function (Hamidi 2016). Therefore vast improvements, reform, and ethical schooling for journalists were demanded for a considerable time. Overall, two mutually dependent phenomena can be observed in the Afghan media development: on the one hand, social and economic change in the direction of democracy and, on the other hand, a strong media market liberalization, which has fragmented ethnic groups public. It has been criticized that international organizations were spending "hundreds of millions of dollars" on the media in Afghanistan (Barker 2008, p. 6), but only a few have considered policies that provide a supportive role for the state-building process (Deane 2013; Page and Siddiqi 2012). The "media boom" initiated an enormous expansion and differentiation of the journalistic professional field. In the last 10 years and more, journalism as a profession has gained popularity dramatically in Afghanistan – especially among the young generation. Notwithstanding the expansion of media offers, the number of "professional" media formats remains relatively small since, although mass media structures have profited from international media funding, lasting projects aimed at the professionalization of media personnel have done so to a lesser extent.

This circumstance has resulted from the fact that in effect a qualitative and long-term oriented good academic journalism education is still missing (see Hamidi 2013, 2016). Just a few fundamental problems of Afghan journalistic education shall be mentioned here. First, university training programs mainly aim to transmit professionally relevant special competences (theoretical knowledge regarding the law and the organization of the media, the journalistic work process, etc.). In contrast, the established and systematic reflection as to the societal and cultural, political, and economic significance and the ethical tenets and challenges of increasingly economical journalism are hardly considered. Second, subject-specific training at state universities can be criticized in so far as their curricula contain no systematic praxis elements but are organized to desist practical work in an editorial office. Third, training programs in journalism schools mainly concentrate on the transmission of individual practical skills in order to "produce" journalists rapidly. Fourth, communication studies research exploring the changing societal demands and needs of journalism in Afghanistan and offering its findings to academic journalism training is equally nonexistent as a subject-specific discussion concerning quality assurance of journalism. Finally, it can be deduced from the research findings as to journalistic education in Afghanistan that there is a need for a model that combines theoretical and praxis simulating training elements by taking into consideration the specific cultural, societal, communicative, and especially confessional scopes and medial structures in Afghanistan while transmitting and reflecting its training aim apart from the mentioned professional competences, especially a mediation competence. Especially these areas have so far been severely neglected in the subject-specific Afghan research. Last but not least, the mediation function according to unanimous views, considering the fragile state structures and the fractured relations between the ethnic groups, has to be the central function of journalism in Afghanistan. It is the task of the media in fragile states like Afghanistan to shape a national identity and deliver compromises between the different fractions and opinions (Deane 2013).

## 25.7 Mediation as a (New) Journalistic Competence

In international comparison, attitudes to intermediary competences in journalist education of respondents in Afghanistan are quite similar to those of the competency models by Nowak (2007), Weischenberg (1998), Hanitzsch (1999), and Meier (2011). The similarity among the interviewees is particularly evident in the fact that professional competence, special competence, presentation and cross-media competence, organizational competence, and social orientation are described as most important for journalistic education. The journalistic competences suggested by Weischenberg (1998), Hanitzsch (1999), Nowak (2007), and Meier (2011), which should be taught in training, are also largely shared by the participants in the study. However, statements made by respondents in this study in Afghanistan suggest a correction of the above developed model of journalistic skills to be necessary. Respondents agreed that in addition to these areas of competence, another area of journalistic training should be added and taught. This competence can be summarized in a memorable way under "mediation competence." This competence field, which emphasizes a journalist's dedication to the public good in fragile state like Afghanistan (national development and harmony), falls under the umbrella of "mediation." Mediational journalism techniques include to be a creator of a "forum," to aim for "consensus," and to act as a "teacher." This involves reporting on stories that foster hope, healing, and resilience, as the interviewed strongly believe that this style of reporting can contribute to the country's positive future. Overall, the respondents' statements on this area of competence can be included in eight clusters or dimensions. So far, mediation competence is not offered in academic journalism education in Afghanistan.

### 25.7.1 Journalist Should Mediate Between State and the People

According to the interviewees, mediation between the state and citizens is seen as one of the most important tasks of journalists. They believe that Afghanistan has a communicative problem between the government and the people. The population does not know what the government is actually doing. As a result, the population is resigned and disappointed with the state's activities. Through mediation, journalists should increase their confidence in state structures so that these structures can consolidate themselves. The journalists should therefore learn how to communicate and explain government programs of the population and how they can act and function as a bridge between the state and the population. But it also means that journalism students learn how to monitor and control government activities "critically" and "transparently" so that the results can then be revealed to citizens. In summary, respondents are expected to learn that not only do students impart knowledge between the state and the population but that state programs also critically monitor and control activities. To train and professionalize students is seen as an important task of the universities.

## 25.7.2 Urban and Countryside

The interviewees also consider that journalists not only produce information for the urban population but also always have an eye on the rural population, because in Afghanistan over 70% of the population live in rural areas. It is emphasized by respondents that journalist students should learn to take the two strata of the population equally serious. They should also learn to act as mediators between the two populations. However, a journalism student should also learn to connect other layers of society, such as children, women, adults, or the elderly (see also Quebral 2012, p. 14).

## 25.7.3 State and Opponents

A journalism student should learn how he later could mediate as a journalist between the state and opponents of the state (Taliban). In this context, she should learn to be "fair" and "objective." He should learn to treat the population living outside of state influence as equal.

## 25.7.4 Social Consensus Between Ethnic Groups

Respondents argue that journalism students should be qualified to work for national consensus. This is explained by the fact that Afghanistan has been in a civil war for 40 years and that the country is a multiethnic society and ultimately a fragmented society. The cause of the civil war, according to the interviewees, is that the different ethnic groups in Afghanistan have problems with each other and these problems are not communicated. That's why journalism should teach students to see the media as a social "forum" in which the various social groups can share and communicate their concerns and problems. Through this communication, long-term peace could be secured. But if this communication does not take place, this conflict can still continue. That's why journalists are very important here to take on this role. Respondents emphasized that journalists should be trained on how to implement this role.

## 25.7.5 Conflict Resolution

Respondents point out that there are long-standing ethnic conflicts in Afghanistan and journalists have the task of finding solutions to these conflicts by providing information about the problems. This could reduce the conflicts in society. In this context, it is central that journalism students learn not only to present the problems but also to look for their constructive solutions. The task of the journalist is not only to offer the solutions themselves, but to seek these solutions in society and show them to society. Especially Afghanistan needs such journalism graduates. These aspects are central and should be taught in journalism studies.

### 25.7.6 Constructive Journalism as an Agent for Change

Respondents agree that most media in Afghanistan engage in "war journalism" and are constantly focused on negative news. As a result, the population is now tired of this "negative reporting." The negative mediatization of society makes the population "depressed." This leads to a great loss of confidence in social developments. Therefore, the respondents expect that journalist training does not only focus on reporting on negative developments but also to engage with positive developments and successful projects in society and to communicate them to society. The role of journalists in Afghanistan should also be that, according to interviewees, to provide guidance and future prospects for society by reporting on positive developments. For this the journalist students should be trained and conditioned. This view has many interfaces with the concept of so-called constructive journalism. Gyldensted (2014, p. 42) described constructive journalism as a more comprehensive form of journalism that accurately portrays the world by covering not only stories about conflict and destruction, but stories about collaboration and progress as well. The concept has been defined and introduced to the academic literature as an emerging form of journalism that involves applying positive psychology and other behavioral science techniques to news processes and production in an effort to create more productive and engaging stories, while remaining committed to journalism's core functions (Gyldensted 2014).

### 25.7.7 Education and Enlightenment

The respondents argue that most of the population in Afghanistan has no education. There the media and journalists should take on an educational intermediary role. They could take on the role of a teacher. For example, in the rural areas of the population, the media could communicate that it is important to send the girls to school. Journalists need to educate people about their problems and solutions. Journalism must not only have the task of communicating information, but should also educate and enlighten. For that the journalism students should be qualified.

## 25.8 Conclusion

The need for mediation skills is seen as urgent by respondents in Afghanistan along with the other areas (professional, special, presentation and cross-media, organizational, as well as social orientation) of competence. They argue that, first, Afghanistan suffers from protracted war; second, that different ethnic groups have been fighting each other for several years; and third, that there is no social understanding among the groups. In this respect, respondents believe that "every journalist" in Afghanistan should have the job of mediating between different social groups and thus possibly foster peace in the long term. Respondents are of the opinion that Afghanistan is undergoing a process of transformation resulting in economic, social, and political problems. For this reason, it is important for respondents that journalists should assume the leadership or orientation role in the transformation process. It is also important that they promote national and cultural identity. They should sensitize people for it.

All respondents agreed that this area of expertise should be systematically integrated into the curriculum of academic journalism education, so that journalists can acquire this qualification or qualify it for later use in professional life. In this respect it is important to integrate the concept of "mediation competence" systematically into journalist training (competence model), since the journalistic responsibility in fragile contexts can only benefit from it, when social cohesion is seen as the most important solution. If these connections are solidifying in basic and advanced training, they could serve a crucial contribution to constructive nation-building. This area of competence represents an additional concept in the competence model of academic journalism education, which offers a basic orientation: How should we report on national development, how should conflicts be reported, and how should interculturally be sensitive in a multicultural and multiethnic society? The aim of the study program with this competence model could be to train journalists who are qualified to work in currently reporting media. The focus should be on a wide range of qualifications. The training should be multimedia, that is, it should cover newspapers, agencies, radio, television, and online journalism. The study of journalism could convey a mixture of knowledge and ability:

- Communication science basic knowledge, so that the students can deal with the media and the journalism, with the functioning and the legal and political conditions of the media system as well as media-ethical questions.
- Basics in journalistic mediation and production for print and electronic media.
- The ability to understand complex issues and make them journalistically accessible to an audience.
- Scientific skills and techniques to enable them to be applied to a research topic by combining, analyzing, and interpreting the methodological-analytical tools, as well as mastering the skills of documentation and presentation.
- Communicative and social skills to work successfully in the editorial offices and mass media editorial offices.
- Knowledge of the social context in order to report competently and critically on current events from different event fields from different perspectives.
- Reflection and ethical competence to work journalistically on the backgrounds and contexts of events and to develop a critical attitude to their own actions in the journalistic profession.
- Mediation literacy skills to enable journalists to live up to their journalistic responsibilities in a development process, thus making an important contribution to building a constructive nation-building.

For reasons of space, many questions could not be addressed in this article. First of all, of course, the mediation system used is seen as a multilevel process involving expertise on different components: *approaches to development and communications*, c*ommunication and conflict*, as well as *intercultural communication* (intercultural dialogue). It cannot be explained here for reasons of space, but this field introduces the students to the issues of development and the specific role played by the media in development support communication, communication and conflict, and intercultural communication. Secondly, it was not possible to differentiate in this article who could study in such a degree program and where the graduates could later work as journalists. Basically, this competency model is best suited for academic education, as it may have the resources to do so. Here, of course, there is the question of whether the economic imperatives in the media even allow the journalists to be able to implement such skills, which are supposed to promote "public good," in everyday professional life. However, it can be assumed that journalist graduates in the fragile states should be sensitized and generally equipped for these six spheres of competence in principle and in general. Whether or not they can implement this competence in everyday professional life would then be a question for another study.

## 25.9 Future Discussions

The subjects of theoretical and practical training correspond to the competences found in international literature, which a journalist has or should acquire during a professional study. The international standard is (1) professional competence, (2) specific competence, (3) cross-media competence, (4) organizational competence, and (5) social orientation. A new and important finding from the project's specific perspective on Islamic multiethnic developing countries and fragile state such as Afghanistan is the recognition that the political, social, and especially cultural and economic framework conditions demand from journalists a high degree of and mediation competence. Referring to these models, the own model for journalist training is divided into six competences, because the competence (6) "mediation competence" was added as a result from the expert interviews. It must be assumed that the societal circumstances in the fragile society (such as Afghanistan) demand a high measure of mediation skills in journalism. The mediation competence plays only a minor role in Western journalism education (e.g., USA, Germany), while it is central to the academic training of journalists in Afghanistan, as stated by the interviewees, and therefore receives an equal position with respect to the areas of competence. The competence area mediation may still be a marginal topic in classical journalistic education, but the main points of contact are inherent among the respondents: They interpret it as a vital goal. They have fewer problems with the concept of mediation than their counterparts in Western industrialized countries, because "they themselves are affected by these crises." This "ultimate goal," which can be called "a constructive contribution to national development," is consistently seen and expected to be an important responsibility and main function of mass media in fragile states. That means that media and journalism have to create more than anything else "nation-building," "national consciousness" and "unity," and the "encouragement of co-operation and peaceful co-existence between diverse and sometimes hostile communities." This meaning implies the culturally sensitive ideology of journalism in Islam too (see Pintak 2014). The goals of this field in journalism education should instill in students a passion for journalism that helps to transform communities, a critical sense of being a mediator of community development, and develop a sense of moral obligation and professional duty to represent the interests of the masses.

## References

Ali OA (2011) Freedom of the press and Asian values in journalism. http://www.pakistanpress foundation.org/userranddwithoutpics.asp?uid=215. Accessed 25 July 2006

Altai Consulting (2015) Afghan Media in 2014. Understanding the Audience. Retrieved from http://www.altaiconsulting.com/wp-content/uploads/2016/05/Altai-Internews-Afghan-Media-in-2014.pdf

Bandyopadhyay PK (1988) Values and concepts of news. J Northwest Commun Assoc 23(1):40–42

Barker MJ (2008) Democracy or polyarchy? US-funded media developments in Afghanistan and Iraq post 9/11. Media Cult Soc 30(1):109–130

Bayuni EM (1996) Asian values in journalism: Do they exist? In: Masterton M (ed) Asian values in journalism. AMIC, Singapore, pp 39–43

Bourdieu P (1998) Homo academicus. 2. Aufl. Suhrkamp. [orig. 1984], Frankfurt a. M.:

Central Intelligence Agency (CIA) (ed) (2015) Afghanistan. Factbook. https://www.cia.gov/library/publi-cations/the-world-factbook/geos/af.html. Accessed 25 Jan 2018

Christians CG, Glasser TL, McQuail D, et al. (2009) Normative Theories of the Media: Journalism in Democratic Societies. Urbana, IL: University of Illinois Press

Collier P (2010) Wars, guns and votes. vintage, New York

Deane J (2013) Fragile states: the role of media and communication. BBC Media action. BBC World Service Trust, London

Grossenbacher R (1988) Journalismus in Entwicklungsländern: Medien als Träger des sozialen Wandels? Böhlau, Köln/Wien

Gunaratne SA (ed) (2000) Handbook of the media in Asia. Sage, New Delhi

Gunaratne SA (1996) Old wine in a new bottle: public journalism versus developmental journalism in the US, Asia Pacific Educator, 1(1):64–75

Gyldensted C (2014) How journalists could be more constructive – and boost audiences. The Guardian. https://www.theguardian.com/media-network/2014/oct/24/constructive-journalism-de-correspondent. Accessed 22 Jan 2018

Hallin DC, Mancini P (2004) Comparing media systems beyond the western world. Cambridge University Press, Cambridge

Hamidi K (2013) Zwischen Information und Mission: Journalisten in Afghanistan. Berufliche Merkmale, Einstellungen und Leistungen. PhD dissertation at the Faculty of Social Sciences and Philosophy of the University of Leipzig

Hamidi K (2016) Transformation der Mediensysteme in fragilen Staaten am Fallbeispiel Afghanistan. Global Media Journal. German Edition. 5(2). http://www.globalmediajournal.de/de/2015/12/18/transformation-der-mediensysteme-in-fragilen-staaten-am-fallbeispiel-afghanistan-2/. Accessed 22 Mar 2018

Hanitzsch T (1999) Journalistenausbildung in Indonesien. Bedingungen, Ziele, Strukturen und Inhalte. Unpublished thesis, Universität Leipzig, Leipzig

Hanitzsch T, Hanusch F, Mellado C (2011) Mapping journalism cultures across nations. Journal Stud 12(3):273–293. https://doi.org/10.1080/1461670X.2010.512502. 04 Apr 2018

Hedebro G (1982) Communication and social change in developing nations: a critical view. Iowarlowa State University Press, Ames
Hippler J (2004) Nation-Building. Ein Schlüsselkonzept für friedliche Konfliktbearbeitung? Dietz, Bonn
Jayaweera N (1987) Rethinking development communication: a holistic view. In: Jayaweera N, Amunugama S (eds) Rethinking development communication. The Asian Mass Communication Research and Information Centre, Singapore, pp 76–94
Kunczik M (1985) Massenmedien und Entwicklungsländer. Böhlau, Köln
Kunczik M (1988) Concepts of journalism; North and South. Friedrich-Ebert-Stiftung, Bonn
Kunczik M (1992) Communication and social change. A summary of theories, policies and experiences for media practitioners in the third world, vol 3. Friedrich- Ebert-Stiftung, Auflage
Lozare BV (2015) Beyond information-sharing: the key roles of communication in national development. https://healthcommcapacity.org/beyond-information-sharing-the-key-roles-of-communication-in-national-development/. Accessed 22 Mar 2018
Massey BL, Chang LA (2002) Locating Asian values in Asian journalism, a content analysis of web newspapers. J Commun 52(4):987–1003
McQuail D (1983) Mass communication theory, an introduction. Sage, London
McQuail D (2000) McQuail's mass communication theory, 4th edn. Sage, Thousand Oaks
Meier K (2011) Journalistik. UVK Verlagsgesellschaft, Konstanz
Menon V (1996) Preface. In: Masterton M (ed) Asian values in journalism. Asian Media Information and Communication Centre, Singapore, pp vii–vix
Mensing D (2010) Rethinking [again] the future of journalism education. Journal Stud 11(4): 511–523
Namra A (2004) Development journalism vs. "envelopment" journalism. http://www.countercurrents.org/hr-namra190404.htm. Accessed 12 Mar 2018
Nowak E (2007) Qualitätsmodell für die Journalistenausbildung. Kompetenzen, Ausbildungswege, Fachdidaktik. PhD dissertation, Journalism of the Faculty of Cultural Studies at the University of Dortmund
Page D, Siddiqi S (2012) The challenges of transition. Policy briefing, vol 5. BBC Media Action, London
Pintak L (2014) Islam, identity and professional values: a study of journalists in three Muslim-majority regions. Journalism 15(4):482–503
Putzel J, van der Zwan J (2005) Why templates for media development do not work in crisis states. Defining and understanding media development strategies in post-war and crisis states. London School of Economics and Political Sciences, London
Quebral NC (2006) Development communication in a borderless world. Global Times (3)
Quebral NC (2012) Development communication primer. Southbound, Penang
Schetter C (2009) Strukturen und Lebenswelten. Stammesstrukturen und ethnische Gruppen. In: Chiari B (ed) Wegweiser zur Geschichte Afghanistan, 3rd edn. Ferdinand Schöningh, Paderborn, pp 123–133
Schulze M, Hamidi K (2016) Frei und fragil: das Dilemma des afghanischen Mediensystems. medienconncret 3(2):66–69
Servaes J (2015) Beyond modernization and the four theories of the press. In: Lee CC (ed) Internationalizing "International communication". University of Michigan Press, Ann Arbor
Shen G (2012) Comparing media systems beyond the Western World. Chin J Commun 5(3): 360–363
von Kaltenborn- Stachau H (2008) The missing link. Fostering positive citizen-state relations in post-conflict environment. Comm Gap-Report. The World Bank. http://siteresources.worldbank.org/EXTGOVACC/Resources/CommGAPMissingLinkWeb.pdf. Accessed 12 Jan 2018
Weischenberg S (ed) (1990) Journalismus und Kompetenz. Qualifizierung und Rekrutierung für Medienberufe Westdeutscher Verlag. Westdeutscher Verlag, Opladen
Weischenberg S (1995) Journalistik. Medientechnik, Medienfunktionen, Medienakteure. Medienberufe Westdeutscher Verlag, Opladen

Weischenberg S (1998) Journalistik. Theorie und Praxis aktueller Medienkommunikation. Bd. 1: Mediensysteme, Medienethik, Medieninstitutionen. Westdeutscher Verlag, Opladen

Wöhrer V, Arztmann G (2017) Partizipative Aktionsforschung mit Kindern und Jugendlichen Von Schulsprachen, Liebesorten und anderen Forschungsdingen. Spriner, Wiesbaden

Xu X (1998) Asian values revisited. The context of intercultural news communication. Media Asia 25(2):37–41

Xu X (2009) Development journalism. In: Wahl-Jorgensen K, Hanitzsch T (eds) The handbook of journalism studies, Routledge, New York and London

Yin J (2008) Beyond the four theories of the press: a new model for the Asian & the World Press and mass communication. The association for education in journalism 10(1). http://journals.sagepub.com/doi/pdf/10.1177/152263790801000101. Accessed 22 Mar 2018

# Glocal Development for Sustainable Social Change

**26**

Fay Patel

## Contents

| | | |
|---|---|---|
| 26.1 | Overview and Discussion of Development Communication | 504 |
| | 26.1.1 Development Communication, Modernity, and Corporate Globalization | 504 |
| | 26.1.2 Technologization and Consumption | 504 |
| | 26.1.3 International Development Communication, Injustice of Modernity, and "Bonded Development" | 505 |
| 26.2 | Glocal Development | 506 |
| 26.3 | The 2030 Sustainable Development Goals Universal Agenda | 507 |
| 26.4 | Adopting and Implementing the 2030 SDGs Universal Agenda | 509 |
| 26.5 | Glocalization of Learning, Glocal Engagement Framework, and Higher Education Social Responsibility (HESR) Strategy | 509 |
| 26.6 | Propositions for Sustainable Social Change | 512 |
| 26.7 | Conclusion | 515 |
| References | | 516 |

### Abstract

Several contested concepts are introduced and critically reviewed to explore the role of glocal (local and global) communities as partners in promoting sustainable social change. In order for Glocal Development (GD) and Glocalization of Learning to contribute to sustainable social change as outlined in the 2030 sustainable development goals (SDGs), it is necessary for international higher education to mobilize as a social responsibility community engagement agent. To this end, it is proposed that a Higher Education Social Responsibility (HESR) Strategy be implemented within and across glocal higher education institutions to generate HESR funds that will enable adoption of one or several of the 2030 SDGs (relevant to their strengths and to their stakeholder community needs).

---

F. Patel (✉)
International Higher Education Solutions, Sydney, Australia
e-mail: dr.fay.patel@gmail.com

© Springer Nature Singapore Pte Ltd. 2020
J. Servaes (ed.), *Handbook of Communication for Development and Social Change*,
https://doi.org/10.1007/978-981-15-2014-3_77

Glocal Development and the Glocal Engagement Framework encourage the building of glocal communities through the respectful co-construction of knowledge from indigenous, local, and global community knowledge perspectives. The GEF provides international higher education opportunity to adopt the 2030 Sustainability Development Goals (SDGs) in an effort to focus on Glocal Development.

The author asserts that international higher education can play a significant role in the delivery of the 2030 Sustainability Development Goals (SDGs) if it adopts the Glocal Engagement Framework to promote the glocalization of learning. In subscribing to the critical paradigm, the author challenges the "old development communication paradigm," which dichotomized glocal community needs and aspirations in the early decades of the last century.

### Keywords

Glocal development · Glocalization of learning · Higher education social responsibility · Sustainable development

### List of Abbreviations

GD     Glocal Development
GED     Glocal Engagement Dimensions
GEF     Glocal Engagement Framework
GL     Glocalization of Learning
PGE     Principles of Glocal Engagement
SDGs     Sustainability Development Goals

In September 2015, when the United Nations General Assembly adopted the 2030 Sustainability Development Goals (SDGs) to enhance the quality of life of future generations, there was also an acknowledgement that the nations of the world failed in their mission to meet the 2015 Millennium Development Goals. In adopting the 17 SDGs, it was noted that the SDGs "seek to build on the Millennium Development Goals and complete what they did not achieve" (General Assembly Resolution 2015, p. 1). Although the Preamble to the Resolution has traces of the previous rhetoric such as eradicating poverty and strengthening universal peace, more importantly the 2015 Resolutions emphasized that the SDGs were not only "to realize the human rights of all" including gender equality of women but also the SGDs "are integrated and indivisible and balance the three dimensions of sustainable development: the economic, social, and environmental" as a "new universal agenda." It is within this universal agenda context that international higher education has a critical role in leading Glocal Development and driving sustainable social change. The author reiterates her assertion that "higher education is a catalyst for social change" (Patel 2017a, p. 1) and that "international higher education institutions have an obligation to embrace and communicate a social change agenda to transform their policy, practice and structures to remain current" (Patel 2017a, p. 2). The United Nations General Assembly has identified that social change agenda, so international higher education and universities in particular, have a responsibility to commit and deliver on the 2030 SDGs "new universal agenda." With the 2030 SDGs Resolution already 3 years old, it is time to take stock of policies, practices, and structures in international higher education that have been implemented to demonstrate their commitment to the universal agenda. It is also necessary to identify how and what international higher education institutions can contribute to the 2030 SDGs moving forward.

Alternate concepts are introduced, defined, and examined in context, as they emerge, through a critical paradigm lens in order to raise pertinent questions about the socioeconomic and political inequities that emerge around the role of social change agency role of higher education institutions internationally.

Glocal Development (GD), which focuses on the enhancement of quality of life of both local and global (glocal) communities as a partnership from their combined perspectives as one humanity, is introduced as an alternative to Development Communication and International Development. Glocalization of Learning, which advocates for the implementation of the Glocal Engagement Framework (GEF), is posited as a glocal community engagement guideline. The GEF promotes equity, diversity, and inclusivity to bridge the divide between glocal communities with a future-oriented international higher education framework for implementation. There are two components in the GEF which include Glocal Engagement Dimensions (GED) to develop and demonstrate desirable attributes for effective social change agency at an individual and team level and the Principles of Glocal Engagement (PGE) that articulate the considerations in and conditions of successful glocal engagement.

The author invites the readership to question inequities in society and demand that international higher education reinstates a values-based education agenda with human-centered priorities that align with the 2030 SDGs. It is through organized collaborative and partner efforts at glocal community engagement and through commitment to higher education social responsibility agendas that the 2030 SDGs will be met on target. As we move toward 2030, the corporate economic agenda should become replaced with a sustainability development agenda at greater speed. International higher education has, unfortunately, become subsumed by the economic or corporate agenda. Cavanagh and Mander (2004, p. 33) note that "Modern globalization is not an expression of evolution. It was designed by human beings with a specific goal: to give primacy to economic- that is, corporate- values above all other values and to aggressively install and codify these values globally."

In the next section, the old paradigm of development communication that rationalized the export of machinery and technology as modernity standards from the West to other nations is discussed. It is presented as an overview on past and present trends in international communication development that continues to impact the political economy of glocal communities.

## 26.1 Overview and Discussion of Development Communication

### 26.1.1 Development Communication, Modernity, and Corporate Globalization

The "old development communication paradigm" dichotomized local and global (glocal) community needs and aspirations about their quality of life. Similarities between the historical origins of the old development communication paradigm of the 1950s and 1960s (Schramm 1964; Lerner 1958) and the current trends are noted: (a) in the export of new communication technologies (e.g., learning management systems, smartphones, multiple options for online learning, chats, collaboration), over the last decade from the West to developing communities in other regions, and (b) in replicating the underlying "economic globalization" or the corporate globalization agenda (Cavanagh and Mander 2004, pp. 34, 38); and (c) which is carefully "design[ed] for corporate rule" (Cavanagh and Mander 2004, pp. 32–54). The ebb and flow of current economic trends globally are monitored by the "three major global institutions (World Bank, the International Monetary Fund, and the World Trade Organization) (p. 55) that create and express the rules of economic globalization" referred to as the "unholy trinity" and which facilitate the ideal of modernity and modernization in international development as a desirable outcome for developing communities.

At the present time, the push toward technological advancement of disadvantaged communities in the "Third World" to adopt the new communication technologies as a modernity standard (McMichael 2004; Thussu 2000; Stevenson 1988) repeats history as McMichael (2004, p. 23) asserts, *"development/modernity* became the standard by which other societies were judged," following President Truman's proclamation in 1949. In the present day, the breadth and depth of one's consumption and possession of technology gadgets have become measures by which *development/modernity* is also judged. "Technology has become the tool of a cult of success implying the dwelling on growth and progress in line with the modernization ideology" (Servaes 2014, p. xiv). Servaes (2014, p. xiv) further maintains that "the technological deterministic approach played a significant role in the so-called modernization perspectives (Servaes 1999). It saw technology as value -free and a politically neutral asset that can be used in every society and historical context as a driving force of social change."

### 26.1.2 Technologization and Consumption

Those unfamiliar with the origins of the field of mass communication and communication studies may not be aware of the hegemonic presence of the West and Western ideal in the assumptions made in the manufacturing and management of the old world order in which development communication dichotomized the needs of the so-called First World and Third World communities when it came to the export of industrial and agricultural machinery and experiments from the Western nations to other "Third World" nations. The current technological revolution in newly emerging communication media in general, in their various generations, and, in particular, the mass export of education communication commodities in international higher education continues to resemble the export market economy of the 1950s and 1960s. Technology consumption is not a new phenomenon as it was a focus of concern decades ago (Lancaster 1966) and continues to be of interest at the present time when the sophistication of technology and consumption of technology have grown considerably (Raven 2013). However, what is of concern is that five decades later, technologization and the consumption of technology are also being regarded as standards of modernity, and judgements of modernity continue to be measured according to the quantity, brand name, and most recent technology device one has acquired. Raven maintains that technology consumption in 1999 was at "6 h of average daily intake," whereas in 2009 it had increased to "10 h and 45 min." In an age where the higher education providers, educators, and their education seeking publics continue to consume technology at an alarming rate (Lancaster 1966; Raven 2013; Servaes 2014) in their daily dose (or overdose?) of technology, it becomes necessary to contest these similar behaviors that resemble the old development paradigm under guise of innovation. In interrogating the observed behaviors, and through critical self-reflection and ongoing research, international higher education stakeholders are expected to action change as they reconstruct their destinies as a sustainable community at harmony with their *people*, *planet*, and *environment*. "Sustainable development is impossible without a change in the way of thinking that mankind can steward nature by using technology" (Servaes 2014, p. xvii).

### 26.1.3 International Development Communication, Injustice of Modernity, and "Bonded Development"

It is evident through a historical glance over the literature in international communication and development of the last century that hegemonic economic agenda of the West is a well-orchestrated corporate globalization strategy that has successfully ensured that the "Third World," "the remaining half of humanity...regarded as impoverished in standard comparative economic terms" (McMichael 2004, p. 22) remain in a state of "bonded development" (Patel et al. 2012, p. 12) which "refers to a form of existence where developing communities (the poor and the starving populations of the world) remain in bondage to their masters in one form or another." The 2030 SDGs are an opportunity to redress "the injustices of modernity" which "requires the revitalization of a social consciousness that invokes socially just frameworks to uplift the quality of life for developing communities within and across developed [First/Second World] and developing world [Third World] contexts" (Patel et al. 2012, p. 13). The 2030 SDGs have set the agenda for pursuing a values-driven humanistic agenda built on partnership, collaboration, and community engagement.

In the section that follows, the author introduces and elaborates on the notion of Glocal Development (GD) as a future-oriented, sustainable development paradigm of development as an alternative to the old development communication paradigm. Next, the 2030 SDGs are examined regarding their strengths and challenges. The examination extends to a discussion about the proposed adoption of the Glocalization of Learning perspective in international higher education which is embedded within a holistic Glocal Engagement Framework (GEF), as an alternative to the internationalization of higher education model (Patel 2017b, UWN, September 2017). The author maintains that in a Glocal Development context and within the Glocal Engagement Framework, international higher education can adopt all of the 2030 SDGs, thereby endorsing these 17 goals as the priority agenda through commitment to a Higher Education Social Responsibility (HESR) Strategy that attracts a social responsibility fund, from among internal and external stakeholders, for social development projects that action change. Finally, propositions are submitted for embedding the 2030 SDGs within a proactive international higher education agenda through the development of policy, practice, and structures that celebrate the diversity, equity, and inclusivity attributes of both the GD and GEF.

In challenging the old development communication paradigm of the 1950s and 1960s as a dominant world order that aspired mainly to uphold Western ideals and which continues to maintain the balance of the current political economy in favor of the West, the author introduces the term Glocal Development as an inclusive, equitable, and diversity embracing paradigm.

## 26.2 Glocal Development

Glocal Development (GD) refers to the simultaneous development of both local and global communities as a partnership in human endeavor rather than as an economic venture where one partner is subordinate to the other. In this way, GD builds glocal communities based on equity, inclusivity, and diversity principles and subscribes to the "third culture" (Lee) building perspective in which "cultural wealth" (Patel) is the "transactional currency." As an alternative to the old development communication paradigm of the last century which continues to divide local and global community resources and threatens the sustainability of their collective futures, GD connects the local and global community development needs as a social change imperative. In GD, there is an imperative for the empowerment of local and global communities who take responsibility for their collective sustainable futures through collaboration, partnerships, and co-construction of knowledge, policy, practice, and structures that will benefit them as a single humanity. Various labels (international development, development communication, international development communication, modernity) have been used to describe the hegemonic influence of Western agendas of the "First World" on the "Third World" (Chomsky 2003; Thussu 2000). For example, Thussu (2000, p. 19) contends that in the 1950s, the world became divided between the First World countries that upheld capitalism led by the United States with the Second World that was made up by communist countries led by Russia. The label Third World emerged referring to those countries that were outside the capitalist and communist divide; other scholars (Chomsky 2003; McMichael 2004) have proffered similar explanations. The author introduced the term "developing communities" as an alternative reference to Third World and developed communities as an alternative to First/Second World (Patel et al. 2012). Within the context of this chapter, the terms are used interchangeably as they appear in international literature. The author also notes that in the last two decades (1998–2018), due to the mass migration of developing communities in search of better quality of life, job security, and human-centered sustainable futures while escaping the atrocities of war and conflict and tyrannical rulers in the different regions of the world, there is a blurry line between developing and developed communities. Many migrants from Third World nations live amidst of First/Second World and are hungry, homeless, and poor. At the same time, among the First/Second World populations, the local indigenous communities also are hungry, homeless, and poor.

Glocal Development is an equalizer of sorts as it attempts to remove the divisional references among the communities from the three worlds and to encourage partnership and collaboration among the developed communities and the developing communities (as noted in Patel 2012, which were referred in the early decades of the twentieth century as the First, Second, and Third World, respectively).

Within the Glocal Development paradigm, there is a linkage between the local, indigenous communities and the global, international communities based on their partnership as a single humanity that aims at building sustainable futures to benefit all people.

It is the partnership, collaborative, and integrated vision of Glocal Development that responds unconditionally to the 2030 SDGs "goals and targets" to "stimulate action over the next 15 years in areas of critical importance for humanity and the planet" (United Nations Resolution, p. 1). Within the Glocal Development paradigm, as with Lee's third culture theoretical approach, one community is not subjugated by the other, and community engagement and dialogue occur in an open, mutually respectful space in which the well-being and prosperity of both the local and global communities are intrinsically interwoven so that, as noted in the 2030 SDGs Resolution, their combined efforts will ensure that among them, they will honor the Resolution (pp. 1–2) so that people, planet, prosperity, peace, and partnership will remain the beacons to move forward as a glocal community.

## 26.3 The 2030 Sustainable Development Goals Universal Agenda

In upholding the 2030 SDGs as fundamental higher education goals that will benefit glocal communities as one humanity, Glocal Development and the Glocalization of Learning Framework will redirect international higher education to its ultimate responsibility as a catalyst for sustainable social change, imbibing human values as critical elements for the future generations.

The 2030 SDGs are listed below (from General Assembly Resolution, 25 September 2015 http://www.un.org/sustainabledevelopment/sustainable-development-goals/) to demonstrate the depth and breadth of the 17 SDGs, followed by a discussion of how the international higher education community can work within the integrated and interlinked community model envisaged by the United Nations General Assembly.

## 26.4 Adopting and Implementing the 2030 SDGs Universal Agenda

The author explores and highlights the challenges for international higher education of pursuing and implementing the SDGs within a Glocal Development paradigm and the Glocal Engagement Framework through the HESR strategy. The 2030 SDGs (all 17 goals) advocate human values necessary for sustainable living as an integrated community through community engagement to combat *poverty, hunger, good health and well-being, quality education, gender equality, clean water and sanitation, affordable and clean energy, decent work and economic growth, industry, innovation and infrastructure, reduced inequalities, sustainable cities and communities, responsible consumption and production, climate action, life below water, life on land, peace, justice and strong institutions*, and *partnerships for the goals*.

The international higher education (IHE) community has vast human resources and a combination of skills and competencies to respond fully to all 17 SDGs through a mutually respectful agenda setting process for community engagement and partnership within and across regions. The community needs to have the will to change (Patel 2011, p. 57) and to adopt the Glocal Engagement Framework with necessary adaptation to local contexts and needs, as relevant, and to be steadfast in implementing the agenda with firm timelines to coincide with the 2030 SDG deadline. The author contends that the 2030 SDGs provide opportunities for creative thought and context-based innovative solutions (Patel 2014) and acknowledges that in some instances it may be acceptable to identify and "follow" best practices. The agenda requires monitoring of progress toward the desired transformation and an ongoing continuous improvement process to evaluate desired progress with the remit to halt abuse of resources and exploitation of glocal communities in the name of the 2030 SGD transformation goals and to redirect misdirected goals.

## 26.5 Glocalization of Learning, Glocal Engagement Framework, and Higher Education Social Responsibility (HESR) Strategy

The author asserts that the adoption of the glocalization of learning perspective requires a mainstream embedded approach for it to be successful in upholding the international higher education institution's commitment to the 2030 SDGs. This means that the institution is expected to adopt the Glocal Engagement Framework and the Higher Education Social Responsibility (HESR) Strategy as a policy and practice with established structures for swift and effective implementation of the 2030 Sustainability Development Goals (SDGs).

The Glocal Engagement Framework has been articulated in greater detail in a previous paper on the deconstruction of internationalization (Patel 2017b). The Framework is briefly explained, and the Figures (Figs. 1, 2, and 3) are reproduced with the permission of the journal editor. The Glocalization Engagement Framework is illustrated in Fig. 1 as a holistic approach in international higher education

**Fig. 1** The Glocal Engagement Framework (GEF). (Permission has been granted by the International Journal of Global Studies Editor Ray Scupin for use in the chapter)

**Fig. 2** The Global Engagement Dimensions (GED). (Permission has been granted by the International Journal of Global Studies Editor Ray Scupin for use in the chapter)

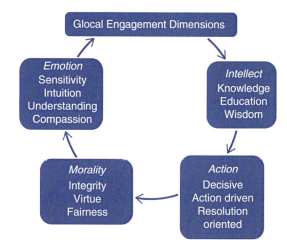

consisting of two critical components (Glocal Engagement Dimensions and Principles of Glocal Engagement). Both components are required for the effective design and delivery of glocalization of learning to be promoted as an equitable, diverse, and inclusive framework of learning in international higher education.

The first component is the Glocal Engagement Dimensions illustrated in Fig. 2 which are a combination of *intellect, emotion, morality, and action* which are critical in the development of compassionate human beings. The hope is that they will engage from a value-driven perspective with local and global communities and respond appropriately to the needs of glocal communities affected by natural and/or human disasters and strife.

The second component is the Principles of Glocal Engagement which are an adaptation of Klyukanov's ten principles of intercultural communication. For an appropriate full response to the 2030 SDGs, glocal community stakeholders would have to be familiar with and informed about the Principles of Glocal Engagement together with the Glocal Engagement Dimensions because all four dimensions (*intellect, emotion, morality, and action*) are required in efforts to drive social change through mutually acceptable and respectful community engagement.

| Number | Intercultural communication principles Klyukanov (2005) | Principles of glocal engagement Patel (2017) |
|---|---|---|
| 1 | Punctuation principle | Draw mutually acceptable boundary lines |
| 2 | Uncertainty principle | Reduce uncertainty through negotiation and sharing of relevant information |
| 3 | Performativity principle | Cultivate new shared meaning |
| 4 | Positionality principle | Position or ground oneself in a context |
| 5 | Commensurability principle | Find common ground among stakeholders |
| 6 | Continuum principle | Consider multiple glocal perspectives |
| 7 | Pendulum principle | Consider ongoing interaction in negotiating shared meaning |
| 8 | Transaction principle | Transaction component of global community building related to exchange of cultural wealth and indigenous knowledge |
| 9 | Synergy principle | Cooperative nature and integration of global community building |
| 10 | Sustainability principle | Long-term mutually respectful relationship |

**Fig. 3** The Principles of Glocal Engagement (PGE) illustrates the adaptation of Klyukanov's intercultural principles

In adopting the Glocalization Engagement Framework as a holistic mainstream policy, practice, and structure, international higher education will prepare its stakeholders (leadership, learners, professionals, administrators, communities, and industry partners) to imbibe a human spirit that transcends all physical and virtual barriers to achieve the 2030 SDGs.

In addition to the GE Framework, the international higher education community is encouraged to adopt the Higher Education Social Responsibility Strategy (HESR) which requires that the institution and all its faculties, schools, and colleges, along with other independent departments and units, contribute a portion of their budgets toward a social responsibility fund that will support research, teaching, and learning within an integrated and interconnected institution-community engagement approach. The Higher Education Social Responsibility Strategy budget is not the same as the Corporate Social Responsibility budget distributed by large and small corporations as a social responsibility contribution to less advantaged communities. HESR budgets are integral to the overall mission and vision of the higher education institution in delivering a transformative higher education agenda that is aligned with the 2030 SGDs. HESR budgets are developed within the institution and distributed equitably across stakeholder groups to deliver on their chosen 2030 SDGs through research, teaching and learning programs, projects, and agendas that impact sustainable social change within local and global (glocal) communities. The Higher Education Social Responsibility Strategy will lead and oversee the gathering of the social responsibility funds and the implementation of social development projects that will benefit the local, regional, national, and international communities to improve their quality of life.

## 26.6 Propositions for Sustainable Social Change

The author subscribes to "context-based innovative solutions" and encourages the seeking of "creative solutions" instead of indiscriminate adoption of "best practices" (Patel 2012, p. 474) and expects guidelines to be adapted to suit the needs of diverse communities so that meaningful, sustainable development is part of actioned social change. In this way, glocal communities will respond meaningfully to the 2030 SDGs and find a myriad context-based innovative solution through community engagement as a partnership in Glocal Development.

So how would international higher education institutions organize themselves around the HESR and the GEF to champion the 2030 SDGs? There are multiple options, opportunities, and permutations that can be explored to creatively design organizational models for glocal community engagement. Glocal communities can adopt the GEF and HESR Strategy as perceived through the author's lens, or they may want to modify existing policies or develop new policies, practices, and structures that are context-based innovative solutions.

Glocal Development and Glocalization of Learning perspectives within which the Glocal Engagement Framework is embedded along with the Higher Education Social Responsibility Strategy are expansive concepts that provide "open spaces for glo[c]al engagement" (Patel et al. 2014). Stakeholder communities are encouraged to explore creative and innovative responses that will in effect be integrated and interlinked among local and global community groups as partnerships to achieve each of the 17 SDGs in ways that have yet to be imagined.

However, the author also submits propositions or critical thought throughout the chapter. As noted in the preceding section, the guidelines must be critically assessed and modified for relevance to the glocal context in which action for change is necessary. In this way, the 2030 SDGs are transformed into the desirable, appropriate innovative solutions among the glocal communities that are affected by any one or more of the 17 SDGs. It is through the glocal efforts of all stakeholders that the 2030 SDGs will become transformed into sustainable living conditions for all.

Within national boundaries, implementation of the 2030 SDGs will require meticulous organization and leadership of 2030 SDG Working Groups (translated into regional indigenous forms of agency for change and transformation) so that the Working Groups work in harmony with indigenous groups to design the level, degree, and area of transformation required in their region. In other words, the GEF and the HESR Strategy ensure that the fundamental mobilization of community, resources, collaborators, and partners is in tandem with the 17 SDGs. The stakeholder groups will decide in a transparent, accountable, and responsible manner and in accordance with principles of equity, diversity, and inclusivity, which SDG/SDGs they will adopt and how they will achieve the 2030 SDG goals, one goal at a time. They will pledge their commitment to the universal agenda so that by 2030, each stakeholder group will achieve some goals through local partnerships building a sustainable community by co-constructing their collective futures together. At the same time, the local partnerships will explore extended mutually respectful partnerships with their global communities to achieve other goals. The 2030 SDG agenda can be set in motion in various ways. However, all variations of agenda setting require strict monitoring measures to assist internal and external quality assurance teams to ensure that (a) the quality of life of all communities meets the highest standard to ensure sustainable growth in their economic prosperity; (b) harmony exists among diverse communities; (c) equity of access and opportunity is paramount in the transformation of the community; and (d) inclusivity is respected of all nationalities, ethnicities, and communities as one human race.

The author highlights creative and innovative context-based solutions that have resulted in the development of policy and practice to support the 2030 SDGs in one country (Malaysia) and in one region of the world (Asia-Pacific Region). The following examples of collaborative partnerships are cited from the Openlearning Conference Keynote Presentation by the first woman vice-chancellor of a Malaysian University (Professor Asma Ismail, November 2017).

The efforts of higher education institutions in Malaysia to adopt the 2030 SDG agenda as a progressive sustainable development agenda may be categorized as Glocal Development in a region of the world where local and global communities are forming partnerships to develop sustainable quality of education to enhance quality of life. Notable exemplars of policy, practice, and structure that resemble variations of the author's proposed Glocal Development, Glocalization of Learning, and Higher Education Social Responsibility Strategy guidelines are presented below through the exemplars from Malaysian Higher Education and the Asia-Pacific Region to support the author's contention throughout the chapter that international higher education institutions have the privilege, authority, and capacity to lead sustainable social change. Malaysia has demonstrated through the Education Blueprints, the CEO Faculty Program, and the various community engagement programs including ACUPEN that learning can be glocalized by beginning with local indigenous knowledge perspectives and becoming harmoniously combined with global community perspectives in areas such as disaster relief and industry engagement to ensure glocal (local and global) impact. The Malaysian model can be modified to other local and global (glocal) community contexts, as relevant to their needs. The Higher Education Social Responsibility Strategy can also be creatively designed utilizing the Malaysian and Asia-Pacific Region projects and programs as a model. These can be further expanded to include mainstream integration within an international education transformation and the 2030 SDGs agenda. In adapting, modifying, and expanding the Malaysian and Asia-Pacific Region initiatives and adopting the 2030 SDGs, Glocal Development for sustainable social change will significantly impact lives among local and global communities.

- Malaysia: Asia-Pacific Region
  At a recent conference (OpenLearning Conference, November 2017), keynote speaker Professor Asma Ismail (Vice Chancellor, USM) presented an overview of the progress made with the Malaysia Education Blueprint 2013–2025 for preschool to post-secondary schools and the 2015–2025 Malaysia Education

Blueprint for universities as a framework for implementing community engagement, collaboration, and a values-based education. The focus was on preparing our society for engagement in a "collaborative economy" in which the collaborative network is key. More importantly, it was stressed that "a collaborative economy is based on trust and trustworthiness without which there is no sharing of information and best practice." Collaborative networking is about building and maintaining trust, so this is an example of the significant impact of a values-based education that is an imperative in the present and future sustainable development goals. The "digital divide" was noted as another area of social inequity, exclusion of those who have no access to technology. Other collaborative community engagement networks (ACUPEN and the CEO and Faculty partnership) were also highlighted as Malaysia's and the Asia-Pacific Region's remarkable capacity to respond creatively and innovatively to the 2030 SDGs through community engagement among national, regional, and international stakeholders.

- ACUPEN: Asia-Pacific Community Engagement Network
  Professor Asma Ismail also highlighted that ACUPEN (Asia-Pacific University Community Engagement Network) was another community engagement initiative among 19 countries and 92 institutions in the region since 2011. From an International Education and a Glocal Development perspective, ACUPEN is an example of a partnership of local communities in the Asia-Pacific Region with global communities in Australia and Germany, for example, that focus on common goals for sustainable social change. The objectives include the promotion of community engagement among staff and students, capacity building, dissemination of information, and collaborative project development.
- CEO and Faculty Program
  In this innovative university-industry partnership (that impact 90,000 staff and students), CEO's tutor selected lecturers for 1500 h over 200 sessions. The CEO Faculty Program is an innovative partnership between industry and faculty in higher education institutions to promote industry participation and integration and exchange of knowledge and expertise among industry experts, university research communities, and students.

The Malaysian and Asia-Pacific collaborative partnership initiatives provide adequate evidence that when glocal communities and institutions of international higher education, in particular, have a will to action change, they make a wide impact on communities near and far. The collective efforts of Malaysia and the relevant stakeholders in the Asia-Pacific Region have collectively begun the rewarding and challenging task of addressing the 2030 SDGs:

1. (Poverty [from an International Education perspective, the region reviews technology access issues within the region, the ACUPEN, and internationally to address issues of access to technology to narrow the divide]).
2. (Good health and well-being [e.g., promoted through a values-based education, student community engagement, and volunteerism in disaster relief programs]).

3. (Quality education [through quality assurance processes to raise standards of education design and delivery to meet international education standards also noted in the 2013–2025 and 2015–2025 Malaysian Education Blueprints].
4. Gender equality [visible efforts are noted to meet international gender equity status, e.g., Professor Asma is the first vice-chancellor of a university, and she was the first woman in a senior leadership position in the Ministry of Education].
5. Decent work and economic growth [in line with international education endeavors, various industry partnerships are forged, and the CEO Faculty Program is an example of enhancing prospects for decent work and economic growth].
6. Innovation and infrastructure [e.g., the CEO Faculty Program is an innovative program to encourage partnerships, capacity building, and industry engagement among staff and students].
7. Reduced inequalities [e.g., the education blueprints are evidence of commitment to education transformation with the intent to bring Malaysian education into the international education arena through knowledge, research, and community engagement programs].
8. Peace, justice, and strong institution [demonstrated through ACUPEN partnerships with countries in the Asia-Pacific Region such as Laos, Bangladesh, and Nepal and multilevel collaborative initiatives, locally, regionally, and internationally with Australia and Germany].

The Malaysian programs and projects to enhance Malaysian higher education, industry partnerships, and community engagement attest Malaysian commitment to the 2030 SDGs to bring Malaysia on par with the international education endeavors, and the Asia-Pacific Region network (ACUPEN) cited above demonstrates both Malaysia's and the Asia-Pacific Region's commitment to enhance higher education to international standards. The blueprints and various community engagement initiatives within Malaysia and the Asia-Pacific Region are evidence of organized efforts to begin a journey of trust, respect, and compassion. As articulated by the Minister of Higher Education 2017 Dato Seri Idris Jusoh, "higher education requires humanizing and that technology should be centered on humanity and public interest" (cited in the Openlearning Conference Keynote Presentation, Professor Asma Ismail, November 2017).

## 26.7 Conclusion

The chapter explored challenges in achieving the 2030 SDGs and recommended that the SDGs be embedded in the Higher Education Social Responsibility Strategy in international higher education within a Glocal Development context. In upholding the 2030 SDGs as fundamental international higher education goals to benefit glocal communities as one humanity, the Glocal Engagement Framework and the Higher Education Social Responsibility Strategy will redirect "international higher education as a catalyst for sustainable social change."

Both Glocal Development and the Glocal Engagement Framework face challenges because change of heart, soul, and spirit is an imperative for success. This is a challenge in the face of adverse reactions from those who wish to maintain their dominance of the corporate globalization agenda. The ultimate goal of GD and GEF is to eradicate the socioeconomic and political inequities facing local and global communities, including those who are local, indigenous communities, and migrant communities crossing through their different worlds (First, Second, and Third Worlds as defined in the communication development literature in the 1950s). Further, GD and GEF, as emerging discourses that aim for equity, also challenge the status quo in international higher education in an era that is fraught with an imbalance of socioeconomic and political power, inadequate visionary leadership, and a lack of commitment to human values. Compassionate human communication is necessary for building glocal communities committed to Glocal Development for sustainable social change.

## References

ACUPEN (2011) Asian Asia-Pacific University – Community Engagement Network. More information https://apucen.usm.my/index.php/en/about-apucen/introduction

Cavanagh J, Mander J (eds) (2004) Alternatives to economic globalization a better world is possible, 2nd edn. BK Publishers, San Francisco

Chib A, May J, Barrantes R (eds) (2015) Impact of information society research in the Global South. Springer Open, Singapore

Chomsky N (2003) Hegemony or survival America's quest for global dominance. Henry Holt Publishers, New York

Lancaster K (1966) Change and innovation in the technology of consumption. Am Econ Rev 56 (1/2):14–23

Lerner D (1958) The passing of traditional society: modernizing the Middle East. Free Press, New York

McMichael P (2004) Development and social change a global perspective, 3rd edn. Pine Forge Press/Sage, California

Nicotra A, Patel F (2016) Contesting the political economy of higher education: educating the good citizen. J Int Glob Stud 7(2):22–39 Lindenwood University Publishers, St. Charles. More information http://www.lindenwood.edu/jigs/docs/volume7Issue2/essays/22-39.pdf

OpenLearning Conference (Social Learning Conference 2017) Sydney, Australia 27–28 Nov 2017 keynote presentation shared through the OpenLearning conference course. More information https://www.openlearning.com/

Patel F (2017a) International higher education as catalyst for social change. Educ Leadersh Action J 4(2). Lindenwood University Publishers, St. Charles. More information http://www.lindenwood.edu/academics/beyond-the-classroom/publications/journal-of-educational-leadership-in-action/all-issues/volume-4-issue-2/faculty-articles/patel/

Patel F (2017b) Deconstructing internationalization: advocating for glocalization as a socially responsible and sustainable framework. J Int Glob Stud 2:64–82. Lindenwood University Publishers, St. Charles. More information http://www.lindenwood.edu/files/resources/64-82-deconstructing-internationlization.pdf

Patel F, Li M, Piscioneri M (2014) Cross-institutional and interdisciplinary dialogue on curriculum for global engagement: emerging perspectives and concerns. J Int Glob Stud 5(2):40–52 Lindenwood University Publishers, St Charles. Retrieved from http://www.lindenwood.edu/jigs/docs/volume5Issue2/essays/40-52.pdf

Patel F, Lynch H (2013) Glocalization as an alternative to internationalization in higher education: Embedding positive 'glocal' learning perspectives. Int J Teach Learn High Educ (IJTLHE) 25 (2). More information: http://www.isetl.org/ijtlhe/past2.cfm?v=25&i=2

Patel F (2012) Whither scholarship in the work of enhancing the quality of teaching and learning? In: Teacher education and practice, Fall 2011 24(4). More information http://www.sfasu.edu/education/departments/secondaryeducation/doctoral/teacheduandpractice.asp

Patel F, Sooknanan P, Rampersad G, Mundker A (2012) Information technology, development and social change. Routledge, New York

Raven M (2013) Our rapidly increasing consumption of technology and data. Retrieved https://medium.com/@micrv/our-rapidly-increasing-consumption-of-technology-and-data-ab363ed71b23/

Servaes J (2014) Technological determinism and social change: communication on a techmad world. Lexington Books, USA

Stevenson RL (1988) Communication, development & the third world the global politics of information. Longman Inc., New York

Sustainable Development Goal Declaration (2015) Transforming our world: the 2030 Agenda for Sustainable Development. Retrieved from https://sustainabledevelopment.un.org/post2015/transformingourworld

Schramm W (1964) Mass media and national development: the role of information in developing countries. Stanford University Press, Stanford

Thussu DK (2000) International communication continuity and change. Arnold, Hodder Headline Group, London

# Communication Policy for Women's Empowerment: Media Strategies and Insights

## 27

Kiran Prasad

## Contents

| | | |
|---|---|---|
| 27.1 | Introduction | 520 |
| 27.2 | Feminist Movement | 522 |
| 27.3 | Silencing Women | 523 |
| 27.4 | Muting Women's Capability | 524 |
| 27.5 | Gender and Development Paradigms | 525 |
| References | | 529 |

### Abstract

Despite rapid economic growth in the post-reform period and all the flagship programs of the government, India ranks very low on narrowing the gender gap in education, health, and economic participation. India is among the few countries of the world where female labor force participation is shrinking with less than 15% in any form of paid work. In this context, the role of the National Mission for Empowerment of Women (NMEW) is crucial for interministerial convergence of gender mainstreaming of programs, policies, and institutional processes of participating ministries which have largely operated independently and in a stand-alone manner. Sustainable development in India can be a reality only when there is gender equality and justice-based development that accords top priority to the welfare of women.

### Keywords

Gender inequality in India · Violence against women · Media strategies · Communication policy · Women's empowerment

K. Prasad (✉)
Sri Padmavati Mahila University, Tirupati, India
e-mail: kiranrn.prasad@gmail.com

© Springer Nature Singapore Pte Ltd. 2020
J. Servaes (ed.), *Handbook of Communication for Development and Social Change*,
https://doi.org/10.1007/978-981-15-2014-3_128

## 27.1 Introduction

> There is no chance for the welfare of the world unless the condition of women is improved. It is not possible for a bird to fly on one wing. - *Swami Vivekananda*

The status of women is one of the most pressing contemporary development concerns at the international as well as the national levels. Historically though women enjoyed a respected position in the ancient cultures of India, Persia, and Greece, over time the situation has been drastically altered, and women were isolated from major developments that may have led to their modernization and autonomy especially in many developing countries. In India, women constitute a population of 586.5 million with 405.2 million (48.6%) in the rural areas and 181.2 million in the urban areas (RGI 2011). It is among the countries with the largest female population which makes it imperative to focus on communication policies that address their advancement and empowerment to foreground sustainable development.

India recorded a high economic growth of 9% per annum during 2005–2006 to 2008–2009. In 2016, India climbed 16 places from 55th place to the 39th rank on the Global Competitiveness Index prepared by the World Economic Forum with the economy characterized by improved business sophistication and goods market efficiency (WEF 2016). This remarkable economic achievement has yet to be translated into human development for half of India's population. While India slipped to 131 among 188 countries ranked in terms of human development, it ranks at 132 out of 146 countries in the Gender Inequality Index, which reflects the dismal status of women (UNDP 2017).

The 2016 Human Development Report (UNDP 2017) red flags the stark reality that the largest gender disparity in development was in South Asia where the female HDI value was 20% lower than for males; India accounts for the largest gender disparity in the region. In rural India, where teenage marriages are common, women face insecurity regarding a regular income, food, shelter, and access to health care. It is an understatement to say that violence against women is multidimensional; it is structural, brutal, and a part of everyday life (Ravi and Sajjanhar 2014). Indian women marry at a median age of just 17 years, and 16% of women aged 15–19 have already started bearing children, according to the 2005–2006 National Family Health Survey (IIPS 2007). With 212 per 100,000 live births, India ranks among the countries with the highest maternal mortality rates (MMR) accounting for one-third of maternal deaths in 2015 worldwide (RGI 2009). It was estimated that every year, 78,000 women die during pregnancy and childbirth even though 75% of these deaths can be prevented by health care (Krishnan 2010). India is ranked 170 out of 185 countries in the prevalence of anemia among women (48% women are anemic) and has the highest rate of malnourished children in the world at 44% and stands at 114/132 in stunted growth of children with 38.7% incidence (UNFPA 2016).

A collaborative study by the National Commission for Protection of Child Rights (NCPCR) and the Young Lives India (NCPCR and YLI 2017) revealed that India stood at 11th rank in the countries worldwide with the highest incidence of child marriages accounting for 47% of all children with 39,000 minor girls being married every day in India. The UNICEF further points out that India accounts for one-third of the global total of over 700 million women married as children leading to high levels of depression among them. The National Crime Records Bureau estimates that over 20,000 young mothers, mainly housewives, commit suicide every year, making them the largest demographic group in India to commit suicide followed by farmers (Bakshi 2016). The triggers for these deaths range from an unplanned pregnancy to an abusive or alcoholic husband, pressures to have a male child, and hormonal changes among others. The Programme for Improving Mental Health Care (PRIME) project in Madhya Pradesh is an intervention program for neonatal depression-related problems, where all expecting mothers are screened (Bakshi 2016). There is hardly any recognition of the pressure and depression of motherhood; many young women are automatically expected to care for their child and their personal health in India.

Gender discrimination and violence against women have had a profound effect on the sex ratio in India. The sex ratio has been dropping steadily for the past 50 years. In 2011 the sex ratio of females stood at 940 females per 1000 males, the lowest ratio after independence (RGI 2011). The Pink Economic Survey (2018) which is the first national data of its kind gives estimates based on the sex ratio of the last child (SRLC) which is heavily male skewed to show that 21 million girls were unwanted by parents in India. These human development indices show that India has failed to convert its economic growth to transform the lives of women and children who are among the most vulnerable population.

In the light of persistent inequality of women, the gender equality survey of the World Economic Forum (WEF 2014) places India at 114 out of 142 countries in the Global Gender Gap Index. While India ranks 126 on educational attainment, it is ranked 134 on economic participation and opportunity and the lowest at 142 on the health and survival of women. Contrary to the popular notion that men are the main breadwinners of the family, according to the 2011 Census, about 27 million households, constituting 11% of total households in the country, are headed by women, often among the poorest. Though many of these women lack formal education and employment skills, their courage to face adversities and steer their families out of poverty goes unrecognized by policy makers and development planners in the country.

There is an avowed strategy of women's empowerment that focuses on women's education and employment. But what is shocking is that studies indicate a link between women's employment and domestic violence. The National Family Health Survey (NFHS-3) data report that there is much higher prevalence of violence against women who were employed at any time in the past 12 months (40%) than women who were not employed (29%) (IIPS 2007). Studies give evidence that women who have more education than their husbands, who earn more, or who are the sole earners in their families have a higher likelihood of experiencing intimate partner violence than women who are not employed or who are less educated than their spouse (Sabarwal et al. 2013). This reality in India contradicts the widely held global perception that better economic status of women lowers their risk to marital violence.

It is astonishing that women are achievers when their very survival is under threat on a daily basis. The enormous physical and psychological pressure of competing in an "equal world" with men who do not experience even a fraction of the constraints is a lived experience for many capable women in leadership positions. The feminist genius lies in the striving for creativity and capability by women who are projected as emblems of modernization and emancipation on one hand but muted by the patriarchal oppression that reinforces male domination that overrides these "progressive" images (Prasad 2009).

The feminist movement grew out of a sense of outrage at such treatment of women, for ending several forms of harassment of women, violence, sexual exploitation, job discrimination, exclusion from public life, and unequal educational opportunities. Society is reluctant to concede that women's rights are an inalienable dimension of human rights. The deliberate eclipsing of women from public affairs led to the struggle for equality, social participation, and legitimate share of autonomy and status enjoyed by their male counterparts. President Barack Obama points out during the US 2016 presidential election campaign that the United States is uncomfortable with powerful women, and that is why the United States has yet to elect a woman President (The Hindu 2016) which further gives insights into the world's deep-seated bias against women. There is a need for restructuring development paradigms to include the perspectives of women.

## 27.2 Feminist Movement

The women's movement has had to contend with several situations and idealized stereotypes dictated by cultural and traditional norms that existed during its early days and continues to persist. Virginia Woolf's writings presented the obstacles that prevented women from attaining self-reliance and a distinct identity. Woolf explains that it was impossible for women to develop creativity without minimum independence and personal space (Woolf 1929). She argued that if young men as inheritors of the family fortunes were encouraged to fend for themselves and carve a niche in the world, women fell behind as servants and possessions, with strict socialization patterns ingrained in them. They had no space or time to call their own and could not write down their thoughts and experiences even if they had a chance at the minimal education accorded to them. Women continue to struggle to seek the freedom to express their choices, however simple, in their daily life and continue to grapple with the question of creating space for themselves within their lives:

> Feminism, however, is not only about achieving social justice, it is also about creating a space which allows women to become something other than how they have been traditionally defined by men. Women, against the odds, are attempting to balance autonomy and dependence; self-fulfilment and a desire and obligation to care for others. In the present climate, as hurdle after hurdle remains in their way, they are encouraged to blame themselves – instead of examining how and why the hurdles were constructed in the first place. (Roberts 2004)

The feminist development strategy often lies in crossing these hurdles at a cost seldom known to the outside world and in many cases unknown to even their own families.

Friedan (1963) adopted a sociological approach and met housewives isolated in their urban apartments and homes striving to live up to the stereotyped and traditional roles and attempting to transform their houses into shining models portrayed by soap and floor wax commercials. Women in the developing countries of Asia and Africa find themselves in a "dichotomizing trap" (Williamson 2006) in which researchers and policy analysts envision women as either "traditional" or "modern" meaning liberated, educated, and independent. These images of women have reinforced simplistic ideas about the nature of society, the interpretation of the status of women, and prescriptions for their future (Williamson 2006: 186). The mass media also actively promote the image of a modern "new woman" as revealed by studies of women in advertising on satellite television in India (Prasad 2005), in China (Birch et al. 2001) and in Thailand (Biggins 2006). Though the mass media under the influence of Western culture represent women as modern, traditional perceptions and expectations of women's roles in society or family remain deep-rooted, which have continued to influence the way society treats women in several developing countries including India.

Greer (1971) alerted women to the reality that in spite of having legal, educational, and political rights, it would fail to raise their status unless women consciously overcome their childhood conditioning and acculturation as these were the forces that kept them quiescent and denied them justice in societies. Feminist scholarship has evolved over time with a focus on sex differences (the traditional approach) to a focus on improving society and making women more like men (the reformist or liberal approach), to the current focus of giving voice to women (the radical feminist approach) (Dervin 1987).

## 27.3 Silencing Women

Women's capability as generators of knowledge in all formal education has been vastly eclipsed, what is taught is men's knowledge; the silencing of active theorizing women takes place in almost all education systems (Kramarae 1989). In a digital age, "if tech is widely seen as a male bastion, it is because women's stories have been deliberately erased" (Sampath 2015). The fact that the world's first programmer and inventor of scientific computing is Ada Lovelace, the Spanning Tree Protocol for network computing was invented by Radia Perlman, and the key designer of Transmission Control Protocol/Internet Protocol was Judith Estrin tells "the untold stories" about women (Shevinsky 2015).

The Chief Operating Officer of Facebook Sheryl Sandberg (2013) believes that women themselves are largely responsible for not accepting leadership roles as they are socially conditioned to be less ambitious, settle for less, and prioritize their homes and families at the cost of their career. Sandberg (2013) calls women to "lean in" to their careers without being guilty about the need to be liked and sacrifice on behalf of the family. Here lies the catch for many creative, capable women geniuses of India. According to the Gender Diversity Benchmark Survey for 2011 and 2014, Indian companies lose 11% of their female workforce every year as women are haunted by "daughterly guilt" and maternal guilt that leads them to prioritize caring for parents, children, and extended family by leaving their careers (Kumar 2016). Such women are also under pressure to support their husbands to fulfil their family aspirations for a better lifestyle by working from home or getting creative by plunging into entrepreneurship which has its own set of challenges.

The creativity of women to develop lay in circumventing the barriers erected in their social environment to continue performing their duties and lead their families and communities daily by silencing their voices and muting their capabilities putting them under enormous strain. Empowerment of women can be achieved only by enabling women to voice their experiences and for society to understand them as human beings and respond to them with sensitivity. With nations (including India) boasting themselves as nuclear powers, "space-age countries," and "information societies," it is unthinkable that any country would burn women alive (for dowry) except in India. But the large-scale brutal violence against girls and women from acid attacks to gang rapes does not seem to touch the hearts of representatives of the people (nearly half of their electorate being women) who remain oblivious to the plight of women. It is estimated that around 30% of the sex workers in India are below 18 years and many women have been pushed into sex work (Saggurti et al. 2011). It is even more shocking that of late several panchayats (local administrative bodies especially in North India) have begun to award punishments to women such as approving their sale to criminals, gang rape, excommunication from the village (along with their families), and social boycott by the community. Such local bodies are not instruments of political empowerment, as presently they seem to be emerging as instruments of oppression against women.

Spender (1985) documents the silencing or threatening of women by the application of deviancy labels. Women who are particularly knowledgeable and witty or those who question or rebel against patriarchy are called aberrations, unnatural, unattractive, unisexual, unnaturally sexed, and man-haters. All forms of violence used to silence women – implicit or explicit – restrict women's capability in all spheres and diminish human development.

## 27.4 Muting Women's Capability

Women are under great social control and scrutiny which has restricted what they can say and where and to whom. They are forced to express their subordination through "feminine" words, voice, and syntax. In this context, Ardener (1975) conceives of women as a "muted" and men as a "dominant" group in relation to language, meaning, and communication. In patriarchal cultures, men determine the general system of meanings for society and validate these meanings through the support received from other men. These meanings, regarded as correct by men, have evolved out of male experiences. The concepts and vocabulary arising out of it are quite different for women to contend with and express themselves in (Prasad 2000). There are several words in almost every language to describe women who are in disrepute for whatever reason. For instance, to be a "prostitute" is to be stigmatized for life, but the men who are the "clients" do not suffer in the least from social mores or sanctions that continue to bestow status on them. They enjoy immunity from stigma and abuse which is strengthened by the fact that there are no words to describe such immoral men. There can be "other women" in extramarital relations but no "other men."

The information and communication media concentrate on fashion, glamor, weight reduction, cookery, and how to sharpen "feminine instincts" to keep men and their in-laws happy rather than on career opportunities, health awareness, entrepreneurship, legal aid, counselling services, childcare services, and financial management that build women's capacity and leadership. Despite the fact that there is no dearth of women leadership in India, it is rarely recalled that the right to information campaign led by Aruna Roy, the struggle for natural resources led by Medha Patkar, the sustainable agriculture movement led by Vandana Shiva, and a host of local agitations led by women like C. K. Janu and Mayilamma have played a stellar role in grounding sustainable development in India (Prasad 2013).

Women in India labor under the brunt of oppressive traditions and exploitation, suffer from lack of self-worth or identity, and are routinely subjected to violence even at home. It is absurd that in a country where even women's dress is dictated by tradition, women must take responsibility for family planning, AIDS, and a host of other maladies affecting society. Patriarchal structures dictate the degrees to which women must be robed or disrobed; women do not even have complete liberty to decide on their dress. The burkini versus bikini debate has ramifications in India where violence against women is attributed to their dress! Women and society in general are in great need of self-introspection and self-conscientization to overcome the downslide in human values and women's rights as an inalienable part of human rights (Prasad 2004).

## 27.5 Gender and Development Paradigms

There has been a coexistence of feminist development approaches worldwide and in India – women in development (WID), women and development (WAD), and gender and development (GAD). WID approach was adopted internationally to achieve women's integration in all aspects of the development process. It is based on liberal feminism which generally treats women as a homogeneous group and assumes that gender roles will change as women gain an equal role to men in the development of education, employment, and health services. The WID approach does not question the existing social structures or explore the nature and sources of women's oppression. It fails to consider the implications of race, class, and gender on women's oppression.

WAD is primarily a neo-Marxist, feminist approach with a strong emphasis on the importance of social class and the exploitation of the "Third World." Affirmative action by the state and proactive approach by the civil society through NGOs and women's groups are advocated by these models for empowerment of women against the forces of patriarchal class society. The major criticism of the WAD approach is that it fails to undertake a full-scale analysis of the relationship between patriarchy and women's subordination. Although work which women do inside and outside homes is central to development, WAD preoccupies itself with the productive aspect at the expense of the reproductive side of women's work and lives.

The GAD approach recognizes that improvements in women's status require analysis of the relations between men and women, as well as the concurrence and cooperation of men. There is recognition that the participation and commitment of men are required to fundamentally alter the social and economic position of women. A gender-focused approach seeks to redress gender inequity from a shift with an exclusive focus on women to an approach that must involve men in the family and the broader sociocultural environment through facilitating strategic, broad-based, and multifaceted solutions to gender inequality. The focus of GAD is also to strengthen women's legal rights.

WID tends to focus on practical needs, whereas GAD focuses on both practical needs and strategic interests. In addition to focusing on everyday problems, GAD is concerned with addressing the root inequalities (of both gender and class) that create many of the practical problems women experience in their daily lives. Practical needs refer to what women perceive as immediate necessities such as water, shelter, and food. Strategic gender interests are long-term, usually not material, and are often related to structural changes in society regarding women's status and equity. They include legislation for equal rights, reproductive choice, and increased participation in decision-making. The strategic interests of women's development and empowerment tend to be long-term which include consciousness-raising, increasing self-confidence, providing education, strengthening women's organizations, and fostering political mobilization. There remains a whole range of women's problems from female feticide, female infanticide, child marriage, sexual abuse, sex trade and trafficking, rising son preference and devaluing daughters, marital rape, unfair burden of population policy and AIDS campaign on women, problems of single women, branding rural women as witches in several parts of the country, to the legal rights of women regarding property, divorce, and succession, which have yet to see concerted action by policy makers and planners. There is a strange silence on heinous crimes of fathers raping daughters especially in Kerala, where women's health, education, and basic socioeconomic indicators match those of the advanced countries in the world (Institute of Applied Manpower Research 2011). Women's activism for gender equality is quite weak even where women enjoy situational advantage. Highly educated and financially independent women are seen succumbing to dowry demands, son-preference, domestic violence, and sexual harassment at the workplace.

Studies on violence against women (VAW) reveal that nearly 1 out of 4 men of 10,000 men in Asia agreed that they had committed rape and sexual entitlement or a belief that men were entitled to sex regardless of consent was the major reason that men gave for committing a rape (Cheng 2013). Crimes against women attract suspicion of women's character (case of love gone sour) and sexual assault and rape in the news media framed in relation to women's freedoms. Basic rights and safety remain out of reach for a vast majority of women in India. The discourse on women's resistance is focused on "get your own safety pin" or teaching self-defense rather than educating boys and men to be humane. Women are under constant surveillance in private and public life. There is the tremendous power of collective punishing mechanism by shaming women and warning against behavior perceived as undesirable by sections of society (khaps, caste councils). There are multiple agencies from family members to criminals regulating or negating women's degrees of freedom. Rarely understood is the trauma of women who become "objects of social ostracism" (due to acid attack, rape, or other violence).

In 2014, the Indian Ordnance Factory, Kanpur, launched a light revolver *Nirbheek* (meaning fearless) for women. This raises the question: if women's safety is all about owning a gun, what about the rapist who lurks at home (Sharma 2014)? There is an even strange justification that if a woman does not reciprocate a man's "love," he can kill or maim her. Instead of strengthening the criminal justice system, improving law and order and policing skills, arranging safety measures like streetlights and transport facilities for women, and creating enabling work environments, the government advocates the use of pepper sprays (widely available in grocery shops), and business is focused on developing apps and various other products that can trigger alarms to emergency contacts if a woman is attacked.

The development of women is not to be perceived as by women, of women, and for women; men as fathers, husbands, brothers, sons, and friends must unite to strengthen the lives of girls and women and for achieving the vision of India as a developed country. Education and the mass media must play a critical role in widening the discourse on gender equality and challenging the social and political order that systematically devalues women.

The Government of India launched the National Policy for Empowerment of Women in 2001 (Department of Women and Child Development 2001) with the following specific objectives:

- Creating an environment through positive economic and social policies for full development of women to enable them to realize their full potential.
- The de jure and de facto enjoyment of all human rights and fundamental freedom by women on equal basis with men in all spheres – political, economic, social, cultural, and civil.
- Equal access to participation and decision-making of women in social, political, and economic life of the nation.
- Equal access to women to health care, quality education at all levels, career and vocational guidance, employment, remuneration, occupational health and safety, social security and public office, etc.
- Strengthening legal systems aimed at elimination of all forms of discrimination against women.
- Changing societal attitudes and community practices by active participation and involvement of both men and women.

- Mainstreaming a gender perspective in the development process.
- Elimination of discrimination and all forms of violence against women and the girl child.
- Building and strengthening partnerships with civil society, particularly women's organizations.

Most of these objectives have largely remained on paper with little action. Nearly after a decade, the National Mission for Empowerment of Women (NMEW) was announced on March 8, 2010, and operationalized during 2011–2012. It works with all the states and union territory governments and has 14 ministries and departments of Government of India as it partners with the Ministry of Women and Child Development as the Nodal Ministry. The mission aims to bring in convergence and facilitate the processes of ensuring economic and social empowerment of women with emphasis on health and education, reduction in violence against women, generating awareness about various schemes and programs meant for women, and empowerment of vulnerable women and women in difficult circumstances (www.nmew.gov.in).

The NMEW has undertaken initiatives to check the declining sex ratio in 12 districts of 7 states to address the sex selective elimination of girls. It is collaborating with various community organizations to further the flagship program of the government *Beti Bachao Beti Padhao* scheme (save the girl child, educate the girl child) and address issues of domestic violence, partnering with Lawyers Collective Women's Rights Initiative (LCWRI) and attempting to bring diverse stakeholders in raising the collective community consciousness on women's issues. The mission must engage in strategic communication for interministerial convergence of gender mainstreaming of programs, policies, institutional arrangements, and processes of participating ministries which have largely hitherto operated independently and in a stand-alone manner. The NMEW is in the process working on a fresh National Empowerment Policy for Women. It is too early to predict the substantial gains in women's empowerment through the NMEW.

There are several social communication campaigns involving men like the *Beti Padhao Beti Bachao* (educate and save the girl child), *Bell Bajao* (Ring the Bell for domestic violence), *What Kind Of Man Are You?* (AIDS awareness campaign), and the *One Billion Rising* campaign to highlight the worldwide violence against women and several community initiatives that attempt to resolve the problems faced by women. The *#MeToo* campaign has also resonated in India with women publicly narrating their experiences on media of violence, exploitation, and humiliation in a gender-unjust society. Positive role modeling through audiovisual communication must be used to break the initial resistance and enable the family and community to participate in empowering women toward health care and improving their status. Family life education and sex education must receive top priority in general education and the mass media. Program modules on parenting must be popularized in schools and colleges to promote greater understanding among the youth about their roles and social responsibility as future parents and create a more responsible dialogue between partners to promote concepts of masculinity that include committed fatherhood. An intensive behavior change communication must be initiated by multi-sectoral agencies to address the masculine norms and behavior that heighten the risk for both men and women and prevent mutually beneficial relationships and inclusive human development. All health information campaigns must stress the importance of inter-spouse communication to improve the lives of girls and women and encourage equitable sharing of household and caregiving responsibilities throughout the life cycle. The compassion of the Buddha and Jesus shows that men are capable of sublime virtues with proper education, guidance, and engagement in personal and social life. The future generation of boys must learn that gender equality will make the world kinder, less violent, and less demanding for men as well (Mander 2014). The message of gender equality given by Nancy Smith (1973) in her powerful poem must become the essence of human sensitization:

*For every woman who is tired of acting weak when she knows she is strong, there is a man who is tired of appearing strong when he feels vulnerable.*

*For every woman who is tired of being called "an emotional female", there is a man who is denied the right to weep and to be gentle.*

*For every woman who is tired of being a sex object, there is a man who must worry about his potency.*

*For every woman who is called unfeminine when she competes, there is a man for whom competition is the only way to prove his masculinity.*

*For every woman who takes a step toward her own liberation, there is a man who finds the way to freedom has been made a little easier.*

## References

Ardener E (1975) Belief and the problems of women. In: Ardener S (ed) Perceiving women. Malaby, London

Bakshi AB (2016) Depression after Childbirth a silent killer in India, 12 July 2016. http://everylifecounts.ndtv.com/depression-childbirth-silent-killer-india-3673?pfrom=home-environment, 17 July 2016

Biggins O (2006) Cultural imperialism and Thai Women's portrayal on mass media. In: Prasad K (ed) Women, globalization and mass media: international facets of emancipation. The Women Press, New Delhi, pp 95–112

Birch D et al (2001) Asia cultural politics in the global age. Allen Unwin, Crow Nest

Cheng M (2013) Study: rape against women 'widespread' in Asia 11 September 2013. http://www.usatoday.com/story/news/world/2013/09/10/rape-asia-mensex/2792811/. 27 Sept 2016

Department of Women and Child Development (2001) National policy for the empowerment of women 2001. Ministry of Human Resources Development, Government of India, New Delhi

Dervin B (1987) The potential contribution of feminist scholarship to the field of communication. J Commun 1987(37):107–120

Friedan B (1963) The feminine mystique. Dell Publishing, New York

Greer G (1971) The female eunuch. McGraw Hill, New York.

IIPS (2007) International Institute for Population Sciences (IIPS) and macro international. In: National Family Health Survey (NFHS-3), 2005–06: India: Volume II, Mumbai, p 39

Institute of Applied Manpower Research (2011) India human development report 2011. Institute of Applied Manpower Research, New Delhi
Kramarae C (1989) Feminist theories of communication. In: Gerbner G, Schramm W, Worth TL, Gross L (eds) International encyclopedia of communication, vol 2. Oxford University Press, New York, pp 157–160
Krishnan J (2010) Dying to be a mother. The Week, 12 Sept 2010, pp 16–24
Kumar KB (2016) Daughterly guilt' haunts Indian working women. The Hindu, 21 Mar 2016, p 14
Mander H (2014) Working with masculinities. The Hindu, 22 Feb 2014
NCPCR and YLI (2017) Report on incidence and intensity of child marriages. NCPCR and YLI, Delhi
Pink Economic Survey (2018) Pink economic survey. Government of India, Niti Aayog
Prasad K (2000) Philosophies of communication and media ethics: theory, concepts and empirical issues. BRPC, New Delhi
Prasad K (2004) Preface. In: Prasad K (ed) Communication and empowerment of women: strategies and policy insights from India. The Women Press, New Delhi
Prasad K (2005) Women's movement and media action: paradoxes and promises. In: Prasad K (ed) Women and media: challenging feminist discourse. The Women Press, New Delhi, pp 213–144
Prasad K (2009) Young women and the modernity project: realities and problems of media regulation in India. J Int Commun 15(1):9–25
Prasad K (2013) Environmental communication from the fringes to mainstream: creating a paradigm shift in sustainable development. In: Servaes J (ed) Sustainable development and green communication: Asian and African perspectives. Palgrave Macmillan, Virginia, pp 95–109
Ravi S, Sajjanhar A (2014) Beginning a new conversation on women. The Hindu, 21 June 2014, p 9
Registrar General of India (2009) Maternal mortality rates. Government of India, New Delhi
Registrar General of India (2011) Census of India, registrar general of India. Government of India, New Delhi
Roberts Y (2004) Pre-feminism is alive and well. The Hindu, 3 Aug 2004
Sabarwal S, Santhya KG, Jejeebhoy S, (2013) Women's autonomy and experience of physical violence within marriage in rural India: evidence from a prospective study. Journal of Interpersonal Violence 29(2):332–347
Saggurti N, Sabarwal S, Verma RK, Halli SS, Jain AK (2011) Harsh realities: reasons for women's involvement in sex work in India. J AIDS HIV Res 3(9):172–179
Sampath G (2015) Yes, there's sexism in science. The Hindu, 24 June 2015, p 9
Sandberg S (2013) Lean in: women, work and the will to Lead. Alfred A. Knopf, New York
Sharma K (2014) In the line of fire. The Hindu, 19 Jan 2014, p 3
Shevinsky E (2015) Lean out: the struggle for gender equality in tech and start-up culture. OR Books, New York/London
Smith NR (1973) For every woman. http://www.workplacespirituality.info/ForEveryWoman.html, 22 July 2016
Spender D (1985) For the record: the making and meaning of feminist knowledge. The Women's Press, London
The Hindu (2016) Our society still feels uncomfortable about powerful women, says Obama. The Hindu, 20 Sept 2016. http://www.thehindu.com/news/International/obama-campaigning-for-hillary-clinton/article9125045.ece, 20 Sept 2016
UNDP (2017) Human development report 2016. UNDP, New York
UNFPA (2016) Global nutrition report 2016. UNFPA, New York
Williamson DA (2006) The promise of change, the persistence of inequality: development, globalization, mass media and women in Sub-Saharan Africa. In: Prasad K (ed) Women, globalization and mass media: international facets of emancipation. The Women Press, New Delhi, pp 183–208
Woolf V (1929) A room of one's own. Available at http://uah.edu/woolf/chrono.html, 15 Aug 2016
World Economic Forum (2016) Global competitiveness report 2016–2017. World Economic Forum, Geneva
World Economic Forum (2014) *Global Gender Gap 2014*, World Economic Forum, Geneva

# Part V
# Strategic and Methodological Concepts

# Three Types of Communication Research Methods: Quantitative, Qualitative, and Participatory

## 28

Jan Servaes

## Contents

| | | |
|---|---|---|
| 28.1 | Introduction | 534 |
| 28.2 | The Context of Research | 535 |
| 28.3 | Quantitative Approaches to Research | 536 |
| | 28.3.1 Objectivity and Subjectivity | 537 |
| | 28.3.2 The "End of Ideology" and "Value-Free" Science | 537 |
| | 28.3.3 Power-Based Research | 538 |
| | 28.3.4 Methodology-Driven Research | 539 |
| 28.4 | Qualitative Approaches to Research | 540 |
| | 28.4.1 Subjectivity and Phenomenology | 540 |
| | 28.4.2 Naturalistic Observation and the Participant Observer | 541 |
| | 28.4.3 Critical Research | 542 |
| | 28.4.4 Validity and Evaluation in Qualitative Research | 543 |
| | 28.4.5 The Role and Place of the Reader, Viewer, and Receiver | 543 |
| 28.5 | Participatory Research | 544 |
| | 28.5.1 Principles of Participatory Communication Research | 545 |
| | 28.5.2 Differences Between Participatory Research and Action Research | 546 |
| | 28.5.3 "Insiders" and "Outsiders" | 547 |
| | 28.5.4 A Definition of Participatory Research (PR) | 547 |
| | 28.5.5 The Process of Participatory Research | 548 |
| | 28.5.6 Evaluation and Validity in Participatory Research | 549 |
| | 28.5.7 The Integration of Different Methodologies | 549 |
| | 28.5.8 A Word of Caution | 550 |
| 28.6 | By Way of Conclusion | 550 |
| References | | 551 |

J. Servaes (✉)
Department of Media and Communication, City University of Hong Kong, Hong Kong, Kowloon, Hong Kong

Katholieke Universiteit Leuven, Leuven, Belgium
e-mail: jan.servaes@kuleuven.be; 9freenet9@gmail.com

© Springer Nature Singapore Pte Ltd. 2020
J. Servaes (ed.), *Handbook of Communication for Development and Social Change*,
https://doi.org/10.1007/978-981-15-2014-3_112

> **Abstract**
>
> This chapter presents a brief overview of the three types of communication research methods being applied in development communication settings: quantitative, qualitative, and participatory. This chapter attempts to outline the relative characteristics and merits of these approaches to research and to emphasize some of the philosophical issues which underpin them. It discusses the strengths and weaknesses of each and highlights the benefits of a more normative approach focused on the "poor" in society.

> **Keywords**
>
> Research methodologies · Quantitative · Qualitative · Participatory · Triangulation · Empowerment

## 28.1 Introduction

All research begins with some set of assumptions which themselves are untested but *believed*. Positivistic research, which comprises the mass of modern communication and development research, proceeds from the presupposition that all knowledge is based on an observable reality and social phenomena can be studied on the basis of methodologies and techniques adopted from the natural sciences. In other words, "reality" exists apart from our interpretation of it, and we can objectively perceive, understand, predict, and control it. Social scientists, enamored by the notion of a predictable universe, therefore concluded that, by applying the methods of positivistic science to study human affairs, it would be possible to predict and ultimately to control human social behavior (McKee et al. 2000). Furthermore, its methodological premises and epistemological assumptions are based almost exclusively on the Western experience and world view, a view which holds the world as a phenomenon to be controlled, manipulated, and exploited (Bhattacherjee 2012).

Quantitative and qualitative researchers today increasingly recognize the weakness of a purely objectivist position. While they recognize that complete objectivity is no longer possible, they nevertheless wish not to replace value freedom with anything of substance. Therefore, many participatory researchers feel that objectivism still resides in the practices of many social scientists.

This chapter attempts to outline the relative characteristics and merits of quantitative, qualitative, and participatory approaches to research and to emphasize some of the philosophical issues which underpin them (Servaes et al. 1996; Jacobson and Servaes 1999; see also ► Chap. 14, "The Relevance of Habermasian Theory for Development and Participatory Communication" by Thomas Jacobson). However, Bryman's (1984; Bryman and Burgess 1994) argument should not be forgotten, i.e., that it is *not* possible to establish a clear symmetry between epistemological positions and associated techniques of social research, and, consequently, they often become confused with each other: "It may be that at the technical level the quantitative/qualitative distinction is a rather artificial one ... At the epistemological level, the distinction is less obviously artificial since the underlying tenets relate to fundamentally different views about the nature of the social sciences, which have resisted reconciliation for a very long time" (Bryman 1984: 88). Or, as stated by Saneh Chamarik (1993: 4): "In social science, approach and methodology are not just a matter of technique and expertise, but essentially represent an attitude of mind, that is to say, a kind of moral proposition."

Therefore, the distinction between the mechanistic diffusion versus organic participatory models (Servaes 1999) is important. Contrary to the scientist who aims for an understanding of the world as it exists "out there," the participatory model takes as the fundamental focus the immediate world of the participants and their analysis of, and subsequent action in, that world. Both models should be regarded as opposites on a continuum.

Moreover, highly rigid methodology serves to limit rather than expand humanistic understanding to the detriment of both the researcher and the "researched." The detachment, the reluctance to "contaminate" one's research design in the dynamic, holistic, and human social context, while claiming to be working for the ultimate benefit of those studied, should be challenged, if not on a methodological, then certainly on a moral basis. Given this aloofness, many so-called quantitative, qualitative, and/or participatory researchers become largely parasitical in character and contribute little to those they research. Jumping from country to country, or community to community, many Western and local researchers alike conduct their "safari" and "airport research," collect the data, and return to the university to write it up, hopefully for publication in a prestigious journal. The subjects of the study do not learn of the results and obtain no benefit, except via a long, distorting cycle of traditional research and top-down social action.

However, the assertion is not that the entirety of empirical research or academic inquiry is of no value. *The intent is not to reject science but to properly define its place within human knowledge.*

## 28.2 The Context of Research

The social group with the greatest interest in the perpetuation of highly complex research practices, which mandate "correct" methodology above all, are the representatives of the scientific community themselves: university teachers and researchers or, in one word, the homo academicus (Servaes 2012). As with bureaucracies, academic institutions strive for results that paint their activities in a concise, easily visible fashion. The reasons for this include peer status as well as personal and organizational sustenance. The rule of "publish or perish" is known to all and those who would go against the flow or approach their task in an unorthodox fashion are jeopardizing their personal careers. It is not the individual's place to question the prevalent paradigms or accepted methodologies. Consequently, most scholars continue to follow the intellectual fashions of the day. In the whole their specialist knowledge prompts little divergence from the prevalent opinion of their social group or class. As a result, institutions of higher education have often stayed aloof to the needs of the majority of the population they are supposed to serve.

New forms of access to and structural changes in the higher education system should be a subject of serious consideration. *The search for truth succumbs to the search for funding, which is subservient to ideology.* Therefore, while academia is supposed to be a "marketplace of ideas" and a forum for free discussion, it has often become part of the market. With time, the parameters under which researchers operate can constitute, for them, a social reality. They see its limits as the limits of the world, not as a more or less arbitrary boundary between what they know and what they do not. Such trends are demeaning to the researcher herself/himself as well.

Due to the widespread use and application of Western models, constructs, and methodologies, intellectuals in the so-called Global South are sadly dependent for their status and acceptance on their links with the West. Mimicking foreign methodologies, as well as striving for more exacting standards, and tighter control, social science research is pulled farther and farther from the dynamic social context which it ostensibly strives to understand.

Unfortunately, this observation is not limited to the academic world only. As a result of bureaucracy and managerialism, more than in the past, many governmental development agencies *outsource* their intellectual capacity to consultants, sometimes NGOs but more often for-profit foundations. These agencies often prefer consultants because consultants are always available on short-term notice, while academics often cannot be easily freed up, or have become "too costly." A range of self-proclaimed experts and consultants fill the gap and provide the background papers and policy recommendations as commissioned by the system. The values of equality and democracy which development agencies claim to promote are often absent within the agency itself. It is clear that power relations are problematic inside of development agencies just as much as inside of academic and scientific institutions (Louw; Van Hemelrijck 2013).

## 28.3 Quantitative Approaches to Research

Logic and common sense tells us that such a thing as *objectivity* exists. It also tells us the bumblebee can fly. But by the laws of physics, the bumblebee is unable to fly. "Common sense" is just that *common sense*.

Social research and developmental research is claimed to be objective, neutral, and thus scientific. In reality, however, it often becomes an instrument in the hands of the powerful elite to rationalize their control of the people. Therefore, as James Halloran (1981: 40), the late President of the International Association for Media and Education Research (IAMCR) claimed, "it is ideology that we are now talking about, and in mass communication research, as elsewhere, there are an increasing number of ideologists who present their work as social science ... it all depends on what one means by ideology, and what one means by social science."

## 28.3.1 Objectivity and Subjectivity

The underlying problem of the importunate posture of "objectivity" is that in assuming that the meanings for the conditions of subject's lives are independent of those subjects – researchers in the human and social studies are assuming they themselves are independent of how they see those subjects or of the phenomena they are examining.

Present discussion rejects this assumption. The idol has feet of clay. Similar to the notion of relative logic, the assertion is "objectivity" is nothing more than subjectivity which a given aggregate of individuals agree upon. It refers to what a number of subjects or judges experience, in short, to phenomena within the public domain. Therefore, as Brenda Dervin (1982: 293), the first female President of the International Communication Association (ICA), observed, "No human being is capable of making absolute observations, and since it is humans that produce that thing we call information, all information is itself constrained." And Jim Anderson and Meyer (1988: 239) adds that "the ultimate source of our knowledge is not 'out there' pressing upon us, but it is in our own consciousness."

Hence, "objectivity" is nothing more than *intersubjectivity*. For example, to ensure coder reliability, the researcher culls out those coders whose subjective coding is incongruent with the mean. The rejected coders' analysis is not intersubjective; it is not "objective." Research defines the categorical constructs, then defines and refines the instruments to measure those constructs. With time (as well as painstaking research and adequate funding), "intelligence" becomes what the tests measure. Science becomes the operationalization of concepts through the development of reliable measures, which, in turn, serve to measure and substantiate the concepts. Reliability and internal validity are perfected. However, whether this cyclical process is related with the social reality outside the lab, with external validity, remains largely unconsidered.

## 28.3.2 The "End of Ideology" and "Value-Free" Science

Roughly corresponding with the rise of communication models and the doctrine of modernization emerged a growing sentiment of the impending "end of ideology" in the social sciences. Superstition and ideology were to yield to reason, positivism, and science. Friberg and Hettne (1985: 209) relate the belief that "controlled experiments and deductive reasoning guarantees an infinite growth of valid knowledge about the mechanisms of nature and this knowledge will lay the foundation of the technological society marked by material abundance."

This new mode of inquiry was claimed to be detached, neutral, and value-free. It was "the noble and final battle of 'knowledge' against 'ignorance'" (Thayer 1983: 89). The belief was that "with just a little more data, with just a few more refined techniques of inquiry, we will indeed 'arrive'" (p. 89). *Not all agree.* Some take a political stand and argue that "neutral" science has led to the dehumanizing and catastrophic utilization of knowledge, nuclear missiles, biological poisons, and psychological brainwashing. Others claim from a more methodological point of view that no responsible social scientist would hold that the results of social science research are comprehensive, free of value premises, or valid for all seasons.

Similar to the majority of efforts towards communication and participation in the development context, research most often derives from, and serves to perpetuate, dominant structures. Tying the notions of intersubjectivity, ideology, and the established order together, Addo (1985: 17) writes that "claims to scientific neutrality and objectivity come easily to the establishment [which] holds fast to a deep-seated common world view ... a theoretical structure of what the world is like and ought to be like; and what it is about and ought to be about."

### 28.3.3 Power-Based Research

The "dominant methodological concern is how to approach the interpretation of the modern world-system in order to make it safe for ... the perpetuation of ... this world-system" (Addo 1985: 20). This methodology reveals its pro-status-quo bias in that it never considers the alternative of the creation of a new system but rather presents "functional" adjustments to the old. It facilitates the functioning of the existing system, without ever questioning its validity, however dangerous that system may be for the future of society and man's and woman's integrity. In this way the research, although often referred to as abstracted empiricism, is certainly not abstracted from the society in which it operates. In sum, science, particularly at the level of social research, is laden with ideology. It serves, and is largely intended to serve, vested power structures. As such, claims of the "end of ideology" or "value-free science" do not hold true either in their premises or in their applications.

Communication is no exception. Robert White (1982: 29) argues that "much of the research tradition in the field of communication – concepts, methodology, and research designs – has been in the service of authoritarian communication." And Halloran (1981: 22) adds that communication research "has developed ... essentially as a response to the requirements of a modern, industrial, urban society for empirical, quantitative, policy-related, information about its operations. ... Research was carried out with a view to ... achieving stated aims and objectives, often of a commercial nature." The goal of inquiry has often been to investigate how commercial or political persuasion could be effectively used. Hence, such research is an "objective" form of intelligence gathering for social control by business firms and governments. The predominant vein of social research, resting on epistemological suppositions, strives to understand reality, with the objectives of prediction and control. However, while claiming detachment and neutrality, research functions to objectify the very "reality" it assumes to be objective. This is done in a manner congruent with, and towards the benefit of, the existent status quo, usually those who pay the bills. In other words, it serves more to create and perpetuate a reality than to understand one.

## 28.3.4 Methodology-Driven Research

In the journals and conferences, debate is more often about the method of investigation than the content. Halloran (1981: 23) argues that "'Scientific' is defined solely or mainly in terms of method, and ... little or no attention is given to ... the nature of the relevant, substantive issues and their relationship to wider sociological concerns." This overemphasis on methodology and techniques, as well as adulation of formulae and scientific-sounding terms, exemplify the common tendency to displace value from the ends to the means. A sociologist or psychologist obsessed with frameworks, jargon, and techniques resembles a carpenter who becomes so worried about keeping his or her tools clean that he or she has no time to cut the wood. Reality as experienced in the lives of people has been sacrificed to methodological rigor. Too often an emphasis upon exact measurement *precludes* the asking of *significant* questions, and the result is fragmented bits of knowledge on "researchable" topics.

As the methodology becomes increasingly complex, the questions posed become ever-more trivial. The parameters of inquiry yield to increasingly sophisticated methods. It is sometimes said that modern scientists are learning so much about so little that they'll eventually know everything about nothing. Therefore, many years ago already, Kuhn (1962: 35) attested that "perhaps the most striking feature of the normal research problems ... is how little they aim to produce major novelties, conceptual or phenomenal."

With the tendency towards reductionism, the social sciences tend to *compartmentalize* the subject matter into separate parts, with only rare attempts at approaching the subject as a whole. This forces nature: "into the preformed and relatively inflexible box that the paradigm supplies. No part of the aim of normal science is to call forth new sorts of phenomena, indeed those that will not fit the box are often not seen at all" (Kuhn 1962: 24).

Categories, taxonomies, terms, and their relations are largely what make a given paradigm. Yet, even though this "terminology is a reflection of reality, by its very nature it is a selection of reality and therefore also functions as a deflection of reality" (Burke 1968: 44). The categorization is often derived from the literature rather than from "concepts that are necessarily real, meaningful, and appropriate from the native point of view" (Harris 2001: 32). This renders invisible the knowledge of the people involved in the real-life activity at which research is aimed.

Prevalent theories, paradigms, methodologies, and taxonomies become tantamount to human understanding; they become the goal and the reality. Adherence to the widely accepted methods produces mostly one-dimensional communication research unable to cope with complex and dynamic social realities. Analysts research isolated variables, thereby removing them from the cultural context which gives them their meaning. But communication cannot be dissected and displaced to increase the convenience of observation without disrupting the process of meaning creation at the same time.

## 28.4 Qualitative Approaches to Research

Qualitative research almost mirrors the quantitative one, and, in this sense, discussion has already addressed the qualitative approaches. However, several points require elaboration (see also Berg and Lune 2014; Denzin and Lincoln 2005).

### 28.4.1 Subjectivity and Phenomenology

Qualitative research accepts subjectivity as *given*. It does not view social science as "objective" in the ordinary and simple understanding of that term but as an active intervention in social life with claims and purposes of its own. As the world is constructed intersubjectively through processes of interpretation, the premise is that "reality does not write itself on the human consciousness, but rather, that human consciousness approaches reality with certain embedded interpretations" (Anderson and Meyer 1988: 238), and those interpretations, in turn, constitute reality. We are thus introduced to a *new principle of relativity* which holds that all observers are not led by the same physical evidence to the same picture of the universe (Bruhn Jensen 2013).

As such, given the idiosyncratic nature of individuals and cultures, social reality has predominantly the character of irregularity, instead of the ordered, regular, and predictable reality assumed in empirical approaches. These implications have profound impact on the manner that we look at communication and the way we strive to understand, and research, social contexts. Further confounding this subjective, idiosyncratic irregularity is the fact of its dynamism. Christians and Carey (1981: 346) assert, "To study this creative process is our first obligation, and our methodology must not reduce and dehumanize it in the very act of studying it." The emphasis should be on discovery rather than applying routinized procedures. Following the humanistic premise that people are not objects, there is no warrant for believing that the social sciences should imitate the natural sciences in form or method. The subjects of research are just that, subjects. We are not objects and cannot be objectified without losing the very humanity that is the focus of inquiry. Needed are intellectual endeavors that reach out to, rather than insulate us from popular experience. We must therefore have the courage to underwrite contextual thinking and contextual research. "Truth" lies not in objective proof, but in the experience of existence or non-existence in the minds of men and women and their mutual affirmation. Therefore, "truth" is relative.

This brings us back to the topic of intersubjectivity. At the risk of being "irrational" and "unscientific," it is argued qualitative and phenomenological approaches, in their broadest sense, advocate a human approach towards understanding the human context. It is based on the human character of the subject matter. As a specifically human approach, it uses lived experience as acts on which to base its findings. Fuglesang (1984: 4) and Ewen (1983: 223) summarize: "If communication is difficult in today's world, it is perhaps because it is difficult to be human. Communication experts do not make it easier when they try to develop science out of something which is essentially an art. The impulse of our analysis ought to respect these voices in their own terms, not continually seek to distort them into faceless data. Such respect may mean our work will be *less* scientifically antiseptic, *less* arrogantly conclusive, *less* neatly tied up, *less* instrumental. Yet, being liberated from such constraints, our work will likely become *more* lyrical *more* speculative, *more* visionary, *more* intelligent and human" (see also Lie 2003).

### 28.4.2 Naturalistic Observation and the Participant Observer

To attempt an understanding of a human being or a social context, to the degree possible, we must lay aside all scientific knowledge of the average man and woman and discard all theories in order to adopt a completely new and unprejudiced attitude. The researcher must set aside pre-derelictions, what she/he "already knows," because if inquiry begins with a preconceived set of interpretive templates, viewpoints will be lost. Through the myopic glasses of rigid theories and constructs, other aspects that may be even more important are typically not noted because they are not looked for. This implies a re-evaluation of conceptions such as "developer," "source," "expert," "neutrality," and "objectivity." In other words, the researcher immerses herself/himself in the context not to verify hypotheses, but as sources for *understanding meaning*. In a very real sense, the most important tool of inquiry becomes the researcher's attitude. Large egos are detrimental when the investigator considers himself or herself far above the group to be investigated and refuses implicitly or explicitly the real encounter with the subject of investigation. Anderson and Meyer (1988: 320) concurs in writing: "The 'us/them' character of ethnographic work and our own socialization as researchers – keepers of the truth – can lead to a sense of superiority about our social action as compared to those we study. Observational distancing can turn people into subjects – objects of study." Therefore, Kennedy (2008), based on extensive research among Alaskan natives, describes the different roles he experienced as a change agent: from social broker to social advocate to social mobilizer.

Quite the opposite of neutrality and detachment, the participant observer involves himself/herself in the natural setting in order to, to the extent possible, obtain an inside view of the social context. This involves "gaining access," but the researcher must also maintain somewhat of an aloof position to observe. The goal is to see the other objectively and, at the same time, to experience his or her difficulties subjectively. For the "outsider," participation is: "a staged effort and observation is a product of routine. At the other extreme, the insider has the 'of-course-it's-true' knowledge of the member. For the insider, participation is effortless, observation difficult" (Anderson and Meyer 1988: 296).

First among the fundamental concepts of qualitative research is the axiom that the study of human life is interpretive, Anderson argues, it is the search for subjective understanding, rather than manipulative, the quest for prediction and control. Qualitative research has replaced prediction with emphatic understanding as its research aim so that social science can be freed from its exploitative uses and enlisted instead in the cause of freedom and justice. Contrary to quantitative inquiry which accepts its premises and focuses on methodological rigor, qualitative study stresses the critical analysis of those suppositions to understand human understanding, thereby restoring the critical and liberating function to intellectual investigation.

In anthropology one usually distinguishes between an "emic" and an "etic" approach: "Emic operations have as their hallmark the elevation of the native informant to the status of the ultimate judge of the adequacy of the observer's descriptions and analyses. The test of the adequacy of emic analyses is their ability to generate statements the native accepts as real, meaningful, or appropriate. In carrying out research in the emic mode, the observer attempts to acquire a knowledge of the categories and rules one must know in order to think and act as a native... Etic operations have as their hallmark the elevation of observers to the status of ultimate judges and concepts used in descriptions and analyses. The test of adequacy of etic accounts is simply their ability to generate scientifically productive theories about the causes of socio-cultural differences and similarities. Rather than employ concepts that are necessarily real, meaningful, and appropriate from the native point of view, the observer is free to use alien categories and rules derived from the date language of science. Frequently, etic operations involve the measurement and juxtaposition of activities and events that native informants may find inappropriate or meaningless" (Harris 1980: 32).

In this context I would like to plead for an "emic" position. However, as Clifford Geertz (1973: 15) warns, this is an extremely difficult approach to accomplish: "We begin with our interpretations of what our informants are up to, or think they are up to, and then systematize ..." Geertz therefore prefers the notions of "experience-near" and "experience-far" above the "emic" and "etic" concepts. The former are internal to a language or culture and are derived from the latter which are posed as universal or scientific.

### 28.4.3 Critical Research

Qualitative research has often been characterized as critical research. The reverse that critical research is often qualitative in nature is also true, especially when that research addresses the dominant societal structures and in particular when the inquiry is directed towards what can be called the academic research complex. Critical research looks at the historical, societal, and dynamic context of its questions, and consequently is not amenable to rigid methodology. Hence, there exist few similarities in approaches deemed critical research. Therefore, Halloran (1981: 26), among others, argues that "the unity of the critical approach lies in its opposition to conventional work rather than in any shared, more positive approach." Lacking in "accepted" methods, critical research is often written off as unscientific or as ideological prattle. However, just as qualitative research accepts its subjectivity, critical analysis admits its ideological premises. It can also be argued the reason critical research is deemed invalid is not because of its disparate methodology but rather because of its more populist stance. It seeks to examine social issues of importance to the generality of the people and not to – say – politicians and media managers. It is attacked "because it challenges the status quo and the vested interests and rejects the conventional wisdom of the service research that supports this establishment" (Halloran 1981: 33). In sum, Halloran (1981: 34) asserts: "It is now suggested that research ... should be shifted away from such questions as 'the right to communicate' to 'more concrete problems.' But what are these 'concrete problems?' They are the same as, or similar to, the safe, 'value-free' micro questions of the old-time positivists who served the system so well, whether or not they intended or understood this. All this represents a definite and not very well disguised attempt to put the clock back to the days when the function of research was to serve the system as it was – to make it more efficient rather than to question it or suggest alternatives."

### 28.4.4 Validity and Evaluation in Qualitative Research

Given this lack of methodological rigor, on what grounds should one accept the qualitative or critical argument? What makes the research worth recognition? Was the research "successful"? To address these questions, reference is made to the assertions above as to the premises of "value-free science" and "objectivity." If we do not accept these – if we see the emperor is indeed naked – then we must accept, and valuate, the conclusions of all inquiry in a different light. Unlike quantitative inquiry, qualitative research does not, as a rule, seek broad generalizability. For the qualitative researcher, there is no possibility of a grand theory of social action, as theory is contextually bound. Nor does it strive for reliability in its tools of analysis. Indeed, its basis of an interpretive, and consequently idiosyncratic, social reality precludes the strict application of such tools.

Whereas the primary concerns of quantitative research are reliability and construct validity, qualitative research seeks external validity, not through "objectivity" but rather through subjectivity. Anderson and Meyer (1988: 239) writes "The closer the analytic effort is to the lifeworld, the more direct its observation; the more it uses the concepts and language of this lifeworld, the greater the validity." Therefore, to the degree it reflects the circumstances, inquiry is valid and reliable even though not based on randomization, repeated and controlled observation, measurement, and statistical inference. In other words, it "is more 'objective' precisely because it can bring in more of the subjective intentionality" (White 1984: 40).

### 28.4.5 The Role and Place of the Reader, Viewer, and Receiver

The primary problem of validity in qualitative inquiry, then, is "of matching the researcher's conclusions and the actor's intention" (Rockhill 1982: 12) and putting those conclusions in form which makes sense to the reader, to go "from understanding to explanation, and from explanation to understanding" (p. 13). In light of cultural variance, indeed, the disparity of realities, this is no simple task.

Ultimately, for both quantitative and qualitative analysis, *validity ultimately rests on the acceptance or rejection of the argument by the reader him/herself.* Whether the statistical analysis was proper, the data clean, the surveyors well trained, etc., innumerable attacks can be brought against any study if the reader does not care to accept the propriety of the argument. On the other hand, if the reader (or policy maker) finds the conclusions amenable, or the author trustworthy, she/he will rarely give the methods a second glance. It comes down to the skill, the honesty, and the ethics of the researcher (Plaisance 2011).

Therefore, in the final analysis, Anderson and Meyer (1988: 355) writes the value of, and criticism toward, the qualitative study must "be appropriate to its character as a personal encounter, documented by a careful record, which results in an interpretation of the social action which rises inductively to a coherent explanation of the scene. To criticize this effort for its lack of objectivity, random sampling, statistical measure of validity, deductive logic, and the like is inane. Such criticism simply established the ignorance of the critic … As with all research, the success of a qualitative study is dependent on reaching the knowledgeable, skeptical reader with a compelling argument. The skeptical reader will give none of these criteria away, demand an honest effort, and consider the weaknesses as well as the insights in reaching a judgement of value. That judgement of value is the bottom line of research."

## 28.5 Participatory Research

If we subscribe to the notion that social research should have beneficial impact on society, it is imperative that we pay more attention to research philosophies that can profitably handle, and indeed stimulate, social change (Chambers 2007; Dervin and Huesca 1999; Jacobson 1993; Huesca 2003; Lennie and Tacchi 2013). Therefore, participatory research, in my opinion, borrows the concept of the interpretive, intersubjective, and human nature of social reality from qualitative research, and the inherency of an ideological stance from critical research combines them and goes one step further. Rather than erecting elaborate methodological facades to mask the ideological slant and purpose of inquiry, the question becomes, "Why shouldn't research have a direct, articulated social purpose?" Instead of relying on participant observation or complex techniques to gain the subjective, "insider's" perspective, it is asked "Why shouldn't the 'researched' do their own research?" Why is it "The poor have always been researched, described and interpreted by the rich and educated, never by themselves?"

In regard to the topic at hand, why is it such a great deal of research has been conducted about participation in a non-participatory fashion? As in the case of participatory communication, the major obstacles to participatory research are anti-participatory, often inflexible structures and ideologies. We cannot be reductionistic about holism, static about dynamism, value-free about systematic oppression, nor detached about participation. Participatory research may not be good social science in positivist terms, but it may be better than positivist social science for many development purposes.

## 28.5.1 Principles of Participatory Communication Research

That the mass of social research is largely guided by the social context in which it operates, and largely does not function to serve those studied, has been argued at length. Participatory research was conceived in reaction to this elitist research bias. It is ideological by intent; it is the research of involvement. It is not only research with the people – it is people's research. As such it largely rejects both the development policies of states and the "objectivity" and "universal validity" claims" of many methodologies in the social sciences. Even if we momentarily assume contemporary research practices are free of ideology and do not constitute a means of oppression, the fact remains they are of little utility to the poor.

"We have moved beyond the whole notion of some of us leading the struggles of others. This shift ... in the control over knowledge, production of knowledge, and the tools of production of knowledge is equally legitimate in our continued struggles towards local control and overcoming dependency. It is here that PR [participatory research] can be an important contribution ... PR is quite the opposite of what social science research has been meant to be. It is partisan, ideologically biased and explicitly non-neutral" (Tandon 1985: 21; see also Tandon 2002). It is the realization that most of the present professional approach to research is in fact a reproduction of our unjust society in which a few decision-makers control the rest of the population that has led many to move away from the classical methods and experiment with alternative approaches. In urging participatory research, we are not speaking of the involvement of groups or classes already aligned with power. These groups already have at their disposal all the mechanisms necessary to shape and inform our explanation of the world.

Therefore, a basic tenet of participatory research is that whoever does the research, the *results must be shared*. They must be available to the people among whom research is conducted and upon whose lives it is based. Data is not kept under lock and key or behind computer access codes, results are not cloaked in obfuscating jargon and statistical symbols (see ▶ Chap. 67, "Fifty Years of Practice and Innovation Participatory Video (PV)" by Tony Roberts and Soledad Muñiz).

Further, and perhaps most importantly, *the inquiry must be of benefit to the community*, and not just a means to an end set by the researcher. This benefit is contrasted to the circuitous theory-research design-data-analysis-policy-government service route which neutralizes, standardizes, dehumanizes, and ultimately functions as a means of social control: "People's voices undergo a metamorphosis into useful data, an instrument of power in the hands of another. Rather than assembling collectively for themselves, political constituencies are assembled by pollsters, collecting fragmentary data into 'public opinion'" (Ewen 1983: 222) (see also ▶ Chap. 65, "Participatory Communication in Practice: The Nexus to Conflict and Power" by Chin Saik Yoon).

Again, participatory research challenges the notion that only professional researchers can generate knowledge for meaningful social reform. Like authentic participation, it believes in the knowledge and ability of ordinary people to reflect on their oppressive situation and change it. To the contrary, in many cases at the local community level, participants have proved to be more capable than "experts" because they best know their situation and have a perspective on problems and needs that no outsider can fully share. This perspective is quite divergent from the abstract concepts, hypothetical scenarios, and macro-level strategies which occupy the minds and consume the budgets of development "experts" and planners (Cooke and Kothari 2001; Midgley 1986; Robson 1995).

### 28.5.2 Differences Between Participatory Research and Action Research

Because of this nature of involvement, participatory research is often known under the rubric of social action or action research (Argyris et al. 1985; Chambers 2008; Fals Borda 1988; Fals and Rahman 1991; Huizer 1989; Whyte 1989, 1991). In numerous respects they are similar, and participatory research is not really new. It is a novel concept only to the extent it questions the domains of the research as well as the economic and political elites (Tacchi et al. 2003).

However, there are fundamental differences between action and participatory research. Chantana and Wun Gaeo (1985: 37) write that action research "can be non-participatory and related to top down development ... whereas participatory research must involve the people throughout the process. Action research can be intended to preserve and strengthen the status quo, whereas participatory research ... is intended to contribute to the enhancement of social power for the hitherto excluded people."

Conversely, participatory research assumes a bias towards the "subjects" involved in the research process rather than the professional. Participatory research is related to the processes of conscientization and empowerment. It was probably Paulo Freire himself who introduced the first version of this approach in his philosophy of conscientization. Rather than agendas being defined by an academic elite and programs enacted by a bureaucratic elite for the benefit of an economic or political elite, participatory research involves people gaining an understanding of their situation, confidence, and an ability to change that situation (see ► Chap. 15, "The Importance of Paulo Freire to Communication for Development and Social Change" by Ana Fernández-Aballí Altamirano). White (1984: 28) writes this is quite divergent from "the functionalist approach which starts with the scientist's own model of social and psychological behavior and gathers data for the purpose of prediction and control of audience behavior. The emphasis is on the awareness of the subjective meaning and organization of reality for purposes of self-determination."

*Participatory research is egalitarian.* Thematic investigation thus becomes a common striving towards awareness of reality and towards self-awareness. It is an educational process in which the roles of the educator and the educated are constantly reversed and the common search unites all those engaged in the endeavor. It immerses the exogenous "researcher" in the setting on an equal basis.

Considering the necessary trust and attitudes as well as cultural differences, the task is not easy and makes unfamiliar demands on researchers/educators.

### 28.5.3 "Insiders" and "Outsiders"

Interaction fosters a pedagogical environment for all participants. The researcher, as a newcomer, contributes in that she/he requires the membership to give an account of how things are done, which fosters an atmosphere where participants may better know themselves, question themselves, and consciously reflect on the reality of their lives and their sociocultural milieu. Through such interaction, a fresh understanding, new knowledge, and self-confidence may be gained. Further, awareness, confidence, and cohesiveness are enhanced not only for group members but also among and between those members and "outsiders" who may participate, thereby increasing their understanding of the context and obstacles under which the people strive. Education goes both ways. This learning process can instill confidence and ultimately empowerment. The intent of participatory research is not latent awareness. Relevant knowledge increases self-respect and confidence and leads to exploration of alternatives towards the attainment of goals, and to *action*. Through this process, the givenness of the group is revealed on which one can build up a superior, higher vision.

### 28.5.4 A Definition of Participatory Research (PR)

The recent popularity of participatory research, the act of labelling it as such, may have implied that it is something special that requires a particular expertise, a particular strategy, or a specific methodology. Similar to participation, there has been great effort towards definitions and models of participatory research to lend an air of "respectability." Also similar to participation, perhaps this is no more than an attempt to claim title or credit for an approach which, by its very nature, belongs to the people involved. As one is dealing with people within changing social relations and cultural patterns, one cannot afford to be dogmatic about methods but should keep oneself open to people. This openness comes out of a trust in people and a realization that the oppressed are capable of understanding their situation, searching for alternatives, and taking their own decisions: "Participatory research is an alternative social research approach in the context of development. It is alternative, because although business, government and the academic also undertake research with development in the end view, little thought and effort go as to how the research project can be used for the benefit of those researched. The central element of Participatory Research is participation... It is an active process whereby the expected beneficiaries of research are the main actors in the entire research process, with the researcher playing a facilitator's role" (Phildhrra 1986: 1).

Because there is no reality "out there" separate from human perception and, as put forth in the multiplicity paradigm, there is no universal path to development, it is maintained each community or grouping must proceed from its own plan in consideration of its own situation. In other words, to the extent the methodology is rigidly structured by the requisites of academia, participatory research is denied.

By its nature, this type of research does not incorporate the rigid controls of the physical scientist or the traditional models of social science researchers. Chantana and Wun Gaeo (1985: 39) state: "There is no magic formula for the methodology of such PR projects ... However, there are common features taking place in the process:

1. It consists of continuous dialogue and discussion among research participants in all stages; [and]
2. Knowledge must be derived from concrete situations of the people and through collaborative reflection ... return to the people, continuously and dialectically."

Therefore we would like to delineate participatory research as an educational process involving three interrelated parts:

1. Collective definition and investigation of a problem by a group of people struggling to deal with it. This involves the social investigation which determines the concrete condition existing within the community under study, by those embedded in the social context;
2. Group analysis of the underlying causes of their problems, which is similar to the conscientization and pedagogical processes addressed above, and;
3. Group action to attempt to solve the problem.

### 28.5.5 The Process of Participatory Research

Therefore, the process of participatory research is cyclical, continuous, local, and accessible. *Study-reflection-action* is the integrating process in this type of research. Kronenburg (1986: 255) gives the following characteristics of participatory research: "[It] rests on the assumption that human beings have an innate ability to create knowledge. It rejects the notion that knowledge production is a monopoly of 'professionals'; [It] is seen as an educational process for the participants ... as well as the researcher; It involves the identification of community needs, augmented awareness about obstacles to need fulfillment, an analysis of the causes of the problems and the formulation and implementation of relevant solutions; The researcher is consciously committed to the cause of the community involved in the research. This challenges the traditional principle of scientific neutrality and rejects the position of the scientist as a social engineer. Dialogue provides for a framework which guards against manipulative scientific interference and serves as a means of control by the community."

### 28.5.6 Evaluation and Validity in Participatory Research

Given a continuous cycle of study-reflection-action, participatory research inherently involves formative evaluation. Indeed, the terms participatory research and participatory evaluation are often used synonymously. Actors are exercising themselves in participatory evaluation by the whole group of the situation of underdevelopment and oppression.

Congruent with the objectives of participatory research, the purpose of evaluation is to benefit the participants themselves. It does not function to test the efficiency of an exogenous program, formulate diffusion tactics or marketing strategies for expansion to a broader level, gather hard data for publication, justify the implementing body, or collect dust on a ministry shelf. In brief, it is an ongoing process as opposed to an end product of a report for funding structures.

Whether participatory research "succeeds" or "fails" is secondary to the interaction and communication processes of participating groups. The success of the research is seen no more in publications in "reputed" journals but in what happens during the process of research. Bogaert et al. (1981: 181) add that "participatory evaluation generates a lot of qualitative data which is rich in experiences of the participants. It may be ... quantitative data is sacrificed in the process. However, what is lost in statistics is more than made up by the enhanced richness of data."

Drawing on innovative and creative research and evaluation approaches, alternative perspectives on development, and significant, rigorous research, Lennie and Tacchi (2013) present a framework for critically thinking about and understanding development, social change, and the evaluation of communication for development.

The book presents a framework for the evaluation of communication for development projects and includes practical ideas and processes for implementing the framework and strategies for overcoming the many challenges associated with evaluating communication for development. The authors highlight the need to take a long-term view of the value of this approach, which can be cost effective when its many benefits are considered.

### 28.5.7 The Integration of Different Methodologies

The implication is not that quantitative or qualitative methods or exogenous collaboration in participatory evaluation are forbidden. Writing of research participants, D'Abreo (1981: 108) states: "While they, as agents of their own programme, can understand it better and be more involved in it, the outside evaluator may bring greater objectivity and insights from other programmes that might be of great use to them. However, the main agents of evaluation, even when conducted with the help of an outside agency or individual, are they themselves."

Turning to the question of validity, Tandon (1981: 22) suggests, on a methodological level, "getting into a debate about reliability and validity of PR is irrelevant because it is quite the opposite shift in understanding what this research is." Its focus is on authenticity as opposed to validity. However, referring to generalizability and validity addressed in relation to qualitative research, it can be argued that validity in its less esoteric sense is participatory research's hallmark. "If ordinary people define the problem of research themselves, they will ensure its relevance" (Tandon 1981: 24), and their involvement "will provide the 'demand-pull' necessary to ensure accuracy of focus" (Farrington 1988: 271).

Finally, the basis of participatory research, *indigenous knowledge* is inherently valid (Tuhiwai Smith 1999). This is not to say conditions are not changing or that this knowledge cannot benefit from adaptation. The argument is that, in most cases, this knowledge is the most valid place from which to begin.

### 28.5.8 A Word of Caution

Participatory research can all to easily be utilized as yet another tool of manipulation by vested interests (Cooke and Kothari 2001). Charges are correctly made that it is often a means of political indoctrination by the right and the left alike. Often organizers have been attacked for manipulating people's minds and managing their actions towards their own ends.

While the approach strives towards empowerment, challenges existing structures, and is consequently ideological, rigidly prescribed ideologies must be avoided. In addition, knowledge and perspective gained may well empower exploitative economic and authoritarian interests instead of local groups. Far from helping the process of liberation, if the researcher is not careful, he or she may only enable the traditional policy makers and vested interests to present their goods in a more attractive package without changing their substance.

Even the best intentioned researcher/activist can inadvertently enhance dependency rather than empowerment. If she/he enters communities with ready-made tools for analyzing reality, and solving problems, the result will likely be that as far as those tools are successful, dependency will simply be moved from one tyrant to another.

In other words, overzealous researchers can easily attempt to compensate for an initial apathy by assuming the role of an advocate rather than a facilitator. "What looks like progress is all too often a return to the dependent client relationship" (Kennedy 1984: 86). This approach is no better than more traditional researchers with hypotheses and constructs to validate, or the diffusionist with an innovation for every ill.

## 28.6 By Way of Conclusion

1. Most contemporary research is based on positivistic assumptions that we can objectively know the social world "out there." It is argued that objectivity is nothing more than intersubjectivity, principles, and parameters which people

agree to agree upon. The danger occurs when one group assumes its constructs and methods are universally valid.
2. Qualitative and quantitative research are largely polar in orientation. The qualitative approach accepts, indeed advocates, subjectivity as a valid avenue of understanding. The emphasis is on how people perceive, understand, and construct reality rather than how they mechanistically react to manipulative stimuli.
3. As such, qualitative research is a human approach to the understanding of a human world. The researcher strives to see the naturalistic environment of those she/he is studying from their perspective. This forbids rigid, preformulated constructs and hypotheses.
4. What makes the qualitative study of value is not its construct validity or reliability but its external validity. This validity arises from its foundations in human perception rather than rigorous methodology. In the final analysis, the relevance of any research rests on the reader's quite subjective acceptance or rejection of the argument advanced.
5. Combining aspects of qualitative and critical research, participatory research asks why the researched shouldn't do their own research? A further question is why is the mass of research on participation done in a non-participatory manner?
6. Like all research, participatory research is ideological. It is biased in the sense it holds research should be guided by, available to, and of direct benefit to the "researched," rather than privileged information for a manipulative elite. It further believes research is not, nor should it be, the domain of a powerful few with the "proper" tools.
7. Participatory research is similar, but not equal, to social action research. It is research of involvement, not of detachment. It includes all parties in a process of mutual and increasing awareness and confidence. It is research of conscientization and of empowerment.
8. As has been argued for participation, and as in qualitative approaches, there can be no strict methodology for participatory research. However, it must actively and authentically involve participants throughout the process, and the general flow is from study to reflection to action.
9. Evaluation is inherent in participatory research. However, it is formative rather than summative evaluation. Its purpose is not for journals, ego-boosting, or to solicit further funding but rather to monitor and reflect on the process as it unfolds. Further, people's involvement in the research assures validity of the inquiry, their validity.

## References

Addo H (1985) Beyond Eurocentricity: transformation and transformational responsibility. In: Addo H, Amin S, Aseniero G et al (eds) Development as social transformation. Reflections on the global problematique. Hodder and Stoughton, London

Anderson J, Meyer T (1988) Mediated communication. A social action perspective. Sage, Newbury Park
Argyris C, Putnam R, Smith D (1985) Action science. Jossey-Bass, San Francisco
Berg B, Lune H (eds) (2014) Qualitative research methods for the social sciences. Pearson, Essex
Bhattacherjee A (2012) Social science research: principles, methods, and practices. Textbooks collection 3. http://scholarcommons.usf.edu/oa_textbooks/3
Bogaert MVD, Bhagat S, Bam NB (1981) Participatory evaluation of an adult education programme. In: Fernandes W, Tandon R (eds) Participatory research and evaluation: experiments in research as a process of liberation. Indian Social Institute, New Delhi
Bruhn Jensen H (ed) (2013) A handbook of media and communication research: qualitative and quantitative methodologies. Routledge, New York
Bryman A (1984) The debate about quantitative and qualitative research: a question of method or epistemology? Br J Sociol 35(1):75–92
Bryman A, Burgess R (eds) (1994) Analyzing qualitative data, Routledge, London
Burke K (1968) Language as symbolic action. University of California Press, Berkeley
Chamarik S (1993) Democracy and development. A cultural perspective. Local Development Institute, Bangkok
Chambers R (2007) Ideas for development. Earthscan, London
Chambers R (2008) Revolutions in development inquiry. Earthscan, London
Chantana P, Wun Gaeo S (1985) Participatory research and rural development in Thailand. In: Farmer's Assistance Board (ed) Participatory research: response to Asian people's struggle for social transformation. Farmer's Assistance Board, Manila
Christians C, Carey J (1981) The logic and aims of qualitative research. In: Stempel G, Weaver D, Westley B (eds) Research methods in mass communication. Prentice Hall, Englewood Cliffs
Cooke B, Kothari U (eds) (2001) Participation. The new tyranny? Zed Books, London
D'Abreo D (1981) Training for participatory evaluation. In: Fernandes, Tandon R (eds) Participatory research and evaluation: experiments in research as a process of liberation. Indian Social Institute, New Delhi
Denzin NK, Lincoln YS (eds) (2005) The Sage handbook of qualitative research, 3rd edn. Sage, Thousand Oaks
Dervin B (1982) Citizen access as an information equity issue. In: Schement JR, Gutierrez F, Sirbu M (eds) Telecommunications policy handbook. Praeger, New York
Dervin B, Huesca R (1999) The participatory communication for development narrative: an examination of meta-theoretic assumptions and their impacts. In: Jacobson T, Servaes J (eds) Theoretical approaches to participatory communication. IAMCR book series. Hampton Press, Creskill, pp 169–210
Ewen S (1983) The implications of empiricism. J Commun 33(3):219
Fals Borda O (1988) Knowledge and people's power: lessons with peasants in Nicaragua, Mexico, and Colombia. Indian Social Institute, New Delhi
Fals Borda O, Rahman HA (eds) (1991) Action and knowledge: breaking the monopoly with participatory action research. Intermediate Technology Publications, London
Farrington J (1988) Farmer participatory research: editorial introduction. Exp Agric 24:269
Friberg M, Hettne B (1985) "The greening of the world", development as social transformation. Westview, Boulder
Fuglesang A (1984) The myth of people's ignorance. In: Development dialogue. The Dag Hammarskjold Foundation, Uppsala, pp 1–2
Geertz C (1973) The interpretation of cultures. Basic Books, New York
Halloran JD (1981) The context of mass communication research. In: Mcanany E, Schnitman J, Janus N (eds) Communication and social structure: critical studies in mass media research. Praeger, New York
Harris M (1980) Cultural Materialism: The Struggle for a Science of Culture, Vintage, New York
Harris M (2001) Cultural materialism: the struggle for a science of culture. Rowman & Littlefield, Lanham/Oxford

Huesca R (2003) Tracing the history of participatory communication approaches to development: a critical appraisal. In: Servaes J (ed) Approaches to development. Studies on communication for development. UNESCO, Paris

Huizer G (1989) Action research and People's participation. An introduction and some case studies. Third World Centre, Nijmegen

Jacobson T (1993) A pragmatist account of participatory communication research for national development. Commun Theory 3(3):214–230

Jacobson T, Servaes J (eds) (1999) Theoretical approaches to participatory communication. IAMCR book series. Hampton Press, Creskill

Kennedy TW (1984) Beyond advocacy: an animative approach to public participation. Dissertation, Cornell University

Kennedy TW (2008) Where the rivers meet the sky: a collaborative approach to participatory development. Southbound, Penang

Kronenburg J (1986) Empowerment of the poor. A comparative analysis of two development endeavours in Kenya. Third World Center, Nijmegen

Kuhn T (1962) The structure of scientific revolutions, 2nd edn, enlarged. University of Chicago, Chicago

Lennie J, Tacchi J (2013) Evaluating communication for development. A framework for social change. Routledge Earthscan, London

Lie R (2003) Spaces of intercultural communication. An interdisciplinary introduction to communication, culture, and globalizing/localizing identities. IAMCR book series. Hampton Press, Creskill

Mckee N, Manoncourt E, Saik YC, Carnegie R (eds) (2000) Involving people, evolving behaviour. Southbound, Penang

Midgley J (ed) (1986) Community participation, social development, and the state. Methuen, London

Plaisance PL (2011) Moral agency in media: toward a model to explore key components of ethical practice. J Mass Media Ethics 26(2):96–113

PPDHRRA – Philippine Partnership for the Development of Human Resources in Rural Areas (1986) Participatory research guidebook. Philippine Partnership for the Development of Human Resources in Rural Areas, Laguna

Robson C (1995) Real world research. A resource for social scientists and practitioner-researchers. Blackwell, Oxford

Rockhill K (Fall, 1982) Researching participation in adult education: the potential of the qualitative perspective. Adult Educ 33(1):3

Servaes J (1999) Communication for development. One world, multiple cultures. Hampton Press, Creskill

Servaes J (2012) Homo academicus: Quo vadis? In: Silvia N-Z, Karyn H (eds) Global academe: engaging intellectual discourse. Palgrave Macmillan, New York, pp 85–98. 230pp

Servaes J, Jacobson T, White S (eds) (1996) Participatory communication for social change. Sage, New Delhi

Tacchi J, Slater D, Hearn G (2003) Ethnographic action research. UNESCO, Paris

Tandon R (1981) Participatory evaluation and research: main concepts and issues. In: Fernandes W, Tandon R (eds) Participatory research and evaluation: experiments in research as a process of liberation. Indian Social Institute, New Delhi

Tandon R (1985) Participatory research: issues and prospects. In: Farmer's Assistance Board (ed) Response to Asian people's struggle for social transformation. Farmer's Assistance Board, Manila

Tandon R (2002) Participatory research: revisiting the roots. Mosaic Books, New Delhi

Thayer L (1983) On 'doing' research and 'explaining' things. In: Gerbner G (ed) Ferment in the Field. J Commun 33(4)

Tuhiwai Smith L (1999) Decolonizing methodologies. Research and indigenous peoples. Zed Books, London

Van Hemelrijck A (2013) Powerful beyond measure? Measuring complex systemic change in collaborative settings. In: Servaes J (ed) Sustainability, participation and culture in communication. Theory and praxis. Intellect-University of Chicago Press, Bristol/Chicago

White R (1982) Contradictions in contemporary policies for democratic communication, Paper IAMCR conference, Paris, September

White R (1984) The need for new strategies of research on the democratization of communication, Paper ICA conference, San Francisco, May

Whyte WF (ed) (1989) Learning from the field. A guide from experience. Sage, Beverly Hills

Whyte WF (ed) (1991) Participatory action research. Sage, Newbury Park

# Visual Communication and Social Change 29

Loes Witteveen and Rico Lie

## Contents

| | | |
|---|---|---|
| 29.1 | Introduction | 556 |
| 29.2 | History | 557 |
| | 29.2.1 Visual Communication as Addressed in Early Textbooks | 557 |
| | 29.2.2 Starting to Respect the Audience | 559 |
| 29.3 | Exploring the Visual | 563 |
| | 29.3.1 Visual Literacy | 563 |
| | 29.3.2 Recognizing the Complexity of Visual Images | 564 |
| | 29.3.3 Valuing the Potential of Visual Images | 567 |
| 29.4 | Contemporary Visual Communication | 568 |
| | 29.4.1 Acknowledging Diverse Professionalisms | 570 |
| | 29.4.2 Centralizing the Importance of Design Thinking | 572 |
| | 29.4.3 Recognizing Social Imaginaries as Options for Visual Communication in Social Dialogues | 574 |
| 29.5 | Conclusion | 576 |
| 29.6 | Cross-References | 577 |
| References | | 577 |

L. Witteveen (✉)
Research Group Communication, Participation and Social Ecological Learning,
Van Hall Larenstein University of Applied Sciences, Velp, The Netherlands

Research Group Environmental Policy, Wageningen University, Wageningen, The Netherlands
e-mail: loes.witteveen@hvhl.nl; loes.witteveen@wur.nl

R. Lie
Research Group Knowledge, Technology and Innovation,
Wageningen University, Wageningen, The Netherlands
e-mail: rico.lie@wur.nl

© Springer Nature Singapore Pte Ltd. 2020
J. Servaes (ed.), *Handbook of Communication for Development and Social Change*,
https://doi.org/10.1007/978-981-15-2014-3_55

**Abstract**

This chapter explores visual communication in the context of communication for social change with a major focus on the field of information and knowledge exchange in agriculture and rural development, formerly termed agricultural and rural extension. The specifics of visual communication, the appropriateness and effectiveness of visual communication in rural development, and social change processes are addressed proceeding from early understandings of visual communication to contemporary views. Reviewing work experiences, the authors searched for an analysis of the praxis and the very realities of visual communication and social change. The chapter concludes by arguing that, in contemporary visual communication, acknowledging diverse professionalisms, centralizing the importance of design thinking, and recognizing social imaginaries as options for visual communication in social dialogues are of crucial importance.

**Keywords**

Visual communication · Visual literacy · Design thinking · Portrayal · Social imaginaries · Social change

## 29.1 Introduction

This chapter explores visual communication in the context of communication for social change with a major focus on the field of information and knowledge exchange in agriculture and rural development, formerly termed agricultural and rural extension. The objective is to address the specifics of visual communication and determine the appropriateness and effectiveness of visual communication in social change processes. To accomplish this, the chapter proceeds from early understandings of visual communication to contemporary views and by doing so provides an insight into continuities and changes in views and praxis of visual communication. The chapter starts by describing the limited and specific attention given to visual communication in early textbooks in the field of agricultural extension. In these textbooks, the voice of the audience was hardly considered by the so-called experts, but gradually the audience seemed to become more respected. At the same time, more attention is paid to visual literacy. Over time, the complexity of the visual image becomes recognized, and the potential of the role of visuals in processes of social change is seen. The chapter concludes by arguing that, in contemporary visual communication, acknowledging diverse professionalisms, centralizing the importance of design thinking, and recognizing social imaginaries as options for visual communication in social dialogues are of crucial importance. The contribution draws on the authors' academic and professional experiences in the field of visual communication for social change. Reviewing work experiences, the authors searched for an analysis of the praxis and the very realities of visual communication and social change.

## 29.2 History

### 29.2.1 Visual Communication as Addressed in Early Textbooks

In an attempt to trace back the development of visual communication and visual literacy in a context of communication for social change, the work of Dutch scholar Anne van den Ban, who focused on agricultural extension, is revisited. One of his first publications was co-authored with Everett Rogers, who is renowned for his work on adoption and diffusion theories. In that article, the term extension was used to describe extension as a process to educate farmers (Rogers and Van Den Ban 1963). Anne van den Ban became famous for the English version of his textbook *Agricultural Extension*, which he wrote with H. Stuart Hawkins. In that textbook, extension is referred to as a process that "involves the conscious use of communication of information to help people form sound opinions and make good decisions" (Van Den Ban and Hawkins 1988, p. 9). Another frequently used textbook in agricultural extension education was *Agricultural Extension in Developing Countries* by M.E. Adams, referring to agricultural extension as "assistance to farmers to help them to identify and analyze their production problems and to become aware of the opportunities for improvement" (1982, p. 9). Both textbooks aimed to contribute to a stronger recognition of extension as an important academic and professional domain, positioning it as an interdisciplinary field between agricultural sciences on the one hand and the field of communication and learning on the other hand.

As the focus here is on visual communication, we find in Adams' book a short section on mass media in the chapter on extension methods. Despite mentioning "special skills" needed for program production, the author quickly seems to move away from conceptual thinking into recommendations like "wherever possible the programmes should have a rural setting. Tricky and artistic effects should be avoided" (p. 39). In the chapter on extension methods, Van den Ban presents sections on mass media, group methods, individual extension, and the use of folk media. The last section, on modern information technology, concludes with the statement that "unfortunately, information technologies seldom are developed in association with the people who are expected to use them" (1988, p. 173). This critique could have remained unnoticed if the chapter had not concluded with a table in which specific functions and characteristics are ascribed to different extension methods. The table is followed by a self-critique that "the nature of the audience and of the message receive only limited treatment in the table" (p. 174). Despite the limited recognition of audiences in the construction of meaning, attention on media is growing and starts to move away from media as mere channels with determined qualities. The aforementioned textbooks depict a time where new media production technologies and increased production of photography and film are on the rise. The term "audio-visual aids" is coming up with the articulation of strong qualities of "AV media" for information transfer, while at the same time, qualities of "making a subject more lively and understandable" (Adams, p. 19) and "develop[ing] or strengthen[ing] the motivation to change as an emotional process" (Van den Ban 1988, p. 161) are mentioned. The latter is somehow read between the lines and not addressed in advanced detail.

Meanwhile, in the adjacent field of development studies, attention was paid to visual literacy from a cross-cultural communication perspective. In 1982, Andreas Fuglesang published his book *About Understanding* to shed a cross-cultural light on the understanding of visual media. *About Understanding* was recognized as a work that articulated common and probably erroneous interpretations of images in development communication. Some of the pictures in his book became iconic; see for example, the repeated figure of the runner who seemed to have 1.5 legs and others in Van Brink and Visser (1983) and Fuglesang (p. 158). See Fig. 1 for an example of such a picture.

Unfortunately, Fuglesang's work was commonly understood as a call for attention on cross-cultural differences in a context of low literacy and otherwise framed to support a lack of insight by third world audiences. His focus on increased understanding and appropriate design of development communication media was less recognized.

Most authors who critically reviewed common extension materials produced in fields like health, agriculture, and sanitation during the 1980s and the 1990s argued that the lack of understanding of visuals or images could be overcome by training, similar to text alphabetization. The upcoming term to describe the limited understanding or "reading" of graphic media outings in extension became *visual literacy*. Although appreciated for its trainable properties, the term also carried a notion of further legitimizing the conventional sender – message – receiver model

**Fig. 1** Example of a picture of a football player with seemingly 1.5 legs (newspaper cutting n.d.)

and therefore talking about receivers as beneficiaries and target groups with a likely positioning as aid-receivers, or referring to absent qualities by using terms like non-westerner, illiterate, and uneducated.

### 29.2.2 Starting to Respect the Audience

In the 1990s, the recognition grew that meaningful media consumption in rural development is also about design and production. FAO increasingly worked on capacity building in media production, referring to the general term of audio-visual media. See for example, the FAO publication *Powerful Images. Planning, Production and Use Guide* (1990). This publication about "slide programmes and filmstrips to inform, motivate and train in developing countries" focuses on production techniques describing why and how "local production by developing countries" can be achieved (p. 2). The booklet reflects conventional views on visual literacy of "rural people" including the man who seems to be missing half a leg (p. 27) while at the same time a consistent focus on testing is recognizable with some well-elaborated statements like "test the material not the audience" (p. 39). Human resources are not yet functionally described, but a short section on production organization presents terms like creative team, field-testing, and postproduction (pp. 48–49).

The Centre for the Study of Education in Developing Countries (CESO) became a renowned institute for studies highlighting cultural, anthropological, and sociological dimensions of communication. CESO author Kees Epskamp articulated the "visual and social conventions, which are widely applied in visual communication" (1995, p. 26). His colleague Ad Boeren (1994, p. 104), referring to "pictorially inexperienced people," is one of the early authors writing about testing and providing examples such as the versions of a drawing of a woman burning a pile of grass, which was not recognized as such by the audience during testing (see Fig. 2).

Testing is further articulated by Epskamp when he states: "The question remains whether the target group should be made visually literate in the language of the designers, or whether the designer should first check whether his future audience recognizes and understands the visual material he has produced. If one intends to accept the target group as they are, one should choose option two: preliminary research and testing" (1995, p. 27). Epskamp widens the debate on visual literacy to media literacy and to a critical media consumption perspective, as he states that training "will create 'media-literate' consumers, by which we mean those consumers who are able to identify the persuasive devices used by media designers" (p. 115). Boeren widens debates on the potential of media, elaborating on the influence of "a number of critical factors" including "apart from the educational characteristics" factors such as the dissemination infrastructure, production facilities, and financial resources (p. 166).

The abovementioned issues are discernible in the field of media for development, as has been reported by the Taller de Comunicaciones Campesinas (TCC) in Juigalpa, Nicaragua, (1989) where it was found that the limited relevance and impact of extension or information material hampered the realisation of the main objectives of their commissioning national clients, such as the Ministry of Agriculture, and international organizations such as UNICEF. Blaming farmers and rural

**Fig. 2** Tested drawings Boeren. (Taken from Gürgen 1987)

communities was not a common interpretation of the TCC staff, as the team of journalists, graphic artists, and media producers had not been trained in the field of agricultural extension or other technical fields. Their backgrounds were media, journalism, sociology, and the arts, and most professional expertise was situated outside the field of extension. In response to disappointing communicative interventions, TCC merged into a new enterprise unifying regional capacities and resources in the Empresa de Comunicaciones (ECOM). ECOM had the ambition to bring together professionals in the field of media, journalism, and the arts with staff members of the extension and public relations departments of diverse ministries and NGOs. This collective approach was considered as a response to urgent and popular demands for extension and information media in wartime in Nicaragua, to align diverse campaigns with the rural communities, and to build capacities for mediated communication in the fifth region (administrative area) of which Juigalpa was the regional capital (ECOM 1990a).

ECOM had difficulty dealing with clients that requested media production and campaign development while most often only indicating the "message" and the budget or resources available. It was soon realized that demarcated positions and mechanisms in the system of rural, health, and sanitation extension and advice were embedded in a system where "experts" sent messages to receivers, but media design and production did not fit as a significant process in the system and was considered a kind of support service that would be carried out at the request of a commissioner or client.

A media design and production system was established with account managers, so common in commercial media. This system, in which production requests were no longer received directly by the ECOM staff from production departments such as radio or print but by the newly established account managers, did not align with existing expectations and working routines of the relevant institutions to "just get the message across." A touch of annoyance crept into the relation between ECOM and commissioning institutes. ECOM's envisaged media design and production of communication processes stood in contrast to the expectations of institutions commonly used to defining messages and ordering media outings. ECOM therefore struggled to recognize the nested roles of extension media as a function to support individual decision making about the adoption of products and service innovations without questioning intended audiences' interest in these innovations. For ECOM working for diverse clients, another issue reported as a hindrance was the limited interest in the alignment of proposed communicative interventions by different organizations focusing on the same rural audiences. ECOM evaluation reports do not dig in full detail into these frictions, but evidence can be found in proposals "to document accumulated experiences and the practice of popular communication in which ECOM participates with other institutions and projects" and "to boost the professionalism of communication professionals" (ECOM 1990b, p. 6). The most concrete strategy to achieve ECOM's formulated media for development objectives was to focus on developing protocols for "appropriate" media design and development while working on productions, thereby allowing prototypes to be tested as a way of conducting applied research. In this context, ECOM conceptualized rural people as "audiences" and accepted accountability for the impact of media developed rather than focusing on rural communities' assumed incapacities. It was realized that, despite major efforts by the national government in its *Cruzada Nacional de Alfabetización* (National Literacy Crusade), which claimed to have reduced the literacy rate from 50% to 12% in five months (UNESCO 2007), a major focus should be on radio and visual media and continuous support for literacy initiatives.

Reference is made to audience participation in a film production developed in Juigalpa in cooperation with ECOM staff for the water sanitation film *Fuente de Amor* (Source of Love). As described in the article by Goris et al. (2015), the film *Fuente de Amor* provides a relevant case study to investigate audience participation in a film production for social change. The film was a concluding activity in a 5-year drinking water project commissioned by the The Hague–Juigalpa twinning committee. The film aimed to address issues of water shortage, which remained unresolved after a technical project for the improvement of the drinking water supply system had been implemented. The initial request had been to produce a documentary with a gender focus, promoting water sanitation for women as main caretakers of children and household tasks like cooking, cleaning, and water storage. As this was considering as blaming the victims, the filmmakers, who included (former) ECOM staff, came up with a counterproposal to request the affected inhabitants to participate in the film and articulate their perceptions on the water shortages. This did not resonate with the commissioner, and the proposal was rejected on the basis that the audience had no capabilities, formulated as "They cannot make a film, they never made a film."

In brief, the script development stage started with the organization of a script contest in Juigalpa, which resulted in writings and drawings indicating that the problem was not (only) about shortages. It was also about distribution and about water problems related to the functioning of the water authority, and it was also about women and about the river (see Fig. 3 for a page of a visual script submitted to the contest). Based on all the contributions and undoubtedly inspired by the submitted drawings, a script was prepared, and in 1993, the film was recorded in Juigalpa in 4 weeks, impelled by the dynamics of community art and a professional core film crew. The main actors, all amateurs, were selected on the basis of screen tests. A city poet wrote a song that was sung by the church choir and recorded in the ECOM radio studio. Over 120 people are included in the credits, and the film was screened in the neighborhoods and accompanied by public debates. Twenty years later, the filmmakers were informed that the film was still being screened.

The *Fuente de Amor* experience is illustrative of continuing debates on audience participation in communicative processes for social change. The film production process represents a rich example of diverse participatory potentials in a complex visual production, positioned as an edutainment communicative intervention. The commissioners took a long time to perceive as realistic the filmmakers' proposal to incorporate the local communities' knowledge and expertise. The two pages with six drawings, each by Julio Madrigal, are acknowledged as a turning point in that lengthy process.

A necessary remark needs to be added here by quoting Buckingham that "the use of such apparently open approaches could be seen to require a greater degree of reflexivity about the relationships of power [...] however 'creative' it might outwardly appear to be." (Buckingham 2009).

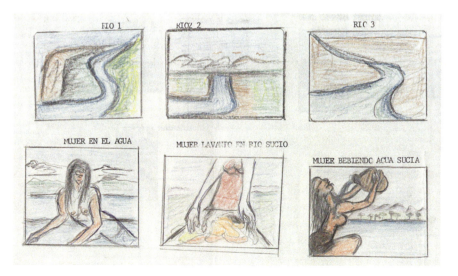

**Fig. 3** Drawings submitted to the script contest. (Courtesy of Julio Madrigal)

## 29.3 Exploring the Visual

### 29.3.1 Visual Literacy

Contemporary views on visual communication resemble those of the 1980s and 1990s, as again technological developments – this time, the rapid increase in digital (mobile) technologies for visual communication – are accompanied with studies on the increasing qualities and options of visual communication (see for example Dondis 1973). Felten (2008, p. 60) elaborates on new qualities or functions as he states that "the 'pictorial turn' means that no longer primarily to entertain and illustrate. Rather they are becoming central to communication and meaning-making." Messaris (2012, p. 106) describes how the increased access to visual tools and the change from analogue to digital photography also implicated new perspectives on visual communication in relation to manipulation of images. The author thereby adds a new aspect to visual literacy arguing that "the awareness of the likelihood of digital manipulation does not necessarily translate into an ability to detect its use in particular instances" (p. 111).

The expression "one picture is worth ten thousand words" is repeatedly quoted and more often contested for its origin than substantiated. A major change is that the search for quantitative "evidence" is replaced by qualitative studies focusing on questions of culturally defined meaning and media production in a societal context. Studying the inherent meaning of particular images is therefore replaced by critical studies, using terms such as portrayal, representation, and visuality to create access to more constructivist views on the creation, circulation, and audiences of images. See, for example, Gillian Rose (2016, p. 18), who states that research on visual culture is concerned "not only with how images look, but also with how they are looked at."

The interest in the visual as another textuality or language is currently articulated by concerns over new types of literacies such as the understanding of emojis (see, for example, Rodrigues et al. 2017). It will be no surprise that authors in the field of visual literacy (Felten 2008; Rose 2016; Serafini 2017) advocate for visual literacy in education curricula referring to arguments of visual culture hegemony and related twenty-first century skills. These suggestions are both applauded as received with resistance as visual literacy may compete with textual literacy in education leading to concerns that conventional writing and reading could possibly become extinct competencies. Such concerns may be viewed as conservative and traditionalist reactions to socio-technological innovations but may also be viewed as critical or as an expression of what has recently been framed as responsible innovation. Decreasing text literacy could be a concern from the perspective of viewing literacy as a transition from oral modes to reading and writing as skills that impact human consciousness, as "not only does it allow for the representation of words by signs, but it gives a linear shape to thought, providing a critical framework within which to think analytically" (UNESCO 2006).

Basic concerns over literacy remain relevant, as it is still considered an important fact that "103 million youth worldwide lack basic literacy skills, and more than 60% of them are women" (UNDP 2018). This fact feeds into the fourth Sustainable Development Goal on Quality Education formulated as "By 2030, ensure that all youth and a substantial proportion of adults, both men and women, achieve literacy and numeracy" (UNDP). This SDG rightly focuses on text and numeric literacy, as it is nowadays more common "to understand literacy as a contextual and societal transformation" (UNESCO 2006, p. 159).

Authors conceiving literacy in such a way refer to literacies and textualities as text based, visual, audio (radio), and increasingly electronic, but fundamentally communicative. Such broader understandings of literacy also provide an entry point to consider the visual as a discourse in its own right. Rose, after Foucault, defines discourse as "a particular knowledge about the world which shapes how the world is understood and how things are done in it" (2016, p. 187), and, quoting Lynda Nead, she refers to discourse as "a particular form of language with its own rules and conventions and the institutions within which the discourse is produced and circulated" (2000, p. 187). Felten argues for the inclusion of visual literacy in the curricula beyond an understanding of the visual as a "linguistic model of communication" and considers that "visual images and representations play a crucial role in many facets of contemporary society" and the author refers to "developing one's identities, shaping worldviews" (2008, p. 16).

The above views create space to look again at visual literacy in close connection with the design and production of visual media to see whether it represents dominant institutional knowledges. We first return in time again to ECOM to demonstrate the complexity of visual images.

### 29.3.2 Recognizing the Complexity of Visual Images

In 1990, ECOM undertook a study on visual literacy using 14 pictures previously used in a study on the visual ability of Kenyan primary school leavers and 10 graphics recently designed by ECOM. The 14 Kenyan pictures were simple line drawings, which had been used to test the school leavers' understanding of perspective, overlap, dimensions, shadow, signs, and symbols. These pictures were selected for their resemblance to the regional context and the probability of appearing in ECOM graphic work. They were included with the same multiple-choice questions as in the original study, translated into Spanish. The 10 ECOM graphics were selected from recent works that had been distributed in large quantities and were considered of high relevance for rural farming communities. These graphics contained dense information and were accompanied by open questions. Answers were recorded by research assistants who were trained in writing down the answers as provided by respondents without any interference.

For the Kenyan drawings, similar quantitative results were obtained as those for the rural respondents in the Kenyan study; this confirmed and refined established protocols for graphic design. These expected results were blown away by outcomes perceived as alarming: the ECOM drawings portraying rural life and life cycles produced unexpected results, countervailing their original intended meaning.

The line drawing of children in front of a home (see Fig. 4) was included in the test, asking "How do these children live?" The answers did not align with the expectation that the drawing would elicit remarks about their unhygienic living conditions as reflected by the mud and the filth. The "correct" answers were

# 29 Visual Communication and Social Change

**Fig. 4** Question on hygiene from the ECOM visual literacy study. (Courtesy of ECOM)

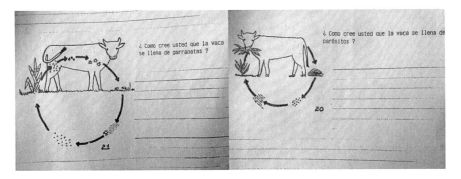

**Fig. 5** Life cycle question from the ECOM visual literacy study. (Courtesy of ECOM)

countered by a similar number of respondents mentioning the wealth of the family, referring to the two turkeys visible in the drawing.

Two ECOM graphics representing the life cycle of ticks and parasites in relation to livestock both scored 91% for incomprehensibility (see Fig. 5). Further investigation clarified the assertion that the drawings ignored the complexity of a life cycle composed of elements that occurred in different time frames but which had been printed over a realistic line drawing of a cow.

A major conclusion was quickly drawn: many pictures used in agricultural extension are not inappropriate because of farmers' lack of visual literacy but rather because of designers' and developers' lack of capacity in combination with insufficient pretesting. The problem of media designers' professional incapacities evolved into a more transformative insight when further consideration was given to the life cycle drawings (Guerra and Witteveen 1991), as the model embodies a characteristic example of modeling in the life sciences. In agricultural extension, farmers or fisherman often receive instructions about recurrent infestations of pests and diseases conveyed through life cycles models. In the ECOM case, discussing this finding with a so-called subject matter specialist became an issue, as these professionals could not recall how they had been taught such models to understand for example a tick infestation. This is not the occasion to further elaborate on more graphic details such as diverse relative sizes and the visibility of the different stages. The basic complexity of a life cycle as a metaphor, portraying a circular evolution, transforming insects into themselves at the closure of the circle rather than portraying the generational evolution over time, provides evidence of how knowledges have been influenced by dominant models and dominant discourses.

When rural communication workers are being trained, it is worth debating the life cycle model with them to exemplify the paradigmatic dilemma of conceptualizing (visual) literacy as an individual quality of audiences rather than seeing the mismatch between that and recognition of the diverse knowledges in the rural communication system. The label "life cycle lies" is still strong enough for participants (students and adult learners) to result in impactful learning outcomes, as exemplified by learners continuously contributing examples of problems with life cycle drawings that they come across in their work. See Fig. 6 showing submitted pictures.

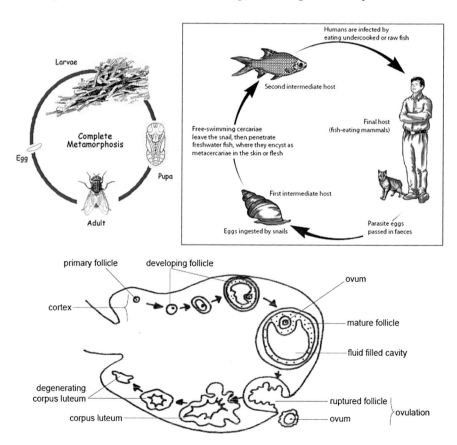

**Fig. 6** Life cycle pictures shared by students in a "life cycle lies" discussion

### 29.3.3 Valuing the Potential of Visual Images

In the following section, the focus on the potential of images and the visual moves deliberately away from less educated audiences whose visual illiteracy could be dismissed as arising from their low formal education. The following examples are positioned in higher education as evidence of the potential of the visual to be seen as transferable, thereby aiming to provide evidence of the impact of images in general. A second reason for looking at images in higher education is to follow-up the conclusion that images become unrecognized parts of dominant discourses, as shown by the life cycle example.

An example of using images in higher education is the increasing use of visual research methods (VRMs) in contemporary research (Rose 2014). VRMs – such as photography or video – are applied as the visual data obtained can disclose aspects of (inter) action, document certain practices (through repeated observation), and further support multidisciplinary analysis. The selection of VRMs is not only grounded in advanced knowledge of the specific qualities of visual data but also results from higher access resulting from technological facilities, both hardware (e.g., mobile phones) and software (e.g., for editing and data management). A VRM course in our university starts from the notion that academia has only recently come to appreciate the full potential of visual data and requires capacity building and due reflection. It is also considered that, in the social sciences, a more critical wait-and-see approach prevails, articulated in debates on ethical implications. It is, however, recognized, as with all methods in all domains, that the assumed richness of visual data also confronts researchers with challenges in the process of collecting, analyzing, and interpreting the visual data.

A more particular case of using visuals in higher education that provides evidence of the strong qualities of images is inspired by the artwork of Ursus Wehrli (2011). This artist provides a series of (often two) photographs, which show an object or a situation as a meaningful whole and as an ordered collection of its constituting elements. To confront students with visual discourse deconstruction, Wehrli's work has been imitated to visualize what dominant models prevail in research and analysis. Pairs of photographs of a bouquet of flowers and lentil soup are presented (see Fig. 7). The soup pictures show the ingredients (see Fig. 7.1) followed by the prepared soup (see Fig. 7.2). The bouquet photograph (see Fig 7.3) is placed beside a photograph showing a neat and tidy ordering of the leaves and petals (See Fig. 7.4). Despite the direct link between the ingredients and the soup, or the bouquet and the petals, each pair of photographs refers to different moments in time. The soup cannot be converted back into its ingredients, and putting the petals and leaves together again cannot restore the bouquet. The soup and bouquet visual analysis are used a visual metaphors to discuss with students underlying assumptions when doing (social science) research by unravelling complex systems into concepts, during technical research design. Unraveling is the action verb used to describe defining concepts in dimensions, aspects, and subaspects (Verschuren and Doorewaard 2010, p. 117). The resulting conceptual framework is often needed for doing research, but it may come with a consequence of ignoring the effects of such modeling process. Once research data are presented in tables, pie diagrams, and convincing analyses,

Figure 7.1 Soup ingredients   Figure 7.2 Lentil soup

Figure 7.3 A bouquet   Figure 7.4 Petals and leaves

**Fig. 7** Visual modeling of research inspired by Wehrli. (Courtesy of Annemarie Westendorp and Loes Witteveen)

they might no longer resemble or represent the societal situation that they attempted to understand. Using the soup and the bouquet visual analysis offers a surprising entry point for students to critically review the assumptions underlying their research. The strength of using pictures here in a context of inducting a critical view on academic research discourses provides an easy link to the current interest in visual research methods.

## 29.4 Contemporary Visual Communication

The previous section aimed to elaborate on the strength of visual communication, stills, and moving images. It also looked at the praxis of media design and media production, trying to move visual literacy away from questions about audiences' capabilities to a perspective on the accountability of media designers, developers, and commissioners. It drew on experiences of praxis in times of Nicaragua's postrevolution decade when educational ideas from Latin-American scholars like

Freire and Illich were explicit adhered to with deliberate text literacy campaigns and other communicative interventions for rural communities. Visual communication was recalled from those early times, evolving around visual literacy in times of international cooperation as a major platform for questioning the relevance and impact of using visual media in rural development, to now create a bridge to contemporary times.

In the following, we move to contemporary times where conventional linear and top-down information flows lose relevance with the increase in complexity and the changing media landscape. Drivers like participation, inspiration, and social learning replace the "message" as the steering element in a process of transformation. The role of communication professionals and innovation intermediaries is therefore changing; accountability and reflexivity become interwoven in new practices as the focus on the visual becomes embedded in contemporary visual culture. Rose describes this from the VRM perspective, stating: "using VRM performs a contemporary visual culture, rather than finding it represented in the images it generates, in the social it performs or in the research participants it involves" (2014, p. 39).

Before pressing the fast-forward button, we refer to contemporary views on analyzing images by using Rose's (2016, p. 24) framework, which distinguishes four sites: the site of the image itself, the site of audiencing, the site of production, and the site of circulation, which the author added in the fourth edition of her book on visual methodologies. These four sites are visualized in a kind of circular matrix with technological, compositional, and social modalities as crosscutting perspectives on every site.

Rose indicates that compositional interpretation refers to compositional elements of the images (discrete and combined aspects of content, colour, spatial organization, montage in the case of moving images, sound, and light) and complicates this by referring to the "good eye" or "visual connoisseurship" (2016, p. 57). As such terms originate from the field of the arts, the focus might seem to be drifting away from exploring visual communication for social science. Yet, why be afraid of art–science hybridization? The site of the image itself forces us to define what constitutes a good look, whereby balancing between aspects of training and talents cannot be ignored. The good look will be supported by (culturally defined) knowledge of colors, graphic techniques, genres, and so on, and this knowledge can be learned and taught. However, following the previous section, which outlined the attribution of meaning to an image considering it as a social construct, implicates making choices. Similar to the example of defining concepts in technical research design, there is no escape from using the compositional elements to interpret images. However, a major difference between interpreting an artwork as an autonomous object and visual communication for social change is that the latter has to align with the circularity embedded in Rose's framework. The analyst is forced to deconstruct the image, taking production, circulation, and the audience into account, and this is not always a happy marriage.

Pereira da Silva sheds light on such dilemmas by elaborating on the frictions resulting from the main purposes of aid campaigns: giving visibility to a cause, giving visibility to a humanitarian agency, and raising funds (2013, p. 41). His article brings the issue of portrayal, and therefore the potential impact or strength of pictures, to a different level, stating that "the act of portrayal itself can be seen as a way of capturing agency," which he further elaborates by stating: "The action that a humanitarian organization performs to portray an individual, a human being in a certain manner serves to create a generic collective image of an *'other,'* a set in which *'their situation'* as a whole becomes a kind of *label* that postulates *a priori* all the needs and aspirations of those under the labeling process. It is important to notice that this process can be intentional or unintentional" (p. 43).

Modeling such a situation for the case of aid campaigns in a post-hurricane situation articulates an almost linear linkage of roles between an aid organization as commissioner, an advertising agency as the portrayer, an audience of potential donors to the aid organization, and somewhere the "portrayed" who in the linear linkage line of thinking would soon convert to aid recipients. Again, as in the days of development aid, the aid recipients are left out of the equation, which should include aspects of inclusion, agency, and participation. Such pessimistic statements do not lose relevance when one considers countervailing actions of portrayal such as those of the photographer Sebastiao Salgado, who aims to show the resilient side of people, always portraying them doing some kind of action rather than framing them as passive victims of drought, war, and other emergencies of others.

It follows that transformative views on design, development, and production are needed to be able to address visual communication for social change. Such quest resonates with the imperatives formulated by Servaes and Lie (2014) to recognize complexity and establish rights-based social dialogues, quoting Slim and Thompson (1993, p. 4) and Witteveen and Lie (2012) "that all people have a right to be heard, especially when the main debates take place in documents which they do not write or in meetings which they do not attend" (p. 19). The focus on "all people" aligns with contemporary interests in informal, everyday communicative interactions among stakeholders. Leeuwis and Aarts (2011) describe this, stating: "meaningful innovation is dependent on changes in discourses, representations and storylines that are mobilized by interacting social actors" (p. 27). Therefore, the remainder of this section reviews three core aspects of designing, developing, and producing visual communication interventions for social change. These three aspects are: acknowledging diverse professionalisms, centralizing the importance of design thinking, and recognizing social imaginaries as options for visual communication in social dialogues.

### 29.4.1 Acknowledging Diverse Professionalisms

To wrap up on the interwoven histories of media design and production at ECOM as an example of traditional views on visual literacy, it is appropriate to state that ECOM continued working on visual media, viewing impact as a shared accountability of commissioners, designers, and developers. ECOM's move to incorporate account managers was a first response towards a framework for operationalizing insights. This organizational configuration was later refined by appointing both an expert team and a communication team for each major media production.

Payan Leiva (1994) reports an interesting example of this configuration, describing the ECOM production process for a cattle health guide (*Guía sanitaria ganado bovino*) as a continuous negotiation between the expert team with vets and zoo technicians, the communication team, and livestock farmers. Figure 8 represents a strong discussion among both production teams over recommendations for dealing with the cadaver of a cow that died of anthrax. The technicians recommended digging a deep hole, which farmers indicated could never happen as the soil was too rocky. The communication team insisted on a more feasible recommendation, as the main goal was to prevent farmers from bringing the highly contagious cadaver to the market. Widening these practices to text treatment in a context of low literacy, it is worth mentioning that the communication team used the Flesch-Douma equation to control readability for newly literate farmers, with the unexpected advantage that this quantitative tool increased the technicians' respect for the communication team, as it was seen as proof of communication professionalism. Again, the silent interpretation by the communication team was the realisation of both teams' very different knowledges and ideas of praxis. The example demonstrates how various professionalisms need to come together to accomplish change.

Another example is provided by Goris et al. (2015), whose article explores the participatory filmmaking production process by explicitly addressing diverse professionalisms needed to deliver a successful production process and product. In examining the collaborative production process, the authors focus on dilemmas encountered in relation to integrating participatory qualities and artistic qualities

**Fig. 8** Fragment for the Guía sanitaria ganado bovino. (Courtesy of ECOM)

and related professionalisms. In doing so, at least three different disciplinary fields and professions are distinguished: filmmakers, facilitators, and researchers. The articulation and understanding of these roles enhance the overall appreciation and outcomes of the production process not only for the community members involved but also for the diverse collaborating professionals.

Haag and Berking (2015) in their publication address gaps in knowledge design as resulting from limited synergy between two key domains of research and practice: learning sciences and human-computer interaction. The authors distinguish between professional educators, instructors, instructional designers and stress the use of learning theories, conceptual frameworks, and user experience design aspects to come to production considerations and design requirements for "mobile learning."

### 29.4.2 Centralizing the Importance of Design Thinking

A recent transformative view in the design, development, and production of visual communication for social change comes from design thinking, which combines with the upcoming influence of user experience (UX) and user interface (UI) design positioned in the work of digital applications. Design thinking is described by the Interaction Design Foundation as "a design methodology that provides a solution-based approach to solving problems. It's extremely useful in tackling complex problems that are ill-defined or unknown, by understanding the human needs involved, by re-framing the problem in human-centric ways, by creating many ideas in brainstorming sessions, and by adopting a hands-on approach in prototyping and testing" (2018, p. 1).

UX refers to how a client's or commissioner's original request is developed into a workable structure, which is tested with prototypes to assess the navigation through the app and other aspects in relation to technological development of the structure. UI is most often used in the design of a digital application focusing on relevance and appreciation as perceived by the user. Framing a UX or a UI designer as an ICT expert versus an artistic designer is a conservative and not a relevant distinction in this context. Rather than defining an exclusive professionalism, it can be stated that UX and UI design are complementary processes with distinct foci contributing equally to better digital development. Focusing on UX and UI or on them together as design thinking means that the user – and thus audience participation – is a serious driver from the very start in the development of a communicative intervention for social change.

An example of design thinking in rural communication is provided by Witteveen et al. (2017, p. 1677), who explain the design process of the Digital Farmer Field School in Sierra Leone and the Digital Herder Service Centre in Mongolia (see Fig. 9). In this example, it is clearly illustrated that a communicative intervention, based on explicitly formulated design principles, is better able to achieve complex outcomes – in this case, developing digital interfaces that make extensive use of visual communication.

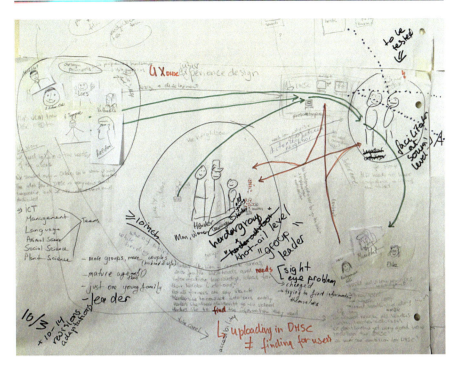

**Fig. 9** Prototype design for a digital herder service center, Munkhbat et al. 2017. (Courtesy of MULS DHSC team)

Design thinking is in fact a way of thinking. It is a way of thinking that values the interaction of diverse professionalisms, but it is also a methodology. It is a participatory methodology in which different disciplinary perspectives embedded in different professions are brought together to interact. It is then the interaction itself that co-creates an environment in which a production design and development process can be shaped. Ivarsson and Nicewonger (2016) elaborate on such a process developing a theory of entropic architecture "building with processes you cannot control" and stated that such a process worked "by setting up a linguistic register," by producing and evaluating prototypes for their material and aspirational properties (2016, p. 76).

Design thinking is a systemic way of thinking. It considers the whole but respects the individual elements. These individual elements are often very diverse, but they are always considered in respect of the whole; all make their own contribution towards the construction of the whole. Design thinking is an interactive process that always centralizes the end-user and respects and actively engages the end-user as an equal partner in the production process of any communicative intervention. In light of the previous section on acknowledging diverse professionalisms, applying design thinking in the field of visual communication for social change encompasses therefore the collaboration of technological experts with visual artists, like painters, photographers, and filmmakers. The dimension anticipation in the framework of responsible innovation (Stilgoe et al. 2013, p. 1573) mentions approaches such as scenario design and vision assessment to impact on existing imaginaries and "participation rather than prediction." With this forward-looking view, in the next section, we will look into social imaginaries in the field of social change.

### 29.4.3 Recognizing Social Imaginaries as Options for Visual Communication in Social Dialogues

As described in the first section, Van den Ban (1996) made a short reference to a quality of visual media to "develop or strengthen the motivation to change as an emotional process." Not to establish short-cut continuities but searching for contemporary views in the field of development for social change, we address social imaginaries as options for visual communication.

Social imaginaries refer to a collection of images, ideas and principles, fantasies, motivations, institutions, rules, which are shared by a group of people, which give meaning and presence in their surroundings and position all members in relation to one another. Social imaginaries refer to the way people imagine their social surroundings expressed in images and stories of popular and artistic expressions. Carpentier (2017) describes social imaginaries as "fantasies that enable an overcoming of the lack generated by the contingency of the social and the structural impossibility of attaining reality" (p. 21). He further refers to Lacan (1999, p. 197) who is said to have conceptualized fantasy as creating protection from the impossibilities of reality, which is eventually counterbalanced by providing subjects with discourse to enable producing, accepting, and dealing with reality (Carpentier, p. 21). Winders et al. describe the use of imaginaries (by scholars) as referring to assumptions "built into and shaping the language and terms used" to discuss certain issues and "their interpretation of the empirical realities that implicitly or explicitly, formed the scaffolding of their arguments."

Social imaginaries may become oppressive grids, defined by arrays of political and commercial influencers. However, we take here an optimistic view by steering towards rethinking dynamic design and communicative qualities of imaginaries for sustainable futures. In this context, the River Flows project provides a case study as it focused on creating social imaginaries, in this case of a new nature conservation area at the northern riverbank of the IJssel river in Gelderland Province, the Netherlands. The project was undertaken with a group of international students from an MSc course, Management of Development, in June 2017. In their home countries, they have functions related to agricultural knowledge and governance systems. In the setting of a Media Design for Social Change course, these adult students engaged in workshops on graphic design, painting, theatre, poetry, and creative processes to support community resilience, participation, and social ecological learning. The experiences were put into practice in a community art project with *Natuurmonumenten*, a Dutch nature foundation, and the students embarked on producing a poetry route to express a sense of place in relation to the new nature conservation area, Koppenwaard, along the IJssel river (see Fig. 10). They were

**Fig. 10** Student writing a poem on location. (Courtesy of Loes Witteveen)

challenged to create their interpretation of the landscape, their appreciation of the natural resources and surroundings in poems and paintings, which were later compiled into a poetry route. Mohamed Jalloh from Sierra Leone stated: *"Prior to this, I had never written a poem and I was wondering how possibly can I be able to do this. After hard thinking and observing things around me (river, forest, etc.), I started composing my poems and ideas kept flowing."*

The ambition of this project was to contribute to positive dissonance in conversations in the public participation process. Using the artworks of outsiders (international students) and yet insiders for their creative production on location meant inducing a positive element of dissonance or disruption to render new openings to the public debate on nature conservation, which became characterized by the same divergent voices heard in similar situations. Challenging people to reflect on a sustainability agenda inspired by the poetry route was meant to change the debate from a merely political debate on vested interests to clashes between present and future wishes and needs. The poetry route was meant to strongly induce social imaginaries, tempting the creation of shared affections and images over ambitions for the newly established conservation area (see Fig. 11).

The ambition to provide the students with an individual and group experience of creating a visual expression of social imaginaries has been successfully achieved. The further trajectory of the poetry route River Flows is more ambiguous. It may happen that audiences look at the cognitive information, but a visual social dialogue is new and therefore requires new processes or conditions for success. The experiences in this project however underline the relevance of including the "site of circulation" in the framework "sites and modalities for interpreting visual materials" by Rose (2016) as the production, the image and the audience do not provide sufficient perspectives on understanding a complex "social imaginary" process such a River Flows. To prevent the poetry project from interpretations of a narrow instrumental use of social imaginary, we take the freedom for a short quotation of Castoriadis that "the power to deliberately shape social thought is always limited" (Gilliard, p. 336).

**Fig. 11** Fragment of a combined poem and painting, part of the resulting poetry route. (Courtesy of Linda Agbotah)

Glimpses of relevance and impact have been observed, pointing in the direction of visual media having a strong potential for creating social imaginaries or affective or emotional inspiration and thereby contributing to opening up debates in processes of social change whereby cognitive arguments tend to have closing or self-affirming qualities. It is, however, recognized that insufficient knowledge and expertise are available to realize this potential. Art–science interfaces are intertwined with issues of social status and resource allocation, understanding of the role and facilitation of positive dissonance in social change debates are limited, and the grammar of visual textualities and discourses in societal debates still leaves much to discover. Yet, a social imaginary as a concept contributing to sustainable futures and social change is a strong driver to dig into these challenges.

## 29.5 Conclusion

As shown by Fuglesang's work, some sparks on specific aspects of visual communication and media design were present in the 1980s and 1990s. Notwithstanding these insights, the field of international development for a long time continued to focus on the "semiotic point of view, analysing and decoding the 'language' of the messages transmitted," and the quest for "systematic study [...] on the evaluation of instructional media" (Epskamp 1978, p. 1) took place in different fields. It was in the next millennium and inspired by the field of the arts that arguments were built to conceive "language" and "text" beyond the understanding of visual literacy as opposed to a merely word-based or written language to recognize visual communication and visual languages.

Even though a final statement on the quality of visual communication will not be formulated, the chapter searches to go beyond the visual turn and to build on a visual future and concludes by arguing that, in contemporary visual communication, acknowledging diverse professionalisms, centralizing diverse professionalism and transdisciplinarity the importance of design thinking, and recognizing social imaginaries as options for visual communication in societal dialogues are of crucial importance.

## 29.6 Cross-References

▶ ICTs for Learning in the field of Rural Communication

## References

Adams ME (1982) Agricultural extension in developing countries. Longman Scientific & Technical, Essex

Boeren A (1994) In other words... the cultural dimension of communication for development. CESO paperback no. 19. Centre for the Study of Education in Developing Countries, The Hague

Buckingham D (2009) 'Creative' visual methods in media research: possibilities, problems and proposals. Media Culture Soc 31(4):633–652. https://doi.org/10.1177/0163443709335280. Sage, Los Angeles/London

Carpentier N (2017) The discursive-material knot. Peter Lang Publishing, New York

Dondis DA (1973) A primer of visual literacy. MIT Press, Cambridge

ECOM (1990a) Evaluación de la empresa de comunicaciones, Agosto – Diciembre 1990. Juigalpa. Nicaragua (Internal report)

ECOM (1990b) Comunicaciones populares en la región V. Juigalpa Nicaragua (Internal report)

Epskamp C (1978) Media, education and development. A bibliography. CESO bibliography no. 3. Centre for the Study of Education in Developing Countries

Epskamp K (1995) On printed matter and beyond: media, orality and literacy. CESO, The Hague

FAO (1990) Powerful images. Slide programme & filmstrips. Planning, production and use guide. Development Support Communication Branch, Information Division, Rome

Felten P (2008) Visual literacy. Change: The Magazine of Higher Learning, 40(6):60–64. https://doi.org/10.3200/CHNG.40.6.60-64

Fuglesang A (1982) About understanding – ideas and observations on cross-cultural communication. Dag Hammarskjöld Foundation, Uppsala

Goris M, Witteveen L, Lie R (2015) Participatory filmmaking for social change: dilemmas in balancing participatory and artistic qualities. J Arts Communities 7(1–2):63–85. https://doi.org/10.1386/jaac.7.1-2.63_1

Gürgen R, Simenyimana B (1987) Visual aids and villagers in Rwanda. Ideas and Action. 1987/5no.176. p.14–18

Guerra E, Witteveen L (1991) Test de Alfabetización Visual. Investigando la percepción visual de sectores populares en la región V, Nicaragua. Empresa de Comunicaciones ECOM, Juigalpa

Haag J, Berking P (2015) Design considerations for mobile learning. The mobile learning research. In: Zhang YA (ed) Handbook of mobile teaching and learning. Springer-Verlag, Berlin/Heidelberg, pp 41–60. https://doi.org/10.1007/978-3-642-54146-9_61

https://www.tandfonline.com/doi/abs/10.1080/10749039.2016.1183131. Accessed 27 Sept 2018

Interaction Design Foundation (2018) 5 stages in the design thinking process. https://www.interaction-design.org/literature/article/5-stages-in-the-design-thinking-process. Accessed 12 July 2018

Ivarsson J, Nicewonger TE (2016) Design imaginaries: knowledge transformation and innovation in experimental architecture. Mind Culture Activity 24(1). https://doi.org/10.1080/10749039.2016.1183131

Lacan J (1999) The Seminar of Jacques Lacan. Book XX, On Feminine Sexuality: the Limits of Love and Knowledge: Encore 1972–1972, Jacques-Alain Miller (ed.) Trans. with notes by Bruce Fink. New York: Norton

Leeuwis C, Aarts N (2011) Rethinking communication in innovation processes: creating space for change in complex systems. J Agric Educ Ext 17(1):21–26

Messaris P (2012) Visual "Literacy" in the digital age. Rev Commun 12(2):101–117. https://doi.org/10.1080/15358593.2011.653508

Munkhbat B, Witteveen L, Erdene-ochir S, Van de Fliert E, Baldan Y, Ochir J, Uyanga N, Tuyagerel G, Ganbold Ya, Bat-erdene E, Batsumber D (2017) Digital herder service Centre prototype design, development and testing. Mongolia University of Life Sciences

Nead L (2000) Victorian Babylon: People, Streets and Images in Nineteenth-Century London. London: Yale University Press

Payan Leiva DM (1994) Communicating technology, making a livestock guide for farmers. Ileia Newsletter March

Pereira da Silva Gama CF, Pellegrino AP, De Rosa F, De Andrade I (2013) Empty portraits – humanitarian aid campaigns and the politics of silencing. Int J Human Soc Sci 3(19)

Rodrigues D, Lopes D, Prada M, Thompson D, Garrido M (2017) A frown emoji can be worth a thousand words: perceptions of emoji use in text messages exchanged between romantic partners. Telematics Inform 34(8):1532–1543

Rogers EM, Van Den Ban AW (1963) Research on the diffusion of agricultural innovations in the United States and the Netherlands. Sociol Rural 3:38–51

Rose G (2014) On the relation between "visual research methods" and contemporary visual culture. Sociol Rev 62:24–46

Rose G (2016) Visual methodologies. An introduction to researching with visual materials, 4th edn. Sage, London

Serafini F (2017, February 27) Visual literacy. Oxford Research Encyclopedia of Education. Ed. From http://education.oxfordre.com/view/10.1093/acrefore/9780190264093.001.0001/acrefore-9780190264093-e-19. Accessed 22 Oct 2018

Servaes J, Lie R (2014) New challenges for communication for sustainable development and social change: a review essay. J Multicult Discourses. https://doi.org/10.1080/17447143.2014.982655

Slim H, Thompson P (1993) Listening for a change: Oral testimony and development. London: Panos

Stilgoe J, Owen R, Macnaghten PP (2013) Developing a framework for responsible innovation. Res Policy 42(9):1568–1580

Taller de Comunicaciones Campesinas (1989) Evaluación del Taller de Comunicaciones Campesinas. Juigalpa. Nicaragua. (Internal report)

UNDP (2018) Goal 4 targets. http://www.undp.org/content/undp/en/home/sustainable-development-goals/goal-4-quality-education/targets/. Accessed 16 July 2018

UNESCO (2006) Education for all global monitoring report. Chapter 6: understandings of literacy. http://www.unesco.org/education/GMR2006/full/chapt6_eng.pdf. Accessed 16 July 2018

UNESCO (2007) La Cruzada Nacional de Alfabetización. http://www.unesco.org/new/es/communication-and-information/memory-of-the-world/register/full-list-of-registered-heritage/registered-heritage-page-6/national-literacy-crusade/. Accessed 15 July 2018

Van Brink L, Visser M (1983) Kijken en …… (niet) begrijpen. Vakgroep voorlichtingskunde. Landbouwhogeschool Wageningen

Van den Ban AW, Hawkins HS (1988) Agricultural extension. Blackwell Science, Oxford

Verschuren PJM, Doorewaard HJACM (2010) Designing a research project. Eleven International Publishing, The Hague

Wehrli U (2011) The art of clean up – life made neat and tidy. Chronicle Books, San Francisco

Witteveen LM, Lie R (2012) Learning about "wicked" problems in the Global South. Creating a film-based learning environment with "Visual Problem Appraisal" MedieKultur. J Med Commun Res 28(52):81–99

Witteveen L, Lie R, Goris M, Ingram V (2017) Design and development of a digital farmer field school. Experiences with a digital learning environment for cocoa production and certification in Sierra Leone Telematics and Informatics https://doi.org/10.1016/j.tele.2017.07.013

# Multidimensional Model for Change: Understanding Multiple Realities to Plan and Promote Social and Behavior Change

## 30

Paolo Mefalopulos

## Contents

30.1 Introduction .................................................................. 580
30.2 The Key Role of C4D in Achieving Change ............................................. 582
30.3 The Multidimensional Model for Change (MMC) ..................................... 584
    30.3.1 Multidimensional Model for Change: Windows of Perceptions .............. 591
30.4 Conclusions .................................................................. 592
References .................................................................. 594

### Abstract

This chapter introduces an innovative multidimensional model to manage change. It starts from a broader overview and assessment of the situation, engaging various groups of stakeholders. It then gradually zooms in critical issues mapping out perceived and real gaps and weaknesses while also identifying strengths and perceived benefits. This provides the inputs to design an effective strategic plan, which includes a C4D strategy to promote the intended social and behavior change. The Multidimensional Model for Change, or MMC, allows to devise relatively simple strategies to promote social and behavior change based on complex analysis that assess the situation in a comprehensive, multi-dimensional, and cross-sectorial way, thus eliminating, or at least greatly reducing, mistakes of the past.

### Keywords

Multidimensional · Change · Communication for development · Interdisciplinary · Cross-sectorial · Participatory

P. Mefalopulos (✉)
UNICEF, Montevideo, Uruguay
e-mail: pmefa@yahoo.com

© Springer Nature Singapore Pte Ltd. 2020
J. Servaes (ed.), *Handbook of Communication for Development and Social Change*,
https://doi.org/10.1007/978-981-15-2014-3_28

## 30.1 Introduction

An in-depth review of the literature and project documentation of organizations operating in the field of development made in 2003 suggested that too many times failures of development initiatives are mostly due to lack of proper planning based on genuine communication among the various stakeholders (Mefalopulos 2003). Since then there have been some progress, even if rather moderate. The challenges posed by the findings of that study are still valid and can be summarized in the following questions: Why sound technical solutions are not enough to achieve the intended change? Why individuals resist changing behaviors even when they have the knowledge that the new behavior will benefit them? These questions provide the benchmark along which this chapter originates and develops, introducing a model for change that considers different dimensions needed to be addressed before a person can be expected to adopt a new behavior.

Communication for Development or C4D has a long tradition in the theory and practices of international development, originally applied "to modernize" the so-called underdeveloped or developing countries. It has been used under various denominations, but the main concept, as it originated soon after World War II, remained substantially consistent through time, i.e., use communication to promote betterment of life conditions. The way this should be achieved has gradually evolved from the one-way communication model of reference into a more horizontal, two-way and participatory model, as currently envisioned by most scholars and practitioners in this area. At first C4D, at the time better known as Development Support Communication, was closely associated with the belief that mass media were all powerful and could be used to assist in the modernization of developing countries. As its conception and applications evolved, there have been several attempts to institutionalize C4D in international organizations, including within the United Nations. In the 1960 Erskine Childers was instrumental in having it introduced in UNDP, and then toward the end of that decade, the Food and Agriculture Organization of the United Nations, or FAO, created its Development Support Communication Branch.

In the 1980s it was UNESCO that put communication at the top of the international agenda, with the New World Information and Communication Order. Subsequently UNICEF established its own Social and Behavior Communication Change (SBCC), currently renamed Communication for Development or C4D section. Even the World Bank at the beginning of the new millennium established a Development Communication Division, which became very active not only within the World Bank programs and operations but also in international fora, taking the lead in organizing the First World Congress on Communication for Development in 2006. It should be mentioned that since 1988 the UN had established the Inter-Agency Roundtables on Communication for Development, a mechanism that, chaired by UNESCO, was convening every 2 years, hosted by a different UN agency on a rotational basis.

The initial objectives of the C4D Roundtables were to exchange ideas and experiences and promote collaboration among United Nations agencies. At the beginning the attendance was mostly limited to agencies that had C4D units and in some cases also saw some NGOs' participation. With time attendance by UN agencies increased and the scope of Roundtables also broadened, including advocacy efforts to institutionalized and strengthened C4D planning and implementation as a key component of most development programs and projects. The case for the need of dedicated funding to C4D was also strongly made using compelling evidence.

UNESCO had the task to compile the Secretary General Biannual Report on C4D that since 1996 was presented to the General Assembly, the UN Roundtables. This was the most important opportunity, if not the only one, to have issues and challenges that could be addressed by C4D presented to a global audience. It was with great surprise that in 2017, the C4D community learned, post-factum, of the UNESCO decision to ask for the suppression of this valuable mechanism. The lack of transparency and consultation leading to this decision was not only controversial but also raised questions of its legitimacy as UNESCO mandate was basically to chair the Inter-Agency Round Tables on Communication for Development and prepare the report for the Secretary General. Hence, many stakeholders saw the decision to suppress it without previous consultation with other UN agencies (some of which would have probably been ready to take up this role) as UNESCO stepping out of the institutional mandate and boundaries.

It is sadly ironic, and shortsighted, that the C4D Round tables were suppressed just when the importance of social and behavior change was being increasingly recognized as a crucial component of development initiatives. The key role of C4D in the successful handling of polio (which in 2008 was certified to be fully eradicated from India) as well as in other health emergencies such as avian flu and Ebola confirmed the value of rigorous social and behavior communication strategies in achieving the intended change. Even the World Bank in its World Development Report of 2015, titled Mind, Society, and Behavior (World Bank 2016), acknowledged that human behavior is not always rationally driven and focused on the importance of understanding the drivers for social and behavior change.

The evolution and growth of C4D went through some ups and down. Unfortunately, despite an increasing acknowledgment by many key actors of the importance of addressing social and behavior determinants to achieve the intended change, within the UN family, C4D is currently thriving only at UNICEF. The model of reference which was increasingly used in UNICEF, starting by the team of the India Country Office and then adopted more systematically at a global level, was an adaptation of the socio-ecological model (SEM), which forms the basis of the Multidimensional Model for Change presented in this chapter.

The Multidimensional Model for Change – MMC – is derived from Bronfenbrenner's ecological framework for human development (Bronfenbrenner 1979). The main differences consist in the framing and the number of the dimensions to be considered, which now are four, each composed by three basic components, devised to assess strengths, weaknesses, and gaps needed to be addressed in the planning of the communication strategy aimed at achieving the intended change. The aim is to steer C4D interventions away from the exclusive focus on individual behavior, which characterized many initiatives of the past, and adopt a broader and intersectoral view of the whole situation, a view that includes a holistic assessment of the factors needed to achieve change, and then consider how, and which ones, communication can address.

## 30.2 The Key Role of C4D in Achieving Change

The main causes for the many failures of development initiatives have been addressed in the presentation, *Why Sound Technical Solutions Are Not Enough: The Role of Communication in Development Initiatives* (Mefalopulos 2010), delivered for the 11th Raushni Deshpande Memorial Oration at the Lady Irwin College in New Delhi, India. If knowledge of solutions that would improve life conditions is not enough to make people change behaviors, what would then be the best way to achieve it? The vast literature evaluating development projects and programs indicate that there are a number of factors that affect the adoption of innovations and new behaviors. Many of such factors have been often neglected to a significant degree.

Too many development initiatives were doomed before they even started being implemented, due to their vertical, techno-driven sectorial planning design, a design that was usually devised by a few experts, often knowing little of the local context and culture, and even more often based on the (flawed) belief that if people knew that the new behavior was beneficial they would embrace it; thus the emphasis was on the production of messages to be disseminated in the media. The role and limited effectiveness of communication were often influenced by such design and the flaws related to it. To be truly effective and loyal to its core values, C4D had to be critical of modernization-driven approaches, which adopted linear and top-down approaches, where communication was mainly used to disseminate information. That is when two-way communication approaches started to increasingly gain ground.

Many publications such as of Jan Servaes (2003), Nobuya Inagaki (2007), and Emile McAnany (2012) help to understand some of the main challenges C4D faced in making the case to prove its value, not only in promoting the needed change but also in assessing the overall situation, as a prerequisite to identify the gaps and drivers for change. The challenge was to rethink theoretical and practical approaches to understand better how change occurs. A broader perspective on how individuals think and act was needed, abandoning the assumption that behaviors are rationally driven and reflecting on what would be the determinants of change shaping human behaviors.

Hence, C4D had not only to adopt a more horizontal and participatory-driven approach, but it had also to broaden its functions, adding assessment and planning to the communicating one. Most important C4D officers in major international organizations had to become advocate for a cross-sectorial approach that should replace the traditional, limited sectorial planning and implementation. The cross-sectorial approach enhances significantly the effectiveness of development initiatives and allows to apply C4D methods and tools at their best. A simple example can further clarify this assertion.

When facing the Ebola crises, the first approach was driven strictly by technical (i.e., medical) considerations. However, that proved to be highly ineffective as the virus was spreading fast due to cultural norms and habits (such as those associated to the burial of the death) that people would not accept changing just based on technically appropriate information, coming from external experts. The situation started to change when C4D specialists went in the affected areas to work with communities in order to reduce the risks of contracting the virus and addressed with them possible alternatives to reduce the risks of contamination.

Participatory Rural Communication Appraisal, or PRCA, is a methodology that was developed and applied along similar lines. It was devised in the 1990s by an FAO team working in Southern Africa, and it combines participatory techniques with communication research, and planning methods were devised. PRCA laid the foundations of an innovative approach, which started to be gradually adopted in projects across the world. It was defined as "a first step in the design of cost-effective and appropriate communication programmes, and materials for development projects ... a flexible, multi-disciplinary and participatory way to conduct communication research" (Anyaebgunam et al. 2004). PRCA was one of the first systematic attempts to bridge and apply participatory approaches with dialogical communication processes and tools, drawing from scholars and practitioners such as Orlando Fals Borda, Luis Ramiro Beltran, Robert Chambers, Paulo Freire, Bella Mody, and many more.

Moreover, PRCA was innovative because it was promoting a broader interdisciplinary approach to sectorial project design. For most areas of work in international development, such as health, nutrition, sanitation, education, etc., working vertically within its own sector has been the norm for a long time. However, cross-sectorial work is an imperative for C4D, because its functions and objectives are intrinsically embedded in those of health, nutrition, education, etc. There is not an ultimate communication goal per se, but there are communication objectives aiming at supporting behavior change to achieve education, health, and other sector goals. That is why C4D can provide the broader assessment, exploring and linking different areas and sectors, which absence is often considered to be the weak element of planning. A single, specific behavior change to occur often requires that several other factors across various areas/sectors be in place.

For instance, let us take the case of a project set up to introduce a new crop, more economically valuable. To increase the income and the living conditions of farmers in a district where farmers are living at subsistence level, the Rural Ministry decided to promote the adoption of the new income-generating crop. The innovation in their cropping system seemed a sound idea, and once an investigation confirmed that the new crop could be grown well in places where staple food is currently grown, the project begins its activities and extensionists are sent to disseminate the message and information about the new crop. To the experts' surprise, several farmers resisted the innovation. In those cases where the new crop has been adopted, problems aroused, such as lack of marketing skills to price and sell the product and lack of transport to access markets; decrease in staple food, which is being substituted by the income-generating crop, led to higher malnutrition of children (Anyaebgunam et al. 2004).

Examples such as this have been quite common in past development initiatives and are due mostly to the lack of two-way communication among project managers and experts on one side and the ultimate stakeholders (in the past labeled beneficiaries) on the other. Another common flow is the limited technical focus on the issue itself, in this case the adoption and growth of a new crop, which neglects other needed elements of the equation, such the knowledge of markets, availability of transport, change in dietary habits, etc. All, or most, of these issues could have been addressed if a genuine two-way communication would have occurred during the assessment and the planning phases of the project (Anyaebgunam et al. 2004).

Before discussing the MMC or Multidimensional Model for Change, it is worthwhile highlighting how C4D scholars and practitioners are becoming increasingly aware of the importance of social pressure in the adoption of rejection of a given behavior. Human behavior is not always guided by "rational" considerations, or maybe rationality does not have a universal applicable standard. What is rational in India might not be rational in Germany, and what is rational for a 20-year-old might not be rational for a 60-year-old.

Social norms, interdisciplinary overview, infrastructure strengths, and institutional capacities are some of the elements that should be considered when planning for behavior change. Hence, a more comprehensive but agile approach is needed to make sure that all key components of a certain situation are taken into consideration and assessed before designing effective strategies leading to social and/or behavior change. This is at the core of the model presented next.

## 30.3 The Multidimensional Model for Change (MMC)

Knowing the rationale behind the development of the Multidimensional Model for Change or MMC helps to understand why and how to use it. A literature review of international experiences plus lessons learned and documented failures about trying to eradicate open defecation in India were instrumental in helping to refine this model. Its main rationale is to ensure that all dimensions and related components that can promote and facilitate change would be considered when developing a C4D strategy. Open defecation usually falls under the domain of the Water and Sanitation sector but affects a number of other sectors such as health, nutrition, and education. Hence adopting a strict sectorial approach would not bear the intended results, while the intersectorial and integrated focus of the MMC can be of much greater value.

MMC combines the insights of participatory communication research and planning with the comprehensive overview and assessment allowed by the contextual framework of the socio-ecological model. This helps to break out from the rigid unilinear approaches of the past. Clearly to successfully eliminate open defecation, the key behavior to be adopted virtually by every individual has to do with the systematic use of latrines. It is of little wonder then to see that the many communication campaigns aimed at promoting the use of latrines both in urban and in rural settings had a very limited effect overall.

This is mostly due to the fact that such communication campaigns have been often devised and implemented with an information-bias approach, that is, focusing on the message to be given to people regardless of other factors of relevance to audiences/stakeholders. To be effective in promoting the behavior of latrines, communication campaigns should be weighed against factors that are likely, if not certain, to affect the adoption of the new behavior. Among these are latrine-friendly policies requiring toilets in public places; number of and access of latrines available; number and qualifications of line workers at community level promoting the building, maintenance, and proper use of latrines; and, last but not least, the presence of social norms, cultural beliefs, and customs that support the use of latrines, as peer pressure could be a crucial factor in accepting or rejecting the adoption of a new behavior.

Many of these "side factors" have been underestimated, if not completely missing, when designing and implementing campaigns aimed at promoting the adoption of the new behavior (something required in virtually any type of innovation or change initiatives), thus explaining to a large degree the reasons for the many failures in the adoption of the latrines by individuals. The MMC has been devised to avoid this kind of shortfalls and promote a comprehensive and cross-sectorial assessment that would then allow to plan and adopt communication approaches, methods, and tools in needed components of each dimension. Hence, the Multidimensional Model for Change is an approach that, starting from an identified situation that needs to be improved, helps to make a comprehensive assessment of the context related to the situation in question. Next, it allows the planning of the overall intervention as well as to devise the parallel communication strategy aimed at ensuring that the intended behaviors will be adopted.

The MMC comprises four dimensions, three of which define the enabling socio-economic environment conducive to change, and the fourth entails the components to be addressed at the individual level for promoting the adoption of the required new behavior. The first dimension is the political one. It has been labeled this way because its components are mostly shaped in the political arena as they are: budget (financial and human resources at disposal), infrastructure (availability and quality), and public policies (laws facilitating or impeding change). The second is the organizational dimension intended to assess the strengths and weaknesses of organized groups and institutions that are linked to the intended change. These organizations or institutions may belong to the state, public, civic society, or private sector. The third dimension is the sociocultural one. Its components are norms, media, and networks of reference. In other words, they are components that can greatly influence decisions taken by individuals, either through media or peer pressure.

In order to ensure that the enabling environment is in place, or to identify what is needed to achieve it, the assessment of the components in these three dimensions needs to precede the last one. If all prerequisites are satisfactorily met, behavior change is much more likely to occur. Hence, the fourth dimension of the MMC is the individual one. Here is where the actual behavior change is expected to occur; hence the components are those directly linked to the individual. The first is about the influential sources in the life of the individual, could be a farmer, a priest, the mother-

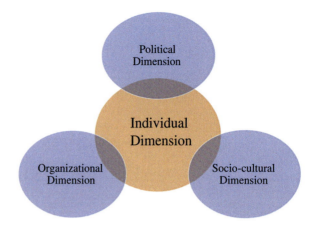

**Fig. 1** MMC dimensions. (Source: Author)

in-law, etc., and the next is about her/his inner circle, including family members and close friends. The last component is a composite one that includes the basic components needed to embrace the new behavior, i.e., AKAP. In much of the communication literature, Awareness, Knowledge, Attitude, and Practice constitute the ladder to achieve and sustain new behaviors, even if recent studies indicate that social norm pressure often has a stronger impact on the adoption of new behaviors. Figure 1 presents the basic MMC dimensions in graphic form.

Before illustrating each dimension in detail, we should mention that the MMC comprises three basic steps or phases. The first phase entails carrying out an assessment of what are the components needed to have the ideal situation (e.g. a polio- free contry), following the four dimensions highlighted in Table 1. It can be achieved reviewing the available literature and complementing it with interviews and/or brainstorming with experts and other key stakeholders. The aim is to identify, as exhaustively as possible, all the elements needed to have the intended situation, that is why is named "ideal mapping." Then, the second phase can begin, which follows the same steps described in Table 1 but now addressing and investigating the situation as it in reality. That is, rather than defining what are the ideal elements needed, it focuses on the hard evidence available to give an accurate picture of the current situation, that is why is named "factual mapping."

Comparing and contrasting the findings of the ideal mapping and the factual mapping provides the basis for planning. This is the third phase of the MMC, where the strategy to achieve the intended change is developed. It is intended to give a clear understanding of what is missing and should be addressed in order to achieve the intended change. The MMC allows to draw a broad strategy covering various sectors of intervention, but it also provides the framework for developing a C4D strategy, which is in sync with the broader one, addressing the social and/or behavior change in question.

1. The political dimension entails components that, to a great extent, are in control or influenced by the state and are needed to facilitate the intended change, namely, policies, infrastructure, and resources available (both financial and human

**Table 1** Ideal v. reality mapping window

| Dimensions | Components | Key questions |
|---|---|---|
| **Political** | *Infrastructure* | What infrastructure would be needed to be able to support the adoption of the innovation/new behavior? |
| | *Public policies/laws* | Which policies or laws should be in place to promote the adoption of the innovation/new behavior? |
| | *Budget/resources* | What is the needed budget and what number and type of human resources are needed to promote effectively the adoption of the innovation/new behavior? |
| **Organizational** | *State/public institutions (mandate and capacity)* | Which type of state institutions would be required to promote the adoption of the innovation/new behavior? And with what mandate and technical capacity? |
| | *Civil society (NGOs, academia, foundations)* | Which CSOs (NGOs, academia, foundations) would be required to promote the adoption of the innovation/new behavior? And with what technical capacity? |
| | *Private sector* | How could the private sector support the adoption of the innovation/new behavior? And which firms could be involved? |
| **Sociocultural** | *Norms* | Which sociocultural norms, beliefs, and habits would facilitate the adoption of the innovation/new behavior? |
| | *Media* | What kind of themes and messages should the media present, discuss, and disseminate? |
| | *Networks of reference* | What kind of themes and messages should be introduced, discussed, and disseminated? |
| **Individual** | *Influential sources* | What themes and messages should influential sources focus on? |
| | *Interpersonal inner circle* | What themes and messages should inner circle individuals focus on? |
| | *Individual AKAP* | What knowledge, which attitudes, and which behaviors are required to achieve the intended change? |

Source: Author

resources). Thus, as stated above, the first step is to sit at a table brainstorming about which of these components would be needed to achieve our objective. Next the factual (that is evidence-based) assessment process can begin in order to compare and contrast how far away are we from such ideal situation and what can be done to address the challenges identified.

To make this clearer, let us refer to the open defecation example. Once it has been decided that the end goal is to eliminate open defecation, the MMC approach assesses the various components of each dimension, starting with the political one. To facilitate the adoption of latrines, there should be laws and policies that strengthen the requirements of building and maintaining toilets in all public places and that give rewards or fiscal incentives to individual building latrines in their homes. Infrastructure wise the state needs to ensure that there are enough quality toilets or latrines available in public spaces as well as in institutions of relevance, such as schools, hospitals, public offices, etc. Finally, the budget needs to be assessed against current availability of toilets and latrines in private and public places and what would be needed. Considerations about the cost of having the needed human resources in place should be addressed as well.

Too many times communication campaigns trying to have people abandon the practice of open defecation have been designed and implemented even though many people did not have access to latrines. Everybody can easily see how this is a sure recipe for disaster. Technical experts have not always taken what seems obvious into account. Nowadays, it is still possible to see nicely crafted media campaigns, appealing to the eye but totally useless for reducing the problem of open defecation.

2. The organizational dimension refers to organizations, institutions, or groups' alliances, either national or international, which are capable and willing to support the required behavior change. The components of this dimension have been labeled as (1) state/public institutions, such as line ministries, departments, or specialized state agencies; (2) civil society organizations, including nongovernmental organizations, academia, and no-profit foundations; and (3) private sector, which basically can include firms and entrepreneurs which are interested in getting actively engaged in activities to promote the achievement of the set objective.

    Hence, in the case of organizations, capacity is not the only factor to be assessed but also the mandate, and the adherence of practice to that mandate, because as Quarry and Ramirez (2009) stated "organizations are half the methodology." Continuing with the open defecation example, the factual assessment would first have to focus on the demographics about the open defecation problem. Where does it happen the most (e.g., urban vs. rural areas)? Is it evenly spread by age, gender, and socioeconomic status? This data will help understand better the specifics of the problem and identify what kind of institutional presence would be needed in the field.

    Starting with the first component, the following questions should be posed to understand what would be needed to put in place an effective behavior change strategy. Which public/state institutions have the mandate to operate in this area? And do they have the financial and human resources to support the needed behavior change? If gaps are identified, they need to be addressed before launching a communication campaign to promote the adoption of latrines. A similar process should be followed for the other components. The one of civil society would require an appraisal of NGOs working in relevant areas as well as of academic centers able and willing to support the strategy to eliminate open defecation.

3. The sociocultural dimension is the one that has been most neglected in the past and is probably the one most effective in promoting or blocking the adoption of

new behaviors. The components of this dimension are (1) social norms, which are the behaviors that people are expected (by other people) to apply in specific social situations, which clearly vary from culture to culture. It is becoming increasingly evident how social norms, through peer pressure, can be the main factor in making individuals accepting or rejecting new behaviors; (2) media, refers to all channels of communication that address messages and themes either in a monologic or dialogic manner. They can be mass media, social media, or even traditional media such as popular theater or flip charts. These are channels that have the main function of mostly passing or exchanging information, even if in some cases they can stimulate dialogue and discussion, e.g., social media; and (3) networks of reference, which refers to networks of people, which can affect individual's decisions, such as a farmers' unions, a national association of pediatricians, or in the case of open defecation an academician network of universities having water and sanitation programs and courses. These three components have one thing in common. They are all effective instruments capable of influencing and shaping the individual's decisions in embracing change and adopting new behaviors.

Recently, the influence of social norms in shaping behaviors has gradually become more and more important. Evidence shows how most individuals are prone to adapt or adopt certain behaviors if they feel others in their community of reference are expecting them to do so and if they are expected to reciprocate. The work of Bicchieri (2017) on the role of expectations in social norms provides interesting insights in this area. The importance of peer pressure in making individuals change their behavior has been documented in many cases, but I find of particular interest the instance described by Rosenberg (2011).

In her book, among the many examples, one stood out as providing a great insight on how to influence behavior change. The author investigates why traditional anti-smoking campaigns that worked with adults (even if after a long time) did not work for teens. It became clear that a key reason was the fact that teens are by nature rebellious and not particularly keen to listen to advice by grown-ups. Things change when the approach was switched to appeal to their rebelliousness, having as the main target not health issues but the big tobacco companies and the way they manipulated marketing message to lure teens to smoke. This approach was made even more effective by the fact that groups of teens themselves were the ones spearheading debates and actions related to this issue in schools and other places where teens would usually convene.

To make clearer the functions of the components in this dimension, let us revert to the open defecation example. The initial step would be to investigate and assess available norms and beliefs related to this practice. If it turns out to be (as it is the case in some places) that open defecation is not only tolerated but also socially acceptable, it becomes clear that before promoting change of individual behaviors, there is the need to address and change perceptions and expectations of the community. In this respect, mapping the media landscape is a useful exercise as it can help to raise awareness about issues and promote change, even if change is usually more effectively achieved through interpersonal communication. Social media can also be quite effective in promoting change given their potential for dialogic communication, as illustrated in the *Development Communication Sourcebook* (Mefalopulos 2008).

Media campaigns can provide valuable contributions toward the goal of eliminating open defecation, but only if research and message design are done in a dialogic and participatory manner. Finally, the third component has a similar function as media, but on a more interpersonal level. Networks of reference are powerful channel through which individuals receive and exchange information while also forming and/or reinforcing their worldviews and opinions; thus they can be effectively used to engage individuals at community level. For instance, in parts of India, C4D strategies have relied successfully on religious leaders in having people accepting to vaccinate their children. The use of religious networks has been of great help in reassuring people who were hesitant or suspicious.

4. The individual dimension is where change is expected to take place, activated by the perceived benefits expected, the pressure of close family and friends, and the inner process of knowing what does it take to adopt the change and have the motivations to act upon it. The three dimensions presented above constitute the whole of the enabling environment, which can greatly facilitate or impede the adoption of the new behavior. This last dimension instead is directly related to the actual decision of embracing the intended change. Hence, the three components of this dimension are meant to be directly capable of achieving the adoption of the new behavior.

The first component is (1) perceived benefits, meant to verify if all stakeholders know about the benefits of the new behavior and if the perceptions of the benefits are consistent among different groups. That is why the Windows of Perception tool is applied again in this step. It ensures that the benefits' perceptions of the experts are shared by other stakeholders, for instance, rural farmers. Too many times projects and programs have failed due to gaps in the understanding and miscommunication among the various parties (Anyaebgunam et al. 2004), resulting in distinct groups of stakeholders having radically different perceptions on the objectives and expected benefits of the development initiative. The other component is (2) the inner circle, which is composed by persons, either family or friends, who are very close to the individual and, thus, have a profound influence in affecting the decision whether to adopt the new behavior. Key persons of the inner circle can be addressed and engaged in cases where other media are not so effective or with individuals who resist to engage in communication with external sources.

Finally, the last component is AKAP, or Awareness, Knowledge, Attitude, and Practice, that is the process leading an individual to adopt a specific innovation or a behavior. Even if its linear approach has been critiqued, the AKAP ladder still provides a good basic reference on individual factors providing useful predictors that change is likely to occur. Awareness is the first step intended to acquire consciousness that change is needed. Knowledge goes a step further bringing detailed information and inputs about the issue, its pro and contra. Most of the channels and tools in the third dimension are used with all relevant stakeholders to ensure awareness and knowledge are properly addressed.

Attitude is a tougher factor to assess and address. It refers to a stance, viewpoint, or state of mind toward a certain issue or situation. When having to shift people's attitudes, usually from a negative stance to a more positive one, evidence suggests that dialogic, interpersonal communication is more effective than media or monologic communication. Practice is the last part of this change ladder. It is about acting upon the decision to adopt the intended change. It should be kept in mind that this ladder of change is only one way, as behavior can also be adopted due to social pressure or other factors, not always easily explainable rationally.

In the open defecation example, the C4D strategy should start to focus on the behavior change element only at this point, while the components of previous dimensions were preparing the ground. In the first dimension, C4D can be used to advocate for ensuring adequate infrastructure, enabling policies and enough resources needed to facilitate the intended change. In the second dimension, C4D can still advocate to have relevant institutions and organizations supporting the actions needed to promote the change. Moreover, C4D experts should also strengthen capacities of these organizations/institutions to effectively use communication and participatory approaches to promote and engage people in the reasons and the need for the change. Finally, they should assist in devising C4D strategies conducive to the adoption of the new behaviors.

In the sociocultural dimension, the last of the "enabling" dimensions, the emphasis is on social norms. Social pressure, recognition, and inclination to conform to the expectations of the rest of the community are elements that have been often underestimated in behavior change. The basic structure of the MMC approach is reflected in the next table, which when used and compared among different groups of stakeholders compose a tool named "windows of perceptions" originally developed as part of the PRCA toolkit (Anyaebgunam et al. 2004). The initial table, named ideal situation mapping, is based on evidence and is expected to be used to map the situation defining the "best case scenario" that is as it should be, i.e., the ideal situation. This can be assessed through rigorous review of available literature eventually complemented by brainstorming sessions with key stakeholders. The other table (which is the same in its format) is filled with inputs from the situation in the field and by consultations with all relevant stakeholders. It is called "reality mapping" and by comparing it and contrasting it with the first table is expected to define the gaps and disjunctures between the situation as is intended to be from the way it is currently. These gaps will form the basis of the strategy leading to the intended change. It should be noted that in the case of the second table, it is not only factual reality that should be assessed but also stakeholders' perceptions of the situation, as perceptions are often the major impediments to achieve change.

## 30.3.1 Multidimensional Model for Change: Windows of Perceptions

See Table 1.

## 30.4 Conclusions

To summarize, let us briefly review the entire process, leading from the situation assessment to the planning of a C4D strategy, keeping in mind that such C4D strategy is part of the broader strategy. Such broader strategy to achieve the intended change can still be devised through the MMC. Thus, the first step is to fill as much as possible the ideal mapping window as per Table 1, which would then be compared and contrasted with the reality of the situation on the ground uncovered in the factual mapping. Hence, the actual review of the evidence-based assessment of the various components of each dimension begins, investigating the issues and questions presented in Table 1. After having reviewed and assessed to what degree adequate public policies, infrastructure, and resources are in place, we can move to the organizational dimension. Mandate and capacities of state and public institutions, as well as of NGOs and other no-profit organizations, should be reviewed and assessed. Private sector opportunities and alliance should also be considered.

In the sociocultural dimension, beliefs and norms of relevance should be thoroughly investigated and assessed to see if they are neutral, if they facilitate, or if they impede the intended change. Media and networks of reference need to undergo the same kind of scrutiny, to assess which can be of value and how they can be used effectively. Finally, in the individual dimension, the perceived benefits as understood by the various groups of stakeholders are assessed and compared. The next component to be investigated is the inner circle, to find out if some persons can be of use in promoting change, e.g., the mother-in-law. The last component is one directly addressing the behavior change at an individual level. Do individuals have the required knowledge about intended the change? And what is their attitude about it? Are they prone to adopt the new behavior? And if not, how can they be engaged in a dialogue about why is the intended change expected to improve life conditions?

The Multidimensional Model for Change has been already applied in practice, and it has shown that by identifying broader dimensions of analysis, it helps to have a more comprehensive and intersectorial approach. It also avoid neglecting major gaps and by comparing the ideal versus the actual it provides helpful insights needed to develop effective strategies, where communication for social and behavior change can be instrumental to achieve the intended results. The following is the summary of its pro and cons, starting with the former. First and foremost, the MMC comprehensive overview of the situation allows considering and assessing all major elements of a given situation, minimizing the risks faced by many past initiatives that failed due to their unilateral sectorial focus. It provides a sort of checklist with all key factors that should be considered when devising a strategy aimed to achieve social and/or behavior change. It also highlights gaps and weaknesses where the strategy needs to focus. The fact that some of these factors are not strictly related to the eventual C4D strategy is another strength, as new behaviors cannot be adopted without a conducive environment. For example, it does not make sense to have a highly appealing media campaign promoting the use of latrines, if there are not enough latrines to respond to the eventual growth to the demand for the use of latrines.

The fact that the MMC allows a broader oversight entailing different areas and sectors is another strength when developing a C4D strategies, since C4D is cross-sectorial and can be used not only across sectors but also for distinct functions, as illustrated clearly in the *Development Communication Sourcebook* (Mefalopulos 2008). The overall C4D strategy can, for instance, be used for advocacy purposes in advocating for certain policies, or to have more resources allocated, it could be used for capacity strengthening of certain institutions or for building alliances between public and private stakeholders, for instance. Naturally, it would be also expected to include media campaigns, participatory approaches, social mobilization, and other approaches with the aim to achieve the intended social and behavior change. The comparison and contrast of the ideal situation vs. the factual one provides valuable inputs in identifying gaps and critical points, thus making it easier where to focus the strategy and how to fit it in a proper timeframe.

Lastly, the MMC allows focusing the strategy on the specific behavior change framing it within the overall multidimensional context, and in this respect other factors become of relevance. The one about social norms is probably the most critical. There is increasing evidence on how social pressure and especially peer pressure are major determinants when deciding if adopting a new behavior, as illustrated effectively by Rosenberg (2011). Social norms have been not fully understood and applied in C4D strategies of the past, but they are now acquiring the relevance they deserve.

On the cons, it should be mentioned the challenges deriving from having to carry out a broader assessment at various levels. Prioritizing these challenges in terms of importance is not always easy, but devising a solid theory of change for the intended change should help in ranking priorities. Time needed to carry out a MMC assessment is another factor that can apparently be a disadvantage. However, the time spent in the initial assessment is recuperated with interest in term not only of time saving but also of planning effectiveness when developing the strategy.

A multidimensional approach, with a strong interdisciplinary connotation, is always challenging. Most of us have been brought up thinking in a linear way, and most academic careers aim at an in-depth specialization on a specific discipline or topic, gradually narrowing the path from broader knowledge to specific areas of specialization. It is only natural that we find challenging to adopt a broader multidimensional and intersectorial approach, having to assess and prioritize among several factors. However, as stated above, devising a sound theory of change helps to guide the design of a multidimensional strategy that can successfully achieve the intended change.

The apparent complexity of the MMC does not necessarily signify that strategies to achieve change must be complicated; on the contrary they could be rather simple and straightforward, once challenges and gaps are properly identified and agreed upon. However, it should be noted that simple solutions often require complex analysis. To ensure that such analysis is done in an effective and empowering manner, MMC should be carried out following the basic C4D principles, having at its core the use of dialogic (i.e., two-way) communication (Mefalopulos 2008), which ensures stakeholder participation throughout all the phases of the change initiative. Applied in this way, MMC has proven to minimize errors in planning the strategy due to misperceptions among stakeholders; errors of neglecting factors not directly linked to the behavior, but still necessary to have the behavior adopted; and errors due to a narrow focus that impedes broader approaches and alliances instrumental to the success of the strategy. In other words, MMC has proven to be an effective tool to develop and design even complex interventions aimed at social and behavior change.

MMC provides an overview of key dimensions constituting the environment of a given situation. Such multidimensional overview allows an interdisciplinary and exhaustive assessment, which provides insights and inputs for a better planning and related strategy, and not just for the C4D strategy. C4D methods and tools are part of the investigation and assessment process of the MMC, using dialogue to engage stakeholders and empower them in the decision-making process. It also ensures that the C4D strategy is closely aligned to the overall strategy, thus greatly enhancing its chances of success. In conclusion, the Multidimensional Model for Change is another powerful approach within the Communication for Development family. It is an approach that can greatly enhance at the same time the overall multidimensional strategy and the C4D strategy to achieve specific social and behavior change, while at the same time engaging and empowering stakeholders. Finally, the MMC provides another vision about the understating of a situation and planning for change: a vision that leaves behind older unilinear strategy design in favor of a multidimensional, cross-sectorial, and interdisciplinary design, taking into full account different realities and their perceptions by the various stakeholders' groups.

## References

Anyaebgunam C, Mefalopulos P, Moetsabi T (2004) Participatory rural communication appraisal: starting with the people, 2nd edn. FAO/SADC, Rome
Bicchieri C (2017) Norms in the wild. Oxford University Press, New York
Bronfenbrenner U (1979) The ecology of human development: experiments by nature and design. Harvard University Press, Cambridge, MA
Inagaki N (2007) Communicating the impact of communication for development. The World Bank, Washington, DC
McAnany EG (2012) Saving the world: a brief history of communication for development and social change. University of Illinois Press, Urbana
Mefalopulos P (2003) Theory and practice of participatory communication: the case of the FAO Project "Communication for Development in Southern Africa." Dissertation. Austin, TX: University of Texas at Austin.
Mefalopulos P (2008) Development communication sourcebook: broadening the boundaries of communication. The World Bank, Washington, DC
Mefalopulos P (2010) Why sound technical solutions are not enough: the role of communication in development initiatives. In the Oration at New Delhi, India: Lady Irwin College, University of Delhi, XX Nov 2010
Quarry W, Ramirez R (2009) Communication for another development: listening before telling. Zed Books, London
Rosenberg T (2011) How peer pressure can transform the world. W. W. Norton & Company, INC, New York
Servaes J (ed) (2003) Approaches to development: studies on communication for development. UNESCO, Paris
World Bank (2016) World development report 2015: mind, society and behavior. The World Bank, Washington, DC

# Broadcasting New Behavioral Norms: Theories Underlying the Entertainment-Education Method

**31**

Kriss Barker

## Contents

| | | |
|---|---|---|
| 31.1 | Introduction | 596 |
| 31.2 | How Does Entertainment-Education Work as a Mechanism for Social Change? | 596 |
| 31.3 | Shannon and Weaver (1949): Communication Model | 599 |
| 31.4 | "Rovigatti" (1981): A Circular Adaptation of Shannon and Weaver and Additional Circuit for a Social Content Serial Drama | 601 |
| 31.5 | Lazarsfeld (1968): Two-Step Flow of Communication | 603 |
| 31.6 | Bentley (1967): Dramatic Theory | 604 |
| 31.7 | Jung (1970): Theory of the Collective Unconscious | 605 |
| 31.8 | Bandura (1977): Social Learning Theory | 606 |
| 31.9 | Sabido (2002): Theory of the Tone | 608 |
| 31.10 | MacLean (1973): Theory of the Triune Brain and Sabido's Theory of the Tone (2002) | 609 |
| 31.11 | Horton and Wohl (1956): Parasocial Interaction | 610 |
| 31.12 | Sood (2002): Audience Involvement | 611 |
| 31.13 | Conclusions and Implications for Future Work | 611 |
| References | | 612 |

### Abstract

The effectiveness of entertainment-education (EE) has been demonstrated time and again in numerous settings to address myriad issues. However, it is still not clear exactly what contributions each of the different theories within the entertainment-education approach make to the achievement of such significant results. The current chapter reviews the theories underlying a specific approach to entertainment-education, referred to as the PMC methodology. The PMC methodology is based on the Sabido methodology, pioneered by Miguel Sabido of Mexico.

---

K. Barker (✉)
International Programs, Population Media Center, South Burlington, VT, USA
e-mail: krissbarker@populationmedia.org

© Springer Nature Singapore Pte Ltd. 2020
J. Servaes (ed.), *Handbook of Communication for Development and Social Change*,
https://doi.org/10.1007/978-981-15-2014-3_67

> **Keywords**
>
> Entertainment-education · Social and behavioral change communication · Behavior change · Theories

## 31.1 Introduction

> Anyone who tries to make a distinction between education and entertainment doesn't know the first thing about either. (Marshall McLuhan)

Entertainment-education (EE) is the process of purposely designing and implementing a media message to both entertain and educate, in order to increase audience members' knowledge about an issue, create favorable attitudes, shift social norms, and change the overt behavior of individuals and communities (Singhal and Rogers 2003, p. 289). The EE strategy in development communication bridges the gratuitous dichotomy that mass media programs must be either entertaining or educational but cannot possibly be both (Singhal and Rogers 1999; Televisa's Institute for Communication Research 1981).

Results show that entertainment-education using mass media channels is able to produce population-wide changes in social norms and individual behavior (Bertrand and Anhang 2006; Figueroa et al. 2001; Khalid and Ahmed 2014; Myers 2002; Nariman 1993; Piotrow et al. 1997; Rogers et al. 1999; Ryerson 2010; Shen and Han 2014; Singhal and Rogers 1999; Singhal et al. 2004; Vaughan et al. 2000a, b; Westoff and Bankole 1997). However, as Singhal and Rogers (2002) suggest: "Theoretical investigations of EE need to move past scholarly research on what effect EE programs have, to better understand how and why EE has these effects" (Singhal and Rogers 2002, p. 120). Papa et al. (2000) echo this observation: "...most studies of EE programs, with a few exceptions (Brown and Cody 1991; Lozano and Singhal 1993; Svenkerud et al. 1995; Storey et al. 1999; and perhaps some others) have not provided an adequate theoretical explanation of how audience members' change as a result of exposure to EE programs" (Papa et al. 2000, p. 33).

## 31.2 How Does Entertainment-Education Work as a Mechanism for Social Change?

This chapter will examine the theories underlying a specific form of EE, applied by Population Media Center (PMC). PMC's methodology is based on the Sabido methodology of behavior change communication via mass media, pioneered by Miguel Sabido of Mexico. To date, PMC has used our specific form of EE to design over 50 dramas in 54 countries worldwide, reaching over 500 million people over a 20-year history.

In order to understand how EE produces such significant behavior change results, we must explore how the different theories underlying EE programming contribute to attitudinal and behavior change. And, more broadly, how does EE work as a mechanism for social change?

We can best understand this question through the use of the following graphics, which show potential trajectories for results when the various theories are differentially applied. If the constellation is all of the theories, what happens when we apply only some of the theories, or apply all of the theories, but to a greater or lesser degree? (Fig. 1)

In the first graphic, all theories are equal. Thus, each incremental addition in application of any of the theories will result in an equal incremental addition in behavior change. The fewer the theories utilized, the poorer the results. In the same vein, the more theories that are applied will generate a proportional increase in impact. In this construct, when all of the principles are ideally applied, this is the purest form of the EE methodology. Of course, this utopian ideal could never be achieved in the real world (Fig. 2).

In the second figure, some theories have more effect than others. We could say that some theories are thus more "important" to achieving attitudinal or behavior change impact than are others. Without these key theories, the EE program will

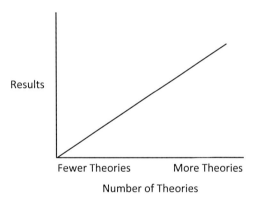

**Fig. 1** All theories are equal

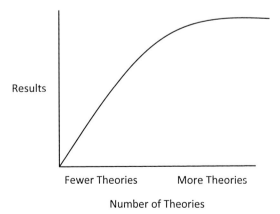

**Fig. 2** Some theories have more impact than others

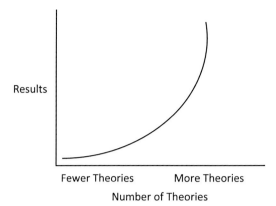

**Fig. 3** "All or nothing"

almost certainly fail. However, additional (although proportionally lesser) impact can be achieved through the addition of other, "less important" theories. That is, you can get some effect with the most basic theories, but the best effect is achieved through application of all of the theories.

For example, Bandura's social learning theory (Bandura 1977) of vicarious role modeling has often been identified as the key theory underlying the EE methodology. But, does that mean that an EE program can achieve equal impact by only applying the principles of vicarious role modeling while ignoring Bentley's dramatic theory (Bentley 1967)?

The third and final graphic contends that application of the theories is an "all or nothing" equation. Thus, in order to achieve any behavior change effect, most or all of the different theories must be rigorously employed in a format that is conducive to their synergistic application (Fig. 3).

The value of this knowledge lies in its potential to create a working model for "best practices" in the application of EE for creating behavior change at the individual and societal levels. Armed with a more complete understanding of the different roles that each of the foundational theories contributes to the effectiveness of EE, we can better design EE programs for the greatest prosocial impact.

As stated above, the current chapter reviews the theories underlying a specific approach to EE, referred to as the PMC methodology. The PMC methodology is based on the Sabido methodology, pioneered by Miguel Sabido of Mexico. Sabido was looking for a theoretical basis for the educational and behavioral effects he created by his *telenovelas*. As Vice President for Research at Televisa, he was in a unique position to explore the impact of application of different communication theories to entertainment-education television.

The Sabido methodology draws from the following theories of communication and behavior change (Nariman 1993):

- Shannon and Weaver's communication model (Shannon and Weaver 1949)
- Rovigatti's circular adaptation of Shannon and Weaver's communication model and additional circuit for a social content serial drama (Rovigatti 1981)

- Lazarsfeld's two-step flow of communication (Lazarsfeld et al. 1968)
- Bentley's dramatic theory (Bentley 1967)
- Jung's theory of archetypes and stereotypes and the collective unconscious (Jung 1970)
- The social learning theory of Albert Bandura (1977)
- Bandura's social cognitive theory (Bandura 1986)
- MacLean's concept of the triune brain (Note: MacLean's concept of the triune brain has been the subject of much recent debate. However, it is included in this list as it was an important concept in Sabido's original methodology.) (MacLean 1973), supplemented by Sabido's own Theory of the tone (Sabido 2002)

The PMC methodology adds to this list Horton and Wohl's concept of parasocial interaction (Horton and Wohl 1956) and the related concept of audience involvement (Sood 2002).

## 31.3 Shannon and Weaver (1949): Communication Model

Shannon and Weaver's communication model (1949) has five basic factors, arranged in a linear format. The basic model sets up a communicator, a message, a medium (channel), a receiver, and a response. These theories, which stem from Aristotle, were initially made explicit in the mathematical model proposed by Shannon and Weaver and later adopted by different authors.

The components in this model are (Shannon and Weaver 1949):

- The information source selects a desired message out of a set of possible messages.
- The transmitter changes the message into a signal that is sent over the communication channel to the receiver.
- The receiver is a sort of inverse transmitter, changing the transmitted signal back into a message and interpreting this message.
- This message is then sent to the destination. The destination may be another receiver (i.e., the message is passed on to someone else), or the message may rest with the initial receiver, and the transmission is achieved.

This model dates back to the Second World War. In the Second World War, communication was of critical importance – battles were won or lost depending on how effectively information was communicated and understood by armed forces on the ground. Shannon and Weaver were engineers, not communication scholars (Barker and Njogu 2010). They were working on communication systems to use on battle lines – thus, they used technical/engineering terms such as "source," "transmitter," "receiver," and "destination" (Shannon and Weaver 1949).

One of the key contributions of Shannon and Weaver's model was the concept of "noise" (Sabido, 10 May 2010, personal communication). Shannon and Weaver posit that, in the process of transmitting a message, certain information that was not intended by the information source is unavoidably added to the signal (or message). This "noise" can be internal (i.e., coming from the receiver's own knowledge, attitudes, or beliefs) or external (i.e., coming from other sources). Such internal or external "noise" can either strengthen the intended effect of a message (if the information confirms the message) or weaken the intended effect (if the information in the "noise" contradicts the original message) (Sabido, 10 May 2010, personal communication).

"Noise" can also be structural – thus, we also have to know if our target audience has access to the channel and medium that we have chosen to use. For example, if we are going to transmit our message via print media, then we need to know that our audience can read. If we are going to use mass media channels like radio or television, we need to know if our audience has radio sets or television sets and whether they listen or watch at the time when our program is being broadcast (Barker and Njogu 2010).

The idea of "noise" suggests that even if there is nothing wrong with the information, it may be that the information does not result in the intended effect because there is something in-between. This can be the lack of our audience's ability to capture the transmission (such as lack of a receiver such as a radio or television or the inability to read a printed/textual message) or lack of the ability to understand the message even if the information is physically received (e.g., if the program is in a language that our audience does not understand) (Barker and Njogu 2010).

Noise can also be cultural. For example, if a person believes that God or Allah or destiny has predetermined the number of children she will have, she is not going to be receptive to messages about family planning. In her view, she doesn't determine the number of children she is going to have – this has already been determined since long before she was born. Or, if a woman is not in control of the decision to use contraceptives (or doesn't perceive that she is in control) because this decision rests with her spouse or extended family, then transmitting information to her about family planning methods will not be effective – she simply does not feel that she controls that decision (Barker and Njogu 2010).

For Sabido, the biggest shortfall in the Shannon and Weaver model is the lack of feedback – there is no way for the transmitter to know if the message has been received and whether the information has been understood. In interpersonal (face-to-face) communication, there is feedback all of the time (nodding of the head, questioning facial expressions) (Sabido, 10 May 2010, personal communication).

In mass media programs (such as radio and television), we use audience feedback (focus groups, randomized telephone surveys, viewers'/listeners' groups, and/or calls and letters from audience members) to get feedback about the program – to see if our messages are being understood by the audience as we intended. To reflect this critical element of feedback, Sabido took Shannon and Weaver's linear model and formed it into a circle (Barker and Njogu 2010).

## 31.4 "Rovigatti" (1981): A Circular Adaptation of Shannon and Weaver and Additional Circuit for a Social Content Serial Drama

Sabido (inspired by the Italian "Rovigatti") (When Miguel Sabido was asked to present his methodology at an international conference on communication in Strasburg, France, in 1981, he was hesitant to use his own name for several of the key components of the nascent approach – so he invented an Italian scholar, Rovigatti, to whom he attributed these ideas, in case they would be shunned by the international academic community.) took the five basic factors in Shannon and Weaver's (1949) communication model (the communicator, message, medium, receiver, and noise) and arranged these factors in a circular model in which factors could interact directly with each other. He then applied this circuit to a commercial serial drama.

In the case of a commercial serial drama on television, the communicator is the manufacturer of a product, the message is "buy this product," the medium is the serial drama, the receiver is the consumer, and the response is the purchase of the product and television ratings (Televisa's Institute for Communication Research 1981) (see Fig. 4).

In the design of a social content serial drama, Sabido left the communication circuit of a commercial serial drama intact; however, he added a second communicator, a second message, a second receiver, and a second response (Fig. 5). These additions to the communication circuit did not impede the function of the first communicator, which is still the product manufacturer, as shown in the figure below (Televisa's Institute for Communication Research 1981).

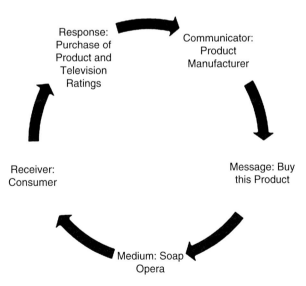

**Fig. 4** Circular adaptation of Shannon and Weaver's model of communication. (Source: Nariman 1993)

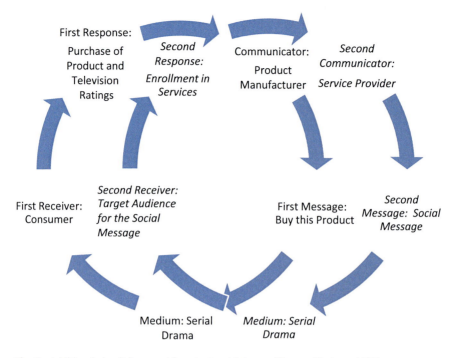

**Fig. 5** Additional circuit for a social content serial drama. (Source: Nariman 1993)

The serial drama continues to be the medium, and in the same way, the audience continues to be its customary receiver. The response, on the other hand, is no longer the sale of the product but also the change and/or reinforcement of opinions and attitudes and actual behavior related to the promoted value, in each individual person and, if possible, in the society (Televisa's Institute for Communication Research 1981).

Social problems can be detected where mass media can contribute to a solution by changing or reinforcing the knowledge, attitudes, or behavior of certain audiences. These solutions can be translated into television language so that, through handling specific theories of commercial television formats, prosocial content may be transmitted while keeping the audience's interest and the entertainment value of the program (Televisa's Institute for Communication Research 1981).

The two circles together are the "70/30" model to which Sabido refers to as a model for a social content serial drama. The 70% should be the commercial serial drama (the entertainment component), and the other 30% should be the educational component (the social component) (Sabido, 10 May 2010, personal communication).

With these modifications, the general process of social use of serial dramas on mass media was established. However, this model did not explain how values flow through society.

## 31.5 Lazarsfeld (1968): Two-Step Flow of Communication

The communication circuit for social content serial dramas established their general communication process, but Sabido still needed to explain how this process actually has an impact upon the society as a whole.

To describe how the social content serial drama works on changing behaviors within a society, Sabido turned to the two-step flow theory of communication described by the sociologist Paul F. Lazarsfeld, which states that values or messages communicated through mass media have the most impact upon a minority of receivers. These people will then communicate the value or message to others, hence a two-step flow. Therefore, although the direct effect of most mass media messages on behavior change is often modest, the indirect effects of the media in encouraging peer communication can be substantial (Lazarsfeld et al. 1968). This process of communication flow has been substantiated by more recent communication theories, most notably Rogers' diffusion of innovations (Rogers 1995).

Sabido distinguishes two types of audience and two types of effects (Televisa's Institute for Communication Research 1981):

1. First, there is the entire audience in which the value is reinforced. Examples of values to be reinforced in the general audience might be: "it is good that adults keep on studying," or "it is good to practice safer sex."
2. Then, there are those members of the audience who have to learn how to use a given service to help them solve a certain problem. For example, adult literacy classes will not be needed by university graduates; the family planning infrastructure will not be used by people who are not sexually active – but the value is reinforced even among these audience members, who in turn might advise others to access the services that are relevant to their needs.

Lazarsfeld's "two-step flow" theory explains the audience effects that Sabido intrinsically recognized. Michael J. Papa and colleagues summarized this effect: "…mass media alone seldom effect individual change, but they can stimulate conversations among listeners, which create opportunities for social learning as people, individually and collectively, consider new patterns of thought and behaviour" (Papa et al. 2000, p. 33).

Thus, conversations about the educational content of a media program can create a socially constructed learning environment in which people evaluate previously held ideas, consider options, and identify steps to initiate social change (Papa et al. 2000).

Rogers' diffusion of innovations further explains this phenomenon: Ideas and information conveyed through the mass media, instead of having a direct impact on the individuals within the audience, often have an impact first upon an "especially receptive" segment of the audience (Rogers 1995). This receptive segment – who Rogers termed "early adopters" – spread ideas and information through interpersonal channels to other individuals who were less interested and less receptive (Rogers 1995).

These secondary discussions are very important in explaining how Sabido-style serial dramas affect behavior change within a community. Based on his experience as a writer, producer, and director of *telenovelas* since 1963, Sabido states that "every *telenovela* is a discussion – both on and off the screen – about what is 'good' and what is 'bad.' Audience members often conduct discussions regarding important social issues with their peers that are similar to those of the characters' 'on the air.' This happens because the characters provide a model on how to discuss issues that are sensitive, or even taboo, and discussions between characters indicate a certain social acceptance of these issues" (Sabido, 10 May 2010, personal communication).

## 31.6 Bentley (1967): Dramatic Theory

Sabido's theoretical work was inspired by Aristotle's *Poetics* (Aristotle 1960) and by Bentley's (1967) theory of five theater genres: tragedy, comedy, tragicomedy, farce, and melodrama (Sabido 2004).

Among these genres, melodrama presents reality in a slightly exaggerated sense in which the moral universes of good and evil are in discord. Melodrama is an emotive genre that confronts (moral) behaviors in discord, emphasizing the anecdote and producing identification between the audience and certain characters (Bentley 1967).

Sabido, originally a dramatic theoretician himself, employed Bentley's structure of the melodrama genre as a basis from which to design characters and plots (Sabido, 10 May 2010, personal communication). "Good" characters in Sabido-style serial dramas accept the proposed social behavior, and "evil" characters reject it. Plots are then constructed around the relationships between good and evil characters as they move closer to or farther away from the proposed social behavior. Their actions encourage the audience to either champion or reject these characters accordingly (Televisa's Institute for Communication Research 1981).

Serial dramas can be defined as a "dramatized account, which confronts the universes of good and evil, through the characters and situation making up an anecdote" (Televisa's Institute for Communication Research 1981, p. 19). The anecdote is divided into several episodes, following the sequence of a story. In the anecdote, the "good" guys are rewarded and the "bad" guys are punished. In the end, the message is conveyed that "goodness" succeeds and "badness" perishes (Televisa's Institute for Communication Research 1981, p. 19).

Thus, melodrama is a moral reflection on what is good and bad for the individual and for society. One melodramatic experience will not change behavior, but it verified existence of certain values. Constant repetition of melodramatic experiences can lead to the reaffirmation of a specific value and/or to the modification of behavior within the social contract. For this reason, a serial drama, that is an iterative melodrama, is an appropriate vehicle to reaffirm the positive values of the social contract within the audience (Televisa's Institute for Communication Research 1981).

The tension between the good and evil characters evoked by the melodrama places the audience between the forces of good and evil. But, in a twist of the typical audience role in melodrama, where audience members simply watch or listen to the battle between good and evil, Sabido inserted the audience into the heart of the action – by representing audience members through a third group, one that is uncertain about the social behavior in question. These "uncertain" characters are intended to be those with whom the target audience most closely identifies. It is also these "transitional" characters who will guide the audience members through their own evolution toward adoption of desired behavior changes.

Although the three groups of characters in Sabido-style serial dramas are exaggerated as is the case in melodrama, they are modeled on real people within the target audience and the perceptions these people might have regarding the social value and behavior being presented. It is critical that the transitional characters, who are the role models for the audience with regard to the attitudinal or behavioral changes in question, be as similar to the target audience as possible, to allow for identification and parasocial interaction with these characters (Barker and Njogu 2010).

## 31.7 Jung (1970): Theory of the Collective Unconscious

The positive and negative characters in Sabido-style dramas are the poles representing the positive and negative values to be addressed in the drama. Jung's representations of archetypes and stereotypes, as described in his theory of the collective unconscious, are extremely useful in developing these positive and negative characters.

Jung's theory states that there are certain scripts or stories with familiar patterns and characters that people play out throughout history. These universal scripts or stories appear in myths, legends, and folktales around the world. Jung posited that these universal scripts or stories are the "archetypes of a collective unconscious" and share common characters such as "the wise old man," "the mother," and "the warrior." Jung further suggests that these archetypes are expressions of a primordial, collective unconscious shared by diverse cultures (Jung 1970).

Probably the most commonly known universal script is the "Cinderella" story. This story describes the rags-to-riches evolution of the archetypical "good" and "pure" Cinderella, abused at the hands of an "evil" stepmother and stepsisters. Aided by her fairy godmother, Cinderella finds her prince and ultimately lives "happily ever after." This is the basic storyline employed by the popular Peruvian *telenovela*, *Simplemente Maria*, which greatly influenced Sabido's work in Mexico (Singhal et al. 1994).

Serial drama characters that imitate a myth represent archetypes, while characters that imitate life represent stereotypes (Sabido 2004). Sabido used the archetypes described in Jung's theory as a basis for developing characters that embody universal psychological and physiological characteristics to address themes within the serial drama. Through these characters, the viewer finds an archetypical essence of himself or herself that interacts with the social message. Sabido portrayed these archetypes as positive or negative stereotypes, representing the societal norms of the target audience (Televisa's Institute for Communication Research 1981).

Sabido-style serial dramas rely on extensive formative research to identify the culture- or country-specific versions of these archetypes and to identify local archetypes that represent the prosocial values (or the antithesis of these values) that will be addressed in the serial drama. If the formative research upon which the serial drama is based is done properly, the scriptwriters will be able to develop archetypical characters with whom audience members will be able to identify. The formative research is used to develop a grid of positive and negative social values that these positive and negative characters will embody (Barker and Sabido 2005).

## 31.8 Bandura (1977): Social Learning Theory

In searching for a theoretical foundation on which to base his first prosocial serial drama, Sabido discovered the research of Stanford University psychologist Alfred Bandura. Bandura's social learning theory (Bandura 1977) reigns as one of the most influential theories in the development of the Sabido methodology (Sabido, 10 May 2010, personal communication). (Bandura's work on the effects of violence in mass media was key to the development of parental content advisories in television and film.) Sabido experimented with the application of social learning theory in developing his social content serial dramas while Vice President of Research at Television in the 1970s and 1980s. As Bandura stated, "My theories have been proven in a laboratory of 80 million people called Mexico" (Poindexter 2009, p. 120).

The major tenets of Sabido's methodology were finally beginning to coalesce. Shannon and Weaver's communication model (Shannon and Weaver 1949) (and Sabido's subsequent restructuring of the model) (Televisa's Institute for Communication Research 1981) outlines how the parts of communication systems are related. Sabido's additional circuit for a social content serial drama shows how serial drama can be used to achieve prosocial impacts (Televisa's Institute for Communication Research 1981). Lazarsfeld's two-step flow theory (Lazarsfeld 1968) describes how information flows through an audience. Bentley's dramatic theory (Bentley 1967) points out how to use melodrama in order to achieve specific effects upon the audience. Jung's theory of the collective unconscious (Jung 1970) directs development of the positive and negative characters. Lastly, Bandura's social learning theory (Bandura 1977) addresses the issue of how the viewers learn from TV serial dramas the models of behavior (values) portrayed in them.

Bandura's social learning theory has tended to dominate most theoretical writing and research about EE. A natural fit exists between Bandura's theory and EE interventions, which often seek to influence audience behavior change by providing positive and negative role models to the audience (Singhal 2004). Sabido often states that social learning theory is the cornerstone theory of his approach (Sabido, 10 May 2010, personal communication).

Sabido determined that three types of characters are fundamental to successful modeling by audience members (Barker and Sabido 2005). The first two types of characters are positive and negative role models, and they will not change during the course of the serial drama but are repeatedly rewarded or punished for their behaviors. As mentioned above, the consequences of these positive or negative behaviors must be directly linked to the behavior in question: for example, a truck driver character that is practicing at-risk sexual behavior should suffer from a sexually transmitted infection or even contract HIV, but should not be the victim of a traffic accident (Barker and Sabido 2005).

The third type of character, the "transitional character," is neither positive nor negative but somewhere in the middle. These transitional characters play the pivotal role in a Sabido-style serial drama and are designed to represent members of the target audience. The transitional characters' evolution toward the desired behavior is what the audience members will use to model their own behavior change (Barker and Sabido 2005).

It should be noted that the positive and negative characters are not intended to be direct role models for the audience – the positive and negative characters are role models for the transitional character. Although audience members are certainly influenced by observing the positive or negative consequences of the behaviors of the positive or negative characters, in a Sabido-style drama, these positive and negative behaviors are generally exaggerated examples of extreme behaviors and are thus more stereotypical than would be found in reality. In a Sabido-style drama, the positive and negative characters are role models for the transitional character within the drama itself – and the transitional character represents the role model for the audience. As such, during the course of the drama, the transitional character will exhibit positive or negative behaviors (following the influence of the positive or negative characters) and will be rewarded or punished accordingly (Barker and Njogu 2010).

For example, in Sabido's first social content serial drama, *Ven Conmigo* ("Come with Me") which addressed adult literacy, transitional characters were specifically chosen from specific subgroups (e.g., the elderly, young adults, homemakers) who represented the key target audiences for the national literacy campaign in Mexico. One of the main transitional characters was a grandfather who struggled to read the many letters he received from his favorite granddaughter. In a cathartic episode, he graduates from literacy training and is finally able to read his granddaughter's letters, albeit with teary eyes. In the year preceding the broadcast of *Ven Conmigo*, the national literacy campaign had registered 99,000 students. Following the broadcast of this episode (and the epilogue which provided information about registration in the literacy campaign), 250,000 people registered for literacy training. By the end of the serial drama, 840,000 people had registered for the literacy program – an increase of almost 750% from the preceding year (Ryerson 2010).

To motivate changes in behavior among the target audience, it is imperative that audience members not only identify with these transitional characters but empathize with these characters as they first experience the suffering that compels them to change negative behaviors and then struggle during the process of change. Thus, the grandfather in *Ven Conmigo* struggles against naysayers who disparage his efforts to learn to read by reasoning that the elderly cannot possibly learn to read and that his efforts to become literate are futile. The grandfather eventually surmounts this wall of cynicism and proves that, in fact, "old dogs can learn new tricks" (Barker and Sabido 2005).

As was demonstrated by this example, the evolution of the transitional characters must be gradual, or the audience will reject the change process as being unrealistic. If the characters' evolution is not gradual and fraught with obstacles, the audience will expect that their own progress toward positive change will be unrealistically rapid and facile (Ryerson 2010).

## 31.9 Sabido (2002): Theory of the Tone

The Sabido methodology is based on conveying a holistic message that is perceived by audience members on several levels of awareness. Sabido, who began his career as a theater director and dramatic theoretician, realized that acting means changing energy within the body to portray convincingly a symbol depicted by an actor/actress (Sabido 2004). Actors change their energies from one place to another in their bodies – that is, they change the locus of their energies. The producer/director uses various nonverbal theories of communication, including facial expressions, body language, lighting, music, sound effects, and tone of voice, to evoke different responses from the audience (Sabido 2004).

Sabido's theory of the tone describes how the various tones that are perceived by humans can be used in drama. In this theory, the producer/director serves almost the same function as an orchestra conductor, who can evoke different tones from each instrument in order to create various harmonies or tones within the body of the music and thereby inspire different moods among the audience. Although the theory is quite complex, it can be summarized by saying that for Sabido, the "tone" is the human communication form to which the receiver gives a tone according to his/her own genetic and acquired repertoire, thus making the "tone" the foundation of human communication. The theory has one main hypothesis: it is possible to change the tone of communication by hierarchically ordering its flow theories in a specific manner. According to Sabido, every human communication has a tone, similar to the one generated by actors on the stage (Sabido 2002).

As EE practitioners, we try to channel this energy: sometimes, we want to talk to people's brains, sometimes we want to talk to their hearts, and sometimes we want to elicit a very visceral reaction. Even with the same information, we might want to channel energy in different areas, to communicate in different ways to the audience (Barker and Njogu 2010).

For example, let us suppose you are watching television, and there is an advertisement for Coca-Cola. All of a sudden you are really thirsty. You weren't thirsty a few minutes ago, but all of a sudden you want something to drink. You don't even realize it, but that advertisement was talking to you in a very visceral way. A good advertisement will follow up with cognitive information about where you can buy a Coke. So, the advertisers have created a desire (thirst) – and they subsequently provide you with a means (intellectual information) that allows you to act on this desire. In a Sabido program, this cognitive information is provided through epilogues, which provide information about where to obtain services (Barker and Njogu 2010).

## 31.10 MacLean (1973): Theory of the Triune Brain and Sabido's Theory of the Tone (2002)

MacLean, in his theory of the triune brain (MacLean 1973), describes these zones physiologically and in terms of their function in the evolution of the human species:

1. The first zone is the reptilian zone of the brain and is common to all animal life – its purpose is self-preservation. It has four functions: feeding, fighting, fleeing, and fornicating.
2. The second zone of the brain is the paleo-mammalian brain. This zone is common to all mammals and is the source of most of memory. It is also the seat of emotions. As such, it is the primary residence of human values.
3. The third zone of the brain is the neo-mammalian brain (neocortex). MacLean posits that this zone is exclusive to the human race and is the center of human cognition.

When energy goes to the reptilian brain, we are blind victims of our drivers, our instincts. Through the area of innate behavior, we can reinforce archetypes and form stereotypes that can be acceptable and beneficial to the social body (Sabido 2002).

When energy goes to the emotional circuits of the mammalian brain, we are moved to identify with a figure (like a *telenovela* character) that we believe is similar to us, and we react emotionally. The emotional area, where emotional processes are located, is linked with communication strategies as well. Those who receive a primarily emotional message in content are more persuaded by it than those who receive a primarily logical message (Sabido 2002).

When energy goes to our third brain (neocortex), we process reality in an intellectual, calculating way (Sabido 2002).

The theory of the triune brain suggests that individuals process messages cognitively (thinking), affectively (emotional), and animalistically (physical) via three separate brain centers (the neocortex, visceral, and reptilian, respectively). MacLean suggested that the type of processing evoked by a message has a great deal to do with the outcome, such that messages processed via the reptilian portion of our brain trigger basic instinctual urges like hunger, need for sex, or aggressiveness. Messages processed viscerally produce intellectual and thoughtful responses (MacLean 1973).

Sabido adopted MacLean's theory and suggested that conventional health education programs failed because they focused on the cognitive process brain centers only (information dissemination). Therefore, Sabido developed serial dramas that prompted intellectual, emotional, and physical responses (Sabido 2002).

To the list of theories underlying the Sabido methodology of EE, PMC's method adds two additional theories: Horton and Wohl's (1956) notion of parasocial interaction and Sood's (2002) concept of audience involvement.

## 31.11 Horton and Wohl (1956): Parasocial Interaction

One of the key concepts in the PMC methodology is identification. It is important to form a relationship with the characters – this is possible in serial drama because the characters are created to be similar to audience members. This was one thing that Sabido learned from *Simplemente Maria*: "Viewers thought that Maria was real – they felt as if they had a relationship with her" (Sabido, 10 May 2010, personal communication).

This "relationship" with fictional characters in film, television, and radio is termed parasocial interaction. An example is when someone is watching a horror movie, and upon seeing that the heroine is about to be terrorized by a monster or villain behind her, the viewer yells out, "Watch out behind you!" The character in the movie obviously cannot hear this sage advice, but as a concerned viewer (and concerned "friend" of the person in the movie), you feel compelled to warn them anyway (Barker and Njogu 2010). Parasocial interaction is also evident in sporting matches – when spectators "advise" the players or coaches (or referees) about what to do to ensure successful play (Barker and Njogu 2010).

Horton and Wohl defined parasocial interaction as a quasi-interpersonal relationship between an audience member and a media personality (Horton and Wohl 1956). The audience member forms a relationship with a character that is analogous to real interpersonal relationships. Some audience members even talk to their favorite characters (i.e., to their TV or radio set) as if the characters were real people (Papa et al. 2000).

Katz et al. (1992) argued that parasocial interaction can prompt referential involvement on the part of audience members. Referential involvement is the degree to which an individual relates a media message to his or her personal experiences. Before audience members consider behavior change as a result of exposure to a media character, they must be able to relate the experiences of the character to their own lives. If such a connection cannot be made, behavior change is less likely (Katz et al. 1992).

Horton and Wohl summarize the important role of parasocial interaction in coping with social interactions on a daily basis: "The function of the mass media is the exemplification of the patterns of conduct one needs to understand and cope with in others as well as of those patterns which one must apply to one's self. Thus the spectator is instructed in the behaviors of the opposite sex, of people of higher and lower status, of people in particular occupations and professions. In a quantitative sense, by reason of the sheer volume of such instruction, this may be the most important aspect of the parasocial experience, if only because an individual's personal experiences are relatively few, while those of the others in his social worlds are very numerous. In this culture, it is evident that to be prepared to meet all the exigencies of a changing social situation, no matter how limited it may be, could – and often does – require a great stream of plays and stories, advice columns, and social how-to-do-it books. What, after all, is serial drama but an interminable exploration of the contingencies to be met with in 'home life'?" (Horton and Wohl 1956, p. 226).

Papa and colleagues (2000) contend that parasocial interaction with characters in EE programs may initiate a process of behavior change for certain audience members by influencing their thinking. Parasocial relationships can prompt role modeling as audience members carefully consider the behavioral choices made by media performers (Papa et al. 2000).

Parasocial interaction can also lead audience members to talk to one another about the media message to which they have been exposed (Rubin and Perse 1987). These conversations can create a social learning environment in which media consumers test new ideas that can spark behavior change. More specifically, a social learning environment exists when a group of people talk about a new behavior, persuade one another of the value of the new behavior, or articulate plans to execute it (Papa et al. 2000; Sood and Rogers 2000).

## 31.12 Sood (2002): Audience Involvement

Audience involvement is the degree to which audience members engage in reflection upon, and parasocial interaction with, certain media programs, thus resulting in overt behavior change. Sood (2002) characterizes audience involvement as being comprised of two dimensions: (1) affective-referential involvement and (2) cognitive-critical involvement. Sood thus describes audience involvement as a multidimensional phenomenon that serves as a mediator motivating behavior change, through increasing self-efficacy and collective efficacy and in promoting interpersonal communication among individuals in the audience. Sood states that audience involvement is a "key factor in the effectiveness of EE interventions" (Sood 2002, p. 154).

## 31.13 Conclusions and Implications for Future Work

This chapter seeks to explore what contribution each of the different theories that underlie the PMC EE method (i.e., the different theories described above) makes to the overall result.

Bandura's social learning theory (Bandura 1977) has often been touted as the foundational theory for all EE. Certainly, numerous findings confirm this premise. However, my work in developing over 50 social content serial dramas using the PMC method over almost 20 years shows that while effective use of social learning theory is necessary, it is not sufficient to achievement of behavior change. Another critical element is audience identification with the characters, specifically the transitional character (as described by Sood 2002 and Horton and Wohl 1956).

In writing this chapter, and reflecting on the question of the relative contribution of different theories to the success of EE in achieving behavior change, the concepts of "necessary" and "sufficient" often arose. In the introduction to this chapter, I propose three possible scenarios:

1. All of the theories have equal effect.
2. Some of the theories have more effect than others.
3. "All or nothing" – that is, all of the various theories must be used in order to achieve any effect.

So, what is the verdict? Overall, my findings show that, although some theories clearly have more impact than others, the best impact is achieved when the creative team brings an understanding of a multitude of behavioral theories and communication strategies to the design "table."

Thus, it appears that the response to the question "What contribution does each of the different theories make to the overall result?" is best represented by Fig. 2 from the introduction to this chapter.

In this depiction, some theories have more effect than others – without these key theories (such as Bandura's social learning theory), the EE program will most certainly fail. The inclusion of additional theories enhances the likelihood of success – with incrementally lesser impact for each added element.

The flaw within this diagram is that it assumes that there is a well-defined constellation of "all" of the theories that can possibly be applied in designing an EE program. Scholars and practitioners do not agree on a definitive list of the various theories that contribute to the success of EE programs, although various lists have been put forward. It seems that the number of theories is longer than earlier scholars may have assumed and will most certainly continue to grow as we learn more about how EE produces its effects.

So, the figure above is a misrepresentation of what is most likely: while the figure levels off and eventually terminates when application of the "entire" constellation of possible theories has been achieved, this would never be possible. In fact, the more theories that can be applied, the better the result (with diminishing – but additional – positive effects). To be completely accurate, the figure would show the number of theories going to infinity. While this concept may seem daunting to communication scholars and practitioners looking for a "magic bullet" to ensure the success of communication campaigns, I contend that this is actually a happy result. As we continue to learn more about what works, our communication programs will achieve even better effects.

## References

Aristotle (1960) The rhetoric of Aristotle (trans: Cooper L). New York
Bandura A (1977) Social learning theory. Prentice-Hall, Englewood Cliffs
Bandura A (1986) Social foundations of thought and action. Prentice-Hall, Englewood Cliffs
Barker K, Njogu K (2010) Scriptwriters' training workshop: PMC/Ethiopia. Addis Ababa
Barker K, Sabido M (eds) (2005) Soap operas for social change to prevent HIV/AIDS: a training guide for journalists and media personnel. Population Media Center (PMC) and United Nations Population Fund (UNFPA), Shelburne
Bentley E (1967) The life of the drama. Atheneum, New York
Bertrand JT, Anhang R (2006) The effectiveness of mass media in changing HIV/AIDS-related behavior among young people in developing countries. WHO technical report series. UNAIDS

Inter-agency Task Team on Young People, Geneva. Preventing HIV/AIDS in young people: a systematic review of the evidence from developing countries. Available http://www.unfpa.org/upload/lib_pub_file/633_filename_preventing.pdf

Brown WJ, Cody MJ (1991) Effects of pro-social television soap opera in promoting women's status. Hum Commun Res 18(1):114–142

Figueroa M, Bertrand JT, Kincaid DL (2001) Evaluating the impact of communication programs: Summary of an expert meeting organized by the MEASURE Evaluation Project and the Population Communication Services Project. University of North Carolina at Chapel Hill, Carolina Population Center, MEASURE Evaluation, Chapel Hill. Available http://www.cpc.unc.edu/measure/publications/pdf/ws-02-09.pdf

Horton D, Wohl R (1956) Mass communication and parasocial interaction: observations on intimacy at a distance. Psychiatry 19:215–229

Jung CG (1970) Archetypes and the collective unconscious. Editorial Paidos, Buenos Aires

Katz E, Liebes T, Berko L (1992) On commuting between television fiction and real life. Q Rev Film Video 14:157–178

Khalid MZ, Ahmed A (2014) Entertainment-education media strategies for social change: opportunities and emerging trends. Rev J Mass Commun 2(1):69–89

Lazarsfeld PF, Berelson B, Gaudet H (1968) The people's choice: how the voter makes up his mind in a presidential campaign, 3rd edn. Columbia University Press, New York

Lozano E, Singhal A (1993) Melodramatic television serials: mythical narratives for education. Commun Eur J Commun 18:115–127

MacLean PD (1973) A triune concept of the brain and behavior, including psychology of memory, sleep and dreaming. In: Kral VA et al (eds) Proceedings of the Ontario Mental Health Foundation meeting at Queen's University. University of Toronto Press, Toronto

Myers M (2002) Institutional review of educational radio dramas. Centers for Disease Control and Prevention, Atlanta. Available http://www.comminit.com/pdf/InstitutionalReviewofRadioDramas.pdf

Nariman H (1993) Soap operas for social change. Praeger, Westport

Papa MJ, Singhal A, Law S, Pant S, Sood S, Rogers EM, Shefner-Rogers C (2000) Entertainment education and social change: an analysis of parasocial interaction, social learning, collective efficacy, and paradoxical communication. J Commun 50:31–55

Piotrow PT, Kincaid DL, Rimon J II, Rinehart W (1997) Health communication: lessons from family planning and reproductive health. Praeger, Westport

Poindexter DO (2009) Out of the darkness of centuries: a forty-five year odyssey to discover the use of mass media for human betterment. Booksurge, Lexington

Rogers EM (1995) Diffusion of innovations, 4th edn. The Free Press, New York

Rogers EM, Vaughan P, Swalehe RMA, Rao N, Svenkerud P, Sood S (1999) Effects of an entertainment-education radio soap opera on family planning behavior in Tanzania. Stud Fam Plann 30(3):193–211

Rovigatti (aka Miguel Sabido) (1981) Cited in: Televisa's Institute of Communication Research. Toward the social use of soap operas. Paper presented at the International Institute of Communication, Strasbourg

Rubin AM, Perse EM (1987) Audience activity and soap opera involvement: a uses and effects investigation. Hum Commun Res 14:246–268

Ryerson WN (2010) The effectiveness of entertainment mass media in changing behavior. Unpublished manuscript, Population Media Center, Shelburne

Sabido M (2002) The tone, theoretical occurrences, and potential adventures and entertainment with social benefit. National Autonomous University of Mexico Press, Mexico City

Sabido M (2004) The origins of entertainment-education. In: Singhal A, Cody MJ, Rogers EM, Sabido M (eds) Entertainment-education and social change: history, research, and practice. Lawrence Erlbaum Associates, Mahwah, pp 61–74

Shannon CE, Weaver E (1949) The mathematical theory of communication. University of Illinois Press, Urbana

Shen F, Han J (2014) Effectiveness of entertainment education in communicating health information: a systematic review. Asian J Commun 24(6):605

Singhal A, Rogers EM (1999) Entertainment-education: a communication strategy for social change. Lawrence Erlbaum Associates, Mahwah
Singhal A, Rogers EM (2002) A theoretical agenda for entertainment-education. Commun Theory 12(2):117–135
Singhal A, Rogers EM (2003) Combating AIDS: communication strategies in action. Sage, New Delhi
Singhal A, Obregon R, Rogers EM (1994) Reconstructing the story of Simplemente Maria, the most popular telenovela in Latin America of all time. Gazette 54:1–15
Singhal A, Cody MJ, Rogers EM, Sabido M (2004) Entertainment-education and social change: history, research and practice. Lawrence Erlbaum Associates, Mahwah
Singhal A (2004) Entertainment-education through participatory theater: Freirean strategies for empowering the oppressed. In: Singhal A, Cody MJ, Rogers EM, Sabido M (eds) Entertainment-education and social change: history, research, and practice. Lawrence Earlbaum Associates, Mahwah, pp. 377–398
Sood S (2002) Audience involvement and entertainment-education. Commun Theory 12(2):153–172
Sood S, Rogers EM (2000) Dimensions of intense para-social interaction by letter-writers to a popular entertainment-education soap opera in India. J Broadcast Electron Media 44(3):386–414
Storey JD, Boulay M, Karki Y, Heckert K, Karmacharya DM (1999) Impact of the integrated Radio Communication Project in Nepal, 1994–1997. J Health Commun 4(4):271–294
Svenkerud P, Rahoi R, Singhal A (1995) Incorporating ambiguity and archetypes and change in entertainment-education programming. Lessons learned from Oshin. Gazette 55:147–168
Televisa's Institute for Communication Research (1981) Toward the social use of soap operas. Paper presented at the International Institute of Communication, Strasbourg
Vaughan PW, Regis A, Catherine ES (2000a) Effects of an entertainment-education radio soap opera on family planning and HIV prevention in St. Lucia. Int Fam Plan Perspect 26(4):148–157
Vaughan PW, Rogers EM, Singhal A, Swalehe RM (2000b) Entertainment-education and HIV/AIDS prevention: a field experiment in Tanzania. J Health Commun 5(Suppl):81–100
Westoff CF, Bankole A (1997) Mass media and reproductive behavior in Africa. Demographic and Health Surveys analytical reports no. 2. Macro International, Calverton. Available http://www.measuredhs.com/pubs/pdf/AR2/AR2.pdf

# Protest as Communication for Development and Social Change

**32**

Toks Dele Oyedemi

## Contents

| | | |
|---|---|---|
| 32.1 | The Prominence of Protest | 616 |
| 32.2 | Protest Culture and Social Change | 617 |
| 32.3 | Limitations and Criticisms of Protest Culture | 619 |
| 32.4 | Theoretical Approaches to Protest as Communication for Social Change | 621 |
| | 32.4.1 Peaceful Assembly | 621 |
| | 32.4.2 Protest as Communication | 622 |
| | 32.4.3 Participation, Power, and Social Change | 623 |
| | 32.4.4 Conceptualizing Social Movement and Collective Action | 624 |
| | 32.4.5 Theoretical Foundations of Social Movement | 625 |
| 32.5 | Protest as Communication for Social Change: A Framework of Guidelines | 627 |
| | 32.5.1 Planning Stage | 627 |
| | 32.5.2 At the Protest | 628 |
| | 32.5.3 Following-Up | 629 |
| 32.6 | Conclusion | 629 |
| 32.7 | Cross-References | 630 |
| References | | 630 |

### Abstract

Protest is an instrument used to agitate for social change globally, however, in spite of the increasing use of protest to address socioeconomic and political grievances, there has been limited critical scholarly study of protest in the communication for development and social change (CDSC) scholarship. Protest is communication. It is a communication strategy for drawing attention to developmental and social issues that affect the well-being of citizens. Communication for development and social change utilizes various communication

T. D. Oyedemi (✉)
Communication and Media Studies, University of Limpopo, Sovenga, South Africa
e-mail: toyedemi@gmail.com

© Springer Nature Singapore Pte Ltd. 2020
J. Servaes (ed.), *Handbook of Communication for Development and Social Change*,
https://doi.org/10.1007/978-981-15-2014-3_132

approaches, such as social marketing, public awareness and information campaign, entertainment-education, media advocacy, social mobilization, and many others. In addition to these approaches, protest action is situated as a communication approach for development and social change. In this chapter, protest culture is engaged critically by drawing theoretically from conceptual discourses of participation, participatory development communication, grassroots bottom-up social change, social movement theory, collective action theory, and critical analysis of power. This chapter also provides a framework for the use of protest as communication for development and social change by identifying types and methods of protest, and ends by offering critical guidelines of how protest can be effectively used as communication strategy for development and social change.

### Keywords

Protest · Mass mobilization · Protest as communication · Participation · Social movement · Collective action

## 32.1 The Prominence of Protest

In 411 BCE, Aristophanes drew a strong attention to the power of protest as a communication tool for social change. In the play, *Lysistrata,* Greek women, mobilized to protest against the Peloponnesian War. Through the withholding of physical intimacy from their husbands, they were able to forge peace and end the war. As frivolous as sex strike may seem, this form of protest has been adopted centuries after *Lysistrata* to mobilize for change. When Funmilayo Ransome-Kuti mobilized thousands of women in protest march against colonial taxation of women in colonial Nigeria, her leadership also highlighted the power of women and the use of protest for social change. From Ghandi's Salt *Satyagraha* against British colonial injunction, the 1963 march on Washington for freedom and economic justice, the 1956 women march to Pretoria against apartheid's laws, to the Treatment Action Campaign's marches for access to drugs for HIV-infected South Africans, protest has been a tool used to communicate grievances and forge social change.

Globally, protest has been a tool used in agitating for social change in economic, health, human rights, politics, and in gender-related issues. In spite of the rising use of protest to address social-economic and political grievances, there has been limited critical scholarly study of protest in the communication for development and social change (CDSC) scholarship. *Protest is communication*. It is a communication strategy used in drawing attention to developmental and social issues that affect the well-being of citizens. A study of protests in "World Protests 2006–2013" shows a worldwide increase in protest actions. Grievances leading to protest actions are related to issue of economic justice, such as unemployment and job creation and elimination of inequality, ensuring affordable food, energy, and housing (Burke 2014). Carothers and Youngs (2015) observe four critical characteristics of the current wave of global protest: First, protest is occurring in every region of the world, and it is affecting every major regime category, including authoritarian countries, semi-authoritarian states as well as democracies. Second, the era of transnational antiglobalization protest has given way to more localized protest triggered by local issues around economic concerns and political decisions. Third, there are long-term enabling causes such as the development in new information and communication technologies that facilitate protest in multiple ways.

There are also global economic trends in many parts of the developing world and post-Communist worlds that have led to a rise in the middle classes clamoring for change beyond material goals. Increasingly protest actions are not only concentrated among the poor but middle- and upper-middle classes have been involved in protest actions galvanized by social issues, unequal economic growth benefits, and stagnation of the middle classes. In addition, there is the growth of civil society organizations, professional associations, informal groups, religious organization, labor unions, village and local cooperatives, women's groups, students group, pro-democracy groups, and so forth who are leading protest against specific grievances. Fourth, Carothers and Youngs (2015) observe that protest are focused and directed at specific agendas. The ideas of "rebels without a cause" is inaccurate in describing protests, they are often anchored in specific aims, agendas and decidedly aimed at producing specific changes.

Many scholars (Rogers 1976; Servaes 1999, 2002; Melkote and Steeves 2015; Wilkins et al. 2014) have provided conceptual and practical characterizations of communication for development and social change. A summation of the many characterizations is that communication for development and social change is about the application of communication research, theory, approaches, and technologies to bring about improvement in human experience towards a sustainable and fulfilling human well-being. It involves the use of communication tools to influence behavioral and attitudinal change as well as to influence policy directions towards a goal of achieving quality human experiences in many areas of the economic, health, education, political, social, and cultural landscapes.

To this effect, actions are required to produce knowledge through research, planning, and to communicate such knowledge to the public and sites of power in order to achieve social change. As a process, communication for development and social change utilizes various communication approaches, such as social marketing, public awareness and information campaign, entertainment-education, media advocacy, information and communication technologies for development, social mobilization, etc. Various communication tools are used in these various approaches, such as broadcast and newsprint media (international, national and community based media), interpersonal communication, public communication, visual media (painting/photography), performances (community drama and theater), and new digital technology platforms (social media tools). Amidst many tools for communication for development and social change, protest action is situated as a communication approach.

## 32.2 Protest Culture and Social Change

Various types of protest action and mass mobilization (see Table 1) have been effective in drawing public and state attention to grievances from ordinary citizens and civil organizations from national to local community levels. Protest culture is an

**Table 1** Types and categories of protest

| Categories of protest | Types and examples |
|---|---|
| Processions and mass gatherings | Marches, silent processions, picket lines, and public events |
| Nonobedience, Noncooperation, and civil disobedience | Sitdowns/sit-ins, civil disobedience of "unjust" ordinances and laws, shutdowns, blockades, occupations, destruction of objectionable symbols and images, walkout, strike, boycott |
| Communication and media | Banners, posters, leaflets, public speeches, newspaper writings, opinion pieces, alternative press, recorded music, graphic arts, and clothing |
| Religious and sacred activities | Vigils, prayers, religious rallies, and spiritual practices |
| Written protest | Petitions, letters (mass letters to people in power), and memorandum of demands |
| Performances | Live music performance, public skits, dramatic performances, and dancing |
| The body | Nudity (e.g., women going topless to draw attention to women issue), inscriptions and slogans on the body, depriving, and hurting the body (e.g., hunger strike) |
| Technology and online protest | Online petitions, mass email to people in power, social media campaigns, mass mobilization through social networking sites and blogs, blocking access to websites, and texting (for coordination and mass texting as petitions to an individual in power) |

attitude and a form of expression that reflects the human traits and desires for reasonable well-being through motivation, agitation, and activities that draw attention of location of societal power to these desires and agitation. As such, protest bears inherent traits that are essential for development and social change:

- They are usually grassroots-oriented by reflecting participation of citizens at local levels who mobilize to make their grievances known.
- Protest and mass mobilization generally tend to be participatory in nature; participants are visibly engaging in protest actions.
- While the collective action of individuals is core element of mass action, individuals can equally single-handedly protest certain issues.
- The optics of mass protest usually draw attention of sites of power – at local, national, and international levels.
- Importantly, protests are emancipatory.
- They reflect human aspiration to maintain or achieve well-being consistent with humane existence.
- They are universal, not peculiar to certain regions or groups.

When *Time Magazine* named "The Protester" as Person of the Year 2011, it reiterated the power of protest and protesters in social change, that the action of individual(s) can incite protest that would lead to major political change, topple dictators, and engender global waves of dissent.

Although protest culture is universal and occurs in all social classes, it tends to occur more among the poor, the oppressed, the deprived, the disadvantaged, those that bear brunt of political and social conditions, and in regions with history and occurrences of political and socioeconomic inequalities, which may result in socioeconomic developmental challenges. These challenges are present in developed, as well as developing countries. Take the protest culture in South Africa as an example. South Africa experiences one of the most frequent rates of protest in the world, in fact some (cf. Runciman 2017) assume the country may be the protest capital of the world. Although there is no dedicated database for protest action in South Africa, there are speculations and debates that there are about 30 protests per day in South Africa (Williams 2009; Bhardwai 2017). Irrespective of the uncertainty of daily protest rate figure, it is uncontested that there is a strong protest culture in South Africa. But protest as tool for social change has always been essential to South Africa's pre and post apartheid eras. The struggle for freedom from the apartheid regime was emboldened by various protest actions and social movements (Solop 1990; Kurtz 2010; Brown and Yaffe 2014).

Many forms of protest actions were adopted by ordinary citizens and activists to fight the apartheid regime, these range from mass demonstration, funeral marches as platform for protest, music – singing and dancing, civil disobedience (such as Sharpville 1960), stay-aways, boycotts, etc. Often these protests were confronted by state violence through the use of the military and the police forces. Internationally, anti-apartheid movements in the UK and the USA were actively using many forms of protest to agitate for the end of apartheid. Take, for instance, the Non-Stop Picket at the South African Embassy in London where the City of London Anti-Apartheid Group maintained a continual presence every day and night at the South African embassy from 1986 until just after the release of Nelson Mandela in 1994 (Brown and Yaffe 2014).

The protest culture has also persisted in postapartheid South Africa. Irrespective of the huge achievements of the postapartheid government in addressing the socioeconomic inequalities and the provision of basic utilities like waters, electricity, and housing, there is still high poverty and unemployment rates, lack of access to municipal and state services, and elective representatives have become self-seeking. All these contribute to a general nature of protest in South Africa, characterized as the *rebellion of the poor* (Alexander 2010). Protest in South Africa is diverse and undertaken by mostly different urban groupings such as labor unions, the unemployed, poor shack dwellers, students, local communities, and ordinary residents with a largely focus on socioeconomic inequalities and justice (Nyar and Wray 2012). Additionally, many social issues around gender abuse and violence, poor state service delivery, access to health services, land and housing, water, electricity, sanitation/waste, and affordable education have been some of the drivers of protest actions.

## 32.3 Limitations and Criticisms of Protest Culture

While evidence abounds in many countries about how protest actions have drawn attention and awareness to social issues, and how authorities have reacted by addressing protesters' concerns at local and national levels, there are challenges, limitations, and questions regarding the efficacy of protest. There is the criticism that beyond visible demonstration of discontent, protests do not achieve much change. As Heller (2017) cautions, "protest is fine for digging in your heals. But work for change needs to be pragmatic and up-to-date" (p. 72). Heller provides practical cases to question the effectiveness of protest: first, the 2003 global protest against American prospective war in Iraq didn't stop the invasion of Iraq; about a decade after Occupy Wall Street protest, no US policies have changed as a result of the protest; in spite of the active Black Lives Matter protests, a majority of the law-enforcement officers were not indicted, of those that were, three were found guilty and only one to date has received a prison sentence; and the international Women's March following Donald Trump's election did not stop the inception of the new administration (Heller 2017).

Another criticism is that protests are not long-lasting or consistent tool of social change. They are ephemeral in nature, and as such they do not leave lasting impression, rather they are fleeting and ritualistic. Srnicek and Williams (2016) note that protest rise easily rapidly through mass mobilization of large people and then fade away to be replaced with a sense of apathy and another protest. They argue that protests have become a symbolic theatrical narrative with people and the police playing their respective roles.

Similar criticism also follows online activism, which is considered a low involvement form of protest. *Slacktivism* describes the low-involvement trend of people protesting or supporting a cause online, it is criticized for being a culture that creates a feel good feeling of being an activist through the clicking of a mouse rather than the hard and arduous work of physically getting involved through education, lobbying, negotiation, mobilizing, and planning. Although such online action tends to boost the morale of those actively participating in a protest and it has huge value in promoting a cause, being able to "like," "share," and "tweet" support for a cause is considered a low-cost involvement with an imagined sense of active engagement.

Another challenge to the use of protest for social change is the occurrence of violence. The destruction of public infrastructure and private properties during protest tends to distract from the genuine discontent that motivated the protest. While the use of violence through destruction of certain public infrastructure and objectionable artifact has been used effectively in pursuit of social change, the sight of protesters throwing bricks, shattering windows, setting public buildings such as libraries and schools on fire have created an image of protesters as unruly mob. But violence is not only perpetrated by protesters, security forces' unabashed use of violence in order to manage protest often tends to instigate violence response. Police use of violence create a huge challenge to protest actions and reflect the high handedness nature of the use of state power ranging from close range pepper spraying of students who refused to move during protest action in California, USA, to the sad and disturbing use of live ammunition against striking miners in South Africa killing 34 mineworkers and leaving 78 seriously injured.

State restriction could also be a challenge to the use of protest for change. Through legislative mechanisms, policies and regulatory frameworks states have developed guidelines to manage protest. These guidelines range from the process requesting permission to protest to requiring adherence to guidelines on what is and not permissible in protests. Take the Regulation of Gatherings Act in South Africa as an example, as a regulatory framework, this Act undeniably locates power to the state to manage protest actions. This generates genuine opposition that the state may use its power not only to avoid violent protest but also to determine the shape of and how people protest. For instance, Chapter 3 section 8 (7) of the Act states: "No person present shall at any gathering or demonstration wear a disguise or mask or any other apparel or item which obscures his facial features and prevents his identification." The state also has the power to impose limitations on the rights of assembly and to protest in cases of public security and use of violence. But state restrictions are equally attempts to balance citizens' rights to protest with national security, public order, health and safety, and the right of nonprotesting citizens.

In the following sections of this chapter, various conceptual framings and theoretical approaches to protest will be engaged; these include the legislative and state constitutional framing of protest, the conceptual assertion of protest as communication, theoretical issues of power and participation and their pragmatic implications for protest in social change activities, and the theoretical conceptualizations of social movement and collective action. The chapter ends by providing a working guideline for effective use of protest as communication for social change.

## 32.4 Theoretical Approaches to Protest as Communication for Social Change

### 32.4.1 Peaceful Assembly

Peaceful Assembly is global framework that describes the rights of citizens to protest. Underlining the concept of peaceful assembly is *freedom of assembly,* which provides legislative consent for citizens' right to protest. The right to freedom of peaceful assembly and association is a fundamental human right that is important in highlighting the power of citizens to participate collectively in shaping opinions, ideas, and processes in society. This concept is enshrined in the human rights declaration and constitutional framework of many states. For example, the article 20 (1) of the UN Declaration of Human Rights captures this right, and "the right of the people peaceably to assemble" is included in the First Amendment of the constitution of the United States. In the South African context, Section 17 of the Bill of Rights makes it explicitly clear: "Everyone has the right, peacefully and unarmed, to assemble, to demonstrate, to picket and to present petitions."

This right to assembly offers three essential qualities for democracies. First, it helps create space for collective politics. A collective voice has a higher chance of getting its message across than a singular voice, especially in systems where power is concentrated in few hands. Second, it permits citizen to meet and deliberate, which allows citizens to share ideas and talk to one another. Third, this right is essential for those who feel that their demands are not being given serious consideration by the state or institutional authorities (Woolman 1998). In addition to the constitutional framework, states may develop regulatory framework for protest. The regulation usually provides conditions and processes for organizing protest and the power of the state, in the most extreme cases, to prohibit a gathering.

Three important theoretical concepts shape the legislative framework of protest actions: assembly, association, and gathering. While gathering and assembly connote similar meaning, there is a difference between freedom of assembly and freedom of association. While freedom of assembly guarantees freedom to protest in public spaces, freedom of association provides the right to join a labor union, clubs, societies, group of people to assembly, or meet with people. As a result, the freedom of association makes the freedom of assembly more effective.

### 32.4.2 Protest as Communication

*Protest is communication.* This assertion is based on the empirical analysis that protest is a tool for communicating grievances and contestation. It is a visible way of communicating support or opposition to certain ideology or practice. Protest as communication can be analyzed from two perspectives. One, as already explained, protest is inherently a communication about certain grievance. Two, protest actions utilize various tools and genres of communication and forms of media in communicating discontent and grievance. *Protest is performance.* The performance may be action of a single person or collective action of individuals. As Ratliff and Hall (2014) note, "protesters, as actors, take the world stage as protagonist (or antagonists, given one's point of view) in performative dialectics, collectively struggling to control the framing of issues and opponents to foment social change, retain challenged power structures, and/or achieve social justice" (p. 270). As performance, *protest is an "art"* in two ways. Protest utilizes artistic practices, such as street drama, paintings, graphics, images, and many forms of artistic displays. Also, protest is metaphorically an art form that involves a skillful and strategic performance that uses various sensory techniques to invoke emotions for social change (Ratliff and Hall 2014).

The use of *music for protest* is a strong example of how protest can invoke people's emotion to "get up, stand up, stand up for your rights!". As Ruhlig (2016) notes, "songs are more than intellectual stimulating texts; they help to create *emotions* stimulating atmosphere of community and solidarity that very often draws on the power of utopia and stimulates people to dream" (p. 60). Globally, songs have been used in protest actions. Protest and socio-political songs aided African-Americans in the trajectory of historical fights for freedom (Trigg 2010). Freedom songs are historical in the USA from the abolitionist era to civil rights period. Freedom songs also known as struggle songs in South Africa are a form of oral arts that are nonfictitious but also performative in nature. They are a form of resistance and persuasion; they are used to resist the injustice of the apartheid system, as a method of self-persuasion, and to rouse others to grow indignant against an unjust system that oppressed them (Le Roux-Kemp 2014). Movement, procession, theater, sacred and religious performances, public speeches, and many symbolic and cultural activities are all elements of performativity and communicative practices that are used in protests.

*Forms of communication media are tools of protest.* Protesters use various forms of communication media and technologies such as broadcast and print media (community media, pamphlets, banners, posters, etc.). The revolution in digital technologies has provided new and advanced ways of using information and communication technologies for protest in two ways: one, ICT tools, such as mobile phones, the internet, social media and web platforms, are critical for coordinating protest actions. Two, these are actual spaces of protest actions, exemplified by virtual/online protest and social media activism.

### 32.4.3 Participation, Power, and Social Change

Participation is an essential theoretical concept relevant to analysis of protest. In democracies, participation is core to citizens' involvement in shaping the functioning of the political system. The theory of participatory democracy reveals that the representative institutions at the national level are not sufficient enough for effective democratic system; as such, citizen's participation in all levels is necessary. As Pateman (1970) notes, for a democratic polity to exist and function, it is paramount for a participatory society to exist. A participatory society is where citizens can participate in decision-making process at macro- and microlevels of society and be able to communicate and exercise their opinions and ideas. Pateman notes that if individuals are to exercise maximum control over their lives and their environment then authority structures must be organized in a way that individuals can participate in decision-making. Jensen et al. (2012) distinguish two forms of political participation: one is representative political participation exemplified by voting and political party activities and the other is extra-representative political participation such as active involvement in activism. Bean (1991) had equally asserted that mass political participation may also take the form of political protest.

*Protest is participation.* It is a form of participation that reflects the actions of individuals in speaking out about or against ideals and processes that have influence in shaping their well-being. It is a form of social involvement in decision-making about principles that shape their lives as citizens. Carpentier (2011) identifies minimalist and maximalist forms of political participation. While minimalist form tends to focus on the role of citizen to participate in electoral processes of representative delegation of power, maximalist participation is not limited to sphere of election of representatives, it is a broader and continuous form of participation. This resonates with Thomas' (1994) macroparticipation (relating to national political spheres of a nation) and microparticipation (relating to local spheres of school, family, workplace, religious formations, and local community). Carpentier (2012) also notes the essence of power in participation, this highlights the distribution of power in society and the power to include or exclude in decision-making. Power is exhibited in the complexities of the dominant political authority organs and in the counter-power of resistance by ordinary citizens (such as in protest).

The concept of participation and its relevance for social change influences the analytical and practical processes of communication for development and social change. The participatory model of development and social change emphasizes the liberating ideology that people should be active agents in decisions and processes for social change that impact their lives, that development involves the active "self" and strategies that stimulate people for action at the grassroots local level rather than a top-down, national, international dictates of what social change means to local agent. This analysis is dominant in the participatory model of development and social change; it is a counter discourse of a bottom-up approach rather than the top-down modernization paradigm of early development discourses and processes (Servaes 1999; Huesca 2008).

*Power* is the over-arching concept in the analytical discourse of participation and social change. To understand the concept of power, as Foucault (1982) suggests, is to comprehend power relations in society. For Foucault, to understand power, and power relations in society, it requires taking forms of resistance against different forms of authority, a form of antagonism to different forms of power. So, "in order to understand what power relations are about, perhaps we should investigate the forms of resistance and attempts made to dissociate these relations" (Foucault 1982, p. 780). For instance, to understand an aspect of the feminist protest will be to question (an opposition to) the power of men over women. Protest is often a form of resistance and oppositional strategies that highlights the power of the individual to participate in society. It provides voice to the people. *Protest is voice as power and participation.* "Voice" is multidimensional. Voice is power to project opinion; it is visibility; it is recognition and communication.

### 32.4.4 Conceptualizing Social Movement and Collective Action

From the introduction of voice as power, social movements can be understood as formations through which collective voice are projected for social change. Snow et al. (2004) offer a conceptual explanation that applies to the current discourse:

> Social movement are one of the principal social forms through which collectivities give voice to their grievances and concerns about rights, welfare, and well-being of themselves and others by engaging in various types of collective action, such as protesting in the street, that dramatize those grievances and concerns and demand that something be done about them.

Protest is thus one of the many tools of collective action that social movements use in demanding change. Snow et al. note further that some conceptual characteristics are identifiable in studying social movement, and these are: collective or joint action, change-oriented goals or claims, extra- or noninstitutional collective action, some degrees of organization, and some degree of temporal continuity. Social movement as a form of collective action overlaps but also is different from other sorts of collective action such as crowd behavior (in spectator events). While crowd action may be utilized by social movement, this is often a result of prior planning, organization, and negotiation.

Snow et al. (2004) also note that social movements in their collective action tactic overlaps with interest groups but are also different to some degrees. Interest groups are typically defined in relation to the government or state, social movement standing in relation to the government is different. Interest groups are embedded within the political arena and pursued their interest through institutionalized means such as lobbying and soliciting campaign donations. Social movements on the other hand are typically outside of the government or state with minimal recognition among political authorities, and they usually pursue their collective ends through protest, marches, boycotts, sit-in, etc. While social movements may occasionally operate within the political realm, their actions are mostly extra-institutional. The goal of social movement is largely to promote and/or resist change in some areas of human lives. Social movements and their activities are typically based on organization and coordination. Small organizing groups can band together to form a social movement for coordinated activities toward social change. They also operate with a flexible temporal variability, while some social movements may disband after achieving change in a selected area, others continue and move to another similar area for change (see Snow et al. 2004).

Social movement is a social process characterized, through collective action, by three distinct features. They are involved in conflictual collective action – that is, they operate on an oppositional relationship with actors who seek or are in control of political, economic, and cultural power in areas where social movement actors are engaged in to promote or oppose social change. Social movements have dense informal networks of individuals and organized groups in pursuit of common good, as a result no single individual can claim ownership of a social movement. Thirdly, the sense of common purpose and shared commitment to a cause leads to a collective identity (Porta and Diani 2006).

## 32.4.5 Theoretical Foundations of Social Movement

Various theoretical foundations shape the understanding of social movement: Resource Mobilization Theory (RMT) suggests access to resources is core to the formation and outcome of social movements. Social movements need critical resources and should be able to mobilize them in order to be successful. This theoretical approach suggests the presence of variety of human, social, political, financial assets, and organizational resources that must be mobilized in the pursuit of the goal of collective action of social movements (Tilly 1977; Jenkins 1983; Edwards and Giliham 2013). These "resources" may include labor, money, knowledge, time, skills, and social support from other organization, media, and political actors, among others.

*Deprivation theory* is premised on the idea that social movements are formed when certain people or groups in society are discontented about their feelings of being deprived of certain resources, services, or goods in society (Morrison 1971; Gurney and Tierney 1982). The sense of the experience of being disadvantaged could be based on absolute deprivation (group deprivation in isolation from the group's position in society), or relative deprivation (a group in a disadvantaged position in relation to some other groups in society (Sen and Avci 2016). *Political process theory* or *political opportunity theory* offers that the success of a social movement is dependent on the political atmosphere. In this sense, a politically positive or accommodating atmosphere must be present for a successful social movement. Developed in the study of civil rights movement in the USA, the core assumption of political process theory is that there must be a political and economic opportunities for change in order for a successful social movement (see McAdam 1982, 2013; Caren 2007). As Sen and Avci (2016) note, if a government's position is entrenched and it is also prone to repressive tendencies, the chance are high that social movement might fail, compared to when a government is more tolerant of dissenting behavior, which augurs well for the success of social movement.

The work of Smelser (1962) on value-added theory has been influential in adopting this theory (also known as *social strain theory*) to the understanding of social movement and collective behavior. Social strain theory argues that a problem may arise in the social system that causes some people to be frustrated and subsequently result to some actions, such as protest, to direct that anger. So a rebellion may develop from a group in society as a reaction to limitation they experience which hindered their pursuit of social goals. Women or people marginalized by discrimination may suffer strain due to the limitation that sexism, patriarchy, racism, or social inequalities caused in the pursuit of equal attainment of social and economic goals. Although originally this theory has been used to explain deviant behavior, such as crime, as a way to achieve certain social and economic goals due to strain that developed from people being unable to attain socioeconomic goals as a result of the social-economic structural system in society or the lack of means to attain these goals which cause strain on people. See Merton's (1938) seminal work on this.

Another theoretical approach to social movement is *New Social Movement Theories* (NSMT). The NSMT approach feature social theories that emerge in reaction to the reductionist nature of classical Marxist approach to collective action, which privileges "proletarian revolution" rooted in the economic reductionism of capitalist production and the marginalization of any other form of social protest (Buechler 1995). Rather than the Marxist economic determinism and class reductionism that see collective action as reaction to economic logic of capitalist production, and conflict of social classes based on economic ideology, while placing socio-cultural logics as secondary, NSMT elevates the social and cultural as basis of collective actions. This tends towards a postmodernist approach to social movement analysis. New social movement theorists locate social actions in other areas such as politics, ideology, culture, and identity as basis of much collective action. As such, varying formations of identity, such as ethnicity, gender, and sexuality are the definers of collective identity and action (Buechler 1995). Issues of human rights and individual rights become prominent in this analysis in contrast to traditional social movement theories that primarily relate to exploitation of one social class by another, usually on economic logic.

## 32.5 Protest as Communication for Social Change: A Framework of Guidelines

Protest actions that have maximum impact are those that are planned and executed well. Organizing an effective protest, march, demonstration, or rally for a social change agenda required careful attention to planning, execution and follow-ups. The following are some of the issues relevant to a successful planning and execution of a protest action.

### 32.5.1 Planning Stage

**Do you really need a protest?** Before deciding to protest, you need to ask if protest is the appropriate medium to communicate a particular grievance. Protest should never be the first action in communicating grievance. Try to communicate grievances, demands, objections, or support through communication and dialogue as a first step. Protest can be organized when other dialogic communicative methods have failed to yield result. However, in certain instances, a protest action can be the first method of action, for example:

- When a law is about to be signed, a government action is about to be implemented or an immediate action is about to take place that leaves no time for other methods of objections.
- If there is a need to make a big impression or visible impact to draw attention of the public or the media to your cause.

**Notification and permits**: There are different procedures involved in allowing a protest to occur in different countries. Check and make sure you notify the responsible state authority before proceeding on a protest. In many instances, the notification procedure is included in a policy or Act regulating protest. Also, keep police well informed and cooperate with them. Policies regulating protests usually provide information on how to notify, when to notify, the role of the protesters and the police, conducts, prohibition, etc.

**Leadership**: Protest actions need a group of coordinators that can form a leadership cohort. They will work together in planning, executing, and following-up processes of the protest. The coordinators will also represent the collective in deliberation and negotiation processes with authority who is the target of the protest action.

**Goals and tone**: Make sure the goal of the protest is clear, succinct, and memorable. It is also important not to combine too many issues into one single protest action. A single broad issue is more powerful when communicating your demands. Consider a theme for your protest. #FeesMustFall as a theme of student protest against high cost of tuition in South Africa is an effective use of a memorable theme and slogan. Whether the protest takes a solemn or upbeat tone depends on the goal of the protest.

**Mobilize support:** Ensure there is a widespread support for your issue. Seek endorsement of civic organization and NGOs working around issue of your protest. A prominent figure in society may also add some credibility to your protest.

**Media and communication strategy:** Contact the news media, provide press releases, create a website, a blog, social media platforms, flyers, posters, and placards to communicate your issue and information on the protest actions. Protest needs visibility and publicity, use all available publicity tools to your advantage, it also helps in garnering community support for your protest.

**Decide on venue**: Choose appropriate and safe venue where your protest can achieve maximum visibility with less disruption. You may lose community support if people perceive your protest to be disruptive of their everyday life. While this is usually unavoidable, the less disruptive the more you gain public support. If it is a march establish a route that provides maximum visibility. Choose appropriate venue for a rally. Students protesting high tuition cost at a shopping mall surely looks odd.

**Funding:** Protest and rallies cost money. Funding is usually difficult for protest activities, contact individuals and organizations interested in your issues. Solicit support for supply of equipment and volunteers for services that cost money.

**Logistics**: Pay serious attention to every small detail, such as:

- *Transportation:* How easy to get people to the venue and to disperse.
- *Time and date:* Appropriate weather condition is important. You do not want to get protesters drenched in rain.
- *Competing publicity:* Make sure there are no conflicting public events in day of your protest, unless your protest is a counter protest to such event.
- *Entertainment:* Music, skit, performances, etc. to keep protesters in high morale.
- *Medical:* Provide first aid supplies, designate transportation to the hospital for emergency cases.
- *Equipment:* Public address system. Depending on the form of protest, it may be mounted in a fixed spot or mounted on a moving vehicle.
- *Refreshment:* Depending on funds, make plans for refreshments, such as water.

### 32.5.2 At the Protest

**No violence**: Protest actions with violence generate negative publicity, police high-handedness and injuries that could be avoided. Students in South Africa have burned building, destroyed lecture theatres, set laboratory on fire, and destroy libraries while protesting lack of resources on university campuses. Choosing appropriate method of protest is paramount to a successful protest action without violence. For example, considering that universities function when students attend classes, students embarking on a sit-in or civil disobedience of nonattendance of classes will have more positive impact than destruction of properties. But there are instances where some forms of violence become necessary part of a protest. For example, when protesters destruct objectionable artifact that represent the motive of the protest action.

**Speeches and performances**: Line up people who would speak at the events and other performances that keep people's attention and boost their morale.

**Ensure communication during protest**: Provide live updates on social media platforms, and blog updates as events are taking place for maximum publicity. Provide access to the news media during protest event.

**Marshals**: Identify marshals who are part of the protest and who will monitor that everyone behaves accordingly while avoiding violence and confrontations.

**Memorandum of demands and petitions:** Have a memorandum of demands that representative of authority or institutions you are protesting against can sign to acknowledge your demands with a promise to address them.

### 32.5.3 Following-Up

**Sustain the energy:** After the protest, there is a need to continue the drive for the issue you protest for/against. Keep pressures on the authority or institution that are the target of the protest.

**Continue media exposure**: Through traditional media and on social media platforms, publicize your success.

**Monitor progress**: Follow-up with intended audience or target of the protest to monitor reaction and progress and keep the protesters informed of the progress.

## 32.6 Conclusion

In this chapter, protest action is centered as a global strategy in the pursuit of development and social change. In the scholarship of communication for development and social change, the use of protest does not feature prominently as a communication tool. The attempt here is to assert protest as communication, and more importantly, to locate protest and mass mobilization as relevant components of communication tools for development and social change.

Protest is a very old approach to show discontent, but it has often been limited to the political sphere. Asserting protest as communication helps to situate it as a viable way, not only to show discontent, but a strategy of giving visibility to causes, promoting certain ideals, drawing attention to various social challenges to the human condition, and demanding that sites of power pay attention. The inherent characteristics of protest, as visible manifestation of citizens' constitutional right to assembly, as communication, as empowering form of collective action and participation, as grassroots bottom-up mechanism for advocating for change, and a form of citizen's power to demand change, make protest a critical tool of communication for development and social change. While acknowledging challenges and limitations of protest actions, there are many instances where protests have yielded results, some immediate and others after long and persistent actions. These results are evidenced in political decisions, economic actions, and improvement to citizens' welfare ranging from changes in political order, access to health services to affordable access to education.

But success of protest action tends to be influenced by political and economic atmosphere. Although they are conditions for protest actions, entrenched state political position in politically repressive regimes and the seemingly embedded nature of neoliberal economic agenda and the naturalized attitude towards exploitative capitalism tend to diminish the success of protest actions. Irrespectively, protest allow members of communities at macro- and microlevels to mobilize for collective action, create visibility for their grievances, demand responses, provoke behavioral change, and request policy formulation in pursuit of a better well-being. All these describe the core essence of communication for development and social change.

## 32.7 Cross-References

▶ Development Communication in South Africa
▶ Empowerment as Development: An Outline of an Analytical Concept for the Study of ICTs in the Global South
▶ Health Communication: Approaches, Strategies, and Ways to Sustainability on Health or Health for All
▶ Importing Innovation? Culture and Politics of Education in Creative Industries, Case Kenya
▶ Millennium Development Goals (MDGs) and Maternal Health in Africa
▶ Protest as Communication for Development and Social Change in South Africa

## References

Alexander P (2010) Rebellion of the poor: South Africa's service delivery protests – a preliminary analysis. Rev Afr Polit Econ 37(123):25–40
Bean C (1991) Participation and political protest: a causal model with Australian evidence. Polit Behav 13(3):253–283
Bhardwai V (Africa Check) (2017) Are there 30 service delivery protests a day in South Africa? Africa Check. Retrieved 8 Oct 2017 from https://africacheck.org/reports/are-there-30-service-delivery-protests-a-day-in-south-africa-2/
Brown G, Yaffe H (2014) Practices of solidarity: opposing apartheid in the centre of London. Antipode 46(1):34–52
Buechler SM (1995) New social movement theories. Sociol Q 36(3):441–464
Burke S (2014) What an era of global protest says about the effectiveness of human rights as a language to achieve social change. SUR: Int J Hum Rights 20:27–33

Carpentier N (2011) Media and participation: A site of ideological-democratic struggle. Intellect, Bristol

Carpentier N (2012) The concept of participation. If they have access and interact, do they really participate? Revista Fronteiras – estudos midiaticos 14(2):164–177

Caren N (2007). Political process theory. In: Ritzer G (ed) Blackwell encyclopedia of sociology. http://www.sociologyencyclopedia.com/public/

Carothers T, Youngs R (2015) The complexities of global protest. Carnegie Endowment for International Peace, Washington, DC

Edwards B, Giliham P (2013) Resource mobilization theory. In: Snow D, Porta D, Klandermans B, McAdam D (eds) The Wiley-Blackwell encyclopedia of social and political movements. Blackwell Publishing Ltd., Oxford

Foucault M (1982) The subject and power. Crit Inq 8(4):777–795

Gurney JN, Tierney KJ (1982) Relative deprivation and social movement: a critical look at the twenty years of theory and research. Sociol Q 23:33–47

Heller N (2017, August 21) Is there any point to protesting? The New Yorker. Retrieved 26 Oct 2017. https://www.newyorker.com/magazine/2017/08/21/is-there-any-point-to-protesting

Huesca R (2008) Tracing the history of participatory communication approaches to development: a critical appraisal. In: Servaes J (ed) Communication for development and social change. Sage, Thousands Oaks

Jenkins JC (1983) Resource mobilization theory and the study of social movements. Annu Rev Sociol 9:527–553

Jensen M, Jorba L, Anduiza E (2012) Introduction. In: Anduiza E (ed) Digital media and political engagement worldwide: a comparative study. Cambridge University Press, New York

Kurtz LR (2010) The anti-apartheid struggle in South Africa. International Center on Nonviolent Conflict. Retrieved 8 Oct 2017 from https://www.nonviolent-conflict.org/the-anti-apartheid-struggle-in-south-africa-1912-1992/

Le Roux-Kemp A (2014) Struggle music: south African politics in song. Law Humanit 8(2):247–268

McAdam D (1982) Political process and the development of black insurgency, 1930–1970. University of Chicago Press, Chicago

McAdam D (2013) Political process theory. The Wiley-Blackwell encyclopedia of social and political movements. Retrieved 19 October 2017 from http://onlinelibrary.wiley.com/doi/10.1002/9780470674871.wbespm160/full

Melkote SR, Steeves HL (2015) Communication for development: theory and practice for empowerment and social justice. Sage, Thousand Oaks

Merton R (1938) Social structure and anomie. Am Sociol Rev 3(5):672–682

Morrison DE (1971) Some notes toward theory on relative deprivation, social movements, and social change. Am Behav Sci 14(5):675–690

Nyar A, Wray C (2012) Understanding protest action: some data collection challenges for South Africa. Transformation 80:22–43

Pateman C (1970) Participation and democratic theory. Cambridge University Press, New York

Porta DD, Diani M (2006) Social movement; an introduction. Blackwell, Malden

Ratliff TN, Hall LL (2014) Practicing the art of dissent: toward a typology of protest activity in the United States. Humanit Soc 38(3):268–294

Rogers E (ed) (1976) Communication and development: critical perspectives. Sage Publications, Beverly Hills

Runciman C (2017, 22 May) SA is the protest capital of the world. Pretoria News. Retrieved 26 Oct 2017 from https://www.iol.co.za/pretoria-news/sa-is-protest-capital-of-the-world-9279206

Ruhlig T (2016) 'Do you hear the people sing' 'Lift your umbrella: Understanding Hong Kong's pro-democratic umbrella movement through YouTube music videos. China Perspective 2016/4

Sen A, Avci O (2016) Why social movements occur: theories of social movements. Bilgi Ekonomisi ve Yönetimi Dergisi 11(1):125–130

Servaes J (1999) Communication for development. one world, multiple cultures. Hampton Press, Cresskill

Servaes J (2002) Approaches to development communication. UNESCO, Paris Available http://www.unesco.org/fileadmin/MULTIMEDIA/HQ/CI/CI/pdf/approaches_to_development_communication.pdf

Smelser N (1962) Theory of collective behavior. The Free Press, New York

Snow DA, Soule SA, Kriesi H (2004) Mapping the terrain. In: Snow DA, Soule SA, Kriesi H (eds) The Blackwell companion to social movements. Blackwell, Malden

Solop FI (1990) Public protest and public policy: the anti-apartheid movement and political innovation. Policy Stud Rev 9(2):307–326

Srnicek N, Williams A (2016) Inventing the future: postcapitalism and a world without work. Verso, Brooklyn

Thomas P (1994) Participatory development communication: philosophical premises. In: White SA, Nair KS, Ascroft J (eds) Participatory communication: working for change and development. Sage, New Delhi

Tilly C (1977) From mobilization to revolution. University of Michigan, Ann Arbor

Trigg C (2010) A change ain't gonna gome: Sam Cooke and the protest song. Univ Tor Q 79(3):992–1003

Wilkins KG, Tufte T, Obregon R (2014) The handbook of development communication and social change. Wiley Blackwell, Malden

Williams JJ (2009) The everyday at grassroots level: poverty, protest and social change in post-apartheid South Africa. Clacso working papers. Retrieved 8 Oct 2017 from http://hdl.handle.net/10566/2036

Woolman S (1998) Freedom of assembly. Retrieved 12 Oct 2017 from http://www.chr.up.ac.za/chr_old/centre_publications/constitlaw/pdf/21-Freedom%20of%20Assembly.pdf

# Political Engagement of Individuals in the Digital Age

## 33

Paul Clemens Murschetz

## Contents

| | | |
|---|---|---|
| 33.1 | Introduction | 634 |
| 33.2 | Civic Engagement in the Digital Age | 636 |
| 33.3 | The Sociology of Engagement | 637 |
| 33.4 | Mediatized Engagement | 640 |
| 33.5 | Future Directions | 642 |
| 33.6 | Cross-References | 643 |
| References | | 644 |

### Abstract

The use of digital media is becoming a prominent feature for individuals engaged as citizen in e-participation processes.

However, while research into e-participation at large has attracted much scholarly attention from various disciplines over the years, the fundamental question of why individuals engage in e-participation and how digital media triggers their engagement in processes of political engagement in the public realm is largely unaddressed.

This chapter reviews two conceptual frameworks for explaining individual engagement in politics: Firstly, the sociological theory of political engagement as promulgated by Laurent Thévenot's work which explores why human agency, rather than social structures, determines the engagement of individuals in the political process and, secondly, the "mediatization" concept as discussed in media and communication studies whereby digital media "mediatize" engagement by shaping and framing the processes of interaction of political communication among citizens and between citizens and government.

P. C. Murschetz (✉)
Faculty of Digital Communication, Berlin University of Digital Sciences, Berlin, Germany
e-mail: murschetz@berlin-university.digital

© Springer Nature Singapore Pte Ltd. 2020
J. Servaes (ed.), *Handbook of Communication for Development and Social Change*,
https://doi.org/10.1007/978-981-15-2014-3_85

> **Keywords**
>
> Agency · Civic engagement · Individual interest · Mediatization · Pragmatic sociology · Social movement · Sociology of engagement

## 33.1 Introduction

Engaged individuals become citizens by using digital media and ICTs that enable their participation in the political communication and interaction process. These digital media help integrating bottom-up decision-making processes, allowing for better informed decisions by all stakeholders in the political communication process. Technically, they support the further development of e-participation in modern democracies.

However, while research into e-participation has attracted much scholarly attention from various disciplines over the years, the fundamental question of why individuals engage in e-participation and how digital media triggers their engagement in processes of political engagement in the public realm is largely unaddressed.

This chapter supports the idea that individual-level models of civic engagement are a logical starting point for understanding why and how individuals participate in the political process in the digital age. As a corollary, investigating how individuals are doing politics would reconcile limited and contradictory findings in research on the concept of "civic engagement" in the digital era. These result from a variety of deficiencies such as:

- Inconsistent definitions as to the subject and scope of the concept of "civic engagement" (e.g., what factors influence the "engagement" of individuals, particularly in new modes of e-participation; what degree of knowledge is required to be competent and effectively engaged, and is there any evidence that this would lead to wider and more sustainable political engagement; Ekman and Amnå 2012, Freeman and Quirke 2013, Macintosh 2004, Mossberger et al. 2008)
- Different units and levels of analysis in identifying the factors predicting civic engagement (e.g., do they predominantly reside in the interests, motives, and preferences of the individuals themselves, that is, in "human agency" understood as the capability of individuals or groups to make free decisions or act, as against the "structure" defined as a patterned influence or limitation derived from rules and resources available to individuals or group actions; Giddens 1984)
- Disagreement on the role of communication processes between individual citizens; governmental (politicians, political parties, ministers, parliamentarians, etc.) and non-governmental stakeholders (NGOs, political lobbies, local communities, etc.) (Dahlgren 2005; Kreiss 2015); political institutions establishing, legitimizing, routinizing, or regularizing certain e-participation practices (Teorell et al. 2007); and citizens' perceived control over the engagement process and outcome (Zimmerman and Rappaport 1998)

- A lack of appreciation of the role of "mediatization" and its role in transforming individual civic engagement into collective action (Collins et al. 2014; Couldry 2014; Dahlgren 2009), given that the change of media communication has fundamental impacts on sociocultural change as part of our everyday communication practices and our communicative construction of reality (Hepp et al. 2015; Hjarvard 2013; Kramp et al. 2016; Hjarvard 2013)
- A largely incoherent base of theory for examining relations between the parameters mentioned above, particularly with regard to civic engagement and e-participation from the point of view of individual engagement and the links that work between different levels of engagement, that is, the "macro-level" of the political and situational environment, the "meso-level" of organizations facilitating political participation, and the "microlevel" of individual citizens who wish to become more or less actively engaged, and their values, motives, attitudes, preferences, and behaviors, respectively (Quintelier and Hooghe 2012)

Hence, the problems are manifold, but the most pressing seems to be the nature of the concept of "civic engagement" and its relationship to "e-participation" in the digital age which has remained complex, multidisciplinary, and difficult to operationalize.

Following this introduction, Sect. 33.2 will explore some more fundamental issues of civic engagement in the digital age, including the presentation of research claims. Section 33.3 will then provide two conceptual frameworks for explaining individual engagement in the e-participation domain. Firstly, some key ideas of Laurent Thévenot's pragmatic sociology of political engagement are described. In doing so, the epistemological value of this part of Thévenot's œuvre is queried to gain new insights into how his thinking could act as a new framework for explaining the dynamic change that is currently blurring boundaries between the individual, the media, culture, and society at large. While Habermas' concept of a "deliberative democracy" emphasizes that public *will* formation is enabled through intersubjective communications in the public sphere so that public interests are commonly created through deliberation rather than aggregated and voted upon, Thévenot's work is an ambitious conceptual attempt to a pragmatic theory of plural "worlds" and "polities" where individuals' interest in the common good plays a constitutional role in how social life is organized and conflicts are resolved. Seen this way, individuals create public spaces for and by themselves, to help their self-development and sensitivity for involvement in political (and sociocultural) issues of the public sphere and to support their political ambitions, learning, knowledge creation, and sharing as well as improving their own decision-making competencies. Section 33.4 then reviews "mediatization" as a theoretical concept from the point of view of media and communication studies and discusses some selected ideas on the issue of "mediatized engagement." Mediatization seems to be constitutional in ways in which (digital and other) media are embedded in processes of civic engagement and e-participation whereby digital media shape and frame the processes and conversation of political communication among citizens and between citizens and government. Section 33.5 sums up key insights gained from research into mediatized engagement and further investigates how a political sociology of engagement, as provided by Thévenot's "grammar of the individual in a liberal public," can deliver a solid theoretical base for analyzing mediatized engagement more profoundly. This means that these insights shall open the way to new departures for debates on how individuals' civic engagement may eventually fertilize into social structures of common political interest. Eventually, this happens when private cognitive models of perception and thinking about various issues of political concern are transformed into specific forms of political action, transformation processes which may well be mediated by digital media, ICTs, and other interactive technologies.

## 33.2 Civic Engagement in the Digital Age

Certainly, the deficiencies mentioned above have mobilized research into several analytical frameworks, which themselves have drawn on different disciplinary roots, such as economics, politics, history, and sociology. This cross-fertilization of ideas has further provided a diversity of intellectual activity and research, with a vigorous program of detailed empirical studies.

However, while research into e-participation has attracted remarkable scholarly attention from these various disciplinary perspectives in recent years, the fundamental question of why individuals engage in e-participation processes as a typical form of civic engagement in political communication and how digital media triggers (or hinders) their engagement in processes of political engagement in the public realm is largely unaddressed.

To answer these questions, the following three research claims are raised: Firstly, it is noteworthy to stress the importance of intensifying research into individuals as political agents with a view to understanding its implications for political argumentation. This is inevitable if we want to understand the emergent digital culture of e-participation in contemporary politics in all its forms, ranging from e-voting to several forms of ICT-supported and ICT-enabled interaction between governments and citizens, including not only direct ones such as consultations, lobbying, and polling but also ones pursued outside of government itself, including campaigning, signing a petition, writing a letter to a public official, blogging about a political issue, donating money to a cause, volunteering for a campaign, joining an activist or interest group, holding a public official position, occupying a building in an act of protest, or committing an illegal or even terrorist act.

Secondly, research needs to draw from disparate academic fields such as media and communication studies and political sociology and thereby systematize, link, and extend familiar definitions, characteristics, types, and dimensions of engagement from extant literature in these fields.

And, thirdly, civic engagement in the digital age is further challenged by the assumption that both the technical effectiveness and moral virtue of conventional forms of civic engagement are called into question by institutional and intellectual transformations that push inexorably toward social fragmentation, political disintegration, and ethical relativism, leaving us behind with many-layered crises that are rendering citizenship and civic engagement ever more difficult (Putnam 2000). Notably, research needs to focus on these developments that seem to threaten the promise of traditional concepts of e-participation: that using online, social, and mobile media would bring a new "political culture" of engagement to the fore, regardless of social backgrounds, material resources, and cultural practices or "repertoires" or "tool kit" of habits, skills, and styles from which "actors select differing pieces for constructing lines of action" (Swidler 1986, p. 277).

Hence, by putting emphasis on e-participation at the micro/individual rather than the meso/organizational and macro/societal level, individuals can be addressed as key actors and central drivers for political action and campaigning in the digital realm. These individuals and their ambitions, motives, and practices create new civic e-participation spaces, new social "spaces for change" (Cornwall and Coelho 2007), that is, publics that establish next to the state and the market and allow for unconventional forms of participation in that they create new and enrich existing digital public spheres. These new spaces enable dialogue via digital communication platforms with a strong "bottom-up" and "do-it-ourselves" ethos that defy "top-down" control by authorities. Being invented by their own agents, these spaces thus develop independently and parallel to invited e-participation spaces, such as government, NGOs, and political parties (Kersting 2013; Bennett and Segerberg 2013). But while standard e-participation practices of invited participation are considered as "top-down" in nature and come as formal government-to-citizen invitation to enter a controlled public sphere, new spaces are "invented" spaces of participation, such as referendums, round tables, or online forums call for the creation of new "mini publics" in the digital realm which are initiated by groups of individuals (Cruickshank et al. 2014) or, as Thimm (2015) defined them: "a group of online users referring to a shared topic in a publicly visible and publicly accessible online space over a period of time, by means of individual activities such as textual or visual contributions" (p. 173).

## 33.3 The Sociology of Engagement

The French sociologists Laurent Thévenot and Luc Boltanski are key proponents of the French school of pragmatic sociology and developed the political and moral "grammar of engagement" approach (Blokker and Brighenti 2011; Boltanski and Thévenot 2006; Thévenot 2013, 2015). Boltanski and Thévenot have been associated with the "pragmatic" school of French sociology, which developed from the 1980s onward, and have also featured leading scholars such as Bruno Latour, Michel Callon, and Michèle Lamont. Pragmatic sociology stands in stark contrast to the critical sociology of Pierre Bourdieu, under whose tutelage Boltanski had initially worked in the early 1970s (Blokker and Brighenti 2011, p. 252, Boltanski 2011). Boltanski came to criticize Bourdieu's critical sociology for presenting social agents as uncritical "cultural dopes," as dominated without knowing it, and for assuming that only sociologists are able to see how people are subjugated by social structures and processes (Boltanski 2011, p. 20). In marked contrast, Boltanski advocates a "pragmatic sociology of critique" which "fully acknowledges actors' critical capacities and the creativity with which they engage in interpretation and action" (Boltanski 2011, p. 43, see Bénatouïl 1999). Thus, the sociologist cannot, and should not, hark back to a priori definitions of power and interest that govern social actions. Instead, the sociologist must reintegrate the capability of critique into the social world and show how actors draw upon different logics in different situations to direct their actions, arguments, and agreements within and through particular disputes. Thévenot, in particular, explored the dimensions of social life "under the public" as a condition to enlarge the scope of public critique to oppressions and to understand the required transformations and obstacles to their exposition in common, to the discord of the political community.

Taking Thévenot's most significant publication *On Justification: Economies of Worth* (2006) as entry point to the present debate (the book was coauthored with Luc Boltanski, a French sociologist and professor at the École des Hautes Études en Sciences Sociales in Paris), this chapter focuses on those "critical moments" when public issues arise and are disputed by different social actors (p. 359–360). At the heart of their analysis are six specific types or "worlds" of justification that social actors draw upon during disputes. The interplay of these worlds produces diverse matrices of conflict, compromise, or collaboration in different situations. Thévenot also suggests that social order arises from the sharing of so-called pragmatic regimes, that is, modes of mutual exchange between actor and the world with a conception of the common good and universal principles that govern human activity.

Consequently, Thévenot theorizes on three pragmatic regimes governing engagement with the world: (1) "publicly justifiable engagement," (2) "familiar engagement," and (3) "engagement in an individual plan." He later transferred these into three implicit social norms – so-called grammars – defining meaningful behavior in a certain situation and enabling agents to evaluate judgments and claims (based on Boltanski's and Thévenot's work *On Justification* of 2006, originally published in 1991): (1) the "grammar of publicly justifiable engagement," (2) the "grammar of personal affinities to a plurality of common places" or "the grammar of close affinities," and (3) the "grammar of individual choices in a liberal public" (Thévenot 2001, 2007, 2013, 2014, 2015).

Firstly, at the level of publicly justifiable engagement, a public discussion would be a typical example and aims at resolving situational conflicts by referring to general-level issues available in the political debate at large. Then, conflict resolution happens by referring to the common good, that is, general principles of what is shared and beneficial for all or most members of a given community or, alternatively, what is achieved by citizenship, collective action, and active participation in the realm of politics and public service. Secondly, the grammar of personal affinities describes familiar attachment and the emotional connections one makes to either close humans (friends, family, etc.) or what Thévenot calls "common-places" that should be understood in its broadest meaning, as referring as much to physical places (e.g., parks, monuments, etc.) as to more "intangible" things (songs, legends, literary characters), whereby agents share experiences and emotions that bring them together on these topics in the public realm (Thévenot 2014).

As these two concepts are all complex and raise difficult questions, not all of which, of course, one can reasonably expect to be addressed fully here, special focus shall be on the third grammar: the grammar of individual choices in a liberal public. This specific grammar refers to legitimacy of individual political claims and rests on the recognition of the rights of individuals and their choice in constructing representative groups: even if actors act as individuals, they rhetorically construct a wider group of people and, in so doing, share their opinion and back their claims. The basic premise of this particular view of Thévenot's approach describes a way of engaging politically in what derives from private interests of individuals as such interests which are, consequently, aggrandized into representative collective preferences and hence structures in democratic politics.

In general, however, there seems to be a conflation of the social or collective *choice* process in these democratic decision-making processes with, say, a *public* voting mechanism whose domain is properly restricted to a "public" subset of alternatives. Liberal democratic theory then presupposes that the set of agents' desires is the basic input which is transformed by the rule into a social choice or output. A typical case of the legitimate interests of subjects is the idea of the referendum: without taking any regard as to the reason or justification behind individual votes, each citizen chooses an idea as a right-carrying subject, and the idea with the most votes (or more usually, the one with over 50% of votes) gets chosen. This means that aggregate social behavior results from the behavior of individual actors, each of whom is making their own individual decisions.

Following Thévenot's grammar of individual choices, the common good is seen to rise out of the private interests of relevant actors – not by finding a single good, but through articulation (and contestation) of preferences by these actors. Even though the general model of the grammar of individual interests is the referendum, negotiations and discussions are also part of the repertoire. These discussions are not deliberations on the relative merits of common goods, but rather trading, haggling, and presentation of individual situations and subjectivities. Such a grammar emphasizes and selects as relevant the differences between individual choices, being granted that they are made public under the form of opinions or interests. In my view, if we wish to better understand people who are engaged with political issues – through electoral processes, city planning, or social media rambling – people who are usually not identified as politicians or activists, nor see projects channeled in their favor through established political communities or organizations, then Thévenot's grammars of engagement and, notably, the "grammar of individual choices in a liberal public" open the way to new theoretical departures in understanding how individuals do politics and create public spheres today.

The next section wishes to mobilize some reasoning about selected ideas on the issue of "mediatized engagement" against the background of communication processes among "digital citizens" and between these citizens and their representatives.

## 33.4 Mediatized Engagement

Now, there is at least one more striking *research desiderate in the present context: to* explore the motivational factors of individuals in "mediated civic engagement" and how this advocates new knowledge and practices that further promote civic engagement, social change, and democratic well-being in the context of e-participation.

Digital media infrastructures create new opportunities for the dissemination of public knowledge. Although the decline in civic participation in established democratic societies has been widely lamented (Putnam 2000), other observers (Dahlgren 2009) have pointed to the growth of new communities online and the growth in quantity and diversity in communication platforms outside of the traditional e-participation platforms, where citizens can exchange information and participate in political debate without any "outside" government influence and control. In fact, "individualization" of civic cultures has emerged in tandem with the growth of mediatized communication processes where individuals use new technologies, with a tendency toward personalization in the public domain (Alvarez and Dahlgren 2016; Bennett and Segerberg 2013). As it looks, social media and networking sites, podcasts, blogs, open-source software, and wikis, have paved the way for an "increasingly individualized civic environment" (Gerodimos 2012, p. 188), with engagement in the public domain being "subjectively experienced more as a personal rather than a collective question" (Dahlgren 2013, p. 52). Here, mediatization research comes as another reminder that political communication and, in its entourage, civic engagement are currently changing.

When seen as a meta-process (Krotz 2011), mediatization, alongside various other "mega-trends" of change in political communication such as digitization of communication technologies, hybridization of communication forms, globalization of communication spaces, or individualization of communication repertoires, comes as another important driver of change to affect individuals in their motivation to engage politically (Vowe and Henn 2016). In theory, mediatization investigates the interrelation between change in media repertoires and usage as drivers for communicative and sociocultural change, understood as a long-term process of change. Naturally, however, (digital) media do not necessarily cause these transformations, but they have become co-constitutive for the articulation of politics, economics, education, religion, etc. (Hepp et al. 2015; Hjarvard 2013; Couldry and Hepp 2016).

Now, these transformations arising from mediatization in its full dimensionality evoke a nexus of research dimensions on the level of the individual citizen. However, if we emphasize that individuals interact with their environments in ways that their interests are voiced through political cultures and commonalities by means of their own individual wills (rather than as a contest of higher principles related to the common good), then models that used to support traditional governmental technologies of e-participation appear not to work any longer. Addressing this void raises the fundamental question of how individuals are doing politics today and how they see new mediated forms of e-participation as a valuable alternative to traditional participation platforms and means of creating political public spheres (Allen et al. 2014).

Essentially, however, understanding of *individuals* engaging in politics and their ways to using digital media technologies within settings of *e-participation* remains elusive. Nonetheless, in view of the observations above, it is about individuals' solicitations and comments about public policies that inspire our understanding of how civic engagement emerges from individual engagement and may eventually fertilize into collective structures of commonality, whereby rather private cognitive models of perception and thinking are transformed into communal and political ways of evaluating (political) arguments. And further, we are well advised to leave behind various traditional perspectives and rhetorics on civic engagement and instead widen its canvas toward a sociological theory of engagement and political action in the digital age as it is promulgated by the works of Laurent Thévenot and his "liberal" notion of "individuals doing politics" (Thévenot 2007, 2014), or what Thévenot himself called "the grammar of the individual in a liberal public," as explained above. And further, if one looks at individuals and their ambitions to engage politically, then research on mediatization advances our thinking about their opportunities for action in new civic e-participation spaces, new social "spaces for change" (Cornwall and Coelho 2007), that is, publics that establish next to the state and the market and allow for unconventional forms of participation in that they create new and enrich existing digital public spheres. Predictors of civic engagement on individual level are reciprocity of participants (Gouldner 1960; Wasko and Faraj 2000), exchange of (symbolic) messages (Rafaeli and Sudweeks 1997), active user control (Rice and Williams 1984), immediacy of feedback (Dennis and Kinney 1998), and trust (van der Meer 2017). Computer-mediated communication (CMC) theory, in particular, stresses the way by which the communicators process social identity and relational cues (i.e., the capability to convey meanings through cues like body language, voice, tones, i.e., basically social information) using different media (Fulk et al. 1990; Walther 1992).

And mediatized engagement is viewed to have more anthropomorphic properties (Quiring 2009). Here, interactivity refers back to the concept of action in the social sciences, whereby action is presupposed to depend on an active human subject intentionally acting upon an object or another subject. Interaction with objects and the creators of these objects modifies their actions and reactions due to the actions by their interaction partner(s). Seen this way, mediated engagement is understood as a subjective mode of perception and cognition and, as interpreted from a communication theory perspective, focuses on how a receiver actively interprets and uses mass and new media messages. In the CMC literature, two more key themes have emerged under this rubric: individual experiential processes of interactivity and perceptions of individual control over both presentation and content. Self-awareness (i.e., the psychological factor that impacts on social interaction as mediated by CMC; Matheson and Zanna 1998), responsiveness (i.e., the degree to which a user perceives a system as reacting quickly and iteratively to user input; Rafaeli and Sudweeks 1997), a sense of presence (discoursed as virtual experience made by humans when they interact with media; Lee 2004), involvement (defined as perceived sensory and cognitive affiliation with media; Franz and Robey 1986), and perceived user control are further constituent psychological activities on this level of discussion (Zimmerman and Rappaport 1998). Furthermore, human agents are not only calculative and rational, nor are they only bound by structures; they are also guided by nonbinding habits that leave room for new engagements and new ways of actions.

Hence, this study should also start an interdisciplinary dialogue and articulate a set of key transformations brought about by ICTs, the media, and individuals as social and political activists (Bakardjieva et al. 2012). These include the realities of "hyper-connectivity" and "mediatization" as facilitated by ICTs and online, social, and mobile media and how research on these facilitating technologies provides insights into barriers and perceived affordances for e-participation as well as the necessary conditions for increased adoption for citizen-led participation.

## 33.5 Future Directions

Theorizing on the political sociology of engagement in politics is a legitimate way of presenting a critique against "top-down" policies of e-participation. It goes beyond these traditional approaches by focusing on "bottom-up" aspects of enriching democratic processes, political discourse, and social change. This opens research to alternative approaches in e-participation that recognize the importance of social movements and other "communities of practice" as these represent a vital and new component of doing politics today "from below" (Della Porta et al. 2017). Hence, fundamentally, this chapter started from the notion that individuals demanding that their idiosyncratic preferences and political interests be recognized and addressed are a key focal point for analyses into civic engagement. Such an individualist "grammar of engagement" emphasizes and selects as relevant the differences between the choices of individuals, being granted that they are made public under the form of opinions or interests. Group building takes place by negotiation among individuals opting for interests that must be publicly recognizable. In this study, consequently, the notion of "methodological individualism" has been supported, an epistemic "individual-first" approach of citizens to politics whereby "structure exists, and has determinant force, but a conscious heuristic decision that what individuals choose to do, or perceive themselves as choosing, is interesting as an object of study – not the individual as a 'case study' of a larger whole, but the individual as exceptional or particular" (Burke 1969, online). Hence, in the present context, it is essential to understand how individuals perceive opportunities of political engagement in order to also grasp how they are influenced by new media and e-participation technologies.

Secondly, political agents of this kind create their own strategies for action by utilizing resources of their own choice. This strategizing is not (only) calculative and maximizing one's own interest but also affected by attitudes, emotions, habits, and social relations. Therefore, when political acts are undertaken, individual actors (residents, techno-politicians, Facebook users, etc.) engage with, utilize, cultivate, and inhabit their cultural repertoires, pooling from their own media resources and using these by turning political strategies into arguments, justifications, memes, protest letters, and other political objects. Moreover, in the present context, we are focusing on a new type of individual, the activist or political campaigner, a human being that is deemed to be free and social at the same time (Dahlgren 2013), whereby "free" is frequently understood as being autonomous, disembodied, rational, well informed, and disconnected: an individual and atomistic self, while that freedom does not occur in a vacuum, but in a space of affordances and constraints (Floridi 2015). In practice, today a political campaigner uses Facebook and other similar tools to create ad hoc campaign groups, utilizes the cultural repertoire of the Internet, and participates in politics when (and only when) he or she sees fit.

And thirdly, the deployment of these strategies and the use of these media tools occur on macro/societal, meso/organizational, micro/individual, and nano/supra-individual levels of social interaction (Bimber 2017; Vowe and Henn 2016). In this context, mediatization comes as a meta-process that blurs the distinction between these levels in many ways, and, as is proposed, it is simultaneously an interactional arena with some of the dynamics from face-to-face encounters, a place for connecting, organizing, and strategizing and a news distribution system or media (Zimmermann 2015).

As for future research, starting from the notion that civic engagement and the challenges arising through e-participation in favor of democratic decision-making of individuals is a contingent, dynamic, and complex social process, some of the paradigms that seem to dominate more traditional research perspectives on e-participation need to be refreshed.

However, while theorizing on civic engagement in the digital age itself is subtle and sophisticated, skepticism as to its value for analyzing changes in e-participation within the digital marketplace prevails. Ultimately, research into *digital* civic engagement needs to confront this challenge because, as it appears, individuals in the resulting digital ecology have to be seen as integral to redefining political participation in ways that may or may not be beneficial to society and democracy at large.

## 33.6 Cross-References

▶ Capacity Building and People's Participation in e-Governance: Challenges and Prospects for Digital India
▶ Communication for Development and Social Change: Three Development Paradigms, Two Communication Models, and Many Applications and Approaches
▶ Communication for Development and Social Change through Creativity
▶ Reducing Air Pollution in West Africa Through Participatory Activities: Issues, Challenges, and Conditions for Citizens' Genuine Engagement
▶ The Importance of Paulo Freire to Communication for Development and Social Change
▶ The Relevance of Habermasian Theory for Development and Participatory Communication
▶ The Theory of Digital Citizenship

# References

Allen D, Bailey M, Carpentier N, Fenton N, Jenkins H, Lothian A, Linchuan Qui J, Schaefer MT, Srinivasan R (2014) Participations: dialogues on the participatory promise of contemporary culture and politics. Part 3: Politics. Int J Commun 8:1129–1151

Alvarez C, Dahlgren P (2016) Populism, extremism and media: Mapping an uncertain terrain. Eur J Commun 31(1):46–57

Bakardjieva M, Svensson J, Skoric MM (2012) Digital citizenship and activism: questions of power and participation online. eJournal eDemocracy Open Gov 4(1):i–iv

Bénatouïl T (1999) A tale of two sociologies: the critical and the pragmatic stance in French contemporary sociology. Eur J Soc Theory 2(3):379–396

Bennett WL, Segerberg A (2013) The logic of connective action: digital media and the Personalization of contentious politics. Cambridge University Press, Cambridge

Bimber B (2017) Three prompts for collective action in the context of digital media. Pol Commun 34(1):6–20

Blokker P, Brighenti A (2011) An interview with Laurent Thévenot: on engagement, critique, commonality, and power. Eur J Soc Theory 14:383–400

Blumler JG (2014) Mediatization and democracy. In: Esser F, Strömbäck J (eds) Mediatization of politics. Palgrave Macmillan, London, pp 31–45

Boltanski L (2011) On critique. A sociology of emancipation. Polity Press, Cambridge

Boltanski L, Thévenot L (2006[1991]) On justification. Economies of worth. Princeton University Press, Princeton/Oxford

Burke K (1969) A grammar of motives. University of California Press, Berkeley (originally published in 1945)

Collins CR, Watling Neal J, Zachary NP (2014) Transforming individual civic engagement into community collective efficacy: the role of bonding social capital. Am J Community Psychol 54(3–4):328–336

Cornwall A, Coelho VS (2007) Spaces for change?: the politics of citizen participation in new democratic arenas. Zed Books, London

Couldry N (2014) The myth of us: digital networks, political change and the production of collectivity. Inf Commun Soc 18(6):608–626

Couldry N, Hepp A (2016) The mediated construction of reality. Polity Press, Cambridge

Couldry N, Livingstone S, Markham T (2007) Media consumption and public engagement: beyond the presumption of attention. Palgrave Macmillan, Basingstoke

Cruickshank P, Ryan B, Smith FC (2014) 'Hyperlocal E-participation'? Evaluating Online Activity by Scottish Community Councils, Conference for E-democracy and Open Government, 21–23 May 2014, Danube University Krems. http://www.donau-uni.ac.at/imperia/md/content/department/gpa/zeg/bilder/cedem/cedem14/cedem14_proceedings.pdf

Dahlgren P (2005) The internet, public spheres, and political communication: dispersion and deliberation. Pol Commun 22(2):147–162

Dahlgren P (2009) Media and political engagement. Polity Press, Cambridge

Dahlgren P (2013) The civic subject and media-based agency. In: Dahlgren P (ed) The political web: media, participation and alternative democracy. Palgrave Macmillan, Basingstoke, pp 133–151

Della Porta D, O'Connor F, Portos M, Subirats Ribas A (2017) Social movements and referendums from below. Direct democracy in the neoliberal crisis. Polity Press, Bristol

Dennis AR, Kinney S (1998) Testing media richness theory in the new media: the effects of cues, feedback, and task equivocality. Inf Syst Res 9(3):256–274

Ekman J, Amnå E (2012) Political participation and civic engagement: towards a new typology. Hum Aff 22(3):283–300

Floridi L (2015) The onlife manifesto. being human in a hyperconnected era. Springer, Cham

Franz CR, Robey D (1986) Organizational context, user involvement, and the usefulness of information systems. Decis Sci 17(4):329–356

Freeman J, Quirke S (2013) Understanding e-democracy government-led initiatives for democratic reform. JeDEM 5(2):141–154

Fulk J, Schmitz J, Steinfield CW (1990) A social influence model of technology use. In: Fulk J, Steinfield C (eds) Organizations and communication technology. Sage, Newbury Park, pp 117–140

Gerodimos R (2012) Online youth attitudes and the limits of civic consumerism: The emerging challenge to the Internet's democratic potential. In: Loader B, Mercea D (eds) Social media and democracy: innovations in participatory politics. Routledge, London, pp 166–189

Giddens A (1984) The constitution of society elements of the theory of structuration. Polity Press, Cambridge

Gouldner AW (1960) The norm of reciprocity: a preliminary statement. Am Sociol Rev 25(2):161–178

Habermas J (1996) Three normative models of democracy. In: Benhabib S (ed) Democracy and difference. Contesting the boundaries of the political. Princeton University Press, Princeton, pp 21–30

Hepp A, Couldry N (2013) Conceptualizing mediatization special issue: mediatization. J Commun Theory 23(3):191–315

Hepp A, Hjarvard S, Lundby K (2015) Mediatization: theorizing the interplay between media, culture and society. Media Cult Soc 37(2):314–322

Hjarvard S (2013) The mediatization of culture and society. Routledge, London

Kersting N (2013) Online participation: from 'invited' to 'invented' spaces. Int J Electron Gov 6(4):270–280

Koc-Michalska K, Lilleker D (2017) Digital politics: mobilization, engagement, and participation. Polit Commun 34(1):1–5

Kramp L, Carpentier N, Hepp A, Tomanić Trivundža I, Nieminen H, Kunelius R, Olsson T, Sundin E, Kilborn R (2016) Media practice and everyday agency in Europe. Various authors on the "dynamics of mediatization". edition lumière, Bremen, pp 33–128

Kreiss D (2015) The problem of citizens: e-democracy for actually existing democracy. Soc Media Soc 1(2). https://doi.org/10.1177/2056305115616151

Krotz F (2011) Mediatisierung als Metaprozess. In: Hagenah J, Meulemann H (eds) Mediatisierung der Gesellschaft? LIT-Verlag, Münster, pp 19–41

Lee KM (2004) Presence, explicated. Commun Theory 14:27–50

Macintosh A (2004) Characterizing e-participation in policy-making. In: Proceedings of the 37th Hawaii international conference on system science (HICSS'04). Computer Society Press, Washington, DC, pp 50117–50126

Matheson K, Zanna MP (1998) The impact of computer-mediated communication on self-awareness. Comput Hum Behav 4(3):221–233

Mossberger K, Tolbert CJ, McNeal RS (2008) Digital citizenship: the internet, society, and participation. The MIT Press, Cambridge, MA

Putnam R (2000) Bowling alone. The collapse and revival of American community. Simon & Schuster, New York

Quintelier E, Hooghe M (2012) Political attitudes and political participation: a panel study on socialization and self-selection effects among late adolescents. Int Polit Sci Rev 33(1):63–81

Quiring O (2009) What do users associate with 'interactivity'? A qualitative study on user schemata. New Media Soc 11(6):899–920

Rafaeli S, Sudweeks F (1997) Networked interactivity. J Comput Mediat Commun 2(4). http://jcmc.huji.ac.il/vol2/issue4/rafaeli.sudweeks.html

Rice RE, Williams F (1984) Theories old and new: the study of new media. In: Rice RE (ed) The new media. Sage, Beverly Hills, pp 55–80

Swidler A (1986) Culture in action: symbols and strategies. Am Sociol Rev 51(2):273–286

Teorell J, Torcal M, Montero JR (2007) Political participation: mapping the Terrain. In: van Deth JW, Montero JR, Westholm A (eds) Citizenship and involvement in European democracies: a comparative analysis. Routledge, London/New York, pp 334–357

Thévenot L (2001) Pragmatic regimes governing the engagement with the world. In: Schatzki TR, Knorr-Cetina K, von Sevigny E (eds) The practice turn in contemporary theory. Routledge, London, pp 56–73

Thévenot L (2007) The plurality of cognitive formats and engagements. Moving between the familiar and the public. Eur J Soc Theory 10(3):413–427

Thévenot L (2013) The human being invested in social forms. Four extensions of the notion of engagement. In: Archer M, Maccarini A (eds) Engaging with the world. Agency, institutions, historical formations. Routledge, London/New York, pp 162–180

Thévenot L (2014) Voicing concern and difference: from public spaces to common-places. Eur J Polit Cult Sociol 1(1):7–34

Thévenot L (2015) Making commonality in the plural, on the basis of binding engagements. In: Dumouchel P, Gotoh R (eds) Social bonds as freedom: revising the dichotomy of the universal and the particular. Berghahn Books, New York, pp 82–108

Thimm C (2015) The mediatization of politics and the digital public sphere. Participatory dynamics in mini-publics. In: Frame A, Brachotte G (eds) Citizen participation and political communication in a digital world. Routledge, London/New York, pp 167–183

van der Meer TWG (2017) Political trust and the "crisis of democracy". Oxford research encyclopedia of politics. https://doi.org/10.1093/acrefore/9780190228637.013.77

Vowe G, Henn P (2016) Political communication in the online world: theoretical approaches and research designs. Routledge, New York

Walther JB (1992) Interpersonal effects in computer-mediated interaction: a relational perspective. Commun Res 19(1):52–89

Wasko M, Faraj S (2000) "It is what one does": why people participate and help others in electronic communities of practice. J Strateg Inf Syst 9(2–3):155–173

Zimmerman MA, Rappaport J (1998) Citizen participation, perceived control, and psychological empowerment. Am J Community Psychol 16(5):725–750

Zimmermann T (2015) Between individualism and deliberation: rethinking discursive participation via social media. Int J Electron Gov 7(4):349–365

# Family and Communities in Guatemala Participate to Achieve Educational Quality

## 34

Antonio Arreaga

## Contents

| | | |
|---|---|---|
| 34.1 | Introduction | 648 |
| 34.2 | Planning and Implementing a Communication for Development Strategy to Achieve Quality Education | 650 |
| 34.3 | Identifying the Problems | 650 |
| 34.4 | Objectives and Communication Approaches | 651 |
| 34.5 | A Communication for Development Strategy for Quality Education | 653 |
| 34.6 | Awareness Campaign on Quality Education in the Classroom | 656 |
| 34.7 | Early Reading Socialization Mechanism | 657 |
| 34.8 | Early Reading Socialization Mechanism Fields of Action | 657 |
| 34.9 | Assessments, Perceptions of Change, and Results of the Communication Intervention | 660 |
| | 34.9.1 Results | 660 |
| | 34.9.2 School Directors | 662 |
| | 34.9.3 Teachers | 662 |
| | 34.9.4 Brief Analysis About the Awareness Campaign Monitoring | 663 |
| 34.10 | Conclusion | 664 |
| References | | 665 |

### Abstract

After a 36-year armed conflict, Guatemala finally signed the Peace Accords in December 1996. Because of these agreements, it is the Educational Reform Design (1998) that promotes different axes and areas of transformation, which include, among other topics, the country achieving educational quality with equity. To achieve this, for example, pedagogical aspects, educational policies, financial topics, and community and family involvement in the education of their children were proposed. Drawing on Educational Reform Design, at the

A. Arreaga (✉)
Juarez & Associates, Inc., Guatemala City, Guatemala
e-mail: aarreaga@juarezassociates.com

beginning of the twenty-first century, USAID/Guatemala begins to develop projects to improve bilingual education and the processes of assessment of the education system, as well as to have expected learning goals to achieve a better education. But it is up to USAID/Education Reform in the Classroom Project (2009–2014), implemented by the firm Juarez & Associates, Inc., to take up the challenge of involving in the educational quality, with an emphasis on reading to municipalities, communities, and families, many of them Mayas, in conditions of poverty, with low rates of scholarship. This approach was called "early reading socialization mechanism," which used communication for development to design and implement strategies that reach their objectives at different local levels. In 2009, engaging these local actors, under these social and economic contexts in quality education focused on reading, was a little known and explored in Guatemala.

**Keywords**

Family involving in educational quality · Communication for development focused on education · Communication for development in Guatemala · Friendly municipalities for reading · Early reading socialization mechanism

## 34.1 Introduction

In Guatemala, for the first decade of the twenty-first century, statistics showed an increase in educational coverage for primary school. In the year 2000, 8 out of 10 children, between 7 and 12 years old, were enrolled in school. And for 2006, 9 of every 10 children of the indicated age were enrolled (PREAL and CIEN 2008). These data were a milestone for a country that in the twentieth century still had very low enrollment coverage for primary school, compared to others in the Latin American region.

Although the coverage was not total, this evidence of school attendance, suggested to authorities and technical staff from the Ministry of Education, tanks of thought and academics related to education, that the great majority of the educational community, especially fathers and mothers, were already convinced that it was important to send their children to school. What was the next step? Communicate and raise awareness that in addition to attending a school center, it was vital to learn, starting with reading and mathematics.

But this step of raising awareness of understanding that more than sending their children to school, they had to learn well gained more relevance knowing the results of the tests of learning in reading and mathematics of the students. In 2005, evaluations for sixth grade students nationwide indicated that in reading only 43% of students achieved a satisfactory level and an excellent 5%. Therefore, a little more than half (52%) of the students evaluated in the country in this last grade of primary school were at levels of "must improve" or "unsatisfactory." In mathematics, the results were very similar. Only 49% achieved a satisfactory level and 6% reached the excellent level (PREAL and CIEN 2008).

The years of these assessments (2005 and 2006) began to link with more intensity in the education system, obtaining better learning by students and having teachers better prepared for teaching reading and mathematics, with the term of achieving educational quality.

However, the goal of achieving quality in education was already clearly presented in the document "Design of Educational Reform," published in 1998 by a specialized education commission. This document is the result of the Peace Accords signed in December 1996, which put an end to Guatemala's internal armed conflict that lasted for 36 years in the country.

In the "Design of Educational Reform," the educational quality is linked from the beginning of the document with one of the great means to achieve a change in education (Diseño de Reforma Educativa 1998, p. 15). And as the document progresses educational quality is linked with key issues as equity, bilingual education, educational policies, structures and organization of the Ministry of Education, as well as having committees of parents and mothers of family, teachers and students, among other topics (Diseño de Reforma Educativa 1998, p. 89).

It is not surprising that educational quality is linked to an active participation of parents. There are studies that report on the involvement in the education of the parents of the students and the levels that this entails (Stevens and King 2004). These authors reported that after the first level of parental involvement, which is minimal, the second stage or level consists of parents arriving to learn about the program to get involved. And they finish this second level with an important milestone: they can act as teachers of their children at home or outside of the regular educational environment.

By the end of 2009, involving parents in Guatemala in the learning of their children, especially in reading, as well as communicating and having them aware about what is educational quality and its importance was an underdeveloped topic. Especially with rural parents, with low schooling or illiterate; and many of them, of Mayan origins and their mother tongue was related to Mayan roots.

On the other hand, communication approaches for the development of education issues with this audiences or target groups were not common in Guatemala either. There existed in the country previous experiences with approaches of social marketing, communication for development, and change of behavior in the health area, supported by international cooperative sources and/or civil society organizations.

In October of 2009, USAID/Guatemala constituted the Education Reform in the Classroom Project (2009–2014). The project aimed to improve access to basic education (including quality, equity, and efficiency) by focusing on increasing teacher effectiveness, improving classroom learning environments, fostering effective first and second (Mayan language [mother tongue] and Spanish) language acquisition and reading, extending access to underserved populations including girls and indigenous groups, and expanding parents', communities', and stakeholders' participation in student learning.

From the design of this project, it was thought that a communication for development approach was necessary to guarantee the successful involvement of parents and stakeholders in student learning, focused in quality education and reading.

In this sense, the project scope of work included the design and implementation of a communication for development strategy, which will include an awareness campaign on quality education in the classroom. The communication strategy for development and the awareness campaign were key tools at the service of the early reading socialization mechanism, focused on parents, promoting reading in rural municipalities.

To carry out the design task for the communication for development strategy and the awareness campaign on quality education, the USAID project formed a communication team, consisting on international and local consultants, technical communication staff from the Ministry of Education, and permanent communication staff from Education Reform in the Classroom Project.

## 34.2 Planning and Implementing a Communication for Development Strategy to Achieve Quality Education

The project communication team followed a theoretical and implementation framework proposed by Ph.D. Jan Servaes, who led the leadership as an international consultant specializing in the design of communication for development strategy. Next, a figure that shows this process (Fig. 1).

Using this work route, built under a communication approach for development, where the key actors of the community take part in the proposal of their problems, a diagnostic was designed that would serve as a basis for the strategy.

In this sense, the communication team of USAID/Education Reform in the Classroom held a participatory assessment process, in which, through interviews and focus groups, representatives of the educational community (students, teachers, principals), local municipal authorities, community leaders, businessmen, and media communication provided their views on the situation of education in their geographical area, specifically in three topics of interest for the project: quality of education, interest in reading, and communications media (Alvarado et al. 2010a). These geographic areas were the municipalities where the project was going to have an impact on education issues: Joyabaj, Quiché; San Pedro Pinula, Jalapa; and Jocotán, Chiquimula. The diagnostic process in those municipalities included the urban center and selected rural areas of each.

## 34.3 Identifying the Problems

After conducting the fieldwork with the participatory diagnostic, the communication team analyzed the findings regarding common problems in the three municipalities.

Table 1 shows these problems, prioritized under the following three aspects:

- The project could solve the problem.
- The project had an impact, but it did not solve the problem.
- It was beyond the scope of the project.

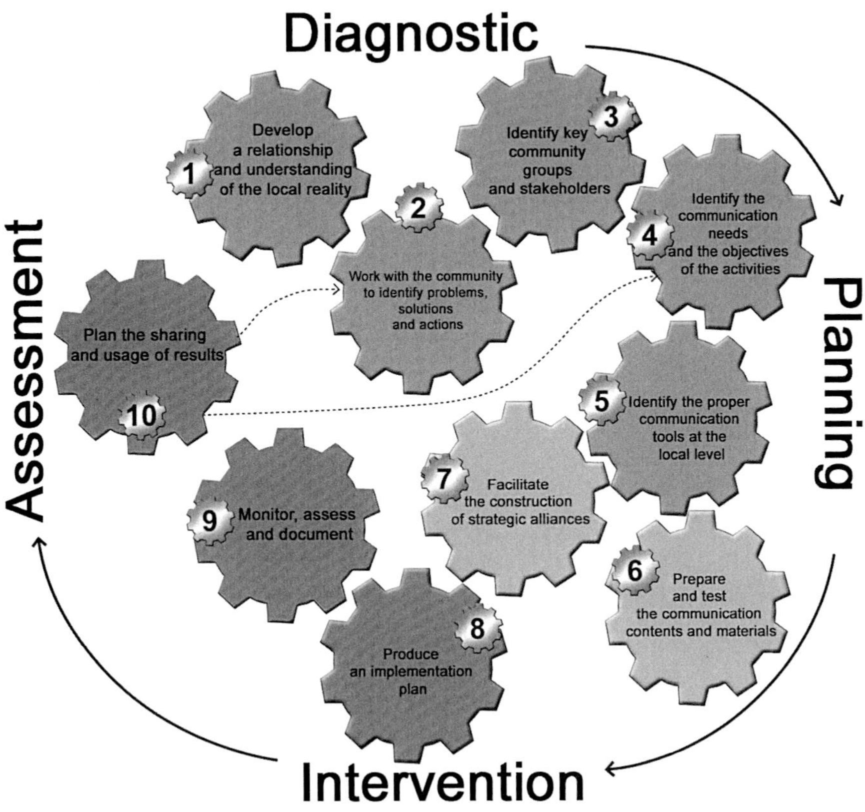

**Fig. 1** The proposed phases in the design of the communication strategy. (Servaes 2011)

Likewise, this table identifies the actors/audiences subject to change and those that must be advocated for the problems that the project is supposed to solve (red color); defines the term for generating changes: short (1 year), medium (2 years), and long (4 years); and finally links the problem with the space where it develops, S (School), F (Family), and C (Community).

## 34.4 Objectives and Communication Approaches

Having the prioritized problems identified in the participatory assessment and splicing the results in the USAID/Guatemala work order required for the USAID/ Education Reform in the Classroom, the communication team defined the following main objectives:
- Promote the involvement of parents, communities, and leaders in student learning.
- Promote the creation of a literate environment in the communities and socialize activities that promote reading as a fundamental instrument of learning.

**Table 1** Problem prioritization, audiences, and terms table. (Alvarado et al. 2010a)

| Audiences | Term | Priority | Problems |
|---|---|---|---|
| Family, teachers and students | Short | 1 F | Language barrier for K'iche' and Ch'orti' Mayan languages |
| Teachers and students | Short | 1 S | Strong difference between urban and rural (student population) |
| Teachers and students | Short | 1 S | Lack of teaching materials |
| Teachers | Short | 1 S | Teacher opposition to quality education campaigns |
| Family | Short | 1 F | Lack of reading material at home, especially in rural areas |
| Community | Medium | 1 C | Little participation of the community in education matters |
| Family, teachers and students | Medium | 1 S | Little participation of the parents in the education of their children |
| Teachers and MINEDUC authorities | Medium | 1 S | Lack of technical training |
| Teachers and MINEDUC authorities | Medium | 1 S | Lack of communication between the teachers and MINEDUC authorities (performance of their students in tests) |
| MINEDUC Education Council | Medium | 1 S | Absence of school libraries and where these exist, little use. |
| MINEDUC, teachers, students | Medium | 1 S | Poor understanding of the contents |
| Donors, national authorities, local community | Large | 1 C | Community leaders and parents think that bilingual education is not practical (Joyabaj, Quiché) |
| Family, teachers and students | Large | 1 F | Poor reading habits |
| Family, teachers and students | Large | 1 F | Education aspirations are very basic (learn reading and writing and numbers) |
| | | 2 S | Frustration by the teachers |
| | | 2 F | Illiteracy of the parents |
| | | 2 F | Lack of values (due to ignorance) |
| | | 2 F | The education of the children is the sole responsibility of the teachers |
| | | 3 C | Lack of support to build an infrastructure for education |
| | | 3 C | Change in local authorities every 4 years which results in lack of continuity |
| | | 3 C | Education is not a priority for the Municipal authorities |
| | | 3 S | Lack of infrastructure (classrooms) |
| | | 3 S | Overcrowded classrooms (number of students, multiple grades) |
| | | 3 S | Outdated education statistics which result in discordant resource allocations |
| | | 3 F | Scarce economic resources |

Communication Media at Editorial Level (*)

Short Term: 1 year. Medium Term: 2 years. Large Term: 4 años (*) Communication media are crosscutting target audiences to assist in social problems. A plan to impact national and local media is an appropriate way to reach and involve this audience.

■ We can solve the problem   ■ We can have an impact, not solve   ■ Out of our reach

- Raise awareness in different audiences about the meaning of quality education and the importance of putting it into practice in the classroom and beyond.

To reach this three main objectives, the following approaches for communication development were proposed:

- Approaches that attempt to *change attitudes* (through information dissemination, public relations)
- *Behavior change approaches* (focusing on changes in individual behavior, interpersonal behavior, and/or behavior of the community and society)
- *Advocacy approaches* (primarily aimed at politicians and decision-making powers at all levels and in all sectors of society)
- *Communication approaches for structural and sustainable social change* (which can be related to the way of the planning design: top-down, horizontal, or bottom-up)

Change attitudes and raise awareness are linked mainly with mass media communication (community media, national media, and TICs). Behavior change is more connected with interpersonal communication. Advocacy approaches are connected with interpersonal communication and mass media communication. Communication approaches for sustainable structural change is a mix of all of them and call for Communication for Structural and Sustainable Social Change (CSSC). Also in this type of communications are the participatory communications, constituted by interpersonal communication and community media (Servaes 2011).

## 34.5 A Communication for Development Strategy for Quality Education

With all the pieces on the table, the team designed a matrix strategy that included the following elements:

- **Specific objectives**, according to the problems presented in the diagnoses, where communication for development can support sustainable social changes. The Table 2 shows only the objectives related with those problems that USAID/Education Reform in the Classroom could support to solve (Table 1).
- **Levels coverage**: Local, regional, national, and international.
- **Audiences**: School (students, parents, principals, school board/education council), family (fathers, mothers, students), and community (municipal and education local authorities, among others).
- **Change sought**: Changing attitudes (information, public relations), behavior change (persuasion), commitment to social policy (incidence), and providing individuals and groups knowledge, values, and skills that lead to actions effective for change (empower) (Table 2).

**Table 2** Communication strategy for the sustainable development of quality education. (Alvarado et al. 2010a)

| Objectives | Levels | Target Audience | Type of "change" | Term | Media |
|---|---|---|---|---|---|
| Harmonize the different concepts of quality education | National | All | Information | Short | Mass communication means, community dialogues (local/national), forums, press (editorial content), lobbying |
| Reduction of the language gap within the education community (Joyabaj and Jocotán) | Local | Family, teachers, students, MDC | Information and Persuasion | Short | Local radio, community meetings, religious actors, alternative means |
| Facilitate access to reading materials for the education community | Local | Family, teachers, students | Incidence | Short | Interpersonal communication |
| Change the image of quality education | Local | Teachers | Information and Persuasion | Short | Local radio, community meetings, religious actors and community leaders (auxiliary majors and COCODES) |
| Facilitate the participation in the education topic | Local | Community | Empowerment | Medium | Local radio, community meetings, religious actors and community leaders (auxiliary majors and COCODES), mobile phones, alternative means |
| Create spaces to improve the communication between the MINEDUC | Local | Teachers and MINEDUC authorities | Information and Empowerment | Medium | MINEDUC web page, cascading communication, workshops, interpersonal |

(continued)

**Table 2** (continued)

| | | | | | |
|---|---|---|---|---|---|
| authorities and the teachers in urban and rural schools | | | | | contact between supervisors and teachers |
| Establish mechanisms to promote the use of libraries at urban and rural schools in Joyabaj | Local | Local | Incidence | Medium | Interpersonal communication |
| Advocacy on parents and local leaders about the importance of bilingual education | National and Local | International cooperation, national authorities, community | Persuasion Incidence | Long | Dialogue and forums, interpersonal contact |
| Awake interest and build reading habits | Local | Family, teachers, students | Persuasion Empowerment | Long | Local radio, community meetings, religious actors, alternative means (storytelling, libraries, reading contests, spelling bees), TV (spill over of the national campaign,) local cable TV |
| Achieve better understanding of the benefits of education to have a better quality of life | Local | Family, teachers, students | Persuasion Empowerment | Long | Local radio, community meetings, religious actors, alternative means TV (spill over of the national campaign,) local cable TV |
| Facilitate the participation of the education community in the education topic | Local | Family, teachers, students | Persuasion Empowerment | Long | Local radio, community meetings, religious actors and community leaders (auxiliary majors and COCODES) |

## 34.6 Awareness Campaign on Quality Education in the Classroom

The communication team design this campaign as a key tactic tool from the communication strategy for development which main objective was to raise awareness on the importance of taking advantage of education as a vehicle for personal, family, and community improvement. It also positioned to the audiences about key conditions to have educational quality in communities. Under the type of change approach in communication, the awareness campaign was the tactic that provided information and persuasion to the defined audiences.

In order of priority, the audience for the campaign were the following: (1) family, (2) community, and (3) school. The main insights of the campaign were as follows: (1) I like to learn. (2) If my children learn, they will have a better future. (3) I have a better attitude toward life when I learn. The discourse included in the pieces and communication tools of the campaign sought to leave this concept in the audience: "Achievement success in my personal, family and community life because I get support to receive education, learn more and be a better person" (Steele 2011).

The main tools of the campaign were:

1. Three different radio spots versions of 60 and 30 s produced in Spanish and Mayan Language K'iche'. The spots addressed different topics about actions to obtain better learning in reading and how to support the educational quality in the family and the community. The spots used dialogues between characters who represented community members (grocer, teacher, municipal councillor), who spoke in colloquial language about the necessary conditions to achieve students better learning in reading and quality of education at municipal levels. The campaign included a media plan with popular local radios (municipal and departmental levels), broadcasting eight radio spots (60 s) per day the first month. The next 3 months, broadcasting 10 radio spots (30 s) per day. The work with the radio included, also only the first broadcasting month, live mentions about the importance of reading and quality education by the announcers (four locutions per day of around 10 s). This media plan with municipal and departmental radios was used in 2011 for municipalities from Quiché, Jalapa, and Chiquimula.
2. Alternative public communication media with mobile publicity. This consisted of a vehicle (car or motorcycle) that carried a banner with images and text of the campaign and broadcasting live the radio spots on speakers. This media actions were used in 2011 (for municipalities of Quiché, Jalapa, and Chiquimula) and 2012 for municipalities from Totonicapán and San Marcos.
3. Large format poster with nine key protocols about the conditions to have quality education at classroom. These nine conditions were developed previously by technical staff from USAID/Education Reform in the Classroom, as well as the design of the poster. But they are included in the campaign because their contents were a conceptual basis for the development of communication tools within the campaign.

## 34.7 Early Reading Socialization Mechanism

According to the objectives set, designing and implementing this mechanism led to include in its tactics and tools all communication approaches for change included on the communication strategy for the sustainable development: of quality education, information, persuasion, incidence, and empowerment.

The communication objectives proposed for the mechanism were:

1. Raise awareness among parents/families whose daughters and sons are enrolled in first, second, and third graders on the importance of being involved in education.
2. Persuade parents/families on the importance of reading.
3. Empower families to recognize the reading skills their children should acquire according to grade level.

To achieve these objectives, the following communication approaches for change were proposed:

1. **Political incidence** generating commitment for policy support and increased public interest in social issues
2. The **social support** that develops alliances and social support systems that legitimize and encourage development-related activities such as social norms
3. The **empowerment** that gives individuals and groups knowledge, values, and skills that promote effective action for change

Table 3 presents the specific problems that the socialization mechanism would try to solve, including objectives and the communication approaches for change.

As indicated above, this mechanism covered all approaches to change, so it was designed to use Communication for Structural and Sustainable Social Change, which integrates interpersonal communication, participatory communication, and mass media communication.

## 34.8 Early Reading Socialization Mechanism Fields of Action

The implementation of the early reading socialization mechanism had three major fields of action:

1. **Text provision**. USAID/Education Reform in the Classroom project made alliances with other USAID projects, private businesses, and education NGOs to provide teaching materials and reading texts to schools where the mechanism would be implemented, located in the departments of Quiché, Chiquimula, Jalapa, Totonicapán, and San Marcos.
   The provision of textbooks and school libraries by the Ministry of Education through its "National Reading Program" was very important in this topic.

**Table 3** Early reading socialization mechanism problems. (Alvarado et al. 2010b)

| Communication Media at Editorial Level (1) | Audiences | Term | Problems | Objectives | Strategies (2) |
|---|---|---|---|---|---|
| | Teachers and students | Short | Lack of teaching materials in the school | Provide teaching materials and reading books | Text provision |
| | Family | | Lack of reading material at home, especially in rural areas | | |
| | Family, teachers and students | Medium | Little participation of the parents in the education of their children | Raise awareness among parents, mothers and community about the importance of getting involved in education | Political incidence (awareness and commitment) |
| | Community MINEDUC. teachers, students | | Little participation of the community in education matters | Empower families to recognize reading skills | Social support and empowerment |
| | | | Absence of school libraries and where these exist, little use | Promote the use of libraries | Political incidence (awareness and commitment) |
| | Family, teachers, students | Large | Poor reading habits | Encourage the habit of reading | Political incidence (awareness and commitment) |
| | Family, teachers, students | | Education aspirations are very basic (learn reading and writing and numbers) | Achieve that the educational community recognizes the benefits of education to have a better quality of life | Political incidence (awareness and commitment) Social support and empowerment |

**Short Term**: 1 year. **Medium Term**: 2 years. **Large Term**: 4 años (1) Communication media are crosscutting target audiences to assist in social problems. A plan to impact national and local media is an appropriate way to reach and involve this audience. (2) **Author's note**: The original source used the term "strategies." In this document, communication approaches for change have been used. For the topic of "text provision" the term "Action" may fit better.

2. **Awareness and commitment**. Through local mass media (especially radio) and alternative media. The tactic of the awareness campaign on quality education was the main arm to fulfill that action, whose general aspects were described above. On the other hand, forums were held on radio and TV where local educational actors talked about the importance of reading, the quality of education, and the conditions to achieve it in their municipalities/communities. In addition to having the audiences aware about the quality of education, this action allowed these actors to appropriate a discourse on the importance of educational quality and initiate a process of social support and commitment in their communities (Arreaga 2014).
It is important to mention that the radio spots of the campaign were used as information and awareness tools in the tactics radio and TV forums and "parent schools" tactic (more information next).
3. **Social support and empowerment**. In this field of action, several tactics and communication actions were designed and implemented.
    (a) **Community dialogues**. Were implemented in the urban municipal centers around the topics reading, educational quality, and the conditions to achieve educational quality, also known as learning opportunities. As well as in the radio and TV forums, people from Guatemala City did not attend the dialogues to lead the dialogues. Local leaders on education issues, teachers and event parents were the leading panelist at the community dialogues. This again generated social support and allowed local actors to participate in actions to improve and change education in their communities and municipalities.
    (b) **Reading Fairs.** These had their scope of action schools and involved local educational and municipal authorities, principals, teachers, volunteers, parents, and students. The main objective of the fairs was to involve them in the development of didactic tools for the local learning communities for the best learning of reading and to raise awareness of the importance of the practice of reading in school and in family settings. The implementation of this tactic of promotion to reading (2011–2012) counted the participation of 694 adults from the zones of the intervention. Among them were volunteers, school directors, teachers, and educational and municipal authorities. If we add the participating children, almost 2000 people participated.
    (c) **"Parent schools."** Were designed and executed to provide parents with skills in recognizing their children's reading skills and to facilitate reading-learning activities at home. The main challenge of implementing such a tactic was that most of the fathers and mothers were of low schooling and/or illiterate and those from the Western Highlands, especially mothers, did not speak Spanish, only the Mayan K'iche' language. In this sense, it was necessary to create, through experts in adult education at rural level with experience in the indicated social and cultural contexts, work guides of around 2 h, which would be implemented through face-to-face sessions, the days and time that the parents agreed in each group. The project implemented this empowerment tactic in 20 schools, with an average of

35 participants per session, through volunteer facilitators in its first phase (2011) and, in the second (2012), through facilitators hired by a project grant with a nongovernmental organization with presence in the highlands of Guatemala. In most of the cases, the facilitators were bilingual (Mayan areas), and where it was not possible, they were supported by a local leader. In the first phase (6 sessions), the focus was reading and in the second (8 sessions) on expected student learning. Of the 20 schools where it was implemented, only 2 participated in the two phases.

For the social support and empowerment field of action, it was necessary to contract, under a grant scheme, organizations with rural and community experience that had more permanence in the work areas. It is key to indicate that the USAID project did not have a full-time regional or local team to implement its tactics and actions.

## 34.9 Assessments, Perceptions of Change, and Results of the Communication Intervention

The awareness campaign of quality education and the reading socialization mechanism in its tactic of "parent schools" contemplated in their design assessment actions regarding perceptions of change and penetration. For the "parent schools," questionnaires were made that explored, at the end of the intervention, the perceptions of change and comments of the fathers and mothers, as well as the teachers of the schools where it was implemented. On the awareness campaign, penetration surveys were designed, included in the post-baseline field studies carried out by USAID/USAID/Education Reform in the Classroom with the educational communities in the geographical areas where the project was implemented. At the end of 2013 and the beginning of 2014, close to the end of the USAID project (March 2014), focus groups with local actors (parents, teachers, school directors, and educational authorities) were designed and implemented to learn about their perceptions of communication actions carried out in 2011 and 2012. Likewise, as part of this final monitoring process, interviews were held with representatives of the organizations that implemented these communication actions locally (Arreaga 2014).

### 34.9.1 Results

**Awareness campaign monitoring (2011).** A couple of months after the campaign was broadcast in its first phase in Quiché and Jalapa (September 2011), USAID/Education Reform in the Classroom conducted a penetration survey in the municipalities where it was broadcasted/disseminated. This survey was conducted with fathers and mothers or managers of students who were in their homes in the areas of opportunity. The sample was taken to a total of 788 people, where 61.8% were women and 38.2% were men. To the question: *"Do you listen to radio?,"* in Quiché,

62% said yes, and 38% did not. In Jalapa, to the same question, 93.6% said they did listen to the radio, and 6.4% did not.

The people who indicated that they did listen to the radio were asked if they remembered any announcement of education through this medium. In Quiché, 13.4% said that they remembered an announcement about education; and 86.6% commented that they did not remember. On the other hand, in the people of Jalapa, 69.6% answered that they did remember announcements about education, and 30.4% indicated that they did not remember.

The education messages that were indicated more than once, as remembered by the people who were surveyed, are:

- Send children to school
- Education is free
- Education is a right
- Do not mistreat children
- On the health and hygiene of children
- There will be no classes for political elections
- **Education begins at home and is also the responsibility of parents**
- **Children must be well educated**
- **Help children with their homework**
- **Education gives a better life**
- **Send children to school so they can learn**
- Educate is to live
- **Reading support from home**
- **Teacher training**
- That there is no educational quality
- How to educate children
- Educate everyone
- Send children to school every day
- **Education is important for life**
- Educate without discrimination
- Educate with love (Arreaga 2014)

The remembered subjects of education that have bold (8) are those that were included as contents in the awareness campaign.

**Awareness campaign monitoring (2012).** Next, the results of the penetration of the campaign by alternative means in 2012 are presented, according to the survey conducted in September 2012, to school directors and teachers of Totonicapán and Quiché (For this department, the campaign was disseminated with these alternative media in 2011).

The survey asked the following two questions to the two target groups:

1. *Have you heard through an automobile, motorcycle with speakers, or speakers an important announcement for the community?*
2. *What do you remember about those ads?*

3. *Have you heard another way or means of communication to talk about education in the municipality?*
4. *What do you remember about those ads?*

### 34.9.2 School Directors

Of a total of 52 directors questioned with question No. 1, 80.8% answered that they had not heard any message by said means of communication, and 19% said they had heard. Of those who affirmed that they had heard messages in alternate media, they remembered about "promotion of the use of libraries," one of the education issues of the project campaign.

About having heard by other media about education, 42.6% said they had heard. These directors said that they had heard these messages on radio, then through educational dialogues, and finally through other media. The main contents heard by the directors on education through these channels were "educational quality," "libraries," "reading, and teacher training." All these contents were part of the project's campaign.

### 34.9.3 Teachers

Of a total of 130 teachers questioned with question No. 1, 72.3% answered that they had not heard any message by these means of communication, and 27.7% said they had heard. Of those who affirmed that they had heard messages in alternative media, they remembered about "reading," "writing," libraries," and "intercultural bilingual education," the first three being education issues of the project campaign.

About having heard by other media about education topics, 46.5% said that they had listened. These teachers stated that they had heard these messages on radio (72%), then educative community dialogues (23%), and other means (5%). The main contents heard by teachers about education through these channels were:

- *Educational quality.*
- *That education is quality and effective for children.*
- *That parents are also responsible for the education of children. Adequate time for reading.*
- *Have a habit of reading.*
- *The importance of reading for personal development.*
- *Orientations to parents on education.*
- *Promotion of reading.*
- *Education is important so that a community can improve every day.*

**About awareness campaign monitoring in focus group (End of 2013–beginning of 2014).** In this period, the USAID project carried out five focus groups (around 9 persons per group) with different representatives of the educational community of the project's intervention areas. In four of the focus groups, one person per group listened to the campaign by radio or mobile publicity.

### 34.9.4 Brief Analysis About the Awareness Campaign Monitoring

The first point that draws attention is the use of radio by audiences in the departments where the project intervened: Quiché (62%) versus Jalapa (96%). And in those who do listen radio, the difference is very high between those departments about the top of mind on the education topic: Quiché (13.6%) and Jalapa (69.6%).

Of 21 education topics remembered, 8 topics (38%) were contents of the campaign. Linked to the previous paragraph is very high probability that this impact and penetration occurred more in Jalapa than in Quiché.

The data recorded in the 2012 monitoring show that mobile publicity did not have a high penetration for school directors (19% heard it) and teachers (27.7% heard it). However, what makes it interesting is that for both audiences, school directors and teachers, almost half said they had heard educational messages in other media: radio, dialogues, and others. The point is that for 2012 there was no awareness campaign by radio in Totonicapán or in Quiché. The radial spots were broadcast only by closed circuit in a station in a market of the municipality of Momostenango, Totonicapán, and in the radio forums described before (early reading socialization mechanism, field of action awareness and commitment). They were also disseminated in the community dialogues, where 23% of teachers said they heard about education topics. But there is not enough evidence, according to the sources consulted for this report, to determine with certainty if these groups received the information through the market closed circuit (Momostenango), through the radio forums, or maybe through other radio campaigns not implemented by the USAID project.

**"Parents schools" monitoring.** For this tactic of the early reading socialization mechanism, field of action social support and empowerment questionnaires were made that explored, at the end of the intervention, the perceptions of change and comments of the parents, as well as the teachers of the schools where it was implemented. Also, exploratory questions were asked about the impact of this tactic, in the focus groups carried out by the project (2013–2014).

We present below the perceptions of parents and teachers regarding changes and impact of the intervention (Arreaga 2014):

Parents perceptions:
1. *"It was to wake up the mind. It is no longer as before when nobody supported us."*
2. *"I have changed my way of thinking about the responsibilities of having a child in school and that he learns well."*
3. *"My son has improved his grades this bimester."*
4. *"It helped us a lot in letting us know what we as parents have to do to help our children learn."*

5. *"I didn't know the library of Joyabaj; now, I take my daughter* [there], *who likes to go."*
6. *"In addition of reading more, I got closer to my daughter."*
7. *"We have to help our children, even if we don't know how to read and write."*
8. *"We learned how to spend time with our children reading stories."*
9. *"We read signs of shops and stores with our children."*
10. *"Now he does his homework and he doesn't bring any reds* [to fail an exam or class]*."*
11. *"Even though we don't know how to read and write, the program gives us great ideas to accompany our children."*
12. *"My son used to fail classes, but this year he passed them all. The teacher says he is improving."*
13. *"We dare to help the children, mainly in reading."*

Teacher perceptions:
1. *"Students are more interested in attending school."*
2. *"*[A] *significant change in the relationship between teachers and parents."*
3. *"I had a student who had problems to read and write. I believe the school for parents has helped him to read more."*
4. *"75% of the children have improved their school performance."*
5. *"Boys and girls are now less shy."*

**Community dialogues and reading fairs monitoring.** At the focus groups, the participants said about the reading fairs that it was very motivating for the boys and girls, awakening interest in them by reading. Likewise, the role of community dialogues was recognized as a space for group reflection to search for strategies at the local level that support educational quality. As indicated above, community dialogues were also a good channel for teachers and principals to listen to the spots designed to be broadcast on radio.

## 34.10 Conclusion

From several angles implementing this communication strategy for development, seeking the involvement of parents to achieve quality education focusing on reading was a model to follow and to take account for different actors who finance, design, and implement education projects. Such is the case of the Ministry of Education, projects of international cooperation agencies, civil society organizations, and even academic and university sectors. Below are concepts that indicate these angles that can serve as a guide to know to this previous experience in education and communication for development in Guatemala and highlight to what was learned from this work in communication.

- According to the perceptions showed previously, implementing an early reading socialization mechanism where parents participated in a tactic like a "parent

schools" demonstrated that it is possible to work in Guatemala with parents with little education or illiterate and many of them speaking Mayan languages.
- This was possible using communication approaches that sought to inform, raise awareness, and empower parents. To achieve these approaches, it was necessary to merge disciplines such as communication and adult education, contextualized to the reality of the participants (language, cultural values, replicability of actions at home according to their educational profile, etc.)
- Implementing communication actions for development with a focus on sustainability implies having the support of local, municipal, or departmental partners. They can be volunteers, hired by a grant, or be part of the permanent staff of a project.
- Facilitating reading fairs, community dialogues, and radio/TV forums where participants are teachers, leaders, and local education authorities can generate sustainability of these communication actions, since members of the educational communities are those who participate in its change. Although the USAID project provided the funding and proposed the action, with a *top-down* approach, local stakeholders, including parents, participated in the development of these communication actions (*bottom-up*).
- The use of mass media, such as radio, is quite heterogeneous in Guatemala as evidenced by the penetration surveys of the awareness campaign in two geographical areas of the country (Jalapa vs. Quiché). This suggests that it is necessary to know in greater detail the use of radio at local levels, by certain audiences.
- According to the surveys and focus group carried out, the advertising mobiles or mobile publicity did not have the penetration and expected impact. Possibly more time was needed for exposure to the hearings, but the USAID project did not have a higher budget to use in these media.

Upon completion of the USAID/Education Reform in the Classroom Project (2014), USAID/Guatemala financed and implemented a new project: USAID Lifelong Learning (2014–2019). This project took this communication model for development, replicating on a larger scale the early reading socialization mechanism, including parents school and reading fairs, and using the conceptual and design basis of the awareness campaign.

## References

Alvarado L, Arreaga A, Bollmann C et al (2010a) Requerimiento 1.7: Estrategia de comunicación para el desarrollo. USAID/Reforma Educativa en el Aula, Guatemala

Alvarado L, Arreaga A, Bollmann C, Rubio F, Servaes J, Steele K (2010b) Requerimiento 3.4: Mecanismo de socialización de Lectura Inicial. USAID/Reforma Educativa en el Aula, Guatemala

Arreaga A (2014) Comunicación para el Desarrollo. Informe final de sistematización. USAID/Reforma Educativa en el Aula, Guatemala

Comisión Paritaria de Reforma Educativa (1998) Diseño de Reforma Educativa. Guatemala. http://www.reaula.org/administrador/files/Dise%C3%B1o%20de%20la%20Reforma%20Educativa%20web.pdf. Accessed Dec 2017

PREAL, CIEN (2008) Educación: un desafío de urgencia nacional. Informe de Progreso Educativo, Guatemala

Servaes J (2011) Communication for Sustainable Development. Indicators for Impact Assessment in USAID Project "Educational Reform in the Classroom in Guatemala". J Lat Am Commun Res 2(2):3–34. http://www.academia.edu/2808398/Communication_for_Sustainable_Development._Indicators_for_Impact_Assessment_in_USAID_Project_Educational_Reform_in_the_Classroom_in_Guatemala_. Accessed Dec 2017

Steele K (2011) Reporte final. Campaña de Concienciación sobre Educación de Calidad en el Aula. USAID/Reforma Educativa en el Aula, Guatemala

Stevens J, King E (2004) Administración de Programas de Educación Temprana y Preescolar. Trillas, México (reimpresión 2004)

# Digital Communication and Tourism for Development

## 35

Alessandro Inversini and Isabella Rega

## Contents

| | | |
|---|---|---|
| 35.1 | Introduction: Tourism and Developing Countries | 668 |
| 35.2 | The Conceptualization of eTourism 4 Development | 669 |
| 35.3 | Community-Based Tourism: The Context of Action | 672 |
| 35.4 | Digital Communication Technologies in Community-Based Tourism: Experts' Opinion | 673 |
| 35.5 | Discussion and Conclusions | 674 |
| References | | 676 |

### Abstract

This book chapter aims to discuss critically the role of digital communication technologies in tourism for development. In doing so, the literature at the intersection of tourism, development studies, and digital communication technologies is presented, with a clear proposal to overcome the current reductionist approach within the field. The chapter presents a holistic model of the role of digital communication technologies in tourism for development, which sees them not only as a tool (i.e., to market the destination) but also as a catalyst and a driver for experience creation and co-creation.

### Keywords

Tourism · Development studies · eTourism 4 Development

---

A. Inversini (✉)
Henley Business School, University of Reading, Greenlands Henley-on-Thames, UK
e-mail: a.inversini@henley.ac.uk

I. Rega (✉)
Centre of Excellence in Media Practice, Faculty of Media and Communication, Bournemouth University, Bournemouth, UK
e-mail: irega@bournemouth.ac.uk

© Springer Nature Singapore Pte Ltd. 2020
J. Servaes (ed.), *Handbook of Communication for Development and Social Change*,
https://doi.org/10.1007/978-981-15-2014-3_82

## 35.1 Introduction: Tourism and Developing Countries

The nature of the contribution of tourism to developing and emerging economies is to date controversial. Both researchers and practitioners have been discussing this subject, debating a series of issues that can be positioned on a spectrum with two clear ends: on the one side, there is the technical and infrastructural development of the destination; on the other side, there is the social development of the communities involved in the tourism initiatives.

Commentators praise the advent of tourism in developing rural contexts endowed by natural beauties: this, in fact, helps the creation of infrastructures and services that are used primarily by the travelers and ultimately by the residents. Apart from infrastructural and eventual economic gains, other benefits attributed to tourism in developing and rural areas include skills development; exposure to the world beyond their immediate locality; better access to education, healthcare, clean water, and transportation; and increased confidence in and sense of ownership of the enterprise and the neighborhood in which the tourism enterprise is situated (Scheyvens 2007). Historically, the growth of tourism destinations in developing countries has been led by private companies, often in conjunction with governments and local authorities; these created the basis for tourism to flourish in the designated area. Governments around the world have started to encourage initiatives for the requalification of rural and developing areas (e.g., Park and Yoon 2009) with specific funding programs. These programs are frequently aimed at transforming the primary source of income for local communities (which are often based on agriculture and breeding) and thereby fostering the culture of tourism and hospitality.

Most of these programs have proved to be successful, often radically transforming and reshaping the landscape and the value chain of the rural/developing areas. However, one of the main challenges of these approaches (both the private and public funding programs) lies in the sustainability of the interventions: community dwellers need to be part of the transformation process; otherwise, as soon as the public funding runs out and/or the destination is considered fully exploited by the private sector, they would not have any ownership of the tourism initiatives, and the community would collapse. Academic literature has often advocated the inclusion and involvement of local communities as they are seen as a key resource for sustainable tourism development. In fact, tourism can facilitate community development and poverty eradication because it is labor-intensive, inclusive of women and the informal sector, and often based on the natural and cultural heritage of the rural communities (Ashley and Roe 2002). However, in this respect, there is the clear possibility of the exploitation of local natural and human resources by international investors (Deller 2010): there is the risk of nurturing the creation of a local working class of poor for the international riches. This attitude clearly reflects a kind of postcolonialist approach to tourism where rich investors are able to drive the change within the destination without really involving the local communities in the process. It is clear that, if there is no feeling of ownership of the tourism projects and of the destination, communities will inevitably breakdown. However, there is also evidence of communities successfully blending the traditional source of income (e.g., farming) with hospitality and tourism practices (funded either by the public or private sponsor); these lucky communities flourished to a new life, and community dwellers are actually the protagonists, and ultimately the added value, of the tourism experience (Park and Yoon 2009).

Additionally, UNWTO – United Nations World Tourism Organization – proclaimed 2017 the International Year of Sustainable Tourism for Development (UNTWO 2017), effectively endorsing the impact of tourism on local socioeconomic development. However, both within the UNTWO documents and in the general academic literature about tourism in developing and emerging economies, there is scant evidence about the role that digital communication technologies could play to facilitate and promote community-based sustainable tourism initiatives, thus underestimating how these technologies changed the global tourism market, which is increasingly relying on the internet for travel search and purchase (Spencer et al. 2012).

## 35.2 The Conceptualization of eTourism 4 Development

The research community is starting to acknowledge the importance of digital communication technologies in tourism for development (e.g., Gössling 2017); however, there remains a long way to go to understand fully the impact of digital communication technologies in this field and their possible evolution. In fact it is recognized that digital communication technologies have reshaped the competitive landscape of tourism (Buhalis 2003) on the one side and that on the other, they can strongly support community-based socioeconomic development (Unwin 2009). Nonetheless, there are few studies investigating the strategic and tactical role of digital communication technologies in community-based tourism for socioeconomic development (Rega and Inversini 2016).

This lack in the literature is peculiar, given (i) the possible contribution that tourism can bring to community development, (ii) the transformation that global tourism has faced in recent decades, and (iii) the recognition of the role played by digital communication technologies in the socioeconomic development processes. Let us first explore the interplay between tourism and community development. The industry is populated and operated by a galaxy of (micro) small and medium tourism enterprises (SMTEs), constituting its "life blood" (Thomas et al. 2011). These businesses operate in synergy with one another and dominate the value chain of the travel and tourism domain. In fact, to deliver a given experience (i.e., a hospitality experience), a series of competencies, products, and services (e.g., taxi service, breakfast, laundry service, etc.) needs to be provided, in addition to accommodation, by a succession of actors within the value chain. It is easy to understand that a popular and successful hospitality establishment in a community can have a waterfall impact on a series of other formal or informal players who provide services to the main business (i.e., hospitality). This is something that academic literature has already acknowledged and studied: Kirsten and Rogerson (2002), for example, focused their study on the contribution of micro and small tourism enterprises

(MSTEs) to the tourism value chain in emerging and developing economies; it transpires that successful interplay between SMTEs and MSTEs may result in a virtuous effect leading to local socioeconomic development for the local communities and for the overall tourism supply chain (Simpson 2008). However, also within these studies, there is no trace of the role of digital communication technologies, and the issue remains scantly investigated.

Let us now take into consideration how digital communication technologies have impacted the tourism industry. Recent decades have witnessed the disruptive rise of information and communication technologies (ICTs), which are playing a pivotal role in travel and tourism and are reshaping the competitive landscape of the industry. Literature acknowledges the impact of ICTs, especially the internet, in the tourism field within a stream of research called eTourism (Buhalis 2003). One of the most interesting issues in the recent development of communication technologies is related to their increasing availability (Rega and Inversini 2016). The rise of smartphones together with the mobile internet is allowing a wider audience to be online and to create and share contents in an extremely quick way. Mobile computing, alongside the unprecedented digitalization of conversations and word-of-mouth, creates an avenue for interactive discussion between communities of interests that were – until few years ago – separate from one another. Thanks to the advancements of technologies, peripheral tourism businesses – that are not included in the popular distribution channels – can now be visible and bookable by travelers.

Finally, let us consider how academics and practitioners have reflected and explored the crucial role of communication technologies to reach the Millennium Developments Goals. In recent decades, a new interdisciplinary field of research has emerged, called ICT4D (Information and Communication for Development – Unwin 2009). ICT4D brings together computer scientists and social scientists tackling a broad range of developmental domain, such as agriculture, health, education, and political participation. Although there are a lack of studies focusing on tourism, an interesting set of works focuses on entrepreneurship and suggests micro-, small-, and medium-sized enterprises as units of analysis (Heeks 2010) to investigate the connection between technologies and development. An effective use of ICTs by SMTEs and MSTEs in developing contexts can enhance businesses' visibility and competitiveness and, therefore, improve the socioeconomic conditions of local communities.

We named the missing piece of research at the intersection between (i) development studies, (ii) tourism studies, and (iii) ICTs eT4D (eTourism 4 Development) (Fig. 1). In order to describe eT4D better, the following diagram will sum up the abovementioned intersections:

*Tourism Studies and Development Studies*: Literature presents an ever-growing body of research about the intersection between development studies and tourism studies (Sharpley and Telfer 2014). The academic debate acknowledges that there could be the risk of an imperialistic and post-colonialist approach of international tourism as outlined above. Exploitative tourism has been historically challenged (Krippendorf 2010), and alternative forms of tourism arose with

**Fig. 1** The conceptualization of eTourism 4 Development

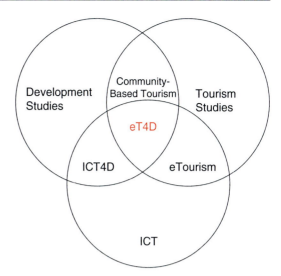

a renewed focus on local sustainability and community development. Alternative tourism is an ideologically different form of tourism that is considered preferable to mass, consumer-driven, and exploitative travel (Wearing 2001), because it takes into consideration the local communities and the local environment.

*Tourism studies and ICT*: ICTs have revolutionized the structure and organization of the tourism industry; the advent of the internet produced a paradigm shift in the industry thanks to the convergence among informatics, communication, and multimedia; technologies can support both (i) marketing, i.e., bringing to the attention of the travelers a given offer (Inversini et al. 2015), and (ii) the travel and tourism value chain, i.e., enabling the effective management of products and services (Zhang et al. 2009).

*Development and ICT*: Within the ICT4D field, ICTs are seen to generate benefits related, for example, to the interaction between customers and suppliers (Donner and Escobari 2010) or the enhancement of labor productivity leading to higher salaries; however, tourism has not been regarded as a research domain by academics and practitioners in the field (Rega and Inversini 2016) resulting in a low number of studies dealing with this issue.

Therefore, merging these three perspectives, tourism can contribute in disrupting its imperialistic and post-colonialist tradition in developing countries (Nash 1989; Pastran 2014) by giving access via the global market to alternative forms of tourism; thanks to digital communication technologies, communities and local players have the unprecedented opportunity to manage their experiential products by themselves and to promote and sell their products online without intermediaries. This can foster the growth of community-based tourism initiatives and enhance the economic and social status of local entrepreneurs and collaborators, thus generating better living conditions in disadvantaged communities.

## 35.3 Community-Based Tourism: The Context of Action

The concept of eTourism 4 Development can be applied to tourism and hospitality initiatives led by local communities (Inversini and Rega, 2016). It is in fact proposed that communities can leverage digital communication technologies to design, deliver, and promote their travel-related products and/or experiences to a wider public (e.g., Inversini et al. 2015). This vision challenges those of Cater (1995) and of Akama (1999) who were describing an oligopolistic scenario in developing and emerging economies dominated by big tour operators and travel agencies with high marketing power. What eTourism for Development proposes is to refocus the location of sustainability within the community, where local dwellers can engage among themselves and with international travelers, thanks to digital communication technologies, and in doing so disrupt the relationship with travel agents and tour operators. In this vision, digital communication technologies are seen as a strategy and tactic to foster both connections between local businesses and between local businesses and travelers.

This approach is facilitating the rise of alternative markets, which include sustainable tourism (Hunter 1997), ecotourism (Cater 1993), ethical tourism (Weeden 2002), and volunteer tourism (Uriely et al. 2003). These alternative forms of tourism are driven by a growing demand for products and services that are more sustainable, pro-poor, and less harmful to local environments and communities (Callanan and Thomas 2005). Community participation is often regarded as one of the most essential tools to drive tourism toward a substantial contribution to the local, regional, and national development of a country. When the community participates fully in tourism activities, there is sustainability, the increased opportunity for local people to benefit from the activity taking place in their vicinity, positive local attitudes, and the conservation of local resources (e.g., Tosun 2006). Murphy (1985) maintained that community-based tourism (CBT) emphasizes the inclusive participation of communities in tourism initiatives to ensure economic returns to the locals and promote the socioeconomic evolution of rural and developing areas (Heeks 2010).

On the same subject, Akama and Kieti (2007) stated that the success of tourism should not merely be measured in terms of the increase in arrivals and revenues but based also on how tourism is integrated with the local and national economies and its contribution to the overall development of the local communities. A bottom-up CBT initiative can induce development for the community, because local spontaneous participation would come with full involvement in production, management and marketing, product development in line with local assets, and reliance on community networks (Zhao and Ritchie 2007). In fact, the key to enhancing the welfare of poor communities is not in expanding the size of the tourism sector but in unlocking opportunities for the poor within the industry (Ashley and Roe 2002). In this way, tourism can be understood as one of many development strategies and could create synergies with other approaches to enhance community development and poverty alleviation (Zhao and Ritchie 2007).

## 35.4 Digital Communication Technologies in Community-Based Tourism: Experts' Opinion

In order to contribute to the discussion related to tourism in developing and emerging economies with the angle of community-based eTourism4 Development, this chapter presents the outcomes of 11 semi-structured interviews which will help in shedding light on the power of digital communication technologies in designing, supporting, and promoting tourism for development at the local level. This chapter presents experts' perspectives and proposes a framework for the adoption of digital communication technologies that would contribute toward the development of a new and transformative learning experience. In particular, the study wanted to understand experts' perspectives about (i) the importance and role of digital technologies in community-based tourism socioeconomic development and (ii) how digital technologies can serve community-based tourism.

Eleven international academic experts working at the crossroads of digital technology, tourism studies, and development studies were interviewed over the telephone. Experts were chosen on the basis of a systematic literature review for the structured keywords "tourism + digital technologies + local development" carried out in two academic databases, namely, EBSCO and Web of Science. Specifically, five of the interviewees had a background and a position related to tourism (marketing and management); five held a position related to development studies, and one had a position related to tourism and development studies. Respondents were asked about the importance and role of digital communication in community-based tourism socioeconomic development. Interviews were recorded, anonymized, and transcribed to be analyzed through thematic analysis.

| Code | Role | Area of research |
|---|---|---|
| #1 | Professor | Tourism |
| #2 | Emeritus Professor | Developmental studies |
| #3 | Senior Lecturer | Tourism and developmental studies |
| #4 | Professor | Developmental studies |
| #5 | Associate Professor | Developmental studies |
| #6 | Professor | Tourism |
| #7 | Professor | Tourism |
| #8 | Professor | Tourism |
| #9 | Assistant Professor | Tourism |
| #10 | Associate Professor | Developmental studies |
| #11 | Professor | Developmental studies |

In the first instance, experts were all supporting the role of tourism for local socioeconomic development together with a series of possible controversial effects, thus confirming literature in the field (Deller 2010).

Engagement with people and developing them personally was one theme stressed by experts: the connection with local communities can hold great importance in motivating them to partake in the tourism development process. Expert #11 said that: "socio development has to do with people looking into themselves in terms of their capability and so on, the financial uplifting of those people. I do think that tourism can serve as a very good tool to foster socio-economic development in local communities."

However, expert #9 said with reference to African tourism that: "some of the local communities in the vicinity of tourism development have resentment towards it as they don't see the benefits of it for them. They might see it as something that only recent people can engage in and as a community they won't have access to it." The issue of helping communities to understand tourism appears to be a hard balance to reach as some projects do not deliver the intended benefits. Expert #6 touched on this, stating: "The return investment just isn't there; there's a lot of evidence on evaluations on why tourism funded projects succeed or fail. When those have happened, the focus has been on looking at the output of the project, not the impacts developed by them."

It is possible that there needs to be a shift in perceptions that moves past a focus on investment and toward assisting communities on a personal level. This is in line with the work of Kleine and Unwin (2009), where the accent is on the development of people's freedom within communities rather than on infrastructural advancement.

The role of digital communication technologies within community-based tourism socioeconomic development was then discussed by interviewees. With mention of the islands in Greece, expert #5 emphasized that "The ICT applications have the ability to redefine distance between places; this impacts the attractiveness of a region for businesses, residences and education." In this way technology is a driving force for development, in that it creates opportunities for communities to be able to make use of it (Inversini et al. 2015). Some experts (e.g., respondent #6) made mention of projects that have failed because of investments not being put to use effectively, and this lack of technological infrastructure in turn restricts the impact that technology can have on a region; expert #9 agreed with this: "I think unfortunately there cannot be socio-economic development without technology but I think in terms of developing countries the problem lies in wasted resources. We might have an idea that technology can make us more complex and efficient but if this is not available then we are stuck. The solution is to find some funding and the drive to focus on that will be most important."

Ultimately communication technologies and tourism can play a critical role in socioeconomic development, but how they are both delivered and implemented is the most important factor. Experts mentioned specific techniques for this operation. For example, expert #1 mentioned the rise of smart tourism, a concept that "enables the collective competitiveness of tourism destination to manage their co-creation of activities."

## 35.5 Discussion and Conclusions

When asked about the impact of digital communication technologies for community-based tourism socioeconomic development, two primary perspectives emerged:

(i) Few experts believed that technology is just one of the possible tools to market and manage tourism at community level but that country/state policies and incentives are the (top-down) driving force for local socioeconomic development; expert #8 stated: "Technology is always only a tool; in order to empower someone, you need to understand the technology, so you need to understand the benefits, operate and handle it and also need a lot of time. If they are used to the proper way then that can empower people but on the other hand it can often be time-consuming and often (especially for smaller communities) it's better for people to communicate face-to-face instead of using ICT."

(ii) Most of the experts discussed innovation (and in particular digital technology innovation) as the prerequisite for local socioeconomic development, claiming a bottom-up vision where community digital empowerment can nurture local socioeconomic development; expert #7 maintained: "Technology is a driving force for local growth and co-operation between different stakeholders at local level. It contributes to overcome physical, social and economic barriers." On the same line of thought, respondent #1 explained: "Technology will provide the infrastructure for the development; interconnectivity will enable communities to put forward their story and attract customers. Technology will be the connector between buyer and seller and will provide the infrastructure to facilitate the co-creation of local experiences."

Taking for granted the infrastructural evolution which is happening in some developing contexts, as asserted by experts (#1, #3, #7, #9, #10), it is possible to overcome the reductionist conception of digital communication technologies as a mere tool within this context to explore a more contemporary role of these as a catalyst for creating and co-creating experiences. Digital communication technologies should be leveraged to empower local communities to create a network of formal and informal business able to cooperate toward the creation and co-creation of sustainable travel experiences; this will allow communities to flourish and to leverage their identity toward the creation of authentic tourism products to be marketed to travelers. The marketing perspective (together with the sales one – eCommerce) is also essential to sustainable community-based tourism; digital communication technologies will, in fact, allow interactive communication between community dwellers and potential visitors toward the co-creation of experiences.

In conclusion, thanks to a detailed discussion of the role of tourism in socioeconomic development coupled with a critical debate on the academic literature about the key role of digital communication technologies in this field, this chapter proposed a new lens with which to study and understand the impact of digital communication technologies in socioeconomic development at the community-based tourism level; the chapter contributes to the body of knowledge of eTourism 4 Development, proposing a pragmatic and conceptual contribution about digital communication technologies to tourism for development.

# References

Akama JS (1999) The evolution of tourism in Kenya. J Sustain Tour 7(1):6–25

Akama JS, Kieti D (2007) Tourism and socio-economic development in developing countries: a case study of Mombasa Resort in Kenya. J Sustain Tour 15:735–748. https://doi.org/10.2167/jost543.0

Ashley C, Roe D (2002) Making tourism work for the poor: strategies and challenges in southern Africa. Dev South Afr 19(1):61–82. https://doi.org/10.1080/0376835022012385

Buhalis D (2003) ETourism: information technology for strategic tourism management. Financial Times Prentice Hall, Harlow

Callanan M, Thomas S (2005) Volunteer tourism: deconstructing volunteer activities within a dynamic environment. In: Niche tourism: contemporary issues, trends and cases. Butterworth-Heinemann, Oxford, pp 183–200

Cater E (1993) Ecotourism in the third world: problems for sustainable tourism development. Tour Manag 14(2):85–90. https://doi.org/10.1016/0261-5177(93)90040-R

Cater E (1995) Consuming spaces: global tourism. In Allen J, Hamnet C (eds) A Shrinking World? Global Unevenness and Inequality (Milton Keynes). Oxford, Oxford University Press, pp 183–231

Deller S (2010) Rural poverty, tourism and spatial heterogeneity. Ann Tour Res 37(1):180–205. https://doi.org/10.1016/j.annals.2009.09.001

Donner J, Escobari MX (2010) A review of evidence on mobile use by micro and small enterprises in developing countries. J Int Dev 22(5):641–658. https://doi.org/10.1002/jid.1717

Gössling S (2017) Tourism, information technologies and sustainability: an exploratory review. J Sustain Tour 25(7):1024–1041

Heeks R (2010) Do information and communication technologies (ICTs) contribute to development? J Int Dev 22(5):625–640. https://doi.org/10.1002/jid.1716

Hunter C (1997) Sustainable tourism as an adaptive paradigm. Ann Tour Res 24(4):850–867. https://doi.org/10.1016/S0160-7383(97)00036-4

Inversini A, Rega I (2016) eTourism for socio-economic development. Symphonya. Emerg Issues Manag 0(1). https://doi.org/10.4468/2016.1.07inversini.rega

Inversini A, Rega I, Pereira IN, Bartholo R (2015) The rise of etourism for development. In: Information and communication technologies in tourism 2015. Springer, Cham, pp 419–431

Kirsten M, Rogerson CM (2002) Tourism, business linkages and small enterprise development in South Africa. Dev South Afr 19(1):29–59

Kleine D, Unwin T (2009) Technological revolution, evolution and new dependencies: what's new about ict4d? Third World Q 30(5):1045–1067. https://doi.org/10.1080/01436590902959339

Krippendorf J (2010) Holiday makers. Taylor & Francis https://www.taylorfrancis.com/books/9781136357558

Murphy PE (1985) Tourism: a community approach. Methuen, London

Nash D (1989) Tourism as a form of imperialism. Hosts and guests: The anthropology of tourism. Hosts and Guests: The Anthropology of Tourism 2:37–52

Park D-B, Yoon Y-S (2009) Segmentation by motivation in rural tourism: a Korean case study. Tour Manag 30(1):99–108. https://doi.org/10.1016/j.tourman.2008.03.011

Rega I, Inversini A (2016) eTourism for development (eT4D): the missing piece in the ICT4D research agenda. Inf Technol Int Dev 12(3):19–24

Scheyvens R (2007) Exploring the tourism-poverty nexus. Curr Issues Tour 10(2–3):231–254

Sharpley R, Telfer DJ (2014) Tourism and development: concepts and issues, 2nd edn. Channel View Publications, Bristol

Simpson MC (2008) Community benefit tourism initiatives – a conceptual oxymoron? Tour Manag 29(1):1–18. https://doi.org/10.1016/j.tourman.2007.06.005

Spencer AJ, Buhalis D, Moital M (2012) A hierarchical model of technology adoption for small owner-managed travel firms: an organizational decision-making and leadership perspective. Tour Manag 33(5):1195–1208. https://doi.org/10.1016/j.tourman.2011.11.011

Thomas R, Shaw G, Page SJ (2011) Understanding small firms in tourism: a perspective on research trends and challenges. Tour Manag 32(5):963–976. https://doi.org/10.1016/j.tourman.2011.02.003

Tosun C (2006) Expected nature of community participation in tourism development. Tour Manag 27(3):493–504. https://doi.org/10.1016/j.tourman.2004.12.004

UNTWO (2017) 2017 is the International Year of Sustainable Tourism for Development. Retrieved 5 May 2017, from http://media.unwto.org/press-release/2017-01-03/2017-international-year-sustainable-tourism-development

Unwin PTH (2009) ICT4D: information and communication technology for development. Cambridge University Press, Cambridge

Uriely N, Reichel A, Ron A (2003) Volunteering in tourism: additional thinking. Tour Recreat Res 28(3):57–62. Special Issue: Volunteer Tourism

Wearing S (2001) Volunteer tourism: experiences that make a difference. CABI, Wallingford

Weeden C (2002) Ethical tourism: an opportunity for competitive advantage? J Vacat Mark 8(2):141–153. https://doi.org/10.1177/135676670200800204

Zhang X, Song H, Huang GQ (2009) Tourism supply chain management: a new research agenda. Tour Manag 30(3):345–358. https://doi.org/10.1016/j.tourman.2008.12.010

Zhao W, Ritchie JRB (2007) Tourism and poverty alleviation: an integrative research framework. Curr Issues Tour 10(2–3):119–143

# Part VI

# Methods, Techniques, and Tools

# A Community-Based Participatory Mixed-Methods Approach to Multicultural Media Research

**36**

Rukhsana Ahmed and Luisa Veronis

## Contents

| | | |
|---|---|---|
| 36.1 | Introduction | 682 |
| 36.2 | Conceptual Framework and Methodological Approaches | 683 |
| 36.3 | Description of the Partnership and Its Development Process | 684 |
| 36.4 | Governance Structure of the Partnership | 684 |
| | 36.4.1 Collaborative Participation in the Intellectual Leadership | 686 |
| 36.5 | Participatory Research Design | 686 |
| | 36.5.1 Design of Research Instruments | 686 |
| | 36.5.2 Data Collection | 687 |
| | 36.5.3 Data Analysis | 688 |
| 36.6 | Participatory Knowledge Mobilization | 688 |
| | 36.6.1 Knowledge Production/Creation | 689 |
| | 36.6.2 Dissemination | 689 |
| 36.7 | Key Learnings and Insights | 690 |
| 36.8 | Cross-References | 691 |
| References | | 691 |

### Abstract

This chapter critically reviews and discusses the methodological approach adopted by the Ottawa Multicultural Media Initiative (OMMI), a team-based, multidisciplinary, and multi-sectoral partnership project, examining the role of Multicultural Media (MCM) in fostering the settlement, integration, and well-being of four immigrant and ethnocultural communities in Ottawa,

---

R. Ahmed (✉)
Department of Communication, University at Albany, SUNY, Albany, NY, USA
e-mail: rahmed4@albany.edu

L. Veronis
Department of Geography, Environment and Geomatics, University of Ottawa, Ottawa, ON, Canada
e-mail: lveronis@uottawa.ca

© Springer Nature Singapore Pte Ltd. 2020
J. Servaes (ed.), *Handbook of Communication for Development and Social Change*,
https://doi.org/10.1007/978-981-15-2014-3_62

Canada – Chinese, Spanish-speaking Latin American, Somali, and South Asian. The project was carried out in accordance with key principles of community-based participatory research in the hopes to influence policy and standards and create a community action project for promoting newcomer well-being and inclusion, as these concepts are understood within specific ethnocultural and immigrant communities. To this end, a mixed-methods approach, including surveys, focus groups, media content analysis, and semi-structured interviews, was adopted, and all research instruments were developed in collaboration with multiple local community and municipal stakeholders. While navigating the challenges of participatory research (e.g., time constraints, competing expectations, and reiterative process), significant opportunities emerged to produce grounded and relevant research and policy outcomes that encompass the diverse needs and practices of the communities under study and are embedded within the local multicultural context. The chapter concludes with a discussion of the key insights gained from the OMMI experience that will help to advance more "holistic" approaches to the study of MCM in a variety of contexts for development and social change.

**Keywords**

Community-based participatory research · Ethnocultural · Knowledge mobilization · Immigrant communities · Mixed methods · Multicultural media · Partnership

## 36.1 Introduction

The Ottawa Multicultural Media Initiative (OMMI) is a team-based, multidisciplinary, and multi-sectoral partnership project, examining the role of Multicultural Media (MCM) in fostering the settlement, integration, and well-being of four immigrant and ethnocultural communities (i.e., Chinese, Spanish-speaking Latin American, Somali, and South Asian) in Ottawa, Canada. OMMI's three main interrelated objectives were (1) to build long-term research capacity and expertise on the intersections between MCM, immigrant integration, and economic prosperity; (2) to forge strong partnerships with the municipal government (the City of Ottawa), targeted communities, and local stakeholders (e.g., settlement agencies, immigrant associations, community groups) through a broad and dynamic network of collaboration; and (3) to cocreate, mobilize, and disseminate knowledge about MCM and their potential to address issues related to Ottawa's economic prosperity and immigrant inclusion and integration. With the aim to co-construct knowledge, the project was carried out in accordance with key principles of community-based participatory research (Israel et al. 1998) in the hopes to influence policy and standards and create a community action project for promoting newcomer well-being and inclusion, as these concepts are understood within specific ethnocultural and immigrant communities (EICs). To this end, the research team adopted a mixed-methods approach (including surveys, focus groups, semi-structured interviews, and media content analysis) and developed all research instruments in collaboration with multiple local community and municipal stakeholders. While navigating the challenges of participatory research (e.g., time constraints, competing expectations, and reiterative process), significant opportunities emerged to produce grounded and relevant research and policy outcomes that encompass the diverse needs and practices of four different EICs and are embedded within the local multicultural context. This chapter focuses on the conceptual, empirical, and methodological contributions of OMMI's approach. The chapter is divided into six sections, with Sect. 36.2 describing the OMMI partnership development. In Sect. 36.3, the conceptual framework and methodological approaches that OMMI adopted are described, while Sect. 36.4 outlines the project governance structure. Section 36.5 describes the participatory design of the project, and Sect. 36.6 provides a discussion of the key insights gained from the OMMI experience that will help to advance more "holistic" approaches to the study of MCM in a variety of contexts for development and social change.

## 36.2 Conceptual Framework and Methodological Approaches

Community-based participatory research – oftentimes equated with action research, participatory research, and participatory action research, among other terms (MacDonald 2012) – is "a collaborative approach to research that equitably involves all partners in the research process and recognizes the unique strengths that each brings" (WK Kellogg Foundation 2013, para 1). The OMMI project was carried out in accordance with seven key principles of community-based participatory research (Israel et al. 1998) that advocate for fruitful research partnerships (WK Kellogg Foundation 2013). These principles include (1) recognizing "community as a unit of identity," (2) building "on strengths and resources within the community," (3) facilitating "collaborative partnership in all phases of the research," (4) integrating "knowledge and action for mutual benefit of all partners," (5) promoting "a co-learning and empowering process that attends to social inequalities," (6) involving "a cyclical and iterative process," and (7) disseminating "findings and knowledge gained to all partners" (Israel et al. 1998, pp. 179–180). These principles are in line with the intended outcome of influencing policy and standards (Brown et al. 2002) and creating a community action project with a view to promote newcomer well-being and inclusion as these concepts are understood within specific EICs. The intention was to engage local stakeholders through the collaborative network and the broader community through focus groups and interviews (Anyaegbunam et al. 2010) in order to benefit from their resources and knowledge in such a way as to produce grounded as well as relevant research, practice, and policy outcomes (Graybill et al. 2010; Mays et al. 2013).

Located within a community-based participatory research framework, the OMMI project incorporated a two-pronged approach that combined corresponding quantitative and qualitative methods. The first approach focused on a Uses and Gratifications approach (Blumer and Katz 1974; Elliott 1974) involving surveys in order to examine MCM consumption behavior of members of EICs (why and how members of EICs consume specific MCM to meet their information needs). The second approach focused on a social capital approach (Bourdieu 1986; Putnam 1993a, b) involving focus groups and interviews in order to examine the potential of MCM for building both community bonding and intercommunity bridging social capital.

Together, these two approaches and their respective methods were adopted and adapted to help identify both opportunities and challenges associated with MCM and, more specifically, to determine the role of MCM in facilitating settlement, integration, and well-being of EICs in Ottawa.

It is also important to note that the OMMI project structure reflected the goal to develop social capital by building and bridging links between the different groups: researchers, the producers and consumers of MCM, and community partners, thus helping to create a rich partnership model that can form the basis for future partnerships and networks of collaboration.

## 36.3 Description of the Partnership and Its Development Process

According to the 2011 census data, 22.6% of Ottawa's population were foreign-born, representing a slightly higher percentage than the national average of 20.6% (Statistics Canada 2013). Through a partnership with key local stakeholders such as service providers, the City of Ottawa launched the Ottawa Immigration Strategy (OIS) (OLIP n.d.) in 2011, a community action plan and strategy to address settlement and integration needs of immigrants at the local level.

The growing ethnocultural diversity of the City of Ottawa has been accompanied by the increasing appeal for MCM. Such demands are linking EICs to diverse media sources, creating an opportunity to foster MCM growth. However, little is known of what challenges and opportunities exist in Ottawa's MCM landscape. Against this backdrop, the OMMI engaged in an important partnership development endeavor to undertake a study of Ottawa's MCM landscape with an in-depth focus on the Chinese, Spanish-speaking Latin American, Somali, and South Asian communities to examine strategies for improving the social inclusion, well-being, and labor market integration of EICs.

The four EICs were selected based on three primary considerations. First, these communities constitute Ottawa's largest visible minority groups and are fast growing. Second, the communities represent a fertile combination of characteristics in terms of population size, group cohesion and diversity (e.g., language, immigrant status, period of settlement, and so on), geographical/national dispersion, and media production/consumption, thus likely leading to rich and nuanced research. Third, the communities were complementary to the research team's expertise and language competencies, thus facilitating fruitful collaborative and participatory engagement.

## 36.4 Governance Structure of the Partnership

The OMMI partnership utilized its governance structure (see Fig. 1) to engage the different actors involved in the project. Specifically, it was able to link the University of Ottawa and the City of Ottawa to a network of MCM producers; leading MCM scholars and practitioners; representatives of the local Chinese, Spanish-speaking Latin American, Somali, and South Asian communities; and other stakeholders (e.g., settlement organizations, schools, and community centers).

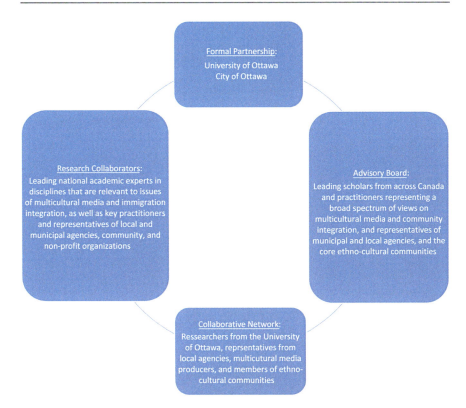

**Fig. 1** Ottawa Multicultural Media Initiative collaborative structure

The governance structure consisted of four main core pillars that created an opportunity for OMMI to develop meaningful network and partnership building. Moreover, it enabled the integration of collaborative expertise in four ways. First, OMMI assembled a strong team of multi- and interdisciplinary researchers specialized in communication, geography, and political science with a unique focus on issues of MCM, immigrant inclusion and integration, diversity, and community building and integration. The project also brought together a group of experienced collaborators comprising academics and practitioners who had a twofold mandate: a) to ensure a rigorous and inclusive methodology and b) to facilitate the goal of cocreating, mobilizing, and disseminating knowledge and understanding about MCM and its potential for addressing the complex, pressing, and interconnected issues of prosperity and newcomer integration. Second, the City of Ottawa, the main partner, involved four municipal departments – Corporate Communications, Community and Social Services, Organizational Development and Performance, and Human Resources – thus bringing expert knowledge about their existing policies, in-kind support, and a willingness to participate in the research to develop improved communications and newcomer integration policy. Third, the project formed an Advisory Board composed of leading scholars from across Canada and practitioners representing a broad spectrum of views on MCM and community integration and representatives of municipal and local agencies and the core ethnocultural communities. The Advisory Board met annually, in conjunction with core team meetings and on a consulting basis, to oversee the direction of the project, monitor progress, and offer advice and guidance as needed. Fourth, the research team developed a collaborative network among the university, local agencies, MCM producers, and the EICs under study. The role of this collaborative network was to bring grounded knowledge of as well as contacts in the four communities in order to recruit study participants, secure access to data sites, and aid in communicating research findings.

### 36.4.1 Collaborative Participation in the Intellectual Leadership

Through their collaboration with the research team, representatives from the City of Ottawa, research collaborators, Advisory Board, and the collaborative network were afforded opportunities to provide input on every aspect of the OMMI project. The research team created a small working group comprised of these representatives in order to ensure more detailed and substantive input on the actual design and direction of the research. For example, in accordance with their needs, policy mandate, and priorities, members of this small working group (1) attended team meetings and workshops; (2) had an opportunity to review and to suggest modifications to the survey questionnaire, focus group materials, and interview guide; and (3) commented on research output drafts as they became available. This process helped to ensure that the research materials reflect the different concerns and needs of the multiple actors involved in the project.

## 36.5 Participatory Research Design

In participatory action research, various methods of data collection, including qualitative, quantitative, and mixed methods, are applied (Pain and Francis 2003; Small and Uttal 2005). This flexibility in data collection is believed to lead to best practices for bringing about social change (McTaggart 1991; Reason 1994; Reason and Bradbury 2001). To meet its objectives, the OMMI project employed a mixed-methods approach in three sequential phases (Creswell 2014) over 3 years (see Fig. 2). At the core of this design was the goal to elicit participatory input from all actors involved in the project (e.g., the research team, the partner, the collaborative network, and the Advisory Board) in order to complement the research findings and offer richer insights (Olson and Jason 2015).

### 36.5.1 Design of Research Instruments

The OMMI adopted four main data collection techniques to meet its research objectives: (1) survey questionnaires with MCM users (i.e., members of the four participating EICs), (2) focus group discussions with MCM users, (3) content

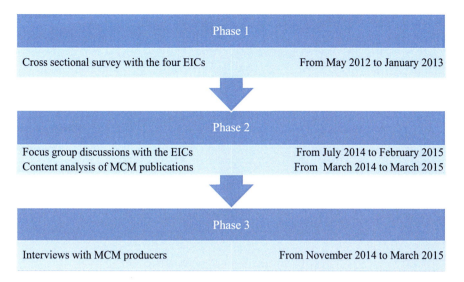

**Fig. 2** Ottawa Multicultural Media Initiative

analysis of MCM publications (press), and (4) semi-structured personal interviews with MCM producers. During the development stage of the main research instruments – the survey questionnaire, focus group guide, and interview protocol – OMMI followed an iterative process, a major principle of participatory action research (Kim 2016; Lewin 1946). As such, a series of consultation meetings were held with the Advisory Board, partners, and community representatives for each phase to collaboratively develop the instruments. In particular, special care was taken to solicit input from the community representatives of each EIC. While this iterative process was time-consuming and involved numerous discussions among various participating actors, it was essential to ensure that the data collection tools reflected the needs and interests of each group.

### 36.5.2 Data Collection

In accordance with the community-based participatory research approach, the collaborative effort of the OMMI research team, community representatives, collaborative network, and local community organizations was leveraged in order to recruit survey, focus group, and interview participants and improve response rates. For example, to recruit survey participants, community partners sent out email messages and letters of invitation to potential survey participants on behalf of the OMMI research team; OMMI researchers attended community events and workshops; with the help of representatives from the collaborative network, language training classes, on and off the university campus, were targeted, where students were provided with information on OMMI research and invited to participate in the survey.

### 36.5.3 Data Analysis

The mixed-methods approach employed by the OMMI project was consistent with a major principle of participatory action research, including flexible research methods, and allowed triangulation of data in consultation with various actors by building on the findings of each stage. Thus survey results were discussed and verified during focus groups; survey and focus groups findings were discussed and verified during interviews with producers; and so on. This process facilitated the amassing of different types of data (quantitative, qualitative) as well as from different types of sources (e.g., media users, media producers, media content). This multiphase, mixed-methods approach led to diverse and complementary perspectives on the use and role of MCM among EICs (i.e., a better picture of the issue from different angles) by engaging researchers, community partners, and collaborators in the analysis.

## 36.6 Participatory Knowledge Mobilization

The knowledge mobilization plan of the OMMI was based on the goal to engage its partners in continued dialogue, cocreation and dissemination of knowledge, and exchange of ideas concerning MCM and community integration. An additional goal was to build bridges among citizens, community leaders, policy makers, and academics. To this end, the research team organized knowledge mobilization activities according to collaborative learning and participatory research principles reflecting the values, expertise, and experience of the team and the nature of the partnership targeted at three broader audiences – academic, policy, and public. In addition to a summary in Table 1, OMMI's collaborative strategies for participatory knowledge production and dissemination are described below.

**Table 1** OMMI collaborative strategies for participatory knowledge production and dissemination

| Audience | Academic | Policy | Public |
|---|---|---|---|
| Knowledge production/ creation | Disciplinary and interdisciplinary research partnership approach Student training | Multi-sectoral team meetings and workshops Participation in research design Participation in preliminary analysis | Broader audience knowledge mobilization strategies Community workshops and feedback sessions Community capacity building |
| Dissemination activities | Conferences Peer-reviewed publication (OA) Project website, online presence, social media | Reports, press releases, policy presentations Project website research data base Participation of partners and collaborators in conferences and professional meetings | Open access website with resources and data bases Interviews in local media Social media |

## 36.6.1 Knowledge Production/Creation

### 36.6.1.1 Academic
A disciplinary and interdisciplinary research partnership approach was employed within the core research team. Comprising five scholars from three different disciplines, the team composition allowed for nuanced, multi-perspective research capable of making a significant contribution to advancing the growing research on MCM in Canada. Students were involved in all aspects of the research process, which allowed the team members and initiative to benefit from the interest, energy, capacity, and knowledge that students brought to the project.

### 36.6.1.2 Policy
The OMMI engaged all its contributors in policy production – i) the City of Ottawa as the main partner; ii) the collaborative network composed of local stakeholders, media producers, and the core ethnocultural communities (Chinese, Spanish-speaking Latin American, Somali, and South Asian); and iii) the Advisory Board composed of leading scholars and practitioners representing a broad spectrum of views on MCM and community integration and the representatives of municipal and local agencies and the four core EICs. Their participation played an important role in knowledge mobilization because of their involvement in every step of the project (some even attended our team meetings while at the proposal stage). In accordance with their needs, policy mandate, and priorities, the Advisory Board members (1) attended team meetings and workshops, (2) offered input in research design (e.g., selection of and access to research sites and groups/issues for study), and (3) commented on research output drafts as they became available.

### 36.6.1.3 Community
The partner, collaborative network, and Advisory Board were actively involved in developing specific knowledge mobilization strategies suitable to community groups for increasing access to and utilization of materials intended for a wider audience. Dissemination workshops were conducted in each of the four core EICs to provide opportunities for community members to offer feedback on the research and the implications of the findings for specific groups. Drawing on the project findings, capacity building activities were created for community members to take part in (e.g., writing and photo contests for students belonging to each of the four EICs).

## 36.6.2 Dissemination

### 36.6.2.1 Academic Dissemination
The OMMI research was presented at regional, national, and international professional meetings (both disciplinary specific and interdisciplinary) to showcase the project's findings while also hosting seminars and organizing conference panels in conjunction with the project. Peer-reviewed articles were published in interdisciplinary and disciplinary journals, including online journals to make project findings

available via open access. A project website was created that served as a key forum for the team, partnership, and research-related communications transcending the academic/policy/public divides, including through the use of social media.

#### 36.6.2.2 Policy Dissemination

In collaboration with the partner and collaborative network, (1) the research team developed and implemented dissemination strategies such as writing, publishing research reports and press releases, and delivering presentations at policy roundtables; (2) the main partner, the City of Ottawa, and the collaborative network had access to the project website research database; and (3) representatives from the partner and collaborative network were invited to present project findings at conferences and regional and national professional meetings.

#### 36.6.2.3 Public Dissemination

For public dissemination of research findings, OMMI adopted three strategies: (1) an open access website was created to archive and document the project findings in publicly accessible databases; (2) the research team engaged in interviews with local mainstream and MCM; and (3) social media fora, such as Facebook and Twitter, were created in order to reach a broader audience.

## 36.7 Key Learnings and Insights

The key learnings and insights gained from the OMMI experience are discussed in light of the challenges that were encountered in applying a community-based participatory research. Overall, the aim is to help advance more "holistic" approaches to the study of MCM in a variety of contexts that can support positive development and social change.

Although the OMMI collaborative effort was successful in closely adhering to and implementing a community-based participatory approach, a few limitations and challenges need to be noted. The main issue was the time and effort invested in the process over a sustained period for all parties involved. Maintaining momentum throughout the project was another significant challenge, even more so due to changes to and turnover among staff and representatives of participating stakeholders. In some cases, there were also changes in the interests and priorities of collaborating actors, which meant that their level of participation, engagement, and involvement fluctuated over time (Gillis and Jackson 2002; MacDonald 2012). The OMMI was able to creatively adapt and find ways to address these challenges, but this required additional efforts, resources, and committed management strategies to ensure progress and completion of the project. Thus, community-based participatory research generally implicates "more than the sum of its parts" both in terms of the process and the outcomes.

Notwithstanding the challenges, the OMMI partnership offered an ideal opportunity for all interested parties, the research team, the City of Ottawa, the collaborative network, and the Advisory Board, to join their efforts in order to pursue relevant research on the pressing questions of well-being and prosperity of the City of Ottawa and its immigrant population. By collaboratively working together, the OMMI research team and all its partners managed to achieve much more than independently. In particular, the project led to the generation of robust and comparable set of empirical data that will advance understanding of the use, role, and potential of MCM in facilitating and promoting newcomer and immigrant integration, inclusion, and participation in Ottawa and in Canadian society more broadly (Ahmed and Veronis 2017; Veronis and Ahmed 2015). More specifically, the research findings have been made available to various local stakeholders (e.g., City of Ottawa, community groups, settlement organizations, etc.) who can apply them to address the complex and interconnected issues of social and economic health of Ottawa's diverse communities.

## 36.8 Cross-References

▶ Capacity Building and People's Participation in e-Governance: Challenges and Prospects for Digital India
▶ Key SBCC Actions in a Rapid-Onset Emergency: Case Study from the 2015 Nepal Earthquakes
▶ Youth Voices from the Frontlines: Facilitating Meaningful Youth Voice Participation on Climate, Disasters, and Environment in Indonesia

**Acknowledgments** This research is part of the Ottawa Multicultural Media Initiative (OMMI) which was funded by the Social Sciences and Humanities Research Council of Canada (SSHRC) (Project Number: 890–2010-0137).

## References

Ahmed R, Veronis L (2017) Multicultural media use and immigrant settlement: a comparative study of four communities in Ottawa, Canada. Journal of International Migration and Integration 18(2):587–612

Anyaegbunam C, Hoover A, Schwartz M (2010) Use of community-based participatory communication to identify community values at a Superfund site. In: World Environmental And Water Resources Congress 2010: challenges of change. ASCE, San Antonio, pp 381–390

Blumler JG, Katz E (eds) (1974) The uses of mass communications: current perspectives on gratifications research. Sage, Beverly Hills

Bourdieu P (1986) The forms of capital. In: Richardson JG (ed) Handbook of theory and research for the sociology of education. Greenwood Press, New York, pp 241–258

Brown D, Howes M, Hussein K, Longley C, Swindell K (2002) Participatory methodologies and participatory practices: assessing PRA use in the Gambia. Overseas Development Institute (ODI), Agricultural Research & Extension Network (AgREN), London

Creswell J (2014) Research design. Qualitative, quantitative, and mixed methods approaches (4th ed.) Thousand Oaks, CA: Sage

Elliott P (1974) Uses and gratifications research: a critique and a sociological alternative. In: Blumler JG G, Katz E (eds) The uses of mass communications: current perspectives on gratifications research. Sage, Beverly Hills, pp 249–268

Gillis A, Jackson W (2002) Research methods for nurses: methods and interpretation. F.A. Davis Company, Philadelphia

Graybill P, Aggas J, Dean RK, Demers S, Finigan EG, Pollard RQ Jr (2010) A community-participatory approach to adapting survey items for deaf individuals and American Sign Language. Field Methods 22:429–448

Israel B, Schulz A, Parker E, Becker A (1998) Review of community-based research: assessing partnership approaches to improve public health. Annu Rev Public Health 19:173–202

Kim J (2016) Youth involvement in participatory action research (par): challenges and barriers. Critical Social Work 17(1):38–53

Lewin K (1946) Action research and minority problems. Journal of Social Issues 2:34–46. https://doi.org/10.1111/j.1540-4560.1946.tb02295.x

MacDonald C (2012) Understanding participatory action research: a qualitative research methodology option. Can J Action Res 13(2):34–50

Mays GP, Hogg RA, Castellanos-Cruz DM, Hoover AG, Fowler LC (2013) Public health research implementation and translation: evidence from practice-based research networks. Am J Prev Med 45:752–762

McTaggart R (1991) Principles for participatory action research. Adult Education Quarterly 41(3):168–187

OLIP (n.d.) Ottawa Local Immigration Partnership (OLIP) (n.d.). Ottawa Immigration Strategy. Retrieved from http://olip-plio.ca/what-we-do/ottawa-immigration-strategy/

Olson BD, Jason LA (2015) Participatory mixed methods research. In: Hesse-Biber SN, Johnson RB (eds) The Oxford handbook of multimethod and mixed methods research inquiry. Oxford University Press, Oxford

Pain R, Francis P (2003) Reflections on participatory research. Area 35(1):46–54.

Putnam RD (1993a) The prosperous community: social capital and economic growth. Current 356:4–10

Putnam RD (1993b) The prosperous community: social capital and public life. Am Prospect 13:35–42

Reason P (1994) Three approaches to participatory inquiry. In: Denzin NK, Lincoln YS (eds) Handbook of qualitative research. Sage, Thousand Oaks, pp 324–339

Reason P, Bradbury H (2001) Handbook of action research: participative inquiry and practice. Sage, London

Small SA, Uttal L (2005) Action-oriented research: Strategies for engaged scholarship. J Marriage Fam 67(4):936–948

Statistics Canada (2013) Immigration and ethnocultural diversity in Canada. Analytical document. National Household Survey, 2011. Ottawa: Statistics Canada. Retrieved from http://www12.statcan.gc.ca/nhs-enm/2011/as-sa/99-010-x/99-010-x2011001-eng.pdf

Veronis L, Ahmed R (2015) The role of multicultural media in connecting local municipal governments with ethnocultural and immigrant communities: the case of Ottawa. Global Media Journal – Canadian Edition 8(2):73–95

WK Kellogg Foundation (2013) Community health scholars program. Community campus partnership for health. Retrieved from http://depts.washington.edu/ccph/commbas.html

# Digital Stories as Data

## 37

Valerie M. Campbell

## Contents

| | | |
|---|---|---|
| 37.1 | Introduction | 694 |
| 37.2 | Digital Storytelling | 694 |
| 37.3 | Creating Digital Stories | 696 |
| 37.4 | Digital Storytelling and Youth | 697 |
| 37.5 | Ethics of Digital Storytelling in Research | 697 |
| 37.6 | Analysis of Digital Stories | 699 |
| 37.7 | Challenges of Digital Storytelling in Research | 700 |
| 37.8 | Conclusion | 701 |
| References | | 702 |

### Abstract

Technology is an integral part of daily life for many young people; much of their communication is mediated by digital and social media. Youth connect with each other through social networking platforms that allow for short, snappy messages, such as Snapchat with images that disappear after a few seconds. Furthermore, YouTube has been steadily growing in popularity as technology changes have made video-making accessible. Anthropological methods suggest that to fully understand your participants, you have to meet them where they are. But perhaps, in addition to that, we also have to adapt our data collection methods to incorporate their means of communication. This chapter describes how digital storytelling can be used as a method for data collection. This chapter is also used to demonstrate the practical application of this method.

---

V. M. Campbell (✉)
University of Prince Edward Island, Charlottetown, PE, Canada
e-mail: vcampbell@upei.ca

**Keywords**

Digital storytelling · Ethics · Youth studies · Technology

## 37.1 Introduction

Technology is ubiquitous in modern life and usage continues to grow in Canada and around the world. In the USA and Europe, the majority of teens own cell phones and use them (and other devices) to participate in social networking sites (Ranieri and Bruni 2013). In Canada, young people are increasingly accessing the Internet using cell phones and other portable devices such as laptops, tablets, and iPods, giving them almost constant access to the Internet (Steeves 2014). Youth in Canada live a significant part of their lives in online spaces; much of their communication is mediated by digital and social media. In the fast-paced world of digital technology, social networks gain and lose popularity on a regular basis. While this change is undoubtedly influenced by a variety of factors, the increasing use of video can account for the growth of networks such as YouTube and Snapchat. A recent Social Media Lab report (Gruzd et al. 2018) provided new information on the social media used most frequently by Canadians. Youth aged 18–24 are most engaged with Facebook (88%) and YouTube (88%), followed by Instagram (65%), Snapchat (62%), and Twitter (47%). Indeed, screen media has become the "preferred means of communication" (Davis and Weinshenker 2012) for today's youth. It follows that scholars in the field of youth studies want to engage with young people in the spaces they inhabit (Hallett and Barber 2014) and through methods that reflect their ways of communicating. This chapter uses the author's dissertation experience (the Project) with using participant-created digital stories as part of the study data to illustrate the incorporation digital storytelling into research design. Despite the relatively wide body of literature about digital storytelling, it is not well documented as a research technique (Walsh et al. 2010, p. 193). The focus is often on the process itself, or the value of engaging participants, or the use of digital storytelling in dissemination of research data. This chapter will begin to fill that gap in knowledge.

## 37.2 Digital Storytelling

The concept of digital storytelling is not new, having been developed in the 1980s by Joe Lambert, cofounder of *The Center for Digital Storytelling* (CDS) in California (now *StoryCenter* https://www.storycenter.org) (Meadows 2003; Robin 2006). Ferrari et al. (2015) describe digital stories as "three- to five-minute videos produced with a mix of voiceover, music, and images to convey first-person narratives" (p. 556). In the StoryCenter context, digital stories are personal expressions about an important person, event, or place in our lives; about what we do; or about a personal journey of love, discovery, or recovery (Lambert 2010). According to Lambert (2010), "we are made of stories" (p. v), and the purpose of the StoryCenter workshops is to "help storytellers find the story they want or need to tell" (p. 8). Thus, digital storytelling, in this context, is intensely personal.

Prior to the introduction of the StoryCenter, digital storytelling existed as the purview of experts in media production who had ultimate control over the completed video (Hartley and McWilliam 2009; Hopkins and Ryan 2014). StoryCenter provided a space and the technology to create exportable videos which enabled "ordinary" people to participate in the practice (Edmonds et al. 2016; Hartley and McWilliam 2009). Today's technology, in particular Web 2.0, has made it possible for practically anyone with access to a computer, or even a smartphone, to create a video (Edmonds et al. 2015; Hopkins and Ryan 2014; Page and Thomas 2011; Ranieri and Bruni 2013). This has opened more avenues for digital storytelling, which is "practiced around the world in increasingly diverse contexts" (Hartley and McWilliam 2009, p. 4). In the StoryCenter model, the emphasis is not on the technology but on the story and the telling of the story (Hartley and McWilliam 2009). As it becomes used in different places and contexts, the way in which digital storytelling is presented varies. However, Hartley and McWilliam (2009) argued that despite this variation, "the practice can usually be directly linked to the CDS" (p. 8).

What makes digital storytelling different from other methods of telling stories? Qiongli (2009, pp. 230–231) listed several characteristics that distinguish digital storytelling:

- Story-oriented: with a focus on the story not the technology
- Disciplined: with a practical framework developed by StoryCenter
- Authentic: storytellers narrate their own personal stories
- Multimedia: weaving together narratives, images, voices, music, and video
- Simple technology: can be done with a computer, microphone, and easy to use software
- Found materials: utilizes preexisting material such as family photo albums
- Collaborative creativity: StoryCenter training workshop model

Authenticity and found materials are key attributes of digital storytelling. Producers narrate their own stories, "prioritizing personal experience and allowing a 'direct voice'" (Hopkins and Ryan 2014, p. 32). The materials they use are also their own, whether preexisting or found/created for the Project, and reflect their connection to the story. Those who engage in the process create, and tell, their own stories and have full editing power. Meadows (2003) lauded this as one of the most exciting aspects of digital storytelling as "no longer must the public tolerate being 'done' by media – that is, no longer must we tolerate media being done *to* [emphasis in original] us" (p. 192). Digital storytelling provides an opportunity to tell your own story and have your voice heard in *your* way (Hopkins and Ryan 2014).

As a pedagogical tool, digital storytelling can be used by educators to deliver content or to engage students in participatory educational practices (Buturian 2017; Clarke and Adam 2012; Robin 2006, 2008). Lowenthal (2009) listed five benefits of using digital storytelling in the classroom: increase student engagement, give access to a global audience, amplify students' voice, leverage multiple literacies, and increase student's emotion (pp. 253–254). While the possibilities and promises of digital storytelling in the pedagogical sense are intriguing, this chapter focuses on digital storytelling as means of sharing personal experiences in a specific research context.

## 37.3  Creating Digital Stories

The creation of a digital story can be seen as a journey (Lambert 2010), often through personal, sometimes emotional, territory. Thus, every story represents a unique experience. However, over time, StoryCenter has developed a seven-step process to guide storytellers in the creation of a digital story. Those steps, as outlined by Lambert (2010), are mentioned in Table 1.

These steps provide a very linear method for the creation of a digital story. In application, the process is much more circular. Each step builds on the one(s) before it, but often the storyteller will revisit a step based on practical considerations as well as insights gained throughout the journey. For example, a storyteller might have an idea of a moment and emotion but, when reviewing images, realize that the emotion generated by the visuals is not what was intended. The choice is then to focus on a different emotional direction or search for new images. Or, once the images and words are put together, there is a discordance – it simply does not work – so, again, new images or new words. And this continues, back and forth between steps, until a story is created. Thus, as in any creative endeavor, the finished product may be

**Table 1**  7 steps of digital storytelling

| Step 1: Owning your insight | Understanding the story behind the story; exploring your motivation for telling this particular story at this particular time |
|---|---|
| Step 2: Owning your emotions | Be aware of the emotions generated within yourself by the story and be honest in conveying that emotion; consider your intended audience when determining how to demonstrate your emotional honesty |
| Step 3: Finding the moment | In the process of developing the story, look for a "moment of change" that will best represent the insight you want to share with your audience |
| Step 4: Seeing your story | Look for the images that come to your mind when thinking about your story. Think about which images best represent the meaning you are trying to convey and then find or create those images for use in the story |
| Step 5: Hearing your story | Think about the script you want to add to your story. Then think about how you want to perform that script. Focus on natural speaking patterns rather than reading. Consider whether your story will be enhanced by the addition of background sounds such as music |
| Step 6: Assembling your story | Now is the time to put it all together – to create a storyboard. In this step, you consider your story as a whole, how you want it to look and sound |
| Step 7: Sharing your story | Consider both the audience you wish to share your story with and the method |

Adapted from Lambert (2010, pp. 9–24)

considerably different from what was conceived. For example, in the Project, participants were asked to create a story about the ethics of online life. Barry (a pseudonym), in the process of preparing his story, was reminded of a bicycle trip he had taken after high school. Although not related to online activity, it recalled an important event and became the story he shared.

## 37.4 Digital Storytelling and Youth

As a means of creative expression, digital stories are opportunities to create visual representations of and give meaning to experiences (Ferrari et al. 2015, p. 556). Davis and Weinshenker (2012) suggested digital storytelling as a method to tap into youth engagement with screen as a "preferred means of communication" (p. 49); taking charge of creating their own messages increases youth agency and empowerment (Clarke and Adam 2012; Davis and Weinshenker 2012; Gubrium et al. 2014; Taub-Pervizpour 2009).

In research with youth, digital storytelling can appear to be a "natural fit" based on young people's engagement with technology. However, we must always beware of making a broad assumption that youths' ubiquitous use of and engagement with digital media translates to skill or interest in all technologies (boyd 2014). The Project incorporated digital storytelling to provide youth participants with the opportunity to engage in a way that resonated with their current practice. Time and resources did not allow the presentation of a three-day digital storytelling workshop. Instead, the concept was discussed in focus groups and ideas brought forward. Youth participants were presented with some information on digital storytelling, including a short presentation on the seven steps outlined above. They were then asked to create a story based on their experiences with ethical issues in digital media. The format and content of the digital story remained the sole responsibility of the participant. There was no training on any specific technology as there are so many different ways to create video, and it was felt that participants would have their own ideas about how to proceed. This proved to be a false assumption demonstrated by Adrienne (a pseudonym) who wanted to tell a story about an experience with a fake Facebook account. However, the video software was unfamiliar to her, and she found it very difficult to understand. She was very confident with the technology she used on a regular basis, but those skills did not translate to this new platform; that became her story.

## 37.5 Ethics of Digital Storytelling in Research

A key element of ethical conduct of research is the informed consent process. Within this process, participants are generally guaranteed that their privacy will be protected, and they will remain anonymous in any publication of study results. The ethical implications of visual methods have been described as "muddy" (Gubrium et al. 2014, p. 1608), particularly if the intention is to share images in dissemination

| |
|---|
| Choose **ONE** of the following options: |
| I agree, or give consent, to provide a limited license to [Author] to use my digital story (as described above) for academic purposes only (including public presentations), at no charge. |
| I agree, or give consent, to provide a limited license to [Author] to use my digital story (as described above) as data for her dissertation only (with no public presentations), at no charge. |
| If you wish to be notified any time your story will be presented, please provide contact information: |

**Fig. 1** Video dissemination options

of results. It is important that the participant understands fully how the images will be used and how anonymity and confidentiality will be limited. Consent should be ongoing, with secondary consent for dissemination (Cox et al. 2014; Gubrium et al. 2014). The Project offered participants several options, formally and informally, regarding dissemination of their digital stories. Formally, they were asked to sign a licensing agreement outlining two options (Fig. 1). This form also allowed the participants to request notification prior to any presentation of their video.

If the content is produced by the participant, consent of nonparticipants who may appear in the images then becomes an issue (Clark 2012; Cox et al. 2014). Indeed "it is often impossible, impractical, or even illogical to maintain the anonymity and confidentiality of individuals in artwork, photographs and film" (Clark 2012, p. 21). Clark further stated that while it may be possible to have the participant make contact with other people who appear in the image(s), it could be time-consuming and difficult to manage. Additionally, "ethical issues may arise when a storyteller publicly identifies people, experiences, or events that others would prefer remain private" (Gubrium et al. 2014, p. 1610). Notwithstanding the ethical challenges mentioned above, participant-created digital stories provide an opportunity to show respect for research participants, a core principle of research ethics (Canadian Institutes of Health Research et al. 2014), by providing them with a forum to present their story, in their words, using methods familiar to them.

Anonymity, another key element of research ethics, is difficult to maintain when using visual methods (Anderson and Muoz Proto 2016). Dissemination of those stories may, and often will, compromise anonymity; however, while that is a key element of research ethics, is it equally important for youth? Cox et al. (2014) argued that anonymity may even be contrary to the intention of participants who create visual data, and Anderson and Muoz Proto (2016) cautioned against paternalism in ethical decision making. The Project storytellers did not attempt to conceal their identity in their videos, and all but one agreed to allow the researcher to use the video in dissemination activities. This suggests that anonymity is not a requirement for youth participation. Indeed, in discussions about research ethics, youth participants were more concerned with transparency; they wanted to know how and why their data would be used.

Despite the participants' willingness to have their videos shared, some of the stories contained images of people other than the storyteller. These stories were received at the end of the Project, after the analysis workshops had been completed. Inclusion of third-party images had not been discussed in the focus groups, and participants had not been asked to ensure images of others were blurred or to garner consent. Consideration was given to have the researcher modify the stories herself and make third-party images blurry, but this created a new ethical dilemma: does the researcher have the right to alter participant-created data? Would the participant be asked to provide consent for this alteration? When presenting direct quotes from participants, researchers will often remove any potentially identifying information – would blurring images in a video be any different? These questions are part of the muddiness (Gubrium et al. 2014) of working with visual data. Kim (2015) discussed "narrative smoothing" as a method used to make a "participant's story coherent, engaging, and interesting to the reader" (p. 192). Although she suggested that this is a method used by many researchers, she also described it as problematic ethically in that it might change the participant's intended meaning and thus not be a faithful account. Kim's concern with this method in narrative inquiry can be considered in the situation above: would blurring the images change the story?

In one example from the Project, a participant submitted a story showcasing the positive side of her social networking sites. This was an important story as much of the discussion in the workshops had been about the negative aspects of social media. However, most of the images in the story included people other than the participant. Although she had given permission to use the digital story in presentations, the only way to do so would be if the researcher modified the video to hide all third-party images. After careful consideration of the ethics of this approach, it was decided to leave the video as submitted and refrain from using it in presentations, although the content would be analyzed as project data.

## 37.6 Analysis of Digital Stories

Analysis of digital stories is complex. The literature on digital storytelling focuses on process or the experiences of participants. There is little discussion on analysis of the stories themselves. Each digital story received in the Project was transcribed verbatim by the researcher and the transcripts read as text, focusing on the content. Reading the stories on paper provides an opportunity to focus on the story without any distraction. However, the text alone offers a limited understanding without the context of the images, voice, and, perhaps, music. Images can be powerful on their own and may illustrate a particular phenomenon in a way which is difficult to explain, however, without the context of the words, may be interpreted in a way contrary to the storyteller's intent. Thus, when digital stories are used as data, both visual and narrative modes of analysis are necessary.

Narrative analysis requires "looking for underlying patterns of meaning" (Shopes 2011, p. 459) within the text. Kim (2015, p. 188) identified the elements of qualitative analysis as codes (identifying concepts), categories (linking codes), patterns (repeated units in categories), and themes (representations of similar patterns). In the narrative context, we are applying these elements to a story, attempting to understand the meanings the participant has given to their experiences (Kim 2015, p. 189). Narratives can be interpreted by focusing on meanings "What does this mean? What does this tell me about the nature of the phenomenon of interest?" (Patton 2002, p. 477). It is necessary to read the words carefully, with a focus on how the words are put together (Daiute 2013) and looking for clues such as a play on words or contradictions in the text (Kim 2015). In other words, if "narrating is a sense-making process" (Daiute 2013, p. 15), analysis of narrative is a process of unpacking how that process unfolded.

Visual research provides a new set of cues for analysis. Visual analysis is similar to narrative in that it is a process of sense-making, developing an understanding of the perception and meanings that can be attributed to a photo or video (Prosser 2011). Whether researcher or participant produced, videos or photos allow the analyst to evaluate more than the words used, for example, the setting, tone of voice (in videos), body language, and other nonverbal behaviors that are captured within an image – "it would take a lot of words to convey the information in the photographs" (Harper 2000, p. 721). However, without the context of the story, there is no framework of analysis for interpretation of the images. Visuals in research can corroborate data, thus improving the trustworthiness of the analysis (Prosser 2011).

Many of the challenges of narrative inquiry (e.g., researcher interpretation of meaning) can be ameliorated by the addition of visual content, "the potential of using visuals in narrative inquiry is enormous" (Kim 2015, p. 149). Kim (2015) provides ten ways in which visuals can enhance narrative inquiry, including the ability of images to express what is difficult to put into words (p. 150). Thus, the digital story provides the best of both worlds; the text explains the images and vice versa. However, that does not mean that there is no space for researcher analysis and interpretation. Key questions to ask include: Why this story at this time? Why those images? Does the connection between the words and images seem authentic? Is there a story behind or beneath the story? Every image in a digital story is the result of a decision-making process by the participant, and analysis of the story includes interpretation of that process.

In addition to narrative and images, digital storytellers often incorporate music into their videos. Music can set the tone for the story, even if only softly playing in the background. Music invokes emotion; it can be romantic, dramatic, invigorating, relaxing, and so much more. While the addition of music adds to the complexity of the analysis, it also can be a cue to aid in interpretation.

## 37.7 Challenges of Digital Storytelling in Research

The common approach to digital storytelling is through a workshop process in which the elements of storytelling and available technology are presented and participants have time/space to develop and create their stories. This process requires time and resources not always available to a researcher. It is also a big ask of participants. Meadows (2003, p. 190) stated that everyone has a story to tell; however, Gubrium and Holstein (2010, p. 256) noted that storytelling takes work "people seldom just 'burst out' in stories." In order to truly understand what is being asked of participants, a researcher should really go through the process first (Buturian 2017). Only then can one fully appreciate the challenge and rewards of creating a digital story.

While it can be argued that the benefits of producing a digital story are many (Hopkins and Ryan 2014), for example, development or enhancement of digital literacy skills, opportunity to explore your own interests and make meaning out of life experiences, ownership of the final product, and so on, do those benefits outweigh the time and effort? How can participants be persuaded to engage in this process? These are important considerations when electing to add digital storytelling to a research project.

Research generally has a specific focus. When incorporating digital storytelling, participants are then asked to tell a particular story – one that meets the requirements of the Project. This is contrary to the agency and voice digital storytelling is meant to provide. The researcher must be careful to ensure that participants understand the nature of the Project well enough to create relevant stories while avoiding being overly directive and imposing their own voice.

Participants will not always produce a story that is in line with the research questions. Our focus as "professional researchers" may not align with the stories participants value (Manning 2010, p. 166), and we have to be prepared to learn from that. Perhaps we are asking the wrong questions? Or have chosen the wrong participants? In the Project, two participants created stories that were not at all about their online lives or experiences with social media. They had other stories to tell. This does not mean that their stories cannot be used. In fact, it is ethically unacceptable to disrespect the data provided by participants in such a way. Even if the content of the story does not align with the research question, the ways in which the stories were told, and even the fact that they chose to tell stories that were off-topic, are worthy of analysis.

## 37.8 Conclusion

Digital storytelling as data is not well researched. The bulk of the literature focuses on the process and the benefits to the participants. Digital storytelling can provide a rich body of data in a research project. It privileges participant voice and can elicit deeper engagement with the research topic. However, as a research tool, digital storytelling requires time, technology, and participant engagement. The use of digital storytelling in youth research is a wonderful opportunity to ask young people to participate in a process congruent with their everyday practices and gives them the agency to tell the stories they want to tell. In selecting this method, researchers must think ethically about the benefits to the participant and consider whether those benefits outweigh the level of involvement required. As research data, digital stories provide a deeper insight than a filmed interview; participants take time to reflect and select the visuals, words, music, and any other elements they incorporate to provide *their* story, in *their* way. And while that story may not always align with the researcher's plan, the process of soliciting and then analyzing digital stories will prove invaluable to a research project.

> Digital Storytelling isn't just a tool; it's a revolution. (Meadows 2003, p. 192)

# References

Anderson SM, Muoz Proto C (2016) Ethical requirements and responsibilities in video methodologies: considering confidentiality and representation in social justice research. Soc Personal Psychol Compass 10(7):377–389
boyd d (2014) It's complicated: the social lives of networked teens. Yale University Press, New Haven
Buturian L (2017) The changing story: digital stories that participate in transforming teaching and learning. University of Minnesota Libraries, Minneapolis
Canadian Institutes of Health Research, Natural Sciences and Engineering Research Council of Canada, Social Sciences and Humanities Research Council of Canada (2014) Tri-council policy statement: ethical conduct for research involving humans 2014 RR4-2/2014E:216
Clark A (2012) Visual ethics in a contemporary landscape. In: Pink S (ed) Advances in visual methodology. Sage, London, pp 17–35
Clarke R, Adam A (2012) Digital storytelling in Australia: academic perspectives and reflections. Arts Humanit High Edu 11(1):157–176
Cox S, Drew S, Guillemin M et al (2014) Guidelines for ethical visual research methods. Visual Research Collaboratory, Melbourne
Daiute C (2013) Narrative inquiry: a dynamic approach. Sage, Thousand Oaks
Davis A, Weinshenker D (2012) Digital storytelling and authoring identity. In: Ching CC, Foley BJ (eds) Constructing the self in a digital world. Cambridge University Press, New York, pp 47–74
Edmonds F, Evans M, McQuire S et al (2015) Digital storytelling, image-making and self-representation: building digital literacy as an ethical response for supporting aboriginal young peoples' visual identities. Vis Methodol 3(2):98–111
Edmonds F, Evans M, McQuire S et al (2016) Ethical considerations when using visual methods in digital storytelling with Aboriginal young people in Southeast Australia. In: Warr D, Guillemin M, Cox S et al (eds) Ethics and visual research methods: theory, methodology and practice. Palgrave Macmillan, New York, pp 171–184
Ferrari M, Rice C, McKenzie K (2015) ACE pathways project: therapeutic catharsis in digital storytelling. Psychiatr Serv 66(5):556
Gruzd A, Jacobson J, Mai P et al (2018) The state of social media in Canada 2017. Ryerson University Social Media Lab, Toronto, Canada, pp 1–18
Gubrium JF, Holstein JA (2010) Narrative ethnography. In: Hesse-Biber SN, Leavy P (eds) Handbook of emergent methods. The Guilford Press, New York, pp 241–264
Gubrium AC, Hill AL, Flicker S (2014) A situated practice of ethics for participatory visual and digital methods in public health research and practice: a focus on digital storytelling. Am J Public Health 104(9):1606–1614
Hallett RE, Barber K (2014) Ethnographic research in a cyber era. J Contemp Ethnogr 43(3):306–330
Harper D (2000) Reimagining visual methods: Galileo to Neuromancer. In: Denzin NK, Lincoln YS (eds) Handbook of qualitative research, 2nd edn. Sage, Thousand Oaks, pp 717–732

Hartley J, McWilliam K (eds) (2009) Story circle: digital storytelling around the world. Wiley, Chichester. Wiley-Blackwell, Oxford

Hopkins S, Ryan N (2014) Digital narratives, social connectivity and disadvantaged youth: raising aspirations for rural and low socioeconomic young people. Int J Widening Partic 1(1):28–42

Kim J (2015) Understanding narrative inquiry: the crafting and analysis of stories as research. Sage, Thousand Oaks

Lambert J (2010) The digital storytelling cookbook. Digital Diner Press, Berkeley

Lowenthal P (2009) Digital storytelling in education: an emerging institutional technology? In: Hartley J, McWilliam K (eds) Story circle: digital storytelling around the world. Wiley-Blackwell, Oxford, pp 252–259

Manning C (2010) 'My memory's back!' Inclusive learning disability research using ethics, oral history and digital storytelling. Br J Learn Disabil 38(3):160–167. https://doi.org/10.1111/j.1468-3156.2009.00567.x

Meadows D (2003) Digital storytelling: research-based practice in new media. Vis Commun 2(2): 189–193

Page R, Thomas B (2011) New narratives: stories and storytelling in the digital age. University of Nebraska Press, Lincoln

Patton MQ (2002) Qualitative research and evaluation methods, 3rd edn. Sage, Thousand Oaks

Prosser J (2011) Visual methodology: toward a more seeing approach. In: Denzin NK, Lincoln YS (eds) Handbook of qualitative research, 4th edn. Sage, Thousand Oaks, pp 479–496

Qiongli W (2009) Commercialization and digital storytelling in China. In: Hartley J, McWilliam K (eds) Story circle: digital storytelling around the world. Wiley-Blackwell, Oxford, pp 230–244

Ranieri M, Bruni I (2013) Mobile storytelling and informal education in a suburban area: a qualitative study on the potential of digital narratives for young second-generation immigrants. Learn Media Technol 38(2):217–235

Robin B (2006) The educational uses of digital storytelling. Technol Teach Edu Ann 1:709

Robin BR (2008) Digital storytelling: a powerful technology tool for the 21st century classroom. Theory Pract 47(3):220–228

Shopes L (2011) Oral history. In: Denzin NK, Lincoln YS (eds) Handbook of qualitative research, 4th edn. Sage, Thousand Oaks, pp 451–466

Steeves V (2014) Young Canadians in a wired world, phase III: online privacy, online publicity. MediaSmarts, Ottawa

Taub-Pervizpour L (2009) Digital storytelling with youth: whose agenda is it? In: Hartley J, McWilliam K (eds) Story circle: digital storytelling around the world. Wiley-Blackwell, Oxford, pp 245–251

Walsh CA, Rutherford G, Kuzmak N (2010) Engaging women who are homeless in community-based research using emerging qualitative data collection techniques. Int J Mult Res Approach 4(3):192–205. https://doi.org/10.5172/mra.2010.4.3.192

# Participatory Mapping

## 38

Logan Cochrane and Jon Corbett

## Contents

| | | |
|---|---|---|
| 38.1 | Introduction | 706 |
| 38.2 | Participatory Mapping: Context, Impacts, and Limits | 706 |
| 38.3 | Power and Empowerment | 708 |
| 38.4 | Understanding Impact | 709 |
| 38.5 | Conclusion | 711 |
| References | | 711 |

### Abstract

Participatory and community mapping has emerged as a key tool for identifying and communicating development needs and been further recognized as a means to support social change. Drawing upon a broad assessment of the literature, more than two decades of experience with mapping initiatives from three continents, and covering a diverse array of applications and issues, this chapter explores both the application of participatory and community mapping and the range of impacts experienced. In so doing, the chapter explores the potential effectiveness for participatory and community mapping to effect positive change. At the same time, we will critically review the assumptions about social change and empowerment, highlighting challenges and limitations to their meaningful usage. This chapter provides practitioners and academics with an overview of participatory and community mapping uses, processes, and impacts and their role as a tool of communication for development and social change.

---

L. Cochrane (✉)
International and Global Studies, Carleton University, Ottawa, ON, Canada
e-mail: logan.cochrane@gmail.com

J. Corbett (✉)
University of British Columbia, Vancouver, BC, Canada
e-mail: jon.corbett@ubc.ca

© Springer Nature Singapore Pte Ltd. 2020
J. Servaes (ed.), *Handbook of Communication for Development and Social Change*,
https://doi.org/10.1007/978-981-15-2014-3_6

**Keywords**

Participatory mapping · Participation · Empowerment

## 38.1 Introduction

The past 25 years have witnessed a rapid increase of participatory mapping initiatives throughout the world (Brown and Kyttä 2014; Cochrane et al. 2014). Robert Chambers wrote that participatory mapping has spread "like a pandemic with many variants and applications" (2006, p. 1). It has become a vibrant area of practice and a well-used research method and is increasingly seen as an area of study in its own right.

Participatory mapping is a mapmaking process that strives to make visible the relationship between a place and local communities through the use of cartography (Aberley 1999; Flavelle 2002). Participatory maps provide a unique visual representation of what a community perceives as their place and identify features of significance within it – both physical and sociocultural (Bird 1995; Tobias 2000). Furthermore, the process of participatory mapping recognizes the intrinsic value of crafting an inclusive environment where all voices have the space to be expressed (Rambaldi et al. 2006).

Participatory mapping employs a range of tools. These include sketch mapping, transect mapping, and participatory three-dimensional modeling. More recently, participatory mapping initiatives have begun to use geographic information technologies including Global Positioning Systems (GPS), aerial photos and remote-sensed images (from satellites), geographic information systems (GIS), and the geospatial web (IFAD 2009; Johnson 2017).

Drawing upon a broad assessment of the literature and informed by more than two decades of experience with mapping initiatives from three continents, this chapter explores both the application of participatory and community mapping and the range of impacts experienced. In so doing, we examine the potential effectiveness of participatory mapping to effect positive change. At the same time, we critically review assumptions made about empowerment and social change, highlighting challenges and limitations to the meaningful usage of these terms in the context of participatory mapping practice. This chapter provides both practitioners and academics with an overview of participatory mapping practice – its uses, processes, and impacts in relation to their role as a tool of communication for development.

## 38.2 Participatory Mapping: Context, Impacts, and Limits

Participatory mapping is a process in which community members, writ large, contribute their own experiences, relationships, information, and ideas about a place to the creation of a map. The practice is usually conducted in an inclusive, or participatory, way. Many examples of participatory mapping initiatives emphasize the process of mapmaking as a transformative agent of change, while others focus upon the product and its use. In this chapter, we utilize the terminology of participatory mapping but in doing so recognize that a diversity of nomenclature and practices have emerged that also fall within the sphere of its practice. Other terms used include community mapping, asset mapping, participatory GIS, bottom-up GIS, community information systems, community-integrated GIS, counter-mapping, cultural mapping, indigenous mapping, participatory 3D mapping, and public participation GIS.

Participatory mapping projects often assume an advocacy role and actively seek recognition for community interests through identifying boundaries, traditional land uses, and place-based issues (Alcorn 2000; Chapin et al. 2005; Hazen and Harris 2007). In this form, participatory maps can play an important role in supporting diverse interests, such as those held by farmers, indigenous peoples, the homeless, recreation groups, youth, and seniors (Bryan 2011; Rocheleau 2005). Mapping initiatives can also challenge dominant worldviews, provide counter-narratives, and be used as a tool of resistance (Cooke 2003; Ghose 2001). The processes and outcomes of participatory mapping are viewed as being more inclusive and democratic and thus have commonly been adopted as a means to facilitate individual empowerment and societal change (Corbett 2003; Lydon 2002; Sieber 2006).

Acknowledgment needs to be given to the role of intermediaries in the practice of participatory mapping; often researchers, governments, NGOs, and community groups play a crucial role as interlocutors, trainers, advocates, and facilitators in participatory mapping initiatives (Alcorn 2000; IFAD 2009). Additionally, participatory mapping can be initiated by outsider groups, and the maps produced will contribute to an outsider's agenda, such as contributing to research, assisting in collaborative spatial planning exercises, ameliorating land and resource conflicts, or assessing local development potential. The levels of community involvement and control over the mapping process vary considerably between projects (Cochrane et al. 2014).

As both a multidisciplinary practice and area of research, the aims and specific objectives, as well as how outcomes are articulated, of participatory mapping initiatives vary significantly. This variation ranges from a focus on the process of participation through to the end use to which these maps are put, which in turn is influenced by the actors that will view and make decisions related to the content of these maps.

A broad body of literature has noted that participatory mapping has been an effective mechanism in amplifying the voice of community concerns. It has also been effective for highlighting and engaging in dialogue on a diverse set of community-relevant issues and themes. Initiatives have contributed to advancing land and resource rights (Stocks 2003) and indigenous title and claims (Parker 2006; Peluso 1995), supporting the revival and recovery of indigenous knowledge (Wilson 2004), improving mobility for individuals living with disabilities and mental health issues (Corbett and Cochrane 2017; Townley et al. 2009), and enhancing spatial information during humanitarian crises (Camponovo and Freundschuh 2014; Harvey 2012; Shekhar et al. 2012). Maps produced through participatory mapping initiatives have been utilized as a communication tool, ranging from climate change issues (Piccolella 2013) to zoning decisions (Zhang et al. 2013) and disease exposure (Keith and Brophy 2004). These examples demonstrate the potential of participatory mapping to contribute to social justice issues and positive societal change. It is often the examples such as those mentioned above that further inspire others to integrate mapmaking into their development activities and processes, with the understanding that they will serve as a pathway to transformational change.

Yet, participatory mapping is not exclusively utilized as a tool to empower individuals and communities. At times, these processes have been used purely to extract information from participants, especially in the realm of academia where participatory mapping is an often-used data acquisition method. Furthermore, they have been adopted as a means to coerce public support, with an understanding that people may be less likely to oppose decisions if they have been involved (even superficially) in the process (Yearley et al. 2003). In other instances, engagement in participatory mapping processes can be cursory or even be used manipulatively (McCall 2004). Mapping processes can also lead to unintended consequences; they can exacerbate conflict, influence land use and ownership, and facilitate the expansion of state control (Anau et al. 2003; Bryan 2011; Corbett 2003; Fox et al. 2003; Pramono et al. 2006; Wright et al. 2009). More fundamentally, there are critiques that these approaches are often not as participatory nor inclusive as claimed (Chapin et al. 2005; Lasker and Weiss 2003).

While online participatory mapping, or geospatial web, technologies have broadened participation in certain contexts, these changes do not equate with equal opportunity and access. Marginalization in many cases is systemically entrenched with continued exclusion. This may manifest itself with the exclusion of individuals of low socioeconomic status, groups living in remote or inner-city areas, indigenous communities, recent migrants, the homeless, people with disabilities or experiencing mental illness, and senior citizens (Beischer et al. 2015; Cochrane et al. 2014). Currently, challenges related to the digital divide reflect caution raised by Harley (1990, pp. 3–4) when commenting on the early development of GIS technologies: "technological progress does not automatically translate into maps that are more relevant in a society."

There are also fundamental challenges with participatory mapping. Based on a review of academic publications, issues of social justice rarely feature in projects related to online mapping and crowdsourcing geographic information (Cochrane et al. 2017). Cochrane et al.'s meta-analysis found that, within the academic literature on participatory mapping since 2005, the practice and analytical findings of participatory mapping have tended toward being overtly technical, despite recognition of the political nature of maps and repeated calls to integrate social, economic, and political components. Limited progress has been made since these issues were first highlighted in the 1990s (Cochrane et al. 2017). Many practitioners, on the other hand, are engaging issues of social justice, but these activities are not well represented in the literature (Brown and Kyttä 2014).

## 38.3 Power and Empowerment

What gets included and excluded on a map, how it is represented, and why a map is made, are all questions linked to power. As Poole (1994, p. 1) notes, maps have always been "both symbols and instruments of power." Harley (1990, p. 16) argued that mapmaking is "never merely the drawing of maps: it is the making of worlds."

The most influential forms of mapmaking in modern history were means to express and exert control. Control over territory was claimed with a map. Taxes were levied with the support of maps. Ruling elite understood the power of maps, and in some cases, mapmaking that challenged official versions was viewed as an act of treason. In the past, mapmaking was primarily undertaken by ruling elite because the task required cartographic proficiency, as well as being time and resource intensive. Maps expressed a "top-down, authoritarian, centrist paradigm" and were produced by experts (Goodchild 2007, p. 29). The emergence of new technologies – from geographic information systems (GIS) to the geospatial web and geo-social media – has enabled more people to engage in mapmaking (Ghose 2001; Harris and Weiner 1998). As these technologies become more accessible, affordable, and user-friendly, new forms of participatory mapping practice are emerging. As a direct result of this shift, the power expressed by and within maps is constantly being challenged (Crampton and Krygier 2005; Wright et al. 2009).

Participatory mapmaking is often claimed to be an empowering process (Bryan 2011; Cochrane et al. 2014; Elwood 2002; Kesby 2005). Increasingly, the literature on this topic has employed empowerment as a simplistic phrase for expressing success or social change, but the majority of these initiatives do not define or measure empowerment (Corbett et al. 2016). While empowering and marginalizing processes are occurring (Harris and Weiner 1998), we know little about the extent and duration of these changes. Participatory and community mapping initiatives have the potential to be empowering; however, with a lack of measurement, we are currently unable to outline how these processes occur, for whom, and for how long, beyond anecdotal experiences.

## 38.4 Understanding Impact

From the literature and practical experience, we know that the outcomes of some mapping initiatives are more instrumental in contributing to social change than others. Cochrane et al. (2014) outline best practices categorized by the stage of the mapmaking initiative. They define three core stages in the participatory mapping process – pre-process, in-process, and post-process. For the pre-process, or dialogue and idea development phase, success includes having clear objectives, ensuring maps are socioculturally and politically contextualized, verifying that the capacity for participation exists, having processes to engage with disagreements, communicating realistic expectations, having champions to promote the initiative, and having clear and transparent communication.

While a mapping initiative is in-process, Cochrane et al. (2014) highlight the importance of feedback mechanisms regarding usability and inclusivity, of celebrating small successes, and of ongoing engagement and the value of quality checks on data accuracy. These design- and process-related features are critical to success, but other factors are less visible and are unpredictable.

Cochrane et al. (2014) further note that impacts are most likely to fall in the post-process phase. The common factors that contribute to impact include participation, empowerment, and ownership. The role of participation in the post-process phase encourages ongoing public awareness raising, which translates into continued use and development of the map (Ganapati 2011). Specifically, with regard to online maps, which may experience enthusiastic participation while in-process, the post-process period may see interaction decline or stop entirely. Some projects have a timeframe or are designed for a specific purpose and period; for those that are ongoing, participation is a critical factor for success. Enabling continued engagement may require creativity, such as using social media and contests, while ensuring that participants see that their interaction is contributing to an objective.

Linked with a feeling of contribution, or of personal benefit, the ability for a project to empower those that engage with it will contribute to its post-process success. Empowerment in this form may be building capacity so that maps and mapmaking can become a feature of the community, whereby maps are revised and reinvented with time (Fox et al. 2008). The complexity both of projects and empowerment, however, makes this task a challenging one.

As Perkins (2008, p. 154) notes "Community empowerment is complex. Projects have different goals...The same project may carry different meanings for different members, who are likely to engage in different ways with the mapping" (Perkins 2008, p. 154).

Ownership of the map and the information that it represents also affect post-process impact. In some cases, the ownership can restrict accessibility and availability of the information and map, which is an outcome that can be purposeful or unintentional. Communities, in some cases, may not have access to information and maps due to ownership by consultants, researchers, or nongovernmental organizations (McCall and Minang 2005). On the other hand, participatory maps that are public, easily accessible, readily available and offer ongoing engagement opportunities can support a larger audience in the post-process period. Regardless of the arrangement and reasons for it, ownership will greatly affect the post-process phase.

Other key factors that contribute to variation in outcomes relate to motivation, organizational capacity, and leadership (Corbett and Cochrane 2017). The less visible factors include issues related to trust in the process, which relates to the person, people, or organization managing the initiative as well as the histories between the various actors. A lack of trust may materialize if the individuals involved have past connections that might be suggestive of bias. Other factors that are beyond strong technical design include the ways in which people interact, the ways in which their contributions are expressed and responded to, and the sociocultural or political incentives that influence actor motivations. Similarly, unpredictable events may alter the impact of a mapmaking process or the impact of a participatory-produced map. An example of this is broad-based re-problematizing of the issue being discussed, spurring widespread engagement and interest, which may translate into political responsiveness. Alternatively, interest may wane due to external forces or the rise of public attention to different issues. As a result, technical design and ideal processes are critical but do not necessarily result in positive cultural, political, economic, legal, or societal changes.

## 38.5 Conclusion

Participatory mapping has been widely utilized, especially in development contexts, as a means to challenge ideas, priorities, and power. These initiatives have contributed to positive social change in a diverse array of spheres, from land rights to language revitalization. However, even with ideal technical design and processes, participatory mapping does not always result in positive outcomes. Sometimes projects fail, and other times there are notable negative impacts. To date the focus of research has been upon the technical and process aspects of mapping, with little measurement and few evaluations occurring on the medium- and long-term impacts. This chapter has highlighted both the potential and the problems of participatory mapping so that academics and practitioners can be better informed of their uses, processes, and impacts. This overview aims to support more informed decision-making about what options might be most suitable and appropriate when considering a range of different tools of communication for development and social change.

## References

Aberley D (1999) Giving the land a voice: mapping our home places. Land Trust Alliance of British Columbia, Salt Spring Island

Alcorn JB (2000) Keys to unleash mapping's good magic. PLA Notes 39(2):10–13

Anau N, Corbett J, Iwan R, van Heist M, Limberg G, Sudana M, Wollenberg E (2003) Do communities need to be good mapmakers? Center for International Forestry Research, Jakarta

Beischer A, Cochrane L, Corbett J, Evans M, Gill M, Millard E (2015) Mapping experiences of injustice: developing a crowdsourced mapping tool for documenting good practices for overcoming social exclusion. Spatial knowledge and information conference proceedings, 27 February – 1 March, Banff

Bird B (1995) The EAGLE project: re-mapping Canada from an indigenous perspective. Cult Survival Q 18(4):23–24

Brown G, Kyttä M (2014) Key issues and research priorities for public participation GIS (PPGIS): a synthesis based on empirical research. Appl Geogr 46:122–136

Bryan J (2011) Walking the line: participatory mapping, indigenous rights, and neoliberalism. Geoforum 42(1):40–50

Camponovo ME, Freundschuh SM (2014) Assessing uncertainty in VGI for emergency response. Cartogr Geogr Inf Sci 41(5):440–455

Chambers R (2006) Participatory mapping and geographic information systems: whose map? Who is empowered and who disempowered? Who gains and who loses? Electron J Inf Syst Dev Ctries 25(2):1–11

Chapin M, Lamb Z, Threlkeld B (2005) Mapping indigenous lands. Annu Rev Anthropol 34:619–638

Cochrane L, Corbett J, Keller P, Canessa R (2014) Impact of community-based and participatory mapping. Institute for Studies and Innovation in community-university engagement. University of Victoria, Victoria

Cochrane L, Corbett J, Evans M, Gill M (2017) Searching for social justice in crowdsourced mapping. Cartogr Geogr Inf Sci 44(6):507–520

Cooke FM (2003) Maps and counter-maps: globalised imaginings and local realities of Sarawak's plantation agriculture. J Southeast Asian Stud 34(2):265–284

Corbett J (2003) Empowering technologies? Introducing participatory geographic information and multimedia systems in two Indonesian communities. Doctoral Dissertation submitted to the University of Victoria

Corbett J, Cochrane L (2017) Engaging with the participatory geoweb: exploring the dynamics of VGI. In: Campelo C, Bertolotto M, Corcoran P (eds) Volunteered geographic information and the future of geospatial data. IGI Global, Hershey

Corbett J, Cochrane L, Gill M (2016) Powering up: revisiting participatory GIS and empowerment. Cartogr J 53(4):335–340

Crampton JW, Krygier J (2005) An introduction to critical cartography. ACME Int E J Crit Geog 4(1):11–33

Elwood S (2002) GIS use in community planning: a multidimensional analysis of empowerment. Environ Plan 34(5):905–922

Flavelle A (2002) Mapping our land: a guide to making maps of our own communities and traditional lands. Lone Pine, Edmonton

Fox J, Suryanata K, Hershock P, Pramono AH (2003) Mapping power: ironic effects of spatial information technology. Spatial information technology and society: ethics, values, and practice papers. East-West Center, Hawaii

Fox J, Suryanata K, Hershock P, Pramono AH (2008) Mapping boundaries, shifting power: the socio-ethical dimensions of participatory mapping. In: Contentious geographies: environmental knowledge, meaning, scale. Ashgate, Burlington, p 203

Ganapati S (2011) Uses of public participation geographic information systems applications in e-government. Public Adm Rev 71(3):425–434

Ghose R (2001) Use of information technology for community empowerment: transforming geographic information systems into community information systems. Trans GIS 5(2):141–163

Goodchild MF (2007) Citizens as sensors: web 2.0 and the volunteering of geographic information. GeoFocus (Editorial) 7:8–10

Harley JB (1990) Cartography, ethics and social theory. Cartographica Int J Geogr Inf Geovisualization 27(2):1–23

Harris T, Weiner D (1998) Empowerment, marginalization, and "community-integrated" GIS. Cartogr Geogr Inf 25(2):67–76

Harvey F (2012) Code/space: software and everyday life by rob kitchen and Martin dodge. Int J Geogr Inf Sci 26(10):1999–2001

Hazen HD, Harris LM (2007) Limits of territorially-focused conservation: a critical assessment based on cartographic and geographic approaches. Environ Conserv 34(4):280–290

IFAD (2009) Good practices in participatory mapping. International Fund for Agricultural Development, Rome

Johnson PA (2017) Models of direct editing of government spatial data: challenges and constraints to the acceptance of contributed data. Cartogr Geogr Inf Sci. 44(2):128–138

Keith MM, Brophy JT (2004) Participatory mapping of occupational hazards and disease among asbestos-exposed workers from a foundry and insulation complex in Canada. Int J Occup Environ Health 10(2):144–153

Kesby M (2005) Retheorizing empowerment-through-participation as a performance in space: beyond tyranny to transformation. Signs 30(4):2037–2065

Lasker RD, Weiss ES (2003) Broadening participation in community problem solving: a multidisciplinary model to support collaborative practice and research. J Urban Health Bull N Y Acad Med 80:14–47

Lydon M (2002) (Re)presenting the living landscape: exploring community mapping as a tool for transformative learning and planning. Master of Arts submitted to University of Victoria

McCall MK (2004) Can participatory-GIS strengthen local-level spatial planning? Suggestions for better practice. In: 7th international conference on GIS for developing countries (GISDECO 2004). University Teknologi, Johor

McCall M, Minang P (2005) Assessing participatory GIS for community-based natural resource management: claiming community forests in Cameroon. Geogr J 171(4):340–356

Parker B (2006) Constructing community through maps? Power and praxis in community mapping. Prof Geogr 58(4):470–484

Peluso NL (1995) Whose woods are these? Counter-mapping forest territories in Kalimantan, Indonesia. Antipode 27(4):383–406

Perkins C (2008) Cultures of map use. Cartogr J 45(2):150–158

Piccolella A (2013) Participatory mapping for adaptation to climate change: the case of Boe Boe, Solomon Islands. Knowl Manag Dev J 9(1):1–13

Poole P (1994) Geomatics: who needs it? Cultural Survival quarterly magazine, December 1994: www.culturalsurvival.org/publications/cultural-survival-quarterly/geomatics-who-needs-it

Pramono AH, Natalia I, Arzimar III J, Bogor I, Janting Y, Kasih PP, Pontianak I (2006) Ten years after: counter-mapping and the dayak lands in west Kalimantan, Indonesia. http://dlc.dlib.indiana.edu/dlc/bitstream/handle/10535/1997/Pramono_Albertus_Hadi.pdf

Rambaldi G, Chambers R, McCall M, Fox J (2006) Practical ethics for PGIS practitioners, facilitators, technology intermediaries and researchers. In: Participatory learning and action mapping for change: practice, technologies and communication, vol 54. IIED, London, pp 106–113

Rocheleau D (2005) Maps as power tools: locating communities in space or situating people and ecologies in place? In: Communities and conservation: histories and politics of community-based natural resource management. AltaMira, Walnut Creek, p 327

Shekhar S, Yang K, Gunturi VM, Manikonda L, Olive D, Zhou X, Lu Q (2012) Experiences with evacuation route planning algorithms. Int J Geogr Inf Sci 26(12):2253–2265

Sieber R (2006) Public participation geographic information systems: a literature review and framework. Ann Assoc Am Geogr 96(3):491–507

Stocks A (2003) Mapping dreams in Nicaragua's Bosawas reserve. Hum Organ 62(4):344–356

Tobias TN (2000) Chief Kerry's moose: a guidebook to land use and occupancy mapping, research design, and data collection. Union of BC Indian Chiefs: Ecotrust Canada, Vancouver

Townley G, Kloos B, Wright PA (2009) Understanding the experience of place: expanding methods to conceptualize and measure community integration of persons with serious mental illness. Health Place 15(2):520–531

Wilson AC (2004) Introduction: indigenous knowledge recovery is indigenous empowerment. Am Indian Q 28(3/4):359–372

Wright DJ, Duncan SL, Lach D (2009) Social power and GIS technology: a review and assessment of approaches for natural resource management. Ann Assoc Am Geogr 99(2):254–272

Yearley S, Cinderby S, Forrester J, Bailey P, Rosen P (2003) Participatory modelling and the local governance of the politics of UK air pollution: a three-city case study. Environ Values 12(2):247–262

Zhang Z, Sherman R, Yang Z, Wu R, Wang W, Yin M, Yang G, Ou X (2013) Integrating a participatory process with a GIS-based multi-criteria decision analysis for protected area zoning in China. J Nat Conserv 21(4):225–240

# Evaluations and Impact Assessments in Communication for Development

**39**

Lauren Kogen

## Contents

| | | |
|---|---|---|
| 39.1 | Introduction | 716 |
| 39.2 | The Logic Behind Evaluations: Accountability and Learning | 716 |
| 39.3 | What We Need to "Learn" About Development | 717 |
| 39.4 | The History of Accountability and Learning Priorities | 717 |
| 39.5 | Conflation of Accountability and Learning | 718 |
| 39.6 | Searching for "Recipes" for Social Change | 719 |
| 39.7 | Literature Reviews as Panacea? | 721 |
| 39.8 | Redefining "Learning" | 723 |
| | 39.8.1 Potential Learning Priority 1: Developing Theory-Based Guidelines for Social Change | 724 |
| | 39.8.2 Potential Learning Priority 2: Social Change Process Evaluation | 726 |
| 39.9 | Summary | 728 |
| References | | 729 |

### Abstract

Within the development field, project evaluations and impact assessments are essential. Donors are increasingly requiring rigorous evaluations in order to (1) ensure that aid dollars are spent on projects that are having positive impacts and not being wasted on projects that are ineffective and (2) promote "evidence-based policy making" in which evaluations contribute to understanding best practices for development aid. These two goals are frequently referred to by the world's major donors as promoting "accountability" and "learning," respectively. However, current conceptions of learning and accountability are problematic – at

---

Adapted from Kogen (2018a). Copyright © 2017 by the Author. Reprinted by permission of SAGE Publications, Ltd.

L. Kogen (✉)
Department of Media Studies and Production, Temple University, Philadelphia, PA, USA
e-mail: lauren.kogen@temple.edu

© The Author(s) 2020
J. Servaes (ed.), *Handbook of Communication for Development and Social Change*,
https://doi.org/10.1007/978-981-15-2014-3_131

times even counterproductive. This chapter provides an overview of the role of evaluations in the CDS field and the concepts of accountability and learning and then describes the problems, contradictions, and ethical dilemmas that arise in the field because of them. The chapter ends with suggestions for how the field might fine tune the concepts of learning and accountability in a way that would better serve both donors and aid recipients.

**Keywords**

Communication for development · Communication for social change · Foreign aid · Monitoring · Evaluation

## 39.1 Introduction

Despite decades of practice, we in the West do not fully understand how to "do" international development. Many development efforts fail, and it is difficult to predict which will succeed (Glennie and Sumner 2014). Some would even say that after more than half a century's worth of efforts, we have done more harm than good (Easterly 2006; Moyo 2009). This is why we are still asking such fundamental questions as "does aid work?" (Burnside and Dollar 2004).

In recognition of the fact that we do not have all the answers, that many aid projects have failed, and that there is a global skepticism about the effectiveness of aid, policymakers have placed increasing importance on formal project evaluations to help them decide which projects to fund and which to discontinue.

## 39.2 The Logic Behind Evaluations: Accountability and Learning

Project evaluations assess whether specific development projects have succeeded or failed and to what degree. Ideally, they serve two primary purposes: first, they ensure that aid dollars are spent on projects that are having positive impacts and not being wasted on projects that are ineffective. Second, they promote what is referred to as "evidence-based policy making," in which evaluations contribute to understanding best practices for development aid. These two goals are frequently referred to by the world's major donors as promoting "accountability" and "learning," respectively. Indeed, USAID cites these two goals as the "primary purposes" of evaluation (2016, p. 8).

*Accountability* typically refers to the question of whether or not a project met the goals it set out to achieve. In order for funders and implementers to be held accountable to taxpayers and beneficiaries, the logic goes, they must be able to show that they are spending money on development programs that are working and not wasting money on those that are not. *Learning* is, in theory, achieved by using data from past projects to improve future projects.

One problem that arises, however, is that these two terms are often used interchangeably by the world's major donors. This stems in part from the fact that, while "accountability" is relatively well established as a concept, "learning" is not. This chapter focuses on the worrisome implications this has for what we can hope to achieve in terms of using evaluations to improve future interventions.

## 39.3 What We Need to "Learn" About Development

While Western donors still lack a fundamental understanding of what works in aid, "learning," as conceptualized by major donors, does not aim to answer fundamental questions. Despite an acknowledgment that we still have much to learn when it comes to aid, present practices reinforce the assumption that our strategies are working reasonably well and that evaluations can, perhaps, provide some finetuning (Power 1997). This means that using current definitions of accountability and learning as starting points for evaluations carries the potential of pushing us *away* from answering foundational questions about how aid works, and ultimately making it harder, not easier, to improve development projects.

Much recent work has focused on the idea that accountability- and learning-based approaches are in many ways incompatible (Armytage 2011). This chapter argues, first, that accountability and learning have been inappropriately conflated in the documentation of two of the world's major donors – USAID and DFID – stemming from the assumption that accountability, by definition, improves projects. This conflation, in turn, makes it much more difficult for project implementers and evaluators to understand which learning-based questions policymakers need answered in order to improve aid priorities. Second, the chapter offers a revised conceptualization of "learning," focused on theories undergirding social change and social change project implementation that would make evaluations more useful.

While this chapter focuses on Communication for Development and Social Change (CDS), current challenges in evaluation practice apply to all development projects that focus on human-centered social change – the kind of complex, messy scenarios that make interventions and evaluations so complicated, that lend themselves to failure, and for which making assumptions about what works is particularly dangerous.

## 39.4 The History of Accountability and Learning Priorities

Placing accountability at the forefront of evaluations is the guiding principle behind results-based management (RBM), the preferred management style of many of the world's major donors (Vähämäki et al. 2011). RBM rose in prominence following the Paris and Accra High Level Forums on Aid Effectiveness convened by the Organization for Economic Cooperation and Development (OECD) in 2005 and 2008, to which 138 countries agreed to adhere. Following increasing global skepticism about the effectiveness of aid, this series of forums take as their raison d'être the notion that "aid effectiveness must increase significantly" (OECD 2008) and that effectiveness must be measurable and visible to the public. The Paris and Accra meetings resulted in a set of principles that together are assumed to make aid more effective. One of these key principles is "managing for results": the need to "manag[e] and implement[] aid in a way that focuses on the desired results..." (OECD 2008, p. 7). The Evaluation Policy Document produced by the U.S. Agency for International Development (USAID) follows suit and cites the OECD's principle of managing for results, stating that projects must monitor "whether expected results are occurring" (2016, p. 9). The organizations rightly acknowledge that it is unethical and nonsensical to oversee billions of dollars of aid without knowing whether or not that aid is doing any good. They conclude that measuring levels of success will ensure that only projects that are truly improving lives will be funded and supported. In this way, the organizations become "accountable" to those the projects are aiming to help, because RBM ensures that money is not being wasted, or failing to reach those in need. RBM also promotes accountability to taxpayers, who are ultimately supporting the bulk of these projects.

Likewise, these large donor organizations tout the importance of *learning* from past interventions in order to improve future interventions. Donors expect evaluation questions to be "explicitly link[ed]" to policymaking decisions (USAID 2011, p. 7) and support evaluations that produce "high quality evidence for learning" (DFID 2014, p. 3).

These conceptualizations of accountability and learning seem logical and straightforward at first glance, but they require parsing. There are thousands of development projects currently underway. What is the best way to take all of these projects and "learn" from them? What precisely are we trying to learn? How do we extract lessons that will make development aid work better for beneficiaries? The evaluation documentation published by USAID and the UK's Department for International Development (DFID) (the world's largest foreign aid donors) do not explicitly address this. Instead, two problematic assumptions are evident regarding how learning occurs through evaluations: first, that accountability and learning are essentially equivalent, and second, that periodic literature reviews help extract lessons across a portfolio of evaluations. Unfortunately, this view of learning creates problems, as explained below, and does not produce adequate high-quality findings.

## 39.5 Conflation of Accountability and Learning

The policy documents produced by USAID and DFID eschew clear definitions of learning. For example, USAID's Evaluation Policy Document defines learning as "systematically generat[ing] knowledge about the magnitude and determinants of project performance" (2011, p. 3). On its face, it is difficult to differentiate this definition from a common understanding of accountability, save for the word "determinants," which implies some attention to *determining why* a project succeeded or failed. DFID's 2014 Evaluation Strategy gives no guidance at all on how it views learning even though, as in the USAID documentation, it emphasizes the importance of it. Both documents incorporate a general sentiment that learning should improve future program planning but provide minimal information regarding what the organizations hope to learn or what type of data would be most useful for learning. The OECD's Principles document hardly mentions learning, except to say that lessons learned should be shared (2008, p. 6). This lack of a clear definition leaves the term open to interpretation.

One implication of the repetitive reference to the importance of "accountability and learning" without a formal definition of learning is the suggestion that accountability and learning are two sides of the same coin: that by addressing one we are necessarily addressing the other. Indeed, when learning *is* referenced in evaluation documentation, the general impression that arises is that learning simply refers to knowing whether projects worked or not and to what degree. But this is precisely the aim of accountability. Accountability is about understanding whether a project has succeeded or failed; learning here suggests that by seeing which projects have succeeded and failed one can *learn* which projects to promote and which to terminate. The implication thus becomes that learning is solely about assessing project success and that it is therefore a natural outcome of accountability. This is the implication of statements such as USAID's claim that learning "represents a continuous effort to... measure the impact on the objectives (results) defined" (2011, p. 3) or DFID's statement that "monitoring results provides us with an incentive to look at the evidence, innovate and learn" (2014, p. 1).

Neither explain what we might learn beyond whether or not to refund or defund the project at hand. But "learning" whether or not an individual project worked does not necessarily help us understand development as a field. The most that can be done with this form of learning is to (a) keep funding the same project or (b) fund an identical project in another location. The dangers of this understanding of learning should be obvious: We cannot improve interventions by copying and pasting. Understanding that intervention A works to improve problem X is much less useful than understanding, for example, what kinds of broad-based strategies work to improve problem X, and why, under what circumstances, etc. *This* understanding of learning would do much more to help practitioners and policymakers improve the world of development interventions.

## 39.6   Searching for "Recipes" for Social Change

To better understand why USAID's and DFID's approaches are problematic, consider the popular website allrecipes.com. The site features 885 different recipes for chicken parmesan. Someone who is trying to make the dish might determine that all recipes with four or more stars (out of five) should be considered "successful" and that all other chicken parmesan recipes should be discarded. However, this criterion still leaves us with hundreds of "successful" recipes. Furthermore, those who are trying to extract lessons in order to teach others how to make chicken parmesan should be able to do more than offer one of the 300 or so ready-made recipes that happen to appear on the website. They should be able to provide overall lessons about the best ingredients and the best cooking techniques. Likewise, policymakers should not be taking an allrecipes.com approach to aid-making because there are no fool-proof instructions for what works to create social change. Instead, policymakers should be trying to extract general principles about how to conduct effective aid-making. If asked what makes for an effective HIV prevention strategy, for example, a well-informed policymaker should be able to do more than advise practitioners to copy one of the hundreds of HIV interventions that have already been conducted around the globe.

As anyone who has worked in the field can attest, copying interventions from one setting to another is nearly impossible. Someone who wants to prevent HIV in New Delhi cannot pluck an intervention plan that was used in rural South Africa from a database, insert it into a vastly different context, and expect it to work in the same way. General guidelines, principles, and theories are needed so that interventions can be designed for the various contexts in which they are implemented. In other words, learning *why* particular programs work in particular contexts is what contributes to improving policy, not simply learning *whether* programs worked.

Indeed, Deaton (2009, p. 4) advocates the formation of "potentially generalizable mechanisms" that help explain why a project worked. Likewise, Stern et al. (2012, p. 27) argue that project findings need to "form[] the basis for more general theory" in order to streamline future projects. This is particularly crucial for projects that address human behavior and social change in which we struggle to understand what leads individuals and institutions to change their thoughts and behaviors and in which well-designed interventions often do not go according to plan. Servaes (2016, p. 2) notes that 75% of recent CDS work has been atheoretical – referencing no theory at all – and that of those pieces that do cite theory, outdated theories stemming from the modernization paradigm are the most common.

A project that provides food to schoolchildren, for example, might succeed in increasing school attendance, but without learning how parents make decisions about when to send their children to school and when to keep them at home, funders cannot extract lessons about what would have made the project even more successful, whether the project could be replicated in another setting, or whether the project is sustainable. It may be the case that in certain contexts, the cost of sending children to school (the danger and length of the commute, the loss in wages, etc.) outweighs the benefits of an additional meal. Alternatively, providing food may provide a short-term increase in attendance but fail to address underlying problems about the local government's funding of schools and teachers. These are but two examples of the many complex factors that affect something as seemingly simple as school attendance. High-quality evaluations may incorporate understanding why an intervention worked or failed, but a lack of emphasis on this kind of learning means that many evaluations do not. It is in part for this reason that many evaluation experts describe "accountability-based" and "learning-based" evaluations as at odds with each other (Armytage 2011; Cracknell 2000; Lennie and Tacchi 2014), despite the insistence by organizations like USAID that both concepts are fundamental and can be "mutually reinforcing" (2016, p. 8). Lennie and Tacchi (2015, p. 27) define learning-based approaches as those that "understand social change as emergent, unpredictable, unknowable in advance." In other words, a focus on learning assumes that we are learning about *how social change works*, not about whether *a particular project worked*.

There is another concern with conflating accountability and learning and using accountability (whether or not a project succeeded) to determine policy (which projects should be supported). Let us assume, for a moment, that donors should continue funding all successful projects. Not only does this not tell funders what makes particular projects successful but it means that they are not being very selective in the projects they are funding. This is because almost all human-centered interventions "succeed." If we define "positive impact" or "success" as achieving the goals the project set out to accomplish, it does not take much for a program to succeed, as long as the organization is savvy and does not overestimate its potential impact. RBM emphasizes quantitative "indicators" and "targets" to ensure projects are truly achieving measurable objectives, such as setting an indicator of "number of children that receive meals at school" and a target of "500 children."

But the very results-based system that was implemented to ensure accountability has led to a system in which program designers are forced to reduce goals to simple, straightforward, low-level, moderately-easy-to-achieve results (Armytage 2011, p. 68; Gumucio-Dagron 2009). They do this, first, so that they are able to translate complex project goals into a number-based spreadsheet that requires everything to be broken down into its component parts, and, secondly, so that organizations are seen as performing well when it comes time for final evaluations. Such evaluation mindsets work better for public relations priorities than for furthering global development (Enghel 2016).

Major funders like USAID and DFID should be able to do better than simply improve semi-successful projects. They should be using evaluations to actively decide which types of activities should be prioritized and which have the highest potential to significantly improve lives. But currently, USAID acknowledges that simple improvement to existing projects is their primary objective. Its documentation suggests that *the* purpose of learning is to allow "those who design and implement projects, and who develop programs and strategies... to refine designs and introduce improvements into future efforts" (2011, p. 3). However, learning how to refine a single project is inadequate. Evaluations that explain why certain projects succeed or fail, build theory, and provide recommendations for overall policies can help refine funders' understanding of development and social change and help improve policy-making decisions. Understanding *why* a project works or fails and tying these findings to policy recommendations therefore needs to be intimately tied to the priorities of evaluations and to the working definition of learning.

## 39.7 Literature Reviews as Panacea?

The evaluation documentation of USAID and DFID does reference the idea of understanding why a project succeeded or failed (e.g., USAID 2016, p. 8), but rarely. Instead, the documentation suggests that real learning, the ability to understand what works across contexts, comes from reviews: periodic syntheses of research, reports, evaluations, etc. that allow policymakers, scholars, and practitioners to produce overall lessons accumulated after analyzing many projects. Through such a process, it is suggested, patterns will emerge, and *learning* will ensue.

Many of these agencies commission literature reviews for this very end. USAID commissions "systematic reviews" to "synthesize the available research on a specific question" (2016, p. 20); DFID similarly assumes that its learning will take place through "thematic [and] sectoral synthesis" of findings (2014, p. 13). Extending from the assumption above that a proper RBM approach will necessarily lead to learning, the idea here is that looking across a portfolio of projects that have succeeded and failed will provide additional insights into what kinds of approaches are worthwhile.

This is an overly restrictive version of learning for two reasons. First, if evaluations focus on accountability (whether a project succeeded or failed), rather than an understanding of why the project succeeded or failed, there is no reason to believe that looking at many evaluations will bring to light new lessons that did not appear in the evaluations of individual projects. Of course, it is possible that patterns will jump out. (It is possible that the vast majority of four-star chicken parmesan recipes, for example, use parmesan cheese and San Marzano tomatoes. If this were the case, a perceptive reviewer could conclude that these ingredients should be considered default strategies for making chicken parmesan.)

Alas, lessons about development are rarely so obvious. The "ingredients" that go into projects are vastly more complex, and those ingredients interact with the setting in which projects are implemented. Even if there is meticulous monitoring of project inputs and outputs, this is usually insufficient to explain why a complicated human-centered project, such as encouraging people to use condoms, succeeds or fails. Without the why, the lessons extracted from a review are limited.

This is exacerbated by the fact that reviewers find it very difficult to deal with disparate data collection methods when trying to synthesize large numbers of studies. Qualitative evaluation data (such as that stemming from in-depth interviews with project beneficiaries) is particularly problematic, yet qualitative data is often how we best answer questions of "why." Thus, reviewers often exclude evaluations that rely on qualitative data, designating them as unhelpful. There are a variety of reasons for which reviewers may exclude studies – that the methods are flawed, the location is mismatched to the needs of the reviewer, the questions asked are not pertinent, etc. – but ultimately many studies are eliminated because they are not conducted in a way that helps reviewers synthesize or extract lessons. USAID explicitly states that its systematic reviews are typically only used for randomized controlled trials (USAID n. d.), which by their very design emphasize answering whether a project worked rather than why (Deaton 2009) and tend to de-emphasize qualitative data.

This is not to say that literature reviews are not useful, or that patterns do not emerge. Sometimes literature reviews reveal important patterns. But reviewers are neither clairvoyants nor magicians; they cannot see, nor draw conclusions about, what is not suggested or addressed in individual reports. If evaluations are focused exclusively on whether or not projects work, it is too easy for evaluations to simply become a series of case studies – disconnected project reports that hint at larger lessons about aid but make it difficult for reviewers to connect the dots.

It is often argued that evaluators can make connecting the dots easier for those conducting literature reviews by using common indicators. For example, if all projects dealing with promoting safe sex were encouraged to measure the "number of people reporting that they practice safe sex," it would be easier to make comparisons across projects. However, this method still does little to explain why particular efforts work in particular contexts. (Instead, a more helpful strategy might be to make discussions of *why* more meaningful and useful – an idea discussed in the next section.)

It is not only literature reviewers that have trouble using evaluations and other studies of development efforts to improve general understanding of development and development policy. A study conducted by Kogen et al. (2012), which surveyed and interviewed policymakers and practitioners working in communication and development, found that overall, policymakers felt that the available literature was impractical. They stated a preference for "research findings [that] are translated into clearly–articulated [*sic*] and replicable recommendations" (2012, p. 3). In survey responses, policymakers scored the "usefulness" of existing evidence relating to the role of media and communication in development at a mere 3.2 out of 5 (close to neutral) and, on average, said they "sometimes" consult it. This does not bode well for a field such as CDS that is trying to achieve greater inroads into policymaking and convince policymakers of its usefulness. (Indeed, many policymakers are not even sure what CDS *is* (Feek and Morry 2009; Lennie and Tacchi 2015).) The inattention to larger policy recommendations is not limited to CDS work. Even when it comes to randomized controlled trials in various fields of development, policymakers often find results to be of little practical use, stating a preference for lessons beyond whether or not an individual intervention worked (Ravallion 2009, p. 2).

## 39.8 Redefining "Learning"

How, then, can we make learning from evaluations more meaningful and useful? The first step is to recognize that accountability and learning are not the same thing and should not be treated as such. A loose use of the word learning is undergirded by a misplaced dependency on accountability and RBM as cure-alls for what ails development funding. Accountability, on its own, does not improve aid, and yet it is clear in rhetoric as well as practice that accountability is being prioritized above learning and that learning is seen as icing on the cake – a perk resulting from good accounting.

But learning is what improves lives, and for this reason, learning must be prioritized (as others have also argued (e.g., Lennie and Tacchi 2014)). Accountability does not easily contribute to a larger conversation about aid, and it does not directly benefit stakeholders, it only provides legitimacy for funders and practitioners to keep pursuing current development projects (Power 1997). Furthermore, learning cannot simply be seen as an automatic byproduct of accountability, as learning requires "different data, methodologies, and incentives" (Armytage 2011, p. 263).

To be sure, funders should be held accountable to their various stakeholders; they should not be funding projects that are failures. But measuring success should be viewed as a secondary purpose of evaluation – as a means to an end, in which the "end" is learning how to effectively promote development.

Meaningful lessons need to come about at the level of individual projects, not at the stage of cumulative literature reviews. In order to answer questions about what works in aid, looking at project success or failure is much less useful than trying to understand why it worked or fell short. Going further, a focus on the "why" could be broken down into two broad fields in which there is currently a large gap in knowledge. The first regards theories about what promotes social change; the second regards understanding how to actually design and implement effective development interventions.

### 39.8.1 Potential Learning Priority 1: Developing Theory-Based Guidelines for Social Change

Even if donors and practitioners agree that answering the question of why and how social change occurs is important, it is still not straightforward how we ought to answer these questions. Imagine a fourth-grader who is working on his long division. If he gets an answer wrong, his teacher can encourage him to figure out exactly where he went wrong when solving the problem. But what if the student gets the question right? How could the student begin to answer the question of where he went... "right"?

The analogy is not simply rhetorical. For a project addressing HIV prevention, for example, an organization's evaluation of why a project was successful could be that the organization focused on addressing social stigma (because it determined this was a key barrier in preventing HIV) or it could be that it hired a top-notch staff that kept the project running efficiently and smoothly. One of these findings is very useful in terms of thinking about what works in HIV prevention, the other far less so.

And yet, project evaluations very often focus on these latter kinds of *site-specific* "why" lessons on how to improve a particular project, such as shifting staff responsibilities, reallocating funds, focusing more on a particular activity, or adding a new activity. Learning "why" a project had particular outcomes or "how" it could be improved then emphasizes recommendations for project refinement, rather than providing lessons that can actually improve aid policies overall. USAID acknowledges this, stating that the most common ways that evaluations are used are to "refocus ongoing activities, including revisions to delivery mechanism work plans, extending activity timelines or expanding activity geographic areas" (2016, p. 16).

Given how much we still need to learn about development, limiting learning efforts to how to improve individual projects is rather narrow. Instead, learning could focus on generalized theory about why and how social change occurs. We should consider development of these theories a key requirement for improving aid effectiveness. This recommendation aligns with the calls of those who have advocated for increased attention to why projects work, such as Deaton (2009) and Stern et al. (2012).

Therefore, evaluators must take this extra step of providing generalizable takeaways, not leave the broader lessons implicitly embedded in the project details for a perceptive literature reviewer to uncover.

Donors could move evaluators in this direction by prioritizing questions about particular genres of social change. These genres need to be broad enough that they do not presume causal pathways to development, but not so broad that they become meaningless. Questions like "How do we use mobile phones to improve government accountability?" or "What are the best methods for opening up media systems?" or "Which works better for persuasion – information or entertainment?" all make assumptions about what works in development. These sorts of questions have their place – and sometimes funders have reasons for asking specific questions. But these very precise questions should not take up the bulk of development funding because they have already missed the boat. They eliminate other potential pathways for change. They assume that we know how to "do" development and that we are at the stage of refining our technique. We are not.

For example, there is currently much research on, and evaluation of, projects that seek to strengthen citizen voice and government accountability through information communication technologies (ICTs) like mobile phones. This work follows an effort by aid policymakers to pursue ICT-related development projects (such as the "Making All Voices Count" development program currently supported by various international governments). Yet, there is a dearth of evidence that increasing access to ICTs is an effective fix for improving governance and government accountability (Gagliardone et al. 2015; Gillwald 2010). Indeed, Shah (2010, p. 17) argues that the obsession with ICTs echoes the modernization paradigm, in which new technologies are "accompanied by determined hope that Lerner's modernization model will increase growth and productivity and produce cosmopolitan citizens." Excitement about the latest innovations in development may be causing policymakers to jump the gun when it comes to choosing the most practical types of programs rather than asking basic questions about how to improve governance.

Questions emphasized by donors should directly address how to improve lives. If large stakeholders could come to an agreement on a handful of these questions, and funders, evaluators, and designers created projects and evaluations that addressed them, this would work to make these disparate, case study projects more comparable, and the job of synthesizing extant data much more productive.

One example of such a question, or genre, relating to social change could be how to reduce gender-based violence (GBV). If projects addressing GBV provided big picture, theory-based, actionable answers to this question, this could help push forward more effective GBV donor policy. For example, one theory about gender-based violence is that for many women, failure to leave abusive relationships is based in large part on financial insecurity, and that if women had their own source of income, this would help them become less financially dependent on men, and therefore less likely to be victims of abuse. If evaluations could provide evidence for policymakers about whether or not this theory is sound, why, and under what circumstances, these insights could easily be incorporated into granting practice and future project designs across regions.

One might argue that this is already being done – that all evaluations of projects regarding GBV address (explicitly or not) broader lessons about how to reduce GBV. Yet this is not the case as often as one might think, and the degree to which these lessons are included is inadequate. As described above, the plethora of development evaluations have not led to as large a knowledge base as one would expect in particular development sub-fields. In the case of GBV, for example, the knowledge base "about effective initiatives is [still] relatively limited" (Bott et al. 2005, p. 3).

This is because many evaluations do not provide actionable insights on what improves lives. Some evaluations focus on whether or not an individual project worked (for instance, reporting whether trauma from GBV was reduced by an intervention, without addressing why); some make assumptions about how to reduce GBV without providing sufficient basis for the assumption (for example, assuming that GBV should be reduced through counseling by local health care providers); some only provide recommendations about how to make the specific project better rather than broader takeaway lessons about how to reduce GBV (for example, by adding more local staff). While there are many evaluations and reports that do provide important insights, there are not nearly enough.

### 39.8.2 Potential Learning Priority 2: Social Change Process Evaluation

Another major assumption of current evaluation policy is that, if we are implementing a project that is based on solid evidence about what promotes positive change, the process of actually designing and implementing the intervention in a particular context is straightforward. The fact that funders do not provide guidelines, or a specific mandate, to evaluate the design and implementation processes suggests that this is the case. However, designing and implementing a development intervention is quite difficult, as anyone who has ever tried to implement a project in unfamiliar territory will attest. Teams need to work with locals, achieve community buy-in, and understand local hurdles to change, among other challenges. Certain project teams have ways of creatively and dynamically engaging with communities; other teams seem to get stuck early on, dealing with communication problems and complex foreign environments they cannot comprehend or negotiate. There are too many projects in which teams paid insufficient attention to local context, did not factor in local institutions or traditions, did not understand ways that the problem was already being addressed, or did not understand what the community truly needed in order to instigate change.

Understanding how to implement projects effectively is crucial, and yet this is not an area that is frequently evaluated or for which project reports and evaluations provide lessons that can inform the development field as a whole. Just because one project distributed menstrual pads to school girls and succeeded in increasing their school attendance does not mean that another organization that attempts the same thing, even in the same community, will be successful.

One major conversation happening right now around project design addresses participatory designs, in which beneficiaries are actively involved in the design, implementation, and even evaluation of projects (see, e.g., Lennie and Tacchi 2013). The idea is certainly not new, but recent conversation stems, in part, from an increased recognition that top-down designs that do not take into account feedback and advice from target beneficiaries are less ethical and less likely to succeed. That said, while many projects engage with their beneficiaries to some degree, at least for feedback on design (Huesca 2003), it is rare for even so-called participatory projects to provide details about how participation was designed, encouraged, and achieved, or how "participatory" a project actually was (Inagaki 2007). So participatory design is at a stage in which many accept its legitimacy, but it is insufficiently clear how to successfully implement such a design, how many practitioners are using it and to what to degree, or how such interventions should be evaluated (Kogen 2018b; Waisbord 2015).

Therefore, with regard to design and implementation evaluation, one specific question evaluations might start answering would be whether deeply integrating stakeholders into the design and implementation process actually improves project outcomes. This is an assumption of much development work (and particularly CDS work and projects that incorporate participatory elements) and the logic seems sound, but we do not have sufficient evidence that this is really the case. To be sure, the fact that so few practice it suggests that most believe it is *not*. There is an underlying normative implication in discussions of participatory research and evaluation that participatory designs are more ethical (Huesca 2003), but this does not mean they produce better development outcomes. In fact, of the limited literature on participatory designs, a significant portion addresses challenges (e.g., Banerjee et al. 2010; Botes and van Rensburg 2000) and some that have tried to implement them have run into problems on the ground (e.g., Campbell and MacPhail 2002; Lennie and Tacchi 2014, p. 16).

In theory, participatory design holds significant advantages over traditional outsider-led project design, in that it is more likely to match the needs of the community and less likely to suffer from major oversights since the community itself is involved. These assumptions have indeed been borne out in many participatory-based projects (e.g., Cashman et al. 2008). These factors would point to participatory projects having greater impact (because they more directly address the needs of the community and avoid missteps) and being more sustainable (because the community has more ownership and buy-in of the project). On the other hand, participatory design is time-consuming, potentially limited to small interventions with a small number of beneficiaries, and therefore potentially too inefficient to promote significant social change (Hornik 1988). Having convincing evidence that participatory project design actually accomplishes development goals in a way that is practical and replicable could create a sea change in the world of development funding. This is an important area for which we need to be providing policymakers with some fundamental evidence that doing projects in a different way is worthwhile. Frameworks for tracking such processes have already been laid out by the evaluation community (Figueroa et al. 2002), but it is as yet unclear to what degree these schema for social change process evaluation have been utilized in the field.

Together, the two broad evaluation priorities defined above – developing theory-based guidelines for social change and evaluating design and implementation processes – can help begin to answer some basic questions about what is going "right" in interventions that succeed and what is going wrong in those that do not. Both can be crucial for improving policy. Not every evaluation will address these two topics, not every project will align with this model, but more need to if we are to begin improving aid policy in a meaningful way.

## 39.9 Summary

The twin pillars of today's development evaluations – accountability and learning – should not be conflated. The terms, as they are constructed by the world's largest donors, assume that donor communities possess a basic competence in development that is simply in need of proof to confirm its benefits, and perhaps some fine tuning. Rather, evaluations need to be reimagined in a way that help us answer more foundational questions about how to "do" development.

Simply knowing whether or not a project was successful (the basic understanding of 'accountability') does not, on its own, do anybody any good, especially if most human-centered social change projects technically "work." When we pledge accountability to taxpayers and beneficiaries, this should mean more than providing accounting mechanisms. Being able to record the "amount" of impact we had on developing regions does not benefit anyone unless it contributes to increasing positive impacts down the line. The only way to increase positive impact is to properly learn from projects. The only way to properly learn from projects is to acknowledge that we do not know much and then to prioritize learning above accounting. The attention to results and impact and the relegation to the back burner of methods for learning and improvement suggests that funders either do not truly believe they need to learn or are at such a loss that they do not even know how to articulate what it is they wish to learn. This is why answering whether a project "worked" is the wrong place to begin an evaluation.

William Easterly (2006) spends the bulk of his book *White Man's Burden* providing convincing arguments that the West does not know what it is doing when it comes to development, that it has had far more failures than successes, and that despite this it evokes a "patronizing confidence" that it knows how to solve development problems (2009, p. 368). A shift in evaluation goals from accountability to learning would signal an acknowledgement that we have much to learn about the basics of aid. This shift would define "learning" as answering foundational questions about aid rather than fine tuning existing programs that themselves may be misguided.

This new form of "learning" could focus on two broad areas: underlying theories about what promotes social change and the process of designing and implementing interventions. Theories that have not been confirmed on the ground should not serve as a backbone for large and ambitious projects, such as using ICTs to promote the economy or government accountability. By skipping the foundational evidence, we risk wasting precious aid dollars on projects that are based on assumptions rather than evidence.

Secondly, evidence on how to effectively design and implement interventions is lacking. Evaluations and reports infrequently analyze the process of conducting development interventions, suggesting that this is an area that does not require analysis because best practices are sufficiently established. They are not.

Many organizations do indeed incorporate useful learning into their evaluations. Good project reports and evaluations include recommendations that are broad and useful to those trying to understand, generally, what works in development. These do not always come in the form of evaluation reports. For example, BBC Media Action often issues "policy briefings" that distill findings from their interventions in order to make them easy to understand and be used by policymakers.

Specific questions could be asked in each of these broad categories. With regard to underlying theories, for example, we need a deep and critical examination of the question of how to reduce gender-based violence. With regard to design and implementation processes, we need explicit evaluations of participatory design, an approach to project development and implementation that has received increasing attention in the literature, seems to offer potential as being more effective and more ethical than traditional, top-down implementation designs, and yet is mostly ignored (except to a superficial degree) by most funders and designers who want more control over the project. The number of projects that fail to take realities on the ground into sufficient account, thus weakening the effectiveness of the project, suggests that this is a question that merits more attention.

Shifting the focus of evaluations away from impact may be anathema to project staff that want to tout the success of their projects. This is understandable. Organizations depend on positive evaluations to receive continued funding. Without evidence that they are doing good work, they risk losing the ability to continue working. But evaluation reports should still address impact. What must change is the mindset around evaluations – the ideology that evaluations are meant to assess impact, that this is their primary function, and that this is what makes development aid accountable and ethical. This frame, I have argued, is *not* adequately ethical because it limits the usefulness of evaluations and of projects. It shows that money was not entirely wasted but it does not efficiently or sufficiently teach us better ways to spend money. In this way, it benefits donors and implementers far more than the beneficiaries they purport to help. Shifting the focus of evaluations to understanding fundamental questions about how aid works may be a better, more ethical, and more just way to improve policy making and to stay accountable to beneficiaries at the same time.

We could simply be happy with the fact that our interventions are working, and continue to fund those that work and reallocate the funds of those that do not. But we should not be satisfied with projects that simply work. We should be seeking to make the most efficient use of money possible and to improve lives to the largest degree that we can. This is what we should mean when we refer to "accountability."

## References

Armytage L (2011) Evaluating aid: an adolescent domain of practice. Evaluation 17(3):261–276. https://doi.org/10.1177/1356389011410518

Banerjee AV, Banerji R, Duflo E, Glennerster R, Khemani S (2010) Pitfalls of participatory programs: evidence from a randomized evaluation in education in India. Am Econ J Econ Pol 2(1):1–30

Botes L, van Rensburg D (2000) Community participation in development: nine plagues and twelve commandments. Community Dev J 35(1):41–58

Bott S, Morrison A, Ellsburg M (2005) Preventing and responding to gender-based violence in middle and low-income countries: a global review and analysis. Policy research working paper no. 3618. World Bank, Washington, DC

Burnside C, Dollar D (2004) Aid, policies, and growth: revisiting the evidence. Policy research paper no. O-2834. World Bank, Washington, DC

Campbell C, MacPhail C (2002) Peer education, gender and the development of critical consciousness: participatory HIV prevention by South African youth. Soc Sci Med 55(2):331–345. https://doi.org/10.1016/S0277-9536(01)00289-1

Cashman SB, Adeky S, Allen AJ III, Corburn J, Israel BA, Montaño J, Rafelito A, Rhodes SD, Swanston S, Wallerstein N, Eng E (2008) The power and the promise: working with communities to analyze data, interpret findings, and get to outcomes. Am J Public Health 98(8):1407–1417. https://doi.org/10.2105/AJPH.2007.113571

Cracknell B (2000) Evaluating development aid: issues, problems and solutions. Sage, London

Deaton A (2009) Instruments of development: randomization in the tropics, and the search for the elusive keys to economic development. Working paper 14690. National Bureau of Economic Research, Cambridge, MA

DFID (2014) DFID evaluation strategy 2014–2019, London

Easterly W (2006) The white man's burden: why the West's efforts to aid the rest have done so much ill and so little good. Oxford University Press, Oxford

Enghel F (2016) Understanding the donor-driven practice of development communication: from media engagement to a politics of mediation. Global Media J: Can Ed 9(1):5–21

Feek W, Morry C (2009) Fitting the glass slipper! Institutionalising communication for development within the UN. Report written for the 11th UN inter-agency round table on communication for development. UNDP/World Bank, Washington, DC

Figueroa ME, Kincaid DL, Rani M, Lewis G (2002) Communication for social change: an integrated model for measuring the process and its outcomes. John Hopkins University Center for Communication Programs, Baltimore

Gagliardone I, Kalemera A, Kogen L, Nalwoga L, Stremlau N, Wairagala W (2015) In search of local knowledge on ICTs in Africa. Stability: Int J Secur Dev 4(1):35

Gillwald A (2010) The poverty of ICT policy, research, and practice in Africa. Inf Technol Int Dev 6:79–88

Glennie J, Sumner A (2014) The $138.5 billion question: when does foreign aid work (and when doesn't it)? Policy paper no. 49. Center for Global Development, Washington, DC

Gumucio-Dagron A (2009) Playing with fire: power, participation, and communication for development. Dev Pract 19(4–5):453–465. https://doi.org/10.1080/09614520902866470

Hornik R (1988) Development communication: information, agriculture, and nutrition in the third world. Longman, New York

Huesca R (2003) Participatory approaches to communication for development. In: Mody B (ed) International and development communication: a 21st-century perspective. Sage, Thousand Oaks, pp 209–226

Inagaki N (2007) Communicating the impact of communication for development: recent trends in empirical research. World Bank working papers. The World Bank, Washington, DC

Kogen L (2018a) What have we learned here? Questioning accountability in aid policy and practice. Evaluation 24(1):98–112. https://doi.org/10.1177/1356389017750195

Kogen L (2018b) Small group discussion to promote reflection and social change: a case study of a Half the Sky intervention in India. Community Dev J. https://doi.org/10.1093/cdj/bsy030

Kogen L, Arsenault A, Gagliardone I, Buttenheim A (2012) Evaluating media and communication in development: does evidence matter? Report written for the Center for Global Communication Studies. Center for Global Communication Studies, Philadelphia

Lennie J, Tacchi J (2013) Evaluating communication for development. Routledge, Oxon/New York

Lennie J, Tacchi J (2014) Bridging the divide between upward accountability and learning-based approaches to development evaluation: strategies for an enabling environment. Eval J Australas 14(1):12. https://doi.org/10.1177/1035719X1401400103

Lennie J, Tacchi J (2015) Tensions, challenges and issues in evaluating communication for development: findings from recent research and strategies for sustainable outcomes. Nordicom Rev 36:25–39

Moyo D (2009) Dead aid: why aid is not working and how there is a better way for Africa. Farrar, Straus and Giroux, New York

OECD Organisation for Economic Cooperation and Development (2008) The Paris declaration on aid effectiveness and the Accra agenda for action, Paris

Power M (1997) The audit society: rituals of verification. Oxford University Press, New York

Ravallion M (2009) Should the randomistas rule? The Economists' Voice 6(2):1–5

Servaes J (2016) How "sustainable" is development communication research? Int Commun Gaz 78(7):701–710. https://doi.org/10.1177/1748048516655732

Shah H (2010) Meta-research of development communication studies, 1997–2006. Glocal Times 15:1–21

Stern E, Stame N, Mayne J, Forss K, Davies R, Befani B (2012) Broadening the range of designs and methods for impact evaluations. Working paper no. 38. Department for International Development, London

USAID (2011) Evaluation: learning from experience. U.S. Agency for International Development, Washington, DC

USAID (2016) Strengthening evidence-based development: five years of better evaluation practice at USAID, 2011–2016. U.S. Agency for International Development, Washington, DC

USAID (n.d.) Meta-analyses, evidence summits, and systematic reviews. Retrieved from http://usaidprojectstarter.org/content/meta-analyses-evidence-summits-and-systematic-reviews

Vähämäki J, Schmidt M, Molander J (2011) Review: results based management in development cooperation. Report written for Riksbankens Jubileumsfond. Riksbankens Jubileumsfond, Stockholm

Waisbord S (2015) Three challenges for communication and global social change. Commun Theory 25:144–165. https://doi.org/10.1111/comt.12068

# Transformative Storywork: Creative Pathways for Social Change

## 40

Joanna Wheeler, Thea Shahrokh, and Nava Derakhshani

## Contents

| | | |
|---|---|---|
| 40.1 | Introduction | 734 |
| | 40.1.1 What This Chapter Will Do | 736 |
| 40.2 | Transformative Storywork | 737 |
| 40.3 | Principles of Transformative Storywork | 738 |
| | 40.3.1 Humanizing | 739 |
| | 40.3.2 Accountable | 740 |
| | 40.3.3 Emancipatory | 740 |
| 40.4 | Core Features of Transformative Storywork in Practice | 741 |
| | 40.4.1 Power of the Personal | 741 |
| | 40.4.2 Creative and Multimodal | 742 |
| | 40.4.3 Role of Iteration | 743 |
| | 40.4.4 Political Listening | 743 |
| | 40.4.5 Relational and Reflexive Learning | 744 |
| 40.5 | How the Principles of Transformative Storywork Are Embodied in Practice and How They Work | 744 |
| | 40.5.1 Group Building | 745 |
| | 40.5.2 Positionalities in Storywork | 746 |
| | 40.5.3 Facilitation | 747 |
| 40.6 | Where and How Transformation Happens | 748 |
| | 40.6.1 Personal | 749 |
| | 40.6.2 Collective | 750 |

---

J. Wheeler (✉)
University of Western Cape, Cape Town, South Africa
e-mail: joannaswheeler@gmail.com

T. Shahrokh (✉)
Centre for Trust Peace and Social Relations at Coventry University, Coventry, UK
e-mail: shahrokt@uni.coventry.ac.uk

N. Derakhshani
Cape Town, South Africa
e-mail: navatree@gmail.com

© Springer Nature Singapore Pte Ltd. 2020
J. Servaes (ed.), *Handbook of Communication for Development and Social Change*,
https://doi.org/10.1007/978-981-15-2014-3_54

| | 40.6.3 Societal | 751 |
|---|---|---|
| 40.7 | Conclusion | 751 |
| References | | 752 |

### Abstract

There is no one way that oppression is experienced in people's lives, and there are also many pathways to undoing exclusion. Understanding the relationships between the lived experiences of exclusion and the possibilities for change requires careful attention. In response, this chapter sets out the case for transformative storywork as an approach that can contribute to social change. Transformative storywork is a complex and multimodal process, which operates on an emotional, creative level. It uses storytelling as a form of inquiry, including the exploration of the self and of daily experiences in connection with life history and social context. Through personal and collective creative expression, transformative storywork builds opportunities to challenge unequal relations of power in our own lives and the lives of others. Creative storytelling can both humanize and politicize learning processes by building new and collective possibilities for social change. This chapter sets out the principles, elements, and practices of transformative storywork. We argue that creative expression through crafting personal stories, within a group-based process, enables a deeper understanding of our lives and, importantly, ourselves in relation to others. This relational learning enables reflection on our own stories and experiences. Building from these story processes, further deconstruction of the power relations within different life stories builds an understanding of the structural injustices that affect people's shared realities. Our analysis is that this relational understanding of structural inequality, built through multimodal creative methods founded in the personal narrative, can support transformative learning and solidarity in shared struggles for social change. This change was not only witnessed but was also experienced both personally and politically by the authors, and the chapter will outline reflexive learnings within these processes to communicate for social change. The chapter draws on examples of previous and ongoing work, including in South Africa.

### Keywords

Creative methods; Storytelling; Intersectionality; Social justice; Emancipatory; Visual; Listening; Transformative storywork

## 40.1 Introduction

Our strategy should be not only to confront the empire but to mock it...with our art...our stubbornness, our joy, our brilliance, our sheer relentlessness. Arundhati Roy

The idea of using storytelling for social change is both old and new. Traditions of storytelling in many cultures and contexts go back hundreds of generations and are part of how knowledge is passed on. Storytelling is part of how traditions are nurtured and transmitted. At the same time, there has been much attention in international development aid, in health communication, and in social movements and campaigns on the potential of stories to communicate messages for change (Matthews and Sunderland 2017; Shahrokh and Wheeler 2014a). This chapter aims to articulate a particular approach to using storytelling for social change: transformative storywork. What we set out here is based on the consolidation of many years of practice, through which this approach has evolved. This chapter intends to make the case for why this approach is important in terms of social change and how storytelling can be connected to transformation at multiple levels.

Transformative storywork refers to a group process that allows participants to use elements of narrative structure to organize life events into sequences and assign these experiences meaning through a story. Transformative storywork is not about compressing one's life story or the story of a group into a linear structure, but instead emphasizes the recognition of moments that have created shifts (both positive and negative). The crafting of stories through this process enables participants to identify these key moments and work around them, in terms of understanding what went into them and what came out of them and where their significance lies. Transformative storywork makes important initial assumptions that the process of how people tell their own stories will be different according to the personal circumstances, nature of the issue, and context; and the stories themselves will also be a reflection of these differences. Importantly, transformative storywork is a process of self- and collective inquiry; it supports a set of processes that build deeper and critical understanding of the world we are a part of and the changes we want to see. Storytellers are at the center of the process of knowledge creation. The development of these stories is intimately connected to allowing storytellers to recognize that the questioning of power relations is valid and valuable. This approach promotes a kind of research that is constructed and directed by the people affected by social injustice.

It is important to situate the "transformative" in storywork: in this approach, the starting point is a deep concern with social justice and how to move toward greater equity. The storytelling focuses explicitly on attention to who and what is oppressed or excluded within a particular context and how this can be changed to be more equitable. It is also informed by the assumption that the personal is political and that the everyday experiences of those living with exclusion need to be expressed and heard. By social change, we are not referring to a generic, political commitment. Instead the focus in transformative storywork is on the importance of surfacing and articulating direct personal experiences of marginality and recognizing and building the strength to challenge them.

Our analysis is that transformative storywork builds a more relational understanding of structural inequality, through multimodal creative methods (Reavey

2011; Clover and Stalker 2007). Through deepening the understanding of the connection between personal experiences and structural inequalities, it can lead to change. Changes emerge at personal, collective, and societal levels through this work (see Sect. 40.6). We have experienced these kinds of changes, personally and professionally, through transformative storywork, and also witnessed as they unfold with groups that we have supported. In our experience, creative storytelling can both humanize and politicize a process of learning for those involved. In some cases, transformative storywork can contribute to social change by building new and collective possibilities for transformation. This chapter draws on our self-reflexive learning over time, including challenges with our own positionality within transformative storywork (see Sect. 40.5 on positionality in transformative storywork).

This chapter represents the synthesis of knowledge gained through experience across many contexts and, over time, through largely collective and collaborative projects. We draw on participatory research we conducted over 8 years in the South African context focused on challenging power inequalities, violence, and injustice and the role of creative and story-based approaches in engaging with and communicating those realities. We also draw on other experience of using storytelling in connecting research with marginalized groups to local, national, and global policy formation processes. Finally, this chapter draws on the experience of storytelling workshops with a wide range of groups, movements, and organizations in Latin America, Europe, Asia, and Africa over the past 15 years. Over these various experiences, the practice of transformative storywork evolved, and this chapter represents an attempt to distill and articulate the approach.

### 40.1.1 What This Chapter Will Do

When researchers and practitioners discuss storytelling approaches, there are diverse ways of interpreting and understanding what they refer to as "storytelling" (cf. Matthews and Sunderland 2017). Storytelling also has different meanings and purposes in different disciplines and from distinct perspectives (psychology, organizational change, corporate, social policy, arts and humanities) (Crossley 2000; Bate 2004; Woodside et al. 2008; Sandercock 2003; Bell 2010; Clover and Stalker 2007). In order to be clear about how this approach is bounded and why it is important for social change, this chapter sets out the parameters of what we mean by transformative storywork.

This chapter defines the approach of "transformative storywork" and positions it in relation to other storytelling methodologies and related processes of communication for social change. It then goes on to articulate a set of principles that inform and guide the approach. These principles have emerged and evolved through an iterative relationship between our own normative assumptions about how to work toward social change and the practice of storytelling. The chapter explores how these principles are embodied in practice and the tensions that emerge. It goes on to describe some of the main elements of using the approach with examples of how these elements function in practice. Finally the chapter turns to the tensions and ethics of our own positionality as researchers and communicators for social change and the implications of this approach for transformation.

We have deliberately chosen not to include a specific example of a story generated through this process and instead draw learning from multiple and diverse stories and storytelling processes. This is because including a story as an example requires substantial contextualization to make its meaning clear in terms of how it relates to the overall approach. The intention of the chapter is to point to an overall area of work and show how it varies significantly by context, individual, group, theme, researchers, and practitioners involved. We hope that by referencing different examples throughout rather than a few in greater depth, the chapter will allow a more critical examination of the principles behind the approach (To view examples of stories produced through transformative storywork, see "My Nightmare Becomes a Reality" [https://vimeo.com/181166061], "Three Broken Hearts" [https://vimeo.com/185294666], "Police Brutality" [https://vimeo.com/192617528], and "We and Them" [https://vimeo.com/206378072] alongside stories responding to gender-based violence [http://interactions.eldis.org/capetown-digital-stories-on-sgbv], policy engagement processes, [http://vimeo.com/participate2015] and wider practice [http://transformativestory.org/case-studies]).

## 40.2 Transformative Storywork

We define transformative storywork in a very specific way. This definition has emerged over time, through many different projects and engagements using story-based methods. Transformative storywork is a complex and multilayered process which operates on an emotional, creative level. It uses storytelling as a form of inquiry, including the exploration of the self and of daily experiences in connection with life history and social context. In doing so, transformative storywork is committed to social justice. Through personal and collective creative expression, transformative storywork builds opportunities to challenge unequal relations of power in our own lives and the lives of others.

Transformative storywork uses a group process to surface and articulate stories of personal and/or collective experience. It does this by connecting a group-learning process to elements of story structure and storytelling that enable freedom, strength, and support in self-expression. Transformative storywork relies on iteration (between versions of stories and modes of expressions), and this iteration allows people to process or work through significant moments in their lives, in turn assigning sequences and meaning to experiences. This iterative process of reflection and renewal allows for shifts in understanding of ourselves, of the group, and of the wider society (Trees and Kellas 2009). These shifts happen, in part, through relational learning and the challenging of assumptions (about ourselves and others and our relationship to society). Transformative storywork is undertaken in order to surface and question power relations, but through a creative, multimodal, and inductive approach, rather than one that is primarily analytical that relies on pre-established categories and classifications.

This work is transformative in the sense that creative expression through crafting personal and/or collective stories, within a group-based process, enables a deeper understanding of our lives and, importantly, ourselves in relation to others. This relational learning enables reflection on our own stories and experiences (Lykes and Scheib 2015). Building from these story processes, further deconstruction of the power relations within different life stories builds an understanding of the structural injustices that affect people's shared realities.

In practical terms, this means that transformative storywork starts with what those involved choose to be their own significant experiences. The "work" is about using the process to deepen an understanding of what these experiences mean and what they mean for others. Storywork, therefore, can lead to stories being produced in different forms and formats. In transformative storywork, we often focus on short-form stories which can be conveyed in different mediums and modes: the group process leads to the production of a story or stories as an artifact of the process. However the process is not driven just by the imperative of creating a particular form of a story or the modality of how to think creatively about a story. It is equally driven by the need to give weight to personal experiences, recognizing their importance and articulating their meaning. From this starting point, storywork is the crafting of those experiences into something legible and powerful in representing and communicating their own meanings.

As researchers, activists, and artists, we have evolved the practice of storywork over time, through many engagements with groups and individuals in different settings and contexts. We draw on existing practice in popular education, art therapy, participatory action research, participatory video, and other story-based methods (such as digital storytelling) (Black et al. 2016; Mills et al. 2015; Shahrokh and Wheeler 2014a, b; Lambert 2013; Milne et al. 2012; Wheeler 2011; Reavey 2011; Wheeler 2009; Boal 2008; Kalmanowitz and Lloyd 2005; Greenwood and Levin 1998; Freire 1996; Heron 1996). While we have been influenced by all of these areas of practice, transformative storywork is bounded in terms of the principles underlying it and the elements of how it is practiced. The next sections explore these principles and elements in greater depth.

## 40.3 Principles of Transformative Storywork

Transformative storywork is an emergent approach. It has grown from our lived experience of learning with the communities and movements that have opened up their worlds, struggles, and aspirations to make new narratives. These narratives make visible and question power inequalities. In doing so, transformative storywork is not simply an exercise in understanding social experience; rather it has emancipatory qualities and provides space to critique forces of domination that participants negotiate in their daily lives and within the world around them. A set of principles have emerged through developing this approach that inform and guide it. These principles reflect the ethics and values of the work: in order to provide space for transformation at personal and collective levels, transformative storywork must be humanizing, accountable, and emancipatory.

## 40.3.1 Humanizing

Systems of oppression that drive exclusions are dehumanizing in that they invisibilize certain people or aspects of people's lives; they are often fed by and reinforce negative labels and stereotypes that reduce people and places to narrow categories. Transformative storywork is humanizing in that it is motivated by how to bring into fuller recognition the people and places that are hidden or made invisible through systemic injustice.

Transformative storywork recognizes that we can grow, in a personal way, through storytelling practice. This approach assumes that, as humans, we are constantly changing and have the potential to change. Systems of exclusion and oppression make personal growth incredibly difficult, but also indicate the potential for strength and resilience (Hernandez-Wolfe et al. 2015). Storytelling practice should therefore enable each person to realize their own creativity, express who they are, and live the freedoms that underpin what it means to be human (Paris and Winn 2014).

A fundamental principle of transformative storywork brings a humanizing approach to how we understand our own identities and experiences. This means that transformative storywork recognizes that people have "mosaic" lives with different aspects coming together in different places and times. This approach considers how structural exclusion is lived, in order to engage with how people experience those exclusions in their everyday experiences as well as in the analysis of the wider structures. Many scholars, practitioners, and activists point to the importance of intersectionality in understanding and addressing the experience of injustice (Hernandez-Wolfe et al. 2015; Crenshaw 1989). Transformative storywork recognizes that intersectionality is present and should be accounted for in the stories that are told (or untold).

The imperative of humanizing, and recognizing intersectionality requires deploying creativity and iteration. Creativity and iteration allow transformative storywork to grow through the textured and mosaic lives and the experiences of exclusion in particular places/people. Creativity is part of a multimodal approach (Reavey 2011; Clover and Stalker 2007). Iteration is moving between different modes (visual, textual, movement, drama, spoken) and different versions of the story. Story (as in narrative form) is therefore fed by creativity and iteration. The central principle, then, is about the importance of creativity in our lives and how this helps us to go beyond limitations and exclusions (Clover and Stalker 2007). This is particularly important when engaging with storywork to transform inequalities.

Our commitment to a humanizing approach challenges the conception of research as an objective process of evidence building or knowledge extraction. Rather, the research process is itself a part of our social world and itself has a role in humanizing and transforming our realities. The knowledge created through transformative storywork is not just a lens onto people's lives; it is actively engaged in creating social change. Through dialogic consciousness raising (the process of becoming more aware of social and political issues in discussion with others), participants and researchers experience change in understanding and actions. Significantly, this means that equality needs to be supported within our research practices and relationships of dignity and care are integral between all involved (Paris 2011).

### 40.3.2 Accountable

Accountability is a central principle to transformative storywork. This entails accountability in a number of aspects. The first form of accountability is from the researcher/practitioner toward the participants in the transformative storywork. This means that we, as researchers and practitioners, must be aware of how we construct and guide the process and the responsibility for working toward transformation. Accountability means that both researchers and participants have a commitment to learning, reflexivity, and change in themselves (Madison 2005). This implies the potential for renewal and an awareness of our own transformation and growth in response to learning from the stories of others, both facilitators and participants. Further, accountability requires an ongoing process of recognizing our own privileges and critically examining how we respond to them in relation to the group we are working with and the tensions that emerge with how we handle those situations. This dimension of accountability in terms of the deep responsibility and commitment of the researcher is important for building trust with the group(s) involved.

Another important aspect to accountability as a principle of transformative storywork is that it is grounded in an "ethics of care." Ethics of care imply shared and contextualized ethical commitments to "care" between facilitators and participants (Edwards and Mauthner 2012). This approach is established from feminist principles and decision-making grounded in care, compassion, attentiveness, and a commitment to acting in the best interests of the individuals or groups involved in the process (Ward and Gahagan 2010). This kind of care is important in transformative storywork as it often involves surfacing painful and difficult emotions, as participants explore their own experiences. As such our humanity must be established and reflected in the relationships held within transformative storywork. Evolving an ethics of care is integral to how the storying work unfolds. Experiences of witnessing each other's experiences and recognizing them provide opportunities for reflection and learning on how our care for and respect of others within the process and within our whole lives can unfold.

### 40.3.3 Emancipatory

Transformative storywork is an epistemological approach that promotes processes of knowledge creation that work in solidarity with marginalized group's own agency.

The approach draws from critical race, decolonizing, feminist, and queer arguments that challenge the privileging of certain groups' social position and in turn ways of seeing and knowing over others. This approach is instantiated through techniques and methods that evolve and change with the group, individual, context, and issues of concern, guided by principles set out here.

Our values and principles affect how we do research and what we value in the results of our research. This means that the objective of transformative storywork as inquiry is emancipatory; it is not simply an exercise in understanding social experience, but rather it works toward transformative social change. Through reflexive, critical, and dialogic learning, transformative storywork aims to effect change at personal, collective, and societal levels. As such, the approach draws on some of the same principles and mechanisms that underpin participatory action research (Fine 2016; Greenwood and Levin 1998; Heron 1996). Importantly however, transformative storywork establishes the power of the storying process in challenging of power inequalities; this happens through the recognition of self and other beyond the categories and labels we are constructed within, enabling representation of who we are and who we as a society can be.

Transformative storywork aims to act as catalyst for social change. This is not to claim that emancipation will be realized through the work but rather that the processes, spaces, and knowledge created through the approach offer a contribution toward wider social and political projects addressing forces of domination such as (neo)-coloniality, (hetero)-patriarchy, and white supremacy. The storywork creates a space for the participants to explore these issues in their own lives. Our efforts as researchers and facilitators within the processes are to support storytellers to expand the possibilities of how they can use their story and the insights they have drawn from the process, to contribute to wider social change.

## 40.4 Core Features of Transformative Storywork in Practice

So what is needed to do transformative storywork? The practice of transformative storywork involves a continuous learning process. There are no set steps to arrive at a transformative story. Rather transformative storywork involves working with a set of core features that must be made responsive to the realities and rhythms of the group of people being engaged. In order to connect to people's complex and changing realities, the researcher/practitioners adapt these features based on the group's histories, aspirations, and desires. As this is an evolving practice, transformative storywork requires constant critical reflecting on how to adapt the approach. This section examines some of the main features of transformative storywork in practice.

### 40.4.1 Power of the Personal

Transformative storywork is grounded in the surfacing of significant stories and reflective practice. Initially this involves creating space in the process for exploring our life stories through the mapping of moments that stand out as having influence in our lives and how we see ourselves. This recognition of the significance of our lives and the moments we have passed through cannot only bring deeper understanding of ourselves but also the personal power to engage with how and why we see these moments the way we do. This brings out opportunities to connect to our own identities as we define them and to question the labels or histories that are constructed by others. From this point of departure, storytellers are supported to identify a moment or set of moments within their larger life history which they want to work through, to process, and to draw insights that can help them connect to the question of who they are and who they want to be.

Connecting to the personal through stories involves an exploration of histories, identities, and perceptions of self. Recognition of personal experiences and valuing of everyday lives can support self-confidence. Enhanced self-confidence through giving recognition to and working through a significant moment in our lives can also strengthen the power to connect to and have empathy for others.

When personal storytelling is embedded within a commitment to social justice, experiences of marginalization and trauma are present in the lives of the people involved. So when people have a space to work through an experience within storytelling, trauma is very likely to emerge, because they may not have previously had the opportunity to explore and understand their own complex histories. As such, a strength-based approach is important for building personal power (Joseph 2015). This strength is often established in the framing of the storywork and how participants are supported to enter into the telling of their story. A statement or question acts as a catalyst for the storywork, and we have learnt that this can help storytellers connect to the past, present, and future. The learning and insights gained from our histories can help build positive pathways forward providing a sense of control and agency within our futures.

### 40.4.2 Creative and Multimodal

Transformative storywork is grounded in an understanding that lived experiences are multimodal, in that we navigate the world not only through narrative but also through space and through our bodies. As articulated in the use of multimodality in visual psychological practice, storywork provides space for the multiple meaning-making resources that we hold within our lived experience to be part of the way we communicate those experiences to others (Reavey 2011). Multilayered forms of expression (through movement, drawing, sculpting, speaking, singing, etc.) in turn enable a different kind of listening, where listening becomes more than hearing. Listening in transformative storywork is a deeper embodied practice that draws on a richer understanding of what is being communicated (see section below on political listening).

The creative and multimodal approach brings unexpected and often obscured knowledge to the fore. For example, the use of the visual through drawings, photography, icons, symbols, story modeling, and drama supports intuitive knowledges to surface. Visual and creative techniques enable reflection on and representation of who we are individually and collectively (Reavey 2011; Clover and Stalker 2007). Creative expression can provide avenues into a nuanced and "unspoken" understanding of people's lives, in particular in relation to difficult issues which can be hidden in the "discursive" (Pink 2004). The visual has both literal and connotative (associated and emotional) meanings created by historical, social, and cultural contexts. Connecting with others through the process of creativity and art-making can allow people to open to each other an intuitive engagement with each other's lives (Kalmanowitz and Lloyd 2005). Creative methods further enable participants to drive the learning experience within storywork, as it is their own creative expression and their narration of that creativity that provides opportunities for dialogue and understanding.

### 40.4.3 Role of Iteration

Iteration forms an important part of transformative storywork. Through iteration, the process helps participants to develop an organized sequence for their stories. Iteration, or the telling and retelling of the story through different modes and versions, allows participants to process significant moments. Iteration also helps participants to assign meaning to particular moments and sequences. Iteration allows the story to unfold and be crafted, rather than to constrain storytelling to a particular mould or version. Iteration allows participants to expand and contract the frame of the story and experiment with different ways of telling it. This kind of iteration helps participants to encounter and establish an organizing sequence that works for their story or stories.

Iteration implies taking time; and time is what is needed for participants to process and express important and often difficult experiences as a story. The process of iterating stories as a group allows participants to build an intimate public. The intimate public refers to the empathy established through the storytelling process with others. The trust created within the intimate public, or group, is also what enables a learning process as the stories are developed. This is the basis for a collective exploration of issues rising through the personal story process.

### 40.4.4 Political Listening

Listening is an essential part of transformative storywork. Listening is cultivated from the outset of the process and is central to how people are able to have the confidence to express their experiences. We treat listening as an active, rather than a passive undertaking – listening is not just about hearing or taking in what is being communicated, but also about recognizing our own position in relation to what is communicated. It is therefore "difficult, emotionally draining cultural work, requiring openness to others, commitment to others and...courage" (Matthews and Sunderland 2017: 8). In political listening, the listener tries to understand the views, perspectives, and experiences of the storyteller and at the same time recognizes their own position in relation to another person and tries to surface the implications of this (Hernandez-Wolfe et al. 2015). Therefore political listening involves making visible power inequalities in the process of listening to the telling of stories. The storyteller therefore experiences this listening as part of the process of articulating their experiences, creating space for shared learning between storytellers in the group. Political listening implies a reciprocity in the way that stories will be heard and in the extent to which members of the group are willing to be open to recognizing their own uses of power. "Political listening" is engaging ourselves in an effort toward social justice. This quality of listening is essential for building trust through the crafting of stories.

### 40.4.5 Relational and Reflexive Learning

Who we are is captured and constructed within our social relationships with others. By working with others and in building an understanding of each other, transformative storywork constitutes a relational process in the construction of the story. Transformative storywork, in similarity to the learning process behind other digital and creative storytelling approaches, emphasizes that storytelling holds the power to establish bonds and relationships through learning from and sharing of narratives of life experience (Poletti 2011).

In dialogue with others, our stories become joint constructions that have the power to shift our interpretations of self and other (cf. Trees and Kellas 2009). This means that transformative storywork goes beyond the personal psychosocial and therapeutic aspects of storytelling to the collective construction of knowledge and the understanding of each other (the relational learning process). This touches on our understanding of ourselves and our own role in witnessing and being heard, seeing and being seen.

For both participants and researcher/facilitators, this relational learning process requires reflexivity. Reflexivity involves finding strategies to question our own thought processes, values, assumptions, and prejudices in order to unpack and question our complex identities and roles in relation to others. Becoming aware of the socially constructed nature of our own gender, race, age, and class builds an understanding of ourselves and our identities that is fluid, multiple, and unfixed. This learning enables a connection to that complexity of self or the textured lives of others. In witnessing the stories of others, we are able to see "mosaic" qualities and in turn review our ways of relating in ways which embody a deeper ethics of care, dignity, and respect.

## 40.5 How the Principles of Transformative Storywork Are Embodied in Practice and How They Work

There is powerful emotional and relational learning within the process of storywork, and this needs to be deliberately supported while leaving enough flexibility for moving forward with telling stories. Our experience is that people from different backgrounds and with different identities can make their own, often surprising connections with one another and can draw strength from this process. At the same time, our different positionalities can cause tensions and constrain the possibilities of transformation. The following section is therefore an attempt to bring transformative storywork to life through sharing our experiences of doing transformative storywork. We share some examples of the careful negotiation involved in upholding the above principles in different contexts. We see the following considerations as essential in carrying out this kind of storywork such that significant care and respect are assigned to the individuals taking part in the process.

### 40.5.1 Group Building

Transformative storywork, when used to negotiate difficult social issues, tends to engage with some form of trauma. Exclusion from human and social rights is in and of itself traumatic and is accompanied by more pronounced and memorable instances of injustice. When working with social change, we hold that the process cannot shy away from the traumatic experiences associated with them, which further contribute to the silencing of such experiences of exclusion. Therefore in order to create a sense of safety for learning and healing, trust must be nurtured in order to respond ethically and adequately to people's experiences.

Trust is an essential element for undertaking transformative storywork. Significant emphasis is therefore placed on building trust between group members themselves and with the facilitators. It is foundational for creating a platform for open sharing within the work and furthermore contributes to building a sense of solidarity within a story group.

Through the sharing of intimate experiences in a structured environment, a strong sense of solidarity and cohesion is established within a story group. The process unites a group beyond the individual storywork and builds a sense of group identity. Participants move beyond participation and can enter into constructing group identity with enduring bonds of friendship and purpose to work together on the issues being addressed. This is not an inconsequential outcome, but must be nurtured and sustained for it to be fruitful. Story group members need to have ownership of their stories and the development of the group and therefore direct a collective approach to solving shared problems.

It is essential for the group members to be clear from the outset on the purpose of the storywork as well as the intended use of the stories articulated through the process. We work on the basis that the stories belong to the individual or group who produced them, and it is their choice to decide how they wish to use the story they produced. They can choose to keep their stories private or to share them with other audiences. But the freedom to choose must remain with the author(s) of the story.

Finally, it is important to note that the formation of trust and group building does not happen quickly or irrevocably. Trust and solidarity require time and effort and can also be lost easily. Our story processes require a minimum of 5 full days, preceded with preliminary meetings and followed by collaborative discussions or by ongoing meetings with group members over a sustained period of time, to work toward a given outcome or intention. Alternately, an established group with its own strong history can take part in transformational storywork (where the process can strengthen bonds and can be used to develop material for a given purpose by the group).

## 40.5.2 Positionalities in Storywork

The starting point or positionality for those initiating storywork is a significant factor which will influence the approach taken. Transformative storywork can be used for various outcomes, but there are also tensions between these outcomes. It is essential to surface these different starting points and assumptions and analyze them critically from the outset. Without this attention, the tensions and contradictions between the different positionalities can work to undermine the process. This section draws on our experience of tensions and conflicts in our positionalities in engaging in transformative storywork.

Based on an analysis of our experience, it is possible to categorize the positionality and intended outcomes of undertaking storywork from different perspectives. This includes empirical research (storytelling as finding out/generating data to answer questions), advocacy or campaigning (storytelling to generate messages aimed at influencing others), and therapeutic interventions (storytelling as a process of healing). Each of these positions implies a focus on a particular kind of outcome, as set out in Table 1.

Recognizing the tensions between these positionalities, we aspire toward a social justice approach (item 4 in Table 1). This approach works to weave a thread between these different positions, continuously questioning and reflecting on the extent to which the storywork is ensuring justice and is more closely realized in the lives of participants. In doing so, researcher/facilitators must engage deeply with the context of the process and who it involves and connect into different spaces accordingly. The process, rather than reflecting primarily the interests of institutions, researchers, or practitioners, must respond to the reality of the group. This often leads to tensions as organizational/professional agendas diverge from those of the group (Black et al. 2016).

Each person and group has a distinct journey in this process; what is possible and the potential involved varies according to their specific characteristics. For example, in working with groups that already self-identify as activists or are working over a longer period with an activist organization, the shift from personal articulation of stories to the communicating of these stories to wider publics can sometimes happen more quickly (sometimes too quickly!). With groups that have no prior group identity or structure, and people living with extreme marginalization, much more time is needed at the stage of allowing people to express and articulate their own stories and start to form a collective identity. In this case, it is not possible to quickly move toward collective action without risking the group formation process or distorting the story content or form. It is integral therefore that the process has value within its own parameters and in relation to the level of change that is possible

**Table 1** Storytelling purposes and outcomes

| Positionality | Empirical research | Advocacy / campaigning | Therapeutic | Social justice |
|---|---|---|---|---|
| Description | Generating evidence/data for specific questions | Conveys particular messages for specific audiences | Focuses on healing for person/group and on individual/group psychological well-being; not on the development of the story as a communicative artifact itself | Combines elements of 1, 2, and 3, but is not wholly aligned with any; starts with assumptions of the importance of individual storytelling through a collective process in contexts of exclusion/marginalization AND importance of rigor in creative practice of storycrafting |
| Intended outcome | Provides information necessary to respond to questions or test hypotheses | Generates stories that can be used to influence behaviors, attitudes, and/or actions | Promotes healing in cases of trauma | Promotes social justice |

for a certain story group (see next section). Importantly, within the storywork, there must be a layered approach that can build toward different kinds of change depending on context.

Furthermore, there should not be a hierarchy of outcomes from storytelling: the measure of success depends on the specific people and group involved, but is fundamentally about the sense of control that the storytellers have over their own experiences – any aspect of research that compromises this sense of control cannot be seen to be legitimate. Yet as researchers, often facing an agenda implies a conflict with the storytelling process (e.g., expectations in a PhD/doctoral research, grant-based funding tied to certain outcomes, pressures within the academy for "credible" research, pressure for stories of a certain artistic quality). Part of the role of the engaged and ethical researcher is to recognize these circumstances and realities and work with them, centering the interests of the storytellers within the process.

## 40.5.3 Facilitation

Skilled and sensitive facilitation plays an important role in fulfilling the principles and ethics of transformative storywork for social change. Effective facilitation of transformative storywork requires particular skills. It involves compassionate accompaniment of participants in articulating deeply personal stories, active and political listening, attention to group building (as discussed above), critical self-awareness, playful creative interaction to enable deeper and alternate ways of understanding individual stories, and an understanding of how to translate elements of stories through group processes into well-crafted stories.

The process is very demanding of facilitators in that it requires us to create a space of safety, freedom, and commitment for group members while also requiring facilitators to be able to share ownership of the process with group members to allow it to develop as it needs to. The role of the facilitator is to support the journey of the group and each individual and to support them in understanding the possibilities of change within the process.

As facilitators there is often a need to cede space to the group itself while also helping the group resolve conflicts that emerge. There is a constant struggle as a facilitator to understand how much input or direction is needed. Ongoing self-reflection within transformative storywork is essential so that facilitators can critically engage with their positionality as the processes unfold.

There is a fine balance of staying accountable to the group and enabling the ongoing longevity and sustainability of a group which is able to continue its work and purpose independently of external input. One dilemma that we have often faced is about how far our commitment as a researcher/facilitator should go in relation to the people and process involved. Ideally the group should be in a position which is strong enough to pursue the ongoing work of social justice autonomously and where the researcher is able to support movements and groups as necessary and not because they/we are needed for the group to continue.

It is important to note that the absence of skilled facilitation can result in things going wrong and can significantly hinder the depth of the process. Bad facilitation can heighten unequal power dynamics within a group which can reinforce silence by those who are extremely marginalized. Insensitive facilitation can undermine genuine trust within a group and therefore curtail the sharing of participants and limit the depth of storywork that can be achieved in the process. The politicized nature of work toward social justice can lead to the colonization of storywork processes by certain sets of interests. This situation of "capture" can happen where there are unequal power dynamics within the group. It is important to recognize that as a social process itself storywork will constitute the politics of change represented in the wider context it is situated in. Facilitators must therefore actively engage with this context and understand the implications for the group so as to be able to prevent wider social inequalities playing out and reinforcing power asymmetries.

## 40.6 Where and How Transformation Happens

Questions of where and how change happens are at the core of understanding how we can create a different kind of world. Understanding processes of transformation can provide us with the power to create lasting shifts in the power inequalities that maintain systems of control and domination. It is pertinent in this work to acknowledge and explore the processes of change that storytellers are a part of and the spaces and moments in which this happens in their lives. With this understanding, transformative storywork can grow and mobilize in order to drive change across the social levels and the spaces that they hold (Shahrokh and Wheeler 2014a). In our analysis of change, we have emphasized personal, collective, and societal levels as holding spaces of transformation. These levels and the spaces within them are integrally linked, and their interconnection or the learning that happens between them is what creates possibilities for lasting social change. Here, we separate these levels of change out in order to articulate their significance; however within transformative storywork, they become deeply connected and are reciprocal in terms of how deeper and lasting change can occur.

### 40.6.1 Personal

At the personal level, there is an impact in sharing one's story and witnessing the stories of others. The process of a group telling stories together has consequences for how we see ourselves and our identities. Lived experiences of exclusion can lead to internalized feelings of invisibility, and self-expression can create a space for being and becoming that allows us not only to question this exclusion, but with others to challenge this power over our own lives. What comes with this is an understanding that working toward social change and on issues related to social justice often involves traumatic experiences for individuals who have faced marginality. Supporting people to voice their difficult experiences, and to be recognized and witnessed by themselves and others, is a powerful element in the process for personal and societal change. It is important to recognize that in personal storytelling what is shared and explored by the storyteller is only one part of a more complex narrative of life (Rose and Granger 2013).

A space of belonging is formed within the story group, enabling an exploration of our own identities and the giving of value to and purpose to life experiences. The dynamics of storytelling are an important part of how transformation can be understood. The telling and retelling of stories highlights that in some way all stories are partial and leave elements untold. Our experience suggests that shifting silences in personal narratives provide insight into when and how possibilities of transformation arise, both at individual and collective levels. For storytellers in the Delft Lives Matters initiative, the surfacing of experiences that had been made invisible by the normalization of violence in a given context enabled a reclaiming of personal struggle as well as collective strength to unpack such a complex and entrenched issue.

The storying process enables participants to identify emotions as they relate to particular moments in their lives and to unpack them. Creative expression enables a connection to emotions through symbolism that can create a way to understand and articulate what these emotions mean and where they come from. For young people with migration-related life experiences in South Africa, visual creativity enabled a representation of their identities through artwork that aimed to challenge assumptions that they felt others had constructed around them. At the same time finding the language to share stories and label experiences gives participants the tools to give a name to what has happened in their lives and redefine or take control of what these experiences mean (Trees and Kellas 2009). Storywork is therefore an active process that has the power to build resilience and strength. This resilience is further built vicariously through understanding which emerges through the experiencing and witnessing of others' stories. This personal transformation is also experienced in the lives of facilitators and researchers. Learning from participant's transformations, storywork can enable a deeper awareness of our own intersectional identities, our silencing of these, and the strength that can come with making visible how others have responded to and survived trauma (Hernandez-Wolfe et al. 2015).

### 40.6.2 Collective

The relational experience of telling and listening to stories creates a sense of freedom that can enable new ways of understanding the world. This understanding stems from a collective consciousness built on relational learning. Participants have shared that transformative storywork has made visible for them that there is no one way that oppression is experienced in people's lives. Rather, a person's life is influenced by many factors, and each of these can influence the way that discrimination or marginalization is felt. In turn transformative storywork can support a deeper understanding of human complexity and how numerous and linked forms of oppression manifest in people's lives.

At the collective level, the process of shared storywork has been shown to contest narrow rigid categories of identity, ways of seeing issues and people, and assumptions about where collective interests and narratives lie. Working with gender justice activists in South Africa has shown that in exploring the intersectionality of identities within groupings that may otherwise be perceived as a 'coherent," there can be a transformation; in this case it was a shift in identifying who and how people are touched by gendered violence. The journey that one storyteller, Vee, took to explore the pain and strength that came with her experience of coming out as a lesbian enabled others to expand their understanding of whose reality counts in terms of representation on the issue of sexual and gender-based violence (watch Vee's story here: http://interactions.eldis.org/capetown-digital-stories-on-sgbv).

By giving space to everyone's story and creating learning relationships that do not privilege certain experiences over others, participants are able to negotiate a "common ground" between themselves. This learning process is partly about finding and describing the common ground. In the example of young refugees in South Africa, this negotiation was critical for their group's understanding and realization of young women's rights to voice and participation both within their joint storywork with young men and more widely in society. These shifts at a group level, as in the case of the gender activists above, can support storytellers to reimagine the world they are in and in turn the possibilities of social justice within it. The storytelling process can promote equality between people and also works to address power hierarchies that can exist within the group.

### 40.6.3 Societal

Within transformative storywork, the support of the group is needed to build strategies for change and ideas about how to put new and shared insights into action. Transformative storywork aims to provide a platform for personal and collective change to moving from being experienced in a more intimate way within the public of the group toward driving wider social change, using the knowledge created to take action for social justice. Seeing the complexities and injustices of structural inequalities through people's stories is one part of transformation.

Recognizing possibilities of change at the societal level grows from an understanding of agency and action which establishes that people have the potential to play a key role in shaping their own lives and relationships within their families, communities, society, authorities, and institutions. This action has multiple forms, and our experience has seen that the significance of this action has been both in the difference one can make in their own life and being expressed as activism in a more public sense. One young man, who was also a storyteller in the process with Vee outlined above, went through an important shift in terms of his own recognition of the exclusions Vee faced and their consequences. This led to him to invite his gay neighbor to walk with him to church each week, both to give recognition to the value of his life and to show his community that their exclusions are unjust.

Practicing agency in the transformation of these relationships involves stepping outside of the constructs, roles, and identities that exclusionary structures have established and creating a change. Expressions of agency such as this involve risk, and it is important that this risk is recognized, analyzed, and accepted before action is taken (Shahrokh and Wheeler 2014a). Importantly the process of transformative storywork supports the development of counternarratives which respond to those that promote discrimination and difference. These counternarratives happen at both the personal and collective level and act as a springboard for conversations across contexts. Our experience however has shown that the possibilities of social change relates deeply to the extent to which transformative storywork is humanizing. In taking action for societal change, retaining a connection to the personal and enabling empathy in understanding supports a commitment to change that is grounded in a shared humanity. Transformative storywork builds this understanding within those driving action, but also the articulation of their stories, and the personal strength and confidence to take that story into the world, can powerfully engage others and build solidarity for change.

## 40.7 Conclusion

This chapter has tried to set out the principles, core features, and elements of practice of transformative storywork. We have outlined some of the possibilities and tensions involved in doing this work. In looking across our experiences with transformative storywork, there are several conclusions that can be drawn about what difference transformative storywork makes and how this happens.

First, it is very clear that transformative storywork must be embedded in social change processes, with specific movements, groups, collectives, and individuals.

When it is driven too much by other agendas and priorities at odds with the real concerns of those living with marginalization and injustice, the integrity of the processes unravels and much of the transformative potential is lost. This more transformative approach works best through accompaniment of these groups/movements/collectives over a longer period of time. One-off workshops and compressed story processes are more vulnerable to collapse and distortion in relation to the principles set out here.

This means that transformative storywork, like social change itself, requires long-term personal commitment from the researcher/practitioner. Undertaking transformative storywork is not only about developing expertise in storytelling and encouraging creativity. It requires that, as researchers and practitioners, we cultivate self-reflexivity and political listening and practice participatory and inclusive facilitation. As transformative storywork often occurs in the context of complex and conflicting agendas, researchers and practitioners of transformative storywork need to develop the ability to negotiate complex organizational contexts and make provisions for coping with emotionally laden and often painful processes. Underlying all of this, the most important characteristic of researchers and practitioners undertaking transformative storywork is respect for the potential of everyone to tell their own story and awareness of the significance of the stories and the process of how the stories emerge. Transformative storywork asks a lot of us as researchers and practitioners, but the responsibility we assume when we enter into this work is also where we find its transformative potential – for ourselves and for wider social justice.

## References

Bate P (2004) The role of stories and storytelling in Organisational change efforts: a field study of an emerging "community of practice" within the UK National Health Service. In: Hurwitz B, Greenhalgh T, Skultans V (eds) Narrative research in health and illness. Blackwell Publishing Ltd, Malden

Bell L (2010) Storytelling for social justice: connecting narrative and the arts in antiracist teaching. In: Routledge. New York/London

Black G, Derakhshani N, Liedeman R, Wheeler J, Members of the Delft Safety Group (2016) What we live with everyday is not right. Sustainable Livelihoods Foundation, Cape Town

Boal A (2008) Theatre of the oppressed, 3rd edn. Pluto Press, London

Brushwood Rose C, Granger CA (2013) Unexpected self-expression and the limits of narrative inquiry: exploring unconscious dynamics in a community-based digital storytelling workshop. Int J Qual Stud Educ 26(2):216–237

Clover D, Stalker J (eds) (2007) The arts and social justice: re-crafting adult education and community cultural leadership. National Institute of Adult Continuing Education, Leicester

Crenshaw K (1989) Demarginalizing the intersection of race and sex: a black feminist critique of antidiscrimination doctrine, feminist theory and antiracist politics. Chicago Legal Forum 140:139–617

Crossley ML (2000) Narrative psychology, trauma and the study of self/identity. Theor Psychol 10(4):527–546

Edwards R, Mauthner M (2012) Ethics and feminist research: theory and practice. In: Miller T, Birch M, Mauthner M, Jessop J (eds) Ethics in qualitative research, 2nd edn. SAGE Publications Ltd, London/California/New Delhi, pp 14–28

Fine M (2016) Participatory designs for critical literacies from under the covers. Lit Res Theor Method Pract 65:47–68
Freire P (1996) Pedagogy of the oppressed. New Revise. Penguin, London/New York
Greenwood DJ, Levin M (1998) Introduction to action research: social research for social change. SAGE Publications, Inc, Thousand Oaks
Hernandez-Wolfe P, Killian K, Engstrom D, Gangsei D (2015) Vicarious resilience, vicarious trauma, and awareness of equity in trauma work. J Humanist Psychol 55(2):153–172
Heron J (1996) Co-operative inquiry: research into the human condition. SAGE Publications Ltd, London/California/New Delhi
Joseph S (2015) Positive therapy: building bridges between positive psychology and person-centred psychotherapy. Routledge, London/New York
Kalmanowitz D, Lloyd B (2005) Epilogue. In: Kalmanowitz D, Lloyd B (eds) Art therapy and political violence: with art, without illusion. Routledge, New York, pp 233–235
Lambert J (2013) Digital storytelling: capturing lives, creating community. Routledge, New York
Lykes MB, Scheib H (2015) The artistry of emancipatory practice: photovoice, creative techniques, and feminist anti-racist participatory action research. In: Bradbury H (ed) The SAGE handbook of action research, 3rd edn. SAGE Publications Ltd, London, pp 131–142
Madison SD (2005) Introduction to critical ethnography. In: Madison SD (ed) Critical ethnography: method, ethics, and performance. SAGE Publications Ltd, London/California/New Delhi/Singapore, pp 1–16
Matthews N, Sunderland N (2017) Digital storytelling in health and social policy. Routledge, London
Mills E, Shahrokh T, Wheeler J, Black G, Cornelius R, Van Den Heever L (2015) Turning the tide: the role of collective action for addressing structural and gender-based violence in South Africa. Institute of Development Studies, Brighton
Milne E-J, Mitchell C, De Lange N (eds) (2012) The handbook of participatory video. AltaMira Press, Lanham/Toronto/New York/Plymouth
Paris D (2011) "A friend who understand fully": notes on humanizing research in a multiethnic youth community. Int J Qual Stud Educ 24(2):137–149
Paris D, Winn MT (2014) Humanizing research: decolonizing qualitative inquiry with youth and communities. SAGE Publications Ltd, London/California/New Delhi
Pink S (2004) Applied visual anthropology social intervention, visual methodologies and anthropology theory. Vis Anthropol Rev 20(1):3–16
Poletti A (2011) Coaxing an intimate public: life narrative in digital storytelling. Continuum: J Media Cult Stud 25(1):73–83
Reavey P (2011) Visual methods in psychology: using and interpreting images in qualitative research. Psychology Press, East Sussex
Sandercock L (2003) Out of the closet: the importance of stories and storytelling in planning practice. Plann Theor Pract 4(1):11–28
Shahrokh T, Wheeler J (eds) (2014a) Knowledge from the margins: an anthology from a global network on participatory practice and policy influence. Institute of Development Studies, Brighton
Shahrokh T, Wheeler J (2014b) Agency and citizenship in a context of gender-based violence, IDS evidence report 73. Institute of Development Studies, Brighton
Trees A, Kellas JK (2009) Telling Tales: enacting family relationships in joint storytelling about difficult family experiences. West J Commun 73(1):91–111
Ward L, Gahagan B (2010) Crossing the divide between theory and practice: research and an ethic of care. Ethics Soc Welf 4(2):210–216
Wheeler J (2009) The life that we don't want': using participatory video in researching violence. IDS Bull 40(3):10–18
Wheeler J (2011) Seeing like a citizen: participatory video and action research for citizen action. In: Shah N, Jansen F (eds) Digital (alter)natives with a cause? Book 2 – To think. Centre for the Internet and Society and Hivos Knowledge Programme, Bangalore/The Hague
Woodside A, Sood S, Miller K (2008) When consumers and brands talk: storytelling theory and research in psychology and marketing. Psychol Mark 25(2):97–145

# Recollect, Reflect, and Reshape: Discoveries on Oral History Documentary Teaching

## 41

Yuhong Li

## Contents

| | | |
|---|---|---|
| 41.1 | Introduction | 756 |
| | 41.1.1 Oral History and Oral History Documentary Filmmaking | 756 |
| | 41.1.2 Research Method and Target | 757 |
| 41.2 | Oral History Documentary Production in College Education | 758 |
| | 41.2.1 Oral History Practice | 758 |
| | 41.2.2 Using Documentary Film to Record Oral History | 758 |
| | 41.2.3 How to Plan and Implement an Oral History Documentary Project | 759 |
| 41.3 | "Our Fathers' Revolution" Oral History Documentary Production | 760 |
| | 41.3.1 Choose Story Theme by Region | 760 |
| | 41.3.2 The Difficulties of Making Oral History Documentaries About the Cultural Revolution in China | 761 |
| | 41.3.3 Accomplish the Target of Liberal Arts Education Through Oral History Documentary Production | 762 |
| 41.4 | Oral History Documentary Series "Family Album" | 763 |
| | 41.4.1 100 Chinese Ordinary Family Stories' Vision Archive | 764 |
| | 41.4.2 Recollect, Reflect, and Reshape | 764 |
| 41.5 | How to Evaluate Oral History Documentary as Historical Data | 765 |
| 41.6 | Conclusion | 766 |
| References | | 766 |

### Abstract

In university education, oral history always hopes to have students involved in a more attractive way, while teaching documentary production one also looks for good stories. Therefore, the combination of oral history and documentary

---

A case study of oral history documentary series "Our Fathers' Revolution" & "Family Album"

Y. Li (✉)
Hong Kong International New Media Group, Hong Kong, China
e-mail: yuhongrush@gmail.com; yuhong@caa.columbia.edu

© Springer Nature Singapore Pte Ltd. 2020
J. Servaes (ed.), *Handbook of Communication for Development and Social Change*,
https://doi.org/10.1007/978-981-15-2014-3_44

production in university education seems to be a necessity. This is especially meaningful for the Chinese young generation since they have a great interest and curiosity about China's modern history and stories about their own family and communities. However, often these stories remained a taboo or untold.

In 2011 the Department of Media and Communication at the City University of Hong Kong launched oral history documentary programs and produced 11 episodes of "Our Fathers' Revolution" and around 100 episodes of "Family Album" short films. This 6-year practice on oral history documentary education turned out oral history and documentary films that combine and accomplish the goals of oral history education. The process of documentary production has been an interesting experience for the students themselves and for their family and friends.

Oral history documentary projects help students foster an intergenerational appreciation and an awareness of the intersection between personal lives and larger historical currents. It could help fulfil the educational target of a liberal arts education – to develop students' critical and creative thinking and inspire the active participation in history and civic life in the future. We can regard the whole process of oral history documentary teaching as a way for students to recollect, reflect, and reshape.

However, there are still many problems to be solved during the production of an oral history documentary. Questions such as: Are the interviewees' subjective feelings consistent with the historical reality? How to carry out trade-offs during the postproduction (video editing) in order to complete a wonderful story? How to complete script writing?

The characters for a documentary require relatively dramatic stories, and it kind of contradicts the relatively boring oral history. How to respect the reality of the history and at the same time satisfy the urge for watching a documentary? How to nurture the students' historical perspective and objective attitudes? All these difficulties have been encountered in the class during the production of oral history documentaries.

**Keywords**

Oral history documentary production · Teaching discovery · Liberal art education · Critical thinking · Recollect · Reflect · Reshape

*The great strength of oral history is its ability to record memories in a way that honours the dignity and integrity of ordinary people.* – Mary Marshall Clark, OHMA

## 41.1 Introduction

### 41.1.1 Oral History and Oral History Documentary Filmmaking

Oral history, by definition, is the collection and study of historical information about individuals, families, important events, or everyday life using audiotapes, videotapes, or transcriptions of planned interviews. Oral history is a research method and a very important way to observe the history as well. It strives to obtain information from different perspectives, and most of these cannot be found in written sources.

In particularly, oral history is conducted with people who participated in these past events. These people could be very important figures in history, but mostly, they are everyday people who tell the details they have witnessed in their daily life. In this way, oral history extends the traditional elites' history to the folk history.

Oral history has been changing with the development of technology. The way to record oral history is getting more and more diversified after the emergence of digital photography and cinematographic technology. Documentary film, as an effective storytelling method, can make good use of oral history.

The unique modern history of China over the past century, especially those hidden, dark, and drifting stories deep in Chinese people's mind, has been buried by so-called national memories controlled by the governmental culture. However, Chinese society has evolved historically since the economic reform in 1976. The middle class is exploding, and they have more concerns about Chinese social and historical issues and are eager to know the truth behind the scenes.

Oral history in Chinese university education has been looking for an attractive way to make more students participate in documentary production and the telling of good stories. Therefore, it is a necessity to combine oral history and documentary production.

### 41.1.2 Research Method and Target

History research need comprehensive information. There are different perspectives about one historical event. Comprehensive messages will help to explain and understand the truth behind the scenes.

In China, oral history has been significant for social and historical research. Due to the limitations of the Chinese political environment, there are major problems about the study of China's history, especially studies on modern Chinese history. Chinese education still focuses on patriotism and nationalism and does not provide a broad vision on history studies and historical issues.

Oral history education provides a better way to fill up these gaps. Oral history documentary production further helps students to explore the stories behind the scenes of their family and what they have suffered or experienced.

Many MA students from Mainland China study in Hong Kong. These students have a great interest and enthusiasm to explore the story that their grandparents and parents' generation experienced and to understand the diversity of the Chinese modern history.

The Department of Media and Communication at City University of Hong Kong initiated the oral history documentary project "Family Album" and aimed to help students foster an intergenerational appreciation and awareness of the intersection between personal lives and larger historical currents.

This chapter will be based on the case study of "Family Album" documentary series and explore the whole process of oral history documentary teaching and how to make it as a way for students to recollect, reflect, and reshape.

## 41.2 Oral History Documentary Production in College Education

### 41.2.1 Oral History Practice

In the United States, since 1966, the Oral History Association (OHA) has served as the principle membership organization for people committed to the value of oral history. OHA has established a set of goals, guidelines, and evaluation standards for oral history interviews and serves a broad and diverse audience including teachers, students, community historians, and filmmakers.

OHA established the Principles and Best Practices for Oral History to guide on how to practice oral history studies. The Practice introduces the three stages of oral history practice, i.e., pre-interview, interview, and post-interview. It emphasized the important role of oral history to develop students' critical and creative thinking and communication and leadership skills and to stimulate young generation's involvement in public affairs and history studies.

China is an authoritarian country with lots of media control; what is on the record may not reflect a complete picture. The voice of the people could present an alternative to the official history. We learn history through documents in archives, but it is difficult for the official, written history to incorporate everyday people's stories in these documents. Oral history is important because it gives a voice to the witnesses of a certain historical event or period.

However, in China, oral history education is still in its early stage. One of China's well-known journalists, Cui Yongyuan, established an oral history center at the Communication University of China and tried to initiate some oral history projects to help the young generation participating in activities related to oral history.

### 41.2.2 Using Documentary Film to Record Oral History

Film shooting technologies have become cheaper, handy cams have become widespread, and digital cameras and mobile telephones provide opportunities of making short film production available to large masses. These developments allow the use of short films in education.

Documentary film can be a great way to record oral history. However, many people are concerned about two aspects about oral history and documentary film. Fist, oral history is about the truth, while documentary film is nonfiction. In other words, will the documentary stories distort the facts or exaggerate the events? Second, in order to attract an audience, documentary films tend to choose more dramatic figures or aspects to focus on. Will this hurt the reliability of oral history?

## 41.2.3 How to Plan and Implement an Oral History Documentary Project

### 41.2.3.1 First, Set Up the Theme

The two documentary series "Our Fathers' Revolution" and "Family Album" were completed by MA students from the City University of Hong Kong. At the Department of Media and Communication, the internship program looked for suitable topics to engage students. Coincidentally, most of these graduate students came from all over Mainland China. Historical themes became the first choice that could attract the student participation and interest.

Students from Mainland China know little about China's history over the past 100 years. For example, the Cultural Revolution (CR) was a social-political movement in China from 1966 until 1976. Launched by Mao Zedong, then Chairman of the Communist Party of China, the movement paralyzed China politically and negatively affected the country's economy and society to a significant degree. However, China's youngsters don't have much access to understand this period of history, except for a couple of general paragraphs in their textbooks at school. The internship project chose the Cultural Revolution as the topic to stimulate students' curiosity. In 2011–2012, around 20 students produced an 11-episode documentary series called "Our Fathers' Revolution." This documentary series focuses on the stories that the students' family suffered then or what happened in their hometown a couple of decades ago. Due to the big success of "Our Fathers' Revolution," another oral history documentary series, "Family Album," was launched in 2013 to explore more personal Chinese everyday people's stories to further understand China's modern history.

These two topics are all based on student's interest and very attractive to the young students from Mainland China. The students took part in the project with great enthusiasm. However, both the Cultural Revolution and family history include an abundance of diverse topics and personalities. It's not easy to choose the topic and character to satisfy the authenticity of oral history and the storylines which documentary films require at the same time. Moreover, the distinction between oral history films and other documentaries is not always clear. The films must use oral history strategies to produce documentaries.

### 41.2.3.2 Second, Training (Pre-production)

For the students who do not have much experience to film a documentary, training is a necessary stage. Usually, the whole process of documentary includes three stages: pre-production, production, and postproduction.

The training stage is about contents training and technical training. It takes one semester (around 3 months). Contents training includes choosing a topic and characters and understanding the sociological significance of the topic and storylines. This part is very important and requires a lot of reading and discussion; especially critical and creative thinking is a must. Contents training will stimulate students' curiosity and interest and help the students conquer many difficulties from the filming and editing.

Since it's about an oral history documentary production, historical literacy training is important. This training was completed through reading and screenings with discussions.

#### 41.2.3.3 Third, Filming (Production)

For both documentary projects, the shooting time was about 1 month and pretty tight since the students need to go back to their hometown during the new year's vacation. However, after 3-month contents and technique training, the students were ready for the filming.

#### 41.2.3.4 Fourth, Video Editing (Postproduction)

It takes around 4 months to complete the video editing. Usually the students would take a lot of footages, and editing is kind of recreating – to further understand the storyline and the character.

## 41.3 "Our Fathers' Revolution" Oral History Documentary Production

The 11-episode Our Fathers' Revolution produced by City University students tried to discover what everyday people experienced during the Cultural Revolution.

### 41.3.1 Choose Story Theme by Region

After deciding the theme of this documentary series, the Cultural Revolution, the next step is to choose a topic or character for individual stories.

The students were from all over China, and finally 11 topics were selected in their regions (Table 1). During the Cultural Revolution, different places in China have different characters and stories. For example, in Chongqing, armed fight was fierce during the CR, and many people were killed or injured. Henan, China, was the place of the Great Famine, and many people died of hunger. The student from Beidahuang, Heilongjiang, tried to explore the stories about youth from big cities who were sent to the countryside for reeducation by the peasants. They found many of them who never got back to their city in their lifetime. One of the students from Hunan province, the hometown of Chairman Mao, went to investigate their memories of Chairman Mao.

For the first time, the students got firsthand stories about the Cultural Revolution and further understood what happened in their hometown during the Cultural Revolution.

**Table 1** 11-Episode "Our Fathers' Revolution" documentary topics and geographical distribution

| Region | Theme | Topic |
| --- | --- | --- |
| Henan | In Memory of the Great Famine | Henan's Great Famine |
| Tianjin | Sorry and Cheer | Encounter of Tianjin's tradition culture during the Cultural Revolution |
| Shijiazhuang, Hebei | The Man Who was God | The story of the old red evolution base |
| Shaoshan, Hunan | Long Live Chairman Mao | How people from his hometown evaluate Mao Zedong |
| Beidahuang, Heilongjiang | Gone Gone Gone | Stories about sent-down youth in China during the CR |
| Fuyang, Anhui | Past of Huizhou | Stories about the experience of grandmothers during the CR |
| Chongqing | Invisible Armed Fight | Victims' experience of armed fight in Chongqing during the CR |
| Chengdu, Sichuan | Story of the Imperial City | The change of Tianfu Square in Chengdu |
| Jingzhou, Hubei | The Silent Jingqian Village | The stories during the CR in a small Chinese village |
| Jingzhou, Hubei | Haven in the Storm | The story of May Seventh Cadre School in Shayang during the CR |
| Hong Kong | Their 1967 | Hong Kong's Cultural Revolution in 1967 |

## 41.3.2 The Difficulties of Making Oral History Documentaries About the Cultural Revolution in China

For most young people in China, the Cultural Revolution is a very unfamiliar historical term. During the training stage, the students need to read many books, especially oral history books about the CR and watch some available movies or documentaries about the CR. To help them know what happened in the past was the first difficult mission since they had very little knowledge about what happened during that period of time.

Students were suggested to read *Mao's Last Revolution* by Harvard professor Roderick MacFarquhar and some other memoirs about the CR which made them understand what happened in order to have a relatively objective and comprehensive picture about that period of history. After they explored their topic, they conducted specific readings. For example, the student who would focus on the Great Famine from 1959 to 1962 would find more related materials to read and research after understanding the whole historical background. This process greatly helped the students prepare the upcoming interviews.

After choosing the topic, the next mission was to select characters. The students could choose some candidates through the Internet or recommendations from their family members and then decide the main character. One of the students, who was from Fuyang, Anhui, was pretty confused to find a proper character. She wanted to know what everyday people experienced during the CR and she talked with her parents. For the first time, she knew her both grandmothers from mother and father sides had suffered a lot due to their family background of being so-called landlord or capitalist. The student therefore filmed a story about her two grandmothers and told the real story how much family background could influence a person's life during the CR.

The biggest challenge at the stage of postproduction was how to visualize those historical stories. Documentary film needs a lot of footage to support, and the audience would feel very bored if there are too many talking-head interviews. When it comes to the topic of the Cultural Revolution, visualization became a big difficulty. There is much footage about the CR; however, the Chinese government does not allow the copyright to be used for a documentary. Family albums and some other resources from the Internet helped to visualize some stories.

How to approach the historical truth at the most? The students were guided to interview different people and ask questions from different perspectives. For example, the story about the Great Famine, the students took great risks (since the Great Famine is still a taboo in China and people were very afraid to talk about it under the government's surveillance) to interview five survivors in a small village, and a scholar who has been doing research about the Great Famine for above 10 years. After talking with these people, the documentary could give an answer about what happened in 1959–1962.

### 41.3.3 Accomplish the Target of Liberal Arts Education Through Oral History Documentary Production

It took almost a year to complete a 15-min short documentary film "Our Fathers' Revolution." What the students have learned from the whole process was much more than making an oral history documentary. Chinese youngsters who are born after the 1980s or 1990s have been called the Me Generation. This oral history documentary project helped them to care about the broader world – the modern history and the Chinese society. Not only the students, even their parents, who were born in the 1960s or 1970s mostly, participated enthusiastically in the whole filming process. Some feedback from the parents indicated that they were eager to know what really happened and why it happened during the Cultural Revolution.

The whole producing process of Our Fathers' Revolution achieved the goal of oral history documentary education – recollect, reflect, and reshape. That means, recollect history facts, reflect the history attitude, and reshape values. The project emphasized critical and creative thinking and required students to remain open-minded and not take anything for granted, for example, not to pre-assume that all the people are victims from the CR and that all the people would say *no* about the CR. The students filmed the documentary with a lot of questions and confusions. For instance, a group of students filmed the story about Chongqing's armed fight during the CR. They assumed many victims would regret about what happed and what they had done. However, when they interviewed a then chief of the armed fighters group, he described his behavior then as a hero and was very proud of what he had done, although he lost one of his fingers in the fight. The students could hardly understand this character and felt even more confused. However, the students got to know that the real world is not only black or white. Human nature is more complicated than we can imagine. In this case, students could further understand why the Cultural Revolution happened in China. It is a process to recognize the diversity of the history.

## 41.4 Oral History Documentary Series "Family Album"

Due to the political sensitive nature of "Our Fathers' Revolution," we could not find major access to broadcast the series in Mainland China, and it was more and more difficult for the students to interview witnesses. Therefore, a new project, "Family Album," was introduced to encourage the students to record family histories in general.

The Family Album documentary series was launched in 2013. By 2017, this project has produced 79 short films. These short films feature the stories of their families, grandparents, parents, and friends. In China, there are major differences between generations. Especially for the young people, most of them have no idea about their family history and even have no idea about their grandparents' name. This project helped them to go back to their own hometown and discover what has shaped their family and themselves.

Since the students came from different family backgrounds, the stories are more related to China's temporary society. Some students tell stories about their grandparents' generation and discover the roots of the whole family. This project drives the students to sit down and talk to their aged grandparents and to explore the old times. For instance, one student produced a short film about her grandparents – how they are losing their memories and express the worries about the issues of an aging society.

The topics of Family Album covered many aspects of social life – the conflicts between generations, the influence of the broken marriage of their parents on the children, China's one-child policy and the discrimination of baby girls in China's rural areas, and so on.

**Table 2** Topic distribution of Family Album project

| Year | Student age | No. of short films | Distribution of topics | |
|---|---|---|---|---|
| | | | Grandparents | Parents |
| 2013–2014 | Born in 1989–1990 | 17 | 14 | 3 |
| 2014–2015 | Born in 1990–1991 | 21 | 11 | 10 |
| 2015–2016 | Born in 1991–1992 | 19 | 10 | 9 |

### 41.4.1 100 Chinese Ordinary Family Stories' Vision Archive

The topics of Family Album are changing with the students' ages and the change in society.

From Table 2, we can see that when the students are getting younger, their focus changes from the grandparent generation to the parent generation. The social background they tried to express was also changing from the Cultural Revolution to the 1980s after China's economic reform. These topics not only care about social change but also focus on their own experience and how to explore their own identity.

One student told a story about her and her father, who divorced her mother when she was only 2 years old. The student moved to her father's home in New Zealand when she was 10, but the relationship between father and daughter never improved until she completed the short film. This short film tries to discuss the process how a young girl experiences her family's hidden history and how she achieve reconciliation with herself and her father.

Another story was about belief and religion. The student told a story about her grandfather, her father, and herself – how did they peruse their own belief and the conflicts between them.

It takes a lot of efforts to choose topics for the Family Album. One of the principles is to emphasize the storyline as well as the truth. Critical thinking is very important during the whole process, especially when the story is about family members. All the topics try to find the meaning of sociology and anthropology behind the story.

During the whole process of film making, the pre-preparation is more important since it is the cornerstone of the film production. The students are required to do a lot of readings and prepare at least 100 questions for the interviews to dig more stories behind the scenes.

### 41.4.2 Recollect, Reflect, and Reshape

Both oral history documentary projects *Family Album* and *Our Fathers' Revolution* aim to help students retell and recognize history; reflect and re-establish universal values; and reflect and reshape their point of view on history and the whole world and eventually reshape themselves.

The *Family Album* project focuses on family history, and the students did actively explore their stories with great enthusiasm. They could start interviews with a great deal of questions and find answers for issues which have confused them for so many years or give themselves a chance to re-recognize the essence of some family stories. The whole process is the way that the students try to recollect, reflect, and reshape consciously.

Whatever it's for retrospective of history or exploration of current society, all these short films recorded the reality of the society. Many years later, these 100 *Family Album* short films will be a valuable visual archive to describe ordinary Chinese families.

## 41.5 How to Evaluate Oral History Documentary as Historical Data

In order to meet the requirement on the dramatic character of stories and the restriction on the length of time, the students were asked to carry out a very complicated postediting. This subjective editing would certainly affect its value in the sense of oral history to some extent.

Moreover, the individual's narrative cannot avoid the composition of the performance. At the same time, it's difficult to judge the veracity or accuracy of the oral narrator without the supporting evidence of more parties. The use of oral history documentaries as historical materials does have to take into account some inherent problems.

Judith Moyer, a historian and educator, raises the following issues to pay attention during the collection of oral history.

1. Is it related to the memory of people's failure?
2. In some cases, people who have little relationship with the event itself will be forced to enhance the structure of the narrative. What is the purpose of doing so?
3. What is the motivation of the storyteller?
4. Between the interviewer and the interviewee, what is the right relationship affecting the reporting of the incident?
5. What's the difference between spoken and written text?
6. What may be inaccuracies in the meaning of words when trying to record a conversation?

Moyer also pointed out that compared to oral history, other forms of historical records have the same problem: they also have personal and social prejudices. Written materials appear in a certain social context. Even historical analysis published by a professional historian still does not reach the goal which is difficult to achieve – a complete "objective" description of the event.

The same problem was encountered during the creative process of *Our Father's Revolution* and *Family Album* film production. Especially for Family Album series, it is difficult not to beautify or exaggerate the possibility of suffering due to the stories of the most intimate family members. Therefore, we hope that by developing the students' critical and creative thinking, we tried our best to overcome these problems.

The students are reminded that they are family members and professional film directors as well; they should keep calm and objective records. A students' history literacy cannot be accomplished by reading a couple of books alone. This process needs to be related to the entire society.

In short, oral history documentaries must be carefully used in the research process as well as other documents and require much analysis and interpretation. But it also provides a new field of information. As a widely disseminated research method and recording process, oral history documentaries democratize historical records – allowing us to have a more inclusive and accurate view of the past.

## 41.6 Conclusion

For education on oral history documentary production, the most fundamental thing is not only to produce a documentary, but especially in the entire creation, students actively participate in the process to recollect, reflect and reshape, and truly realize the critical and creative thinking that our education has been pursuing to cultivate students' awareness and concern for social and public issues.

Oral history documentary teaching requires a lot of patience and sincere input.

## References

Fang F (2003) The history of Chinese documentary film development. Theatrical Publishing House of China, Beijing
Guixin J (2006) Research on literature documentary film. North-east Normal University Journal, Changchun
https://en.wikipedia.org/wiki/Oral_history
https://www.oralhistory.org/education/
Jiajing F (2009) The construction of the realism of oral history documentary film. Hebeijing University Journal, Baoding
Nichols B (2001) Introduction to documentary. Indiana University Press, Bloomington
Qingfu W (2005) A type of documentary film. Television Research, Beijing
Shijun Z (2003) The research on narrative of oral history film in the context of postmodern culture. Guangzhou, Jinan Journal, (Philosophy & Social Science Edition)
Wenguang W (2001) The lens is like your own eyes – documentary and people. Shanghai Literature and Art Publishing House, Shanghai
Yunhui Y (2011) Narrative TV documentary research. Fuzhou Normal University Journal, Fuzhou

# Differences Between Micronesian and Western Values

**42**

Tom Hogan

## Contents

| | | |
|---|---|---|
| 42.1 | Small Communities and Geographical Isolation | 769 |
| 42.2 | Youth's Dependence on Elders | 771 |
| 42.3 | The Nonliterary Nature of Micronesian Culture | 777 |
| 42.4 | Differences in Gender Roles | 782 |
| 42.5 | Micronesians as Information Sharers | 785 |
| 42.6 | Micronesian Concepts of Time | 786 |
| 42.7 | Micronesian Notions of Politeness | 788 |
| References | | 790 |

### Abstract

In 1986 an organisation called the Micronesian Institute was set up in Washington, DC. It followed the mutual signing of the Compact of Free Association between the US government and the Federated States of Micronesia (FSM). The Institute was assisted by the US Department of the Interior.

A small group of expatriate Micronesian college students, calling themselves the Student Program Committee, saw a need to explain some of the sociocultural differences between young Micronesians and Americans. They wrote the script of a radio play, calling it *An Adventure Far from Home*.

The students recorded the play and sent a master open-reel tape to Ezekiel Lippwe, the Broadcast Chief of the FSM, with instructions to dub the tape to

---

This chapter first appeared in Understanding Micronesia: A cultural guide for researchers and visitors, published by Southbound Publishing, Penang (ISBN 978–983–9054–49–1). Springer has obtained the rights to re-publish the work in English language only.

---

T. Hogan (✉)
Sydney, NSW, Australia
e-mail: tehogan@optusnet.com.au

audio cassettes and distribute them to the four radio stations of the FSM states as well as those of the Marshall Islands and Palau.

I first heard the tape when it was broadcast one evening on Truk's (now Chuuk's) radio the following year. It seemed a very handy *expository device* upon which to base the ethnographic observations made in the following article. It is not used as ethnographic or anthropological material *per se.*

## Keywords

Bilingualism · Chuuk · Kosrae · Micronesians · Yap State

One of the remarkable aspects of the dramatized radio feature, *An Adventure Far From Home*, is the willingness of the young Micronesian producers to engage in sociological generalizations. I found it simple enough to check the validity of these with a suitably wide range of informed people from Micronesia. (In the course of this study's fieldwork, all points of cultural difference raised in the dramatized radio feature program *An Adventure Far From Home* were used as points of reference and checked with a cross section of Chuukese, Pohnpeian, Yapese, Kosraean, Marshallese, and Palauan interviewees, as well as with relevant ethnographic material. All points raised by the program-makers were regarded as valid by all Micronesian interviewees.) Micronesians might be reluctant to explain aspects of their culture, but I've noticed that normally they're quick to point out mistakes or to verify the truth. So the investigator is spared the sort of soul-searching that's typified by David Nevin's apologetic recourse to a generalized account of aspects of Micronesian society.

> Yet I will generalize about them and their culture, for their differences are not germane to my subject, which is the part the United States has played in the social malaise gripping them all... I will try to deal only with the parts of Micronesian culture that seem to exacerbate the malaise — for understanding that exacerbation is critical to understanding the malaise itself. The views that follow are anthropologically sound, but they are based on my own observation and interviews. (Nevin 1977, p. 46)

That's a bold statement for a person who isn't a practicing anthropologist, but I've found in general that Micronesians find Nevin's views to be culturally sound enough. Sociological and cultural observations made public by Micronesians about themselves, however, are obviously preferable, but it wasn't David Nevin's fault that no such revelations existed when he was researching his book almost 10 years before the production of *An Adventure Far From Home*.

In the first section of the radio play, "Cultural Barriers," Mike, a graduate of an American college, nostalgically visits his alma mater and finds Arkangkel from Pohnpei. The unfortunate Pohnpeian is homesick and confused, so Mike advises him particularly on matters relating to social aspects of college life. They're joined by Maria, a very worried young woman from Saipan, the capital of the Commonwealth of the Northern Mariana Islands. Maria's problems form the basis of further discussion and of more advice from Mike. It's within this imaginary context that Mike, Maria, and Arkangkel bring forward and discuss openly a significant range of sociocultural issues that relate to differences between them and Americans in particular. It's important to this work to consider them all and relate them to considerations of what Americans or others contemplating developing relationships with Micronesians probably most need to know. I've checked all the claims made by the Micronesian students against pertinent ethnographic material. And please remember, I'm using the radio play as *an expository device*, not as a source of ethnographic information in itself.

It will be helpful to you if I first list the actual generalizations raised by the Micronesian students. You'll find all the quotations I've selected to illustrate them in the Appendix to this work which is a full transcription of the relevant first part of the play. These broad generalizations should be thought of as "macro-values," each of which simply states a cultural area of belief and practice within which there's a collection of specific attitudes and approaches to existence. I'll examine the general issues raised in the order in which they occurred in the program, as the students who prepared the instructional radio play obviously saw a certain logic in that arrangement of ideas:

- Unlike the majority of Americans, Micronesians are from small communities on small islands, and the social rules governing behavior are very different.
- Young Micronesians are not encouraged to use their own initiative.
- Micronesian culture has no literary roots.
- Sex roles are more clearly differentiated in Micronesia than in the United States.
- Micronesians are not ready sharers of information.
- There are considerable differences in concepts of time between Micronesians and Americans.
- Micronesians generally avoid contradicting others.

## 42.1  Small Communities and Geographical Isolation

ARKANGKEL   I have found that other foreign students have problems too. But I think for us it's harder because we come from tiny tropical islands and atolls, where our chances of seeing the outside world are very small. Our lives are simple, and we are brought up by family and traditional rules which don't fit into Western society.

From a sociological perspective, it's not surprising that most Micronesian island groups developed similar ideas about society and the place of the individual within it. For David Nevin "The key cultural point in today's dilemma is the pressure of life on a small island on the individual and on the community" (Ibid.). Fr Hezel prefers to emphasize the smallness of the community rather than the habitat, pointing out that this is not a chicken-or-egg genesis argument, for small and isolated communities may be found even in very big countries. In the course of one of our six or so discussions of this and other issues, he told me that his experience tells him that within small, mainly self-contained communities, unique sets of social pressures arise which, in turn, generate rituals of behavior which determine the lifestyles of all acceptable members of the group. When the balance is affected by the ingress of powerful outsiders who have different codes of social behavior and are persistent in their attempts to bring about change, then change will inevitably occur. Fr Hezel, however, takes issue with the "anthropological zoo" theorists, pointing out that the notion that irreparable damage is necessarily caused to developing states by potent change agents is naïve and simplistic:

> Years ago it was fashionable for cultural anthropologists to regard societies or cultures as complex bits of machinery, like the old-fashioned spring watch, in which every part was inter-related. An alteration in one of the parts would invariably change and often damage the functioning of the entire machine. Lately we have come to realize that societies are as organic as the people who make them up. Like the human body, a society can adapt to stresses and changes in the environment and even to the viruses and bacteria that assail its inner workings. (Hezel 1987, p. 12)

"Micronesia" means, literally, "small islands" and is aptly named. William Alkire notes that there are more "low" islands or coral atolls, than "high" islands of volcanic origin in the total collection of islands and archipelagos. "The societies of the central Caroline atolls, Truk, Ponape, and Kusaie constitute the major cultural groups of central Micronesia... The islands of the central atolls are all small; few have as much as one square mile of land area" (Alkire 1972, p. 24).

Many of the atolls of the Federated States of Micronesia, even today, contain fewer than 100 people. On larger islands, such as Pohnpei, Kosrae, Yap proper, and Moen (the largest in Chuuk State), several thousand people will be found. "But there the society tends to break down into communal villages that are nearly as isolated and territory-conscious as are atoll communities," writes David Nevin. "I believe...the differences between the two are slight" (Nevin, David, op. cit., p. 47).

Today's Micronesian states have evolved from pure subsistence societies which were largely static by nature. Work habits weren't dictated by the passage of the sun but rather by the need for human sustenance within a self-sustaining ecosystem. As a consequence, observes Nevin, there was never a need for constant toil. "Rather it was done as needed, often on impulse, often in groups as in fishing or digging taro, so that it had a social and even a sporting note. There was ample time for doing as one pleased, which in the humid tropics often meant resting. It was not a life to implant a burning work ethic or to make Western work patterns attractive" (Ibid., pp. 47–48). You might recall that Marshall Sahlins pointed out that Western societies were underpinned by economic forces and that David Hanlon pointed to quite a different cultural institution fundamental to Micronesian societies, a web of social forces which make up the social and family scene. It's to David Nevin's credit that although he didn't have access to the analyses of the two social scientists, he got it right anyway.

Part of the "simplicity" of Micronesian societies mentioned by Arkangkel is no doubt related to a pace of life and a speed at which things normally get done which are attuned to the physical demands of a tropical maritime climate and the psychological demands of a well-defined lifestyle. Day follows day in Micronesia without much variation. Why rush toward the predictable and inevitable?

The key members of the Trust Territory administration were alerted to this difference between American and Micronesian attitudes to the pace of existence, as well as to a number of others – such as static Micronesian and fluid, dynamic American visions of the immediate and long-term future – as early as August 1955. During that month, the Trust Territory district administrators' conference was held at the Guam headquarters. Heading the agenda was a keynote paper from the Trust Territory staff anthropologist, Dr A.H. Smith, who was seconded from Washington State College. Most of the paper appeared in Micronesia Monthly, the headquarters newsletter, early in 1956. It was one of few items written by an anthropologist on cross-cultural issues to be circulated by the Trust Territory administration.

Dr Smith began with a brief outline of his subject. "Mr Nucker [the High Commissioner] has requested me to discuss with you briefly the broad and difficult question of American-Micronesian attitudes and relationships" (Smith 1956, p. 26). On the subject of the complexities accompanying the sheer pace at which most Americans appeared to lead their lives, Dr Smith issued the district administrators a caution that "Speed is an important American goal and ideal, so deep-seated in our cultural background and our psychological world-view that it appears in such seemingly irrelevant contexts as our intelligence tests. For this reason it's often difficult for us to become reconciled to the fact that many other societies do not place the same value upon speed as we do, and, indeed, that speed may not be so fundamentally important after all" (Ibid., p. 31). Dr Smith led the administrators step by step through a comprehensive analysis of Micronesians' negative attitudes to a number of cherished American goals. There's no evidence whatsoever in the history of the United States administration of the Trust Territory since the mid-1950s that his advice had any significant impact.

The "simplicity" of life and the "traditional rules" referred to by Arkangkel appear to be directly related to the subsistence origins of small and geographically isolated Micronesian societies. As Fr Hezel has noted, the static nature of a confined community of food-gatherers, whose peaceful coexistence was forced on them by circumstance, necessitated the development and perpetuation, by word of mouth, of ritualized survival techniques. Most of these, naturally enough, didn't work well on the American mainland or, for that matter, anywhere else in the West.

## 42.2   Youth's Dependence on Elders

ARKANGKEL    Another problem is that we aren't trained to make the independent decisions that are expected of us here.

There's a good deal more to this deceptively simple statement than is apparent at first. In fact, it's the key to a range of cultural differences between Micronesians and Westerners. In much of contemporary Micronesia, a baby is born into a kin group, clan, or community in which elders – male or female – or chiefs are still the recognized repositories of knowledge and wisdom.

In the more tradition-conscious societies of Yap and Pohnpei, hereditary leaders still pass on the rules for regulating the use of available resources and for coexisting gainfully with others in a close-knit community, and all are expected to obey. It's significant that Arkangkel is from Pohnpei, for it's there that the most complex sociopolitical organization of central Micronesia, involving stratification and complexity of ranks and titles, is still intact today.

By the same token, many of Yap's traditions remained unscathed. American anthropologist Dr Sherwood Lingenfelter records that "much of the traditional political life remains. All of the major statuses and institutions for villages and regional nets persist today... [But] Yapese mix and match traditional and new alternatives to plan the most promising strategies for political success and to find the most satisfying solutions to their contemporary problems" (Lingenfelter 1975, p. 193). Lingenfelter here is also pointing to one of all Micronesian societies' strengths in the process of acculturation: the willingness and capability to adopt and adapt. There's a lesson here for those who wish to introduce innovations to Micronesian societies: don't expect them to be taken on board wholesale. Rather, if they're not rejected outright, expect adaptations and accept them without rancor. And remember at all times that the real power doesn't necessarily reside with governors and the dignitaries who inhabit the state legislatures. Throughout Micronesia, the strings are pulled by traditional leaders or, in the case of Kosrae, the senior clergy.

The most striking example of the power of traditional leadership can be observed in Yap State. Here, the governmental structure shares with the other Micronesian states and republics the typical executive, legislative, and judicial branches of government that are the salient features of most constitutional democracies. There's one difference, however. In Yap State, there's a fourth branch of government, originally acknowledged in a charter issued by the former Congress of Micronesia and now enshrined in the State's constitution. Yap's constitution acknowledges the two councils of traditional and hereditary leaders in the legislative process. The councils of Pilung and Tamol represent, respectively, the collective voices of the chiefs of the main island, Yap Proper, and of the outer islands. I've found from experience that if an outsider wishes to do serious business of some sort in Yap, it's imperative to make contact with the relevant council of traditional leaders at the earliest possible opportunity.

In Kosrae, the roles of traditional leaders have not been abandoned with the passing of traditional chiefly leadership. Interestingly, they've been taken over by the Protestant Christian leaders within the four municipalities of the island state. Kosrae's leadership is indicative of the ways in which Micronesian peoples borrow, change, and adapt as organic social entities yet, in the process, retain significant components of their traditional ways, as William Alkire has observed:

> Primarily as a consequence of missionary influence the importance of titles began to decrease in the middle and late nineteenth century. By 1960 only two title holders survived and their titles had lost political significance. Nevertheless, rank is still important to a

Kusaien, but prestige is now measured either though control of church affairs or, more recently, through positions established in the...administration of the island... The church and the American municipal system have offered mobility to many who would have had fixed status in the old system. (Alkire, William H., op. cit., p. 37)

Fundamentalist Protestant Christianity is the single most powerful influence in Kosrae. No work of any kind is permitted on Sunday, which is set aside totally for worship and rest. The Kosraean radio station, WTFL, starts every day with at least a half hour of religious music. The young are by no means repressed, but neither are they encouraged to develop initiative and individuality in the American sense. The reason is simple enough; the majority still live in villages where the traditional need for consensus operates as strongly as ever. Alkire notes that "Although the old hierarchical, highly centralized system of old no longer functions, Kusaien society at the village level is still dependent on subsistence cultivation of breadfruit, taro and bananas. Post marital residence is usually patrilocal unless an alternative choice offers a couple greater potential gain. Membership in descent groups is still defined by rights to particular land parcels" (Ibid.).

Dr Pedro Sanchez believes that young Micronesians in general are far more dependent on their elders than young Americans and are often bewildered by the typical Western emphasis on the development of initiative and individuality in the young. They come into contact with notions of competition and the promotion and assertion of "self" in the American-oriented school curricula and their associated literature and often find that these ideas are reinforced and actively promoted by American-trained teachers, Peace Corps volunteers, and others who have derived their values from the United States:

> Young Micronesians are told by expatriate teachers and by a lot of Westernized Micronesian instructors to "run with the ball", to strive for the top marks, to push themselves forward. This can lead to severe tensions in the home and the village if the child behaves in this way, because this sort of conduct is the opposite of the traditional ways in which children are taught to behave by their parents and other elders. So it's normally the home which wins, despite the typical Micronesian parents' desire for a "modern" education for their children. And then there can be tensions in the classroom. (Sanchez, P., interview, 1987)

In Chuuk, the rate of social change from the 1960s to the present appears to have been more rapid than in the other three states of the Federated States of Micronesia. Since his arrival in 1967, Fr Hezel has observed the acceleration of a process involving "the breakdown of the traditional extended family into what we can call standard packaged families — that is, nuclear families composed of father, mother and children, often with a few spare relatives added to the household" (Hezel, Francis X., "The Dilemmas of Development", p. 14).

Fr Hezel maintains that in traditional terms, the basic unit of Chuukese life has always been the lineage. Arkangkel pointed out that young Micronesians were not encouraged to make decisions on their own. This is still true of Chuukese youth but for reasons that are more closely aligned with those which hold good in Kosrae. In Chuuk, there has been, well within living memory, a dramatic shift in leadership responsibilities. Until quite recently, writes Fr Hezel, "The lineage group might be made up of three or four households, each with a single nuclear family and perhaps some other relatives added. The members of these households, especially the younger ones, were subject to the authority of the lineage chief, the senior man in the lineage. He was empowered to oversee the lineage land, assign work responsibilities to other members of the lineage, and supervise the distribution of the food among them" (Ibid., pp. 14–15).

More recently, however, as the economic influence of the American administration has inexorably led Chuuk from a subsistence to a cash economy, the major area of occupational growth has become the public service, so jobs and cash income have come to represent, as Fr Hezel put it to me, the first wedge driven into the economic monopoly of the lineage. From the point at which this trend commenced, he says, traditional leaders have lost much power. As the lineage head lost more of his hold over the economy of the lineage, he suffered a corresponding loss of authority. More and more, children were expected to remain under the supervision of their own biological parents, even after adolescence when they would have normally come under the authority of the lineage seniors. Here's a clear case of acculturation and traditional behavior working in tandem. The young in Chuuk are still controlled well into adulthood, but not in the traditional way.

Fr Hezel has raised an issue which needs to be emphasized at this point, as it concerns a social process which has affected and continues to affect all Micronesia's traditional institutions and folkways, and thus is highly relevant to other points of cultural difference raised in this study. Just as the introduction of American practices has led to shifts in Chuukese family power structures, so other traditional aspects of life in Chuuk and, indeed, throughout Micronesia have undergone similar sorts of modifications. When confronted with new Western models, Micronesians frequently become involved in a form of two-way adaptation.

Some of the principles governing traditional approaches are preserved and some discarded, and some of the new ways are taken on and some also discarded. Just as changes are occurring in the form of the extended family and methods of communal government, so other areas of traditional Micronesian life, including communications, medicine, and even jurisprudence, are undergoing similar adaptive processes. Some commentators, in fact, see this phenomenon – which I term selective acculturation – as typical of developing states and nations on a Pacific-wide scale. Drs Daniel Hughes and Stanley Laughlin, from Hawaii's Institute of Polynesian Studies, for example, are prepared to include Polynesia, Melanesia, and Micronesia in their observations on the particular theme of the administration of justice:

> In every Pacific country law and justice have been influenced by introduced principles and practices, but in every country some aspects of the traditional order remain — in some cases very strongly. Some years ago many people assumed that the traditional elements would soon be totally replaced, but it has not been so or likely to be so. Each Pacific country is evolving a unique amalgam of local and foreign precedents in creating its own system of justice.

Little is known or recorded about these systems or about their needs — individually and for the Pacific as a whole. (Pacific Courts and Justice. No. vii, 1977, p. 77) Hughes and Laughlin claim that the point made on the adaptive acculturation process "can validly be applied to political as well as to legal principles" (Ibid.). In fact, it could validly be said to have a universal application to the developing world.)

Arkangkel's point relating to social pressures which aim at discouraging initiative in young Micronesians may be further illustrated by reference to traditions related to land use and disposal, understandably still very powerful in a part of the world where land is genuinely a scarce resource, and to certain checks and constraints which have to be surmounted by the young if they wish to take part in their state's political life. Most American and Australian attitudes to land ownership are at odds with traditional Micronesian value systems. In the West, for example, the major impediment to a young person's wish to own land is the availability of money. If the young person is of age, has the urge to buy, and can raise the capital, land is normally readily available. In Micronesia, the situation is not as simple, as a committee chaired by ex-Harvard professor of International Business Relations, Anthony Solomon – sent to investigate and report on the Trust Territory on President John F. Kennedy's orders in 1963 – observed:

> Large areas of land, indeed, sometimes whole islands, often classified as in "private ownership", are in reality controlled by a clan, with use rights being apportioned through historical practice, communal agreement and the intervention of the elders or chiefs of the clan. (A curious and important factor tending to perpetuate the power of the traditional chiefs in many communities where terribly complicated structures of land rights exist is their uniquely authoritative knowledge of boundaries and rights.) Land is the tie that binds families and clans together; the right to occupy and use land, guaranteed by family membership, is the security of young and old, and the index of hierarchical patterns. (Solomon, Anthony Morton (Chairman) 1963, Part 1, p. 18 [This document has become known popularly as *The Solomon Report.*])

By virtue of his intimate knowledge of family and clan land rights, the traditional Micronesian chief had an encyclopedic knowledge of the social ranking of each of the members of the area in which he held sway. In Kosrae, social status, as I've noted, is still considered very important, but the role of arbiter these days has been taken over by religious and political leaders. In Chuuk, says Fr Hezel, the chiefs' prerogatives stemmed from the land originally. Land tenure has changed, yet the status of chiefs remains high. This situation, however, might be expected to change soon as the concept of the lineage itself weakens with the breakup of the extended family and land continues to decline in importance in a cash economy. Fr Hezel is sure the lineage head still has a strong say over the disposal of lineage land, but he also recognizes that the latter has been devalued in recent years as an ever greater proportion of the family's resources is provided by cash income.

In Pohnpei and Yap, the traditional leaders in the villages and municipalities still perform a number of tasks which have important social and political ramifications.

I've mentioned that in Yap, chiefs and elders participate directly and openly in the workings of the overall sociopolitical system. In Pohnpei, there's still participation in politics by hereditary chiefs, but it's covert, especially at times when a Pohnpeian decides to run for political office. Secretary to the governor of Pohnpei State, ex-Peace Corps American Marialice Eperiam, wrote to me in 1987 that "Whoever decides to run for an office will go to the traditional leader of his municipality and inform that person of his decision. By doing this, the candidate is seeking the approval of the chief and asking for his support. The chief at some time in the campaigning process will announce who he supports. He will also tell the people in his municipality to vote for that person." (Marialice Eperiam. Personal letter to the author, October 19, 1987) Mrs Eperiam was then secretary to the governor of Pohnpei State, the Honourable Resio Moses. She was an expatriate American journalism graduate and ex-Peace Corps volunteer, married to Emensio Eperiam, Pohnpeian by birth. They are now with the US State Department posted to Mexico.)

Indeed, some of the American politicians who took a personal interest in the development of Micronesia's political relationships with the United States during the 1960s and 1970s viewed with utter dismay the apparent mutual exclusiveness of traditional Micronesian chiefly functions and American concepts of democracy. Donald F. McHenry, senior Department of State employee during this period and author of the most authoritative work on US-Micronesian relationships at the time, *Micronesia: Trust Betrayed*, recorded dryly in 1975 that:

> Congressman Thomas S. Foley's primary concern is that the United States might be making permanent the hierarchy of Micronesians who are presently in power. Traditional chiefs have retained a great deal of power in modern Micronesia. Senator (James A.) McClure, travelling with Foley in 1969, tells of asking a senator of the Congress of Micronesia whether he would run for re-election. The Micronesian replied that he did not know; he had not yet asked the Chief. "Traditional culture," says Foley, "violates almost every principle the American people have ever known." (McHenry 1975, p. 2)

This is the most authoritative available source of information on matters to do with the political relationships between the United States, the United Nations, and Micronesia in the 1960s and 1970s. McHenry has an intimate knowledge and understanding of the affairs of the State Department, having served as Officer in Charge of Dependent Area Affairs and Special Assistant to the Counsellor of the Department of State from 1963 to 1971.

When I mentioned Congressman Foley's concerns to Fr William McGarry in one of our discussions in Pohnpei, he was thoughtful for a moment then replied with a smile: "I guess that's reasonable since American culture violates almost every principle Micronesians have ever known."

A ready recognition of hierarchical status is still of considerable importance to Micronesians, not for reasons related directly to social competition but on such grounds as being able to identify the social rank of a member of the community and, as a consequence, being able to provide the appropriate responses to his or her presence. The importance attached to status can be seen in Pohnpei where its caretakers and arbiters are the traditional chiefs. Iris Falcam has told me that Pohnpeians call each other by their titles, not by their given and family names. Often, she says, an office worker, say, won't know the "real" name of a colleague.

Some visiting American scholars have noticed this phenomenon. One, Dr J.L. Fischer, analogized it to common Western academic practice so that "a Ponapean without a title is much like a professor without a PhD: there can be such people but it's likely to be awkward and embarrassing for most" (Fischer 1974, p. 171).

It's not difficult to imagine the difficulties and tensions faced by a young Micronesian such as Arkangkel, denied ready access in his home state to the economic and social mobility that would often follow the exercise of intelligent initiative in a Western culture, returning to his traditionally structured Micronesian homeland with a mind full of American egalitarian ideals and values. As Pohnpeian Emensio Eperiam says, "You've got problems when you get back from the mainland. In my case, I just don't bother any more getting permission to do this or to do that. I'll give you an example. I really wanted to start a shop, so Marialice and I started our shop. We didn't ask permission as I guess we should've. I just don't listen to our chiefs any more. And you know what? They come and buy booze from us!" (Interview with Pohnpei's Historic Preservation Officer, Emensio Eperiam, March 9, 1989).

## 42.3 The Nonliterary Nature of Micronesian Culture

ARKANGKEL   We are very different from most foreign students... there is a lot to read, and many of us don't read English very well when we arrive. We haven't had a reading background for newspapers and libraries. Ours isn't a reading society.

Western societies, because of the scale of operations within them, and the bureaucracies which have developed as they've become more complex, depend heavily on the generation, dissemination, storage, and retrieval of written, typed, and printed information. This information documents and schedules the processes of the institutions fundamental to the effective functioning of the component groups within those societies. The documentation, furthermore, is normally perceived as part of an orderly, sequential and cumulative set of events which assists in the organization of group behavior and its culmination in predetermined ends.

If Micronesia's American-style governments are moving, albeit slowly, in this direction, there are few other aspects of Micronesian society which seem to be doing so. David Nevin saw little evidence of any rapidity in the rate of change:

> Since nothing was expected to change, there was little reason for planning, and even today Micronesians seem to give little consideration to the future... This attitude also relates to why there is so little institutionalization all across the Pacific. Relationships tend to be personal and to rely on oral contact. Even among literate Micronesians, for example, letters count for little and often are not answered, nor is what they say accepted —let alone integrated — into an ongoing situation. (Nevin, D., op. cit., p. 49)

During the period he spent researching his subject, as a consultant for the Ford Foundation examining education in Micronesia and preparing a background report that would help the Foundation evaluate a grant it was contemplating there, Nevin spent some time with a political scientist from the University of Hawaii, Dr Norman Meller. Meller helped organize the Congress of Micronesia which first met on September 28, 1964. It consisted of elected legislators from the Trust Territory districts of the Northern Mariana Islands, Palau, Yap, Chuuk, Pohnpei (Kosrae was still the part of Pohnpei known as Kusaie), and the Marshall Islands. He was also used as a consultant in the formulation of the constitution of Pohnpei State. Meller appeared to be well aware of the nonliterary nature of the Micronesian societies:

> Well, for instance, one of the key differences between American administrators and Micronesians is that the American tends to deal in formalized ways, with emphasis on written materials. He doesn't feel something is settled until it's written down. And he's time-conscious, schedule-oriented... [In Micronesia] there is no institutionalized follow-through, no way to get something in motion and expect it to follow an orderly, agreed upon, scheduled path. And this frustrates lots of Americans. (Meller, N., quoted in Nevin, D., op. cit., p. 49)

In a society in which, historically, written or printed records have played no part, it's significant that so much traditional culture has survived. But Micronesians are used to deploying their memories in directions they see as relevant to their own domestic existences. The "secrets" which have to be memorized concern matters which are traditionally circumscribed and, after all, in most parts of Micronesia chiefs, elders, and religious leaders are at hand when information is needed about such complex matters as land rights and social status. Like most Westerners, they're not adept at remembering the "administrivia" of bureaucratic activity, yet, unlike Westerners, they generally avoid systems of documentation, storage, and retrieval. It's obvious that such predilections are at odds with many of the work practices within the more bureaucratized societies of contemporary Micronesia, the most "Americanized" of which, in my experience, is the Republic of the Marshall Islands.

Even today, the majority of people who need a good working knowledge of English – professional broadcasters, for example – experience difficulties with the language, and yet this is the lingua franca of most of the Pacific. The extent of the problem is often not readily apparent, even to expatriates who have accumulated a good deal of experience in Micronesian societies. Marialice Eperiam is such a person. Owing to a background in journalism and her experience as media secretary to the Congress of Micronesia as well as the governor of Pohnpei, she was asked to work as co-trainer in a Micronesian subregional radio news workshop, run in Kolonia, Pohnpei, in May 1988, by PACBROAD Coordinator, Hendrik Bussiek. Even the acculturated Mrs Eperiam, who speaks Pohnpeian well, was not prepared for some of the linguistic surprises she encountered when teaching professional broadcasters from Pohnpei, Yap, Chuuk, Kosrae, and the Republics of the Marshall Islands and Palau. "One thing I did notice in almost all the trainees," she reported, "was their limited understanding of English. This came as a shock to me. I have been on Pohnpei for many years and felt like most people could hear, speak and write English fairly well. I have now discovered this is not true. The trainees found it very difficult to read something [in English] and comprehend correctly what it said. This lack of understanding can only limit their success." (Eperiam, Marialice. Report to Hendrik Bussiek of May 26, 1988, on the discharge of her responsibilities and final results of the subregional workshop on radio news gathering and news presentation held for Micronesian journalists in Kolonia, Pohnpei, May 1988.)

It's worthwhile remembering that there's no history of the use of English as a second language in Micronesia prior to the end of the Second World War. Between the wars, Japanese had become the working language of Micronesian administration, and many older Micronesians are still fluent in the language of their former colonial overlords. I've found no traces left today of the languages of the previous colonizers, Germany and Spain.

I've had the opportunity of working for a number of years with a distinguished Australian linguist, Dr George Kanarakis, who speaks English and his native Greek with equal proficiency. Thus I've referred some of the linguistic problems I've encountered in Micronesia – at least in principle – to him. Kanarakis regards "bilingualism" as the practice of using two languages with similar levels of proficiency (This is the definition favored in Kanarakis 1983, p. 7). My experience tells me that most Micronesians have not reached this linguistic status. Generally, they use English only at need and favor their vernaculars in normal communication. Almost exclusively, Micronesian broadcasters are "dependent" rather than "coordinate" bilingual, in Kanarakis's vocabulary. Whereas the dependent bilingual person uses the vernacular consistently as a point of reference for encoding and decoding a second language, "In the coordinate bilingual the two languages function like independent and separate systems without fusing either during the encoding or during the decoding of the message" (Ibid., p. 8).

Dr Kanarakis believes that "Dependent bilingualism carries with it quite serious [negative] personal, social and educational consequences" (Ibid., p. 13). Related to this observation, I've found that a major difficulty can easily emerge, say, in a training program where Micronesians actively wish to learn some of the experiences of the West in an area of knowledge relevant to their needs. In many training and development courses with which I've been intimately associated or which I've observed in operation, there have been difficulties experienced in communication between Micronesian participants and English-speaking overseas trainers, and this has been accompanied by a slowing of progression from task to task in the training curriculum and, on occasions, an accompanying frustration all round. This can explain why accurate consultants' reports of training missions submitted to the assistance agencies which sponsored them can actually look so threadbare.

Added to the problems associated with dependent bilingualism are problems Micronesians encounter with different English accents, the pace of speech, and idioms. On the subject of accent, Micronesians readily admit difficulties. Paulino Pablo, Program Director of Radio Pohnpei, told me in 1986 that "I went to Sydney, Australia — to the International Training Institute — in 1984. For a while, I thought I was in a country where they didn't talk English. I'm used to the way Americans talk. It took me a week to get used to the Australian accent and, with one instructor, I just didn't understand a word he said the entire program" (Interview with Paulino Pablo, Program Director Radio Pohnpei, WSZD, Kolonia, Pohnpei, April 30, 1986).

Yap State's Information Officer, Henry Muthan, told me in 1988, during a workshop I was running at Yap State Radio, about a course in radio news production he'd attended about 2 months previously. "In Pohnpei, the course was run by Mr [Hendrik] Bussiek and Mrs Eperiam. I could not understand Mr Bussiek. I am sure his English is good, but I think he is German, and I had problems with the way he said words. So I had to sneak up to Mrs Eperiam sometimes and get a translation. And, you know," he added with a wry smile, "I am a print and information [public relations] person these days. I really went to the course to help improve my English" (Interview with Henry Muthan, Information Officer for Yap State, Colonia, Yap, July 11, 1988).

And a couple of days later, one of the Chuukese participants, Kemeterio Joseph, said to me quietly: "You will think I am rude for saying this, but can you slow down your English so I can understand?" (Kemeterio Joseph, Announcer Radio Chuuk, WSZC. Request made to me at a subregional workshop in radio magazine programs, Colonia, Yap, July 14, 1988.) Native English speakers – and I include myself among the guilty parties – are often unaware of an acceleration of pace, especially when they're warming to a subject.

It was Paulino Pablo of Pohnpei who told me the story of his Micronesian friend who worked as an assistant at the office of Pohnpei's governor. On one occasion, apparently, an expatriate economic adviser asked him about a message he was supposed to deliver from the lieutenant governor, and Mr Pablo's friend read it to him from notes he'd pencilled on a piece of paper. "What?" said the American. "Sorry, I guess my head was out to lunch. Would you run that by me once again?" So the Pohnpeian walked backward about 30 paces and, waving the paper over his head, ran furiously past the expat. The American, according to Paulino Pablo's grinning account, was mystified. This story genuinely amused a group of Yapese with whom I've worked often. They're used to having to contend with the vagaries of American idiom. I told it to them early in 1987, and the last time I worked there, in December 1989, the Yapese Media Chief, Peter Garamfel, told me he still chuckled about "the Pohnpeian who ran the message past the American once again. I remember it because it's so funny. I can see it in my head."

On the subject of idiom, anthropologist, Dr A.H. Smith, offered the district administrators' conference of August 1955 some practical advice when he addressed the assemblage in Guam:

> One often hears that Micronesians in our service must be reminded over and over to carry out actions of the simplest and most obvious kind. Perhaps this is frequently the case. If so, it's not because of any deficiency in Micronesian intelligence but rather because of a failure through language difficulty to comprehend precisely what is wanted... In our interpersonal relations with Micronesians we should make every effort to be certain that the Micronesians understand our wishes and directions.

> How excessively idiomatic our everyday conversational speech is cannot easily be imagined by one who has not been repeatedly faced with the knotty problem of translation into a non-European language. (Smith, A. H., op. cit., pp. 26–27)

Micronesians understand the difference between "levels" of language well. In Pohnpei and Yap, notes William Alkire, more formal and refined modes of speech are reserved for speaking with traditional leaders (Cf. Alkire, William H., op. cit., p. 34). Thus, if we use an example from my field of expertise, Micronesians understand easily that a friendly, conversational style of English is favored for radio by professional Western broadcasters, in lieu of the comparatively more formal and stilted form generally used in written English, but find it difficult to use the style because of bilingual dependency. This problem, says Henry Muthan of Yap, has caused a deal of friction between American broadcast trainers and Micronesian trainees in the past. "Some American trainers have got upset and they ask us why we do not do what they say. We do not say much but I can tell you they have made many of us ashamed of our English" (Interview with Henry Muthan, op. cit.).

Many Micronesians are sensitive about what they feel are inadequacies in their spoken and written English. On occasions, in conversations with English speakers, some make attempts to justify their status as dependent bilinguals by resorting to rationalizations that might appear obvious to a Western researcher who's gained some insights into their cultures but are regarded as truth by the respondents. In Pohnpei, writes Fr McGarry, "The power of the word is extremely important... Words or speeches can change the nature of something which is going on... the very fact that I am able to explain it convincingly makes it what it is" (McGarry, William F., op. cit., p. 48).

The same concept relating to the relativity of truth operates in Yap. Henry Muthan, who is expected to perform well in English and Yapese as a disseminator of government information, confesses that he has difficulties with English but attempts to minimize the effects of these by claiming some advantages in dependent bilingualism:

> When I speak English, I think first in Yapese and then find the right words in English. There are often ideas which have to be treated very carefully. Like, if I want to say: "I love my sister," that's OK in English, but I could be kicked out of my village for saying it like that in Yapese. Yapese is a much simpler language, and when you translate that thought, the closest you get means "make love to", and that's taboo if you are talking about your sister. So I find it best to think in Yapese all the time. Then I would think of our word "liyor" [lee-or] and say: "I care for, or respect, my sister," and that is an idea that is OK in both languages. (Interview with Henry Muthan, op. cit.)

This statement does not stand close inspection from the standpoint of Western concepts of conventional logic. In fact, from the visitor's viewpoint, it's strangely put. A typical Western response might be: "If you do most of your thinking in Yapese, surely that's a safeguard against making embarrassing mistakes." From the Micronesian point of view, what the Yapese man is saying is: "If I thought and spoke in English just as easily as I do in Yapese, I might mix up my expressions and make embarrassing mistakes." Further inspection of the statement reveals an obliqueness in making the point that is foreign to a Westerner but typically Micronesian.

Fr McGarry provides more insight into a circumstance which you can notice throughout Micronesia. It relates to the references made already to the power of the word. In its most obvious form, Micronesians will explain a custom, event, or course of action to a Westerner or to a fellow Micronesian on one occasion in some depth and on another occasion will offer a different set of reasons for precisely the same subject under discussion. This is not seen by Micronesians as inconsistent, but this form of verbal and intellectual behavior can cause considerable confusion and frustration to a visitor who's seeking the truth in Western terms:

> One particular characteristic which stands out in Ponape is a type of casuistry which is characterized by sloganizing and ambivalence. This is not ambivalence as used in psychology. This is rather an ambivalent standard for judging action. By casuistry I mean the justification of almost any action by some principle or other. On the one hand something is condemned while on the other hand, by the enunciation of a particular saying or proverb, I can justify my behavior or non-action on any particular occasion. What counts is getting out the statement on time.
>
> There is a saying in Ponape which I do not suspect is very ancient, but it's common enough today: *Nan awatail me kitail kin pitkihla*. A rough translation: Our clever tongue helps us to escape. We can get out of any number of difficulties, we can justify our actions, we can escape from condemnation by being good talkers. (McGarry, William F., op. cit., p. 49)

Again, here's food for thought for the researcher in Micronesia. I'm sure you can see that if you gather information empirically, you normally need to check it rigorously from a number of sources before you can reach a reasonable conclusion.

## 42.4 Differences in Gender Roles

| MARIA | ...so far those American men are just marvellous. |
|---|---|
| ARKANGKEL | What do you mean? You like them better than us? |
| MARIA | Well, no – but they treat us better than you guys do! They don't order us around but do what we ask... |
| MIKE | Yes, but Micronesian girls aren't used to being treated so well by men. It's a cultural difference we should be aware of. |

Even in those parts of Micronesia where descent and kinship are traced through the female line, men are expected to act as the dominant members of their social groups. Visitors to Micronesia witness women apparently occupying socially inferior roles in relation to men. It's rare, for example, for an overseas consultant to any of the governments of Micronesia to have dealings with anyone but males. I've seen women arranging and preparing food for domestic purposes and for public feasts, shopping, looking after children, and weaving artifacts for the household. But in the public services, I've noticed they generally occupy minor positions. Fr Hezel has observed that appearances, as the aphorism goes, can be deceptive. "Many of us today are misled into believing that the function of men was to rule while that of women was merely to obey. Appearances to the contrary, this was certainly not the case in Micronesia. While women generally were expected to avoid 'center stage' positions and were barred from speaking in public, they were very often the real movers behind the scenes when it came to allocating resources and even initiating political intrigues." (Hezel, Francis X., op. cit., p. 17).

The strong traditional allocation of gender roles, and the rate of recent change, can be detected in observations made on Chuukese society by William Alkire:

> Gardening and preparation of breadfruit for consumption are responsibilities of men — perhaps two or three working together for several hours two or three days a week. Until recent times the women provided most of the fish consumed, taken from the fringing reefs of the islands. Women did not, however, fish from canoes (a practice forbidden to women throughout Micronesia). Today the women are rarely seen with their small hand nets working the reefs; most fish are either caught by the men or taken from cans purchased at local stores with proceeds from copra sales. (Alkire, William H., op. cit., p. 29)

Fr McGarry has noted that, in Pohnpei, women appear publicly to occupy "a lower position than men" but perform, nonetheless, highly significant functions in their society. "There are phrases [proverbs] which give some indication of the importance of women... At the same time, a man will be laughed at if he shows [in front of others] too much consideration for his wife or if he lets her run his life in any way... the husband is supposed to be in charge and may express this in-chargeness from time to time by a few blows about the head and shoulders" (McGarry, William F., op. cit., p. 69). I might add that these "blows" are tokenistic and aren't intended to hurt. And Micronesian women, I've noticed, frequently smack their men folk playfully and affectionately (but quite hard) on the arms or shoulders when sharing a joke or pointing out the ineptitude of a man who has dared to try his hand at what they see as "women's work."

Throughout Micronesia, men and women, at least in public, appear to behave in a dominant/submissive relationship. Men seem to treat their womenfolk in the sort of cavalier way referred to by Maria in the radio play. As is almost always the case in Micronesia, there is a good deal more occurring behind the scenes, as Fr Hezel has noticed in Chuuk:

> In gender relations as in other aspects of traditional life there was a strong note of reciprocity. Just as women were required to show certain kinds of deference to men, especially to male relatives, men were also required to practice respect behavior towards women. Men were prohibited from using certain kinds of language in their presence, and this to a far greater degree than was practiced in Victorian Europe. What we might call "women's rights", limited as they may have been, were well protected in traditional society. It's true that these "rights" fell considerably short of today's standards. Women, for instance, might be beaten by their husbands. Even so, however, the woman's family kept a close watch over her and were ready to intervene on her behalf in the case of excess. Women in traditional Micronesian societies surely did not enjoy equality with men, but they were not without a considerable measure of security and even power in those societies. (Hezel, Francis X., op. cit., p. 17)

Recently, however, says Fr Hezel, the circumstances of Micronesian men and women have undergone considerable change, and it seems that the changes "have had a far more unsettling effect on men than on women" (Ibid., p. 18).

The spread throughout Micronesia of the American-induced cash economy and the development of American-style government institutions have led to a change and devaluation of men's traditional roles. For a period of time, it seemed that new roles had been created for men. "Indeed there are a variety of salaried jobs as well as new political positions that offer money and influence," says Fr Hezel. "Volunteer organizations, churches, and even athletic associations provide new avenues of status. All these are attractive alternatives to the traditional roles that men have lost with the advent of the modern society" (Ibid., p. 18).

The educational opportunities for Micronesian women, however, have advanced to the point at which women are successfully competing for positions in overseas colleges and universities. In recent years, educated women have become frustrated by the fact that men appear to be monopolizing the new roles in the changing societies and have begun to act and, as Fr Hezel notes, "herein lies the problem. Men see women scrambling for these... positions and intruding in a domain they regard as rightfully theirs. Women work in government agencies, they drive cars, they play basketball and volleyball, and they even run for elected political office. Hence, women are perceived as competitors rather than partners. With the old restrictions on male and female roles giving way, women become a hostile rather than a complementary force" (Ibid.).

My observations throughout contemporary Micronesia bear out Fr Hezel's. Men generally are attempting to hold on to a rigorous differentiation of sex roles, and women are trying to break them down. There's little doubt in my mind that in the workplace Micronesian women will probably see the scales balanced more equitably only when more of them move into positions of managerial responsibility. Of course, this is also still true of the Western workplace, but it's fair to state that most Western societies have evolved in the direction of sexual equality to a significantly greater degree than have Micronesian nations.

An example of a victim of the sort of discrimination which is still deeply rooted in Micronesia is a hardworking and intelligent woman, and very experienced broadcaster, Emiko Boaz (Emiko Robert when I first met her in 1986). Near the end of 1988 – in her early forties and she told me with some pride, newly remarried to a handsome 25-year-old – she took me aside at the end of the first day of my fourth visit to the station since 1984. She was visibly upset: "I put in for the job of News Editor, and what happens? Nothing! I've had 18 years broadcasting experience, and who do they appoint? A man. A man with *no experience*. You know my problem? I'm a woman. That's a real problem in Pohnpei. And my mother was Kosraean, and that's a real problem, too, because here you inherit your clan from your mother, so I don't belong to a Pohnpeian clan."(Interview with Emiko Boaz, News and Program Producer, Radio WSZD, Kolonia, Pohnpei, September 28, 1988) Emiko Boaz, in my professional opinion, is a talented broadcaster. The difficulty with such discrimination, of course, is akin to the problems related to privilege by birth. Often individuals of genuine merit miss out on the sort of employment their talents best fit them for, and their society misses out on the potential quality of their contribution to its general welfare.

## 42.5  Micronesians as Information Sharers

MIKE   ...we and the Westerners have different ideas about ownership. They give away their knowledge but don't want people to take their things.

We let anyone take our things, but don't give away our knowledge...You'll also find that when Americans ask us personal questions, usually they mean no harm. They're just being friendly, and aren't trying to learn our secrets.

MARIA   Yes. They think we're unfriendly if we don't share our information...

This pervasive aspect of Micronesian behavior is a fundamental part of the value systems of all Micronesian social groups. This phenomenon whereby the "secrets," or special knowledge, of individuals or groups leads to a virtual taboo on sharing information with *anybody* certainly makes life difficult for researchers.

The second point made by Mike that Micronesians are very generous with their personal property is worthy of comment. Certainly, Micronesians are genuinely free with their possessions. Frequently, if something one person owns is admired by another, the object is presented as a gift. This can present something of a problem to a Micronesian manager such as Sam Jordan from the Marshall Islands who told me that "I'm the Program Director of our State-run radio station at Majuro and I hardly ever know the exact time. It's impossible to keep a wristwatch. I've given up buying even the cheap digital ones. What happened to the last one I bought was typical. It was admired by one of my cousins and so I had to give it to her. It's our custom."(Interview with Sam Jordan, Program Director of Radio Marshalls, WSZO, Majuro, Republic of the Marshall Islands, February 12, 1987) Sam Jordan is in his 50s and is wise in the ways of the West, as well as his native culture. But traditional pressures are brought to bear on even the most sophisticated Micronesians. Some such as Emensio Eperiam of Pohnpei, a member of the younger generation, decide to open a shop without the traditional seeking of permission. A member of the older generation, Sam Jordan, says he "just rolls with the punches."

Sam Jordan referred to one of his "cousins." Westerners are inclined to think of the term quite literally, but it's often not intended as such throughout Micronesia. A "cousin" is more frequently not a blood relative but rather someone from the same village or geographical area or even a long-standing friend of either sex.

Micronesians genuinely are not overconcerned about possessions. I recall having to placate an Australian tourist, a man in his thirties, after a party at the home of a Peace Corps volunteer in Palau. It's Pacific-wide etiquette to leave your footwear at the door when entering someone's house, and there gets to be quite a pile of items, normally "flip-flops," "sandals," "slippers," "thongs," "jandals," or "zoris" (referring to the same variety of footwear but depending on where you're from) at the entrance. Micronesians, incidentally, use the last term for the now universal flat footwear with the support that goes between the first two toes. "Zori" is just one of many linguistic legacies of the Japanese colonial period.

A few moments after the tourist had left the gathering of locals and expatriates, there was a roar of: "Some bastard's pinched me bloody thongs!" I went to the door with the American hostess and discovered the man peering at the pile of footwear in the gloom of late evening. "Got a torch?" he asked of the hostess. "What do you want a torch for? You want to weld something or crack a safe?" she replied. "He means a flashlight," I put in.

In the light of the flashlight, it was clear that the visitor's "zoris," apparently purchased only the day before, had disappeared. "No problem," said one of the Palauan guests. "You have big feet like me. Take mine." And he fished out a dilapidated pair, very much the worse for wear.

"Great!" replied the Australian. "And what are *you* going to do for thongs?"

"Take someone else's, of course," replied the Palauan. After the Australian had left, somewhat disgruntled, the Palauan guest turned to me with a wide smile. "You know, mine were someone else's yesterday, anyway."

## 42.6 Micronesian Concepts of Time

MARIA — I have learnt another Western custom and that is that they really expect people to be on time – Western time! To them the clock is important, but to us it's finishing what we are doing. I was late for a job interview because I was finishing a paper, and the man was gone when I got there and I didn't get the job.

ARKANGKEL — Yeah! There's a beautiful girl who won't speak to me now because I was an hour late for a date with her. She still doesn't understand that I first had to finish what I was doing.

Micronesians make no apologies for their general lack of respect for "Western time." Frequently, differences in concepts of time form the basis of wry or whimsical comments made by Micronesians to Western visitors. They often enjoy a joke at their own expense, as well, on matters that relate to some of the more trying aspects of Micronesian existence. Their physical environment is a case in point. At times, they're among the most rained-upon people in the world. Some of the "high" islands, the mountainous ones of volcanic origin, are characterized by tropical maritime climates generating very high annual rainfall. The rain forests covering the mountains of Pohnpei, for example, receive about 400 inches a year, among the highest in the world. Most Micronesians make light of the drenchings they so often receive by making fun of their climate, just as they "laugh continually at themselves and their own shortcomings," as Glenn Petersen observes (Petersen 1985, p.19). They also make no attempt to hide a sense of time which is not just non-Western but is seen even by themselves as distinctly idiosyncratic. On one occasion, I arrived at Pohnpei Airport in drenching rain and was greeted as follows: "Kaselehlie! Welcome to Ponape's liquid sunshine! I hear you have a workshop schedule to prepare. You're Australian, so you know about Western time. You've worked in the Pacific, so you know Pacific time. But now you're in Micronesia, so you've got to learn to think in Micronesian time. We invented it a long time ago, and we still use it. I hope you will learn to use it, too, after you have been with us a little while" (Greeting me on my first visit to Pohnpei [and indeed to Micronesia north of the equator] from Ezikiel Lippwe, then Federated States of Micronesia Information Office Broadcast Chief, May 2, 1984).

It's undoubtedly true that, as David Nevin recognized, "Most Pacific Islanders aren't time conscious. The more Westernized they become, the more conscious of time they get — you can see the change" (Meller, N., cited in Nevin, D., op. cit., p. 49). It's also true that there are very few Micronesians who appear to have adopted wholesale the ways of the West. The influence of the traditional folkways, in the area of chronology, is still strong.

Nevin's interpretation of "Micronesian time" provides part of an explanation. "[It] is a phrase Americans throughout the islands use constantly, at first in despair and later in comfort, since it also relieves them of the necessity for promptness. It means simply that things will be done when the doer gets around to it"(Nevin, D., op. cit., p. 49).

There's a good deal more to the Micronesians' idea of time than this. The concept is fleshed out in Arkangkel's and Maria's treatment of the subject in *An Adventure Far From Home*. There, the idea of "finishing what we are doing" before attending to other matters is stated as part of a value system. One has to be very careful, however, in interpreting this phrase. Indeed, it would be very easy to fall into the misconception, as I originally did, that Micronesians are task-oriented. Maria's reference to her lateness for a job interview being caused by her "finishing a paper" appears to point in the direction of task-orientation, a concept which seems, on reflection, to coincide more with aspects of the typical Western work ethos. Pohnpeian Anita Bryan told me that "We are taught that if a job is started it must be finished otherwise what is the point of beginning it? If it's something big that could take more than a day, such as a large sleeping mat, or even many days, such as some of the woodcarvings the men do, we have ways of thinking about the parts. So a part at a time is started and finished until they all add up and the job is done" (Interview with Anita Bryan, April 8, 1987, Kolonia, Pohnpei). Mrs Bryan, wife of expatriate American Information Adviser to the FSM Information Office, 1983–1987, Tom Bryan, was born into a Mokilese family residing on Pohnpei Island. Mokil is a small outer island of the Pohnpei group and is culturally akin to Pohnpei).

Fr Hezel was responsible for pointing out that the behavior described by Anita Bryan was atypical of Micronesians. "I'm very sceptical about what she's told you," he mentioned in one of our discussions. "I've a hunch she's working to a non-local agenda. Micronesians just aren't so conscious of finishing tasks within a given timeframe as she says. I'd check back on that if I were you. What Micronesians are really very conscious of bringing to a satisfactory conclusion are personal interactions. This is what most often holds them up and makes them late getting home or to appointments or whatever" (Interview with Fr Hezel, op. cit.).

Careful checking did indeed confirm Fr Hezel's doubts. Mrs Bryan had observed the work habits of her American journalist husband, invariably involving finishing what he started to tight deadlines, and, according to principles of Pohnpeian-Mokilese etiquette, had anticipated the answer she thought I probably wanted to hear. Micronesians, too, have also discovered that pleading task completion as a reason for lateness is much more acceptable to Westerners than an excuse such as: "I had to finish talking to my cousin, and it made me late." They believe, I'm sure you'll recall, that truth necessarily resides in whichever words seem most appropriate to social lubrication. I guess you could call them "white lies," but they're no strangers to our culture either. Generally, Micronesians like to carry out and finish their relationships with others in a courteous and unhurried way. It's much more likely that Maria's paper had to be finished in order to augment her relationship with her teacher rather than as a simple task *per se.*

One final point related to Micronesians' consciousness of time has been raised by Glenn Petersen who's researched Pohnpeian oral history for a number of years. It's interesting for a Westerner to think of life without a calendar, a watch, or a clock. I invite you to consider that prospect for a moment. Imagine ending up not knowing the year, the month, the week, the day, or the time of day. The risk of psychological problems resulting from temporal disorientation would be undoubtedly high unless you recorded scrupulously, in some fashion, the passing of the days from the moment you lost contact with our system of time-tellers. We've all seen the classic cartoon prisoner who scratches off the days on the wall of his cell. Micronesians simply can't see the point in this. The Western consciousness of the passage of time is often so acute that an individual will actively choose a watch with an analogue rather than a digital face. I've heard friends say things like "I just can't organize my life around a digital watch."

In Petersen's view, such temporal orientation appears of no account to Pohnpeians, despite the fact that they probably came into contact with European devices for measuring time some hundreds of years ago. "Ponapean accounts of their own history seem to emphasize distinctions in space over temporal chronology. Individuals, events, and changes seem to be linked together by variations in spatial organization... Ponapeans continue to conceptualize events in terms of where they take place, and thus remain acutely aware of local variations in historical process" (Petersen, Glenn, op. cit., pp. 16–17).

Be warned. If a Micronesian arranges to meet you, or you arrange to meet a Micronesian, at a particular time at a particular place, the encounter mightn't take place immediately. "Hang loose," as the Hawaiians say, try again, and take no offense if times aren't observed punctiliously.

## 42.7  Micronesian Notions of Politeness

MARIA   Something else I'm learning is that they don't understand our sense of politeness. They think we are uncertain, or dumb, or even rude if we don't say "no" but just keep quiet or say "yes" but don't mean it because we don't want to hurt their feelings or humiliate them in public.

There is, in fact, more to this aspect of Micronesian behavior than mere politeness. The values discussed here are associated with the "secretiveness" noticed by

David Nevin and others which "permeates the culture... No one lets others know exactly what he is thinking or how he really feels. Face is all important and no man, high or low, may be contradicted or denied" (Nevin, D., op. cit., p. 53).

The avoidance of contradiction and denial is closely allied to the pressures of consensus forced upon small groups of people living on small areas of available land. And the concept of not saying "no" is further related to the Micronesian propensity for telling their peers and visitors to their islands what they think the listeners want to hear. This characteristic is often commented upon by expatriate Americans. Tom Bryan, expatriate American Information Adviser to the FSM government, told me that "If you go to a store and they're out of potatoes, say, you might ask when they'll be getting some. And the shop assistant will look at you and think: 'Now what does he want to hear?' And she'll say: 'Maybe tomorrow.' And you know the ship won't be in for two weeks! It's sort of nice in its way — they want you to feel better about things. But you've got to be real careful of this if you're doing serious research in Micronesia" (Interview with Tom Bryan, Kolonia, Pohnpei, April 29, 1986). Stories about earnest foreign students, dutifully recording other people's interpretations of what they themselves are thinking, abound in Micronesia.

There's an element of Micronesian behavior which many visitors find puzzling, and that's the lack of animation on the typical Micronesian face in everyday contact. "Don't these people ever smile?" is a remark made often by Westerners. In fact, after the mandatory warm-up period, Micronesians smile a good deal with visitors, but the fact remains that in their day-to-day relationships with one another, there's a good deal less overt expression of beliefs or feelings than one encounters in most Western countries. Glenn Petersen, as I've mentioned, has identified the phenomenon in Pohnpei as *kanengamah*, which implies the avoidance of expressing "one's sentiments, feelings or beliefs. It can mean, as I have heard it put, constructing and keeping a blank countenance" (Petersen, Glenn, op. cit., pp. 38–39).

One source of frustration and irritation to Americans working in Micronesia is the impact the "sense of politeness," referred to by Maria in the radio play, has on Micronesian administrative systems. Locals who work in them are brought up to avoid opposing others, or hurting their feelings, at any cost. According to Fr McGarry, "It's hard for such people to give or to accept orders. It's hard to demand specific performance at a specific time or to evaluate another person or to punish failure or to fire someone. Micronesians strenuously avoid such actions, which is one reason government is so huge; people are transferred, not fired" (Nevin, David, op. cit., p. 54). He has identified, in the Pohnpeian context, a further reason for an absence of what the Westerner calls "discipline" in the Micronesian workplace:

> There's another very widespread characteristic of Ponapeans. It's contained in the word *mahk*. The word is hard to translate. It might be translated as the embarrassment experienced in confronting a person who has done something which is displeasing to yourself. I am embarrassed at the embarrassment of someone else when caught in the act... There are a number of people in authority who are quite unwilling to correct the mistakes of another person or to dismiss an employee simply because of this *mahk*. They do not want to be embarrassed at having to embarrass another person. (McGarry, William F., op. cit., p. 58)

It's not difficult to imagine a possible end result of a process whereby incompetent or otherwise inadequate people are not disciplined but, to avoid further problems, are transferred to positions of equivalent rank. At some time, there's no longer anywhere they can possibly be transferred to. In Yap, Henry Muthan talks of "the situation where a senior State Government official has been replaced by a younger, better educated man. This younger man comes to the office every day and does his work, and the older man comes in, too, and just sits at the other end of the office. He has nothing to do but they both pick up their pay checks each week. The Governor won't fire the old man. It would upset the Governor too much. They have known each other for a long time" (Interview with Henry Muthan, op. cit.).

The same principles apply in Chuuk. Fr Hezel suggests a scenario in which a Chuukese man, a provider of food for his family for much of his life, makes the transition to a job in the state civil service:

> For failing to provide [food] for kin and others to whom he is especially obligated he can undergo the humiliation of being looked on as niggardly and suffer a corresponding loss of status among those closest to him. On the other hand, the retribution for neglecting his job or abusing his position can be uncertain at best. Supervisors in many parts of Micronesia would be reluctant to fire him or even impose pay deductions and other penalties because of their fear of incurring the lasting enmity of the employee and his family. (Hezel, Francis X., op. cit., p. 13)

## References

Alkire WH (1972) An introduction to the peoples and cultures of Micronesia. Addison-Wesley Modular Publications, Reading. Module 18

*An Adventure Far From Home* (radio play script) (1986) The Student Program Committee of the Micronesian Institute, Washington, DC

Ashby GA (1987) Guide to Pohnpei: an Island argosy. Rainy Day Press, Eugene

Caughey JL (1977) Fa'a'nakkar: cultural values in a Micronesian society, vol 2. University of Pennsylvania Publications in Anthropology, Philadelphia

Fischer JL (1974) The role of the traditional chiefs on Ponape in the American period. In: Hughes DT, Lingenfelter SG (eds) Political development in Micronesia. Ohio State University Press, Columbus

Geertz C (1973) The interpretation of cultures. Basic Books, New York

Hanlon DL (1982) Understanding our misunderstanding of Micronesians. The Micronesian Seminar, Chuuk, January

Hezel FX (1987) The dilemmas of development: the effects of modernization on three areas of Island life. J Pacific Soc

Hezel FX (1995) Strangers in their own land: a century of colonial rule in the caroline and Marshall Islands. University of Hawaii Press, Honolulu

Hezel FX (2001) The new shape of old Island cultures. University of Hawaii Press, Honolulu

Hezel FX (2003) Chuuk: caricature of an Island. http://www.micsem.org/pubs/articles/historical/frames/chuukcarfr.htm

Hezel FX (2006) Is that the best you can do? A tale of two Micronesian economies. The Micronesian Seminar. http://www.micsem.org/pubs/articles/economic/frames/taleoftwofr.htm

Hogan T (1988) Micronesia and the West: major socio-cultural differences and their implications for broadcast training and development. PACBROAD, Suva

Kanarakis G (1983) Bilingualism and language learning, Monograph series. Mitchell College, Bathurst

Lingenfelter SG (1975) Yap: political leadership and culture in an Island society. The University Press of Hawaii, Honolulu

Mason L (1968) The ethnology of Micronesia. In: Vayda AP (ed) Peoples and cultures of the Pacific. The Natural History Press, Garden City

McGarry WF SJ (1982) West From Katau. The Micronesia-Pacific Collection, Community College of Micronesia, Kolonia

McHenry DF (1975) Micronesia: trust betrayed. Carnegie Endowment for International Peace, Washington, DC

Micronesia and Australian Foreign Policy (1987) Proceedings of a symposium held at the Australian National University, Canberra, October, 1987. The Australian Development Studies Network, Canberra

Nevin D (1977) The American touch in Micronesia. W.W. Norton & Company Inc, New York

Pacific Courts and Justice. No. vii (1977) Cited by Hughes DT, Laughlin SK (1982) Key elements in the evolving political culture of the Federated States of Micronesia, Pacific Studies, The Institute for Polynesian Studies, Honolulu

Petersen G (1985) A cultural analysis of the Ponapean independence vote in the 1983 plebiscite. Pacific Studies 9(1):13

Sahlins M (1976) Culture and practical reason. University of Chicago Press, Chicago

Smith AH (1956) Attitudes and relationships. Micronesian Monthly. IV(1). Trust Territory Headquarters. Agana, January–February

Solomon AM (Chairman) (1963) A report by the US Government Survey Mission to the Trust Territory of the Pacific Islands. 9 Oct 1963

The Micronesia Seminar. http://www.micsem.org